A Level Chemistry for OCR A

Rob Ritchie
Dave Gent

OXFORD
UNIVERSITY PRESS

Great Clarendon Street, Oxford, OX2 6DP, United Kingdom

Oxford University Press is a department of the University of Oxford. It furthers the University's objective of excellence in research, scholarship, and education by publishing worldwide. Oxford is a registered trade mark of Oxford University Press in the UK and in certain other countries

© Rob Ritchie and Dave Gent 2015

The moral rights of the authors have been asserted

First published in 2015

All rights reserved. No part of this publication may be reproduced, stored in a retrieval system, or transmitted, in any form or by any means, without the prior permission in writing of Oxford University Press, or as expressly permitted by law, by licence or under terms agreed with the appropriate reprographics rights organization. Enquiries concerning reproduction outside the scope of the above should be sent to the Rights Department, Oxford University Press, at the address above.

You must not circulate this work in any other form and you must impose this same condition on any acquirer

British Library Cataloguing in Publication Data
Data available

978-0-19-835197 9

14

Paper used in the production of this book is a natural, recyclable product made from wood grown in sustainable forests. The manufacturing process conforms to the environmental regulations of the country of origin.

Printed in China by Golden Cup

This resource is endorsed by OCR for use with specification H032 AS Level GCE Chemistry A and H432 A Level GCE Chemistry A. In order to gain endorsement this resource has undergone an independent quality check. OCR has not paid for the production of this resource, nor does OCR receive any royalties from its sale. For more information about the endorsement process please visit the OCR website www.ocr.org.uk

Acknowledgements
Cover: EYE OF SCIENCE/SCIENCE PHOTO LIBRARY **p2-3**: Tischenko Irina/Shutterstock; **p6-7**: Science Photo Library; **p9**: Author; **p.10**: Charles D. Winters/Science Photo Library; **p13**: Mauro Fermariello/Science Photo Library; **p16**: Nagydodo/Shutterstock; **p20**: Martyn F. Chillmaid/Science Photo Library; **p22**: Apttone/iStockphoto; **p23**: Nadezda Boltaca/Shutterstock; **p24**(T): Fablok/Shutterstock; **p24**(B): Martyn F. Chillmaid/Science Photo Library; **p26**: Haveseen/Shutterstock; **p28**: David Hay Jones/Science Photo Library; **p40**: Author; **p41**(T): Martyn F. Chillmaid/Science Photo Library; **p41**(B): Andrew Lambert Photography/Science Photo Library; **p43**(T): Author; **p44**(B): Author; **p43**(B): Author; : Martyn F. Chillmaid/Science Photo Library; **p44**(T): Author; **p50**: Martyn F. Chillmaid/Science Photo Library; **p51**: Martyn F. Chillmaid/Science Photo Library; **p60**: Usas/iStockphoto; **p62**(L): Author; **p62**(R): Author; **p79**: Claude Nuridsany & Marie Perennou/Science Photo Library; **p80**: Andrew Lambert Photography/Science Photo Library; : Science Photo Library; : Ria Novosti/Science Photo Library; **p87**: Bizroug/Shutterstock; **p88-89**: Elena Moiseeva/Shutterstock; **p90**: Science Photo Library; **p94**(T): Arhip4/Shutterstock; **p94**(B): Stuart Monk/Shutterstock; **p81**: Darren Begley/Shutterstock; **p106**: Charles D. Winters/Science Photo Library; **p107**: Charles D. Winters/Science Photo Library; **p108**: Charles Brutlag/Shutterstock; **p109**(L): Author; **p109**(R): Author; **p110**: Andrew Lambert Photography/Science Photo Library; **p113**(B): Petegar/iStockphoto; **p111**(T): Martyn F. Chillmaid/Science Photo Library; **p111**(B): Andrew Lambert Photography/Science Photo Library; **p115**(T): Andrew Lambert Photography/Science Photo Library; **p115**(B): Charles D. Winters/Science Photo Library; **p116**: Andrew Lambert Photography/Science Photo Library; **p117**: Hong Xia/Shutterstock; **p101**: Denis Burdin/Shutterstock; **p102**: Mathier/Shutterstock; **p113**(T): Author; **p130**: Martyn F. Chillmaid/Science Photo Library; **p143**(T): Blvdone/Shutterstock; **p143**(B): Trevor Clifford Photography/Science Photo Library; **p126**: Author; **p148**(B): Dorling Kindersley/Uig/Science Photo Library; **p153**: Charles D. Winters/Science Photo Library; **p142**(T): Gary718/Shutterstock; **p142**(B): Jean Morrison/Shutterstock; **p145**(T): Andrew Lambert Photography/Science Photo Library; **p145**(BL): Author; **p145**(BR): Author; **p148**(T): Ssuaphotos/Shutterstock; **p155**: Charles D. Winters/Science Photo Library; : Vipavlenkoff/Shutterstock; **p157**: Hacohob/Shutterstock; **p163**: Jg Photography; **p185**: Claffra/Shutterstock; **p187**(T): Author; **p188**: Author; **p177**(L): Daniel Korzeniewski/Shutterstock; **p177**(R): Nito/Shutterstock; **p166**: Piccia Neri/Shutterstock; **p195**: Author **p196**(T): Zixian/Shutterstock; **p187**(B): Ekaterina Baranova/Shutterstock; **p202**(B): Author; **p202**(T): Stanzi/Shutterstock; **p203**: Martyn F. Chillmaid/Science Photo Library; **p209**: Hasnuddin/Shutterstock; **p210**: Gyvafoto/Shutterstock; **p212**(T): FLPA/Alamy; **p211**(T): Irin-K/Shutterstock; **p211**(C): Maxal Tamor/Shutterstock; **p211**(B): Vankad/Shutterstock; **p196**(BL): Thinkstock; **p196**(BR): Ingram; **p220**: Andrew Lambert Photography/Science Photo Library; **p197**(T): Fabio Sacchi/Shutterstock; **p197**(B): Pavla/Shutterstock; : Gayvoronskaya_Yana/Shutterstock; **p206**: Author; **p213**: Satit_Srihin/Shutterstock; **p212**(C): Spwidoff/Shutterstock; **p212**(B): Author; **p216**: Christian Draghici/Shutterstock; **p227**: Andrew Lambert Photography/Science Photo Library; **p218**: Jorg Hackemann/Shutterstock; **p230**: Anna Omelchenko/Shutterstock; **p235**(T): Author; **p235**(C): Author; **p235**(B): Author; **p236**(T): Author; **p236**(B): Jerry Mason/Science Photo Library; **p242**(L): David Nunuk/Science Photo Library; **p242**(R): Molekuul.Be/Shutterstock; **p251**: Fenton One/Shutterstock; **p254**: Jim Varney/Science Photo Library; **p246**: Simon Fraser/Science Photo Library; **p249**: Helene Wiesenhaan/Getty Images; **p164-165**: Mopic/Shutterstock; **p231**: Africa Studio/Shutterstock; **p259**: Lanych/Shutterstock; **p277**(T): Martyn F. Chillmaid/Science Photo Library; **p299**: Robert Boesch/Corbis; **p300**: Hacohob/Shutterstock; **p313**: Costi Iosif/Shutterstock; **p315**: Photong/Shutterstock; **p321**: Africa Studio/Shutterstock; **p327**: Charles D. Winters/Science Photo Library; **p337**: Sherry Yates Young/Shutterstock; **p340**: Science Photo Library; **p444**: Andrew Lambert Photography/Science Photo Library; **p454**(T): Rikkert Harink/Shutterstock; **p454**(B): Dusan Jankovic/Shutterstock; **p456**: Andrew Lambert Photography/Science Photo Library; **p461**: Andrew Lambert Photography/Science Photo Library; **p463**: Artem Furman/Shutterstock; **p468**: Brian Lasenby/Shutterstock; **p475**: S Duffett/Shutterstock; **p480**: LittleStocker/Shutterstock; **p432**(R): Dionisvera/Shutterstock; **p432**(L): Tim UR/Shutterstock; **p438**: Gannet77/iStockphoto; **p440**: Filipw/Shutterstock; **p442**(C): Jon Le-Bon/Shutterstock; **p450**: Sovfoto/UIG/Getty Images; **p510**(d): Andrew Lambert Photography/Science Photo Library; **p469**: Andrew Lambert Photography/Science Photo Library; **p477**: Optimarc/Shutterstock; **p484**(T): Anukool Manoton/Shutterstock; **p485**: Tommaso lizzul/Shutterstock; **p357**: No_limit_pictures/iStockphoto; **p360**(L): Impactimage/iStockphoto; **p360**(R): Byjeng/Shutterstock; **p363**(T): Valentyn Volkov/Shutterstock; **p363**(CT): Science photo/Shutterstock; **p363**(CB): CAN BALCIOGLU/Shutterstock; **p366**: Mangojuicy/Shutterstock; **p369**: Africa Studio/Shutterstock; **p372**: Peticolas/Megna/Fundamental Photos/Science Photo Library; **p373**(L): Martyn F. Chillmaid/Science Photo Library; **p373**(R): Martyn F. Chillmaid/Science Photo Library; **p378**: Avarand/Shutterstock; **p384**(C): Marco mayer/Shutterstock; **p384**(B): Gyvafoto/Shutterstock; **p389**: Andrew Lambert Photography/Science Photo Library; **p395**(T): Claus Lunau/Science Photo Library; **p395**(B): Martin Bond/Science Photo Library; **p403**(C): Andrew Lambert Photography/Science Photo Library; **p415**(T): Power and Syred/Science Photo Library; **p491**: Melinda Fawver/Shutterstock; **p498**(T): Monika Wisniewska/Shutterstock; **p498**(B): Maks Narodenko/Shutterstock; **p508**: Tanewpix/Shutterstock; **p510**(a): Martyn F. Chillmaid/Science Photo Library; **p510**(b): Andrew Lambert Photography/Science Photo Library; **p511**: Andrew Lambert Photography/Science Photo Library; **p512**: Mauro Fermariello/Science Photo Library; **p513**(B): Colin Cuthbert/Science Photo Library; **p270-271**: Eye of Science/Science Photo Library; **p384**(T): Andrew Lambert Photography/Science Photo Library; **p442**(B): Science Photo Library; **p443**: Martin Bond/Science Photo Library; **p455**(L): AntoinetteW/Shutterstock; **p455**(R): Maksimilian/Shutterstock; **p467**: Africa Studio/Shutterstock; **p481**(L): Imageman/Shutterstock;

AS/A Level course structure

This book has been written to support students studying for OCR AS Chemistry A and OCR A Level Chemistry A. It covers all of the modules from the OCR A Level Chemistry A specification, with modules 2, 3, and 4 also part of the OCR AS Chemistry A specification. The modules covered are shown in the contents list, which also shows you the page numbers for the main topics within each module. There is also an index at the back to help you find what you are looking for. If you are studying for OCR AS Chemistry A, you will only need to know the content in the blue box.

AS exam

Year 1 content
1. Development of practical skills in chemistry
2. Foundations in chemistry
3. Periodic table and energy
4. Core organic chemistry

Year 2 content
5. Physical chemistry and transition elements
6. Organic chemistry and analysis

A level exam

A Level exams will cover content from Year 1 and Year 2 and will be at a higher demand. You will also carry out practical activities throughout your course.

---

p481(R): Valentina Proskurina/Shutterstock; **p513**(T): Bibiphoto/Shutterstock; **p334**: Charles D. Winters/Science Photo Library; **p355**: MarcelClemens/Shutterstock; **p429**(T): Abramova Elena/Shutterstock; **p429**(B): Remik44992/Shutterstock; **p502**: Science Photo/Shutterstock; **p430-431**: Eye of Science/Science Photo Library; **p541**: Olha Rohulya/Shutterstock; **p510**(e): Andrew Lambert Photography/Science Photo Library; **p419**(B): Charles D. Winters/Science Photo Library;

Author Photos: p277(B), p280(R), p296, p301, p460, p464, p465, p442(T), p510(c), p510(f), p484(B), p363(B), p376(L), p376(R), p381(L), p381(C), p381(R), p382(L), p382(C), p382(R), p382(B), p394, p400(L),p400(C), p400(R), p402, p403(T), p403(B), p405(T), p405(B), p407, p413, p414(T), p414(C), p414(B), p415(C), p415(B), p416, p418, p419(T), p420, p494, p495(TL), p495(TC), p495(TR), p495(BL), p495(BC), p495(BR), p496(T), p496(BL), p496(BC), p497(T), p497(B), p506(T), p506(C), p506(B), p507, p382(B), p486, p496(C), p496(BR), p280(L), p291, p316; lithium battery: Author;

Artwork by Q2A media

Thank you to St John Rigby college, Wigan, for the use of their laboratory facilities in the production of photographs.

Although we have made every effort to trace and contact all copyright holders before publication this has not been possible in all cases. If notified, the publisher will rectify any errors or omissions at the earliest opportunity.

Links to third party websites are provided by Oxford in good faith and for information only. Oxford disclaims any responsibility for the materials contained in any third party website referenced in this work.

Contents

How to use this book   vii
Kerboodle   x

Module 1 Development of practical skills in chemistry   2

Module 2 Foundations in chemistry   6

Chapter 2 Atoms, ions, and compounds   8
2.1 Atomic structure and isotopes   8
2.2 Relative mass   12
2.3 Formulae and equations   15
Practice questions   19

Chapter 3 Amount of substance   20
3.1 Amount of substance and the mole   20
3.2 Determination of formulae   22
3.3 Moles and volumes   26
3.4 Reacting quantities   32
Practice questions   38

Chapter 4 Acids and redox   40
4.1 Acids, bases, and neutralisation   40
4.2 Acid–base titrations   43
4.3 Redox   48
Practice questions   52

Chapter 5 Electrons and bonding   54
5.1 Electron structure   54
5.2 Ionic bonding and structure   59
5.3 Covalent bonding   63
Practice questions   68

Chapter 6 Shapes of molecules and intermolecular forces   70
6.1 Shapes of molecules and ions   70
6.2 Electronegativity and polarity   74
6.3 Intermolecular forces   77
6.4 Hydrogen bonding   81
Practice questions   84
Module 2 summary   86
Module 2 practice questions   88

Module 3 Periodic table and energy   90

Chapter 7 Periodicity   92
7.1 The periodic table   92
7.2 Ionisation energies   96
7.3 Periodic trends in bonding and structure   101
Practice questions   106

Chapter 8 Reactivity trends   108
8.1 Group 2   108
8.2 The halogens   112
8.3 Qualitative analysis   117
Practice questions   121

Chapter 9 Enthalpy   124
9.1 Enthalpy changes   124
9.2 Measuring enthalpy changes   129
9.3 Bond enthalpies   135
9.4 Hess' law and enthalpy cycles   138
Practice questions   142

Chapter 10 Reaction rates and equilibrium   144
10.1 Reaction rates   144
10.2 Catalysts   149
10.3 The Boltzmann distribution   152
10.4 Dynamic equilibrium and le Chatelier's principle   154
10.5 The equilibrium constant K_c – part 1   160
Practice questions   162
Module 3 summary   164
Module 3 practice questions   166

Module 4 Core organic chemistry   170

Chapter 11 Basic concepts of organic chemistry   172
11.1 Organic chemistry   172
11.2 Nomenclature of organic compounds   174
11.3 Representing the formulae of organic compounds   179

11.4 Isomerism	182
11.5 Introduction to reaction mechanisms	184
Practice questions	187

Chapter 12 Alkanes — 190
- 12.1 Properties of the alkanes — 190
- 12.2 Chemical reactions of the alkanes — 193
- Practice questions — 198

Chapter 13 Alkenes — 200
- 13.1 The properties of the alkenes — 200
- 13.2 Stereoisomerism — 203
- 13.3 Reactions of alkenes — 207
- 13.4 Electrophilic addition in alkenes — 211
- 13.5 Polymerisation in alkenes — 215
- Practice questions — 220

Chapter 14 Alcohols — 222
- 14.1 Properties of alcohols — 222
- 14.2 Reactions of alcohols — 226
- Practice questions — 229

Chapter 15 Haloalkanes — 230
- 15.1 The chemistry of the haloalkanes — 230
- 15.2 Organohalogen compounds in the environment — 235
- Practice questions — 238

Chapter 16 Organic synthesis — 240
- 16.1 Practical techniques in organic chemistry — 240
- 16.2 Synthetic routes — 244
- Practice questions — 250

Chapter 17 Spectroscopy — 252
- 17.1 Mass spectrometry — 252
- 17.2 Infrared spectroscopy — 256
- Practice questions — 262
- **Module 4 summary** — 264
- **Module 4 practice questions** — 266

Module 5 Physical chemistry and transition elements — 270

Chapter 18 Rates of reactions — 272
- 18.1 Orders, rate equations, and rate constants — 272
- 18.2 Concentration–time graphs — 277
- 18.3 Rate–concentration graphs and initial rates — 282
- 18.4 Rate-determining step — 287
- 18.5 Rate constants and temperature — 289
- Practice questions — 292

Chapter 19 Equilibrium — 294
- 19.1 The equilibrium constant K_c – part 2 — 294
- 19.2 The equilibrium constant K_p — 298
- 19.3 Controlling the position of equilibrium — 302
- Practice questions — 307

Chapter 20 Acids, bases, and pH — 310
- 20.1 Brønsted–Lowry acids and bases — 310
- 20.2 The pH scale and strong acids — 315
- 20.3 The acid dissociation constant K_a — 319
- 20.4 The pH of weak acids — 322
- 20.5 pH and strong bases — 326
- Practice questions — 330

Chapter 21 Buffers and neutralisation — 332
- 21.1 Buffer solutions — 332
- 21.2 Buffer solutions in the body — 337
- 21.3 Neutralisation — 340
- Practice questions — 344

Chapter 22 Enthalpy and entropy — 346
- 22.1 Lattice enthalpy — 346
- 22.2 Enthalpy changes in solution — 352
- 22.3 Factors affecting lattice enthalpy and hydration — 358
- 22.4 Entropy — 362
- 22.5 Free energy — 365
- Practice questions — 370

Chapter 23 Redox and electrode potentials — 372
- 23.1 Redox reactions — 372
- 23.2 Manganate(VII) redox titrations — 376

Contents

23.3 Iodine/thiosulfate redox titrations	381
23.4 Electrode potentials	386
23.5 Predictions from electrode potentials	391
23.6 Storage and fuel cells	394
Practice questions	397
Chapter 24 Transition elements	**400**
24.1 d-block elements	400
24.2 The formation and shapes of complex ions	405
24.3 Stereoisomerism in complex ions	409
24.4 Ligand substitution and precipitation	413
24.5 Redox and qualitative analysis	418
Practice questions	422
Module 5 summary	**424**
Module 5 practice questions	**426**

Module 6 Organic chemistry and analysis — 430

Chapter 25 Aromatic chemistry	**432**
25.1 Introducing benzene	432
25.2 Electrophilic substitution reactions of benzene	437
25.3 The chemistry of phenol	442
25.4 Directing groups	446
Practice questions	451
Chapter 26 Carbonyls and carboxylic acids	**454**
26.1 Carbonyl compounds	454
26.2 Identifying aldehydes and ketones	460
26.3 Carboxylic acids	463
26.4 Carboxylic acid derivatives	466
Practice questions	472
Chapter 27 Amines, amino acids, and polymers	**474**
27.1 Amines	474
27.2 Amino acids, amides, and chirality	478
27.3 Condensation polymers	483
Practice questions	488
Chapter 28 Organic synthesis	**490**
28.1 Carbon–carbon bond formation	490

28.2 Further practical techniques	494
28.3 Further synthetic routes	498
Practice questions	504
Chapter 29 Chromatography and spectroscopy	**506**
29.1 Chromatography and functional group analysis	506
29.2 Nuclear magnetic resonance (NMR) spectroscopy	512
29.3 Carbon-13 NMR spectroscopy	515
29.4 Proton NMR spectroscopy	520
29.5 Interpreting NMR spectra	525
29.6 Combined techniques	530
Practice questions	534
Module 6 summary	**536**
Module 6 practice questions	**534**

Unifying concepts — 542

Analysing and answering a synoptic question	542
Practice questions	547

Reference	
Glossary	552
Answers	560
Index	593
Periodic table	603

How to use this book

Learning outcomes
→ At the beginning of each topic, there is a list of learning outcomes.
→ These are matched to the specification and allow you to monitor your progress.
→ A specification reference is also included in the topic header.

This book contains many different features. Each feature is designed to support and develop the skills you will need for your examinations, as well as foster and stimulate your interest in chemistry.

Terms that you will need to be able to define and understand are highlighted by **bold text**.

Application features

These features contain important and interesting applications of chemistry in order to emphasise how scientists and engineers have used their scientific knowledge and understanding to develop new applications and technologies. There are also practical application features, with the icon , to support further development of your practical skills.

1 All application features have a question to link to material covered with the concept from the specification.

Study Tips
Study tips contain prompts to help you with your understanding and revision.

Synoptic link
These highlighted the key areas where topics relate to each other. As you go through your course, knowing how to link different areas of chemistry together becomes increasingly important. Many exam questions, particularly at A Level, will require you to bring together your knowledge from different areas.

Extension features

These features contain material that is beyond the specification. They are designed to stretch and provide you with a broader knowledge and understanding and lead the way into the types of thinking and areas you might study in further education. As such, neither the detail nor the depth of questioning will be required for the examinations. But this book is about more than getting through the examinations.

1 Extension features also contain questions that link the off-specification material back to your course.

Summary Questions

1 These are short questions at the end of each topic.

2 They test your understanding of the topic and allow you to apply the knowledge and skills you have acquired.

3 The questions are ramped in order of difficulty. Lower-demand questions have a paler background, with the higher-demand questions having a darker background. Try to attempt every question you can, to help you achieve your best in the exams.

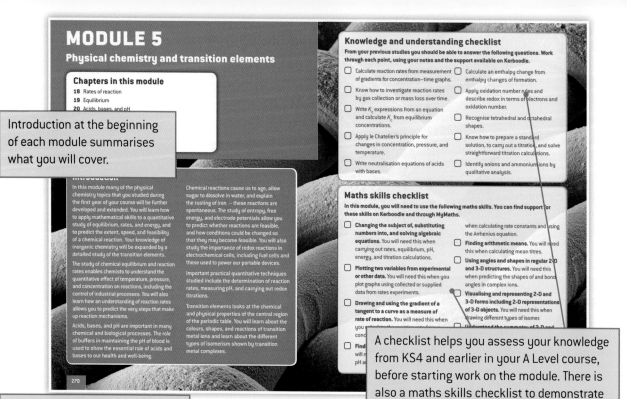

Introduction at the beginning of each module summarises what you will cover.

A checklist helps you assess your knowledge from KS4 and earlier in your A Level course, before starting work on the module. There is also a maths skills checklist to demonstrate the skills you will learn in that module.

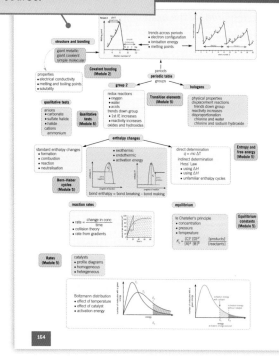

Visual summaries show how some of the key concepts of that module interlink with other modules, across the entire A Level course.

Application task brings together some of the key concepts of the module in a new context.

Extension task bring together some key concepts of the module and develop them further, leading you towards greater understanding and further study.

Practice questions at the end of each chapter and the end of each module, including questions that cover practical and math skills.

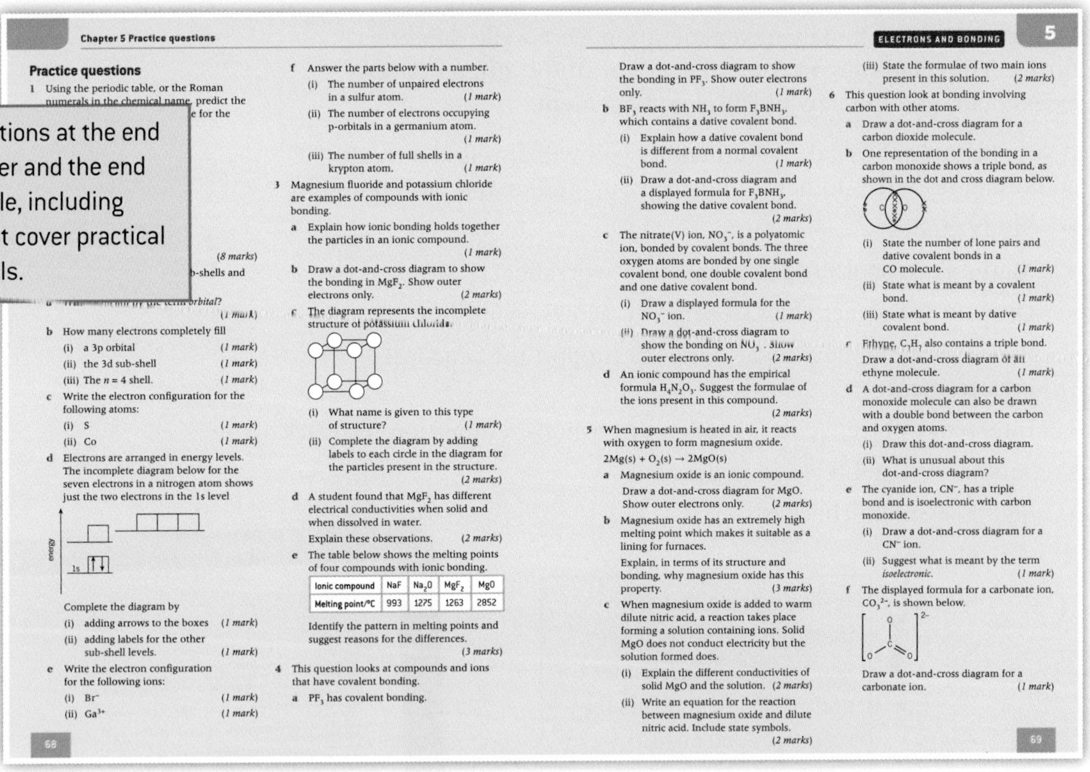
Chapter 5 Practice questions

ELECTRONS AND BONDING 5

Practice questions

1 Using the periodic table, or the Roman numerals in the chemical name, predict the ... e for the ...

(8 marks)

... b-shells and ... orbital?

(1 mark)

b How many electrons completely fill

(i) a 3p orbital (1 mark)

(ii) the 3d sub-shell (1 mark)

(iii) The $n = 4$ shell. (1 mark)

c Write the electron configuration for the following atoms:

(i) S (1 mark)

(ii) Co (1 mark)

d Electrons are arranged in energy levels. The incomplete diagram below for the seven electrons in a nitrogen atom shows just the two electrons in the 1s level

Complete the diagram by

(i) adding arrows to the boxes (1 mark)

(ii) adding labels for the other sub-shell levels. (1 mark)

e Write the electron configuration for the following ions:

(i) Br^- (1 mark)

(ii) Ga^{3+} (1 mark)

f Answer the parts below with a number.

(i) The number of unpaired electrons in a sulfur atom. (1 mark)

(ii) The number of electrons occupying p-orbitals in a germanium atom. (1 mark)

(iii) The number of full shells in a krypton atom. (1 mark)

3 Magnesium fluoride and potassium chloride are examples of compounds with ionic bonding.

a Explain how ionic bonding holds together the particles in an ionic compound. (1 mark)

b Draw a dot-and-cross diagram to show the bonding in MgF_2. Show outer electrons only. (2 marks)

c The diagram represents the incomplete structure of potassium chloride.

(i) What name is given to this type of structure? (1 mark)

(ii) Complete the diagram by adding labels to each circle in the diagram for the particles present in the structure. (2 marks)

d A student found that MgF_2 has different electrical conductivities when solid and when dissolved in water.

Explain these observations. (2 marks)

e The table below shows the melting points of four compounds with ionic bonding.

Ionic compound	NaF	Na_2O	MgF_2	MgO
Melting point/°C	993	1275	1263	2852

Identify the pattern in melting points and suggest reasons for the differences. (3 marks)

4 This question looks at compounds and ions that have covalent bonding.

a PF_3 has covalent bonding.

Draw a dot-and-cross diagram to show the bonding in PF_3. Show outer electrons only. (1 mark)

b BF_3 reacts with NH_3 to form F_3BNH_3, which contains a dative covalent bond.

(i) Explain how a dative covalent bond is different from a normal covalent bond. (1 mark)

(ii) Draw a dot-and-cross diagram and a displayed formula for F_3BNH_3, showing the dative covalent bond. (2 marks)

c The nitrate(V) ion, NO_3^-, is a polyatomic ion, bonded by covalent bonds. The three oxygen atoms are bonded by one single covalent bond, one double covalent bond and one dative covalent bond.

(i) Draw a displayed formula for the NO_3^- ion. (1 mark)

(ii) Draw a dot-and-cross diagram to show the bonding on NO_3^-. Show outer electrons only. (2 marks)

d An ionic compound has the empirical formula $H_4N_2O_3$. Suggest the formulae of the ions present in this compound. (2 marks)

5 When magnesium is heated in air, it reacts with oxygen to form magnesium oxide.

$2Mg(s) + O_2(s) \rightarrow 2MgO(s)$

a Magnesium oxide is an ionic compound. Draw a dot-and-cross diagram for MgO. Show outer electrons only. (2 marks)

b Magnesium oxide has an extremely high melting point which makes it suitable as a lining for furnaces.

Explain, in terms of its structure and bonding, why magnesium oxide has this property. (3 marks)

c When magnesium oxide is added to warm dilute nitric acid, a reaction takes place forming a solution containing ions. Solid MgO does not conduct electricity but the solution formed does.

(i) Explain the different conductivities of solid MgO and the solution. (2 marks)

(ii) Write an equation for the reaction between magnesium oxide and dilute nitric acid. Include state symbols. (2 marks)

(iii) State the formulae of two main ions present in this solution. (2 marks)

6 This question look at bonding involving carbon with other atoms.

a Draw a dot-and-cross diagram for a carbon dioxide molecule.

b One representation of the bonding in a carbon monoxide shows a triple bond, as shown in the dot and cross diagram below.

(i) State the number of lone pairs and dative covalent bonds in a CO molecule. (1 mark)

(ii) State what is meant by a covalent bond. (1 mark)

(iii) State what is meant by dative covalent bond. (1 mark)

c Ethyne, C_2H_2 also contains a triple bond. Draw a dot-and-cross diagram of an ethyne molecule. (1 mark)

d A dot-and-cross diagram for a carbon monoxide molecule can also be drawn with a double bond between the carbon and oxygen atoms.

(i) Draw this dot-and-cross diagram.

(ii) What is unusual about this dot-and-cross diagram?

e The cyanide ion, CN^-, has a triple bond and is isoelectronic with carbon monoxide.

(i) Draw a dot-and-cross diagram for a CN^- ion.

(ii) Suggest what is meant by the term *isoelectronic*. (1 mark)

f The displayed formula for a carbonate ion, CO_3^{2-}, is shown below.

Draw a dot-and-cross diagram for a carbonate ion. (1 mark)

68 69

Practice questions at the end of each chapter, with multiple choice questions and synoptic style questions, also covering the practical and math skills. The questions at the end of the AS modules are also labelled according to the AS exam structure.

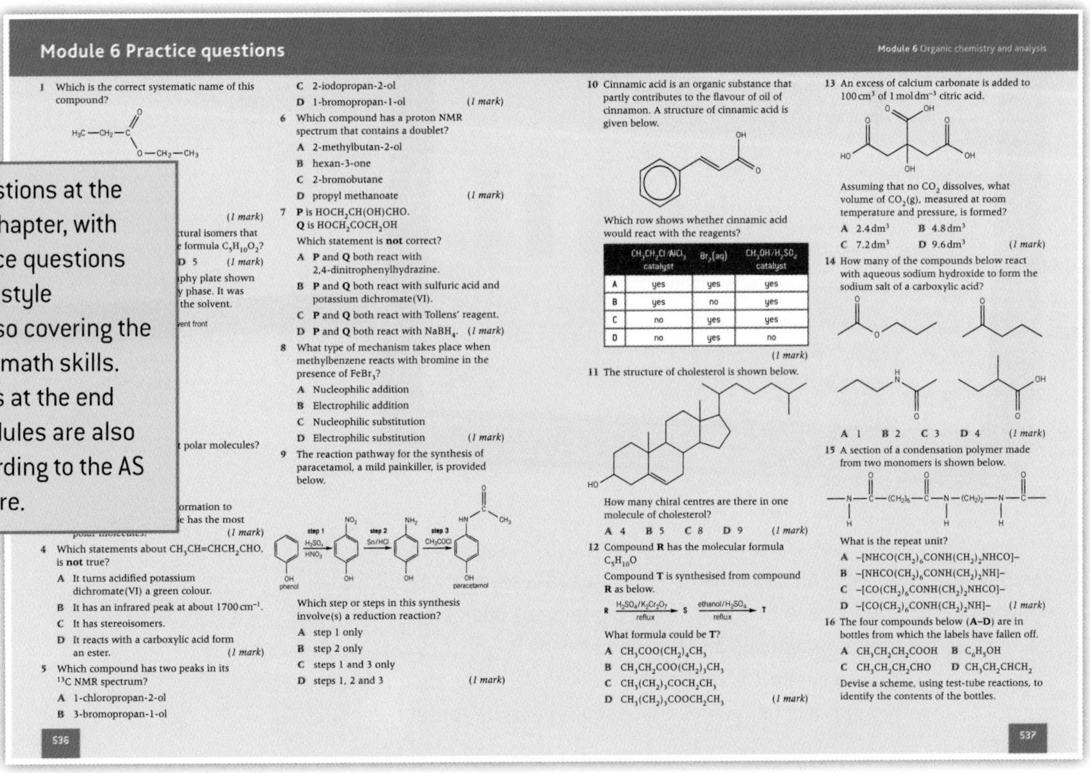
Module 6 Practice questions

Module 6 Organic chemistry and analysis

1 Which is the correct systematic name of this compound?

(1 mark)

... tural isomers that ... formula $C_5H_{10}O_2$?

D 5 (1 mark)

... phy plate shown ... phase. It was ... the solvent.

... polar molecules?

... ormation to ... e has the most polar molecules?

(1 mark)

4 Which statements about $CH_3CH{=}CHCH_2CHO$, is **not** true?

A It turns acidified potassium dichromate(VI) a green colour.

B It has an infrared peak at about 1700 cm^{-1}.

C It has stereoisomers.

D It reacts with a carboxylic acid form an ester. (1 mark)

5 Which compound has two peaks in its ^{13}C NMR spectrum?

A 1-chloropropan-2-ol

B 3-bromopropan-1-ol

C 2-iodopropan-2-ol

D 1-bromopropan-1-ol (1 mark)

6 Which compound has a proton NMR spectrum that contains a doublet?

A 2-methylbutan-2-ol

B hexan-3-one

C 2-bromobutane

D propyl methanoate (1 mark)

7 P is $HOCH_2CH(OH)CHO$.
Q is $HOCH_2COCH_2OH$

Which statement is **not** correct?

A P and Q both react with 2,4-dinitrophenylhydrazine.

B P and Q both react with sulfuric acid and potassium dichromate(VI).

C P and Q both react with Tollens' reagent.

D P and Q both react with $NaBH_4$. (1 mark)

8 What type of mechanism takes place when methylbenzene reacts with bromine in the presence of $FeBr_3$?

A Nucleophilic addition

B Electrophilic addition

C Nucleophilic substitution

D Electrophilic substitution (1 mark)

9 The reaction pathway for the synthesis of paracetamol, a mild painkiller, is provided below.

Which step or steps in this synthesis involve(s) a reduction reaction?

A step 1 only

B step 2 only

C steps 1 and 3 only

D steps 1, 2 and 3 (1 mark)

10 Cinnamic acid is an organic substance that partly contributes to the flavour of oil of cinnamon. A structure of cinnamic acid is given below.

Which row shows whether cinnamic acid would react with the reagents?

	$CH_3CH_2Cl/AlCl_3$ catalyst	$Br_2(aq)$	CH_3OH/H_2SO_4 catalyst
A	yes	yes	yes
B	yes	no	yes
C	no	yes	yes
D	no	yes	no

(1 mark)

11 The structure of cholesterol is shown below.

How many chiral centres are there in one molecule of cholesterol?

A 4 B 5 C 8 D 9 (1 mark)

12 Compound R has the molecular formula $C_5H_{10}O$

Compound T is synthesised from compound R as below.

What formula could be T?

A $CH_3COO(CH_2)_4CH_3$

B $CH_3CH_2COO(CH_2)_3CH_3$

C $CH_3(CH_2)_3COCH_2CH_3$

D $CH_3(CH_2)_3COOCH_2CH_3$ (1 mark)

13 An excess of calcium carbonate is added to 100 cm^3 of 1 mol dm^{-3} citric acid.

Assuming that no CO_2 dissolves, what volume of CO_2(g), measured at room temperature and pressure, is formed?

A 2.4 dm^3 B 4.8 dm^3

C 7.2 dm^3 D 9.6 dm^3 (1 mark)

14 How many of the compounds below react with aqueous sodium hydroxide to form the sodium salt of a carboxylic acid?

A 1 B 2 C 3 D 4 (1 mark)

15 A section of a condensation polymer made from two monomers is shown below.

What is the repeat unit?

A $-[NHCO(CH_2)_6CONH(CH_2)_2NHCO]-$

B $-[NHCO(CH_2)_6CONH(CH_2)_2NH]-$

C $-[CO(CH_2)_6CONH(CH_2)_2NHCO]-$

D $-[CO(CH_2)_6CONH(CH_2)_2NH]-$ (1 mark)

16 The four compounds below (A–D) are in bottles from which the labels have fallen off.

A $CH_3CH_2CH_2COOH$ B C_6H_5OH

C $CH_3CH_2CH_2CHO$ D $CH_3CH_2CHCH_2$

Devise a scheme, using test-tube reactions, to identify the contents of the bottles.

536 537

Kerboodle

This book is supported by next generation Kerboodle, offering unrivalled digital support for independent study, differentiation, assessment, and the new practical endorsement.

If your school subscribes to Kerboodle, you will also find a wealth of additional resources to help you with your studies and with revision.

- Study guides
- Maths skills boosters and calculation worksheets
- On your marks activities to help you achieve your best
- Practicals and follow up activities to support the practical endorsement
- Interactive objective tests that give question-by-question feedback
- Animations and revision podcasts
- Self-assessment checklists

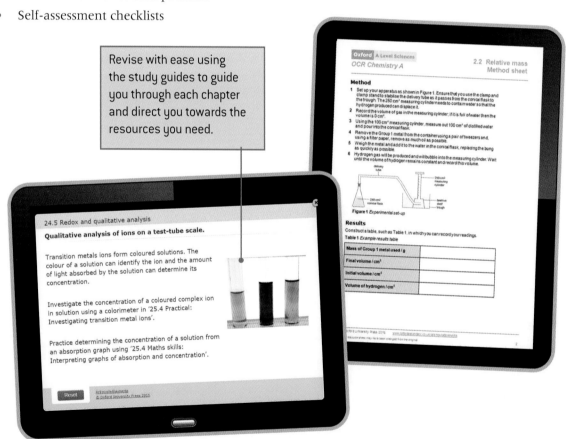

Revise with ease using the study guides to guide you through each chapter and direct you towards the resources you need.

For teachers, Kerboodle also has plenty of further assessment resources, answers to the questions in the book, and a digital markbook along with full teacher support for practicals and the worksheets, which include suggestions on how to support and stretch students. All of the resources are pulled together into teacher guides that suggest a route through each chapter.

MODULE 1
Development of practical skills in chemistry

Chemistry is a practical subject and experimental work provides you with important practical skills, as well as enhancing your understanding of chemical theory. You will be developing practical skills by carrying out practical and investigative work in the laboratory throughout both the AS and the A level Chemistry course. You will be assessed on your practical skills is two different ways:
- written examinations (AS and A level)
- practical endorsement (A level only)

Practical coverage throughout this book

Practical skills are a fundamental part of a complete education in science, and you are advised to keep a record of your practical work from the start of your A level course that you can later use as part of your practical endorsement. You can find more details of the practical endorsement from your teacher or from the specification.

In this book and its supporting materials practical skills are covered in a number of ways. By studying Application boxes and Exam-style questions in this student book, and by using the Practical activities and Skills sheets in Kerboodle you will have many opportunities to learn about the scientific method and carry-out practical activities.

1.1 Practical skills assessed in written examinations

In the written examination papers for AS and A level, at least 15% of the marks will be from questions that assess practical skills. The questions will cover four important skill areas, all based on the practical skills that you will develop by carrying out experimental work during your course.

- Planning – your ability to solve a chemistry problem in a practical context.
- Implementing – your understanding of important practical techniques and processes.
- Analysing – your interpretation of experimental results set in a practical context and related to the experiments that you would have carried out.
- Evaluating – your ability to develop a plan that is fit for the intended purpose.

1.1.1 Planning
- Designing experiments
- Identifying variables to be controlled
- Evaluating the experimental method

Skills checklist
- [ ] Selecting apparatus and equipment
- [ ] Selecting appropriate techniques
- [ ] Selecting appropriate quantities of chemicals and scale of working
- [ ] Solving chemical problems in a practical context
- [ ] Applying chemistry concepts to practical problems

1.1.2 Implementing

- Using a range of practical apparatus
- Carrying out a range of techniques
- Using appropriate units for measurements
- Recording data and observations in an appropriate format

Skills checklist

- ☐ Understanding practical techniques and processes
- ☐ Identifying hazards and safe procedures
- ☐ Using SI units
- ☐ Recording qualitative observations accurately
- ☐ Recording a range of quantitative measurements
- ☐ Using the appropriate precision for apparatus

1.1.3 Analysis

- Processing, analysing, and interpreting results
- Analysing data using appropriate mathematical skills
- Using significant figures appropriately
- Plotting and interpreting graphs

Skills checklist

- ☐ Analysing qualitative observations
- ☐ Analysing quantitative experimental data, including
  - calculation of means
  - amount of substance and equations
- ☐ For graphs,
  - selecting and labelling axes with appropriate scales, quantities, and units
  - drawing tangents and measuring gradients

1.1.4 Evaluation

- Evaluating results to draw conclusions
- Identify anomalies
- Explain limitations in method
- Identifying uncertainties and errors
- Suggesting improvements

Skills checklist

- ☐ Reaching conclusions from qualitative observations
- ☐ Identifying uncertainties and calculating percentage errors
- ☐ Identifying procedural and measurement errors
- ☐ Refining procedures and measurements to suggest improvements

1.2 Practical skills assessed in practical endorsement

You will also be assessed on how well you carry out a wide range of practical work and how to record the results of this work. These hands-on skills are divided into 12 categories and form the practical endorsement. This is assessed for A level Chemistry qualification only.

The endorsement requires a range of practical skills from both years of your course. If you are taking only AS Chemistry, you will not be assessed through the practical endorsement but the written AS examinations will include questions that relate to the skills that naturally form part of the AS common content to the A level course.

1.2.1 Practical skills

By carrying out experimental work through the course, you will develop your ability to:

- design and use practical techniques to investigate and solve problems
- use a wide range of experimental and practical apparatus, equipment, and materials, including chemicals and solutions
- carry out practical procedures skillfully and safely, recording and presenting results in a scientific way
- research using online and offline tools.

Along with the experimental work, these skills are covered in practical skills questions throughout the book.

1.2.2 Use of apparatus and techniques

To meet the requirements for the practical endorsement, you will be assessed in at least 12 practical experiments to enable you to experience a wide range of apparatus and techniques. These practical experiments are incorporated throughout the book in practical application boxes.

This will help to give you the necessary skills to be a competent and effective practical chemist.

Practical Activity Group (PAG) overview and Application features

The PAG labels are opportunities for activities that could count towards the practical endorsement. The table below shows where these PAG references are covered throughout this course.

PAG1–PAG3 and PAG5 will be covered in Year 1, PAG6–8 and PAG10–11 in Year 2, and PAG4 and PAG9 throughout the two-year course. There are a wide variety of opportunities to assess PAG12 throughout the specification.

Specification reference	Topic reference
PAG1 Moles determination 2.1.3(d); 2.1.3(i)	3.2, 3.4, 4.2, 9.2, 10.1
PAG2 Acid–base titration 2.1.4(d)	4.2
PAG3 Enthalpy determination 3.2.1(e)	9.2
PAG4 Qualitative analysis of ions 3.1.4(a) 5.3.2(a)	8.3
PAG5 Synthesis of an organic liquid 4.2.3(a)	16.1
PAG6 Synthesis of an organic solid 6.2.5(a) 6.3.1(a)	28.2, 29.1
PAG7 Qualitative analysis of organic functional groups 6.3.1(c)	13.3, 15.1, 26.2, 29.1
PAG8 Electrochemical cells 5.2.3(g)	23.4
PAG9 Rates of reaction – continuous monitoring method 3.2.2(e) 5.1.1(h)	10.1, 18.2
PAG10 Rates of reaction – initial rates method 5.1.1(h)	18.3
PAG11 pH measurement 5.1.3(o)	21.3
PAG12 Research skills	

Maths skills and How Science Works across Module 1

Maths is a useful tool for scientists and as you study your course you will learn maths techniques and equations that support the development of your science knowledge. Each module opener in this book has an overview of the maths skills that relate to the theory in the chapter. There are also questions using maths skills throughout the book that will help you practice.

How Science Works skills help you to put science in a wider context, and to develop your critical and creative thinking skills and help you solve problems in a variety of contexts. How Science Works is embedded throughout this book, particularly in application boxes and practice questions.

You can find further support for maths and how science works on Kerboodle.

MODULE 2
Foundations in Chemistry

Chapters in this module
2 Atoms, ions, and compounds
3 Amount of substance
4 Acids and redox
5 Electrons and bonding
6 Shapes of molecules and intermolecular forces

Introduction

Chemistry is about all matter and the chemical reactions that take place between atoms. This makes chemistry a vast subject, whether you are studying the 34 elements making up a human body, synthesising a new medicine, or discovering nanoparticles for a use yet to be found. This module studies the foundations of chemistry that form the building blocks for all the other modules in your A level course.

Atoms, ions, and compounds looks at some of the essential language of chemistry. You will learn about the atomic masses that you see on your periodic table. You will also learn about the special code of chemistry – the formulae and equations that allow chemists to communicate.

Amount of substance, and its unit the mole, provides chemists with an ability to convert between mass, concentration, and volume to predict how much product can be made in a chemical reaction.

Acids and redox are two important topics. Central to acids is the analysis of solutions by titration using pipettes and burettes the basis of quality testing for many materials, from washing powders to medicines. In redox you learn about oxidation numbers another essential part of the chemists toolkit for describing chemical change.

Electrons and bonding looks at the role of electrons in atoms and in chemical bonding. A good understanding of bonding and structure is essential for all further topics.

Shapes of molecules and intermolecular forces is all about the molecule. You will see how electrons determine the shape and polarity of molecule. You will learn about how intermolecular forces explain many properties of molecular compounds such as why ice floats and why water is a liquid.

Knowledge and understanding checklist

From your Key Stage 4 study you should be able to answer the following questions. Work through each point, using your Key Stage 4 notes and the support available on Kerboodle.

- ☐ Recall relative charges and approximate relative masses of protons, neutrons, and electrons.
- ☐ Calculate numbers of protons, neutrons, and electrons in atoms, given atomic number and mass number.
- ☐ Write formulae and balanced chemical equations.
- ☐ Calculate relative formula masses of species separately and in a balanced chemical equation.
- ☐ Use a balanced equation to calculate masses of reactants or products.
- ☐ Recall that acids react with some metals and with carbonates and write equations predicting products from given reactants.
- ☐ Describe neutralisation as an acid reacting with an alkali to form a salt and water.
- ☐ Explain reduction and oxidation in terms of gain or loss of electrons, identifying which species are oxidised and which are reduced.
- ☐ Construct dot-and-cross diagrams for simple ionic and covalent substances.

Maths skills checklist

In this module, you will need to use the following maths skills. You can find support for these skills on Kerboodle and through MyMaths.

- ☐ **Working with standard form and significant figures, and using appropriate units**, for carrying out all calculations in this chapter.
- ☐ **Changing the subject of an equation**, for carrying out structured and unstructured mole calculations.
- ☐ **Using ratios, fractions, and percentages**, for working with moles and equations using ratios, calculating percentage yields, and calculating atom economies.
- ☐ **Finding arithmetic means**, for calculating weighted means when determining an atomic mass and when calculating mean titres.
- ☐ **Using angles and shapes in regular 2-D and 3-D structures**, for predicting the shapes of and bond angles in molecule and ions.

MyMaths.co.uk
Bringing Maths Alive

2 ATOMS, IONS, AND COMPOUNDS

2.1 Atomic structure and isotopes

Specification reference: 2.1.1

Learning outcomes

Demonstrate knowledge, understanding, and application of:

→ atomic structure
→ isotopes.

Protons, neutrons, and electrons

The nuclear atom

At GCSE, you learnt about the nuclear model of the atom (Figure 1).

- The atom consists of a **nucleus** made up of two types of subatomic particle – **protons** and **neutrons**.
- A third type of subatomic particle, called an **electron**, occupies a region outside the nucleus. Electrons are arranged around the nucleus in **shells**.

Properties of protons, neutrons, and electrons

Mass

Atoms and their subatomic particles have tiny masses. Instead of working in grams, chemists compare the masses of subatomic particles using **relative masses** (Figure 2).

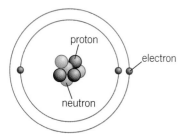

▲ Figure 1 The nuclear atom

- A proton has virtually the same mass as a neutron.
- An electron has negligible mass, about $\frac{1}{1836}$th the mass of a proton.

Accurate measurements show that a neutron has a slightly greater mass than a proton, by a factor of 1.001375. This is so close to 1 that chemists usually assume that protons and neutrons have the same mass.

▲ Figure 2 Relative masses of protons, neutrons, and electrons

Charge

- A proton has a positive charge.
- An electron has a negative charge.
- The charge on a proton is equal but opposite to the charge on an electron. The charges balance.
- A neutron, as its name suggests, is neutral and has no charge.

The actual charge on a single proton is tiny: $+1.60217733 \times 10^{-19}$ C (coulombs). The charge on a single electron must balance the charge on a proton and is $-1.60217733 \times 10^{-19}$ C. It is much easier to use relative charges of 1+ for a proton and 1– for an electron.

Building the atom

Table 1 summarises the relative charges and masses of protons, neutrons, and electrons.

- Nearly all of an atom's mass is in the nucleus.
- Atoms contain the same number of protons as electrons.

ATOMS, IONS, AND COMPOUNDS — 2

▼ Table 1 *Charges and masses of some subatomic particles, relative to the proton*

Particle	Abbreviation	Relative charge	Relative mass
proton	p^+	1+	1
neutron	n	0	1
electron	e^-	1−	$\frac{1}{1836}$

- The total positive charge from protons is cancelled by the total negative charge from electrons.
- The overall charge of an atom is zero – an atom is neutral.

Neutrons can be thought of as providing the glue that holds the nucleus together despite the electrostatic repulsion between its positively charged protons.

- Most atoms contain the same number of, or slightly more, neutrons than protons.
- As the nucleus gets larger, more and more neutrons are needed.

Atomic number – the identity of an element

The number of protons in an atom identifies the element. As of 2014, the existence of 114 elements has been confirmed, and others have been tentatively reported.

- Every atom of the *same* element contains the *same* number of protons.
- *Different* elements contain atoms that have *different* numbers of protons.
- The periodic table lists elements in order of the number of protons in the nucleus. Each element is shown with the number of protons as its **atomic number** (or proton number).

Figure 4 shows the first 18 elements in the periodic table with their atomic numbers.

1 H							2 He
3 Li	4 Be	5 B	6 C	7 N	8 O	9 F	10 Ne
11 Na	12 Mg	13 Al	14 Si	15 P	16 S	17 Cl	18 Ar

▲ Figure 4 *Atomic numbers (proton numbers) for the first 18 elements*

Isotopes

Every atom of an element has the same number of protons.

- Every atom of *nitrogen*, atomic number 7, contains *7 protons*.
- Every atom of *oxygen*, atomic number 8, contains *8 protons*.
- … And so on.

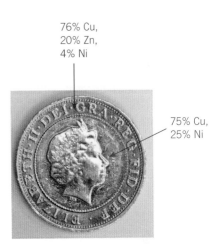

▲ Figure 3 *One coin, two pounds, and three elements – next to one another in the periodic table*

28 Ni	29 Cu	30 Zn

- all Ni atoms have 28 protons
- all Cu atoms have 29 protons
- all Zn atoms have 30 protons

Study tip

Every periodic table shows each element labelled with its atomic number. You will always have access to a copy of the periodic table. Using the periodic table, you will always be able to work out the number of protons (and electrons) in an atom.

2.1 Atomic structure and isotopes

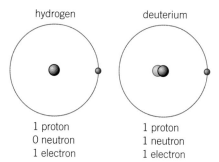

▲ Figure 5 *Two isotopes of hydrogen with the same number of protons but different numbers of neutrons*

Unlike protons, the number of neutrons in the atoms of an element can be different, usually within a narrow range.

- **Isotopes** are atoms of the same element with different numbers of neutrons and different masses (Figure 5).
- Most elements are made up of a mixture of isotopes.

Representing isotopes

Isotopes are represented using the chemical notation shown in Figure 6.

- **Mass number** (nucleon number) A

  A = number of protons + number of neutrons

- **Atomic number** (proton number) Z

  Z = number of protons

You can use this notation to work out the number of protons, neutrons, and electrons in different isotopes of an element. Table 2 shows the atomic structures of three isotopes of oxygen.

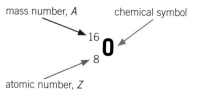

▲ Figure 6 *Isotope notation*

> **Study tip**
>
> To work out the number of neutrons in an atom, simply subtract the atomic number, Z from the mass number A:
>
> number of neutrons = $A - Z$

▼ Table 2 *Atomic structures for isotopes of oxygen*

Isotope	Protons, p^+	Neutrons, n	Electrons, e^-
$^{16}_{8}O$	8	8	8
$^{17}_{8}O$	8	9	8
$^{18}_{8}O$	8	10	8

Other ways of representing isotopes

Chemists refer to isotopes in different ways, so you may see an isotope of oxygen written as $^{16}_{8}O$, ^{16}O, or simply as oxygen-16. All oxygen atoms contain eight protons, so if the '8' is omitted, as in ^{16}O and oxygen-16, you still know how many protons the isotope contains.

Isotopes and chemical reactions

Chemical reactions involve the electrons surrounding the nucleus.

- Different isotopes of the same element have the same number of electrons.
- The number of neutrons has no effect on reactions of an element.
- Different isotopes of an element therefore react in the same way.

There may be small differences in physical properties – with higher-mass isotopes of an element having a higher melting point, boiling point, and density – but the chemical reactions are the same.

▲ Figure 7 *Both glasses contain water. The left-hand glass contains ice cubes made from heavy water. The right-hand glass contains normal ice. Solid D_2O is denser than liquid water and so D_2O ice cubes sink in water*

 Heavy water

You may have heard of heavy water, used to control processes in nuclear reactors. The H_2O molecules in normal water nearly all contain the $^{1}_{1}H$ isotope of hydrogen. In heavy water, all molecules of H_2O contain the $^{2}_{1}H$ isotope of

ATOMS, IONS, AND COMPOUNDS

hydrogen. The $^{2}_{1}H$ isotope is often referred to as deuterium and even given its own symbol, D. The formula for heavy water is often written simply as D_2O. The chemical properties of heavy water are almost identical to those of normal water. However, it has slightly different physical properties, shown in Table 3.

▼ **Table 3** *Properties of D_2O and H_2O*

Physical property	Normal water, H_2O	Heavy water, D_2O
melting point /°C	0.00	3.80
boiling point /°C	100.00	101.40
density / g cm^{-3}	1.00	1.11

The greater density of D_2O gives heavy water its name. If all water were heavy water, you would see ice far more often, as the water would freeze at a higher temperature.

> Tritium, T, is a third isotope of water containing two neutrons in the nucleus. Tritium forms an oxide called super-heavy water.
> a What is the relative mass of a molecule of (i) H_2O, ii D_2O, and (iii) T_2O?
> b Predict how the melting point, boiling point, and density of T_2O would be different from H_2O and D_2O.

Atomic structure of ions

An **ion** is a charged atom. The number of electrons is *different* from the number of protons.

- Positive ions, or **cations**, are atoms with *fewer* electrons than protons. Cations have an overall positive charge.
- Negative ions, or **anions**, are atoms with *more* electrons than protons. Anions have an overall negative charge.

Ions are always shown with their overall relative charge. Take the two ions $^{24}_{12}Mg^{2+}$ and $^{35}_{17}Cl^{-}$.

- Mg^{2+} has two fewer electrons than protons.
- Cl^- has one more electron than protons.

Table 4 shows the number of protons, neutrons, and electrons in the ions $^{24}_{12}Mg^{2+}$ and $^{35}_{17}Cl^{-}$.

▼ **Table 4** *Atomic structures of ions*

Ion	Protons	Neutrons	Electrons	Overall relative charge
$^{24}_{12}Mg^{2+}$	12	12	10	(12+) + (10−) = 2+
$^{35}_{17}Cl^{-}$	17	18	18	(17+) + (18−) = 1−

Ions and atoms of an element have the *same* number of protons but a *different* number of electrons.

Summary questions

1 State the number of protons, neutrons, and electrons in the following isotopes:
 a $^{12}_{6}C$ *(1 mark)*
 b $^{13}_{6}C$ *(1 mark)*
 c $^{14}_{6}C$ *(1 mark)*

2 Iron contains a mixture of four different isotopes: ^{54}Fe, ^{56}Fe, ^{57}Fe, ^{58}Fe.
 a State how these isotopes differ. *(1 mark)*
 b State the similarity between these isotopes. *(1 mark)*

3 State the number of protons, neutrons, and electrons in the following isotopes:
 a $^{15}_{7}N$ *(1 mark)*
 b $^{109}_{47}Ag$ *(1 mark)*
 c $^{207}_{82}Pb$ *(1 mark)*

4 State the number of protons, neutrons, and electrons in the following ions:
 a $^{41}_{19}K^+$ *(1 mark)*
 b $^{34}_{16}S^{2-}$ *(1 mark)*
 c $^{53}_{24}Cr^{3+}$ *(1 mark)*

5 State the difference in the number of protons, neutrons, or electrons of the following:
 a ^{6}Li and ^{7}Li *(1 mark)*
 b ^{18}O and $^{18}O^{2-}$ *(1 mark)*
 c $^{39}K^+$ and $^{40}Ca^{2+}$ *(1 mark)*

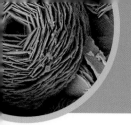

2.2 Relative mass

Specification reference: 2.1.1

Learning outcomes

Demonstrate knowledge, understanding, and application of:

→ relative isotopic mass and relative atomic mass

→ mass spectrometry.

This topic looks at how chemists use a mass system based on **relative mass** to compare the masses of atoms. Later you will see that this idea is extended to all chemicals.

Carbon-12

Table 1 shows the relative mass of protons, neutrons, and electrons.

To find the relative mass of an isotope, it might seem sensible to add together the relative masses of the protons, neutrons, and electrons – but things are not that simple. In fact, the strong nuclear force holding together protons and neutrons comes at the expense of the loss of a fraction of their mass. This astonishing fact was worked out by Albert Einstein over 100 years ago. The small amount of mass lost is called the mass defect. If you are studying A Level physics, you may learn more about the mass defect and its importance to the nucleus.

▼ **Table 1** *Masses of three subatomic particles*

Particle	Symbol	Relative mass
proton	p^+	1
neutron	n	1.001 375
electron	e^-	0.000 544

So how do chemists calculate the mass of atoms if some mass is lost to hold the nucleus together?

First, a standard isotope is needed on which to base all atomic masses. This role is taken by the carbon-12 isotope, which is the international standard for the measurement of atomic masses. One atom of carbon-12 has a mass of $1.992\,646\,538 \times 10^{-26}$ kg. Working in kg would be very awkward so instead a new unit called the atomic mass unit u is used.

- The mass of a carbon-12 isotope is defined as exactly 12 atomic mass units (12 u).
- The standard mass for atomic mass is 1 u, the mass of $\frac{1}{12}$th of an atom of carbon-12.
- On this scale, 1 u is approximately the mass of a proton or a neutron.

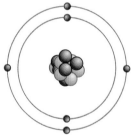

▲ **Figure 1** *Carbon-12 contains six protons (black), six neutrons (blue), and six electrons (red). An atom of carbon-12 has a mass of exactly 12 u*

Relative isotopic mass

Relative isotopic mass is the mass of an isotope relative to $\frac{1}{12}$th of the mass of an atom of carbon-12. Table 2 shows the relative isotopic mass of several isotopes relative to the mass of carbon-12. Relative isotopic mass has no units because it is a ratio of two masses.

Study tip

You need to learn the definition for relative isotopic mass.

▼ **Table 2** *Relative isotopic masses*

Isotope	Mass number	Relative isotopic mass	
		Accurate	One decimal place
^{12}C	12	12 exactly	12.0
^{14}N	14	14.003 074 005 29	14.0
^{16}O	16	15.994 914 635	16.0
^{19}F	19	18.998 403 221 5	19.0

ATOMS, IONS, AND COMPOUNDS

For A Level chemistry, you will be working with masses to one decimal place. In most cases, you can assume that the relative isotopic mass is the same as the mass number A of the isotope (number of protons and number of neutrons)

Relative atomic mass

Most elements contain a mixture of isotopes, each with a different relative isotopic mass. **Relative atomic mass** A_r is the weighted mean mass of an atom of an element relative to $\frac{1}{12}$th of the mass of an atom of carbon-12.

The weighted mean mass takes account of:

- the percentage abundance of each isotope
- the relative isotopic mass of each isotope.

In the periodic table, in addition to the atomic number, each element is shown with its relative atomic mass, A_r. Figure 2 shows six elements of the periodic table, together with their atomic numbers and relative atomic masses.

Determination of relative atomic mass

The percentage abundances of the isotopes in a sample of an element are found experimentally using a **mass spectrometer**.

Different types of mass spectrometer exist but all work to the same basic principle.

1. A sample is placed in the mass spectrometer.
2. The sample is vaporised and then ionised to form positive ions.
3. The ions are accelerated. Heavier ions move more slowly and are more difficult to deflect than lighter ions, so the ions of each isotope are separated.
4. The ions are detected on a mass spectrum as a mass-to-charge ratio m/z. Each ion reaching the detector adds to the signal, so the greater the abundance, the larger the signal.

$$\text{mass-to-charge ratio } \frac{m}{z} = \frac{\text{relative mass of ion}}{\text{relative charge on ion}}$$

For an ion with one positive charge, this ratio is equivalent to the relative isotopic mass, which is recorded on the x-axis of the spectrum.

Figure 4 shows part of the mass spectrum obtained from a sample of chlorine, with the percentage abundances for each isotope shown by each peak. The mass spectrum reveals two isotopes:

- 75.78% of chlorine-35
- 24.22% of chlorine-37.

> **Study tip**
> You may be provided with relative isotopic masses, but if not you can use the mass number (the sum of the numbers of protons and neutrons).

> **Study tip**
> You should learn the definition for relative atomic mass.

14	15	16
Si	**P**	**S**
silicon	phosphorus	sulfur
28.1	31.0	32.1
32	33	34
Ge	**As**	**Se**
germanium	arsenic	selenium
72.6	74.9	79.0

▲ **Figure 2** *Elements in the periodic table. The smaller number at the top is the atomic number Z, the larger number underneath is the relative atomic mass A_r*

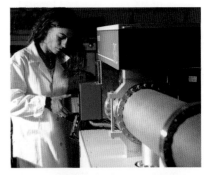

▲ **Figure 3** *A scientist using a mass spectrometer*

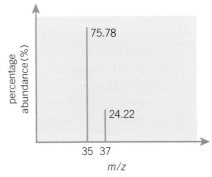

▲ **Figure 4** *Mass spectrum of chlorine*

2.2 Relative mass

You can work out the relative atomic mass using the method in the worked example below.

> **Worked example: Relative atomic mass of chlorine**
>
> As accurate relative isotopic masses have not been provided, the mass number for each isotope is used.
>
> $$\text{Relative atomic mass} = \frac{\overbrace{75.78 \times 35}^{\text{contribution from }^{35}\text{Cl}} + \overbrace{24.22 \times 37}^{\text{contribution from }^{37}\text{Cl}}}{100}$$
>
> Relative atomic mass = 35.4844 = 35.5 to one decimal place

Determination of relative isotopic mass

The mass spectrometer can also record the accurate m/z ratio for each isotope so that accurate values of relative isotopic mass can be measured.

> **Relative atomic masses – time for change?**
>
> Every two years, the International Union of Pure and Applied Chemistry (IUPAC) reviews values for relative atomic masses for use across the world. The review usually results in some very small adjustments, but the 2011 review made a more fundamental change. It has been known for many years that the isotopic abundances of an element may vary slightly depending on where the sample originates. In 2011, IUPAC published a new periodic table to take into account this variation by showing the relative atomic mass of some elements as a range rather than a single value. Figure 4 shows an extract from this periodic table. Compare the relative atomic masses of silicon and sulfur in Figure 2 and Figure 5. This change does not really affect the values used at A Level, but atomic masses shown to greater accuracy are affected, and future reviews may affect other elements.
>
> | 5
B
boron
[10.80, 10.83] | 6
C
carbon
[12.00, 12.02] | 7
N
nitrogen
[14.00, 14.01] | 8
O
oxygen
[15.99, 16.00] | 9
F
fluorine
19.00 |
> |---|---|---|---|---|
> | 13
Al
aluminium
26.98 | 14
Si
silicon
[28.08, 28.09] | 15
P
phosphorus
30.97 | 16
S
sulfur
[32.05, 32.08] | 17
Cl
chlorine
[35.44, 35.46] |
>
> ▲ **Figure 5** *Extract from 2011 IUPAC Periodic Table*
>
> Boron occurs naturally as a mixture of two isotopes, ^{10}B and ^{11}B. The relative isotopic masses are ^{10}B 10.00 and ^{11}B 11.00.
>
> Calculate the percentage abundances by mass of ^{10}B and ^{11}B in samples of boron with relative atomic masses of (a) 10.80 and (b) 10.83, the limits of the range in the new IUPAC Periodic Table.

Summary questions

1. Define the terms
   a. relative isotopic mass *(1 mark)*
   b. relative atomic mass *(1 mark)*

2. Calculate the relative atomic mass of the following elements. Give your answers to two decimal places.
   a. A sample of potassium consisting of 93.20% of ^{39}K, 0.07% of ^{40}K, and 6.73% of ^{41}K. *(1 mark)*
   b. A sample of antimony consisting of 56.87% of ^{121}Sb and 43.13% of ^{123}Sb. *(1 mark)*
   c. A sample of neon consisting of 91.07% of ^{20}Ne and 8.93% of ^{22}Ne. *(1 mark)*

3. The accurate relative isotopic masses for the isotopes chlorine-35 and chlorine-37 are 34.968 852 721 69 (^{35}Cl) and 36.965 902 621 1 (^{37}Cl).
   a. Calculate the relative atomic mass of chlorine, as shown in the worked example, using these accurate isotopic masses. *(2 marks)*
   b. Comment on whether the difference from use of mass numbers is significant. *(1 mark)*

2.3 Formulae and equations
Specification reference: 2.1.2

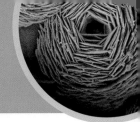

Ionic charges

To study chemistry successfully at any level, you need to be able to write chemical formulae and construct balanced chemical equations. In this topic, you will review how to write the formula of an ionic compound from ionic charges and how to balance chemical equations.

Simple ions from the periodic table

You should remember from GCSE that many atoms lose or gain electrons to achieve the same electron structure as the nearest noble gas, helium (He) to radon (Rn).

- Atoms of metals on the left of the periodic table *lose* electrons to form cations (positive ions)
- Atoms of non-metals on the right of the periodic table *gain* electrons to form anions (negative ions).

For many elements you can use the element's position in the periodic table to work out the likely charge on the ion, as shown in Figure 1.

Learning outcomes

Demonstrate knowledge, understanding, and application of:

→ writing formulae of ionic compounds
→ prediction of ionic charge from the periodic table
→ names and formulae of ions
→ chemical equations.

Synoptic link

Electron structure and ionic bonding will be developed later in Topic 5.1, Electron structure and Topic 5.2, Ionic bonding and structure.

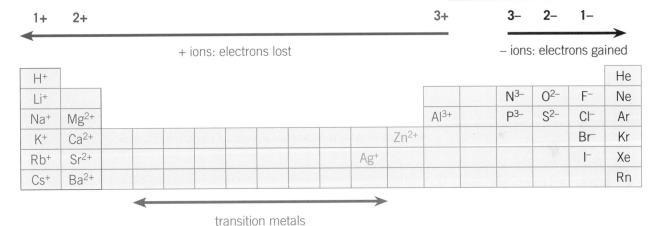

▲ **Figure 1** *The charges of some simple ions can often be deduced from their position in the periodic table*

Some metals, mostly transition metals (Figure 1), can form several ions with different charges. The ionic charge is then shown with a Roman numeral in the name of the ion. For example:

- Copper forms two ions – copper(I), Cu^+, and copper(II), Cu^{2+}.
- Iron forms two ions – iron(II), Fe^{2+}, and iron(III), Fe^{3+}.

Binary compounds

A **binary compound** contains *two* elements only.

- To name a binary compound, use the name of the first element but change the ending of the second element's name to -ide.
- For ionic compounds, the metal ion always comes first.

For example, sodium and oxygen form sodium ox**ide**.

Study tip

You are expected to know the charges on all the ions shown in Figure 1. Notice that Zn^{2+} and Ag^+, shown in green, do not fit into this pattern and you will need to learn these ionic charges.

Synoptic link

You will learn more about using Roman numerals in names in Topic 4.3, Redox.

15

2.3 Formulae and equations

▲ Figure 2 Cu^{2+} ions are responsible for the blue colour of copper(II) sulfate crystals, $CuSO_4$

Study tip

You are expected to know the names and formulae of the ions shown in blue in Table 1, but you will find it useful to learn them all. Be warned — there is no easy way to work out these formulae!

Polyatomic ions

Sometimes, an ion may contain atoms of more than one element bonded together. These ions are called **polyatomic ions**. Table 1 shows some common polyatomic ions and their names.

▼ Table 1 Common polyatomic ions and their names

1+	1−	2−	3−
ammonium NH_4^+	hydroxide OH^-	carbonate CO_3^{2-}	phosphate PO_4^{3-}
	nitrate NO_3^-	sulfate SO_4^{2-}	
	nitrite NO_2^-	sulfite SO_3^{2-}	
	hydrogencarbonate HCO_3^-	dichromate(VI) $Cr_2O_7^{2-}$	
	manganate(VII) (permanganate) MnO_4^-		

Writing formulae from ions

An ionic compound contains a cation and an anion. The formula can be worked out from the charge on each ion.

In a correct formula:

- the overall charge is zero so the ionic charges must balance
- sum of positive charges = sum of negative charges.

Worked example: Ionic formulae

Compound name	Ions present	Balance charges	Formula
zinc chloride	Zn^{2+} and Cl^-	1 Zn^{2+} ions balances **2** Cl^- ions	$ZnCl_2$
aluminium sulfate	Al^{3+} and SO_4^{2-}	**2** Al^{3+} ions balance **3** SO_4^{2-} ions	$Al_2(SO_4)_3$

The charges must balance, but it's just a matter of multiplication tables.

1 Zn^{2+} (1 × 2+ = 2+) is balanced by 2 Cl^- (2 × 1− = 2−)
2 Al^{3+} (2 × 3+ = 6+) is balanced by 3 SO_4^{2-} (3 × 2− = 6−)

Writing the formula

- The number of each ion present is shown as a subscript *after* the ion.
- The ionic charges are usually omitted in the completed formula.
- Brackets are used if there is more than one polyatomic ion.

Aluminium sulfate contains 2 Al^{3+} ions and 3 SO_4^{2-} ions and so the formula is $Al_2(SO_4)_3$.

ATOMS, IONS, AND COMPOUNDS

Writing equations

You will have practised balancing equations at GCSE. For A Level chemistry, you will come across many more equations, but you will find that balancing them quickly becomes second nature. By the end of the course you should be able to write equations for all the reactions you have studied and also for some unfamiliar reactions.

Representing elements and compounds in equations

Elements

In equations, elements are shown simply as their symbol *except* for the few elements that exist as small molecules. Most of these elements exist as **diatomic molecules**, containing two atoms bonded together – H_2, N_2, O_2, F_2, Cl_2, Br_2, and I_2. The only other elements that exist as small molecules are phosphorus, P_4, and sulfur, S_8. (However, it is normal practice to write sulfur simply as S in equations – otherwise every formula in the equation has to be multiplied up by a factor of 8.)

Compounds

Covalent compounds do not contain ions. Most covalent compounds exist as molecules with a small number of atoms bonded together, for example, CO_2 and H_2O. In equations, the formula of the molecule is used.

For ionic compounds, the formula worked out from the ionic charges is used in equations. This is called the **formula unit**.

State symbols in chemical equations

State symbols are shown in brackets after a formula to indicate the physical state. There are four state symbols:

- (g) – gas
- (l) – liquid
- (s) – solid
- (aq) – dissolved in water (aqueous)

Balancing equations

To balance an equation, you multiply each formula by a balancing number until the number of atoms of each element is the same on each side of the equation. Balancing numbers are written in front of each formula.

2 Na_2O means **two** Na_2O formula units giving **2 × 2 = 4** Na^+ and **2 × 1 = 2** O^{2-} ions.

- When balancing an equation you *must not* change any chemical formula.
- Balancing numbers go *in front of* chemical formulae and *on the line* (not subscripted).
- The equation is balanced when there are the same number of atoms of each element on each side of the equation.

> **Study tip**
>
> Remember to use brackets. Students often lose marks by omitting brackets when writing the formula of a hydroxide.
>
> Magnesium hydroxide contains Mg^{2+} and OH^- ions.
>
> $MgOH_2$ is incorrect as it means one oxygen and two hydrogen atoms! The correct formula is $Mg(OH)_2$.
>
> Brackets are added if there is more than one polyatomic ion in a formula.

> **Synoptic link**
>
> You will find out more about these different formulae in Topic 3.2, Determination of formulae.

> **Study tip**
>
> When balancing equations, it is a common mistake to change a formula.
>
> The formula for water is H_2O. Do NOT change the formula to H_2O_2, this is hydrogen peroxide!

2.3 Formulae and equations

> **Study tip**
> Aluminium oxide is an ionic compound, so you will need to use ionic charges.

> **Study tip**
> Aluminium is shown simply as its symbol, Al.
> Oxygen exists as diatomic molecules and so is shown as O_2.

> **Study tip**
> Start with the formulae of *compounds* and count the number of atoms of each element.

> **Study tip**
> You are allowed to use fractions when balancing equations. If you are comfortable using fractions, this can be easier than whole numbers. For the worked example, this would give
> $2Al + 1\frac{1}{2}O_2 \rightarrow Al_2O_3$

> **Study tip**
> Don't forget to add state symbols!

 Worked example: Constructing a balanced equation

Aluminium reacts with oxygen to form aluminium oxide.

Step 1: Work out the formulae.

Aluminium oxide contains Al^{3+} and O^{2-} ions, so the formula is Al_2O_3.

Step 2: Write an equation using formulae for all reactants and products.

Formulae only: $Al + O_2 \rightarrow Al_2O_3$

Step 3: Balance the equation by placing balancing numbers on the line, in front of formulae.

The key here is to get the oxygen atoms equal on both sides.

$4Al + 3O_2 \rightarrow 2Al_2O_3$

As a final check,

- Left-hand side has **4** Al and **3** × 2 (= 6) O
- Right-hand side has **2** × Al_2O_3 = **2** × 2 (= 4) Al and **2** × 3 (= 6) O

The equation is balanced.

Step 4: Finally, add state symbols to complete the equation.

$4Al(s) + 3O_2(g) \rightarrow 2Al_2O_3(s)$

Balancing formulae with brackets

Take care when balancing formulae with brackets. Remember that the balancing number multiplies the entire formula.

3 $Zn(NO_3)_2$ means **3** × $Zn(NO_3)_2$

This gives **3** Zn, **3** × 2 = 6 N, and **3** × 3 × 2 = 18 O

Summary questions

1. Write formulae for the following:
   a. potassium oxide *(1 mark)*
   b. magnesium iodide *(1 mark)*
   c. calcium phosphide *(1 mark)*
   d. iron(III) hydroxide *(1 mark)*
   e. ammonium carbonate *(1 mark)*
   f. iron(II) nitrate *(1 mark)*

2. Name each compound from the formula.
   a. AlN *(1 mark)*
   b. $(NH_4)_3PO_4$ *(1 mark)*
   c. $Fe_2(SO_4)_3$ *(1 mark)*

3. Balance the following equations:
   a. $NH_3(g) + O_2(g) \rightarrow NO(aq) + H_2O(l)$ *(1 mark)*
   b. $C_6H_{14}(l) + O_2(g) \rightarrow CO_2(g) + H_2O(l)$ *(1 mark)*
   c. $Al_2O_3(s) + H_3PO_4(aq) \rightarrow AlPO_4(aq) + H_2O(l)$ *(1 mark)*
   d. $Zn(s) + HNO_3(aq) \rightarrow Zn(NO_3)_2(aq) + NO(g) + H_2O(l)$ *(1 mark)*

4. Write balanced equations with state symbols for the following reactions:
   a. Magnesium reacts with solid phosphorus to form solid magnesium phosphide. *(2 marks)*
   b. Iron reacts with aqueous copper(II) nitrate to form copper and aqueous iron(III) nitrate. *(2 marks)*
   c. Lead(II) nitrate decomposes to form solid lead(II) oxide and two gases, nitrogen dioxide and oxygen. *(2 marks)*

ATOMS, IONS, AND COMPOUNDS

Practice questions

1. This question refers to species **A–D**. For each part, select the correct species.

   A $^{31}_{15}P$ **B** $^{19}_{9}F^-$ **C** $^{16}_{8}O^{2-}$ **D** $^{23}_{11}Na^+$

   a. The number of protons and neutrons are the same. *(1 mark)*
   b. The number of neutrons and electrons are the same. *(1 mark)*
   c. The number of protons, neutrons, and electrons are all different. *(1 mark)*

2. The answer to each part of this question is a number.

   a. How many neutrons are in an atom of zinc-68? *(1 mark)*
   b. What is the total number of electrons in a CO_3^{2-} ion? *(1 mark)*
   c. What is the total number of ions in one formula unit of chromium(III) sulfate? *(1 mark)*

3. This question looks at isotopes of three elements.

   a. An isotope of element **A** contains the same number of neutrons as are found in an atom of ^{51}V. The isotope of **A** also contains 26 protons.
      (i) How many protons in an atom of ^{51}V? *(1 mark)*
      (ii) Write the symbol, including the mass number and the atomic number, of this isotope of **A**. *(2 marks)*
   b. An isotope of element **B** has half as many protons and half as many neutrons as an atom of ^{48}Ti.
      Write the symbol, including the mass number and the atomic number, of this isotope of **B**. *(2 marks)*
   c. An isotope of element **C** has three more protons and four more neutrons than an atom of ^{81}Br.
      Write the symbol, including the mass number and the atomic number, of this isotope of **C**. *(2 marks)*

4. Neon exists as a mixture of isotopes.

   a. What is meant be the term *isotopes*? *(1 mark)*
   b. Define the term *relative isotopic mass*. *(2 marks)*
   c. A sample of gallium, $A_r = 69.7$, was analysed and was found to consist of 65% ^{69}Ga and one other isotope.
      Determine the mass number of the other isotope in the sample of gallium. *(1 mark)*
   d. Complete the table below for two ions that have the same number of electrons as a neon atom. *(2 marks)*

Species	Protons	Neutrons	Electrons	Charge
		9		2–
		14		3+

5. A sample of sulfur, $Z = 16$, was analysed in a mass spectrometer to give the following composition of isotopes.

Isotope	Abundance (%)
^{32}S	94.93
^{33}S	0.76
^{34}S	4.29
^{36}S	0.02

   From the results, the relative atomic mass of the sulfur sample can be calculated.

   a. Define the term *relative atomic mass*. *(3 marks)*
   b. Calculate the relative atomic mass of the sample of sulfur. Give your answer to **two** decimal places. *(2 marks)*
   c. How many of each sub-atomic particle are there in a ^{33}S atom and an $^{34}S^{2-}$ ion? *(1 mark)*

6. Write equations, with state symbols, for the following reactions.

   a. Magnesium reacts with nitrogen to form magnesium nitride. *(2 marks)*
   b. Calcium reacts with water to form a solution of calcium hydroxide and hydrogen. *(2 marks)*
   c. Sodium hydroxide solution reacts with iron(III) sulfate solution to form an iron(III) hydroxide precipitate and sodium sulfate solution. *(2 marks)*

3 AMOUNT OF SUBSTANCE

3.1 Amount of substance and the mole

Specification reference: 2.1.3

Learning outcomes

Demonstrate knowledge, understanding, and application of:

→ the amount of substance, the mole, and the Avogadro constant

→ molar mass

→ calculations involving masses and moles.

Synoptic link

You learnt about carbon-12 in Topic 2.1, Relative mass.

Study tip

The mass of 1 mole of atoms of an element equals the relative atomic mass in grams

▲ **Figure 1** *Molar quantities for chemical elements, clockwise from top left*
- *12.0 g carbon, C*
- *32.1 g sulfur, S*
- *55.8 g iron, Fe*
- *63.5 g copper, Cu*
- *24.3 g magnesium, Mg.*

Each sample contains the same number of atoms, but their masses are different

Counting and weighing atoms

Chemicals are usually measured by mass or volume. Because reactions take place on an atomic scale, chemists need a method for converting a measured mass or volume into the actual number of particles involved in reactions.

Amount of substance and the mole

Chemists use a quantity called **amount of substance** n to count the number of particles in a substance, measured in a unit called the **mole** *mol*. One mole is the amount of a substance that contains 6.02×10^{23} particles. The **Avogadro constant** N_A is $6.02 \times 10^{23}\,\text{mol}^{-1}$, the number of particles in each mole of carbon-12.

The choice of 6.02×10^{23} particles per mole may seem strange, but is directly linked to the mass of carbon-12, the standard for the measurement of relative atomic masses. 12 g of carbon-12 contains 6.02×10^{23} atoms. You can easily find the mass of one mole (1 mol) of atoms of any element – it is the relative atomic mass in grams.

- One mole of carbon, C, atoms has a mass of 12.0 g
- One mole of hydrogen, H, atoms has a mass of 1.0 g
- One mole of magnesium, Mg, atoms has a mass of 24.3 g
- One mole of iron, Fe, atoms has a mass of 55.8 g

So if you have a sample of an element and know its mass, you now have a way of knowing the number of atoms. This is a very important idea for chemistry – it offers an easy way of counting something that cannot be seen, just by measuring the mass.

The Avogadro constant – amazingly large

It is difficult for us to comprehend the size of very large numbers. If 6.02×10^{23} (1 mol) pennies were shared evenly between all humans on Earth, every person could spend £1m every hour for their whole life.

Particles matter

Amount of substance and moles can refer to anything, not just atoms. When you work in moles, it is important to use the formula or unambiguous name of a substance for clarity.

- 1 mol of H: 1 mol of hydrogen atoms
- 1 mol of H_2: 1 mol of hydrogen molecules

AMOUNT OF SUBSTANCE 3

Molar mass

Molar mass, M, gives a convenient way of linking moles with mass for *any* chemical substance.

- $M(C)$ = 12.0 g mol^{-1}.
- $M(NO_2)$ = 14.0 + 16.0 × 2 = 46.0 g mol^{-1}.
- $M(Na_2CO_3)$ = 23.0 × 2 + 12.0 + 16.0 × 3 = 106.0 g mol^{-1}.

Molar mass gives the mass in grams in each mole of the substance.

- Molar mass is the mass per mole of a substance.
- The units of molar mass are g mol^{-1}.

Amount of substance n, mass m, and molar mass M are linked by the equation below.

$$\text{amount } n = \frac{\text{mass } m}{\text{molar mass } M} \quad \text{or more simply,} \quad n = \frac{m}{M}$$

$n = \frac{m}{M}$ is a key equation for working out n, m, or M.

Worked example: Amount of substance, mass, and molar mass

1 Calculate the amount of substance, in moles, in 96.0 g of carbon, C.

$$n = \frac{m}{M} = \frac{96.0}{12.0} = 8.0 \text{ mol}$$

2 Calculate the mass, in g, of 0.050 mol of NO_2.

rearrange $n = \frac{m}{M}$ to $m = n \times M = 0.050 \times 46.0 = 2.3$ g

3 Calculate the molar mass when 2.65 g contains 0.025 mol of a substance.

$n = \frac{m}{M}$ rearrange: $M = \frac{m}{n} = \frac{2.65}{0.025} = 106.0$ g mol^{-1}

Study tip

Learn $n = \frac{m}{M}$ and make sure that you are comfortable rearranging the equation so that you can work out any of n, m, and M from the other two. You may find this format useful to remember:

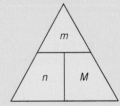

If you cover n, you are left with $\frac{m}{M}$. If you cover m, you are left with $n \times M$. If you cover M, you are left with $\frac{m}{n}$.

Summary questions

1 Calculate the amount of substance, in mol, in the following. Use relative atomic masses to one decimal place.
   a 6.00 g HF (*1 mark*)
   b 220 g N_2O (*1 mark*)
   c 1.14 g Cr_2O_3 (*1 mark*)
   d 0.0150 g $C_6H_{12}O_6$ (*1 mark*)
   e 3.45 × 10^{-2} g $Ca(OH)_2$ (*1 mark*)

2 Calculate the mass, in g, of the following. Use relative atomic masses to one decimal place.
   a 280 mol BeO (*1 mark*)
   b 0.150 mol HNO_3 (*1 mark*)
   c 0.0500 mol H_3PO_4 (*1 mark*)
   d 1.25 × 10^{-2} mol Na_2CO_3 (*1 mark*)
   e 4.55 × 10^{-3} mol $Ca(NO_3)_2$ (*1 mark*)

3 Calculate the molar mass, in g mol^{-1}, of the following substances. Use relative atomic masses to one decimal place.
   a 5.00 mol **A** has a mass of 140 g (*1 mark*)
   b 0.125 mol **B** has a mass of 9.25 g (*1 mark*)
   c 4.50 × 10^{-2} mol **C** has a mass of 3.825 (*1 mark*)

3.2 Determination of formulae

Specification reference: 2.1.1, 2.1.3

Learning outcomes

Demonstrate knowledge, understanding, and application of:
→ empirical and molecular formula
→ formula determination
→ hydrated salts
→ practical techniques for measuring mass.

▼ **Table 1** *Examples of compounds with their empirical and molecular formulae*

Molecular formula	Empirical formula
N_2O_4 ⟶	NO_2
H_2O ⟶	H_2O
CO_2 ⟶	CO_2
C_2H_6 ⟶	CH_3
P_4O_6 ⟶	P_2O_3
$C_9H_{12}O_3$ ⟶	C_3H_4O

▲ **Figure 1** *A diamond crystal and the giant structure of carbon atoms in a diamond*

Synoptic link

You learnt about relative isotopic mass, relative atomic mass, the important role of carbon-12 in atomic mass measurements in Topic 2.2, Relative atomic mass.

Chemical formulae

In this topic, you will see how to use the results from chemical experiments to work out a chemical formula.

Molecular formulae

Some compounds are made up of small units called molecules — two or more atoms held together by covalent bonds. The **molecular formula** is the number of atoms of each element in a molecule.

In Topic 2.3, Formulae and equations, you looked at the elements that exist as molecules. In equations, these elements are shown as their molecular formulae – H_2, N_2, O_2, F_2, Cl_2, Br_2, I_2, P_4, and S_8.

Many compounds also exist as molecules, and again the molecular formula is used in equations.

Empirical formulae

The **empirical formula** is the simplest whole-number ratio of atoms of each element in a compound.

The empirical formula is important for substances that do not exist as molecules. This includes metals, some non-metals (e.g., carbon, C, and silicon, Si), and ionic compounds (e.g., sodium chloride, NaCl).

These substances form giant crystalline structures of atoms or ions.

It would be impossible to base a formula on the actual number of atoms or ions — the numbers would go into billions of billions and would vary depending on the size of the crystals.

The empirical formula is the *ratio* of atoms or ions in the structure and will *always* be the same.

Figures 1 and 2 show giant crystalline structures of carbon and sodium chloride. You will find out more about these structures later in the course.

More relative masses

Some compounds exist as simple molecules (e.g., water, H_2O, and carbon dioxide, CO_2). Other compounds exist as giant crystalline structures (e.g., all ionic compounds). Two terms are needed for relative mass, one for simple molecules and another for giant structures.

Relative molecular mass

Relative molecular mass M_r compares the mass of a molecule with the mass of an atom of carbon-12. You can easily calculate a relative molecular mass by adding together the relative atomic masses of the elements making up a molecule. The examples below show how to work out the relative molecular mass of molecules of water, H_2O, carbon dioxide, CO_2, and glucose, $C_6H_{12}O_6$.

AMOUNT OF SUBSTANCE 3

- $M_r(H_2O) = (1.0 \times 2) + 16.0 = 18.0$
- $M_r(CH_4) = 12.0 + (1.0 \times 4) = 16.0$
- $M_r(C_6H_{12}O_6) = (12.0 \times 6) + (1.0 \times 12) + (16.0 \times 6) = 180.0$

Relative formula mass

Relative formula mass compares the mass of a formula unit with the mass of an atom of carbon-12. It is calculated by adding together the relative atomic masses of the elements in the empirical formula, as shown in the examples below.

- NaCl = 23.0 + 35.5 = 58.5
- $Ca(NO_3)_2$ = 40.1 + (14.0 + 16.0 × 3) × 2 = 164.1

Finding formulae by experiment

You can predict the formula of an ionic compound from its ions, but if you do not know which ions are in a compound, the formula can be worked out from the results of experiments. Investigating the chemical composition of a substance is called **analysis**.

These worked examples show two common ways to calculate empirical and molecular formulae from experimental mass readings. Notice the central role of the mole in these calculations.

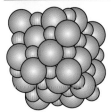

▲ **Figure 2** *Rock salt (sodium chloride) crystals and the giant structure with one sodium ion for every chloride ion*

Worked example: Empirical formula from mass

In an experiment, 1.203 g of calcium combines with 2.13 g of chlorine to form a compound [A_r: Ca, 40.1; Cl, 35.5].

Step 1: Convert mass into moles of atoms using $n = \frac{m}{M}$

$n(Ca) = \frac{1.203}{40.1} = 0.030 \text{ mol} \qquad n(Cl) = \frac{2.13}{35.5} = 0.060 \text{ mol}$

Step 2: To find the smallest whole-number ratio, divide by the smallest whole number.

$n(Ca) : n(Cl) = \frac{0.030}{0.030} : \frac{0.060}{0.030} = 1 : 2$

Step 3: Write the empirical formula: $CaCl_2$

Study tip

Students sometimes wrongly round ratios of moles that appear quite close to a whole-number ratio. If you calculate a ratio of 1 : 1.67, don't be tempted to round it up to 1 : 2. Instead see whether you can convert the ratio into a whole-number ratio by multiplying both sides by the same factor, in this case 3 to give 3 : 5.

Worked example: Determination of a molecular formula

Chemical analysis of a compound gave the percentage composition by mass C: 40.00%; H: 6.67%; O: 53.33% [A_r: C, 12.0; H, 1.0; O, 16.0]. The relative molecular mass of the compound is 180.0.

Step 1: Convert % by mass into moles of atoms using $n = \frac{m}{M}$

$n(C) = \frac{40.00}{12.0} = 3.33 \text{ mol} \qquad n(H) = \frac{6.67}{1.0} = 6.67 \text{ mol} \qquad n(O) = \frac{53.33}{16.0} = 3.33 \text{ mol}$

Step 2: Find smallest whole-number ratio and empirical formula.

$n(C) : n(H) : n(O) = \frac{3.33}{3.33} : \frac{6.67}{3.33} : \frac{3.33}{3.33} = 1 : 2 : 1 \qquad$ empirical formula = CH_2O

Step 3: Write the relative mass of the empirical formula CH_2O: 12.0 + 1.0 × 2 + 16.0 = 30.0

Step 4: Find number of CH_2O units in one molecule: $\frac{180}{30.0} = 6$

Step 5: Write the molecular formula: $CH_2O \times 6 = C_6H_{12}O_6$

3.2 Determination of formulae

Hydrated salts

Many coloured crystals are **hydrated** – water molecules are part of their crystalline structure. This water is known as **water of crystallisation**. When blue crystals of hydrated copper(II) sulfate are heated, bonds holding the water within the crystal are broken and the water is driven off, leaving behind white **anhydrous** copper(II) sulfate. The equation below represents the change when water is removed. The water of crystallisation is shown in the formula of hydrated copper(II) sulfate with a large dot • between the compound formula and the five water units.

$$CuSO_4 \cdot 5H_2O(s) \rightarrow CuSO_4(s) + 5H_2O(l)$$
$$\text{hydrated} \qquad\qquad \text{anhydrous}$$

Without water, the crystalline structure is lost and a white powder remains. It is difficult to remove the last traces of water, as you can see from the very pale blue colour of the anhydrous copper(II) sulfate in Figure 4.

▲ Figure 4 Hydrated (blue) and anhydrous (white) copper(II) sulfate

Formula of a hydrated salt

The method below describes how you could carry out an experiment to determine the water of crystallisation in hydrated crystals. The calculation is similar to the method described for an empirical formula.

The experiment uses hydrated copper(II) sulfate but the method would be suitable for any hydrated salt.

Step 1: Weigh an empty crucible.

Step 2: Add the hydrated salt into the weighed crucible. Weigh the crucible and the hydrated salt.

Step 3: Using a pipe-clay triangle, support the crucible containing the hydrated salt on a tripod (Figure 5). Heat the crucible and contents gently for about one minute. Then heat it strongly for a further three minutes.

Step 4: Leave the crucible to cool. Then weigh the crucible and anhydrous salt.

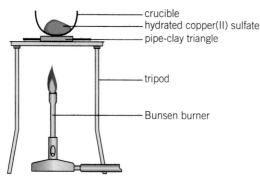

▲ Figure 5 Apparatus for heating crucible

The results are shown below.

mass of crucible / g	18.742	← Reading A
mass of crucible + hydrated salt / g	28.726	← Reading B
mass of crucible + anhydrous salt / g	25.126	← Reading C

Step 1: Calculate the amount, in mol, of anhydrous $CuSO_4$.
mass m of $CuSO_4$ formed = C – A = 25.126 – 18.742 = 6.384 g

$$n(CuSO_4) = \frac{m}{M} = \frac{6.384}{159.6} = 0.0400 \text{ mol}$$

Step 2: Calculate the mass and amount, in mol, of water.
mass m of H_2O formed = B – C = 28.726 – 25.126 = 3.600 g

$$n(H_2O) = \frac{3.600}{18.0} = 0.200 \text{ mol}$$

Step 3: Find the smallest whole-number ratio.

$$n(CuSO_4) : n(H_2O) = 0.0400 : 0.200 = 1 : 5$$

Step 4: Write down the value of x and the formula of hydrated copper sulfate.

$x = 5$ so the formula is $CuSO_4 \cdot 5H_2O$

Use the student results below to determine the value of x and the formula of $CoCl_2 \cdot xH_2O$.

mass of crucible / g	17.265
mass of crucible + hydrated salt / g	18.438
mass of crucible + anhydrous salt / g	17.906

24

AMOUNT OF SUBSTANCE 3

How accurate is an experimental formula?

The application above gives a perfect formula for hydrated copper(II) sulfate. Some assumptions have been made, and real experiments may not always work out as well.

Assumption 1 – All of the water has been lost

If the hydrated and anhydrous forms have different colours, you can be fairly sure when all water has been removed. However you only see the surface of the crystals and some water could be left inside. If the hydrated and anhydrous forms are similar colours, it is not as easy. A good solution is to heat to constant mass – the crystals are reheated repeatedly until the mass of the residue no longer changes, suggesting that all water has been removed.

Assumption 2 – No further decomposition

Many salts decompose further when heated; for example, if heated very strongly, copper(II) sulfate decomposes to form black copper(II) oxide. This can be very difficult to judge if there is no colour change.

Synoptic link

As part of the practical skills required for your course, you need to know how to measure mass, volumes of solutions, and volumes of gases.

This practical application box tells you how to measure masses.

Measuring mass, volumes of solutions, and volumes of gases are also covered in:

- Topic 3.4, Reacting quantities, for how to measure the volumes of gases
- Topic 4.2, Acid–base titrations, for how to measure volumes of solutions
- Topic 9.2, Measuring enthalpy changes, for how to measure mass
- Topic 10.1, Reaction rates, for how to measure volumes of gases.

Summary questions

1. **a** Determine the relative molecular mass of the following: *(3 marks)*
   i SO_2
   ii P_4O_{10}
   iii $HClO_4$
   b Determine the relative formula mass of the following: *(3 marks)*
   i $MgBr_2$
   ii NH_4NO_3
   iii $Al_2(SO_4)_3$

2. A nickel compound has the formula $Ni(NO_3)_2 \cdot xH_2O$ and a molar mass of 290.7 g mol^{-1}. Calculate the value of x. *(3 marks)*

3. Determine all possible molecular formulae and M_r values less than 120 for the empirical formula CH_2. *(1 mark)*

4. A student carried out an experiment to calculate the water of crystallisation of two hydrated salts. For the first salt, the student did not remove all the water of crystallisation. For the second salt, the student removed all the water of crystallisation, but unfortunately the salt decomposed further.

   For each salt, explain whether the student's calculated value of x, the number of water molecules in the formula unit, would be greater or smaller than the actual value of x in data books. *(4 marks)*

3.3 Moles and volumes

Specification reference: 2.1.3

Learning outcomes

Demonstrate knowledge, understanding, and application of:

→ concentration, solution volumes, and the mole
→ molar gas volumes
→ the ideal gas equation.

Using volume for measuring amount of substance

Liquids and gases are measured by volume. As with mass, the volume of a solution or a gas can be converted into amount of substance, in moles, giving us a way to count the particles present.

The volume measurements commonly used in chemistry are:

- the cubic centimetre (cm^3) or millilitre (ml): $1\,cm^3 = 1\,ml$
- the cubic decimetre (dm^3) or litre (l):
  $1\,dm^3 = 1000\,cm^3 = 1000\,ml = 1$ litre, 1 l.

You will be expected to use cm^3 and dm^3. In practical work, you will use glassware graduated in ml and l, and you should record these readings in cm^3 and dm^3.

Moles and solutions

To work out the amount, in moles, of a measured **volume** of the solution, you need to know the **concentration** ($mol\,dm^{-3}$) of the solute (the dissolved compound). The concentration of a solution is the amount of solute, in moles, dissolved in each $1\,dm^3$ ($1000\,cm^3$) of solution.

A $1\,mol\,dm^{-3}$ solution contains 1 mol of solute dissolved in each $1\,dm^3$ of solution.

▲ Figure 1 *Laboratory glassware*

Converting between moles and solution volumes

For a solution, amount n (mol) and volume V (dm^3) are linked by the concentration c ($mol\,dm^{-3}$).

$$n = c \times V$$

You will usually measure volumes in cm^3 so you will need to convert into dm^3 by dividing by 1000. The equation then becomes:

$$n = c \times \frac{V\,(cm^3)}{1000}$$

Study tip

You will use these equations many times, so you will need to learn them.

 Worked example: Converting between solution volumes and moles

1 Calculate the amount of NaCl, in mol, in $30.0\,cm^3$ of a $2.00\,mol\,dm^{-3}$ solution.

$$n(NaCl) = c \times \frac{V\,(cm^3)}{1000} = 2.00 \times \frac{30.0}{1000} = 0.0600\,mol$$

2 Calculate the volume of a $0.160\,mol\,dm^{-3}$ solution that contains 3.25×10^{-3} mol of NaCl.

$$n = c \times \frac{V\,(cm^3)}{1000} \quad so,\ V = \frac{1000 \times n}{c} = \frac{1000 \times 3.25 \times 10^{-3}}{0.160}$$
$$= 20.3\,cm^3$$

Standard solutions

A **standard solution** is a solution of known concentration. In practical work, you will have seen bottles of standard solutions labelled with their concentration, often as $1\,\text{mol}\,\text{dm}^{-3}$.

Standard solutions are prepared by dissolving an exact mass of the solute in a solvent and making up the solution to an exact volume. Using your understanding of the mole, you can work out the mass required to prepare a standard solution.

> **Synoptic link**
>
> Standard solutions and their preparation are discussed in detail in Topic 4.2, Acid–base titrations.

Worked example: Standard solutions

Calculate the mass of Na_2CO_3 required to prepare $100\,\text{cm}^3$ of a $0.250\,\text{mol}\,\text{dm}^{-3}$ standard solution.

Step 1: First work out the amount in moles required.

$$n(Na_2CO_3) = c \times \frac{V\,(\text{in cm}^3)}{1000} = 0.250 \times \frac{100}{1000} = 0.0250\,\text{mol}$$

Step 2: Then work out the molar mass of Na_2CO_3.

$$M(Na_2CO_3) = 23.0 \times 2 + 12.0 + 16.0 \times 3 = 106.0\,\text{g}\,\text{mol}^{-1}$$

Step 3: Rearrange $n = \frac{m}{M}$ to calculate the mass of Na_2CO_3 required.

$$m = n \times M = 0.0250 \times 106.0 = 2.65\,\text{g}$$

Other ways of showing concentrations

You will often see mass concentrations with units of $\text{g}\,\text{dm}^{-3}$. For the solution of Na_2CO_3 in the worked example above, the concentration is $0.250\,\text{mol}\,\text{dm}^{-3}$. To work out the mass concentration, you need to convert between moles and grams.

- $n = \frac{m}{M}$ so, $m = n \times M = 0.250 \times 106.0 = 26.50\,\text{g}$
- mass concentration of $Na_2CO_3 = 26.5\,\text{g}\,\text{dm}^{-3}$.

Moles and gas volumes

In Topic 3.1, you saw how to convert between mass in grams and amount of substance in moles. It is difficult to measure the mass of a gas but easy to measure gas volumes.

At the same temperature and pressure, equal volumes of different gases contain the same number of molecules.

So when you measure a gas volume, you are indirectly counting the number of gas molecules (or the amount of gas molecules in moles).

Molar volume

The **molar gas volume** V_m is the volume per mole of gas molecules at a stated temperature and pressure.

The volume of a gas depends on the pressure and temperature, but many experiments are carried out at room temperature and pressure (RTP).

3.3 Moles and volumes

- RTP is about 20 °C and 101 kPa (1 atm) pressure
- At RTP, 1 mole of gas molecules has a volume of approximately 24.0 dm³ = 24 000 cm³.
- Therefore, at RTP, the molar gas volume = 24.0 dm³ mol⁻¹.

Converting between amount in moles and gas volumes

Using the following equation, you can convert between the amount in moles of a gas, and the volume of the gas, V.

$$\text{amount } n \text{ (mol)} = \frac{\text{volume } V}{\text{molar gas volume } V_m}$$

At RTP, $V_m = 24.0 \text{ dm}^3\text{mol}^{-1}$, so

- when V is in dm^3 $\quad n = \dfrac{V(\text{dm}^3)}{24.0}$
- when V is in cm^3 $\quad n = \dfrac{V(\text{cm}^3)}{24\,000}$

▲ Figure 2 *A helium weather balloon is released by a meteorologist. A sensor attached to the balloon will measure ozone distribution in and beyond the ozone layer, which is 20–30 km above the Earth's surface. So helium balloons are far less dense than air and they are able to rise up to the ozone layer*

 Worked example: Converting between gas volume and amount in moles

1 Calculate the amount (mol) of hydrogen, $H_2(g)$, in 480 cm³ at RTP.

$$n = \frac{V(\text{in cm}^3)}{24\,000} = \frac{480}{24\,000} = 0.0200 \text{ mol of } H_2$$

2 Calculate the volume, in dm³, of 0.150 mol $O_2(g)$ at RTP.
Rearrange to give $V(\text{dm}^3) = n \times 24.0$
so, $V = 0.150 \times 24.0 = 3.60 \text{ dm}^3$

The ideal gas equation

Room temperature and pressure will always be approximate, chosen to match the typical conditions that experiments are carried out in. So what do you do when carrying out experiments where the gases are at different temperatures or pressures, or if you need to be more accurate? The ideal gas equation provides a solution.

You will have come across the following assumptions for the molecules making up an ideal gas:

- random motion
- elastic collisions
- negligible size
- no intermolecular forces.

The ideal gas equation is shown below.

$$pV = nRT$$

The ideal gas constant R is a $pV = nRT$ constant and always has the same value of 8.314 J mol⁻¹ K⁻¹. Temperature is in units of K (Kelvin), which starts at absolute zero (−273 °C). Each 1 K rise in temperature is the same as a 1 °C rise in temperature.

AMOUNT OF SUBSTANCE   3

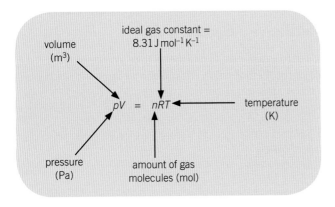

Synoptic link
If you have not met the Kelvin scale of temperature before, you can find out more detail in Topic 9.2, Measuring enthalpy changes.

As long as you know three of p, V, n, and T, you can always find out the unknown variable using the ideal gas equation.

Before using the ideal gas equation, you need to convert any quantities into the correct units Pa, K, and m^3. The conversions from measurements likely to be made when carrying out experiments are shown below.

- cm^3 to m^3 $\times 10^{-6}$
- dm^3 to m^3 $\times 10^{-3}$
- °C to K   + 273
- kPa to Pa $\times 10^3$

> **Worked example: What is room temperature and pressure?**
>
> You can use the ideal gas equation to find out the conditions that give a molar gas volume of $24.0\,dm^3\,mol^{-1}$. Assume that the pressure is 1 atm = 101 kPa and use this to calculate room temperature.
>
> **Step 1:** Convert all quantities to match the ideal gas equation.
>
> $p = 101\,kPa = 101 \times 10^3\,Pa$
> $V = 24.0\,dm^3 = 24.0 \times 10^{-3}\,m^3$
> $n = 1\,mol$
> $T = $ unknown
>
> **Step 2:** Use the ideal gas equation to calculate the unknown.
>
> $pV = nRT$ rearranges to, $T = \dfrac{pV}{nR}$
>
> $T = \dfrac{(101 \times 10^3) \times (24.0 \times 10^{-3})}{1 \times 8.314} = 292\,K = 19\,°C$

Study tip
The hardest part of calculations using the ideal gas equation, $pV = nRT$, is making sure that you are working in units of Pa, m^3, and K. Learn the conversion rules.

Many people make the incorrect assumption that room temperature is 25 °C (298 K), the temperature often regarded as standard temperature for chemistry. Using the ideal gas equation, you can show that the molar gas volume at 25 °C and an atmospheric pressure of 101 kPa is actually equal to $24.5\,dm^3\,mol^{-1}$.

3.3 Moles and volumes

Finding a relative molecular mass

You can use the ideal gas equation to find the relative molecular mass of a volatile liquid. Using the method below, the unknown compound would need to be a liquid at room temperature but have a boiling point below 100 °C so that it vaporises.

1. Add a sample of the volatile liquid to a small syringe via a needle. Weigh the small syringe.

2. Inject the sample into a gas syringe through the self-sealing rubber cap (Figure 3). Reweigh the small syringe to find the mass of the volatile liquid added to the gas syringe.

3. Place the gas syringe in a boiling water bath at 100 °C, as shown in Figure 3. The liquid vaporises producing a gas. The pressure is recorded.

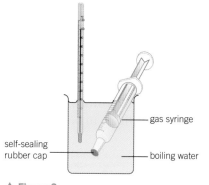

▲ Figure 3

Results

mass of volatile liquid = 0.2245 g
volume of gas in gas syringe = 81.0 cm³
atmospheric pressure = 100 kPa

Follow the steps below to calculate the relative molecular mass of the volatile liquid.

Step 1: Convert all quantities to match the ideal gas equation.

$V = 81.0 \text{ cm}^3 = 81.0 \times 10^{-6} \text{ m}^3$

$T = 100 \text{ °C} = 100 + 273 \text{ K} = 373 \text{ K}$

$p = 100 \text{ kPa} = 100 \times 10^3 \text{ Pa}$

$n = \text{unknown}$

Step 2: Use the ideal gas equation to calculate the unknown.

$pV = nRT$ rearranges to, $n = \dfrac{pV}{RT}$

$n = \dfrac{(100 \times 10^3) \times (81.0 \times 10^{-6})}{8.314 \times 373} = 0.00261 \text{ mol of X}$

Step 3: Find the molar mass.

$n = \dfrac{\text{mass } m}{\text{molar mass } M}$

$M = \dfrac{m}{n} = \dfrac{0.2245}{0.00261} = 86.0 \text{ g mol}^{-1}$

relative molecular mass, $M_r = 86.0$

A 0.320 g sample of a volatile liquid was heated until it vaporised. The resulting vapour then occupied 61.5 cm³ at 101 kPa and 100 °C.

Calculate the relative molecular mass of the volatile liquid.

AMOUNT OF SUBSTANCE 3

> ### ➕ Real gases
>
> The ideal gas equation relies on two key assumptions:
> - forces between molecules are negligible
> - gas molecules have negligible size compared to the size of their container.
>
> These assumptions hold at low pressures and high temperatures when the gas molecules are far apart and moving fast.
>
> When gas molecules are close together, the volume of the molecules compared with the volume of the container starts to become significant. Also if gas molecules move comparatively slowly, they have less energy and intermolecular forces may become significant.
>
> Scientists have developed several improvements to the ideal gas equation for real gases. In the real gas equation, corrections have been made to take into account the volume of gas molecules and intermolecular forces.
>
> real gas equation: $\left(p + \dfrac{n^2 a}{V^2}\right)(V - nb) = nRT$
>
> $\dfrac{n^2 a}{V^2}$ accounts for intermolecular forces
>
> nb accounts for volume of gas molecules
>
> Predict the conditions of pressure and temperature that cause the ideal gas equation to break down. Explain your answer.

Summary questions

1. Calculate the amount of substance, in moles, in:
   a. 250 cm³ of a 1.00 mol dm⁻³ solution *(1 mark)*
   b. 10.0 cm³ of a 0.200 mol dm⁻³ solution. *(2 marks)*

2. Calculate the concentration, in g dm⁻³, for:
   a. 2.00 mol of NaOH in 4.00 dm³ of solution *(2 marks)*
   b. 0.500 mol of HNO_3 in 200 cm³ of solution. *(2 marks)*

3. Calculate the amount of substance, in mol, in the following gas volumes at RTP:
   a. 1440 dm³ O_2(g) *(2 marks)*
   b. 720 cm³ He(g) *(2 marks)*
   c. 34.0 cm³ H_2(g) *(2 marks)*

4. Calculate the volume (dm³) of one mole of a gas at:
   a. 10 °C and 100 kPa *(2 marks)*
   b. 35 °C and 92.0 kPa *(2 marks)*

5. Calculate the volume in cm³ at RTP of:
   a. 0.136 g NH_3(g) *(2 marks)*
   b. 0.088 g CO_2(g) *(2 marks)*
   c. 0.0175 g N_2(g) *(2 marks)*

6. 0.1565 g of X occupies 80.0 cm³ at 101 kPa and 100 °C. Calculate the relative molecular mass of X. *(3 marks)*

3.4 Reacting quantities

Specification reference: 2.1.3

Learning outcomes

Demonstrate knowledge, understanding, and application of:

→ stoichiometry
→ quantities of reactants and products from equations
→ percentage yield
→ atom economy
→ practical techniques for measuring the volume of a gas.

Stoichiometry

In a balanced equation, the balancing numbers give the ratio of the amount, in moles, of each substance. This ratio is called the **stoichiometry** of the reaction.

equation	$2H_2(g)$	+	$O_2(g)$	→	$2H_2O(l)$
amount	2 mol		1 mol	→	2 mol

Chemists use balanced equations to find:

- the quantities of reactants required to prepare a requried quantity of a product
- the quantities of products that should be formed from certain quantities of reactants.

These quantities can then be changed to adjust the scale of a preparation.

Quantities from amounts and equations

The two worked examples show how unknown information about a substance can be obtained using amounts and an equation together. Each example follows the same basic method:

- **Step 1:** Work out the amount in moles of whatever you can.
- **Step 2:** Use the equation to work out the amount in moles of the unknown chemical.
- **Step 3:** Work out the unknown information required.

Study tip

For most problems, steps 1 and 2 will follow the same method. The processing for step 3 can vary.

> **Worked example: Reacting masses**
>
> Calculate the mass of aluminium oxide, Al_2O_3, formed when 8.10 g of aluminium completely reacts with oxygen.
>
> **Step 1:** Calculate the amount, in moles, of Al that reacts.
>
> $n(Al) = \dfrac{m}{M} = \dfrac{8.10}{27.0} = \mathbf{0.300\,mol}$
>
> **Step 2:** Use the equation to find the amount of Al_2O_3, in moles, that forms.
>
> | equation | 4 Al(s) | + | 3 O_2(g) | → | 2 Al_2O_3(s) |
> |---|---|---|---|---|---|
> | moles | 4 mol | + | 3 mol | → | 2 mol |
> | amounts | 0.300 mol | | | → | 0.150 mol |
>
> **Step 3:** Calculate the mass of Al_2O_3 formed.
>
> $n(Al_2O_3) = \dfrac{m}{M}$ so, $m = n \times M = 0.150 \times (27.0 \times 2 + 16.0 \times 3)$
> $= 0.150 \times 102.0 = \mathbf{15.3\,g}$

Study tip

You need the ratios:

4 mol Al → 2 mol Al_2O_3

Then halve the moles of Al to get the moles of Al_2O_3:

0.300 mol Al → 0.150 mol Al_2O_3

AMOUNT OF SUBSTANCE 3

 Worked example: Reacting mass, gas volumes, and concentration

0.552 g of lithium reacts with water to form 125 cm³ of a solution of lithium hydroxide and hydrogen gas. Calculate the concentration of the lithium hydroxide and the volume of hydrogen formed at room temperature and pressure (RTP).

Step 1: Calculate the amount, in mol, of lithium that reacts.

$$n(Li) = \frac{m}{M} = \frac{0.552}{6.9} = 0.0800 \text{ mol}$$

Step 2: Use the equation to find the amounts of LiOH and H_2 formed.

equation	$2\,Li(s)$	+	$2\,H_2O(l)$	→	$2\,LiOH(aq)$	+	$H_2(g)$
moles	2 mol			→	2 mol	+	1 mol
amounts	0.0800 mol			→	0.0800 mol		0.0400 mol

Step 3: Calculate the concentration of LiOH(aq) and volume of $H_2(g)$ in cm³ at RTP.

$$n(LiOH) = c \times \frac{V \text{ (in cm}^3)}{1000}$$

so, $c = \dfrac{1000 \times n}{V} = \dfrac{1000 \times 0.0800}{125}$

$= 0.640 \text{ mol dm}^{-3}$

$$n(H_2) = \frac{V \text{ (in cm}^3)}{24\,000}$$

so, $V = n \times 24\,000 = 0.0400 \times 24\,000 = 960 \text{ cm}^3$

 Identifying an unknown metal

The method below shows how you could carry out an experiment to identify an unknown Group 2 metal **X**. The results can then be analysed using the set method for reacting quantities.

1. Set up the apparatus shown in Figure 1.
2. Weigh a sample of the metal and add to the flask.
3. Using a measuring cylinder, add 25.0 cm³ 1.0 mol dm⁻³ HCl(aq) (an excess) to the flask and quickly replace the bung.
4. Measure the maximum volume of gas in the syringe.

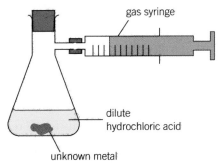

▲ **Figure 1** *Apparatus for determination of an unknown metal*

Synoptic link

As part of the practical skills required for your course, you need to know how to measure mass, volumes of solutions, and volumes of gases.

This practical application box tells you how to measure the volumes of gases.

Measuring mass, volumes of solutions, and volumes of gases are also covered in:

- Topic 3.2, Determination of formulae, for how to measure mass
- Topic 4.2, Acid–base titrations, for how to measure volumes of solutions
- Topic 9.2, Measuring enthalpy changes, for how to measure mass
- Topic 10.1, Reaction rates, for how to measure volumes of gases.

3.4 Reacting quantities

Results
mass of unknown metal = 0.14 g
volume of H_2 collected = 84 cm^3

Step 1: From the experimental results, the amount of $H_2(g)$ can be calculated.

$$\text{Assuming RTP, } n(H_2) = \frac{V \text{ (in cm}^3\text{)}}{24\,000}$$

$$= \frac{84}{24\,000}$$

$$= 0.00350 \text{ mol}$$

Step 2: From the equation, and the result from **Step 1**, the amount of metal **X** can be determined.

$$\begin{array}{ccc}
X(s) + 2\,HCl(aq) & \rightarrow & XCl_2(aq) + H_2(g) \\
1 \text{ mol} & & 1 \text{ mol} \\
0.00350 \text{ mol} & & 0.00350 \text{ mol}
\end{array}$$

Step 3: Work out the unknown information.
You now know the amount, in moles, and mass of **X**, so you are nearly there!

$$n(X) = \frac{m}{M}$$

$$\text{so, } M(X) = \frac{m}{n(X)}$$

$$= \frac{0.14}{0.00350}$$

$$= 40 \text{ g mol}^{-1}$$

From the periodic table, Ca has a relative atomic mass of 40.1.

The unknown metal **X** is calcium

> The experiment was repeated with another Group 2 metal. 0.064 g produced 63 cm^3 of hydrogen. Analyse the results to identify the metal.

Percentage yield

So far, all our calculations have assumed that *all* of the reactants are converted into products. This maximum possible amount of product is called the **theoretical yield**. Unfortunately this is difficult to achieve for several reasons, including:

- the reaction may not have gone to completion
- other reactions (side reactions) may have taken place alongside the main reaction
- purification of the product may result in loss of some product.

The **actual yield** obtained from a reaction is usually lower than the theoretical yield. The conversion of starting materials into a desired product is expressed by the **percentage yield**.

$$\text{percentage yield} = \frac{\text{actual yield}}{\text{theoretical yield}} \times 100\%$$

AMOUNT OF SUBSTANCE 3

> **Worked example: Percentage yield**
>
> 1.15 g of sodium reacts with an excess of chlorine, forming 1.872 g of sodium chloride. What is the percentage yield of sodium chloride?
>
> **Step 1:** Calculate the amount, in moles, of Na that reacts.
>
> $$n(\text{Na}) = \frac{m}{M} = \frac{1.15}{23.0} = \mathbf{0.0500\,mol}$$
>
> **Step 2:** Use the equation to find the theoretical yield of NaCl, in moles.
>
> | equation | 2 Na(s) | + | Cl_2(g) | → | 2 NaCl(s) |
> |---|---|---|---|---|---|
> | moles | 2 mol | | | → | 2 mol |
> | amounts | 0.0500 mol | | | → | 0.0500 mol |
>
> **Step 3:** Calculate the actual yield of NaCl, in moles.
>
> $$n(\text{NaCl}) = \frac{m}{M}$$
> $$= \frac{1.872}{58.5}$$
> $$= 0.0320\,\text{mol}$$
>
> **Step 4:** Calculate the percentage yield of NaCl.
>
> $$\%\text{ yield} = \frac{\text{actual yield}}{\text{theoretical yield}} \times 100\,\%$$
> $$= \frac{0.0320}{0.0500} \times 100 = 64.0\,\%$$

The limiting reagent

In the example above, you have used two reactants with one reactant in excess. The reactant that is *not* in excess will be completely used up first and stop the reaction – it is called the **limiting reagent**.

If you do not know which reactant is in excess, you need to find out by working out the amount in moles of each reactant and comparing with the equation. Calculations *must* be based on the limiting reagent.

For example, when hydrogen and oxygen gases react to form water, 2 mol of hydrogen are required for every 1 mol of oxygen:

$$2H_2(g) + O_2(g) \rightarrow 2H_2O(l)$$
2 mol   1 mol

If equal amounts of hydrogen and oxygen are allowed to react, hydrogen will be used up first, and half the oxygen will be unreacted. The limiting reagent is hydrogen, so calculations must be based on hydrogen.

Atom economy

The **atom economy** of a chemical reaction is a measure of how well atoms have been utilised.

3.4 Reacting quantities

Reactions with high atom economies:

- produce a large proportion of desired products and few unwanted waste products
- are important for sustainability as they make the best use of natural resources.

Atom economy is based solely on the balanced chemical equation for a reaction and assumes a 100% yield.

$$\text{atom economy} = \frac{\text{sum of molar masses of } desired \text{ products}}{\text{sum of molar masses of } all \text{ products}} \times 100\%$$

The idea of atom economy has been developed alongside awareness of dwindling finite resources and environmental concerns about processing or disposing of harmful waste. Improving atom economy makes industrial processes more efficient, preserves raw materials, and reduces waste. In an ideal chemical process, a use would be found for all products and thus the atom economy would be 100%.

> **Study tip**
>
> Atom economy is worked out from the balanced equation. Unlike percentage yield, no experimental results are needed.

> **Worked example: Atom economy**
>
> Hydrogen is an important raw material and is produced from the reaction of carbon with steam.
>
> What is the atom economy of this reaction?
>
> **Step 1:** Write the equation and the molar masses of the products.
>
> $$C(s) + 2H_2O(g) \rightarrow 2H_2(g) + CO_2(g)$$
> $$\rightarrow 2 \times 2.0 \quad 12.0 + 16.0 \times 2 = 44.0$$
>
> **Step 2:** Calculate the atom economy.
>
> $$\text{atom economy} = \frac{\text{sum of molar masses of } desired \text{ products}}{\text{sum of molar masses of } all \text{ products}} \times 100$$
>
> $$= \frac{2 \times 2.0}{2 \times 2.0 + 44.0} \times 100 = 8.3\%$$

> **Study tip**
>
> Remember to take account of the balancing numbers when accounting for the sum of all molar masses.

How sustainable?

The worked example reveals a poor atom economy, especially as the calculation assumes a 100% yield of hydrogen – the overall yield will be even worse. Furthermore, the undesired product is carbon dioxide, one of the gases that causes global warming. So how sustainable is this process?

Atom economy only provides part of the answer.

- The process uses reactants that are readily available, carbon from coal and steam from water. Energy will be needed to produce the steam, but costs for obtaining starting materials are low.
- Other reactions may have a much larger atom economy but poor percentage yields. Efficiency will depend on both factors.

AMOUNT OF SUBSTANCE 3

Summary questions

1. **a** Balance the equation
   $Ca(s) + O_2(g) \rightarrow CaO(s)$. *(1 mark)*
   b State the masses of Ca and O_2 that completely react together to form 4.488 g CaO. *(2 marks)*
   c Calculate the volume of $O_2(g)$ at RTP that reacts with 2.807 g Ca. *(2 marks)*

2. Calculate the atom economy for the following industrial processes.
   a NH_3 production:
   $N_2 + 3H_2 \rightarrow 2NH_3$ *(2 marks)*
   b C_2H_5OH production:
   $C_6H_{12}O_6 \rightarrow 2C_2H_5OH + 2CO_2$ *(2 marks)*

3. 35 g of HF was prepared by reacting 112 g of CaF_2 with an excess of H_2SO_4:

   $CaF_2 + H_2SO_4 \rightarrow 2HF + CaSO_4$

   Calculate the percentage yield of HF and the atom economy of this process. *(3 marks)*

4. **a** Balance the equation:
   $SO_2(g) + O_2(g) \rightarrow SO_3(g)$ *(1 mark)*
   b State the volumes of SO_2 and O_2, at RTP, that would produce 180 cm³ of SO_3. *(1 mark)*
   c 150 cm³ $SO_2(g)$ and 100 cm³ $O_2(g)$ react together. *(2 marks)*
     i State which reactant is in excess.
     ii State the volume of $SO_3(g)$ that could form using these quantities.

5. Hydrogen can be prepared from the reaction of methane in natural gas with steam. The other product is carbon monoxide.
   a Write an equation for this reaction and calculate the atom economy. *(2 marks)*
   b 100 g of methane react with an excess of steam, forming 324 dm³ of hydrogen at RTP. Calculate the percentage yield of hydrogen. *(1 mark)*

6. **a** Balance the equation:
   $Cr(s) + HCl(aq) \rightarrow CrCl_2(aq) + H_2(g)$ *(1 mark)*
   b Calculate the volume of H_2 at RTP formed by the complete reaction of 1.17 g Cr with excess HCl(aq). *(2 marks)*
   c The volume of HCl(aq) used is 150 cm³. What is the minimum concentration of HCl(aq) needed to react with all of the Cr? *(2 marks)*

7. 0.054 g of an unknown metal **X** reacts with an excess of sulfuric acid to form 72 cm³ of $H_2(g)$.
   The equation is: $2X(aq) + 3H_2SO_4 \rightarrow X_2(SO_4)_3 + 3H_2(g)$
   Identify metal **X**. *(3 marks)*

Chapter 3 Practice questions

Practice questions

1. A compound has the percentage composition by mass: Ca, 30.35%; N, 21.20; O, 48.45%
   What is the empirical formula of the compound? *(2 marks)*

2. When heated, potassium chlorate(VII), $KClO_3(s)$ decomposes to form $KCl(s)$ and $108\,cm^3$ $O_2(g)$, measured at RTP. The unbalanced equation is shown below.
   $$KClO_3(s) \rightarrow KCl(s) + O_2(g)$$
   a. Balance the equation. *(1 mark)*
   b. What is the amount, in mol, of O_2 produced? *(1 mark)*
   c. What is the mass of $KClO_3$ required to produce $108\,cm^3$ of O_2 at RTP? Show your working *(3 marks)*

3. $1.893\,g$ of hydrated zinc sulfate, $ZnSO_4 \cdot xH_2O$ is heated to remove all water of crystallisation. The mass of anhydrous $ZnSO_4$ formed is $1.061\,g$.
   What is the formula of the hydrated zinc sulfate?
   Show your working *(4 marks)*

4. On heating, sodium hydrogencarbonate, $NaHCO_3$ decomposes, forming $2.48\,g$ of sodium carbonate, Na_2CO_3.
   $$2NaHCO_3 \rightarrow Na_2CO_3 + CO_2 + H_2O$$
   The percentage yield of Na_2CO_3 is 65%.
   What is the mass of $NaHCO_3$ that was heated?
   Show your working *(5 marks)*

5. Iron ore contains iron(III) oxide. In a blast furnace, the iron(III) oxide reacts with carbon monoxide to form iron and carbon dioxide.
   a. Write an equation for this reaction. *(1 mark)*
   b. Each day, a blast furnace typically reacts 10 000 tonnes of Fe_2O_3.
   Calculate the typical mass of iron produced each day from a blast furnace. *(3 marks)*

6. A nitrogen fertiliser, **A**, has the composition by mass Na, 27.1%; N, 16.5%; O, 56.4%.
   On heating, $3.40\,g$ of **A** decomposes into sodium nitrite, $NaNO_2$, and oxygen gas.
   a. Calculate the empirical formula of **A**. *(2 marks)*
   b. Write an equation for the decomposition of **A**. *(1 mark)*
   c. Calculate the volume of oxygen gas formed at RTP. *(4 marks)*

7. A chemist reacts $0.0250\,mol$ of sodium metal with water, forming $50.0\,cm^3$ of sodium hydroxide solution and hydrogen gas.
   a. What mass of sodium was reacted? *(1 mark)*
   b. Write an equation, including state symbols, for the reaction. *(1 mark)*
   c. Calculate the volume of hydrogen gas formed at RTP. *(2 marks)*
   d. Calculate the concentration of sodium hydroxide solution formed in (i) $mol\,dm^{-3}$; (ii) $g\,dm^{-3}$ *(3 marks)*

8. Chlorine gas is prepared by the electrolysis of brine, a concentrated solution of sodium chloride in water.
   $$2NaCl + 2H_2O \rightarrow 2NaOH + Cl_2 + H_2$$
   The concentration of brine can be assumed to be $4.00\,mol\,dm^{-3}$.
   a. Calculate the mass of sodium chloride dissolved in $250\,cm^3$ of brine. *(1 mark)*
   b. Each day, the UK produces $2.5 \times 10^9\,dm^3$ of chlorine gas, at RTP, from brine.
   Calculate the volume of brine required for chlorine production each day in the UK. *(3 marks)*

9. a. An organic compound **X**, contains carbon, hydrogen and oxygen only. Compound X has the following percentage composition by mass: 54.55%, C; 9.09% H.
   (i) Calculate the empirical formula of **X**. *(2 marks)*
   (ii) In an experiment, $0.2103\,g$ of a vaporised sample of **X** was shown to occupy $72.0\,cm^3$ at $100\,°C$ and $103.0\,kPa$.
   Calculate the relative molecular mass of **X**. *(4 marks)*
   (iii) Deduce the molecular formula for **X**. *(1 mark)*

10. Hydrated aluminium sulfate, $Al_2(SO_4)_3 \cdot xH_2O$, and chlorine, Cl_2, are used in water treatment.

3 AMOUNT OF SUBSTANCE

a A student attempts to prepare hydrated aluminium sulfate by the following method.

The student heats dilute sulfuric acid with an excess of solid aluminium oxide.

The student filters off the excess aluminium oxide to obtain a colourless solution of $Al_2(SO_4)_3$.

(i) State the formulae of the two main ions present in the solution of $Al_2(SO_4)_3$. *(2 marks)*

(ii) Write an equation for the reaction of aluminium oxide, Al_2O_3, with sulfuric acid. Include state symbols. *(2 marks)*

(iii) What does '·xH_2O' represent in the formula $Al_2(SO_4)_3 \cdot xH_2O$? *(1 mark)*

(iv) The student heats 12.606 g of $Al_2(SO_4)_3 \cdot xH_2O$ crystals to constant mass. The anhydrous aluminium sulfate formed has a mass of 6.846 g.

Use the student's results to calculate the value of x. The molar mass of $Al_2(SO_4)_3 = 342.3\,g\,mol^{-1}$. *(3 marks)*

F321 June 2013 (2a)

11 a Borax, $Na_2B_4O_7 \cdot 10H_2O$, can be used to determine the concentration of acids such as dilute hydrochloric acid.

A student prepares 250 cm³ of a 0.0800 mol dm⁻³ solution of borax in water in a volumetric flask.

Calculate the mass of borax crystals, $Na_2B_4O_7 \cdot 10H_2O$, needed to make up 250 cm³ of 0.0800 mol dm⁻³ solution. *(3 marks)*

b The student found that 22.50 cm³ of 0.0800 mol dm⁻³ $Na_2B_4O_7$ reacted with 25.00 cm³ of dilute hydrochloric acid.

$Na_2B_4O_7 + 2HCl + 5H_2O \rightarrow 2NaCl + 4H_3BO_3$

(i) Calculate the amount, in mol, of $Na_2B_4O_7$ used. *(1 mark)*

(ii) Calculate the amount, in mol, of HCl used. *(1 mark)*

(iii) Calculate the concentration, in mol dm⁻³, of HCl. *(1 mark)*

F321 Jan 2013 5(c)(d)

12 Lithium carbonate, Li_2CO_3 is added to an excess of dilute hydrochloric acid.

$Li_2CO_3(s) + 2HCl(g) \rightarrow 2LiCl(g) + CO_2(g) + H_2O(l)$

A student adds 1.845 g Li_2CO_3 to 125 cm³ of 0.500 mol dm⁻³ HCl. The volume of the solution formed is 125 cm³.

a Predict **two** observations that you would expect to see during this reaction. *(2 marks)*

b Explain what is meant by 0.500 mol dm⁻³ HCl. *(1 mark)*

c (i) Calculate the amount, in moles, of Li_2CO_3 and HCl that were reacted. *(3 marks)*

(ii) Calculate the amount, in mol of HCl that was in excess. *(2 marks)*

d (i) Calculate the volume of $CO_2(g)$ that would be expected, measured at RTP. *(1 mark)*

(ii) Suggest why the volume of CO_2 produced at RTP is likely to be less than your answer to (i). *(1 mark)*

e Calculate the concentration, in mol dm⁻³, of LiCl in the solution formed. *(1 mark)*

13 a A factory makes ethyne gas, C_2H_2 from calcium carbide, CaC_2. One of the waste products is calcium hydroxide.

$CaC_2 + 2H_2O \rightarrow Ca(OH)_2 + C_2H_2$

Each day 1.00×10^6 grams of calcium carbide are used and $3.60 \times 10^5\,dm^3$ of ethyne gas, measured at room temperature and pressure, is manufactured.

(i) Calculate the atom economy for this process using the relative formula masses in the table below. *(2 marks)*

Compound	Relative formula mass
CaC_2	64.1
H_2O	18.0
$Ca(OH)_2$	74.1
C_2H_2	26.0

(ii) Calculate the amount, in moles, of CaC_2 used each day. *(1 mark)*

(iii) Calculate the amount, in moles, of C_2H_2 made each day. *(1 mark)*

(iv) Calculate the percentage yield of C_2H_2. *(1 mark)*

(v) Comment on the percentage yield and the atom economy of this process in terms of sustainability. *(2 marks)*

F322 Jun 2010 1(e)

4 ACIDS AND REDOX
4.1 Acids, bases, and neutralisation

Specification reference: 2.1.4

Learning outcomes

Demonstrate knowledge, understanding, and application of:
- acids and bases
- neutralisation.

Acids

All acids contain hydrogen in their formulae (Table 1). When dissolved in water, an acid releases hydrogen ions as protons, H^+, into the solution. In the equation below, hydrogen chloride gas releases H^+ ions as it dissolves in water.

$$HCl(g) + aq \rightarrow H^+(aq) + Cl^-(aq)$$

In this equation + aq has been included to show that an excess of water is present. The equation is essentially hydrogen chloride gas dissolving to form an aqueous solution.

Strong and weak acids

A **strong acid**, such as hydrochloric acid, HCl, releases all its hydrogen atoms into solution as H^+ ions and **completely dissociates** in aqueous solution.

$$HCl(aq) \rightarrow H^+(aq) + Cl^-(aq)$$

A **weak acid**, such as ethanoic acid, CH_3COOH, only releases a small proportion of its available hydrogen atoms into solution as H^+ ions. A weak acid **partially dissociates** in aqueous solution.

$$CH_3COOH(aq) \rightleftharpoons H^+(aq) + CH_3COO^-(aq)$$

The equilibrium sign $\rightleftharpoons$ indicates that the forward reaction is incomplete.

It is important to realise that not all compounds that contain hydrogen atoms are acids. Each molecule of ethanoic acid contains four hydrogen atoms, but only the hydrogen atom on the COOH group is released as H^+. Even then, only about one molecule in every hundred dissociates, so ethanoic acid is a weak acid. Most organic acids, like ethanoic acid, are weak acids.

Bases and alkalis

Metal oxides, metal hydroxides, metal carbonates, and ammonia, NH_3, are classified as **bases**. A base **neutralises** an acid to form a **salt**. Table 2 and Figure 2 show some common bases.

An **alkali** is a base that dissolves in water releasing hydroxide ions (OH^-) into the solution. The equation below shows the alkali sodium hydroxide releasing hydroxide ions as it dissolves in water.

$$NaOH(s) + aq \rightarrow Na^+(aq) + OH^-(aq)$$

▼ Table 1 Common acids

Acid	Formula
hydrochloric acid	HCl
sulfuric acid	H_2SO_4
nitric acid	HNO_3
ethanoic acid (vinegar)	CH_3COOH

Synoptic link

You will learn more and equilibrium and weak acids in Chapter 20, Acids, bases, and pH.

▲ Figure 1 Some common acids – vinegar contains ethanoic acid, orange and lemon juice and even sink cleaner contains citric acid, and apples contain malic acid

ACIDS AND REDOX 4

Table 2 Common bases

Metal oxides	Metal carbonates	Alkalis
MgO	Na_2CO_3	NaOH
CaO	$CaCO_3$	KOH
CuO	$CuCO_3$	NH_3

▲ Figure 2 Three bases – calcium carbonate, $CaCO_3$, copper oxide, CuO, and sodium hydroxide, NaOH

Neutralisation

In neutralisation of an acid, $H^+(aq)$ ions react with a base to form a salt and neutral water. The H^+ ions from the acid are replaced by metal or ammonium ions from the base. Table 3 shows salts of common acids.

- Notice the link between the name of the acid and the salt, in blue.
- To form the salt, the hydrogen (shown in purple) in the acid is replaced by a metal or ammonium ion to form the salt.

Table 3 Acids and their salts

Acid			Salt	
Name	Formula	Type	Name	Formula
hydrochloric acid	HCl	chloride	sodium chloride	NaCl
sulfuric acid	H_2SO_4	sulfate	sodium sulfate	Na_2SO_4
nitric acid	HNO_3	nitrate	calcium nitrate	$Ca(NO_3)_2$
ethanoic acid (vinegar)	CH_3COOH	ethanoate	ammonium ethanoate	CH_3COONH_4

Neutralisation of acids with metal oxides and hydroxides

An acid is neutralised by a metal oxide or metal hydroxide to form a salt and water only. The equations below show the neutralisation of sulfuric acid and hydrochloric acid by copper(II) oxide to form a salt and water only. Figure 3 shows solutions of the salts formed.

$$CuO(s) + H_2SO_4(aq) \rightarrow CuSO_4(aq) + H_2O(l)$$
$$CuO(s) + 2HCl(aq) \rightarrow CuCl_2(aq) + H_2O(l)$$

Alkalis

With alkalis, the reactants are in solution. As with metal oxides, the overall reaction forms a salt and water only:

$$acid + alkali \rightarrow salt + water$$

The ionic equation, shown below, is much simpler than the overall equation: neutralisation of $H^+(aq)$ ions by $OH^-(aq)$ ions to form neutral water, $H_2O(l)$.

Full equation: $HCl(aq) + NaOH(aq) \rightarrow NaCl(aq) + H_2O(l)$
Ionic equation: $H^+(aq) + OH^-(aq) \rightarrow H_2O(l)$

Study tip

The reactions for the neutralisation of an acid by a metal oxide or hydroxide are all essentially the same:

acid + metal oxide/hydroxide → salt + water

Learn this to help you write the equation for any neutralisation by a metal oxide or hydroxide.

▲ Figure 3 Solutions of two copper salts prepared by neutralisation of two different acids with the base copper(II) oxide:
- sulfuric acid has been neutralised to form blue copper(II) sulfate, $CuSO_4$ (left)
- hydrochloric acid has been neutralised to form green copper(II) chloride, $CuCl_2$ (right)

4.1 Acids, bases, and neutralisation

Study tip

Remember the equation for the neutralisation of acids by metal carbonates.

acid + metal carbonate →
   salt + water + carbon dioxide(g)

It will help you write the equation for any neutralisation of this type.

Neutralisation of acids with carbonates

Like metal oxides, carbonates neutralise acids to form a salt and water. There is also a third product, carbon dioxide gas. The equations show neutralisation of two carbonates by sulfuric acid and hydrochloric acid.

$$ZnCO_3(s) + H_2SO_4(aq) \rightarrow ZnSO_4(aq) + H_2O(l) + CO_2(g)$$
$$MgCO_3(s) + 2HCl(aq) \rightarrow MgCl_2(aq) + H_2O(l) + CO_2(g)$$

Dissociation in sulfuric acid

Sulfuric acid, H_2SO_4, is a strong acid, but this is true only for one of the two hydrogen atoms. When sulfuric acid is mixed with water each H_2SO_4 molecule dissociates, releasing just one of its two hydrogen atoms as an H^+ ion:

$$H_2SO_4(aq) \rightarrow H^+(aq) + HSO_4^-(aq)$$

The resulting $HSO_4^-(aq)$ ions then only partially dissociate:

$$HSO_4^-(aq) \rightleftharpoons H^+(aq) + SO_4^{2-}(aq)$$

- Sulfuric acid first behaves as a strong acid.
- The HSO_4^- ions formed behave as a weak acid.

Other strong acids containing more than one hydrogen atom behave similarly.

> Write equations to show the dissociation of the three hydrogen atoms in phosphoric acid, H_3PO_4.

Summary questions

1. Explain what is meant by the following terms.
   a. strong acid *(1 mark)*
   b. weak acid *(1 mark)*

2. Write equations to show the dissociation of the following acids when dissolved in water.
   a. nitric acid, HNO_3 (a strong acid) *(2 marks)*
   b. propanoic acid, CH_3CH_2COOH (a weak acid) *(2 marks)*

3. Write equations for the following neutralisation reactions. For each reaction, name the salt formed.
   a. The reaction of MgO(s) with hydrochloric acid *(2 marks)*
   b. The reaction of NaOH(aq) with sulfuric acid *(2 marks)*
   c. The reaction of $ZnCO_3(s)$ with nitric acid *(2 marks)*
   d. The reaction of aqueous sodium hydroxide with ethanoic acid *(2 marks)*

4. Baking powder usually contains an organic acid and sodium hydrogencarbonate, $NaHCO_3$, also known as bicarbonate of soda. Sodium hydrogencarbonate is an acid salt, formed by partial neutralisation of carbonic acid, H_2CO_3, with sodium hydroxide.
   a. Write equations for the partial and complete neutralisation reactions of carbonic acid with sodium hydroxide. *(4 marks)*
   b. Why is sodium hydrogencarbonate called an acid salt? *(1 mark)*
   c. Baking powder is used to make cakes. When it is mixed into a batter or dough containing water and heated, the sodium hydrogencarbonate reacts with the acid in the powder in the same way as a carbonate. The small quantity of baking powder does not affect the taste, so why is it used? *(2 marks)*

4.2 Acid–base titrations

Specification reference: 2.1.4, 2.1.3

Titrations

A **titration** is a technique used to accurately measure the volume of one solution that reacts exactly with another solution. Titrations can be used for:

- finding the concentration of a solution
- identification of unknown chemicals
- finding the purity of a substance.

Checking purity is an important aspect of quality control, especially for compounds manufactured for human use such as medicines, food, and cosmetics. It is essential that pharmaceuticals have a high level of purity – just a tiny amount of an impurity in a drug could cause a great deal of harm to a patient.

Preparing a standard solution

A standard solution is a solution of known concentration. A **volumetric flask** is used to make up a standard solution very accurately. Volumetric flasks are manufactured in various sizes and can measure volumes very precisely. The volumetric flasks that you will use are manufactured to the typical tolerances below:

- a $100\,cm^3$ volumetric flask: $\pm 0.20\,cm^3$
- a $250\,cm^3$ volumetric flask: $\pm 0.30\,cm^3$.

> **Learning outcomes**
> Demonstrate knowledge, understanding, and application of:
> → preparation of a standard solution
> → carrying out a titration
> → analysing titration results by calculation
> → practical techniques for measuring the volume of solutions.

> **Synoptic link**
> Standard solutions were introduced in Topic 3.3, Moles and volumes.

Preparing standard solutions

1. The solid is first weighed accurately.
2. The solid is dissolved in a beaker using less distilled water than will be needed to fill the volumetric flask to the mark.
3. This solution is transferred to a volumetric flask. The last traces of the solution are rinsed into the flask with distilled water.
4. The flask is carefully filled to the graduation line by adding distilled water a drop at a time until the *bottom* of the meniscus lines up exactly with the mark (Figure 1). Care at this stage is essential – if too much water is added, the solution will be too dilute and must be prepared again. You should view the graduation mark and meniscus at eye level for accuracy.
5. Finally, the volumetric flask is slowly inverted several times to mix the solution thoroughly. If this stage is omitted, titration results are unlikely to be consistent. You will be able to see the solution mixing when you invert the flask as the more dense original solution moves through the solution.

Explain the effect on the titre of the following errors.
1. The flask is filled with water above the graduation line.
2. The flask is not inverted.

▲ **Figure 1** *The volumetric flask is filled so that bottom of the meniscus just touches the graduation line*

▲ **Figure 2** *The volumetric flask is slowly inverted several times to ensure that the solution is mixed evenly*

4.2 Acid–base titrations

Acid–base titrations

Apparatus

In an acid–base titration, a solution of an acid is titrated against a solution of a base using a **pipette** and a **burette**, which are typically manufactured to the uncertainty below:

- a 10 cm³ pipette: ±0.04 cm³
- a 25 cm³ pipette: ±0.06 cm³
- a 50 cm³ burette: ±0.10 cm³.

A burette reading is recorded to the nearest half division, with the bottom of the meniscus on a mark or between two marks. Each burette reading is measured to the nearest ±0.05 cm³ so the reading always has *two* decimal places, the last place being either 0 or 5, for example, 25.40 cm³ or 26.25 cm³.

▲ Figure 3 *When filling a burette, run excess solution out through the tap to remove any air bubbles. If a bubble is left in the neck of the burette, the air could be released during the titration, leading to an error in the titre*

▲ Figure 4 *Take burette readings from the bottom of the meniscus, with your eye is at the level of meniscus. Read the burette to the nearest 0.05 cm³. This burette reading is 25.55 cm³*

The acid–base titration procedure

1. Add a measured volume of one solution to a conical flask using a pipette.
2. Add the other solution to a burette, and record the initial burette reading to the nearest 0.05 cm³.
3. Add a few drops of an **indicator** to the solution in the conical flask.
4. Run the solution in the burette into the solution in the conical flask, swirling the conical flask throughout to mix the two solutions. Eventually the indicator changes colour at the **end point** of the titration. The end point is used to indicate the volume of one solution that exactly reacts with the volume of the second solution.
5. Record the final burette reading. The volume of solution added from the burette is called the titre, which is calculated by subtracting the initial from the final burette reading.
6. A quick, trial titration is carried out first to find the approximate **titre**.
7. The titration is then repeated accurately, adding the solution dropwise as the end point is approached. Further titrations are carried out until two accurate titres are **concordant** – agreeing to within 0.10 cm³.

The readings from the titration are recorded in a table such as Table 1.

▼ Table 1 *Table for recording titration results*

Trial	1	2	3
final burette reading / cm³			
initial burette reading / cm³			
titre / cm³			
mean titre / cm³			

Explain the effect on the titre of the following errors.

1. The pipette has an air bubble inside.
2. The burette readings are taken from the top, rather than the bottom, of the meniscus.

The mean titre

When working out the mean titre, it is important to use only your *closest accurate titres*.

- By repeating titres until two agree within 0.10 cm³, you can reject inaccurate titres.
- If you were to include all the titres in the mean, you have lost the accuracy of the titration technique.

Titration calculations

From the results of a titration, you will know the following:

- *both* the concentration c_1 and the reacting volume V_1 of one of the solutions
- *only* the reacting volume V_2 of the other solution.

The method for analysing the results follows a set pattern.

Step 1: Work out the amount, in mol, of the solute in the solution for which you know *both* the concentration c_1 and volume V_1.

Step 2: Use the equation to work out the amount, in mol, of the solute in the other solution.

Step 3: Work out the unknown information about the solute in the other solution.

Worked example: Determination of an unknown concentration

Pipette: 25.00 cm³ of 0.100 mol dm⁻³ KOH(aq)

Mean titre from burette: 25.70 cm³ of H_2SO_4(aq)

Unknown information: The concentration of H_2SO_4(aq)

Step 1: From the titration results, calculate the amount of KOH.

$$n(KOH) = c \times \frac{V}{1000} = 0.100 \times \frac{25.00}{1000} = \mathbf{0.00250}\ mol$$

Step 2: From the equation and Step 1, determine the amount of H_2SO_4.

$$2\ KOH(aq) + H_2SO_4(aq) \rightarrow K_2SO_4(aq) + 2\ H_2O(l)$$

2 mol   **1 mol** (balancing numbers)

0.00250 mol   **0.00125 mol**

Step 3: Work out the unknown information.

$$n(H_2SO_4) = c \times \frac{V(cm^3)}{1000}$$

$$c = \frac{1000 \times n}{V} = \frac{1000 \times 0.00125}{25.70} = 0.0486\ mol\ dm^{-3}$$

concentration of H_2SO_4(aq) is 0.0486 mol dm⁻³

Synoptic link

As part of the practical skills required for your course, you need to know how to measure mass, volumes of solutions, and volumes of gases.

The practical application boxes in this topic tell you how to measure the volume of solutions.

Measuring mass, volumes of solutions, and volumes of gases are also covered in:

- Topic 3.2, Determination of formulae, for how to measure mass
- Topic 3.4, Reacting quantities, for how to measure the volumes of gases
- Topic 9.2, Measuring enthalpy changes, for how to measure mass
- Topic 10.1, Reaction rates, for how to measure volumes of gases.

Study tip

This method is essentially the same as the one used to calculate unknown quantities using the mole in Topic 3.4, Reacting quantities.

Study tip

It's all done with the ratios of the balancing numbers and moles.

4.2 Acid–base titrations

Identification of a carbonate

You can use the titration technique to work out some unknown information about a substance, for example, you can identify an unknown carbonate, X_2CO_3.

The key steps are shown below, together with results.

1. Prepare a solution of an unknown carbonate, X_2CO_3, in a volumetric flask.
2. Using a pipette, measure 25.00 cm³ of your prepared solution into a conical flask.
3. Using a burette, titrate this solution using 0.100 mol dm⁻³ hydrochloric acid.
4. Analyse your results to identify the carbonate.

Mass measurements

mass of weighing bottle / g	11.41
mass of weighing bottle + X_2CO_3 / g	12.60
mass of X_2CO_3 / g	1.19

Titration readings

	Trial	1	2	3
final burette reading / cm³	24.10	22.35	44.95	22.45
initial burette reading / cm³	1.00	0.00	22.35	0.00
titre / cm³	23.10	**22.35**	22.60	**22.45**
mean titre / cm³			**22.40**	

Analysis

Step 1: Calculate the amount of HCl that reacted.

Use the mean titre V and the concentration of the hydrochloric acid c.

$$n(HCl) = c \times \frac{V}{1000} = 0.100 \times \frac{22.40}{1000} = \mathbf{0.00224\ mol}$$

Step 2: Determine the amount of X_2CO_3 that reacted.

Use the equation and $n(HCl)$.

$$X_2CO_3(aq) + 2\ HCl(aq) \rightarrow 2XCl(aq) + H_2O(l) + CO_2(g)$$

1 mol   **2 mol** (balancing numbers)

0.00112 mol   **0.00224 mol HCl**

Step 3: Work out the unknown information. There are several stages.

1. Scale up to find the amount of X_2CO_3 in the 250 cm³ solution that you prepared.

   $n(X_2CO_3)$ in 25.00 cm³ used in the titration = 0.00112 mol

   $n(X_2CO_3)$ in 250.0 cm³ solution = 0.00112 × 10 = **0.0112 mol**

2. Find the molar mass of X_2CO_3.

   Use the amount, $n(X_2CO_3)$, in the 250 cm³ solution and the mass m of X_2CO_3 used to prepare this solution.

   $$n(X_2CO_3) = \frac{m}{M}$$

   $$M(X_2CO_3) = \frac{m}{n} = \frac{1.19}{0.0112} = \mathbf{106.25\ g\ mol^{-1}}$$

Study tip

You might be given a *structured calculation* — you will be helped through the calculation with prompts similar to the bullet points in this example.

But you might also be given an *unstructured calculation* with no help. You should practise both structured titration calculations and the harder unstructured problems.

Study tip

The scaling here is obvious because 250 is 10 times larger than 25. If you were scaling up a burette reading (say 29.55 cm³), the process is the same, but you would need to find the scaling factor by dividing 250 by the titre.

ACIDS AND REDOX 4

3 Finally use $M(X_2CO_3)$ to identify **X** in the formula X_2CO_3.

$$106.25 = M(X) \times 2 + 12.0 + (16.0 \times 3) = 2M(X) + 60.0$$

$$M(X) = \frac{106.25 - 60.0}{2} = \mathbf{23.125} \text{ g mol}^{-1}$$

From the periodic table, 23.125 most closely matches Na ($A_r = 23.0$).

Unknown carbonate is sodium carbonate, Na_2CO_3.

Use the results to determine the molar mass of an unknown acid HA.

Mass readings: For preparation of a 250.0 cm³ solution of HA

mass of weighing bottle / g	9.64
mass of weighing bottle + HA / g	12.51

Titration readings: Titration of 25.0 cm³ volumes of the solution of HA with 0.0600 mol dm⁻³ Na_2CO_3

	Trial	1	2	3
final burette reading / cm³	27.40	25.05	29.55	24.65
initial burette reading / cm³	2.00	0.00	5.00	0.00

Summary questions

1 **a** 25.00 cm³ of 0.110 mol dm⁻³ NaOH(aq) reacts exactly with 23.30 cm³ of HNO_3(aq). *(5 marks)*
   i Calculate the amount, in moles, of NaOH(aq) used.
   ii Write the equation for the reaction and calculate the amount, in moles, of HNO_3(aq) in the titre.
   iii Calculate the concentration of the HNO_3(aq) solution.
   b 25.00 cm³ of 0.125 mol dm⁻³ KOH(aq) reacts exactly with 26.60 cm³ of H_2SO_4(aq). Write the equation and find the unknown concentration. *(3 marks)*

2 1.96 g of an unknown hydroxide $X(OH)_2$ was dissolved in water, and the solution was made up to 250.00 cm³ in a volumetric flask. In a titration, 25.00 cm³ of this solution of $X(OH)_2$ required 21.20 cm³ of 0.250 mol dm⁻³ HCl(aq) to reach the end point.

Equation: $2HCl(aq) + X(OH)_2(aq) \rightarrow XCl_2(aq) + 2H_2O(l)$
Identify the unknown hydroxide $X(OH)_2$. *(4 marks)*

3 1.654 g of a hydrated acid $H_2C_2O_4 \cdot xH_2O$ was dissolved in water and the solution was made up to 250.00 cm³ in a volumetric flask. 23.80 cm³ of this solution required 25.00 cm³ of 0.100 mol dm⁻³ NaOH(aq) to reach the end point.

Equation: $H_2C_2O_4(aq) + 2NaOH(aq) \rightarrow Na_2C_2O_4(aq) + 2H_2O(l)$
Identify the value of x and hence the formula of the hydrated acid $H_2C_2O_4 \cdot xH_2O$. *(4 marks)*

4.3 Redox

Specification reference: 2.1.5

Learning outcomes

Demonstrate knowledge, understanding, and application of:
- oxidation number
- oxidation and reduction
- redox reactions.

Oxidation number

Oxidation number is based on a set of rules that apply to atoms, and can be thought of as the number of electrons involved in bonding to a different element. Use of oxidation numbers helps when writing formulae and balancing electrons as a check that all electrons have been accounted for.

Rules for elements

The oxidation number is *always* zero for elements.

- In a pure element, any bonding is to atoms of the same element.
- So in H_2, O_2, P_4, S_8, Na, and Fe the oxidation number of each atom of the element is 0.

Rules for compounds and ions

- Each atom in a compound has an oxidation number.
- An oxidation number has a sign, which is placed *before* the number.

Table 1 shows examples of oxidation numbers of atoms in compounds and ions, including some special cases. The oxidation number of an ion of an element is numerically the same as the ionic charge but the sign comes before the number.

Study tip

It is a common mistake to confuse oxidation numbers of elements and ions. The oxidation number of oxygen in O_2 is zero but in H_2O is −2.

Study tip

All oxidation numbers, except zero (0), have a sign, + or −.

The *sign* of an oxidation number is placed *before* the number: the oxidation number of Ca in a Ca^{2+} ion is +2, *not* 2+.

▼ Table 1 Oxidation number rules in compounds and ions

Combined element	Oxidation number	Examples
O	−2	H_2O, CaO.
H	+1	NH_3, H_2S.
F	−1	HF
Na^+, K^+	+1	NaCl, K_2O
Mg^{2+}, Ca^{2+}	+2	$MgCl_2$, CaO
Cl^-, Br^-, I^-	−1	HCl, KBr, CaI_2
Special cases:		
H in metal hydrides	−1	NaH, CaH_2
O in peroxides	−1	H_2O_2
O bonded to F	+2	F_2O

Working out oxidation numbers

In addition to the rules in Table 1, there is a further rule for combined atoms:

- sum of the oxidation numbers = total charge.

48

ACIDS AND REDOX 4

> **Study tip**
> Remember: the sum of the oxidation numbers must equal the overall charge.

Worked example: Oxidation number in compounds

What is the oxidation number of sulfur in sulfuric acid, H_2SO_4?

Step 1: Assign any oxidation numbers from the rules.

$$H_2SO_4$$

Total H = +1 × 2 = +2 $\begin{Bmatrix} +1 & -2 \\ +1 & -2 \\ & -2 \\ & -2 \end{Bmatrix}$ Total O = −2 × 4 = −8

Step 2: What is the sum of oxidation numbers?

sum of oxidation numbers = total charge = 0

Step 3: Work out the unknown oxidation numbers.

sum of oxidation numbers = (+2) + (X) + (−8) = 0

H_2SO_4

Oxidation number of sulfur in H_2SO_4 = +6

Worked example: Oxidation numbers in ions

What is the oxidation number of nitrogen in NO_3^-?

Step 1: Assign any oxidation numbers from the rules.

NO_3^-

$\begin{Bmatrix} -2 \\ -2 \\ -2 \end{Bmatrix}$ Total O = −2 × 3 = −6

Step 2: What is the sum of oxidation numbers?

sum of oxidation numbers = total charge = −1

Step 3: Work out the unknown oxidation numbers

sum of oxidation numbers = (X) + (−6) = −1

NO_3^-

Oxidation number of nitrogen in NO_3^- = +5

Using Roman numerals in naming

Roman numerals are used in the names of compounds of elements that form ions with different charges. The Roman numeral shows the oxidation state (oxidation number) of the element, without a sign. The sign of the oxidation state is obvious from the overall charge:

- iron(II) represents Fe^{2+} with oxidation number +2
- iron(III) represents Fe^{3+} with oxidation number +3.

You have already seen that polyatomic ions containing oxygen, such as NO_2^- and NO_3^-, are sometimes named using –ite and –ate. Although still in common usage, use of –ite and –ate in naming is old-fashioned and modern names use oxidation numbers shown as Roman numerals.

▼ Table 2 *Naming of polyatomic ions*

Ion	Common name	Oxidation number of nitrogen	Modern name
NO_2^-	nitrite	+3	nitrate(III)
NO_3^-	nitrate	+5	nitrate(V)

In common usage, the Roman numeral is often omitted for the common ion, usually with more oxygen atoms:

- nitrate is assumed to be NO_3^-
- sulfate is assumed to be SO_4^{2-}.

Redox reactions

Reduction and oxidation

Originally the terms *oxidation* and *reduction* were used solely for reactions involving oxygen.

- Oxidation is addition of oxygen.
- Reduction is removal of oxygen.

The reaction below shows oxidation and reduction:

$$CuO(s) + H_2(g) \rightarrow Cu(s) + H_2O(l)$$

- Copper(II) oxide has *lost oxygen* and has been *reduced*.
- Hydrogen has *gained oxygen* and has been *oxidised*.

Redox reactions involve **red**uction and **ox**idation. If one process happens, so must the other – if something is reduced, something else must be oxidised.

The terms oxidation and reduction are now applied to many reactions that do not involve oxygen. The modern definitions are in terms of either electrons or oxidation number.

Redox in terms of electrons

- Reduction is the gain of electrons.
- Oxidation is the loss of electrons.

The redox reaction below does not involve oxygen but does involve gain and loss of electrons.

$$2Fe(s) + 3Cl_2(g) \rightarrow 2FeCl_3(s)$$

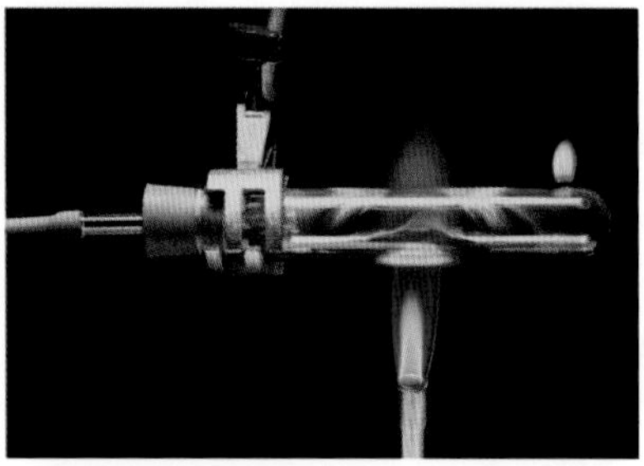

▲ Figure 1 *Reduction of copper(II) oxide with hydrogen – the green flame is excess hydrogen burning off*

Study tip

Remember **OILRIG**:

OIL **O**xidation **I**s **L**oss of electrons

RIG **R**eduction **I**s **G**ain of electrons.

$FeCl_3$ contains positive and negative ions, Fe^{3+} and Cl^-.

- Iron loses electrons and is oxidised $\quad 2Fe \rightarrow 2Fe^{3+} + 6e^-$
- Chlorine gains electrons and is reduced $\quad 3Cl_2 + 6e^- \rightarrow 6Cl^-$

The electrons gained and lost balance:

- 2 Fe in 2Fe   each Fe loses 3e⁻   total of 6 electrons lost
- 6 Cl in $3Cl_2$ each Cl gains 1e⁻   total of 6 electrons gained

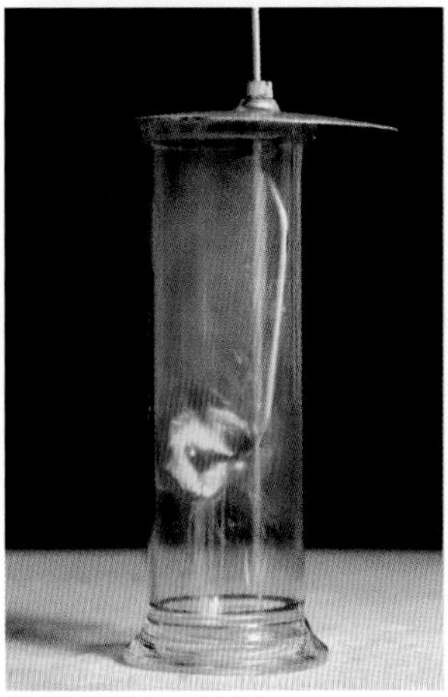

Redox in terms of oxidation number

- Reduction is a decrease in oxidation number.
- Oxidation is an increase in oxidation number.

The reaction below shows oxidation and reduction in terms of oxidation number. Each atom is assigned an oxidation number using the oxidation number rules:

$$Cu(s) + 2AgNO_3(aq) \rightarrow 2Ag(aq) + Cu(NO_3)_2(aq)$$

0 → +2   oxidation
+1 → 0   reduction

The changes in oxidation number apply to each atom and the total changes in oxidation number balance:

- 1 Cu in Cu   Cu increases by +2   total increase = +2
- 2 Ag in $2AgNO_3$ each Ag decreases by −1   total decrease = −2

Redox reactions of acids

In Topic 4.1, you saw how acids produce salts in neutralisation reactions. Dilute acids also undergo redox reactions with some metals to produce salts and hydrogen gas.

metal + acid → salt + hydrogen

Reaction of zinc with dilute hydrochloric acid

$$Zn(s) + 2HCl(aq) \rightarrow ZnCl_2(aq) + H_2(g)$$

0 → +2   oxidation
+1 → 0   reduction

- 1 Zn in Zn   Zn increases by +2   total increase = +2
- 2 H in 2HCl   each H decreases by −1   total decrease = −2

The overall decrease in oxidation number of 2H = 2 × −1 = −2 balances the increase of Zn by +2.

Reaction of aluminium with dilute sulfuric acid

$$2Al(s) + 3H_2SO_4(aq) \rightarrow Al_2(SO_4)_3(aq) + 6H_2(g)$$

0 → +3   oxidation
+1 → 0   reduction

- 2 Al in 2Al   each Al increases by +3   total increase = +6
- 6 H in $3H_2SO_4$ each H decreases by −1   total decrease = −6

▲ Figure 2 Iron reacts with chlorine gas in a gas jar, forming iron(III) chloride, $FeCl_3$

Study tip

The oxidation number applies to each atom of an element. In this example, each atom of Ag decreases from +1 to 0. The total change is −2 because there are 2 Ag, each decreasing by −1.

Summary questions

1 State the oxidation state of the species in the following:
   a Ag^+ (1 mark)
   b F_2 (1 mark)
   c $NaClO_3$ (1 mark)

2 State the oxidation number of sulfur in the following:
   a H_2S (1 mark)
   b SO_4^{2-} (1 mark)
   c $Na_2S_2O_3$ (1 mark)

3 The following reaction is a redox process:
   $Mg + 2HCl \rightarrow MgCl_2 + H_2$
   a Identify the changes in oxidation number. (2 marks)
   b Explain which species is being oxidised and which is being reduced. (2 marks)

Chapter 4 Practice questions

Practice questions

1. 25.0 cm³ sample of an aqueous H_2SO_4 solution of unknown concentration is titrated with 0.125 mol dm⁻³ NaOH. 22.40 cm³ of NaOH are required to reach the end point.

   $H_2SO_4 + 2NaOH \rightarrow Na_2SO_4 + 2H_2O$

   a Calculate the amount, in mol of NaOH used. *(1 mark)*

   b Calculate the amount, in mol of H_2SO_4 used. *(1 mark)*

   c Calculate the concentration in mol dm⁻³ of the H_2SO_4. *(1 mark)*

2. The reaction below is a redox reaction.

   $3CuO + 2NH_3 \rightarrow 3Cu + 3H_2O + N_2$

   Explain in terms of oxidation number, what has been oxidised and what has been reduced. *(2 marks)*

3. This question looks at two redox reactions.

   a $3Mg + 2Fe(NO_3)_3 \rightarrow 3Mg(NO_3)_2 + 2Fe$

   (i) Explain in terms of electrons, what has been oxidised and what has been reduced. *(2 marks)*

   (ii) What is the systematic name for $Fe(NO_3)_3$? *(1 mark)*

   b $MnO_2 + 4HCl \rightarrow MnCl_2 + Cl_2 + 2H_2O$

   (i) Explain, in terms of oxidation numbers, what has been reduced. *(2 marks)*

   (ii) Use oxidation numbers to show that chlorine has only partly been oxidised. *(2 marks)*

4. a A solution of calcium chloride can be prepared by neutralisation reactions of dilute hydrochloric acid with solid calcium carbonate, with solid calcium oxide and with aqueous calcium hydroxide.

   (i) Write equations, with state symbols, for these methods of preparing calcium chloride. *(6 marks)*

   (ii) Why are these reactions all neutralisation reactions? *(1 mark)*

   (iii) Write an ionic equation for the reaction of aqueous calcium hydroxide and hydrochloric acid. *(1 mark)*

   b A solution of calcium chloride can be also prepared by the redox reaction of dilute hydrochloric acid and calcium metal.

   (i) Explain reduction and oxidation in terms of electrons and oxidation numbers *(2 marks)*

   (ii) Write an equation, with state symbols, for this reaction. *(2 marks)*

   (iii) For this redox reaction, determine what had been oxidised and what has been reduced and identify the changes in oxidation numbers. *(2 marks)*

5. Tungsten ore contains WO_3. Tungsten can be extracted from WO_3 present in its ore in a redox reaction with hydrogen. The unbalanced equation is shown below.

   $WO_3 + H_2 \rightarrow W + H_2O$

   a Balance the equation. *(1 mark)*

   b What is meant by oxidation and reduction in terms of electrons? *(1 mark)*

   c Using oxidation numbers show that oxidation and reduction have taken place in this reaction. *(3 marks)*

   d Some ore contains 2% of WO_3 by mass. Calculate the maximum mass of tungsten that could be obtained from the processing of 100 tonnes of ore. *(4 marks)*

6. A household cleaner containing dissolved ammonia is analysed as follows.

   25.0 cm³ of the cleaner is diluted to 250.0 cm³. 25.0 cm³ of the resulting solution is titrated with 0.125 mol dm⁻³ H_2SO_4(aq) and 24.40 cm³ were required to reach the end point. The equation is shown below.

   $2NH_3(aq) + H_2SO_4(aq) \rightarrow (NH_4)_2SO_4(aq)$

   a Calculate the amount, in mol, of H_2SO_4 used in the titration. *(1 mark)*

   b Calculate the amount, in mol, of NH_3 used in the titration. *(1 mark)*

   c Calculate the concentration of ammonia in the household cleaner

   (i) in mol dm⁻³; *(1 mark)*

   (ii) in g dm⁻³. *(1 mark)*

7. a The reaction between magnesium and sulfuric acid is a redox reaction.

   $Mg(s) + H_2SO_4(aq) \rightarrow MgSO_4(aq) + H_2(g)$

(i) Use oxidation numbers to identify which element has been oxidised. Explain your answer. *(2 marks)*

(ii) Describe what you would see when magnesium reacts with an excess of sulfuric acid. *(2 marks)*

b Epsom salts can be used as bath salts to help relieve aches and pains. Epsom salts are crystals of hydrated magnesium sulfate, $MgSO_4 \cdot xH_2O$.

A sample of Epsom salts was heated to remove the water. 1.57 g of water was removed leaving behind 1.51 g of anhydrous $MgSO_4$.

(i) Calculate the amount, in mol, of anhydrous $MgSO_4$ formed. *(2 marks)*

(ii) Calculate the amount, in mol, of H_2O removed. *(1 mark)*

(iii) Calculate the value of x in $MgSO_4 \cdot xH_2O$. *(1 mark)*

F321 June 2009 1(b) (c)

8 A student carries out experiments using acids, bases and salts.

a Calcium nitrate, $Ca(NO_3)_2$, is an example of a salt.

The student prepares a solution of calcium nitrate by reacting dilute nitric acid, HNO_3, with the base calcium hydroxide, $Ca(OH)_2$.

(i) Why is calcium nitrate an example of a salt? *(1 mark)*

(ii) Write the equation for the reaction between dilute nitric acid and aqueous calcium hydroxide. Include state symbols. *(2 marks)*

(iii) Explain how the hydroxide ion in aqueous calcium hydroxide acts as a base when it neutralises dilute nitric acid. *(1 mark)*

b A student carries out a titration to find the concentration of some sulfuric acid.

The student finds that 25.00 cm³ of 0.0880 mol dm⁻³ aqueous sodium hydroxide, NaOH, is neutralised by 17.60 cm³ of dilute sulfuric acid, H_2SO_4.

$H_2SO_4(aq) + 2NaOH(aq) \rightarrow Na_2SO_4(aq) + 2H_2O(l)$

(i) Calculate the amount, in moles, of NaOH used. *(1 mark)*

(ii) Determine the amount, in moles, of H_2SO_4 used. *(1 mark)*

(iii) Calculate the concentration, in mol dm⁻³, of the sulfuric acid. *(1 mark)*

c After carrying out the titration in (b), the student left the resulting solution to crystallise. White crystals were formed, with a formula of $Na_2SO_4 \cdot xH_2O$ and a molar mass of 322.1 g mol⁻¹.

(i) What term is given to the '$\cdot xH_2O$' part of the formula? *(1 mark)*

(ii) Using the molar mass of the crystals, calculate the value of x. *(2 marks)*

F321 Jan 10 Q2

9 Compound **A** is an organic acid containing C, H and O only.

A student analyses compound **A** to find its molar mass as follows.

The student dissolves 2.6432 g of compound **A** in water and dilutes the solution to 250 cm³ in a volumetric flask. The student fills a burette with this solution.

The student adds 25.0 cm³ of 0.224 mol dm⁻³ KOH to a conical flask.

31.25 cm³ of compound **A** was required for reach the end point.

The formula of compound **A** can be simplified as H_2X. The equation in the titration is:

$H_2X(aq) + 2KOH(aq) \rightarrow K_2X(aq) + 2H_2O(l)$

a Calculate the amount, in mol, of compound **A** used in the titration. *(2 marks)*

b Calculate the molar mass of compound **A**. *(2 marks)*

c The percentage composition by mass of compound A is C, 40.68%; H, 5.08%; O, 54.24%

Calculate the empirical and molecular formulae of compound **A**. *(3 marks)*

d Write an equation for the reaction in the titration of compound **A** with KOH. *(1 mark)*

5 ELECTRONS AND BONDING
5.1 Electron structure
Specification reference: 2.2.1

Learning outcomes
Demonstrate knowledge, understanding, and application of:
→ electrons and shells
→ atomic orbitals
→ filling of orbitals
→ electron configurations.

Electrons and shells
You already know that the nuclear atom contains electrons in shells surrounding the nucleus. In this topic you will learn more about how electrons are arranged within shells.

Shells
For GCSE, you learnt the 2,8,8 rule for the number of electrons that can fill each shell.

For A Level, you need to know the number of electrons that fill the first four shells ($n = 1$ to $n = 4$) shown in Table 1.

What are shells?
In an atom, electron shells make up a model that helps us to visualise something that cannot be seen.
- Shells are regarded as energy levels.
- The energy increases as the shell number increases.
- The shell number or energy level number is called the **principal quantum number n**.

▼ Table 1 Maximum number of electrons in the first four shells

Shell number n	Number of electrons
1	2
2	8
3	18
4	32

Electrons in shells

The current model for an electron is based the idea that an electron has properties both of a wave and a particle — wave–particle duality. If you take A level physics, you will learn more about wave–particle duality.

Around an atom, electrons can only fit into energy levels defined by the *wave* nature of an electron.

The maximum number of electrons in a shell n is given by the formula:
- number of electrons = $2n^2$

1. Use the $2n^2$ formula to check the number of electrons in Table 1.
2. Find out the maximum number electrons for the energy levels $n = 5, 6,$ and 7.

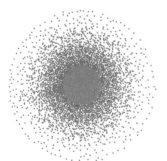

▲ Figure 1 *An atomic orbital shown as a cluster of dots for the probability of finding an electron in a given space*

Atomic orbitals
Shells are made up of **atomic orbitals**. An atomic orbital is a region around the nucleus that can hold up to two electrons, with opposite spins.

Models visualise an atomic orbital as a region in space where there is a high probability of finding an electron (Figure 1). An electron can be thought of as a negative-charge cloud with the shape of the orbital, referred to as an electron cloud.

- An orbital can hold one or two electrons, but no more.
- There are different types of orbitals: s-, p-, d- and f-orbitals.
- Each type of orbital has a different shape.

This may seem bewildering, but you will soon see patterns emerge.

ELECTRONS AND BONDING

s-orbitals

In an s-orbital the electron cloud is within the shape of a sphere (Figure 2). As with all orbitals, an s-orbital can hold one or two electrons.

- Each shell from $n = 1$ contains *one* s-orbital.
- The greater the shell number n, the greater the radius of its s-orbital.

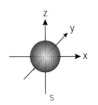

▲ Figure 2 Shape of an s-orbital

p-orbitals

In a p-orbital, the electron cloud is within the shape of a dumb-bell. As with an s-orbital, one orbital can contain one or two electrons. There are three separate p-orbitals at right angles to one another (Figure 3). These orbitals are referred to as p_x, p_y, and p_z.

- Each shell from $n = 2$ contains *three* p-orbitals.
- The greater the shell number n, the further the p-orbital is from the nucleus.

d-orbitals and f-orbitals

The next orbitals, d- and f-orbitals, are more complex.

- Each shell from $n = 3$ contains *five* d-orbitals.
- Each shell from $n = 4$ contains *seven* f-orbitals.

Sub-shells

You may have noticed that a new type of orbital is added for each additional shell. Within a shell, orbitals of the same type are grouped together as **sub-shells**. The orbitals and sub-shells in the first four shells are shown in Table 2.

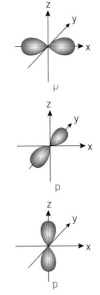

▲ Figure 3 Shapes of the three p-orbitals in a shell

> **Study tip**
>
> You need to know about the existence of each type of orbital, but you only need to know the shapes of s- and p-orbitals.

▼ Table 2 Shells, sub-shells, and orbitals

Shell	Number of orbitals				Sub-shells present	Number of electrons in sub-shells	Number of electrons in shell
	s	p	d	f			
1	1				1s	2	2
2	1	3			2s + 2p	2 + 6	8
3	1	3	5		3s + 3p + 3d	2 + 6 + 10	18
4	1	3	5	7	4s + 4p + 4d + 4f	2 + 6 + 10 + 14	32

- Each new shell gains a new type of orbital.
- The number of orbitals increases with each new type of orbital
  s, 1   p, 3   d, 5   f, 7
- two electrons fit into each orbital, so the number of electrons in each sub-shell also increases
  s, 2   p, 6   d, 10   f, 14

5.1 Electron structure

Filling of orbitals

There is a set of rules for how orbitals are occupied by electrons.

Orbitals fill in order of increasing energy

The sub-shells that make up shells have slightly different energy levels. Within each shell, the new type of sub-shell added has a higher energy. For example, in the second shell, the 2p sub-shell is the new type, and is at a higher energy than the 2s sub-shell.

- in the $n = 2$ shell, the order of filling is 2s, 2p.
- in the $n = 3$ shell, the order of filling is 3s, 3p, 3d.
- in the $n = 4$ shell, the order of filling is 4s, 4p, 4d, 4f.

Figure 4 shows the relative energy levels of the sub-shells making up the first four shells. The *highest* energy level in the third shell overlaps with the *lowest* energy level in the fourth shell.

- The 3d sub-shell is at a *higher* energy level than the 4s sub-shell.
- The 4s sub-shell therefore fills *before* the 3d sub-shell.
- The order of filling is therefore 3p, 4s, 3d.

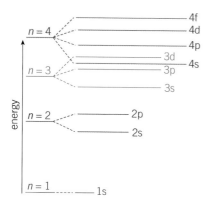

▲ **Figure 4** *Energy levels of the sub-shells*

Electrons pair with opposite spins

Each orbital can hold up to two electrons. Rather than drawing different orbital shapes, it is convenient to use an electrons-in-box model.

- Electrons are negatively charged and repel one another.
- Electrons have a property called *spin* – either up or down.
- An electron is shown as an arrow indicating its spin, either ↑ or ↓.
- The two electrons in an orbital must have *opposite spins*, as shown in Figure 5. The opposite spins help to counteract the repulsion between the negative charges of the two electrons.

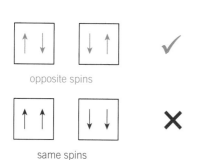

▲ **Figure 5** *'Electrons-in-box' model showing allowed and unallowed spins*

Orbitals with the same energy are occupied singly first

Within a sub-shell, the orbitals have the same energy. One electron occupies each orbital before pairing starts. This prevents any repulsion between paired electrons until there is no further orbital available at the same energy level.

Figure 6 shows how four electrons occupy the p-orbitals of a p-sub-shell.

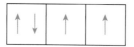 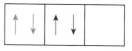

- With four electrons, one electron occupies each p-orbital.
- Only then can pairing take place.
- The paired electrons have opposite spins, ↑ and ↓.

- Not allowed.
- One p-orbital has been left empty and a second p-orbital has been paired.

▲ **Figure 6** *Allowed spins when filling p-orbitals*

5 ELECTRONS AND BONDING

Electron configuration

Electron configuration of atoms

For A Level, you are expected to be able to work out the **electron configuration** of atoms up to atomic number Z = 36 (krypton, Kr,) at the end of Period 4 in the periodic table). The electron configuration of an atom shows how sub-shells are occupied by electrons.

An atom of krypton, Z = 36, contains 36 electrons. In Figure 4, you can see the energy levels of the sub-shells in the first four shells. Table 3 shows how the 36 electrons in a krypton atom fill the sub-shells from the lowest energy level upwards.

▼ Table 3 *Filling the sub-shells of a krypton atom (Z = 36)*

sub-shell	1s →	2s →	2p →	3s →	3p →	4s →	3d →	4p
36 electrons	2	2	6	2	6	2	10	6

The 4s sub-shell fills before the 3d sub-shell because 4s has a lower energy level.

The electron configuration of krypton is therefore written as $1s^22s^22p^63s^23p^64s^23d^{10}4p^6$.

Figure 7 shows the meaning of the numbers in an electron configuration, using nitrogen as an example.

Table 4 shows the electron configurations of the elements across Period 2 of the periodic table, together with the much simpler electron structures from GCSE showing shells only.

> **Study tip**
>
> Electron configurations should be shown in shell order rather than the order of filling. So the electron configuration of krypton *should* show the 3d sub-shell before the 4s:
>
> $1s^22s^22p^63s^23p^63d^{10}4s^24p^6$
>
> instead of
>
> $1s^22s^22p^63s^23p^64s^23d^{10}4p^6$

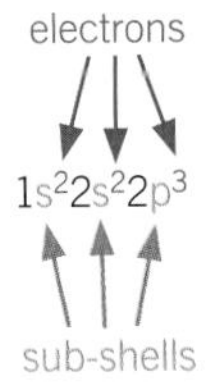

▲ Figure 7 *Electron configuration for the seven electrons in a nitrogen atom*

> **Synoptic link**
>
> You will learn more about the link between electron configurations and the periodic table in Topic 7.1, The periodic table.

▼ Table 4 *Electron configurations across Period 2*

	Li	Be	B	C	N	O	F	Ne
A level	$1s^22s^1$	$1s^22s^2$	$1s^22s^22p^1$	$1s^22s^22p^2$	$1s^22s^22p^3$	$1s^22s^22p^4$	$1s^22s^22p^5$	$1s^22s^22p^6$
GCSE	2,1	2,2	2,3	2,4	2,5	2,6	2,7	2,8

Shorthand electron configurations

Electron configurations can be expressed more simply in terms of the previous noble gas in the periodic table plus the outer electron sub-shells. Table 5 shows this shorthand notation for elements in Group 1 of the periodic table. This notation is useful for emphasising similarities in the electron configurations of the outer shell.

▼ Table 5 *Shorthand notation for electron configuration*

	Electron configuration	Shorthand notation
Li	$1s^22s^1$	[He]$2s^1$
Na	$1s^22s^22p^63s^1$	[Ne]$3s^1$
K	$1s^22s^22p^63s^23p^64s^1$	[Ar]$4s^1$

5.1 Electron structure

Synoptic link
You will find out more about blocks in Topic 7.1, The periodic table

Synoptic link
You will learn more about the electron configuration of d-bock elements in Topic 24.1, d-block elements.

Study tip
With the 4s and 3d sub-shells, it is a case of first in, first out.
- The 4s electrons are first in.
- The 4s electrons are first out.

Summary questions

1 State how many electrons the following can hold.
   a 3s orbital   *(1 mark)*
   b 4p sub-shell   *(1 mark)*
   c $n = 4$ shell   *(1 mark)*
   d 4d orbital   *(1 mark)*

2 Write the electron configuration for the following atoms:
   a C   *(1 mark)*
   b S   *(1 mark)*
   c K   *(1 mark)*
   d Co   *(1 mark)*
   e As   *(1 mark)*

3 Write the electron configuration for the following ions:
   a Mg^{2+} *(1 mark)*
   b P^{3-} *(1 mark)*
   c Br^- *(1 mark)*
   d Fe^{2+} *(1 mark)*

4 Write the shorthand electron configurations for
   a Si   *(1 mark)*
   b Cl^- *(1 mark)*
   c Mn^{2+} *(1 mark)*
   d Ga   *(1 mark)*

Electron configuration of ions
- Positive ions or **cations** are formed when atoms *lose* electrons.
- Negative ions or **anions** are formed when atoms *gain* electrons.

Blocks and the periodic table
The periodic table can be divided into **blocks** corresponding to their highest energy sub-shell.

- s-block
  highest energy electrons in the s-sub-shell (left block of two groups)
- p-block
  highest energy electrons in the p-sub-shell (right block of six groups)
- d-block
  highest energy electrons in the d-sub-shell (centre block of 10 groups)

Ions of s-block and p-block elements
When forming ions, the highest energy sub-shells lose or gain electrons. Table 6 compares the electron configurations of calcium and oxygen atoms and their common ions.

▼ Table 6 *Electron configurations of calcium and oxygen atoms and ions*

	Number of electrons	Electron configuration	
Ca	20	$1s^2 2s^2 2p^6 3s^2 3p^6 4s^2$	↓
Ca^{2+}	18	$1s^2 2s^2 2p^6 3s^2 3p^6$	2 electrons *lost* from 4s sub-shell
O	8	$1s^2 2s^2 2p^4$	↓
O^{2-}	10	$1s^2 2s^2 2p^6$	2 electrons *gained* by 2p sub-shell

Ions of d-block elements
For atoms of d-block elements, the 4s sub-shell is at a lower energy than the 3d sub-shell, so is filled first. The energies of the 4s and 3d sub-shells are very close together and, once filled, the 3d energy level falls below the 4s energy level. The consequence is that:

- the 4s sub-shell *fills* before the 3d sub-shell
- the 4s sub-shell also *empties* before the 3d sub-shell.

The d-block element nickel, $Z = 28$, forms the ion Ni^{2+}.

- For a nickel atom, the electron configuration of the 28 electrons is $1s^2 2s^2 2p^6 3s^2 3p^6 3d^8 4s^2$
- To form the nickel ion, Ni^{2+}, two electrons are lost from the 4s sub-shell: $1s^2 2s^2 2p^6 3s^2 3p^6 3d^8$

5.2 Ionic bonding and structure
Specification reference: 2.2.2

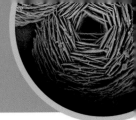

Ionic bonding

Ionic bonding is the electrostatic attraction between positive and negative ions. It holds together cations (positive ions) and anions (negative ions) in ionic compounds.

Common cations include:

- metal ions (e.g., Na^+, Ca^{2+}, Al^{3+})
- ammonium ions (NH_4^+).

Common anions include:

- non-metal ions (e.g., Cl^-, O^{2-})
- polyatomic ions (e.g., NO_3^-, SO_4^{2-}).

Ionic compounds

Dot-and-cross diagrams

The simplest ionic compounds contain metal ions and non-metal ions.

At GCSE, you used a model for ionic bonding based on electron transfer.

- Outer-shell electrons from a metal atom are transferred to the outer shell of a non-metal atom.
- Positive and negative ions are formed.
- The ions formed often have outer shells with the same electron configuration as the nearest noble gas.

In dot-and-cross diagrams, the electrons in the original atoms are shown as either dots or crosses. It is then easy to work out the charge on each ion and to account for all electrons.

Example – potassium fluoride, KF

Figure 1 shows electron transfer of the one outer-shell electron in a potassium atom, K, to the outer shell of a fluorine atom, F, forming a K^+ ion and an F^- ion. The square brackets show that the charge is spread over each ion and that the ions are separate entities. Only the outer-shell electrons are shown because the inner shells are full and not involved in bonding. The electron structures of the K^+ and F^- ions formed are now the same as the nearest noble gas.

> **Learning outcomes**
>
> Demonstrate knowledge, understanding, and application of:
>
> → ionic bonding
> → giant ionic lattice structures
> → properties of ionic compounds.

> **Synoptic link**
>
> In Topic 2.3, Formulae and equations, you looked at how ionic charges are used for writing the formulae of ionic compounds.

> **Study tip**
>
> Dot-and-cross diagrams often show the outer-shell electrons only. In the examples, each cation has been shown with an empty shell. This circle can be omitted for even greater clarity.

	K atom	F atom	K^+ ion	F^- ion
Electron structure	19 p⁺, 19 e⁻	9 p⁺, 9 e⁻	19 p⁺, 18 e⁻	9 p⁺, 10 e⁻
	2,8,8,1	2,7	2,8,8 (argon)	2,8 (neon)
Charge	neutral	neutral	1+	1−

◀ **Figure 1** Dot-and-cross diagram of potassium fluoride, KF

5.2 Ionic bonding and structure

Example – magnesium chloride, MgCl$_2$

Figure 2 shows electron transfer of the two outer-shell electrons in a magnesium atom to two chlorine atoms to form an Mg^{2+} ion and two Cl$^-$ ions. The electron structures of the Mg^{2+} and Cl$^-$ ions are the same as the nearest noble gas.

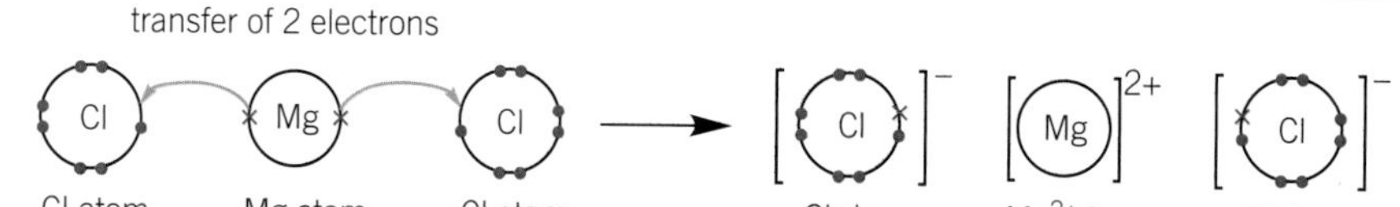

	Cl atom	Mg atom	Cl atom	Cl$^-$ ion	Mg^{2+} ion	Cl$^-$ ion
	17 p$^+$, 17 e$^-$	12 p$^+$, 12 e$^-$	17 p$^+$, 17 e$^-$	17 p$^+$, 18 e$^-$	12 p$^+$, 10 e$^-$	17 p$^+$, 18 e$^-$
electron structure	2,8,7	2,8,2	2,8,7	2,8,8 (argon)	2,8 (argon)	2,8,8 (argon)
charge	neutral	neutral	neutral	1–	2+	1–

▲ **Figure 2** *Dot-and-cross diagram of magnesium chloride, MgCl$_2$*

Structure of ionic compounds

Although it is convenient to look at ionic bonding acting between a small number of ions, each ion attracts oppositely charged ions in *all* directions.

The result is a **giant ionic lattice** structure containing billions of billions of ions, the actual number only determined by the size of the crystal. The giant ionic lattice of sodium chloride is shown in Figure 4. A giant lattice is a key structural feature of all ionic compounds.

- Each Na$^+$ ion is surrounded by 6 Cl$^-$ ions.
- Each Cl$^-$ ion is surrounded by 6 Na$^+$ ions.
- Each ion is surrounded by oppositely charged ions, forming a giant ionic lattice.

▲ **Figure 3** *Crystals of rock salt (sodium chloride) with a cubic shape*

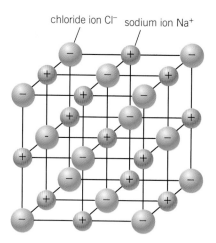

◀ **Figure 4** *Part of the sodium chloride lattice – the regular cubic arrangement of Na$^+$ and Cl$^-$ ions in the giant ionic lattice structure gives the crystal its cubic shape*

Properties of ionic compounds

The physical properties of ionic compounds can be explained in terms of the giant ionic lattice structure and ionic bonding.

Melting and boiling points

Almost all ionic compounds are solids at room temperature. At room temperature, there is insufficient energy to overcome the strong electrostatic forces of attraction between the oppositely charged ions in the giant ionic lattice. High temperatures are needed to provide the large quantity of energy needed to overcome the strong electrostatic attraction between the ions. Therefore most ionic compounds have *high* melting and boiling points.

Table 1 compares the melting points of several ionic compounds.

The melting points are higher for lattices containing ions with greater ionic charges, as there is stronger attraction between ions. Ionic attraction also depends on the size of the ions, but in this example the Na^+ and Ca^{2+} ions are similar sizes so this is not a factor here.

▼ **Table 1** *Melting points of ionic compounds*

Ionic compound	Ions	Melting point / °C
NaF	Na^+ and F^-	993
CaF_2	Ca^{2+} and F^-	1423
Na_2O	Na^+ and O^{2-}	1275
CaO	Ca^{2+} and O^{2-}	2614

Solubility

Many ionic compounds dissolve in **polar** solvents, such as water. Polar water molecules break down the lattice and surround each ion in solution.

In a compound made of ions with large charges, the ionic attraction may be too strong for water to be able to break down the lattice structure. The compound will not then be very soluble. Table 2 compares the solubility in water of several ionic compounds. The most soluble compound.

▼ **Table 2** *Solubility of ionic compounds in water*

Ionic compound	Ions	Solubility at 20 °C / mol dm^{-3}
NaCl	Na^+ and Cl^-	6.1
$CaCl_2$	Ca^{2+} and Cl^-	0.67
Na_2CO_3	Na^+ and CO_3^{2-}	2.0
$CaCO_3$	Ca^{2+} and CO_3^{2-}	1.3×10^{-4}

> **Synoptic link**
>
> You will find out more about polarity in Topic 6.2, Electronegativity and polarity.

A word of caution

Solubility requires two main processes:

- the ionic lattice must be broken down
- water molecules must attract and surround the ions.

The solubility of an ionic compound in water therefore depends on the relative strengths of the attractions within the giant ionic lattice and the attractions between ions and water molecules. For the ionic compounds in Table 2, the attractions in the giant ionic lattice have the greater effect, and solubility decreases as ionic charge increases. But predictions of solubility should be treated with caution.

Electrical conductivity

In the solid state, an ionic compound does *not* conduct electricity. But once melted or dissolved in water, the ionic compound does conduct electricity (Figure 5).

5.2 Ionic bonding and structure

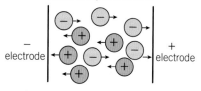

- ions fixed in position in the lattice
- ions cannot move
- no conductivity

- ions are not fixed in a lattice
- ions are now free to move
- electricity is conducted

▲ **Figure 5** *Electrical conductivity of an ionic compound in solid, liquid, and aqueous states*

In the *solid* state:

- the ions are in a fixed position in the giant ionic lattice
- there are no mobile charge carriers.

An ionic compound is a *non-conductor* of electricity in the solid state.

When *liquid* or *dissolved in water*:

- the solid ionic lattice breaks down
- the ions are now free to move as mobile charge carriers.

An ionic compound is a *conductor* of electricity in liquid and aqueous states.

Summary of properties

Most ionic compounds:

- have high melting and boiling points
- tend to dissolve in polar solvents such as water
- conduct electricity only in the liquid state or in aqueous solution.

 Ionic bones and teeth

You rely on ionic compounds for the skeleton framework of your body and for your teeth. So what are bones and teeth made out of? The chemistry is complex but the main ionic compound is calcium hydroxyapatite, which can be represented with the formula $Ca_5(PO_4)_3OH$. This compound is also the main constituent in tooth enamel.

Unfortunately ions in tooth enamel are removed in acid conditions. Once broken down, gaps in enamel can allow tooth decay to develop beneath. Saliva helps to neutralise acidic food and also to replace ions but this may not be enough. Fluoride ions help to replace lost ions by forming fluoropatite, $Ca_5(PO_4)_3F$, which is stronger than hydroxyapatite and more resistant to acid conditions.

Most toothpastes contain fluoride as sodium fluoride. Your water supply may also contain fluoride depending on where you live.

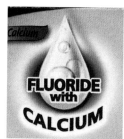

▲ **Figure 6** *Fluoride toothpaste*

1. Show that the ionic charges of the ions in $Ca_5(PO_4)_3OH$ and $Ca_5(PO_4)_3F$ balance.
2. Suggest why $Ca_5(PO_4)_3OH$ is very insoluble in water.
3. Many people confuse fluoride with fluorine. What is the difference and why would it be very strange to find fluorine in your toothpaste?

Summary questions

1. Draw dot-and-cross diagrams, with outer shells only, for the following:
   a Na_2O *(1 mark)*   b MgS   *(1 mark)*   c AlF_3 *(1 mark)*

2. Explain why ionic compounds have high melting and boiling points. *(2 marks)*

3. Explain why ionic compounds dissolve in water. *(2 marks)*

5.3 Covalent bonding

Specification reference: 2.2.2

Covalent bonding

Covalent compounds and molecules

Covalent bonding is the strong electrostatic attraction between a shared pair of electrons and the nuclei of the bonded atoms. Covalent bonding occurs between atoms in:

- non-metallic elements, for example, H_2 and O_2
- compounds of non-metallic elements, for example, H_2O and CO_2
- polyatomic ions, for example, NH_4^+.

For covalent bonding, the key feature is the sharing of a pair of electrons between the two atoms. The atoms are bonded together in a single unit – a small molecule (e.g., H_2), a giant covalent structure (e.g., SiO_2), or a charged polyatomic ion (e.g., NH_4^+). This is very different from the model of electron transfer to form separate ions in ionic bonding.

Learning outcomes

Demonstrate knowledge, understanding, and application of:

→ single covalent bonding
→ multiple covalent bonding
→ dative covalent (coordinate) bonding
→ average bond enthalpy.

The covalent bond

Orbital overlap

A covalent bond is the overlap of atomic orbitals, each containing one electron, to give a shared pair of electrons (Figure 1).

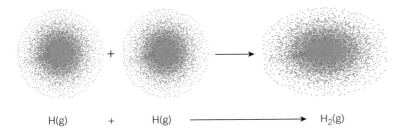

H(g)   +   H(g)   → H_2(g)

▲ **Figure 1** *Overlap of the two 1s orbitals of two hydrogen atoms to form a hydrogen molecule*

- The shared pair of electrons is attracted to the nuclei of both the bonding atoms.
- The bonded atoms often have outer shells with the same electron structure as the nearest noble gas.

A covalent bond is localised

In Topic 5.2, you saw that an ion attracts oppositely charged ions in all directions, resulting in a giant ionic lattice structure containing many billions of ions.

A covalent bond is very different. The attraction is *localised*, acting solely between the shared pair of electrons and the nuclei of the two bonded atoms. The result can be a small unit called a **molecule**, consisting of two or more atoms, such as H_2 and H_2O. A molecule is the smallest part of a covalent compound that can exist whilst retaining the chemical properties of the compound.

ionic
Na^+ ion attracts in all directions in three dimensions

covalent
In H_2, the attraction is solely between the shared pair of electrons in the covalent bond and the nuclei of the bonding atoms

▲ **Figure 2** *Comparison of attraction from an ion and in a covalent bond*

5.3 Covalent bonding

Single covalent bonds

Dot-and-cross diagrams

As with ionic bonding, dot-and-cross diagrams are used to account for electrons in covalent bonding.

- In covalent bonding electrons are shared.
- In ionic bonding electrons are transferred.

Figure 3 shows dot-and-cross diagrams for some simple molecules. Use of dots and crosses allows the origin of each electron to be shown clearly. Each bonding atom now has the electron structure of the nearest noble gas.

Figure 3 also shows a second structure for each molecule called a **displayed formula**.

- A displayed formula shows the relative positioning of atoms and the bonds between them as lines.
- Paired electrons that are not shared are called **lone pairs**. These can also be added to displayed formulae, as in water, H_2O, and ammonia, NH_3, in Figure 3.

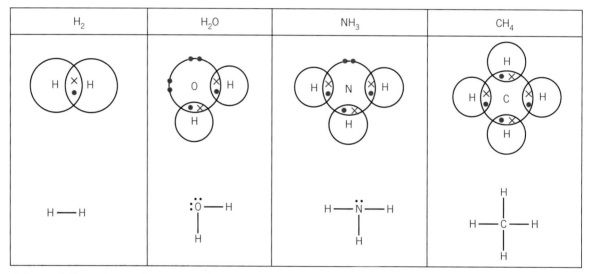

▲ **Figure 3** *Dot-and-cross diagrams and displayed formulae for hydrogen, H_2, water, H_2O, ammonia, NH_3, and methane, CH_4*

Number of covalent bonds

Most of the covalent compounds that you will meet during the course are compounds of hydrogen, carbon, nitrogen, and oxygen. You can see in Figure 3 the number of covalent bonds usually formed by atoms of these elements:

- carbon forms 4 bonds
- nitrogen forms 3 bonds
- oxygen forms 2 bonds
- hydrogen forms 1 bond.

> **Study tip**
>
> You will meet covalent compounds of carbon, nitrogen, oxygen, and hydrogen frequently throughout the course, particularly in organic chemistry.

ELECTRONS AND BONDING

What about other elements?

Boron

Boron, in Period 2 of the periodic table, has the electron configuration $1s^2 2s^2 2p^1$, so only three outer-shell electrons can be paired. Boron forms covalent compounds, such as boron trifluoride, BF_3 (Figure 4), in which its three outer-shell electrons are paired. So a molecule of BF_3 only has six electrons around the boron atom.

BF_3 shows that predictions for bonding cannot be based solely on the noble gas electron structure.

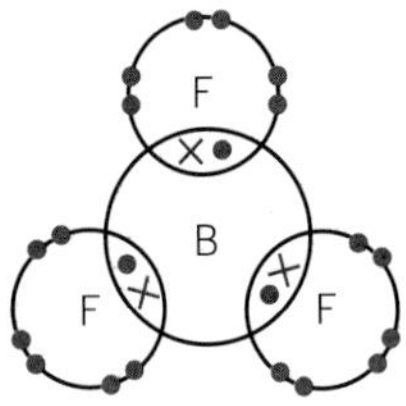

▲ **Figure 4** Dot-and-cross diagram of boron trifluoride, BF_3, showing how the three boron electrons are paired

Phosphorus, sulfur, and chlorine

Table 1 shows the formulae of the fluorides of the non-metals phosphorus, sulfur, and chlorine in Period 3 of the periodic table.

▼ **Table 1** Number of covalent bonds formed by phosphorus, sulfur, and chlorine

Element	phosphorus	sulfur	chlorine
Electron structure	$[Ne]3s^2 3p^3$	$[Ne]3s^2 3p^4$	$[Ne]3s^2 3p^5$
Outer-shell electrons	5	6	7
Formula of fluoride	PF_3 PF_5	SF_2 SF_4 SF_6	ClF ClF_3 ClF_5 ClF_7

> **Synoptic link**
>
> You can use Table 1 to work out the oxidation numbers of phosphorus, sulfur, and chlorine in each fluoride. You studied oxidation numbers in Topic 4.3, Redox.

Phosphorus trifluoride, PF_3, sulfur difluoride, SF_2, and chlorine monofluoride, ClF, follow the expected pattern of formulae, with the bonded atoms having a noble gas electron structure. But how can atoms of phosphorus, sulfur, and chlorine bond with more fluorine atoms to give the other fluorides in Table 1?

For the elements in Period 2, the $n = 2$ outer shell can hold just eight electrons. But for phosphorus, sulfur, and fluorine, the $n = 3$ outer shell can hold 18 electrons, so more electrons are available for bonding. Figure 5 shows how different arrangements for the six outer-shell electrons of sulfur and the different numbers of bonds possible to bond with fluorine.

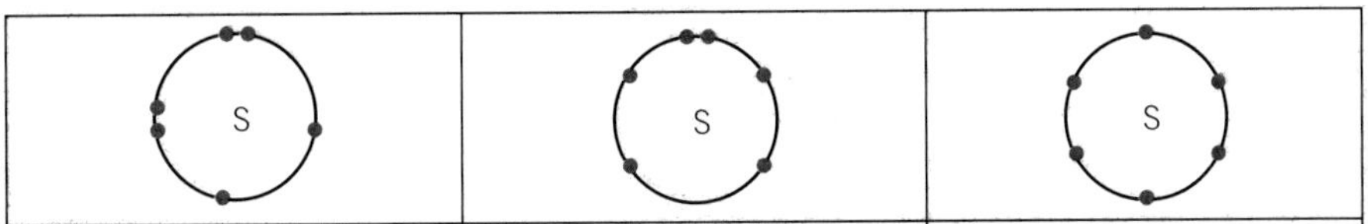

two unpaired electrons	four unpaired electrons	six unpaired electrons
two bonds possible	four bonds possible	six bonds possible
SF_2	SF_4	SF_6

◀ **Figure 5** Different numbers of unpaired electrons lead to different possibilities for covalent compounds of sulfur

5.3 Covalent bonding

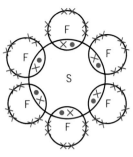

▲ Figure 6 *Dot-and-cross diagram of sulfur hexafluoride, SF_6, showing how the six sulfur electrons all paired forming six covalent bonds*

Figure 6 shows the dot-and-cross diagram of sulfur hexafluoride, SF_6.

- In SF_6, six unpaired electrons from sulfur are paired.
- The outer shell of sulfur now contains 12 electrons, far more than the nearest noble gas, argon, Ar.

This is called expansion of the octet and is possible only from the $n = 3$ shell, when a d-sub-shell becomes available for the expansion.

Multiple covalent bonds

A multiple covalent bond exist when two atoms share more than one pair of electrons.

Double covalent bonds

- In a *double* bond, the electrostatic attraction is between *two* shared pairs of electrons and the nuclei of the bonding atoms.
- Figure 7 shows dot-and-cross diagrams of double bonds in molecules of oxygen (O=O) and carbon dioxide (O=C=O).
- All atoms have eight electrons in their outer shell and the electron structure of the nearest noble gas.
- C=C and C=O double bonds are very important in organic chemistry.

> **Synoptic link**
>
> You will learn more about the nature of a single and double covalent bond when you study organic chemistry in Chapter 12, Alkanes, and Chapter 13, Alkenes.

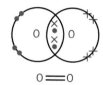 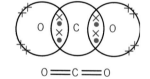

▲ Figure 7 *Double covalent bonds in oxygen, O_2, and in carbon dioxide, CO_2*

Triple covalent bonds

- In a *triple* bond, the electrostatic attraction is between *three* shared pairs of electrons and the nuclei of the bonding atoms.
- Figure 8 shows dot-and-cross diagrams of triple bonds in molecules of nitrogen (N≡N) and hydrogen cyanide (H—C≡N).
- Again, all atoms have the electron structure of the nearest noble gas.

▲ Figure 8 *Triple covalent bonds in nitrogen, N_2, and in hydrogen cyanide, HCN*

Dative covalent bonds

A **dative covalent** or **coordinate bond** is a covalent bond in which the shared pair of electrons has been supplied by one of the bonding atoms only. In a dative covalent bond the shared electron pair was originally a *lone pair* of electrons on one of the bonded atoms. For example, Figure 9 shows formation of an ammonium ion, NH_4^+, by reaction of ammonia, NH_3, and a H^+ ion.

- An ammonia molecule donates its lone pair of electrons to a H^+ ion.
- The dative covalent bond in NH_4^+ is shown by a bond with an arrowhead, →, to show that the nitrogen atom provides both electrons to the covalent bond.
- In an NH_4^+ ion, all four bonds are equivalent and you cannot tell which is the dative covalent bond. The arrow for the dative covalent bond (Figure 10) just helps with accounting for all electrons.

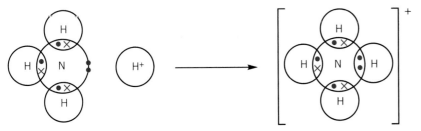

▲ Figure 9 *Formation of a dative covalent bond in NH_4^+*

▲ Figure 10 *Displayed formula of the NH_4^+ ion*

Average bond enthalpy

Average bond enthalpy serves as a measurement of covalent bond strength. The larger the value of the average bond enthalpy, the stronger the covalent bond. Table 2 shows some average bond enthalpies.

▼ Table 2 *Average bond enthalpies*

Bond	Average bond enthalpy / kJ mol^{-1}	Relative strength
Br–Br	193	Increasing bond strength
C–Br	290	
C–O	358	
O–H	464	

Summary questions

1. State what is meant by the term covalent bond. *(1 mark)*

2. Draw dot-and-cross diagrams and displayed formula for the following.
   a F_2O *(2 marks)*
   b PH_3 *(2 marks)*
   c CS_2 *(2 marks)*

3. Draw dot-and-cross diagrams and displayed formulae for the following:
   a C_2H_4 *(1 mark)*
   b H_2CO (carbon is the central atom) *(1 mark)*
   c H_3O^+ (dative bond) *(1 mark)*
   d CO (triple bond, one dative) *(1 mark)*

4. Draw dot-and-cross diagrams and displayed formulae for the following:
   a PF_5 *(1 mark)*
   b ClF_3 *(1 mark)*
   c SO_2 (two double bonds) *(1 mark)*
   d SO_3 (three double bonds) *(1 mark)*

5. Draw displayed formulae for:
   a HNO_3 (nitrogen is bonded to three oxygens only) *(2 marks)*
   b H_2SO_4 (sulfur is bonded to four oxygens only) *(2 marks)*

Synoptic link

Average bond enthalpies and related calculations are covered in detail in Topic 9.3, Bond enthalpies.

In Topic 15.1 Chemistry of the haloalkanes, you will see how average bond enthalpies can explain different rates of reaction.

Chapter 5 Practice questions

Practice questions

1 Using the periodic table, or the Roman numerals in the chemical name, predict the ionic charges and write formulae for the following:

   a lithium phosphide
   b lithium phosphate
   c chromium(III) hydroxide
   d iron(III) selenide
   e titanium(III) nitride
   f barium sulfate
   g barium sulfite
   h nickel(II) manganate(VII) *(8 marks)*

2 This question looks at shells, sub-shells and orbitals.

   a What is meant by the term *orbital*? *(1 mark)*
   b How many electrons completely fill
      (i)   a 3p orbital *(1 mark)*
      (ii) the 3d sub-shell *(1 mark)*
      (iii) The $n = 4$ shell. *(1 mark)*
   c Write the electron configuration for the following atoms:
      (i)   S *(1 mark)*
      (ii) Co *(1 mark)*
   d Electrons are arranged in energy levels. The incomplete diagram below for the seven electrons in a nitrogen atom shows just the two electrons in the 1s level

   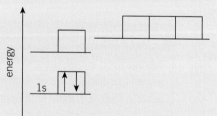

   Complete the diagram by
      (i)   adding arrows to the boxes *(1 mark)*
      (ii) adding labels for the other sub-shell levels. *(1 mark)*
   e Write the electron configuration for the following ions:
      (i)   Br⁻ *(1 mark)*
      (ii) Ga³⁺ *(1 mark)*

   f Answer the parts below with a number.
      (i)   The number of unpaired electrons in a sulfur atom. *(1 mark)*
      (ii) The number of electrons occupying p-orbitals in a germanium atom. *(1 mark)*
      (iii) The number of full shells in a krypton atom. *(1 mark)*

3 Magnesium fluoride and potassium chloride are examples of compounds with ionic bonding.

   a Explain how ionic bonding holds together the particles in an ionic compound. *(1 mark)*
   b Draw a dot-and-cross diagram to show the bonding in MgF_2. Show outer electrons only. *(2 marks)*
   c The diagram represents the incomplete structure of potassium chloride.

   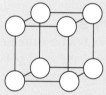

   (i) What name is given to this type of structure? *(1 mark)*
   (ii) Complete the diagram by adding labels to each circle in the diagram for the particles present in the structure. *(2 marks)*

   d A student found that MgF_2 has different electrical conductivities when solid and when dissolved in water.

   Explain these observations. *(2 marks)*

   e The table below shows the melting points of four compounds with ionic bonding.

Ionic compound	NaF	Na_2O	MgF_2	MgO
Melting point/°C	993	1275	1263	2852

   Identify the pattern in melting points and suggest reasons for the differences. *(3 marks)*

4 This question looks at compounds and ions that have covalent bonding.

   a PF_3 has covalent bonding.

Draw a dot-and-cross diagram to show the bonding in PF$_3$. Show outer electrons only. *(1 mark)*

b BF$_3$ reacts with NH$_3$ to form F$_3$BNH$_3$, which contains a dative covalent bond.

  (i) Explain how a dative covalent bond is different from a normal covalent bond. *(1 mark)*

  (ii) Draw a dot-and-cross diagram and a displayed formula for F$_3$BNH$_3$, showing the dative covalent bond. *(2 marks)*

c The nitrate(V) ion, NO$_3^-$, is a polyatomic ion, bonded by covalent bonds. The three oxygen atoms are bonded by one single covalent bond, one double covalent bond and one dative covalent bond.

  (i) Draw a displayed formula for the NO$_3^-$ ion. *(1 mark)*

  (ii) Draw a dot-and-cross diagram to show the bonding on NO$_3^-$. Show outer electrons only. *(2 marks)*

d An ionic compound has the empirical formula H$_4$N$_2$O$_3$. Suggest the formulae of the ions present in this compound. *(2 marks)*

5 When magnesium is heated in air, it reacts with oxygen to form magnesium oxide.

$2Mg(s) + O_2(s) \rightarrow 2MgO(s)$

a Magnesium oxide is an ionic compound.

Draw a dot-and-cross diagram for MgO. Show outer electrons only. *(2 marks)*

b Magnesium oxide has an extremely high melting point which makes it suitable as a lining for furnaces.

Explain, in terms of its structure and bonding, why magnesium oxide has this property. *(3 marks)*

c When magnesium oxide is added to warm dilute nitric acid, a reaction takes place forming a solution containing ions. Solid MgO does not conduct electricity but the solution formed does.

  (i) Explain the different conductivities of solid MgO and the solution. *(2 marks)*

  (ii) Write an equation for the reaction between magnesium oxide and dilute nitric acid. Include state symbols. *(2 marks)*

  (iii) State the formulae of two main ions present in this solution. *(2 marks)*

6 This question look at bonding involving carbon with other atoms.

a Draw a dot-and-cross diagram for a carbon dioxide molecule. *(1 mark)*

b One representation of the bonding in a carbon monoxide shows a triple bond, as shown in the dot and cross diagram below.

  (i) State the number of lone pairs and dative covalent bonds in a CO molecule. *(1 mark)*

  (ii) State what is meant by a covalent bond. *(1 mark)*

  (iii) State what is meant by dative covalent bond. *(1 mark)*

c Ethyne, C$_2$H$_2$ also contains a triple bond. Draw a dot-and-cross diagram of an ethyne molecule. *(1 mark)*

d A dot-and-cross diagram for a carbon monoxide molecule can also be drawn with a double bond between the carbon and oxygen atoms.

  (i) Draw this dot-and-cross diagram. *(1 mark)*

  (ii) What is unusual about this dot-and-cross diagram? *(1 mark)*

e The cyanide ion, CN$^-$, has a triple bond and is isoelectronic with carbon monoxide.

  (i) Draw a dot-and-cross diagram for a CN$^-$ ion. *(1 mark)*

  (ii) Suggest what is meant by the term *isoelectronic*. *(1 mark)*

f The displayed formula for a carbonate ion, CO$_3^{2-}$, is shown below.

Draw a dot-and-cross diagram for a carbonate ion. *(2 marks)*

6 SHAPES OF MOLECULES AND INTERMOLECULAR FORCES

6.1 Shapes of molecules and ions

Specification reference: 2.2.2

Learning outcomes

Demonstrate knowledge, understanding, and application of:
→ electron-pair repulsion theory
→ shapes of molecules and ions.

Electron-pair repulsion theory

An electron has a negative charge, so electron pairs repel one another. The electron-pair repulsion theory is a model used in chemistry for explaining and predicting the shapes of molecules and polyatomic ions.

- The electron pairs surrounding a central atom determine the shape of the molecule or ion.
- The electron pairs repel one another so that they are arranged as far apart as possible.
- The arrangement of electron pairs minimises repulsion and thus holds the bonded atoms in a definite shape.
- Different numbers of electron pairs result in different shapes.

Shapes of molecules

Representing molecules in three dimensions

A molecule of methane, CH_4, is symmetrical with *four* C—H covalent bonds.

- *Four* bonded pairs of electrons surround the central carbon atom.
- The *four* electron pairs repel one another as far apart as possible in three-dimensional (3D) space.

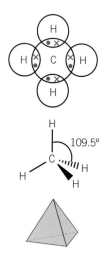

▲ Figure 1 *The four electron pairs and tetrahedral shape of a methane, CH_4, molecule*

The result is a *tetrahedral* shape with *four* equal H—C—H bond angles of 109.5°, shown in Figure 1.

Wedges

It is difficult to show a three-dimensional shape on a flat sheet of paper, so chemists use wedges to help visualise structures in three dimensions:

- a solid line ——— represents a bond in the plane of the paper
- a solid wedge ◢ comes out of the plane of the paper
- a dotted wedge ⋮⋮⋮ goes into the plane of the paper

Wedges are especially useful for representing bonds in organic molecules.

Synoptic link

The shapes of molecules and ions are extremely important, especially for organic chemistry and transition element chemistry, Chapter 24.

Bonded-pair and lone-pair repulsions

A lone pair of electrons is slightly closer to the central atom, and occupies more space, than a bonded pair. This results in a lone pair

SHAPES OF MOLECULES AND INTERMOLECULAR FORCES 6

repelling *more strongly* than a bonding pair. The relative repulsions between lone pairs and bonding pairs are shown below.

bonded-pair/bonded-pair < bonded-pair/lone-pair < lone-pair/lone-pair

increasing repulsion

Molecular shapes from four electron pairs

Methane, CH_4, ammonia, NH_3, and water, H_2O all have four electron pairs surrounding the central atom, but in ammonia and water the electron pairs are a mixture of bonded pairs and lone pairs.

- The four electron pairs around the central atom repel one another as far apart as possible into a tetrahedral arrangement.
- Lone pairs repel more strongly than bonded pairs.
- Therefore, lone pairs repel bonded pairs slightly closer together, decreasing the **bond angle** – the angle between the bonded pairs of electrons.
- The bond angle is reduced by about 2.5° for each lone pair.

Figure 2 compares the shapes of methane, ammonia, and water molecules, and their different bond angles.

Bonded pairs	4	3	2
Lone pairs	0	1	2
Name of shape	tetrahedral	pyramidal	non-linear
Shape and bond angle	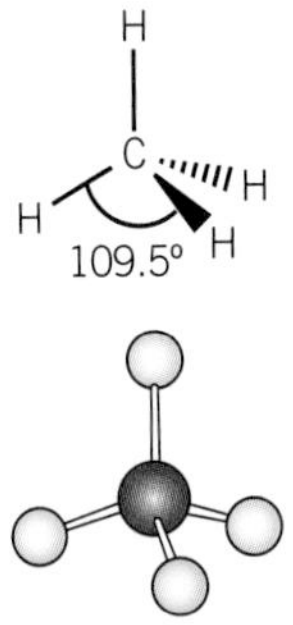 109.5°	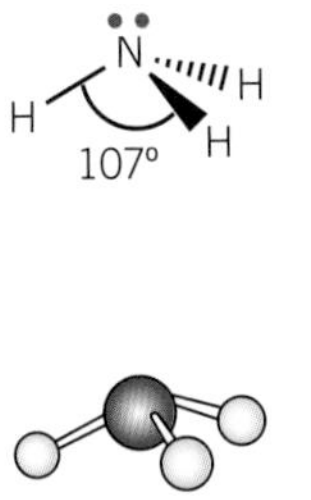 107°	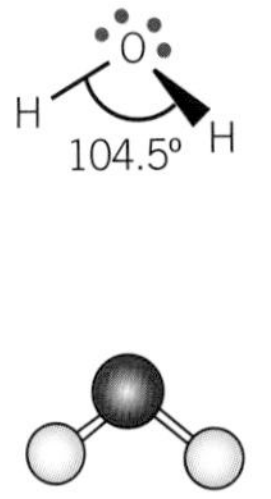 104.5°

▲ **Figure 2** *Shapes of methane, CH_4, ammonia, NH_3, and water, H_2O*

> **Study tip**
>
> Make sure that you learn the shapes and bond angles in these molecules.

Molecular shapes from multiple bonds

In molecules containing multiple bonds, each multiple bond is treated as a bonding region. For example, the bonding and shape of a carbon dioxide, CO_2, molecule is shown in Figure 3.

Molecule	Dot-and-cross diagram	Number of bonding regions	Shape and bond angle	Name of shape
CO_2	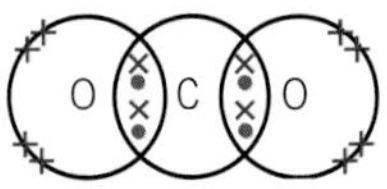	2	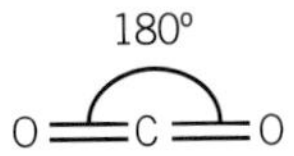 180° O=C=O	linear

▲ **Figure 3** *Dot-and-cross diagram and shape of a carbon dioxide, CO_2 molecule*

6.1 Shapes of molecules and ions

- The 4 bonded pairs around the central carbon atom are arranged as *two* double bonds, which count as *two* bonded regions.
- The two bonded regions repel one another as far apart as possible. This gives the carbon dioxide molecule a linear shape with all three atoms aligned in a straight line.

Molecular shapes from other numbers of electron pairs

The principles of electron-pair repulsion theory can be applied to any number of electron pairs surrounding the central atom.

- Electron pairs around the central atom repel each other as far apart as possible.
- The greater the number of electron pairs, the smaller the bond angle.
- Lone pairs of electrons repel more strongly than bonded pairs of electrons.

Three electron pairs

Boron trifluoride, BF_3, has only *three* bonded pairs around the central boron atom. Electron-pair repulsion gives a trigonal planar shape with equal bond angles of 120° (Figure 4).

Six electron pairs

Sulfur hexafluoride, SF_6, has *six* bonded pairs of electrons around the central sulfur atom. Electron-pair repulsion gives an octahedral shape with equal bond angles of 90° (Figure 4).

> **Synoptic link**
>
> The octahedral shape resulting from six bonding pairs is extremely important in transition metal chemistry, which you will study in Chapter 24, Transition elements.

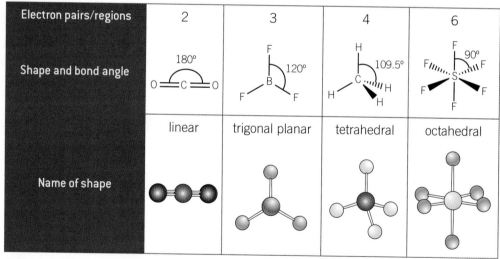

▲ Figure 4 *Shapes and bond angles for molecules with different numbers of electron pairs*

An octahedral shape from six bonded pairs

It may seem strange that SF_6, with *six* bonded pairs, forms a molecule that is an *octa*hedral shape. The reason lies with what is meant by the shape of a molecule. The *six* fluorine atoms are positioned at the *corners* of an octahedron. From the diagrams in Figure 5, you should be able to see that an octahedral shape with eight sides is obtained by joining together the six corners occupied by fluorine atoms.

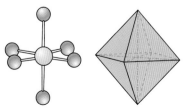

▲ Figure 5 *In an SF_6 molecule, six bonded pairs give an octahedral shape*

SHAPES OF MOLECULES AND INTERMOLECULAR FORCES 6

Shapes of ions

Electron-pair repulsion theory can also be used to explain and predict the shapes of ions.

The ammonium ion

The ammonium ion, NH_4^+, has four bonded pairs surrounding the central nitrogen atom (Figure 6).

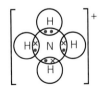

▲ Figure 6 *The four bonded pairs and tetrahedral shape of an NH_4^+ ion*

- An NH_4^+ ion has the same number of bonded pairs of electrons around the central atom as a methane molecule.
- NH_4^+ has the same tetrahedral shape and bond angles (109.5°) as a methane molecule.

Carbonate, nitrate, and sulfate ions

Figure 7 shows the shapes of carbonate, CO_3^{2-}, nitrate, NO_3^-, and sulfate, SO_4^{2-}, ions.

- CO_3^{2-} and NO_3^- ions have three regions of electron density surrounding the centre atom. So they have the same shape as a BF_3 molecule.
- SO_4^{2-} ions have four centres of electron density around the central sulfur atoms and have the same shape as a methane molecule.

trigonal planar CO_3^{2-} ion   trigonal planar NO_3^- ion   tetrahedral SO_4^{2-} ion

▲ Figure 7 *Shapes of carbonate, nitrate, and sulfate ions*

Predicting molecular shapes and bond angles

You should be able to predict the shapes of, and bond angles in, molecules and ions with different numbers of electron pairs, whether bonded pairs or lone pairs. The principles of electron-pair repulsion enable you to predict the arrangement of electron pairs around a central atom of unfamiliar molecules and ions.

Study tip

Remember that dot-and-cross diagrams can be very useful for working out the arrangement of electron pairs and therefore predicting molecular shapes.

Summary questions

1. Name the shapes and give the bond angles around the central atom in the following molecules:
   a. BeI_2 (1 mark)
   b. BCl_3 (1 mark)
   c. SiH_4 (1 mark)
   d. $H_2C=O$ (1 mark)
   e. CS_2 (1 mark)

2. Name the shapes and bond angles in the following:
   a. CH_3OH (around C and O) (1 mark)
   b. SO_3 (1 mark)
   c. SO_2 (1 mark)
   d. H_3O^+ (1 mark)

3. BF_3 reacts with NH_3 to form the compound F_3BNH_3. The bond angles in BF_3 and NH_3 are different from the bond angles in F_3BNH_3.
   a. State the bond angles in BF_3 and NH_3. (2 marks)
   b. Predict, with reasons, the bond angles in F_3BNH_3. (2 marks)

6.2 Electronegativity and polarity

Specification reference: 2.2.2

Learning outcomes

Demonstrate knowledge, understanding, and application of:
→ electronegativity
→ Pauling electronegativity values
→ bond polarity.

Electronegativity

In a covalent bond, the nuclei of the bonded atoms attract the shared pair of electrons. In molecules of elements, such as hydrogen, H_2, oxygen, O_2, nitrogen, N_2, and chlorine, Cl_2, the atoms are the same element and the bonded electron pair is shared evenly.

This changes when the bonded atoms are different elements:

- the nuclear charges are different
- the atoms may be different sizes
- the shared pair of electrons may be closer to one nucleus than the other.

The shared pair of electrons in the covalent bond may now experience more attraction from one of the bonded atoms than the other.

The attraction of a bonded atom for the pair of electrons in a covalent bond is called **electronegativity**.

▲ Figure 1 *Attraction causing a covalent bond*

How is electronegativity measured?

The Pauling scale is used to compare the electronegativity of the atoms of different elements. Figure 2 shows how **Pauling electronegativity values** depend on an element's position in the periodic table.

Across the periodic table

- the nuclear charge increases
- the atomic radius decreases.

A large Pauling value indicates that atoms of the element are very electronegative. You can see that electronegativity increases across and up the periodic table. Consequently, fluorine is the most electronegative atom and is given a Pauling value of 4.0. The noble gases are not included as they tend not to form compounds.

- The non-metals nitrogen, oxygen, fluorine, and chlorine have the most electronegative atoms.
- The Group 1 metals, including lithium, sodium, and potassium have the least electronegative atoms.

electronegativity increases →

H 2.1						
Li 1.0	Be 1.5	B 2.0	C 2.5	N 3.0	O 3.5	F 4.0
Na 0.9	Mg 1.2	Al 1.5	Si 1.8	P 2.1	S 2.5	Cl 3.0
K 0.8						Br 2.8

▲ Figure 2 *Pauling electronegativity values in the Periodic Table*

Ionic or covalent?

If the electronegativity difference is large, one bonded atom will have a much greater attraction for the shared pair than the other bonded atom. The more electronegative atom will have gained control of the electrons and the bond will now be ionic rather covalent.

Electronegativity values can be used to estimate the type of bonding as shown in Table 1.

▼ Table 1 *Ionic or covalent*

Bond type	Electronegativity difference
covalent	0
polar covalent	0 to 1.8
ionic	greater than 1.8

Bond polarity

Non-polar bonds

In a **non-polar bond**, the bonded electron pair is *shared equally* between the bonded atoms. A bond will be non-polar when:

- the bonded atoms are the same, or
- the bonded atoms have the same or similar electronegativity.

In molecules of elements such as hydrogen, oxygen, and chlorine, the bonded atoms come from the same element and the electron pair is shared equally. The bond is a **pure covalent bond** (Figure 3). Carbon and hydrogen atoms have very similar electronegativities and form non-polar bonds. Hydrocarbon liquids such as hexane, C_6H_{14}, are non-polar solvents and do not mix with water.

non-polar
electron pair is attracted
equally to each bonded atom

▲ Figure 3 *Two non-polar molecules, hydrogen, H_2, and chlorine, Cl_2. Only the bonding electrons have been shown for chlorine*

Polar bonds

In a polar bond, the bonded electron pair is shared unequally between the bonded atoms. A bond will be polar when the bonded atoms are different and have different electronegativity values, resulting in a **polar covalent bond**.

Example: Hydrogen chloride

Hydrogen chloride, HCl, has atoms of different elements.

- From Figure 2, hydrogen has an electronegativity of 2.1 and chlorine has an electronegativity of 3.0.
- The chlorine atom is *more* electronegative than the hydrogen atom.
- The chlorine atom has a greater attraction for the bonded pair of electrons than the hydrogen atom, resulting in a polar covalent bond.

polar
bonded electron pair is
attracted closer to Cl atom

▲ Figure 4 *A polar molecule of hydrogen chloride, HCl. Only the bonding electrons have been shown for chlorine*

The H—Cl bond is **polarised** with a small partial positive charge (δ+) on the hydrogen atom and a small partial negative charge (δ−) on the chlorine atom (Figure 4). The delta δ sign means small. The two charges, δ+ and δ− are partial charges, and are much smaller than a full + and − charge.

- This separation of opposite charges is called a **dipole**.
- The hydrogen chloride molecule is **polar**, with δ+ and δ− charges at different ends of the H—Cl bond.

A dipole in a polar covalent bond does not change and is called a **permanent dipole** to distinguish it from an *induced* dipole, which you will meet in the next topic.

Study tip

If you are provided with Pauling electronegativity values:

- the atom with the larger electronegativity value has the δ− charge
- the atom with the smaller electronegativity value has the δ+ charge.

Polar molecules

Hydrogen chloride is a polar molecule, as the H—Cl bond has one permanent dipole acting in the direction of the H—Cl bond. For molecules with more than two atoms, there may be two or more polar bonds. Depending on the shape of the molecule, the dipoles may reinforce one another to produce a larger dipole over the whole molecule, or cancel out if the dipoles act in opposite directions.

6.2 Electronegativity and polarity

A water, H_2O molecule is polar.

- The two O—H bonds each have a permanent dipole.
- The two dipoles act in different directions but do not exactly oppose one another.
- Overall the oxygen end of the molecule has a δ− charge and the hydrogen end of the molecule has a δ+ charge (Figure 5).

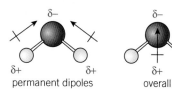

▲ **Figure 5** *The water, H_2O, molecule is polar*

A carbon dioxide, CO_2, molecule is non-polar.

- The two C=O bonds each have a permanent dipole.
- The two dipoles act in opposite directions and exactly oppose one another.
- Over the whole molecule, the dipoles cancel and the overall dipole is zero (Figure 6).

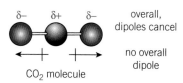

▲ **Figure 6** *The carbon dioxide, CO_2, molecule is non-polar*

Polar solvents and solubility

Figure 7 shows a sodium chloride lattice being dissolved by water molecules to form aqueous sodium and chloride ions:

$$NaCl(s) + aq \rightarrow Na^+(aq) + Cl^-(aq)$$

- Water molecules attract Na^+ and Cl^- ions.
- The ionic lattice breaks down as it dissolves.
- In the resulting solution, water molecules surround the Na^+ and Cl^- ions.

You can see that:

- Na^+ ions are attracted towards the oxygen of water molecules (δ−)
- Cl^- ions are attracted towards the hydrogen of water molecules (δ+).

> **Synoptic link**
>
> The solubility of ionic compounds by polar solvents was introduced in Topic 5.2, Ionic bonding and structure.

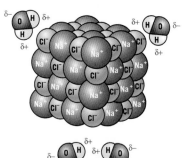

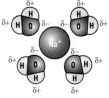

▲ **Figure 7** *Solid sodium chloride, NaCl, (top) dissolving in water to give sodium and chloride ions (bottom) surrounded by water molecules*

Summary questions

1. Define:
   a. electronegativity *(1 mark)*
   b. polar covalent bond *(1 mark)*
   c. dipole *(1 mark)*

2. This question refers to the following compounds:
   Br_2, NO_2, Na_2O, PH_3, Al_2O_3, SiO_2, KF, and LiBr
   Predict the type of bonding in each compound using the electronegativity values from Figure 2 in order to list the compounds from most ionic to most covalent. *(2 marks)*

3. This question refers to the following compounds, all with polar bonds:
   H_2S, $BeBr_2$, NH_3, BF_3, $SiCl_4$, $CHCl_3$, $H_2C=O$, PF_5, SF_6
   a. State which way round the dipole is in each bond. (Refer to the electronegativity values in Figure 2.)
   b. Classify the molecules as polar and non-polar, explaining how you have made your decisions. (You may find it useful to refer to Topic 6.1 to work out the shapes of the molecules.) *(9 marks)*

6.3 Intermolecular forces

Specification reference: 2.2.2

Forces between molecules

Covalent bonding is strong and holds the atoms in a molecule together. **Intermolecular forces** are weak interactions between dipoles of *different* molecules. Intermolecular forces fall into three main categories.

- induced dipole–dipole interactions (**London forces**)
- permanent dipole–dipole interactions
- hydrogen bonding.

Intermolecular forces are largely responsible for physical properties such as melting and boiling points, whereas covalent bonds determine the identity and chemical reactions of molecules.

Table 1 compares the strengths of intermolecular forces with covalent bonds.

Learning outcomes

Demonstrate knowledge, understanding, and application of:

→ induced dipole–dipole interactions (London forces)
→ permanent dipole–dipole interactions
→ structure and properties of simple molecular lattices.

▼ **Table 1** Strengths of intermolecular forces and covalent bonds

Type of bond	Bond enthalpy / kJ mol^{-1}
London forces	1–10
permanent dipole–dipole interactions	3–25
hydrogen bonds	10–40
single covalent bonds	150–500

Induced dipole–dipole interactions (London forces)

London forces are weak intermolecular forces that exist between *all* molecules, whether polar or non-polar. They act between induced dipoles in different molecules. The origin of induced dipoles is described in Figure 1.

- Movement of electrons produces a changing dipole in a molecule.
- At any instant, an **instantaneous** dipole will exist, but its position is constantly shifting

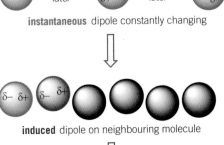

instantaneous dipole constantly changing

- The **instantaneous** dipole **induces** a dipole on a neighbouring molecule

induced dipole on neighbouring molecule

- The **induced** dipole induces further dipoles on neighbouring molecules, which then attract one another.

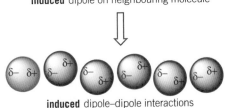

induced dipole–dipole interactions

▲ **Figure 1** *Origin of induced dipole–dipole interactions (London forces)*

6.3 Intermolecular forces

Induced dipoles are only temporary. In the next instant of time, the induced dipoles may disappear, only for the whole process to take place amongst other molecules.

The strength of induced dipole–dipole interactions (London forces)

Induced dipoles result from interactions of electrons between molecules. The more electrons in each molecule:

- the larger the instantaneous and induced dipoles
- the greater the induced dipole–dipole interactions
- the stronger the attractive forces between molecules.

Table 2 compares the boiling points of the first three noble gases, helium to argon.

- Larger numbers of electrons mean larger induced dipoles.
- More energy is then needed to overcome the intermolecular forces, increasing the boiling point.

Synoptic link

The shapes of molecules also affect the strength of London forces. This is especially important in organic chemistry (see Topic 12.1, the properties of the alkanes).

▼ Table 2 *London forces and boiling points of the noble gases helium to argon*

Noble gas	Number of electrons	Boiling point / °C	Relative strength
helium, He	2	−269	stronger London forces ↓
neon, Ne	10	−246	
argon, Ar	18	−186	

What ever happened to van der Waals' forces?

The term van der Waals, forces has sometimes been used to describe induced dipole–dipole interactions. The International Union of Pure and Applied Chemistry (IUPAC), which publishes guidelines for chemical terminology, recommends that van der Waals' forces be used for both permanent and induced dipole–dipole interactions, and London forces for induced dipole–dipole interactions. So the term van der Waals' forces is ambiguous.

London forces is the correct term to use when describing induced dipole–dipole interactions. In other sources, you may see the term dispersion forces. Whoever said the language of science is straightforward?

Permanent dipole–dipole interactions

In Topic 6.1, you looked at polarity in bonds and molecules. Permanent dipole–dipole interactions act between the permanent dipoles in different polar molecules.

Figure 2 shows intermolecular forces arising from permanent dipole–dipole interactions in hydrogen chloride, HCl, molecules.

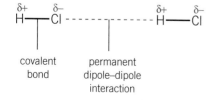

▲ Figure 2 *Permanent dipole–dipole interactions between hydrogen chloride, HCl, molecules*

Table 3 compares hydrogen chloride with fluorine, F_2. Molecules of hydrogen chloride and fluorine have the same number of electrons and the same shape, so the strength of the London forces in hydrogen chloride and fluorine should be very similar.

SHAPES OF MOLECULES AND INTERMOLECULAR FORCES

▼ Table 3 *Induced and permanent dipoles*

Molecule	Dipole	London forces	Permanent dipole–dipole interactions	Number of electrons	Boiling point / °C
F—F	none	yes	no	9 × 2 = 18	−220
H—Cl	$\delta+ \ \delta-$ H—Cl	yes	yes	1 + 17 = 18	−85

Study tip
It is a common mistake to forget that polar molecules have induced dipole interactions *as well as* permanent dipole–dipole interactions.

- Fluorine molecules are non-polar and only have London forces between molecules.
- Hydrogen chloride molecules are polar and have London forces *and* permanent dipole–dipole interactions between molecules.
- Extra energy is needed to break the additional permanent dipole–dipole interactions between hydrogen chloride molecules.
- The boiling point of hydrogen chloride is therefore higher than fluorine.

Simple molecular substances

A **simple molecular substance** is made up of simple molecules – small units containing a definite number of atoms with a definite molecular formula, such as neon, Ne, hydrogen, H_2, water, H_2O, and carbon dioxide, CO_2.

In the solid state, simple molecules form a regular structure called a **simple molecular lattice**. In the simple molecular lattice:

- the molecules are held in place by *weak* intermolecular forces
- the atoms within each molecule are bonded together *strongly* by covalent bonds.

Figure 3 shows the different forces in the simple molecular lattice of iodine, I_2.

Properties of simple molecular substances

Low melting point and boiling point
All simple molecular substances are covalently bonded. At room temperature, they may exist as solids, liquids, or gases. All simple molecular substances can be solidified into simple molecular lattices by reducing the temperature.

- In a simple molecular lattice, the weak intermolecular forces can be broken even by the energy present at low temperatures.
- Simple molecular substances have *low* melting and boiling points.

When a simple molecular lattice is broken apart during melting,
- only the weak intermolecular forces break
- the covalent bonds are strong and do *not* break.

Solubility
Covalent substances with simple molecular structures fall into two categories – polar and non-polar.

The solubility of non-polar substances is easier to predict.

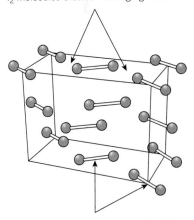

weak intermolecular interactions *between* I_2 molecules break on changing state

strong covalent bonds *between atoms* in I_2 molecule *do not* break on changing state

▲ Figure 3 *Iodine, I_2, is a solid at room temperature, but can easily be turned into a purple vapour. A small Bunsen flame provides enough energy to break the weak intermolecular forces between I_2 molecules in the simple molecular lattice*

6.3 Intermolecular forces

Solubility of non-polar simple molecular substances

- When a simple molecular compound is added to a *non-polar* solvent, such as hexane, intermolecular forces form between the molecules and the solvent.
- The interactions weaken the intermolecular forces in the simple molecular lattice. The intermolecular forces break and the compound dissolves.

Therefore, non-polar simple molecular substances tend to be *soluble* in *non-polar* solvents.

- When a simple molecular substance is added to a polar solvent, there is little interaction between the molecules in the lattice and the solvent molecules.
- The intermolecular bonding within the polar solvent is too strong to be broken.

Therefore simple molecular substances tend to be *insoluble* in *polar* solvents.

Figure 4 compares the solubility of iodine in water (polar solvent) and in cyclohexane (non-polar solvent).

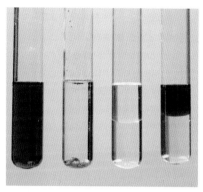

▲ Figure 4 *Solubility of iodine (purple when dissolved) in polar and non-polar solvents. From left to right – cyclohexane and iodine, water and iodine, water and cyclohexane, and layers of water (bottom) and cyclohexane (top) with iodine. Iodine dissolves readily in cyclohexane, a non-polar solvent, but not in water, a polar solvent*

Solubility of polar simple molecular substances

Polar covalent substances may dissolve in polar solvents as the polar solute molecules and the polar solvent molecules can attract each other. The process is similar to dissolving of an ionic compound. For example, sugar dissolves in water, a polar solvent. Sugar is a polar covalent compound with many polar O—H bonds, which attract and bond with polar water molecules. This solubility can extend to liquids and gases. Hydrogen chloride is a gas with a polar H—Cl bond that is extremely soluble in water, forming hydrochloric acid.

The solubility depends on the strength of the dipole and can be hard to predict. Some compounds such as ethanol, C_2H_5OH, contain both polar (the O—H) and non-polar (the carbon chain) parts in their structure and can dissolve in both polar and non-polar solvents.

Some biological molecules have hydrophobic and hydrophilic parts. The hydrophilic part will be polar and contain electronegative atoms (usually oxygen) that can interact with water. The hydrophobic part will be non-polar and comprised of a carbon chain.

Electrical conductivity

- There are no mobile charged particles in simple molecular structures.
- With no charged particles that can move, there is nothing to complete an electrical circuit.

Therefore simple molecular structures are *non-conductors* of electricity.

Summary questions

1. Explain how an induced dipole forms. *(2 marks)*

2. Explain why simple molecular compounds:
   a. have low melting and boiling points
   b. do not usually dissolve readily in water
   c. have poor electrical conductivity. *(3 marks)*

3. For each of the following, state the structure. Predict and explain the following physical properties: melting and boiling points, electrical conductivity, solubility.
   a. KBr *(4 marks)*
   b. CCl_4 *(4 marks)*

6.4 Hydrogen bonding
Specification reference: 2.2.2

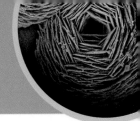

Hydrogen bonds

A **hydrogen bond** is a special type of permanent dipole–dipole interaction found between molecules containing:

- an electronegative atom with a lone pair of electrons, for example, oxygen, nitrogen, or fluorine
- a hydrogen atom attached to an electronegative atom, for example, H—O, H—N, or H—F.

The hydrogen bond acts between a lone pair of electrons on an electronegative atom in one molecule and a hydrogen atom in a different molecule. Hydrogen bonds are the strongest type of intermolecular attractions.

The hydrogen bond is shown by a dashed line. The lone pair of electrons on the oxygen in water, and the nitrogen in ammonia, play a key role in a hydrogen bond. The shape around the hydrogen atom involved in the hydrogen bond is linear.

Anomalous properties of water

Hydrogen bonding has a significant influence on the properties of many molecules but none more so than water. Hydrogen bonding gives water some unique and anomalous (unusual) properties that support the existence of life on Earth.

The solid (ice) is less dense than the liquid (water)

- Hydrogen bonds hold water molecules apart in an open lattice structure.
- The water molecules in ice are further apart than in water.
- Solid ice is less dense than liquid water and floats.

Ice floating on water may seem obvious, but water is one of the few substances in which the solid is less dense than water. The consequence is that ice floats instead of sinking in ponds and lakes, forming an insulating layer and preventing the water from freezing solid – good news if you are a fish. With *two* lone pairs on the oxygen atom and *two* hydrogen atoms, each water molecule can form *four* hydrogen bonds. The hydrogen bonds extend outwards, holding water molecules slightly apart and forming an open tetrahedral lattice full of holes (Figure 3). The bond angle about the hydrogen atom involved in the hydrogen bond is close to 180°.

The holes in the open lattice structure decrease the density of water on freezing. When ice melts, the ice lattice collapses and the molecules move closer together. So liquid water is denser than solid ice.

> **Learning outcomes**
> Demonstrate knowledge, understanding, and application of:
> → hydrogen bonding
> → anomalous properties of water.

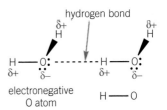

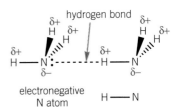

▲ **Figure 1** *Hydrogen bonding between molecules of water, H_2O, and ammonia, NH_3*

> **Study tip**
> When you draw a diagram to show hydrogen bonding, remember to show lone pairs and the atom polarities. Hydrogen is always $\delta+$!

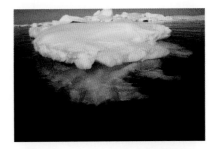

▲ **Figure 2** *At 0 °C, the density of ice is 0.917 g cm^{-3} compared to 1.029 g cm^{-3} for seawater. Only about $\frac{1}{9}$ of the volume of the iceberg is above the surface of the water*

6.4 Hydrogen bonding

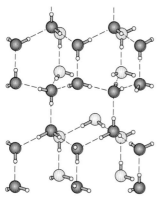

▲ **Figure 3** *The open lattice structure of ice, in which the linear H···O—H arrangement holds the water molecules apart*

Water has a relatively high melting point and boiling point

As with all molecules, water has London forces between molecules.

- Hydrogen bonds are extra forces, over and above the London forces.
- An appreciable quantity of energy is needed to break the hydrogen bonds in water, so water has much higher melting and boiling points than would be expected from just London forces.
- When the ice lattice breaks, the rigid arrangement of hydrogen bonds in ice is broken. When water boils, the hydrogen bonds break completely.

Without hydrogen bonds, water would have a boiling point of about −75 °C and would exist as a gas at room temperature and pressure. There would be no liquid water in most places on Earth and there would be no life as we know it.

Boiling points of hydrides and intermolecular forces

Figure 4 shows the boiling points of the hydrides of Groups 14–17 (Groups 4–7) in the periodic table.

1. Water, hydrogen fluoride, and ammonia do not follow the trend shown by the other hydrides in each group.
   a. Estimate what the boiling points of water, hydrogen fluoride, and ammonia would be if they were to follow the group trends.
   b. Explain why water, hydrogen fluoride and ammonia do not follow the group trends.
2. Explain why all groups show an increase in boiling point from Period 3 to Period 6.
3. What conclusions can be drawn about the relative strengths of London forces and permanent dipole–dipole interactions for the hydrides of Groups 14–17 (Groups 4–7)?

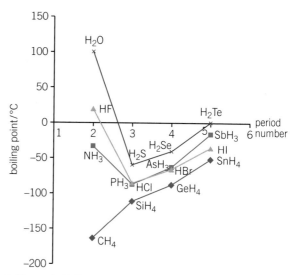

▲ **Figure 4** *Boiling points of hydrides of Groups 14–17 (Groups 4–7)*

Synoptic link

Hydrogen bonding is very important in organic compounds containing O—H, N—H, and C=O groups, such as alcohols, carboxylic acids, carbonyl compounds, amines, and amino acids. You will learn about all these groups when you study organic chemistry later in the course (Chapter 11 onwards).

Other anomalous properties of water

The extra intermolecular bonding from hydrogen bonds also contributes to many more unusual properties of water. Other examples are a relatively high surface tension and viscosity, properties than result in droplets that are 'not wet' and allow insects to walk on pond surfaces. Detergents reduce the surface tension, making water 'wetter'. Water has dozens of anomalous properties, so you can be relieved that you are only required to know about density, melting points, and boiling points.

6 SHAPES OF MOLECULES AND INTERMOLECULAR FORCES

Hydrogen bonding in DNA

The double helix structure of DNA (Figure 5) is held together by hydrogen bonds which enable a single DNA strand to create a perfect copy of itself in a process called replication.

The replication process depends on four bases adenine *A*, thymine *T*, cytosine *C*, and guanine *G* being present in the correct order.

In Figure 5, you can see that:

- A and T pair by forming two hydrogen bonds
- C and G pair by forming three hydrogen bonds

So why do the bases always pair up correctly?

The chemical structure and shape of these four bases ensure correct pairing.

Adenine and guanine are both purine bases with two-ringed structures.

Thymine and cytosine are both pyrimidine bases with single-ringed structures.

- Hydrogen bonding in the double helix can only take place between a purine and a pyrimidine base.
- The bases must fit together so that a hydrogen atom from one molecule and an electronegative atom (either oxygen or nitrogen) from the other molecule are aligned correctly to maximize hydrogen bonding.

Figure 6 shows the pairing of A with T and C with G.

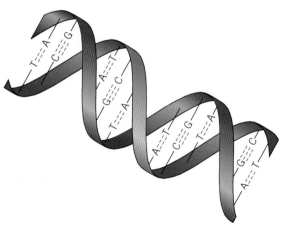

▲ **Figure 5** *Hydrogen bonds in DNA hold the double helix structure together and enable DNA to replicate by ensuring that the correct bases are paired.*

1. Suggest why pairing does not take place between two purine bases or between two pyrimidine bases.
2. Human DNA typically contains just over three billion base pairs in a definite sequence. During replication, this sequence must be copied perfectly. If replication took place randomly, how many possible sequences of the four bases could there be in human DNA?

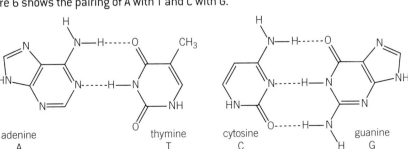

▲ **Figure 6** *Base pairing in DNA by hydrogen bonding*

Summary questions

1. State and explain two anomalous properties of water. *(4 marks)*

2. State which of the following compounds have hydrogen bonding H_2O, H_2S, CH_4, CH_3OH, NO_2. *(2 marks)*

3. Draw diagrams showing two ways that hydrogen bonding can form between:
    a. One molecule of ammonia and one molecule of water
    b. One molecule of water and one molecule of ethanol, CH_3CH_2OH. *(4 marks)*

83

Chapter 6 Practice questions

Practice questions

1. This question looks at polarity and shapes of molecules and ions
   a. Arrange the molecules below in order of increasing polarity.
      (i) HCl, HBr, HF, HI *(1 mark)*
      (ii) CH_2Br_2, CH_3I, $CHCl_2F$, CF_4 *(1 mark)*
   b. The list of ten molecules and ions below can be arranged into five pairs, each with a different shape.
      H_2O, HCN, $AlCl_4^-$, BF_3, SO_3, SCl_2, H_3O^+, NH_4^+, NH_3, CO_2
      (i) Select the five pairs and state the shape of each pair. *(5 marks)*
      (ii) Which of the molecules and ions are planar? *(1 mark)*

2. Complete the table below as follows.
   State the bond angle(s). *(5 marks)*
   b. Name the shape of each molecule. *(5 marks)*

Molecule	Bond angle	Shape
CO_2		
SF_6		
CH_4		
PF_3		
BF_3		

3. $SbCl_3$, exists as simple covalent molecules. A 'dot-and-cross' diagram of $SbCl_3$ is shown below.

   a. Predict the shape of a molecule of $SbCl_3$. Explain your answer. *(3 marks)*
   b. $SbCl_3$ molecules are polar. Explain why. *(2 marks)*

   F321 Jan 2014 Q1(d)

4. This question is about the molecular fluorides, F_2O, CF_4, NF_3, and OF_2.
   a. Complete the table as follows.
      (i) Add the number of bonded pairs and lone pairs of electrons around the atom in bold.
      Ignore any inner shells. *(4 marks)*
      (ii) Draw a 3D diagram for the shape of each molecule. Include any lone pairs *(4 marks)*

Molecule	Bonded pairs	Lone pairs	Shape
BF_3			
CF_4			
NF_3			
OF_2			

   b. In their solid structures, the four molecular fluorides all have London forces.
      Explain the origin of London forces. *(3 marks)*
   c. Electronegativity explains why all of the molecules have polar bonds.
      (i) Explain the term *electronegativity*. *(1 mark)*
      (ii) Show all the dipoles on a molecule of CF_4. *(1 mark)*
   d. Explain which of the molecules in the table are polar and which are non-polar. *(3 marks)*

5. Much of the chemistry of water is influenced by its polarity and its ability for form hydrogen bonds.
   a. What is meant by a hydrogen bond? *(1 mark)*
   b. Explain why water molecules are polar. *(2 marks)*
   c. Draw a diagram showing hydrogen bonding between two molecules of water. Include dipoles and lone pair of electrons. *(2 marks)*
   d. State the bond angle in a water molecule. *(1 mark)*
   e. State and explain two properties of ice that are a direct result of hydrogen bonding. *(4 marks)*

SHAPES OF MOLECULES AND INTERMOLECULAR FORCES

6 Linus Pauling was a Nobel Prize winning chemist who devised a scale of electronegativity. Some Pauling electronegativity values are shown in the table.

Element	Electronegativity
B	2.0
Br	2.8
N	3.0
F	4.0

   a What is meant by the term *electronegativity*? (2 marks)

   b Show, using δ+ and δ– symbols, the permanent dipoles on each of the following bonds. N—F; N—Br (1 mark)

   c Boron trifluoride, BF_3, ammonia, NH_3, and sulfur hexafluoride, SF_6, are all covalent compounds. The shapes of their molecules are different.

   (i) State the shape of a molecule of SF_6. (1 mark)

   (ii) Using outer electron shells only, draw 'dot-and-cross' diagrams for molecules of BF_3 and NH_3.
   Use your diagrams to explain why a molecule of BF_3 has bond angles of 120° and NH_3 has bond angles of 107°. (5 marks)

   (iii) Molecules of BF_3 contain polar bonds, but the molecules are non-polar.
   Suggest an explanation for this difference. (2 marks)

   F321 Jan 11 Q3

7 Simple molecules are covalently bonded.

   a State what is meant by the term *covalent bond*. (1 mark)

   b Chemists are able to predict the shape of a simple covalent molecule from the number of electron pairs surrounding the central atom.

   (i) Explain how this enables chemists to predict the shape. (2 marks)

   (ii) The 'dot-and-cross' diagram of the simple covalent molecule, H_3BO_3, is shown below.

   Predict the O—B—O and B—O—H bond angles in a molecule of H_3BO_3. (2 marks)

   c Give an example of a simple covalent molecule which has all bond angles equal to 90°. (1 mark)

   F321 Jan 2013 Q2

8 Hydrogen chloride is a colourless gas which forms white fumes in moist air.

   a Molecules of hydrogen chloride, HCl, and molecules of fluorine, F_2, contain the same number of electrons. Hydrogen chloride boils at –85 °C and fluorine boils at –188 °C.

   Explain why there is a difference in the boiling points of HCl and F_2. In your answer you should refer to the types of force acting between molecules and the relative strength of the forces between the molecules.

   In your answer, you should use appropriate technical terms, spelled correctly. (4 marks)

   b Hydrogen chloride reacts with water to produce an ion with the formula H_3O^+. An H_3O^+ ion has one dative covalent bond.

   Draw a 'dot-and-cross' diagram to show the bonding in H_3O^+. Show outer electrons only. (2 marks)

   F321 Jan 13 Q5(a)(b)

9 The compounds, NH_3, PF_3 and SF_6 are all gases at room temperature and pressure.

   a What intermolecular forces are present in liquid samples of each compound? (3 marks)

   b The boiling point of NH_3 is much higher than PF_3 and SF_6.
   Explain why. (2 marks)

   c Name the shapes and give the bond angles in molecules of NH_3 and SF_6? (4 marks)

Module 2 summary

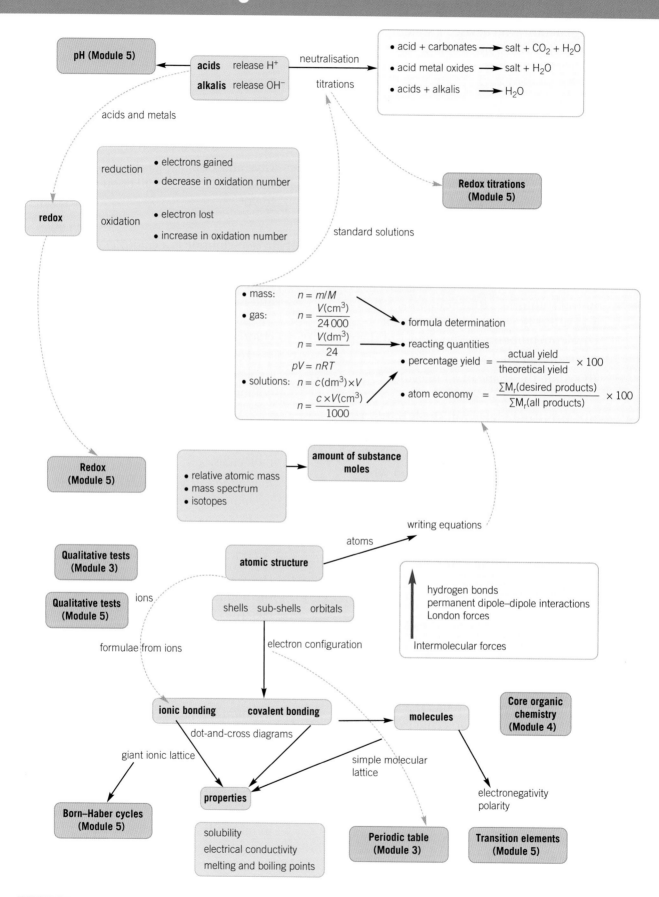

Module 2 Foundations in Chemistry

Krypton

Krypton, Kr, is an element with atomic number 36. It is a colourless gas at room temperature and is used in energy-saving fluorescent lights.

Krypton is in Group 18(0) of the periodic table. It was thought that krypton was completely unreactive. However, in 1963 it was discovered that krypton could react with fluorine, under extreme conditions, to form the molecule krypton difluoride, KrF_2.

$$Kr + F_2 \rightarrow KrF_2$$

Krypton difluoride is unstable and very reactive – it is the most powerful oxidising agent so far discovered.

1 The main isotope of krypton is $^{84}_{36}Kr$. State the number of protons, neutrons, and electrons in this isotope and write down the electron configuration of a krypton atom. Use your answer to suggest why krypton is usually very unreactive.
2 What is the oxidation state of krypton in KrF_2?
3 What mass of krypton would be needed to form 1.0 g of KrF_2? What would the volume in dm^3 of this mass of krypton (at RTP)?
4 Draw out a dot-and-cross diagram of the molecule KrF_2. Show outer electron shells only.

Extension

1 Research the molecule diborane, B_2H_6 and draw a dot and cross diagram of this molecule. Comment on what is unusual about its bonding and structure.
2 Produce a summary for how to deduce whether a molecule has an overall dipole. Include details of the factors that affect bond polarity and the effect of molecular shape. In your summary, include examples of several molecules, with and without overall dipoles
3 There are two isotopes of bromine atoms. Bromine in its standard state is a liquid consisting of Br_2 molecules. When a sample of bromine vapour is analysed using a mass spectrometer, three peaks are seen, with m/z values of 158, 160, and 162.
   a Use ideas about isotopes to explain why three peaks are seen and state the mass number of the two isotopes of bromine
   b The percentage abundances of these peaks are as follows:
      158 : 25.52%, 160 : 50.00%, 162 : 24.48%
      Use these data to calculate the percentage abundance (to 4 s.f.) of the isotopes of bromine

Module 2 Practice questions

1. What is the atomic structure of $^{80}Se^{2-}$?

	Protons	Neutrons	Electrons
A	34	46	36
B	34	80	36
C	36	44	34
D	46	34	48

   AS Paper 1 style question (1 mark)

2. A sample of ethanol, C_2H_6O, contains 3.00 mol of carbon atoms.

   How many hydrogen atoms are in the sample of ethanol?

   A 4.98×10^{-24} B 1.50×10^{-23}
   C 1.81×10^{24} D 5.42×10^{24} (1 mark)

   AS Paper 1 style question

3. A compound has the percentage composition by mass Ag: 71.03%, C: 7.90%, and O: 21.07%.

   What is the empirical formula of the compound?

   A $AgCO_2$ B $AgCO_3$ C Ag_2C_2O D Ag_3C_3O

   AS Paper 1 style question (1 mark)

4. 14.3 g of hydrated sodium carbonate, $Na_2CO_3 \cdot 10H_2O$ is dissolved in water and made up to 250 cm³ of solution.

   What is the concentration of sodium ions in the solution, in mol dm⁻³?

   A 0.0500 B 0.100 C 0.200 D 0.400

   AS Paper 1 style question (1 mark)

5. What is the oxidation number of Mn in $KMnO_4$?

   A +2 B +4 C +6 D +7 (1 mark)

   AS Paper 1 style question

6. The atoms of an element contain seven full orbitals and two singly-occupied orbitals.

   What is the element?

   A Si B P C S D Cl (1 mark)

   AS Paper 1 style question

7. What is the F—B—F bond angle in boron trifluoride, BF_3?

   A 90° B 107° C 120° D 180° (1 mark)

   AS Paper 1 style question

8. This question looks at some chemistry of phosphorus and its compounds.

   a State the full electron configuration of phosphorus showing sub-shells. (1 mark)

   b Phosphoric acid, H_3PO_4, can be made as the only product by reacting the oxide of phosphorus, P_4O_{10}, with water.

   (i) Write a balanced equation for this reaction. (1 mark)

   (ii) What is the oxidation number of phosphorus in phosphoric acid, H_3PO_4? (1 mark)

   c A student prepares a sample of the water softener Na_3PO_4 from 0.960 mol dm⁻³ NaOH and 0.500 mol dm⁻³ H_3PO_4. The equation is shown below.

   $3NaOH(aq) + H_3PO_4(aq) \rightarrow Na_3PO_4(aq) + 3H_2O(l)$

   (i) The student measures out 150 cm³ of 0.960 mol dm⁻³ NaOH.

   Calculate the amount, in mol, of NaOH that the student uses. (1 mark)

   (ii) Calculate the minimum volume 0.500 mol dm⁻³ H_3PO_4 that the student needs to add to completely react with the NaOH. (2 marks)

   (iii) How could the student obtain a solid sample of Na_3PO_4 from their reaction mixture? (1 mark)

   (iv) What is the maximum mass of Na_3PO_4 that the student could obtain from this experiment? (2 marks)

   d Calcium phosphate, $Ca_3(PO_4)_2$, contains two different ions. Suggest the formulae of the two ions. (2 marks)

   AS Paper 1 style question

9. This question is about the simple molecular compounds water, ammonia and boron trichloride.

   a Complete the table below to show the numbers of bonded or lone pairs of electrons surrounding the central atom per molecule.

Molecule	BCl_3	NH_3	H_2O
Number of bonded pairs			
Number of lone pairs			

   (3 marks)

b Ammonia and water form hydrogen bonds. Draw a diagram, including relevant dipoles to show a hydrogen bond between one molecule of water and one molecule of ammonia. *(2 marks)*

c Boron trichloride reacts vigorously with water forming a mixture of two acidic products. One of the products has a relative molecular mass of 61.8 and the following percentage composition by mass B: 17.48%, H: 4.85%, and O: 77.67%.

(i) Calculate the molecular formula of compound **C**. *(2 marks)*

(ii) Suggest an equation for the reaction between boron trichloride and water.
AS Paper 1 style question *(1 mark)*

10 Nitrogen oxides are emitted as pollutant gases from car exhausts. The mixture of nitrogen oxides is commonly referred to as NO_x and typically consists of a mixture of nitrogen monoxide, NO, and nitrogen dioxide, NO_2.

a (i) Nitrogen monoxide is formed from nitrogen and oxygen in the high temperatures in a car engine. Write an equation for this reaction. *(1 mark)*

(ii) Nitrogen dioxide is formed when nitrogen monoxide reacts with oxygen as the emissions cool. Write an equation for this reaction. *(1 mark)*

b On leaving the car's exhaust system, NO_2 dissolves in water to form a mixture of acids:

$$2NO_2(g) + H_2O(l) \rightarrow HNO_2(aq) + HNO_3(aq)$$

This is a disproportionation reaction. Explain what is meant by disproportionation and, using oxidation numbers, show that disproportionation has taken place. *(3 marks)*

c In urban traffic, a car releases 150 cm^3 of NO_x per kilometre.

Calculate the number of molecules of NO_x emitted each kilometre by this car. Give your answer to two significant figures and in standard form. *(2 marks)*

d The mass of NO_x released per kilometre in urban traffic is 0.250 g.

Using your answer to **(c)**, calculate the relative molecular mass of the NO_x.

Explain whether there are more NO_2 molecules or more NO molecules in this sample of NO_x. *(3 marks)*
AS Paper 2 style question

11 A student was asked to carry out an acid–base titration to find the concentration of some sulfuric acid. The student was supplied with a solution of sodium hydroxide with a concentration of 0.106 mol dm^{-3}.

The equation for the reaction is given below.

$$2NaOH(aq) + H_2SO_4(aq) \rightarrow Na_2SO_4(aq) + 2H_2O(l)$$

a The aqueous sodium hydroxide was to be added to a conical flask for the titration. The student was supplied with a 25.00 cm^3 pipette which may have been left unclean from a previous experiment. Describe how the pipette should be prepared for use in the titration. *(1 mark)*

b The student carried out the titration using a burette measuring to an accuracy of 0.05 cm^3.

- The student filled the burette with the sulfuric acid and the burette reading was 0.00 cm^3.
- She carried out a first titration until the burette reading was 21.55 cm^3.
- The student carried out a second titration, starting at 21.55 cm^3 and continuing until the burette reading was 42.25 cm^3.
- She then added more acid to the burette so that the burette read 10.00 cm^3.
- She carried out a third titration after which the burette reading was 30.75 cm^3.

Calculate the mean titre to one decimal place. Use this value to calculate the concentration of the sulfuric acid.
(4 marks)

c The pipette and burette were both labelled to show the accuracy of measurements:
pipette: ± 0.06 cm^3 burette: ± 0.05 cm^3

Calculate the percentage error in the volume of NaOH(aq) delivered from the pipette and the volume of H_2SO_4 in the titre.
(2 marks)
AS Paper 2 style question

MODULE 3
Periodic table and energy

Chapters in this module
7 The periodic table
8 Reactivity trends
9 Enthalpy
10 Equilibrium and reaction rates

Introduction

The legacy left by Mendeleev in producing the first periodic table and the work of many of other scientists have led to the development of the modern periodic table. Classifying the elements as both metal and non-metal and having a way or predicting the properties and reactivity of the elements is of upmost importance to scientific research. This module will provide the important chemical ideas for inorganic and physical chemistry.

The periodic table looks at physical trends and extends your understanding of structure and bonding.

Reactivity trends looks at group properties using Group 2 as a typical metal group and the halogens as a typical non-metal group. Redox reactions are a common theme.

Enthalpy focuses on enthalpy changes and their determination from experimental results and data tables. You will discover how to calculate energy changes from both your own experiments and from given data.

Reaction rates and equilibrium focuses on how changing conditions affect the rate of a reaction and the position of equilibrium. The integrated roles of enthalpy changes, rates, catalysts and equilibria are important for industrial processes, increasing yield ,and reducing energy demand whilst improving sustainability

Knowledge and understanding checklist

From your Key Stage 4 study you should be able to answer the following questions. Work through each point, using your Key Stage 4 notes and the support available on Kerboodle.

- ☐ Describe metals and non-metals and explain the differences between them on the basis of their characteristic physical and chemical properties.
- ☐ Explain how observed simple properties of Groups 1, 7, and 0 depend on the outer shell of electrons of the atoms and predict properties from given trends down the groups.
- ☐ Distinguish between endothermic and exothermic reactions on the basis of the temperature change of the surroundings.
- ☐ Calculate energy changes in a chemical reaction by considering bond making and bond breaking energies.
- ☐ Interpret rate of reaction graphs.
- ☐ Describe the effect of changes in temperature, concentration, pressure, and surface area on rate of reaction.
- ☐ Describe the characteristics of catalysts and their effect on rates of reaction.
- ☐ Recall that some reactions are reversible.

Maths skills checklist

In this module, you will need to use the following maths skills. You can find support for these skills on Kerboodle and through MyMaths.

- ☐ **Changing the subject of an equation**, for carrying out enthalpy change calculations.
- ☐ **Substituting numbers into algebraic equations**, for carrying out enthalpy change calculations.
- ☐ **Solving algebraic equations**, for carrying out Hess' law calculations.
- ☐ **Plotting two variables from experimental or other data**, for plotting graphs from collected or supplied data to follow the course of a reaction.
- ☐ **Drawing and using the gradient of a tangent to a curve as a measure of rate of reaction**, for when you calculate the reaction rate from a concentration–time graph.

MyMaths.co.uk
Bringing Maths Alive

7 PERIODICITY
7.1 The periodic table
Specification reference 3.1.1

Learning outcomes
Demonstrate knowledge, understanding, and application of:
→ the arrangement of elements
→ periodicity
→ groups
→ trend in electron configuration.

The periodic table

The periodic table – then

Just over 60 elements were known when Mendeleev arranged them in order of atomic mass (nothing was known about subatomic particles and atomic number at the time). He also lined up the elements in groups with similar properties. If the group properties did not fit, Mendeleev swapped elements around and left gaps, assuming that the atomic mass measurements were incorrect and that some elements were yet to be discovered. He even predicted properties of the missing elements from group trends – a remarkable insight at the time. It was not until protons were discovered in the early 1900s that the real reason for the order in Mendeleev's table was revealed.

Focus on ekasilicon

Mendeleev was so confident in his periodic law that he predicted the properties of elements that would fill the gaps in his periodic table. In 1871, he predicted the properties of an undiscovered element that would fit in the gap below silicon. He called this element 'ekasilicon'.

Ekasilicon was discovered in 1886 and was named germanium. Mendeleev's predictions are not far off, considering that he had never seen the element (Table 1).

▼ **Table 1** Mendeleev's predictions for ekasilicon, Eka-Si

Property	Eka-Si	Ge
atomic mass	72.00	72.61
density (g cm^{-3})	5.50	5.35
formula of oxide	EO$_2$	GeO$_2$
oxide density (g cm^{-3})	4.70	4.70

Look up the properties of aluminium and indium, and then predict the following properties for eka-aluminium **a** atomic mass **b** formula of chloride.

▲ **Figure 1** Dimitri Mendeleev is one of the main chemists credited with the periodic table

The periodic table – now

As of 2014, the periodic table has 114 elements arranged in seven horizontal periods and 18 vertical groups (Figure 2). The periodic table is the most important organisational tool in chemistry. It is the first point of reference for chemists everywhere and most chemistry laboratories have a periodic table prominently placed on the wall.

You do not need to memorise the periodic table (although some people have done so). It is nevertheless helpful to know where the common elements are positioned, and most students know the atomic numbers and relative atomic masses of common elements such as hydrogen, carbon, nitrogen, and oxygen. It is essential that you are able to *use* the periodic table.

Key
atomic number
Symbol
relative atomic mass

(1)	(2)											(3)	(4)	(5)	(6)	(7)	(0)
1	2	3	4	5	6	7	8	9	10	11	12	13	14	15	16	17	18
1 **H** 1.0																	2 **He** 4.0
3 **Li** 6.9	4 **Be** 9.0											5 **B** 10.8	6 **C** 12.0	7 **N** 14.0	8 **O** 16.0	9 **F** 19.0	10 **Ne** 20.2
11 **Na** 23.0	12 **Mg** 24.3											13 **Al** 27.0	14 **Si** 28.1	15 **P** 31.0	16 **S** 32.1	17 **Cl** 35.5	18 **Ar** 39.9
19 **K** 39.1	20 **Ca** 40.1	21 **Sc** 45.0	22 **Ti** 47.9	23 **V** 50.9	24 **Cr** 52.0	25 **Mn** 54.9	26 **Fe** 55.8	27 **Co** 58.9	28 **Ni** 58.7	29 **Cu** 63.5	30 **Zn** 65.4	31 **Ga** 69.7	32 **Ge** 72.6	33 **As** 74.9	34 **Se** 79.0	35 **Br** 79.9	36 **Kr** 83.8
37 **Rb** 85.5	38 **Sr** 87.6	39 **Y** 88.9	40 **Zr** 91.2	41 **Nb** 92.9	42 **Mo** 95.9	43 **Tc**	44 **Ru** 101.1	45 **Rh** 102.9	46 **Pd** 106.4	47 **Ag** 107.9	48 **Cd** 112.4	49 **In** 114.8	50 **Sn** 118.7	51 **Sb** 121.8	52 **Te** 127.6	53 **I** 126.9	54 **Xe** 131.3
55 **Cs** 132.9	56 **Ba** 137.3	57–71	72 **Hf** 178.5	73 **Ta** 180.9	74 **W** 183.8	75 **Re** 186.2	76 **Os** 190.2	77 **Ir** 192.2	78 **Pt** 195.1	79 **Au** 197.0	80 **Hg** 200.6	81 **Tl** 204.4	82 **Pb** 207.2	83 **Bi** 209.0	84 **Po**	85 **At**	86 **Rn**
87 **Fr**	88 **Ra**	89–103	104 **Rf**	105 **Db**	106 **Sg**	107 **Bh**	108 **Hs**	109 **Mt**	110 **Ds**	111 **Rg**	112 **Cn**		114 **Fl**		116 **Lv**		

lanthanides	57 **La** 138.9	58 **Ce** 140.1	59 **Pr** 140.9	60 **Nd** 144.2	61 **Pm** 144.9	62 **Sm** 150.4	63 **Eu** 152.0	64 **Gd** 157.2	65 **Tb** 158.9	66 **Dy** 162.5	67 **Ho** 164.9	68 **Er** 167.3	69 **Tm** 168.9	70 **Yb** 173.0	71 **Lu** 175.0
actinides	89 **Ac**	90 **Th** 232.0	91 **Pa**	92 **U** 238.1	93 **Np**	94 **Pu**	95 **Am**	96 **Cm**	97 **Bk**	98 **Cf**	99 **Es**	100 **Fm**	101 **Md**	102 **No**	103 **Lr**

▲ **Figure 2** *The periodic table*

Arranging the elements

The arrangement, pattern, and shape of the periodic table reveal trends among the elements. The positions of the elements in the periodic table are linked to their physical and chemical properties. This makes the periodic table essential for predicting the properties of elements and their compounds.

Atomic number

Reading from left to right, the elements are arranged in order of increasing atomic number. Each successive element has atoms with one extra proton – H 1, He 2, Li 3, Be 4, and so on.

Groups

The elements are arranged in vertical columns called **groups**. Each element in a group has atoms with the same number of outer-shell electrons and similar properties.

Periods and periodicity

The elements are arranged in horizontal rows called **periods**. The number of the period gives the number of the highest energy electron shell in an element's atoms.

7.1 The periodic table

Across each period, there is a repeating trend in properties of the elements, called **periodicity**. The most obvious periodicity in properties is the trend from metals to non-metals. The topics in this chapter look at periodicity of several properties:

- electron configuration
- ionisation energy
- structure
- melting points.

Periodic trend in electron configuration

The chemistry of each element is determined by its electron configuration, particularly the outer, highest energy electron shell. Figure 3 shows the electron configuration for elements in the first three periods.

> **Synoptic link**
>
> This topic builds upon the earlier work covered in Topic 5.1, Electron structure, on electron shells, sub-shells, and energy levels. You may find it useful to look back to Topic 5.1.

	1	2	13	14	15	16	17	18
1	1 H $1s^1$							2 He $1s^2$
2	3 Li $[He]2s^1$	4 Be $[He]2s^2$	5 B $[He]2s^22p^1$	6 C $[He]2s^22p^2$	7 N $[He]2s^22p^3$	8 O $[He]2s^22p^4$	9 F $[He]2s^22p^5$	10 Ne $[He]2s^22p^6$
3	11 Na $[Ne]3s^1$	12 Mg $[Ne]3s^2$	13 Al $[Ne]3s^23p^1$	14 Si $[Ne]3s^23p^2$	15 P $[Ne]3s^23p^3$	16 S $[Ne]3s^23p^4$	17 Cl $[Ne]3s^23p^5$	18 Ar $[Ne]3s^23p^6$

s-sub-shell fills ← | → p-sub-shell fills

▲ **Figure 3** *Electron configuration in Periods 1–3*

Trend across a period

Each period starts with an electron in a new highest energy shell.

- Across Period 2, the 2s sub-shell fills with two electrons, followed by the 2p sub-shell with six electrons.
- Across Period 3, the same pattern of filling is repeated for the 3s and 3p sub-shells.
- Across Period 4, although the 3d sub-shell is involved, the highest shell number is $n = 4$. From the $n = 4$ shell, only the 4s and 4p sub-shells are occupied.

For each period, the s- and p-sub-shells are filled in the same way – a periodic pattern.

> **Study tip**
>
> A repeating, periodic pattern is called *periodicity*. There is a periodicity in electron configuration in the periodic table. This is where the periodic table gets its name.

Trend down a group

You will know from GCSE that elements in each group have atoms with the same number of electrons in their outer shell. Elements in each group also have atoms with the same number of electrons in each sub-shell. This similarity in electron configuration gives elements in the same group their similar chemistry.

> **Synoptic link**
>
> In Topic 8.1, Group 2, and Topic 8.2, The halogens, you will study the chemistry of a metal group and a non-metal group.

Blocks

The elements in the periodic table can be divided into **blocks** corresponding to their highest energy sub-shell. This gives four distinct

7 PERIODICITY

blocks, s, p, d, and f. You can follow the order of sub-shell filling by looking at the periodic table (see Figure 4).

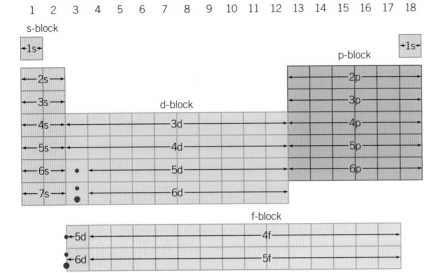

▲ Figure 4 *Sub-shell blocks in the Periodic Table*

> **Synoptic link**
>
> You first met the idea of blocks in the periodic table in Topic 5.1 Electron structure

> **Study tip**
>
> You should know the names halogens and transition elements, but you do not need to know the names of the other groups. However, you may come across them in reference books or on the Internet.

Names and numbers for groups

Some groups have names. You will recognise some names such as halogens, but others such as chalcogens may be unfamiliar.

Two ways of numbering groups are in use (Table 3).

- The old numbers are the numbers that you used at GCSE. Groups 1–7 and then 0. This system is based on the s- and p-blocks. The advantage of the old numbering is that the group number matches the number of electrons in the highest energy electron shell.
- The new numbers run from 1–18, numbering each column in the s-, d-, and p-blocks sequentially. The new numbers were approved for use by IUPAC in 1988, but it can take many years for old practices to change. Your periodic table uses both numbers with the old numbers bracketed.

▼ Table 3 *Group numbers and names*

Group number		Name	Elements
Old	New		
1	1	alkali metals	Li, Na, K, Rb, Cs, Fr
2	2	alkaline earth metals	Be, Mg, Ca, Sr, Ba, Ra
	3–12	transition elements	
5	15	pnictogens	N, P, As, Sb, Bi
6	16	chalcogens	O, S, Se, Te, Po
7	17	halogens	F, Cl, Br, I, At
0	18	noble gases	He, Ne, Ar, Kr, Xe, Rn

Summary questions

1. State the chemical symbol of:
   a. the second element in Period 5 *(1 mark)*
   b. the third element in Group 15 (5). *(1 mark)*

2. State the outer-shell electron configuration of elements in Group 15 (5). Use *n* for the shell number. *(1 mark)*

3. Mendeleev left gaps in his periodic table that were filled later when, for example, scandium, gallium, and germanium were discovered. In addition, there is an entire group in the modern periodic table that Mendeleev omitted. State which group was missing and suggest why Mendeleev was unaware of it. *(2 marks)*

7.2 Ionisation energies

Specification reference: 3.1.1

Learning outcomes

Demonstrate knowledge, understanding, and application of:

→ ionisation energy
→ successive ionisation energies
→ making predictions from successive ionisation energies
→ trends in first ionisation energies down a group
→ trends in first ionisation energies across a period.

Synoptic link

You saw that electron sub-shells are filled from the lowest energy level upwards in Topic 5.1, Electron structure.

▲ Figure 1 *An ionised gas is called a plasma. Lightning is an example of ionisation. Neon lights work by ionising neon gas sealed in a tube at low pressure. The plasma gives out light when electricity passes through it, just as it does in lightning*

What is ionisation energy?

Ionisation energy measures how easily an atom loses electrons to form positive ions.

The **first ionisation energy** is the energy required to remove *one electron* from each atom in one mole of *gaseous atoms* of an element to form one mole of gaseous 1+ ions. For example:

$$Na(g) \rightarrow Na^+(g) + e^- \quad \text{first ionisation energy} = +496 \text{ kJ mol}^{-1}$$

Factors affecting ionisation energy

Electrons are held in their shells by attraction from the nucleus. The first electron lost will be in the *highest* energy level and will experience the *least* attraction from the nucleus. For example, the first electron lost from a sodium atom ($1s^2 2s^2 2p^6 3s^1$) is from the 3s sub-shell.

Three factors affect the attraction between the nucleus and the outer electrons of an atom, and therefore, the ionisation energy.

Atomic radius

The greater the distance between the nucleus and the outer electrons, the less the nuclear attraction. The force of attraction falls off sharply with increasing distance, so atomic radius has a large effect.

Nuclear charge

The more protons there are in the nucleus of an atom, the greater the attraction between the nucleus and the outer electrons.

Electron shielding

Electrons are negatively charged and so inner-shell electrons repel outer-shell electrons. This repulsion, called the **shielding effect**, reduces the attraction between the nucleus and the outer electrons.

Successive ionisation energies

An element has as many ionisation energies as there are electrons. For example, helium has two electrons and two ionisation energies:

$$He(g) \rightarrow He^+(g) + e^- \quad \text{first ionisation energy}$$
$$He^+(g) \rightarrow He^{2+}(g) + e^- \quad \text{second ionisation energy}$$

The second ionisation energy of helium is greater than the first ionisation energy. In a helium atom, there are two protons attracting two electrons in the 1s sub-shell. After the first electron is lost, the single electron is pulled closer to the helium nucleus. The nuclear attraction on the remaining electron increases and more ionisation energy will be needed to remove this second electron.

Successive ionisation energies are defined in the same way as the first ionisation energy. Just be careful with the species losing the electron.

The **second ionisation energy** is the energy required to remove *one electron* from each ion in *one mole* of *gaseous 1+ ions* of an element to form one mole of gaseous 2+ ions.

Successive ionisation energies and shells

Successive ionisation energies provide important evidence for the different electron energy levels in an atom. Figure 2 shows a graph for the successive ionisation energies of fluorine.

> **Study tip**
>
> Don't get your ionisation energies muddled. The number of the ionisation energy is the same as the charge on the ion produced. For helium, the *second* ionisation energy produces a 2+ ion from a 1+ ion.

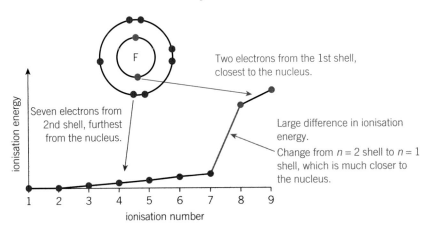

▲ **Figure 2** *Successive ionisation energies of fluorine*

> **Study tip**
>
> The key to analysing graphs of successive ionisation energies is to look for any large jumps in ionisation energy. This marks a change from one shell to another.

The large increase between the seventh and eighth ionisation energies suggests that the eighth electron must be removed from a different shell, closer to the nucleus and with less shielding.

- The first shell ($n = 1$, closer to the nucleus) contains two electrons.
- The second shell ($n = 2$, the outer shell) contains seven electrons.

Making predictions from successive ionisation energies

Successive ionisation energies allow predictions to be made about:

- the number of electrons in the outer shell
- the group of the element in the periodic table
- the identity of an element.

The ionisation energies shown in Table 1 steadily increase but then there is a large increase between the third and fourth ionisation energies. This shows that the fourth electron is being removed from an inner shell. Therefore there are three electrons in the outer shell and the element must be in Group 13 (3). Since it is in Period 3, the element must be aluminium.

Trends in first ionisation energies

Periodic trends in first ionisation energies provide important evidence for the existence of shells and sub-shells.

> **Study tip**
>
> Make sure that you get the shells the right way round.
>
> The electrons with the *largest* ionisation energy are from the shell *closest* to the nucleus.

▼ **Table 1** *Successive ionisation energies for an element from Period 3*

Ionisation number	Ionisation energy / kJ mol^{-1}
1st	578
2nd	1817
3rd	2745
4th	11 577
5th	14 842
6th	18 379

7.2 Ionisation energies

Figure 3 shows the first ionisation energies for the first 20 elements in the periodic table. There are two key patterns:

- a general *increase* in first ionisation energy across each period (H → He, Li → Ne, Na → Ar)
- a sharp *decrease* in first ionisation energy between the end of one period and the start of the next period (He → Li, Ne → Na, Ar → K).

These trends can be explained in terms of atomic radius, electron shielding, and nuclear charge.

Trend in first ionisation energy down a group

First ionisation energies *decrease* down a group. You can see this trend in Figure 3 and Table 2 by comparing the three noble gases helium, neon, and argon.

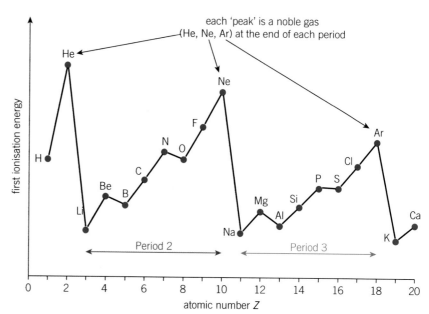

▲ **Figure 3** *Trend in first ionisation energy*

> **Synoptic link**
>
> You will see how the trend in ionisation energy down a group affects the reactivity in Topic 8.1, Group 2.

> **Study tip**
>
> *Down a group*, the *increased atomic radius* and *shielding* are the important factors for the decrease in first ionisation energy.

Table 2 explains the decrease in first ionisation energies down a group. First ionisation energy decreases down every group in the periodic table for the same reasons. Although the nuclear charge increases, its effect is outweighed by the increased radius and, to a lesser extent, the increased shielding.

▼ **Table 2** *Trend in first ionisation energy down a group*

Noble gas	Atomic radius	Number of inner shells	Trend
helium, He	○	0	• atomic radius increases • more inner shells so shielding increases • nuclear attraction on outer electrons decreases • first ionisation energy decreases
neon, Ne	◎	1	
argon, Ar	◉	2	

Trend in first ionisation energy across a period

Figure 3 shows the general increase in first ionisation energy across the first three periods. Table 3 explains the general increase in first ionisation energies of the elements across Period 2.

▼ **Table 3** *General trend in first ionisation energy across Period 2*

Atomic radius	◯	◯	◯	◯	◯	◯	◯	◯
Element	Li	Be	B	C	N	O	F	Ne
Protons	3 p⁺	4 p⁺	5 p⁺	6 p⁺	7 p⁺	8 p⁺	9 p⁺	10 p⁺
Trend	• Nuclear charge increases • Same shell: similar shielding • Nuclear attraction increases • Atomic radius decreases • First ionisation energy increases							

Study tip

Across a period, the increased nuclear charge is the most important factor for the general increase in first ionisation energy.

Sub-shell trends in first ionisation energy

Although first ionisation energy shows a general increase across both Period 2 and Period 3, it does fall in two places in each period. The drops occur at the same positions in each period, suggesting that there might be a periodic cause. The reason is linked to the existence of sub-shells, their energies, and how orbitals fill with electrons.

Across Period 2, the first ionisation energy graph (Figure 4) shows three rises and two falls:

- a rise from lithium to beryllium
- a fall to boron followed by a rise to carbon and nitrogen
- a fall to oxygen followed by a rise to fluorine and neon.

Figure 4 links the pattern to the filling of the s- and p-sub-shells.

Synoptic link

Review Topic 5.1, Electron structure, for details about the electron structure, shells, sub-shells, and orbitals.

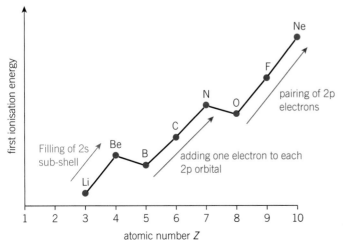

◀ **Figure 4** *Filling the sub-shells across Period 2*

7.2 Ionisation energies

Comparing beryllium and boron
The fall in first ionisation energy from beryllium to boron marks the start of filling the 2p sub-shell (Figure 5).

The 2p sub-shell in boron has a higher energy than the 2s sub-shell in beryllium. Therefore, in boron the 2p electron is easier to remove than one of the 2s electrons in beryllium. The first ionisation energy of boron is less than the first ionisation energy of beryllium.

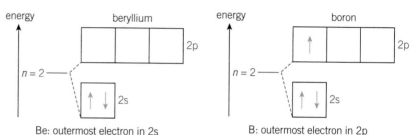

▲ Figure 5 Outer-shell electrons in beryllium and boron atoms

Comparing nitrogen and oxygen
The fall in first ionisation energy from nitrogen to oxygen marks the start of electron pairing in the p-orbitals of the 2p sub-shell (Figure 6).

- In nitrogen and oxygen the highest energy electrons are in a 2p sub-shell.
- In oxygen, the paired electrons in one of the 2p orbtals repel one another, making it easier to remove an electron from an oxygen atom than a nitrogen atom.
- Therefore the first ionisation energy of oxygen is less than the first ionisation energy of nitrogen.

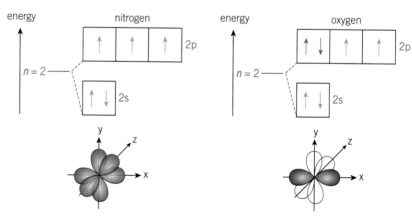

Three 2p electrons: $2p_x^1 2p_y^1 2p_z^1$
- one electron in each 2p orbital
- spins are at right angles – equal repulsion as far apart as possible

Four 2p electrons: $2p_x^2 2p_y^1 2p_z^1$
- two electrons in one 2p orbital
- 2p electrons start to pair
- the paired electrons repel

▲ Figure 6 Outer-shell electrons in nitrogen and oxygen atoms

Summary questions

1. Write the equations to represent the first two ionisation energies of sulfur. *(2 marks)*

2. Write an equation, with state symbols, to represent the first ionisation energy of aluminium. *(1 mark)*

3. Explain why successive ionisation energies always increase. *(2 marks)*

4. a Explain the general trend in first ionisation energy from Na to Ar. *(3 marks)*
   b Explain the sharp drop in first ionisation energy between Ne and Na. *(3 marks)*
   c Explain the trend in first ionisation energy shown by He, Ne, and Ar. *(3 marks)*

5. The first six ionisation energies of an element in Period 3 are 787, 1577, 3232, 4356, 16 091, 19 805 kJ mol^{-1}. Identify the element and explain your reasoning. *(3 marks)*

6. a Explain why Al has a lower first ionisation energy than Mg. *(2 marks)*
   b Explain why S has a lower first ionisation energy than P. *(2 marks)*

7.3 Periodic trends in bonding and structure

Specification reference: 3.1.1

Metals and non-metals

One of the key trends in the periodic table is the change from metals to non-metals from left to right across each period. The changeover from metal to non-metal takes place on a diagonal line from the top of Group 13 (3) to the bottom of Group 17 (7) (Figure 1). Elements near to the metal/non-metal divide (e.g., boron, B, silicon, Si, germanium, Ge, arsenic, As, and antimony, Sb) can show in-between properties and are called semi-metals or metalloids.

Going down these groups, there is a trend from non-metal to metal. The divide is clearest in Group 14 (4), carbon (non-metal) to lead (metal). The two elements on either side of the divide, silicon and germanium, are semi-metals.

There are far more metallic than non-metallic elements – 92 metals to 22 non-metals. However, despite being in a minority in the periodic table, non-metals are extremely important, especially the elements carbon, hydrogen, nitrogen, and oxygen in organic chemistry and biochemistry.

Metallic bonding, structures, and properties

At room temperature, all metals except mercury are solids. The 92 known metals have a wide range of properties – some are strong and hard like tungsten, W, some soft like lead, Pb, some light like aluminium, Al, and some very heavy like osmium, Os, which is twice as dense as lead.

The one constant property of all metals is their ability to conduct electricity. This is a remarkable property for a solid, as charge must be able to move within a rigid structure for conduction to take place.

Metallic bonding and structure

You have already met two main types of chemical bonding – ionic and covalent. **Metallic bonding** is a special type of bonding for metals.

- In a solid metal structure, each atom has donated its negative outer-shell electrons to a shared pool of electrons, which are **delocalised** (spread out) throughout the whole structure.
- The positive ions (cations) left behind consist of the nucleus and the inner electron shells of the metal atoms.

Metallic bonding is the strong electrostatic attraction between cations (positive ions) and delocalised electrons (see Figure 2).

- The cations are *fixed* in position, maintaining the structure and shape of the metal.
- The delocalised electrons are *mobile* and are able to move throughout the structure. Only the electrons move.

Learning outcomes

Demonstrate knowledge, understanding, and application of:

→ metallic bonding and structure
→ giant covalent structures
→ periodic trend in melting points.

(3)	(4)	(5)	(6)	(7)	(0)
13	14	15	16	17	18
					2 He
5 B	6 C	7 N	8 O	9 F	10 Ne
13 Al	14 Si	15 P	16 S	17 Cl	18 Ar
31 Ga	32 Ge	33 As	34 Se	35 Br	36 Kr
49 In	50 Sn	51 Sb	52 Te	53 I	54 Xe
81 Tl	82 Pb	83 Bi	84 Po	85 At	86 Rn
	114 Fl		116 Lv		

▲ Figure 1 *The division between metals and non-metals in the p-block of the periodic table. The blue elements are metals and the yellow elements are non-metals*

7.3 Periodic trends in bonding and structure

▲ Figure 2 Metallic bonding

In Figure 2, there are 12 cations, each with a 1+ charge, and 12 electrons, each with a 1− charge. This balances the charge. For metals containing 2+ cations, twice as many negatively charged electrons are present to balance the charge.

In a metal structure, billions of metal atoms are held together by metallic bonding in a **giant metallic lattice**.

Properties of metals

Most metals have:

- strong metallic bonds – attraction between positive ions and delocalised electrons
- high electrical conductivity
- high melting and boiling points.

The physical properties of metals can be explained in terms of the giant structure of the lattice and metallic bonding.

Electrical conductivity

Metals conduct electricity in solid and liquid states. When a voltage is applied across a metal, the delocalised electrons can move through the structure, carrying charge, as shown in Figure 3. Contrast this ability with the conductivity of ionic compounds, which have no mobile charge carriers in the solid state.

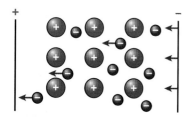

▲ Figure 3 Metal conductivity – all the delocalised electrons will move towards the positive terminal whilst the negative terminal donates electrons that move into the structure

> **Synoptic link**
>
> Review Topic 5.2, Ionic bonding for details about the conductivity of ionic compounds.

Melting and boiling points

Most metals have high melting and boiling points. In Figure 4 coloured shading is used to compare the melting points of the transition metals. Tungsten, W, has the highest melting point at 3422 °C, which is why it is used in the filaments of halogen lamps; other metals would melt. In contrast, mercury, Hg, melts at −39 °C. Other metals that melt at low temperatures include those in Group 1 of the periodic table, which all have melting points below 200 °C.

Sc	Ti	V	Cr	Mn	Fe	Co	Ni	Cu	Zn
Y	Zr	Nb	Mo	Tc	Ru	Rh	Pd	Ag	Cd
Lu	Hf	Ta	W	Re	Os	Ir	Pt	Au	Hg

▲ Figure 4 Comparison of the melting points of transition metals

The melting point depends on the strength of the metallic bonds holding together the atoms in the giant metallic lattice.

- For most metals, high temperatures are necessary to provide the large amount of energy needed to overcome the strong electrostatic attraction between the cations and electrons.
- This strong attraction results in most metals having high melting and boiling points.

Solubility

Metals do not dissolve. It might be expected that there would be some interaction between polar solvents and the charges in a metallic lattice, as with ionic compounds, but any interactions would lead to a reaction, rather than dissolving, as with sodium and water.

Giant covalent structures

You already know that many non-metallic elements exist as simple covalently bonded molecules. In the solid state, these molecules form a simple molecular lattice structure held together by weak intermolecular forces. These structures therefore have low melting and boiling points.

The non-metals boron, carbon, and silicon have very different lattice structures. Instead of small molecules and intermolecular forces, many billions of atoms are held together by a network of strong covalent bonds to form a **giant covalent lattice**.

Carbon and silicon are in Group 14 (4) of the periodic table and their atoms have four electrons in the outer shells. Carbon (in its diamond form) and silicon use these four electrons to form covalent bonds to other carbon or silicon atoms. The result is a tetrahedral structure, as shown in Figure 5 for carbon (diamond).

- Figure 5 shows the tetrahedral arrangement of atoms in the diamond form of carbon.
- The bond angles are all 109.5° by electron-pair repulsion.
- The dot-and-cross diagram shows part of the covalently bonded network of carbon atoms.

Properties

Typical properties of substances with a giant covalent lattice structure are shown below. The properties are dominated by the strong covalent bonds, which make for very stable structures that are very difficult to break down.

Melting and boiling points

Giant covalent lattices have high melting and boiling points. This is because covalent bonds are strong. High temperatures are necessary to provide the large quantity of energy needed to break the strong covalent bonds.

Solubility

Giant covalent lattices are insoluble in almost all solvents. The covalent bonds holding together the atoms in the lattice are far too strong to be broken by interaction with solvents.

Electrical conductivity

Giant covalent lattices are non-conductors of electricity. The only exceptions are graphene and graphite, which are forms of carbon.

- In carbon (diamond) and silicon, all four outer-shell electrons are involved in covalent bonding, so none are available for conducting electricity.
- Carbon is special in forming several structures in which one of the electrons is available for conductivity. Graphene and graphite are able to conduct electricity.

> **Synoptic link**
>
> See Topic 6.3, Intermolecular forces, if you need to review the simple molecular lattice structures of covalently bonded molecules.

▲ **Figure 5** *Structure and bonding in carbon (diamond)*

> **Synoptic link**
>
> Review electron-pair repulsion in Topic 6.1, Shapes of molecules and ions.

▲ **Figure 6** *Sand dunes in the Sahara desert. Sand is mainly silicon dioxide, SiO_2, which has a giant covalent lattice structure similar to diamond. The strong covalent bonds make giant covalent substances stable and generally unreactive*

7.3 Periodic trends in bonding and structure

Graphene and graphite

Apart from diamond, carbon forms giant covalent structures based on planar hexagonal layers.

You can see from the dot-and-cross diagram in Figure 7 that only three electrons of the four outer-shell electrons are used in covalent bonding. The remaining electron is released into a pool of delocalised electrons shared by all atoms in the structure. Structures of carbon containing planar hexagonal layers are therefore good electrical conductors.

Graphene and graphite are both giant covalent structures of carbon based on planar hexagonal layers with bond angles of 120° by electron-pair repulsion.

Graphene

Graphene is a single layer of graphite, composed of hexagonally arranged carbon atoms linked by strong covalent bonds (Figure 8). Graphene has the same electrical conductivity as copper, and is the thinnest and strongest material ever made.

Graphene was discovered in 2004 by Andre Geim and Konstantin Novoselov from the University of Manchester. They were awarded a Nobel prize in 2010. Geim famously made graphene by using sticky tape to pull single layers of carbon atoms from the surface of graphite. Such a simple idea to start work that won a Nobel prize!

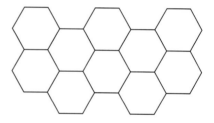
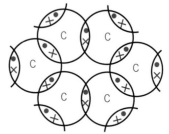

▲ **Figure 7** *Planar hexagonal layer (top) and dot-and-cross diagram (bottom) in graphene and graphite layers*

Graphite

Graphite is composed of parallel layers of hexagonally arranged carbon atoms, like a stack of graphene layers (Figure 9). The layers are bonded by weak London forces.

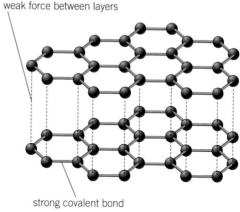

▲ **Figure 9** *Structure of graphite*

The bonding in the hexagonal layers only uses three of carbon's four outer-shell electrons. The spare electron is delocalised between the layers, so electricity can be conducted as in metals.

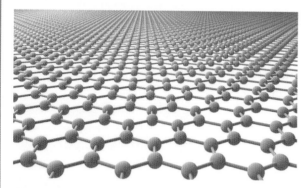

▲ **Figure 8** *Graphene*

> Carry out some research to find these answers.
> Many materials are made out of carbon fibre.
> a How is carbon fibre linked to the structures of graphite and graphene.
> b How are carbon fibre materials strengthened?

PERIODICITY 7

Periodic trend in melting points

The melting points of the elements in Period 2 and Period 3 are shown in Figure 10.

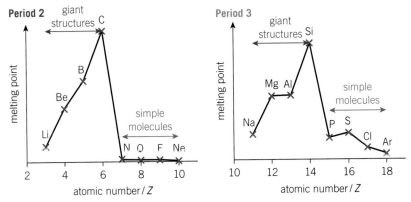

▲ **Figure 10** *Trend in melting points across Periods 2 and 3*

Across Period 2 and Period 3,
- the melting point increases from Group 1 to Group 14 (4)
- there is a sharp decrease in melting point between Group 14 (4) and Group 15 (5)
- the melting points are comparatively low from Group 15 (5) to Group 18 (0).

The sharp decrease in melting point marks a change from giant to simple molecular structures, shown in Figure 11. You can also see the start of the diagonal divide between metals and non-metals.

On melting, giant structures have strong forces to overcome so have high melting points. Simple molecular structures have weak forces to overcome, so have much lower melting points.

The trend in melting points across Period 2 is repeated across Period 3, and continues across the s- and p-blocks from Period 4 downwards.

Giant metallic structure		Giant covalent structure		Simple molecular structure			
Strong metallic bonds between cations and delocalised electrons		Strong covalent bonds between atoms		Weak London forces between molecules			
Li	Be	B	C	N_2	O_2	F_2	Ne
Na	Mg	Al	Si	P_4	S_8	Cl_2	Ar

▲ **Figure 11** *Trend in structure across Periods 2 and 3*

> **Study tip**
> Substances with giant structures have *high* boiling points as strong forces are broken on boiling.

> **Study tip**
> Substances with simple molecular structures have low boiling points as weak forces are broken on boiling.

Summary questions

1. Explain what is meant by metallic bonding and why this type of bonding enables metals to conduct electricity. *(3 marks)*

2. Explain how the bonding in a simple molecular lattice differs from that in a giant covalent lattice. *(2 marks)*

3. Across Period 4, the trend in properties is not exactly the same as across Periods 2 and 3. Suggest explanations for the following.
    a. Germanium is a good conductor of electricity. *(2 marks)*
    b. Arsenic has a much higher melting point than nitrogen and phosphorus. *(2 marks)*

Chapter 7 Practice questions

Practice questions

1. This question looks at ionisation energies
   a. For the third ionisation energy of Mg,
      (i) write an equation, with state symbols for the third ionisation energy of magnesium. *(1 mark)*
      (ii) which sub-shell loses this electron? *(1 mark)*
   b. The 1st to 8th successive ionisation energies, in kJ mol^{-1}, of an element in Period 3 are listed below.

      1012, 1903, 2912, 4957, 6274, 21269, 25398, 29855

      What is the element?
      Explain your reasoning. *(3 marks)*
   c. Suggest, with a reason, which successive ionisation energy is being described below.

      The energy required to remove one electron from each ion in one mole of gaseous 4+ ions. *(1 mark)*

2. This question looks at trends across a period the periodic table. Four sequences of elements across Period 3 are shown below.

   Na, Mg, Al   Al, Si, P   Si, P, S   P, S, Cl

   a. Which sequence shows the melting point increasing across Period 3? *(1 mark)*
   b. Which sequence shows the **first** ionisation energy increasing across Period 3? *(1 mark)*
   c. Which sequence shows the **second** ionisation energy increasing across Period 3? *(1 mark)*
   d. Two sequences contain elements that have the same structure in the solid state.
      Identify these two sequences and state the structure for each. *(2 marks)*

3. Solid graphite and iodine contain covalent bonds but their melting points and electrical conductivities are very different.
   a. State and explain the difference in melting points. *(4 marks)*
   b. State and explain the difference in electrical conductivity. *(3 marks)*

4. Ionisation energies of the elements H to K are shown below.

   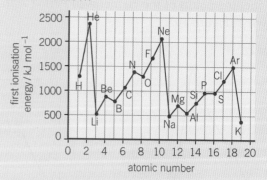

   a. Define the term *first ionisation energy*. *(3 marks)*
   b. Explain why the first ionisation energies show a general increase across Period 3 (Li to Ne). *(3 marks)*
   c. Explain why the first ionisation energy of B is less than that of Be. *(2 marks)*
   d. Explain why the first ionisation energy of O is less than that of N. *(2 marks)*
   e. State and explain the trend in first ionisation energies shown by the elements with atomic numbers 2, 10, and 18. *(3 marks)*
   f. The 1st to 3rd ionisation energies, in kJ mol^{-1} of nitrogen are 1402, 2856, and 4578.
      (i) Write an equation, with state symbols, to represent the **fourth** ionisation energy of nitrogen. *(1 mark)*
      (ii) Suggest why the successive ionisation energies of nitrogen increase in value. *(1 mark)*

5. This question is about aluminium oxide, Al_2O_3.
   a. Successive ionisation energies provide evidence for the arrangement of electrons in atoms. The graph below shows 8 successive ionisation energies of oxygen.

PERIODICITY — 7

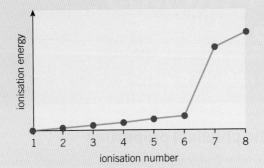

(i) Write an equation, including state symbols, to represent the second ionisation energy of oxygen. *(2 marks)*

(ii) How does this graph provide evidence for the existence of two electron shells in oxygen? *(2 marks)*

b (i) Write the electron configuration for an aluminium atom. *(1 mark)*

(ii) Sketch a graph to show the thirteen successive ionisation energies of aluminium. *(3 marks)*

OCR 2811 Jun 2002 Q3(a)(b)

6 Solids exist as lattice structures.

a Giant metallic lattices conduct electricity. Giant ionic lattices do not. If a giant ionic lattice is melted, the molten ionic compound will conduct electricity. Explain these observations in terms of bonding, structure, and particles present. *(3 marks)*

b The solid lattice structure of ammonia, NH_3, contains hydrogen bonds.

(i) Draw a diagram to show hydrogen bonding between two molecules of NH_3 in a solid lattice. Include relevant dipoles and lone pairs. *(2 marks)*

(ii) Suggest why ice has a higher melting point than solid ammonia. *(2 marks)*

c Solid SiO_2 melts at 2230 °C. Solid $SiCl_4$ melts at –70 °C. Neither of the liquids formed conducts electricity. Suggest the type of lattice structure in solid SiO_2 and in solid $SiCl_4$ and explain the difference in melting points in terms of **bonding** and **structure**. *(5 marks)*

OCR F321 Jun 11 Q5

7 The table below shows the melting points of the elements Na to Cl in Period 3.

	Na	Mg	Al	Si	P	S	Cl
Melting point °C	98	639	660	1410	44	113	–101
Structure							
Bonds/forces broken on boiling							

a (i) Complete the structure row of the table using **M** for giant metallic, **C** for giant covalent, and **S** for simple molecular. *(3 marks)*

(ii) Complete the final row by using **MB** for metallic bonds, **CB** for covalent bonds, **LF** for London forces *(1 mark)*

b State what is meant by metallic bonding. Use a diagram as part of your answer. *(2 marks)*

c Suggest why the melting point increases from Na to Al. *(2 marks)*

d Explain why the melting point of phosphorus is much lower than that of silicon. *(3 marks)*

e Explain why the melting point of sulfur is higher than that of chlorine *(2 marks)*

8 This question compares the electrical conductivity of sodium, sodium chloride, and chlorine in their solid and molten states.

a State the structure and particles making up the structure in the solid state for

(i) sodium *(2 marks)*

(ii) sodium chloride *(2 marks)*

(iii) chlorine *(2 marks)*

b Compare and explain the electrical conductivities of sodium, sodium chloride and chlorine in the solid state. *(3 marks)*

c Compare and explain any differences in electrical conductivities in the liquid state compared with the solid state. *(1 mark)*

8 REACTIVITY TRENDS
8.1 Group 2

Specification reference: 3.1.2

Learning outcomes
Demonstrate knowledge, understanding, and application of:
→ redox reactions of Group 2 elements
→ trends in reactivity and ionisation energies
→ action of water on Group 2 oxides
→ uses of some Group 2 compounds.

▼ **Table 1** *Outer-shell electron configurations of Group 2 atoms and Group 2 ions*

Group 2 atom		Group 2 ion	
Be	[He] $2s^2$	Be^{2+}	[He]
Mg	[Ne] $3s^2$	Mg^{2+}	[Ne]
Ca	[Ar] $4s^2$	Ca^{2+}	[Ar]
Sr	[Kr] $5s^2$	Sr^{2+}	[Kr]
Ba	[Xe] $6s^2$	Ba^{2+}	[Xe]
Ra	[Rn] $7s^2$	Ra^{2+}	[Rn]

▲ **Figure 1** *Group 2 elements, from left to right – calcium, beryllium, and magnesium*

Synoptic link
If you are unsure about redox reactions and oxidation numbers, look back at Topic 4.3, Redox.

Characteristic physical properties
The elements in Group 2 of the periodic table are metals, sometimes named the alkaline earth metals. The name comes from the alkaline properties of the metal hydroxides. The elements are reactive metals and do not occur in their elemental form naturally. On Earth, they are found in stable compounds such as calcium carbonate, $CaCO_3$.

Redox reactions and reactivity
Reducing agents
Each Group 2 element has two outer shell electrons, two more than the electron configuration of a noble gas. The two electrons are in the outer s sub-shell.

Redox reactions are the most common type of reaction of Group 2 elements. Each metal atom is oxidised, losing two electrons to form a 2+ ion with the electron configuration of a noble gas.

$$Ca \rightarrow Ca^{2+} + 2e^- \quad \text{Ca is oxidised}$$
$$[Ar]4s^2 \quad\quad [Ar]$$

- Another species will gain these two electrons and be reduced.
- The Group 2 element is called a **reducing agent** because it has reduced another species.

Table 1 compares the outer shell electron configurations of Group 2 atoms with Group 2 ions.

Redox reactions with oxygen
The Group 2 elements all react with oxygen to form a metal oxide with the general formula MO, made up of M^{2+} and O^{2-} ions.

You will have seen the reaction of magnesium with oxygen in the air in the laboratory. The magnesium burns with a brilliant white light and forms white magnesium oxide.

$$2Mg(s) + O_2(g) \rightarrow 2MgO(s)$$
$$0 \quad\quad\quad \rightarrow \quad +2 \quad\quad \text{oxidation}$$
$$\quad\quad 0 \quad \rightarrow \quad -2 \quad\quad \text{reduction}$$

The total changes in oxidation number balance:

- two Mg in 2Mg   each Mg increases by +2   total increase = +4
- two O in O_2 each O decreases by –2   total decrease = –4

8 REACTIVITY TRENDS

Redox reactions with water

The Group 2 elements react with water to form an alkaline hydroxide, with the general formula $M(OH)_2$, and hydrogen gas. Water and magnesium react very slowly, but the reaction becomes more and more vigorous with metals further down the group – reactivity *increases* down the group.

$$Sr(s) + 2H_2O(l) \rightarrow Sr(OH)_2(aq) + H_2(g)$$

0 → +2   oxidation
+1 → 0   reduction

Not all the hydrogen atoms are reduced:

- 1 Sr in Sr   Sr increases by 2   total increase = +2
- 4 H in $2H_2O$ two H decrease by 1 forming H_2 total decrease = –2
  two H do not change forming $Sr(OH)_2$

Redox reactions with dilute acids

Many metals take part in redox reactions with dilute acids to form a salt and hydrogen gas.

metal + acid → salt + hydrogen

All Group 2 elements react in this way. Again, the reactivity *increases* down the group.

Figure 2 shows the reaction of magnesium with dilute hydrochloric acid.

$$Mg(s) + 2HCl(aq) \rightarrow MgCl_2(aq) + H_2(g)$$

0 → +2   oxidation
+1 → 0   reduction

The total changes in oxidation number balance:

- 1 Mg in Mg   Mg increases by 2   total increase = +2
- 2 H in 2HCl   each H decreases by 1   total decrease = –2

Trend in reactivity and ionisation energy

When the redox reactions above are carried out with each Group 2 element, the reactivity increases down Group 2 (Figure 2).

So why does the reactivity increase?

The atoms of Group 2 elements react by losing electrons to form +2 ions. The formation of +2 ions from gaseous atoms requires the input of two ionisation energies:

- $M(g) \rightarrow M^+(g) + e^-$ first ionisation energy
- $M^+(g) \rightarrow M^{2+}(g) + e^-$ second ionisation energy

Figure 4 shows the first and second ionisation energies for the Group 2 elements. The ionisation energies decrease down the group because the attraction between the nucleus and the outer electrons decreases as a result of increasing atomic radius and increasing shielding.

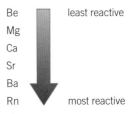

▲ **Figure 2** *The reactivity of Group 2 elements*

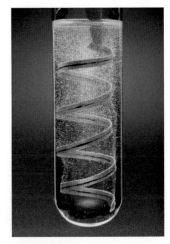

▲ **Figure 3** *Magnesium ribbon reacts with dilute hydrochloric acid to give off tiny bubbles of hydrogen*

Synoptic link

You can review ionisation energies in Topic 7.2, Ionisation energy.

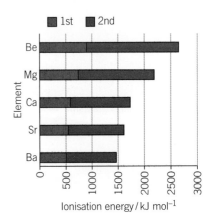

▲ **Figure 4** *First and second ionisation energies of Group 2 elements*

8.1 Group 2

Although other energy changes take place when Group 2 elements react, the first and second ionisation energies make up most of the energy input. From Figure 4, it is clear that the total energy input from ionisation energies to form 2+ ions decreases down the group.
The Group 2 elements become more reactive and stronger reducing agents down the group.

Reactions of Group 2 compounds

Group 2 oxides

Reactions with water

The oxides of Group 2 elements react with water, releasing hydroxide ions, OH^-, and forming alkaline solutions of the metal hydroxide.

$$CaO(s) + H_2O(l) \longrightarrow Ca^{2+}(aq) + 2OH^-(aq)$$

The Group 2 hydroxides are only slightly soluble in water. When the solution becomes saturated, any further metal and hydroxide ions will form a solid precipitate:

$$Ca^{2+}(aq) + 2OH^-(aq) \longrightarrow Ca(OH)_2(s)$$

Solubility of hydroxides

The solubility of the hydroxides in water *increases* down the group, so the resulting solutions contain more $OH^-(aq)$ ions and are more alkaline.

- $Mg(OH)_2(s)$ is only very slightly soluble in water.
  The solution has a low $OH^-(aq)$ concentration and a pH ~ 10.
- $Ba(OH)_2(s)$ is much more soluble in water.
  The solution has a greater $OH^-(aq)$ concentration and a pH ~ 13.

The trend is shown in Table 2.

▼ **Table 2** *Trend in alkalinity*

Hydroxide	Trend
$Mg(OH)_2$	• solubility increases
$Ca(OH)_2$	
$Sr(OH)_2$	• pH increases
$Ba(OH)_2$	• alkalinity increases

You can easily show this trend by carrying out the following experiment.

1. Add a spatula of each Group 2 oxide to water in a test tube.
2. Shake the mixture. On this scale, there is insufficient water to dissolve all of the metal hydroxide that forms. You will have a saturated solution of each metal hydroxide with some white solid undissolved at the bottom of the test-tube.
3. Measure the pH of each solution. The alkalinity will be seen to increase down the group.

Uses of Group 2 compounds as bases

The Group 2 oxides, hydroxides, and carbonates have many uses related to their basic properties and ability to neutralise acids.

Group 2 compounds in agriculture

Calcium hydroxide, $Ca(OH)_2$, is added to fields as lime by farmers to increase the pH of acidic soils. You may have seen the white lime powder on fields (Figure 5). The calcium hydroxide neutralises acid in the soil, forming neutral water:

$$Ca(OH)_2(s) + 2H^+(aq) \rightarrow Ca^{2+}(aq) + 2H_2O(l)$$

▲ **Figure 5** *Spreading lime $(Ca(OH)_2)$ on a field to reduce soil acidity*

REACTIVITY TRENDS 8

Group 2 compounds in medicine

Group 2 bases are often used as antacids for treating acid indigestion. Many indigestion tablets use magnesium and calcium carbonates as the main ingredients, whilst 'milk of magnesia' is a suspension of white magnesium hydroxide, $Mg(OH)_2$, in water. Remember that magnesium hydroxide is only very slightly soluble in water.

The acid in your stomach is mainly hydrochloric acid and the equations below show the neutralisation reactions that take place with $Mg(OH)_2$ and $CaCO_3$.

$$Mg(OH)_2(s) + 2HCl(aq) \longrightarrow MgCl_2(aq) + 2H_2O(l)$$

$$CaCO_3(s) + 2HCl(aq) \longrightarrow CaCl_2(aq) + H_2O(l) + CO_2(g)$$

> **Synoptic link**
>
> If you are unsure about neutralisation reactions look back at Topic 4.1, Acids, bases, and neutralisation.

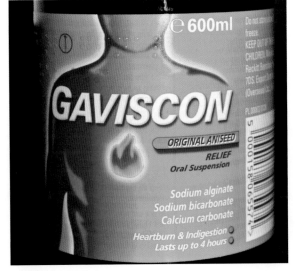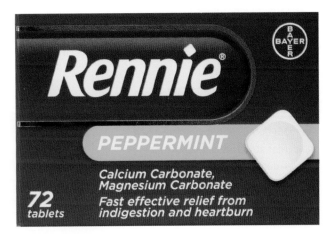

▲ Figure 6 *Common indigestion remedies made with Group 2 compounds, such as calcium carbonate, $CaCO_3$*

Summary questions

1. Explain why Group 2 elements are reducing agents. *(2 marks)*

2. The following reaction is a redox process: $Mg + 2HCl \longrightarrow MgCl_2 + H_2$
   a. Identify the changes in oxidation number. *(2 marks)*
   b. State which species is being oxidised and which is being reduced. *(1 mark)*

3. Explain why the Group 2 elements become more reactive down the group. *(4 marks)*

4. State and explain the trend in alkalinity of the solution formed when Group 2 oxides are added to water. *(3 marks)*

8.2 The halogens
Specification reference: 3.1.3

Learning outcomes
Demonstrate knowledge, understanding, and application of:
→ characteristic physical properties of halogens
→ redox reactions and reactivity of halogens
→ the trend in reactivity of the halogens Cl_2, Br_2, and I_2
→ disproportionation
→ the use of chlorine in water purification and bleach.

Characteristic physical properties
The halogens, Group 17 (7) of the periodic table, are the most reactive non-metallic group. The elements do not occur in their elemental form in nature. On Earth, the halogens occur as stable halide ions (Cl^-, Br^-, and I^-) dissolved in sea water or combined with sodium or potassium as solid deposits, such as in salt mines containing common salt, NaCl.

Trends in boiling points
At room temperature and pressure (RTP), all the halogens exist as diatomic molecules, X_2. The group contains elements in all three physical states at RTP, changing from gas to liquid to solid down the group (Figure 1). In their solid states the halogens form lattices with simple molecular structures. Table 1 explains the trend in boiling points of the five halogens – fluorine to astatine.

Synoptic link
For more details about simple molecular structures and induced dipole–dipole interactions (London forces) look back at Topic 6.3, Intermolecular forces.

◀ **Figure 1** *Chlorine is a pale green gas at RTP. Bromine liquid is extremely toxic, and vaporises readily at room temperature, as can be seen from the orange gas above the red-brown liquid. Iodine is a solid with grey-black crystals*

▼ **Table 1** *Boiling points of the halogen. Astatine only exists as short-lived isotopes and its boiling point has been estimated from the group trend*

Halogen molecule	Number of electrons	Boiling point / °C	Appearance and state at RTP	Trend
F_2	18	−188	pale yellow gas	• more electrons • stronger London forces • more energy required to break the intermolecular forces • boiling point increases
Cl_2	34	−34	pale green gas	
Br_2	70	59	red-brown liquid	
I_2	106	184	shiny grey-black solid	
At_2	170	230	never been seen	

Redox reactions and reactivity of halogens
Redox reactions
Each halogen has seven outer-shell electrons, just one electron short of the electronic configuration of a noble gas. Two electrons are in the outer s sub-shell and five in the outer p sub-shell – s^2p^5.

REACTIVITY TRENDS 8

Redox reactions are the most common type of reaction of the halogens. Each halogen atom is reduced, gaining one electron to form a 1− halide ion with the electron configuration of the nearest noble gas (see Table 2).

$$Cl_2 + 2e^- \rightarrow 2Cl^-$$ chlorine is reduced

Another species loses electrons to halogen atoms – it is oxidised. The halogen is called an **oxidising agent** because it has oxidised another species.

Halogen–halide displacement reactions

Displacement reactions of halogens with halide ions can be carried out on a test-tube scale. The results of the displacement reactions show that the reactivity of the halogens *decreases* down the group.

A solution of each halogen is added to aqueous solutions of the other halides. For example, a solution of chlorine (Cl_2) is added to two aqueous solutions containing bromine (Br^-) and iodine (I^-) ions. If the halogen added is more reactive than the halide present

- a reaction takes place, the halogen displacing the halide from solution
- the solution changes colour.

Solutions of iodine and bromine in water can appear a similar orange-brown colour, depending on the concentration. To tell them apart, an organic non-polar solvent such as cyclohexane can be added and the mixture shaken. The non-polar halogens dissolve more readily in cyclohexane than in water. In cyclohexane their colours are much easier to tell apart, with iodine being a deep violet. The colours are shown in Table 3.

The results and conclusions for these displacement reactions of aqueous solution of halogens and halides are shown in Table 4.

▼ **Table 2** *Comparison of outer-shell electron configurations of halogen atoms and halide ions*

Halogen atom		Halide ion	
F	$2s^22p^5$	F^-	$2s^22p^6$
Cl	$3s^23p^5$	Cl^-	$3s^23p^6$
Br	$4s^24p^5$	Br^-	$4s^24p^6$
I	$5s^25p^5$	I^-	$5s^25p^6$
At	$6s^26p^5$	At^-	$6s^26p^6$

▼ **Table 3** *Halogen solutions in water and cyclohexane*

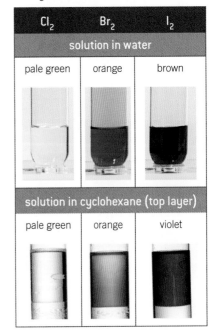

▼ **Table 4** *Halogen displacement reactions*

Halogen	Cl_2(aq)	Br_2(aq)	I_2(aq)
Cl^-(aq)		no reaction	no reaction
Br^-(aq)	orange colour from Br_2 formation Cl_2(aq) + 2Br^-(aq) → 2Cl^-(aq) + Br_2(aq)		no reaction
I^-(aq)	violet colour from I_2 formation Cl_2(aq) + 2I^-(aq) → 2Cl^-(aq) + I_2(aq)	violet colour from I_2 formation Br_2(aq) + 2I^-(aq) → 2Br^-(aq) + I_2(aq)	

From the results:

- chlorine has clearly reacted with *both* Br^- and I^-
- bromine has reacted with I^- *only*
- iodine has *not* reacted at all.

8.2 The halogens

> **Synoptic link**
>
> If you are unsure about redox reactions and oxidation numbers, look back at Topic 4.3, Redox.

Reaction of chlorine with bromide ions

The equation and oxidation number changes are shown in full for the redox reaction between aqueous solutions of chlorine and sodium bromide.

full equation: $Cl_2(aq) + 2NaBr(aq) \rightarrow 2NaCl(aq) + Br_2(aq)$

ionic equation: $Cl_2(aq) + 2Br^-(aq) \rightarrow 2Cl^-(aq) + Br_2(aq)$

$$0 \rightarrow -1 \quad \text{reduction}$$
$$-1 \rightarrow 0 \quad \text{oxidation}$$

- 2 Br in 2Br⁻ — each Br increases by +1 — total increase = +2
- 2 Cl in Cl₂ — each Cl decreases by −1 — total decrease = −2

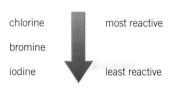

▲ Figure 2 *The order of reactivity of chlorine, bromine, and iodine*

Figure 2 shows the order of reactivity of the three halogens chlorine, bromine, and iodine.

What about fluorine and astatine?

Fluorine is a pale yellow gas, reacting with almost any substance that it comes in contact with.

Astatine is extremely rare because it is radioactive and decays rapidly, and the element has never actually been seen. It is predicted to be the least reactive halogen.

Trend in reactivity

In redox reactions, halogens react by gaining electrons. Down the group, the tendency to gain an electron decreases and the halogens become less reactive (Table 5).

▼ Table 5 *Trend in reactivity of the halogens*

Halogen molecule	Atomic radius	Number of inner shells	Trend
F_2	○	1	• Atomic radius increases
Cl_2	○	2	• More inner shells so shielding increases
Br_2	○	3	
I_2	○	4	• Less nuclear attraction to capture an electron from another species
At_2	○	5	• Reactivity decreases

In the halogens, fluorine is the strongest oxidising agent, gaining electrons from other species more readily than the other halogens. The halogens become weaker oxidising agents down the group.

Disproportionation

Disproportionation is a redox reaction in which the same element is both oxidised and reduced. The reaction of chlorine with water and with cold, dilute sodium hydroxide are two examples of disproportionation reactions.

The reaction of chlorine with water

You will know that chlorine is used in water purification. Chlorine began to be widely used as a disinfectant for drinking water treatment over 100 years ago, revolutionising public health by reducing the incidence of waterborne diseases by killing harmful bacteria.

When small amounts of chlorine are added to water, a disproportionation reaction takes place. For each chlorine molecule, one chlorine atom is oxidised and the other chlorine atom is reduced.

$$Cl_2(aq) + H_2O(l) \rightarrow HClO(aq) + HCl(aq)$$
$$0 \rightarrow -1 \quad \text{reduction}$$
$$0 \rightarrow +1 \quad \text{oxidation}$$

The two products are both acids, chloric(I) acid, HClO, and hydrochloric acid, HCl. The bacteria are killed by chloric(I) acid and chlorate(I) ions, ClO$^-$, rather than by chlorine. Chloric(I) acid also acts as a weak bleach. You can demonstrate this by adding some indicator solution to a solution of chlorine in water. The indicator first turns red, from the presence of the two acids. The colour then disappears as the bleaching action of chloric(I) acid takes effect.

The reaction of chlorine with cold, dilute aqueous sodium hydroxide

The reaction of chlorine with water is limited by the low solubility of chlorine in water. If the water contains dissolved sodium hydroxide, much more chlorine dissolves and another disproportionation reaction takes place.

$$Cl_2(aq) + 2NaOH(aq) \rightarrow NaClO(aq) + NaCl(aq) + H_2O(l)$$
$$0 \rightarrow -1 \quad \text{reduction}$$
$$0 \rightarrow +1 \quad \text{oxidation}$$

The resulting solution contains a large concentration of chlorate(I), ClO$^-$, ions from the sodium chlorate(I), NaClO, that is formed. This solution finds a use as household bleach, which is made by reacting chlorine with cold dilute aqueous sodium hydroxide (See Figure 3).

▲ **Figure 3** *Household bleach contains sodium chlorate(I), NaClO, made by reacting chlorine with sodium hydroxide*

> **Synoptic link**
>
> You will learn about the reaction of chlorine and methane in Topic 12.2, Chemical reactions of alkanes.

Benefits and risks of chlorine use

Although chlorine is beneficial in ensuring that our water is fit to drink and that bacteria are killed, chlorine is also an extremely toxic gas. Chlorine is a respiratory irritant in small concentrations, and large concentrations can be fatal.

Chlorine in drinking water can react with organic hydrocarbons such as methane, formed from decaying vegetation. Chlorinated hydrocarbons are formed, which are suspected of causing cancer. However, the overall risk to health of not adding chlorine to the water supply is far greater than the risk posed by the chlorinated hydrocarbons. The quality of drinking water would be compromised and diseases such as typhoid and cholera might break out. Before safeguarding against what might be a minimal risk, you should also consider why chlorine is added to drinking water in the first place. After a natural disaster one of the very first, life-saving tasks is to ensure that the survivors have a safe water supply (Figure 4).

▲ **Figure 4** *In water, chlorine tablets release chlorine in low concentrations at a steady rate. The tablets are used to purify water for drinking and in swimming pools and can be used to purify water supplies after a natural disaster*

8.2 The halogens

Tests for halide ions

Precipitation reactions with aqueous silver ions

Aqueous halide ions react with aqueous silver ions to form precipitates of silver halides, as shown by the general equation below. X^-(aq) represents an aqueous solution of any halide.

$$Ag^+(aq) + X^-(aq) \rightarrow AgX(s)$$

This reaction forms the basis for a test for the presence of halides. Halide tests are discussed in detail in Topic 8.3, Qualitative analysis.

>
>
> ### Halide ions as reducing agents
>
> In the displacement reactions between halogen and halide ions, the halogen gained electrons and the halide lost electrons. So halogens are oxidising agents and halide ions are reducing agents.
>
> The reducing ability of halide ions can be shown by their reactions with sulfuric acid, H_2SO_4, which is a strong oxidising agent. If a reaction takes place, the halide ions will be oxidised to form the halogen.
>
> Chloride ions are not powerful enough to reduce H_2SO_4.
>
> Bromide ions are more powerful and can reduce H_2SO_4 to sulfur dioxide, SO_2:
>
> $$2H^+ + H_2SO_4 + 2Br^- \rightarrow SO_2 + Br_2 + 2H_2O$$
>
> Notice that the both the atoms and the charges balance.
>
> Iodide ions are even more powerful and reduce the sulfur dioxide formed to sulfur, S, which is then reduced further to hydrogen sulfide, H_2S.
>
> > The unbalanced equations below are for the stepwise reduction, by I^- ions, of H_2SO_4 to H_2S.
> >
> > Step 1 $H^+ + H_2SO_4 + I^- \rightarrow SO_2 + I_2 + H_2O$
> > Step 2 $H^+ + SO_2 + I^- \rightarrow S + I_2 + H_2O$
> > Step 3 $H^+ + S + I^- \rightarrow H_2S + I_2$
> >
> > **a** Balance the equations
> > **b** What are the oxidation number changes in each stage?
> > **c** Write an overall equation for the conversion of H_2SO_4 to H_2S

Summary questions

1. State and explain the trend in boiling points of the halogens fluorine to iodine. *(3 marks)*

2. Write full and ionic equations for displacement reactions of Cl_2(g) with
   a KBr(aq)   **b** MgI_2(aq). *(4 marks)*

3. Chlorine reacts with hot concentrated NaOH(aq) as below.
   $$3Cl_2(aq) + 6NaOH(aq) \rightarrow NaClO_3(aq) + 5NaCl(aq) + 3H_2O(l)$$
   Show that this is a disproportionation reaction. *(3 marks)*

8.3 Qualitative analysis
Specification reference: 3.1.4

Qualitative analysis of ions

You have already learned about titration as a **quantitative analysis** technique (a technique with numerical results). **Qualitative analysis** relies on simple observations rather than measurements, and can often be carried out quickly on a test-tube scale. The observations may be gas bubbles, precipitates, colour changes, or identification of gases.

Tests for anions

Tests based on gases

Carbonate test

Carbonates react with acids to form carbon dioxide gas. The equation below shows the reaction of dilute nitric acid with aqueous sodium carbonate.

$$Na_2CO_3(aq) + 2HNO_3(aq) \rightarrow 2NaNO_3(aq) + CO_2(g) + H_2O(l)$$

This reaction forms the basis for a test for the carbonate ion, CO_3^{2-}.

1. In a test tube, add dilute nitric acid to the solid or solution to be tested.
2. If you see bubbles, the unknown compound could be a carbonate.
3. To prove that the gas is carbon dioxide, use the test that you will remember from GCSE.
   - Bubble the gas through lime water – a saturated aqueous solution of calcium hydroxide, $Ca(OH)_2$.
   - Carbon dioxide reacts to form a fine white precipitate of calcium carbonate, which turns the lime water cloudy (milky) (Figure 1).
   $$CO_2(g) + Ca(OH)_2(aq) \rightarrow CaCO_3(s) + H_2O(l)$$

Tests based on precipitates

Sulfate test

Most sulfates are soluble in water, but barium sulfate, $BaSO_4$, is very insoluble. The formation of a white precipitate of barium sulfate is the basis for the sulfate test, in which aqueous barium ions are added to a solution of an unknown compound. The ionic equation is shown below.

$$Ba^{2+}(aq) + SO_4^{2-}(aq) \rightarrow BaSO_4(s)$$

Usually the $Ba^{2+}(aq)$ ions are added as aqueous barium chloride or barium nitrate. If you intend to carry out a halide test afterwards, use barium nitrate – with barium chloride, you are introducing chloride ions to your solution.

Learning outcomes

Demonstrate knowledge, understanding, and application of:
→ qualitative analysis of ions
→ test-tube tests
→ analysing unknown compounds.

Synoptic link

All carbonates react with acids in a similar way, as described in Topic 4.1, Acids, bases, and neutralisation.

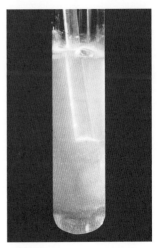

▲ Figure 1 Lime water test for carbon dioxide

▲ Figure 2 A dense white precipitate forms when $Ba^{2+}(aq)$ ions are added to a solution containing sulfate ions, SO_4^{2-} – the heavy precipitate of $BaSO_4$ sinks to the bottom of the test tube

117

Halide tests

Most halides are soluble in water, but silver halides are insoluble. Aqueous silver ions react with aqueous halide ions to form precipitates of silver halides, as shown by the general equation below. X⁻(aq) represents an aqueous solution of any halide.

$$Ag^+(aq) + X^-(aq) \rightarrow AgX(s)$$

This reaction forms the basis for a halide test, which was briefly introduced in Topic 8.2, The halogens.

1 Add aqueous silver nitrate, $AgNO_3$, to an aqueous solution of a halide.

2 The silver halide precipitates are different colours – silver chloride is white, silver bromide is cream-coloured, and silver iodide is yellow (Figure 3).

3 Add aqueous ammonia to test the solubility of the precipitate. This stage is very useful because the three precipitate colours can be difficult to tell apart.

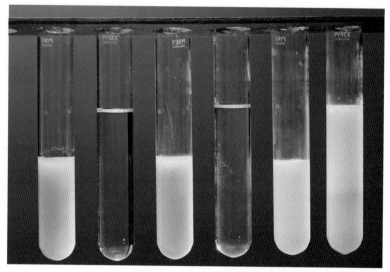

▲ Figure 3 *From left to right, the test tubes contain – a precipitate of silver chloride, silver chloride after addition of dilute aqueous ammonia, a precipitate of silver bromide, silver bromide after addition of concentrated aqueous ammonia, a precipitate of silver iodide, and silver iodide after addition of concentrated aqueous ammonia, which fails to dissolve the silver iodide precipitate*

The reactions with Ag^+(aq) ions and the solubilities of the silver halide precipitates in aqueous ammonia are summarised in Table 1.

▼ Table 1 *Halide tests using aqueous silver ions, Ag^+(aq)*

Halide	Ionic equation	Colour of precipitate	Solubility in NH_3(aq)
chloride, Cl⁻	$Ag^+(aq) + Cl^-(aq) \rightarrow AgCl(s)$	white	soluble in dilute NH_3(aq)
bromide, Br⁻	$Ag^+(aq) + Br^-(aq) \rightarrow AgBr(s)$	cream	soluble in conc. NH_3(aq)
iodide, I⁻	$Ag^+(aq) + I^-(aq) \rightarrow AgI(s)$	yellow	insoluble in conc. NH_3(aq)

REACTIVITY TRENDS 8

A barium meal – making use of precipitation reactions

You may have heard of barium meals – not some kind of food, but an application of precipitation in medicine.

Barium meals are used to enable doctors to see the outline of the gullet, stomach, and upper small intestine in order to identify abnormalities such as ulcers or tumours. The patient swallows water that has been shaken with barium sulfate, the insoluble compound that forms during the sulfate test. The white precipitate coats the inner lining of the gut. An X-ray image is then taken that displays the barium sulfate coating the gut.

> Barium ions in solution are extremely toxic. Why are patients not poisoned by this treatment?

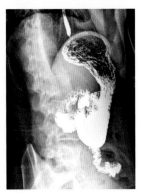

▲ **Figure 4** *X-ray image of a patient who has taken a barium meal to coat the upper gastrointestinal tract with barium sulfate, which shows up white in the picture*

Sequence of tests

If you are asked to analyse an unknown inorganic compound, you will need to carry out the tests for anions in the correct order. Otherwise you could obtain confusing results and make an incorrect identification. For anions, the correct order for tests is:

1. carbonate, CO_3^{2-}
2. sulfate, SO_4^{2-}
3. halides, Cl^-, Br^-, and I^-

Why is there a correct order?

To understand the reasons behind the correct order, you need to think about the chemistry involved in each test.

Carbonate test

In the carbonate test, you add a dilute acid and are looking for effervescence from carbon dioxide gas.

Neither sulfate nor halide ions produce bubbles with dilute acid. The carbonate test can be carried out without the possibility of an incorrect conclusion. If the test produces no bubbles, then no carbonate is present and you can proceed to the next test.

Sulfate test

In the sulfate test, you add a solution containing $Ba^{2+}(aq)$ ions and are looking for a white precipitate of $BaSO_4(s)$.

Barium carbonate, $BaCO_3$, is white and insoluble in water. So if you carry out a sulfate test on a carbonate, you will get a white precipitate too. Therefore it is important to carry out the carbonate test first and only proceed to the sulfate test when you know that *no* carbonate is present.

Halide test

In the halide test, you add a solution containing $Ag^+(aq)$ ions, as $AgNO_3(aq)$, and are looking for a precipitate.

Silver carbonate, Ag_2CO_3, and silver sulfate, Ag_2SO_4, are both insoluble in water and will form as precipitates in this test. It is therefore important to carry out the halide test last, after carrying out carbonate and sulfate tests to rule out those possibilities.

What about a mixture of ions?

If you are asked to analyse a mixture of chemicals, you carry out the tests in the same sequence and on *the same* solution.

1 Carbonate test
- If you see bubbles, continue adding dilute nitric acid until the bubbling stops.
- All carbonate ions will then have been removed and there will be none left to react in the next tests.

If you intend to test for sulfate or halide ions, it is important to use dilute nitric acid, HNO_3, for this test. Sulfuric acid contains sulfate ions and hydrochloric acid contains chloride ions, which will show up in the sulfate and halide tests.

2 Sulfate test
- To the solution left from the carbonate test, add an excess of $Ba(NO_3)_2(aq)$.
  Any sulfate ions present will precipitate out as barium sulfate.
- Filter the solution to remove the barium sulfate.

If you intend to test for halide ions, it is important not to use $BaCl_2(aq)$, because the chloride ions will show up in the halide test.

3 Halide test
- To the solution left from the sulfate test, add $AgNO_3(aq)$.
- Any carbonate or sulfate ions initially present have already been removed. Therefore any precipitate formed must involve halide ions.
- Add $NH_3(aq)$ to confirm which halide you have.

Tests for cations

Test for ammonium ion, NH_4^+

When heated together, aqueous ammonium ions and aqueous hydroxide ions react to form ammonia gas, NH_3.

$$NH_4^+(aq) + OH^-(aq) \rightarrow NH_3(g) + H_2O(l)$$

This reaction forms the basis for a test of the ammonium ion.

1. Aqueous sodium hydroxide, NaOH, is added to a solution of an ammonium ion.
2. Ammonia gas is produced. You are unlikely to see gas bubbles as ammonia is very soluble in water.
3. The mixture is warmed and ammonia gas is released.
4. You may be able to smell the ammonia, but it is easy to test the gas with moist pH indicator paper. Ammonia is alkaline and its presence will turn the paper blue.

Summary questions

1. How could you distinguish between NaCl, NaBr, and NaI by a simple test? *(3 marks)*

2. Explain why it is important to carry out the carbonate test before carrying out a sulfate test on an unknown chemical. *(2 marks)*

3. Explain why, if you are testing a mixture, it is important to use dilute nitric acid, rather than sulfuric or hydrochloric acid, for the carbonate test? *(2 marks)*

Practice questions

1 The equation for the redox reaction between chlorine and hot concentrated NaOH(aq) is shown below.

$$Cl_2(g) + 6NaOH(aq) \rightarrow 5NaCl(aq) + NaClO_3(aq) + 3H_2O(l)$$

This redox reaction is an example of disproportionation.

a Explain what is meant by disproportionation. *(1 mark)*

b Using oxidation numbers, show that this reaction is an example of disproportionation. *(2 marks)*

c Chlorine reacts with hot iron to form iron(III) chloride.

(i) Write the full equation, with state symbols, for the reaction. *(2 marks)*

(ii) Write a half equation to show the reduction of chlorine in this reaction. *(1 mark)*

(iii) The reaction is repeated using bromine instead of chlorine.

What difference would you expect in the rate of the reaction. Explain your answer. *(1 mark)*

2 In Group 2, reactivity increases down the group.

a Explain, in terms of first ionisation energy, this trend in reactivity. *(4 marks)*

b The second ionisation energy is important when explaining trends in reactivity in Group 2.

(i) Write an equation, with state symbols, to represent the second ionisation energy of calcium. *(1 mark)*

(ii) Why is the 2nd ionisation enthalpy important when explaining Group 2 reactivity? *(1 mark)*

3 The Group 2 element barium was first isolated by Sir Humphrey Davy in 1808.

Barium has a giant metallic structure and a melting point of 725 °C.

a Describe, with the aid of a labelled diagram, the structure and bonding in barium and explain why barium has a high melting point.

Include the correct charges on the metal particles in your diagram.

In your answer, you should use appropriate technical terms, spelled correctly. *(3 marks)*

b A chemist reacts barium with water. A solution is formed which conducts electricity.

(i) Write the equation for the reaction of barium with water. Include state symbols. *(2 marks)*

(ii) Predict a value for the pH of the resulting solution. *(1 mark)*

(iii) Give the **formula** of the negative ion responsible for the conductivity of the solution formed. *(1 mark)*

c Heartburn is a form of indigestion caused by an excess of stomach acid.

State a compound of magnesium that could be used to treat heartburn. *(1 mark)*

d In an experiment, a student makes a solution of strontium chloride, $SrCl_2$, by adding excess dilute hydrochloric acid to strontium carbonate.

(i) Describe what the student would observe and write the equation for the reaction. *(2 marks)*

(ii) Draw a dot-and-cross diagram to show the bonding of strontium chloride. Show **outer** electrons only. *(2 marks)*

e In another experiment, a student attempts to make a solution of strontium chloride by adding chlorine water to aqueous strontium bromide.

(i) Describe what the student would observe. *(1 mark)*

(ii) Write the ionic equation for the reaction which takes place. *(1 mark)*

(iii) Chlorine is more reactive than bromine. Explain why. *(4 marks)*

OCR F321 Jan 2013 Q4

Chapter 8 Practice questions

4 Group 2 elements react with the halogens
   a Describe and explain the trend in reactivity of Group 2 elements with chlorine as the group is descended.
      In your answer you should use appropriate technical terms, spelled correctly. *(5 marks)*
   b A student was provided with an aqueous solution of calcium iodide. The student carried out a chemical test to show that the solution contained iodide ions. In this test, a precipitation reaction took place.
      (i)   State the reagent that the student would need to add to the solution of calcium iodide. *(1 mark)*
      (ii) What observation would show that the solution contained iodide ions? *(1 mark)*
      (iii) Write an ionic equation, including state symbols, for the reaction that took place. *(1 mark)*
      (iv) The student is provided with an aqueous solution of calcium bromide that is contaminated with calcium iodide. The student carries out the same chemical test but this time needs to add a second reagent to show that iodide ions are present.
            State the second reagent that the student would need to add. *(1 mark)*
            F321 June 2013 Q4

5 A student used the internet to research chlorine and some of its compounds.
   a They discovered that sea water contains chloride ions. The student added aqueous silver nitrate to a sample of sea water.
      (i)   What would the student see? *(1 mark)*
      (ii) Write an ionic equation, including state symbols, for the reaction that would occur. *(2 marks)*
      (iii) After carrying out the test in (i), the student added dilute aqueous ammonia to the mixture.
            What would the student see? *(1 mark)*

   b The student also discovered that chlorine, Cl_2, is used in the large-scale treatment of water.
      (i)   State one benefit of adding chlorine to water. *(1 mark)*
      (ii) Not everyone agrees that chlorine should be added to drinking water.
            Suggest one possible hazard of adding chlorine to drinking water. *(1 mark)*
   c The equation for the reaction of chlorine with water is shown below.
      $Cl_2(g) + H_2O(l) \rightarrow HCl(aq) + HClO(aq)$
      (i)   State the oxidation number of chlorine in
            Cl_2; HCl; HClO *(1 mark)*
      (ii) The reaction of chlorine with water is a *disproportionation* reaction.
            Use the oxidation numbers in (i) to explain why. *(2 marks)*
      (iii) Chlorine reacts with sodium hydroxide to form bleach in another disproportionation reaction.
            Write an equation for this reaction. *(1 mark)*
   d Two other compounds of chlorine are chlorine dioxide and chloric(V) acid.
      (i)   Chlorine dioxide, ClO_2, is used as a bleaching agent in both the paper and the flour industry. When dry, ClO_2 decomposes explosively to form oxygen and chlorine.
            Construct an equation for the decomposition of ClO_2. *(1 mark)*
      (ii) Chloric(V) acid has the following percentage composition by mass: H, 1.20%; Cl, 42.0%; O, 56.8%.
            Using this information, calculate the empirical formula of chloric(V) acid. Show **all** of your working. *(2 marks)*
      (iii) What does (V) represent in chloric(V) acid? *(1 mark)*
            F321 Jan 2009 Q3

6. This question is about properties of the halogens
   a Describe and explain the trend in boiling points in the halogens. *(4 marks)*
   b Bromine reacts with calcium to form an ionic compound.
      State the electron configuration, in terms of sub-shells, for:
      (i) Ca and Br atoms. *(2 marks)*
      (ii) the ions formed in the reaction. *(2 marks)*
   c Iodine and strontium are reacted together.
      Explain why it is difficult to predict whether this reaction is more or less reactive than the reaction of bromine with calcium. *(3 marks)*

7. The Group 2 element barium, Ba, is silvery white when pure but blackens when exposed to air. The blackening is due to the formation of both barium oxide and barium nitride. The nitride ion is N^{3-}.
   a Predict the formula of:
      (i) barium oxide
      (ii) barium nitride *(2 marks)*
   b A 0.11 g sample of pure barium was added to 100 cm³ of water.
      $Ba(s) + 2H_2O(l) \rightarrow Ba(OH)_2(aq) + H_2(g)$
      (i) Show that 8.0×10^{-4} mol of Ba were added to the water. *(1 mark)*
      (ii) Calculate the volume of hydrogen, in cm³, produced at room temperature and pressure. *(1 mark)*
      (iii) Calculate the concentration, in mol dm⁻³, of the $Ba(OH)_2(aq)$ solution formed. *(1 mark)*
      (iv) State the approximate pH of the $Ba(OH)_2(aq)$ solution. *(1 mark)*
   c A student repeated the experiment in (b) using a 0.11 g sample of barium that had blackened following exposure to the air.
      Suggest why the volume of hydrogen produced would be slightly less than the volume collected using pure barium. *(1 mark)*

   d Describe and explain the trend, down the group, in the reactivity of the Group 2 elements with water. *(5 marks)*

   F321 June 2009 Q5

8. A student is provided with a solution containing two different sodium compounds. The student carries out the series of tests below on the mixture.

   Test 1 Add dilute nitric acid.
   Observation: Effervescence.

   Test 2 Add aqueous barium nitrate to the resulting mixture. If a precipitate forms, filter off the precipitate before carrying out Test 3.
   Observation: No observable change

   Test 3 Add aqueous silver nitrate
   Observation: Cream precipitate which dissolves in concentrated aqueous ammonia.

   a What conclusions can be drawn from the tests?
      Include the formulae of any ions identified and equations for any reactions that take place. *(4 marks)*
   b A second student carries out the same sequence but added dilute sulfuric acid in the first step rather than dilute nitric acid. The student continues with the series of tests and obtains a different result and comes to an incorrect conclusion.
      State and explain the second student's different result and conclusion. Include an equation for any different reaction. *(3 marks)*

9. You are provided with unlabelled solutions of sodium bromide, ammonium bromide, sodium iodide, and ammonium iodide.
   Describe test-tube tests that would allow you to identify which solution is which. For each test, state the observation for a positive result and write an equation to illustrate a positive test. *(7 marks)*

9 ENTHALPY

9.1 Enthalpy changes

Specification reference: 3.2.1

Learning outcomes

Demonstrate knowledge, understanding, and application of:

→ exothermic and endothermic changes
→ enthalpy profile diagrams
→ activation energy
→ standard enthalpy changes.

Enthalpy and enthalpy change

Enthalpy

Enthalpy H is a measure of the heat energy in a chemical system. The chemical **system** refers to the atoms, molecules, or ions making up the chemicals.

Enthalpy is sometimes thought of as the energy stored within bonds. Enthalpy cannot be measured, but enthalpy changes can.

Enthalpy change

In a chemical reaction, the reactants and products are likely to have different enthalpies. The difference in the enthalpies is the **enthalpy change** ΔH:

$$\Delta H = H(\text{products}) - H(\text{reactants})$$

ΔH can be positive or negative, depending on whether the products contain more or less energy than the reactants.

Conservation of energy

The law of **conservation of energy** is one of the fundamental rules of science and states that energy cannot be created or destroyed.

When a chemical reaction involving an enthalpy change takes place, heat energy is transferred between the system and the **surroundings**.

- The system is the chemicals – the reactants and products.
- The surroundings are the apparatus (e.g., the thermometer and apparatus), the laboratory, and everything that is not the chemical system.
- The universe is everything, and includes both system and surroundings, as shown in Figure 1.

▲ **Figure 1** *The system and the surroundings*

Heat in or heat out?

Unlike enthalpy, an enthalpy change ΔH *can* be determined experimentally by measuring the energy transfer between the system and the surroundings.

Energy transfer can be in either of two directions:

- *from* the system *to* the surroundings – an **exothermic** change
- *from* the surroundings *to* the system – an **endothermic** change.

In Figures 2 and 3, the diagrams on the left are called **enthalpy profile diagrams**. Each diagram shows the relative enthalpies of the reactants and products and the enthalpy change ΔH.

Exothermic – heat out of the system

The conservation of energy means that:

energy transferred *from* the system ⟶ energy transferred *to* the surroundings.

- The chemical system releases heat energy to the surroundings.
- Any *energy loss* by the chemical system is balanced by the same *energy gain* by the surroundings.
- ΔH is negative.
- The temperature of the surroundings increases as they gain energy.

> **Study tip**
>
> Remember the directions of energy transfer:
>
> Exothermic
> - ΔH is negative
> - chemical system loses energy
> - surroundings gain energy
> - temperature of surroundings increases.

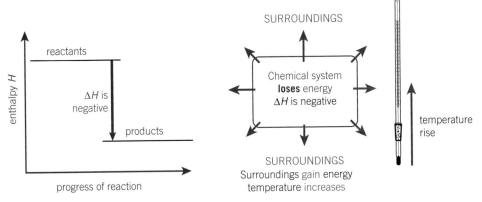

▲ **Figure 2** *Exothermic energy change*

Endothermic – heat into the system

The conservation of energy means that:

energy transferred *to* the system ⟵ energy transferred *from* the surroundings.

- The chemical system takes in heat energy from the surroundings.
- Any energy *gain* by the chemical system is balanced by the same *energy loss* by the surroundings.
- ΔH is positive.
- The temperature of the surroundings decreases as they lose energy.

> **Study tip**
>
> Remember the directions of energy transfer:
>
> Endothermic
> - ΔH is positive
> - chemical system gains energy
> - surroundings lose energy
> - temperature of surroundings decreases.

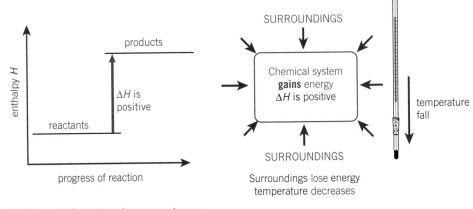

▲ **Figure 3** *Endothermic energy change*

9.1 Enthalpy changes

Activation energy

Atoms and ions are held together by chemical bonds. During chemical reactions, the bonds in the reactants need to be broken by an input of energy. New bonds in the products can then form to complete the reaction. The energy input required to break bonds acts as an energy barrier to the reaction, known as the **activation energy** E_a. Activation energy is the minimum energy required for a reaction to take place.

Figures 4 and 5 show full enthalpy profile diagrams for exothermic and endothermic reactions. Note that these diagrams show the reactants and products, together with labels for ΔH and E_a.

In general, reactions with small activation energies take place very rapidly, because the energy needed to break bonds is readily available from the surroundings. Very large activation energies may be present such a large energy barrier that a reaction may take place extremely slowly or even not at all.

> **Synoptic link**
> You will learn more about activation energy and its relevance to catalysis in Chapter 10, Reaction rates and equilibrium.

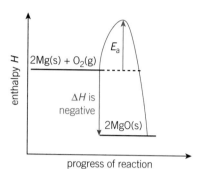

▲ **Figure 4** *Exothermic enthalpy profile diagram*

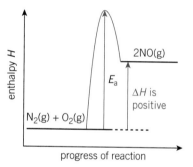

▲ **Figure 5** *Endothermic enthalpy profile diagram*

> **Synoptic link**
> You will learn about the Kelvin scale of temperature in Topic 9.2, Measuring enthalpy changes.

Standard enthalpy changes

The enthalpy change for a reaction can vary slightly depending on the conditions used. Chemists use standard conditions for physical measurements such as enthalpy changes, close to typical working conditions of temperature and pressure. Tables of data always include values taken under standard conditions.

Standard conditions

A standard physical value, such as an enthalpy, is shown in data tables using a special standard sign $^\ominus$. A standard enthalpy change $\Delta H^\ominus$ refers to an enthalpy H change Δ under standard conditions $^\ominus$.

Units are usually kJ mol^{-1}, with mol^{-1} referring to the amount in mol given by the balancing numbers of the chemicals in a stated equation for the reaction (see enthalpy change of reaction below).

- **Standard pressure** is 100 kPa. This is very close to a pressure of one atmosphere, 101 kPa.
- **Standard temperature** is a stated temperature, usually 298 K (25 °C). In this book, standard temperature refers to 298 K.
- **Standard concentration** is 1 mol dm^{-3} (this is relevant for solutions only).
- **Standard state** is the physical state of a substance under standard conditions. Most data tables show the standard state at 100 kPa and 298 K.

Enthalpy change of reaction

The **standard enthalpy change of reaction** $\Delta_r H^\ominus$ is the enthalpy change that accompanies a reaction in the *molar quantities* shown in a chemical *equation* under standard conditions, with all reactants and products in their standard states.

$\Delta_r H^\ominus$ always refers to a stated equation, and its value depends on the balancing numbers. For example, the equation for the reaction of magnesium with oxygen to form magnesium oxide can be written using a fraction to balance O_2.

$$Mg(s) + \tfrac{1}{2}O_2(g) \rightarrow MgO(s) \qquad \Delta_r H^\ominus = -602\,kJ\,mol^{-1}$$

1 mol $\tfrac{1}{2}$ mol 1 mol

If the equation is balanced with whole numbers, the amounts are doubled and the enthalpy change is doubled.

$$2Mg(s) + O_2(g) \rightarrow 2MgO(s) \qquad \Delta_r H^\ominus = -1204\,kJ\,mol^{-1}$$

2 mol 1 mol 2 mol

Both $\Delta_r H^\ominus$ values are correct, but only for the quantities given in each equation written for each value.

Enthalpy change of formation

The **standard enthalpy change of formation** $\Delta_f H^\ominus$ is the enthalpy change that takes place when *one* mole of a compound is *formed from its elements* under standard conditions, with all reactants and products in their standard states.

> **Study tip**
> Learn the definition for $\Delta_f H^\ominus$.

Compounds

The fractional equation for the reaction of magnesium and oxygen above gives the enthalpy change for the formation of 1 mol of a compound. By definition, this is $\Delta_f H^\ominus$ for MgO(s).

$$\underbrace{Mg(s) + \tfrac{1}{2}O_2(g)}_{\text{elements}} \rightarrow \underbrace{MgO(s)}_{\rightarrow \text{1 mol}} \qquad \Delta_f H^\ominus = -602\,kJ\,mol^{-1}$$

The equation could have been written using whole balancing numbers, but it would then not match the definition for $\Delta_f H^\ominus$, which requires formation of *one mole* of MgO.

> **Study tip**
> When balancing equations for enthalpy changes of formation, you must *not* add a balancing number in front of the product that has formed. Balance the equation to give one mole of the product.
>
> Remember that $\Delta_f H^\ominus$ is for the formation of *one mole* of a substance.

Elements

From its definition, $\Delta_f H^\ominus$ for an element refers to the formation of one mole of an element from its element. This is clearly no change, so all elements have an enthalpy change of formation of $0\,kJ\,mol^{-1}$.

Enthalpy change of combustion

The **standard enthalpy change of combustion** $\Delta_c H^\ominus$ is the enthalpy change that takes place when *one* mole of a substance *reacts completely with oxygen* under standard conditions, with all reactants and products in their standard states.

> **Study tip**
> Learn the definition for $\Delta_c H^\ominus$.

When a substance reacts completely with oxygen the products are the oxides of the elements in the substance.

The equation for the combustion of one mole of butane, C_4H_{10}, is shown below.

$$C_4H_{10}(g) + 6\tfrac{1}{2}O_2(g) \rightarrow \underbrace{4CO_2(g) + 5H_2O(l)}_{\rightarrow \text{combustion products}} \qquad \Delta_c H^\ominus = -2877\,kJ\,mol^{-1}$$

1 mol

9.1 Enthalpy changes

The equation could have been balanced without fractions, but it would then not match the definition, which requires combustion of *one mole* of C_4H_{10}.

Enthalpy change of neutralisation

The **standard enthalpy change of neutralisation** $\Delta_{neut}H^\ominus$ is the energy change that accompanies the reaction of an acid by a base to form one mole of $H_2O(l)$, under standard conditions, with all reactants and products in their standard states.

$$H^+(aq) + OH^-(aq) \rightarrow H_2O(l) \quad \Delta_{neut}H^\ominus = -57 \text{ kJ mol}^{-1}$$
$$\text{acid} \qquad \text{base} \qquad \rightarrow \text{1 mol}$$

The neutralisation of hydrochloric acid by sodium hydroxide to form one mole of $H_2O(l)$ is shown below:

$$HCl(aq) + NaOH(aq) \rightarrow H_2O(l) + NaCl(aq)$$
$$\text{acid} \qquad \text{base} \qquad \rightarrow \text{1 mol}$$

For $\Delta_{neut}H^\ominus$, neutralisation involves the reaction of $H^+(aq)$ with $OH^-(aq)$ to form one mole of $H_2O(l)$. The value of $\Delta_{neut}H^\ominus$ is the same for all neutralisation reactions.

> **Study tip**
>
> Learn the definition for $\Delta_{neut}H^\ominus$.

> **Summary questions**
>
> 1. Sketch enthalpy profile diagrams, including E_a, for the following reactions:
>   a. $N_2(g) + 3H_2(g) \rightarrow 2NH_3(g)$
> $\Delta H = -92 \text{ kJ mol}^{-1}$
>   *(2 marks)*
>   b. $N_2O_4(g) \rightarrow 2NO_2(g)$
> $\Delta H = +58 \text{ kJ mol}^{-1}$
>   *(2 marks)*
>
> 2. Write equations, including state symbols, that give $\Delta_f H^\ominus$ for:
>   a. CH_4 *(1 mark)*
>   b. NO_2 *(1 mark)*
>
> 3. Write equations, including state symbols, that give $\Delta_c H^\ominus$ for:
>   a. $H_2S(g)$ *(1 mark)*
>   b. Al *(1 mark)*

What about calories?

Calories are in the news and the importance of a calorie-controlled diet is constantly talked about. You may know that calories have something to do with energy but how are they related to joules? The calorie content of many foods is stated on the packaging in the nutrition label, usually on the back or side of packaging. This information will appear under the Energy heading. Strangely the food calorie is really a kcal.

Figure 6 shows the label from a chocolate bar. You can see the energy information but how many people read this? The label also shows how much this single bar contributes to daily energy requirement!

▲ **Figure 6** *Chocolate bar with energy information*

1. A calorie is equal to 4.18 kJ.
   a. Where have you seen this number before?
   b. How many kJ of energy are available in a Mars bar?
2. Use kitchen scales to find the mass of a cup of tea. Then calculate how many cups of tea could be made from the energy available in a chocolate bar. Assume that the water from the tap is at 15 °C

9.2 Measuring enthalpy changes

Specification reference: 3.2.1, 2.1.3

Measuring energy changes

In Topic 9.1, you looked at the distinction between the chemical system and the surroundings. When you carry out experiments to determine enthalpy changes, the thermometer is part of the surroundings and you will be measuring the temperature change of the surroundings.

The Kelvin scale of temperature

The Kelvin scale of temperature is commonly used in science. It starts at absolute zero, 0 K and is equivalent to −273 °C. The Kelvin scale is part of the International System of Units (SI units). On the Kelvin scale, ice melts at 273 K (0 °C) and water boils at 373 K (100 °C). So a 1 K rise in temperature is the same as a 1 °C rise in temperature. If you record temperatures using a thermometer graduated in °C, the value of the temperature *change* is exactly the same in °C and K.

Calculating an energy change

The energy change of the surroundings is calculated from three quantities – mass, specific heat capacity, and temperature change.

The mass of the surroundings *m*

The mass is measured simply by weighing. You have to identify the materials that are changing temperature. Mass is usually measured in grams (g) to match the scale often used in experiments.

The specific heat capacity of the surroundings *c*

Different materials require different quantities of energy to produce the same temperature change. The **specific heat capacity** *c* is the energy required to raise the temperature of 1 g of a substance by 1 K.

Every substance has a specific heat capacity. Good conductors of heat, such as metals, have small values of *c*. Insulators of heat such as foam plastic, have large values of *c*. In most experiments, you will be measuring the temperature change of water or aqueous solutions. For water, $c = 4.18\,\text{J}\,\text{g}^{-1}\,\text{K}^{-1}$.

The temperature change of the surroundings ΔT

The temperature change ΔT is determined from the thermometer readings:

$\Delta T = T(\text{final}) - T(\text{initial})$.

Calculating an energy change

Heat energy is given the symbol *q*. Once you have values for *m*, *c*, and ΔT from an experiment, it is very easy to calculate the energy in joules (J) using the simple equation below.

$$q = mc\Delta T$$

Learning outcomes

Demonstrate knowledge, understanding, and application of:

→ calculating energy changes from experiments

→ determination of enthalpy changes directly

→ practical techniques for measuring mass.

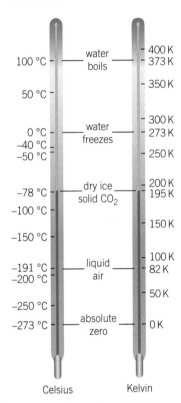

▲ **Figure 1** *Comparison of Celsius and Kelvin scales of temperature*

Study tip

It is easy to convert between Kelvin and Celsius. Add 273 to a Celsius reading to get the Kelvin reading. Subtract 273 from a Kelvin reading to get the Celsius reading.

9.2 Measuring enthalpy changes

Study tip

Learn $q = mc\Delta T$.

Remember that m is the mass that changes temperature *not* the mass of the reactants.

Determination of an enthalpy change of combustion $\Delta_c H$

Combustion is simply a reaction of a substance with oxygen (see Topic 9.1) and is one of the easiest enthalpy changes to determine experimentally. The equation for the combustion of methanol is shown below.

$$CH_3OH(l) + 1\tfrac{1}{2}O_2(g) \rightarrow CO_2(g) + 2H_2O(l)$$

Spirit burners

Liquid fuels, such as methanol, can easily be burnt using small spirit burners. The experimental method and the calculation are outlined below.

1. Using a measuring cylinder, measure out 150 cm³ of water. Pour the water into the beaker. Record the initial temperature of the water to the nearest 0.5 °C.
2. Add methanol to the spirit burner. Weigh the spirit burner containing methanol.
3. Place the spirit burner under the beaker as shown in Figure 2. Light the burner and burn the methanol whilst stirring the water with the thermometer.
4. After about three minutes extinguish the flame. Immediately record the maximum temperature reached by the water.
5. Re-weigh the spirit burner containing the methanol. Assume that the wick has not been burnt.

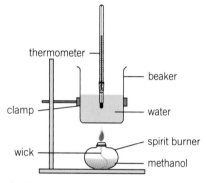

▲ **Figure 2** *Apparatus for determination of $\Delta_c H$ – of a liquid fuels such as methanol*

 Worked example: Determination of $\Delta_c H$ of methanol

Results

Mass of spirit burner and methanol before burning	= 196.97 g
Mass of spirit burner and methanol after burning	= 195.37 g
Mass of fuel burnt	= 1.60 g
Initial temperature of water	= 21.5 °C
Final temperature of water	= 62.5 °C
Temperature change ΔT of water	= 41.0 °C

For water, density = 1.00 g cm⁻³, $c = 4.18 \text{ J g}^{-1}\text{ K}^{-1}$

Calculation

Step 1: Calculate the energy change q of the water in kJ.

Density = 1.00 g cm⁻³ so 150 cm³ of water has a mass of 150 g.

$$q = mc\Delta T = 150 \times 4.18 \times 41.0 = 25\,707 \text{ J} = 25.707 \text{ kJ}$$

Step 2: Calculate the amount, in mol, of CH_3OH burnt.

$$n(CH_3OH) = \frac{m}{M} = \frac{1.60}{32.0} = 0.0500 \text{ mol}$$

Study tip

For Step 1, you are calculating the energy change in the surroundings – where the thermometer is.

You always use $q = mc\Delta T$ Joules

Finally divide by 1000 to convert to kJ

Synoptic link

You learnt about this equation, $n = \frac{m}{M}$, in Topic 3.1, Amount of substance and the mole.

ENTHALPY 9

Step 3: Calculate $\Delta_c H$ in kJ mol^{-1}.

$\Delta_c H$ is the enthalpy change for the complete combustion of 1 mol CH$_3$OH.

In the experiment, 0.0500 mol CH$_3$OH transfers 25.707 kJ of energy to the water.

The water gained 25.707 kJ of energy from the combustion of 0.050 mol CH$_3$OH

1 mol CH$_3$OH has lost $\dfrac{25.707}{0.0500}$ = 514.14 kJ of energy on combustion $\Delta_c H(\text{CH}_3\text{OH}) = -514$ kJ mol^{-1}.

In the calculation, rounding was left until the final answer. This will always give the most accurate result. Work out the error that is introduced in the final answer by rounding the answer in **Step 1** to:
a three significant figures
b two significant figures.

How accurate is the experimental $\Delta_c H$ value?

The data book value for $\Delta_c H$ of methanol is -726 kJ mol^{-1}. This seems very different from the experimental value of -514 kJ mol^{-1} shown above. Data book values are obtained using much more sophisticated apparatus than a spirit burner and a beaker of water. Clearly much less heat was transferred to the water in our experiment than expected. Possible reasons are listed below.

- Heat loss to the surroundings other than the water.
  This includes the beaker but mainly the air surrounding the flame.
- Incomplete combustion of methanol.
  There may be some incomplete combustion, with carbon monoxide and carbon being produced instead of carbon dioxide. You would see carbon as a black layer of soot on the beaker.
- Evaporation of methanol from the wick.
  The burner must be weighed as soon as possible after extinguishing the flame. Otherwise some methanol may have evaporated from the wick. Spirit burners usually have a cover to reduce this error.
- Non-standard conditions.
  The data book value is a standard value. The conditions for this experiment are unlikely to be identical to standard conditions.

All but the last of these reasons would lead to a value for $\Delta_c H$ that is less exothermic than expected.

Study tip

ΔH only makes an appearance right at the end. This is the time to decide on the *sign*. As water has gained energy, the combustion of methanol must have lost the same quantity of energy. So ΔH is negative and the reaction is exothermic.

Synoptic link

As part of the practical skills required for your course, you need to know how to measure mass, volumes of solutions, and volumes of gases.

This practical application box tells you how to measure masses.

Measuring masses, volumes of solutions, and volumes or gases are also covered in:

- Topic 3.2, Determination of formulae, for how to measure mass
- Topic 3.4, Reacting quantities, for how to measure the volumes of gases
- Topic 4.2, Acid–base titrations, for how to measure volumes of solutions
- Topic 10.1, Reaction rates, for how to measure volumes of gases.

Study tip

Be very careful when comparing negative numbers.

The experimental and data book values for $\Delta_c H$ are -514 kJ mol^{-1} and -726 kJ mol^{-1}. Correct comparisons for the experimental value would be less exothermic or less negative. Do not be tempted to state simply that the experimental value is less.

-514 is more than -726 as -514 is less negative!

9.2 Measuring enthalpy changes

Use of draught screens and an input of oxygen gas could minimise errors from heat loss and incomplete combustion.

Determination of an enthalpy change of reaction $\Delta_r H$

Many reactions take place between two solutions, or between a solid and a solution. The enthalpy change of these reactions can be determined using plastic cups made of polystyrene foam. These are cheap, waterproof and light weight, and offer some insulation against heat loss to the surroundings.

When carrying out reactions between aqueous solutions, the solution itself is the immediate surroundings. The chemical particles within the solutions may react when they collide, and any energy transfer is between the chemical particles and water molecules in the solution. A thermometer in the solution will record any temperature change, allowing the heat energy change to be calculated using $mc\Delta T$.

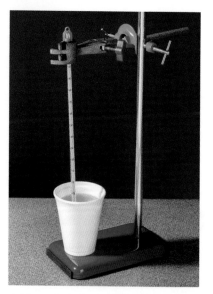

▲ **Figure 3** *A simple experiment to determine the enthalpy change of a reaction*

> **Worked example: Determination of $\Delta_r H$ for a solid and a solution**
>
> An excess of zinc powder is added to 50.0 cm³ of 1.00 mol dm⁻³ copper(II) sulfate. The mixture is stirred until a maximum temperature is obtained.
>
> Find $\Delta_r H$ for Zn(s) + CuSO$_4$(aq) → Cu(s) + ZnSO$_4$(aq)
>
> **Results**
> Initial temperature of solution = 22.5 °C
> Final temperature of solution = 60.5 °C
> Temperature change ΔT of solution = 38.0 °C
>
> For the solution, density and specific heat capacity are close to those of water (density = 1.00 g cm⁻³; c = 4.18 J g⁻¹ K⁻¹)
>
> **Calculation**
> **Step 1:** Calculate the energy change q in the solution in kJ.
>
> Density = 1.00 g cm⁻³ so 50.0 cm³ of the solution has a mass of 50.0 g.
>
> $q = mc\Delta T = 50.0 \times 4.18 \times 38.0 = 7942\,\text{J} = 7.942\,\text{kJ}$
>
> **Step 2:** Calculate the amount, in mol, of CuSO$_4$ that reacted (Zn is in excess).
>
> $n(\text{CuSO}_4) = c \times \dfrac{V\,(\text{in cm}^3)}{1000} = 1.00 \times \dfrac{50.0}{1000} = 0.0500\,\text{mol}$
>
> **Step 3:** Calculate $\Delta_r H$ in kJ mol⁻¹.
>
> $\Delta_r H$ is for the reaction   Zn(s) + CuSO$_4$(aq) → Cu(s) + ZnSO$_4$(aq)
> Balancing numbers give amounts: 1 mol   1 mol   → 1 mol   1 mol

ENTHALPY 9

In the experiment, 0.0500 mol $CuSO_4$ transfers 7.942 kJ of energy to the solution.

1 mol $CuSO_4$ has lost $\frac{7.942}{0.0500} = 159$ kJ of energy to the solution.

$\Delta_r H = -159$ kJ mol^{-1}.

Cooling curves

This experiment measures the enthalpy change for the reaction as on the previous page but the method has been adapted to correct for heat loss by use of a cooling curve correction.

This method can be used to correct for heat loss in other similar enthalpy experiments.

Method

1. Pipette 25.0 cm^3 of 1.00 mol dm^{-3} $CuSO_4$ into a polystyrene cup. Weigh out an excess of zinc powder.
2. Start a stop-clock and take the temperature of the solution every 30 s until the temperature stays constant.
3. Add the zinc to the solution and stir the mixture. Record the temperature every 30 seconds until the temperature has fallen for several minutes.
4. Plot a graph of temperature against time (Figure 4).

To correct for cooling, extrapolate the cooling curve section of the graph back to the time when the zinc was added. Draw a vertical line from the time that the solutions were mixed to the extrapolated cooling curve (Figure 4).

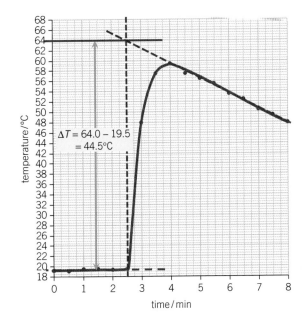

▲ **Figure 4** *Cooling curve method for heat loss*

Using the results shown on Figure 4, $\Delta_r H = -186$ kJ mol^{-1}. This compares with -159 kJ mol^{-1} for the method on the previous page. You should be able to check this.

The table below gives results for the reaction of excess Mg with 100.0 cm^3 0.500 mol dm^{-3} $CuSO_4$(aq)

Time / min	0.0	0.5	1.0	1.5	2.0	2.5	3.0	3.5	4.0	4.5	5.0	5.5	6.0
Temperature / °C	20.0	20.0	20.0	20.0	20.0	63.0	72.5	74.5	72.5	71.5	70.5	68.5	67.5	65.5

Plot a graph of temperature against time and make a correction for heat loss based on the cooling curve. Hence obtain a corrected value for ΔT.

From your results, calculate the enthalpy change for the reaction of 1 mol $CuSO_4$(aq) with Mg.

9.2 Measuring enthalpy changes

Determination of an enthalpy change of neutralisation $\Delta_{neut}H$

This procedure is very similar to the previous example. The only difference is that two solutions react, rather than a solution and a solid.

Worked example: Determination of $\Delta_{neut}H$

A student measures out and mixes 35.0 cm³ of 2.40 mol dm⁻³ NaOH and 35.0 cm³ of 2.40 mol dm⁻³ HCl. The temperature rises by 16.5 °C.

Specific heat capacity of the mixture is 4.18 J g⁻¹ K⁻¹. The density of the mixture is 1.00 g cm⁻³.

Calculate the enthalpy change of neutralisation, in kJ mol⁻¹.

Step 1: Calculate the energy change q in the solution in kJ.

Total volume of solution changing temperature = 35.0 + 35.0 = 70.0 cm³.

Density of mixture is 1.00 g cm⁻³, so 70.0 cm³ has a mass of 70.0 g.

$$q = mc\Delta T = 70.0 \times 4.18 \times 16.5 = 4827.9 \text{ J} = 4.8279 \text{ kJ}$$

Step 2: Calculate the amount, in mol, of NaOH and HCl that reacted.

$$n(\text{NaOH}) = n(\text{HCl}) = c \times \frac{V \text{ (in cm}^3\text{)}}{1000} = 2.40 \times \frac{35.0}{1000} = 0.0840 \text{ mol}$$

Step 3: Calculate $\Delta_{neut}H$ in kJ mol⁻¹.

$\Delta_{neut}H$ is defined as the enthalpy change required for the neutralisation of an acid by an alkali to form 1 mol H₂O(l).

Reacting quantities	NaOH(aq)	+ HCl(aq)	→ NaCl(aq)	+ H₂O(l)
	1 mol	1 mol	→ 1 mol	1 mol
Experiment	0.0840 mol	0.0840 mol →		0.0840 mol

Formation of 0.0840 mol H₂O transfers 4.8279 kJ of energy to the solution.

Formation of 1 mol H₂O loses $\frac{4.8279}{0.0840}$ = 57.5 kJ of energy to the solution.

$\Delta_{neut}H = -57.5$ kJ mol⁻¹.

Summary questions

1. Calculate the energy change, in kJ, of the following (c = 4.18 J g⁻¹ K⁻¹):
   a. 50 cm³ of water decreases in temperature from 62 °C to 19 °C; *(2 marks)*
   b. 50 cm³ of one aqueous solution is mixed with 75 cm³ of a second solution and the temperature increases from 23 °C to 39 °C. *(1 mark)*

2. Combustion of 1.29 g C_6H_{14} releases 34.2 kJ of energy. Calculate Δ_cH for C_6H_{14}. *(2 marks)*

3. Combustion of 1.656 g of ethanol, C_2H_5OH, raised the temperature of 150 g of water from 22.5 °C to 74.5 °C.
   a. Write the equation to represent the enthalpy change of combustion of ethanol. *(2 marks)*
   b. Calculate the enthalpy change of combustion of ethanol to three significant figures. *(2 marks)*

9.3 Bond enthalpies

Specification reference: 3.2.1

Average bond enthalpy

In this topic you will look at how average bond enthalpies can be used to calculate enthalpy changes of reaction without carrying out any experiments, as well as the limitations of this approach.

Average bond enthalpy is the energy required to break one mole of a specified type of bond in a gaseous molecule.

- Energy is always required to break bonds.
- Bond enthalpies are always endothermic.
- Bond enthalpies always have a positive enthalpy value.

Table 1 lists average bond enthalpies for some of the more common covalent bonds. You will need to refer to this table during this topic.

Limitations of average bond enthalpies

The actual bond enthalpy can vary depending on the chemical environment of the bond. Figure 1 shows examples of the actual bond enthalpy of a C—H bond in different environments.

C—H 439 kJ mol^{-1} C—H 420 kJ mol^{-1} C—H 422 kJ mol^{-1}
                                                      C—H 411 kJ mol^{-1}

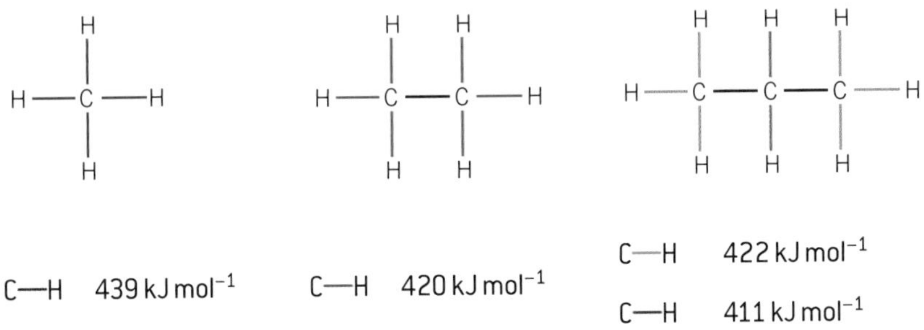

▲ **Figure 1** *Bond enthalpies of C—H bonds in different environments*

An *average* bond enthalpy is calculated from the actual bond enthalpies in different chemical environments. In calculations, you will usually be provided with average bond enthalpies, but sometimes you may be provided with the actual bond enthalpy of an individual bond.

Bond breaking and bond making

In chemical reactions, bonds break and new bonds are formed.

- Energy is required to break bonds
  bond breaking is endothermic ΔH is positive
- Energy is released when bonds form
  bond making is exothermic ΔH is negative

The difference between the energy required for bond breaking and the energy released by bond making determines whether an overall reaction is exothermic or endothermic (Figure 2).

Learning outcomes

Demonstrate knowledge, understanding, and application of:
→ average bond enthalpy
→ calculating enthalpy changes from bond enthalpies.

Synoptic link

In Topic 5.3, Covalent bonding, you learnt that average bond enthalpy is a measurement of covalent bond strength – the larger the value of the average bond enthalpy, the stronger the covalent bond.

▼ **Table 1** *Average bond enthalpies*

Bond	Average bond enthalpy (kJ mol^{-1})
C—H	413
C—C	347
C—O	358
O—H	464
O=O	498
N≡N	945
C=C	612
C=O	805
N—H	391
H—H	436
Cl—Cl	243
Br—Br	193
I—I	151
H—Cl	432
H—Br	366
H—I	298

9.3 Bond enthalpies

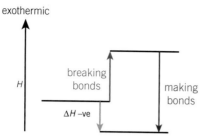
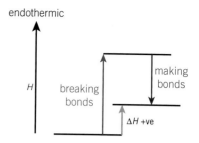

The energy released when making bonds is *greater* than the energy required when breaking bonds.

The energy required when breaking bonds is *greater* than the energy released when making bonds.

▲ **Figure 2** *Enthalpy profile diagrams for bond breaking and bond making in exothermic and endothermic reactions*

Calculating enthalpy changes from average bond enthalpies

The enthalpy change of reaction $\Delta_r H$ can be found by calculating the bond enthalpies of the bonds in the reactants and the products.

For a reaction involving gaseous molecules of covalent substances:

$$\Delta_r H = \Sigma(\text{bond enthalpies in reactants}) - \Sigma(\text{bond enthalpies in products})$$

> **Study tip**
>
> Σ (a capital Greek letter sigma) is shorthand for sum of.

🖩 Worked example: Combustion of propane

Using average bond enthalpies, calculate $\Delta_r H$ for the reaction of propane with oxygen.

$$C_3H_8(g) + 5O_2(g) \rightarrow 3CO_2(g) + 4H_2O(g)$$

▼ **Table 2** $\Delta_r H$ *calculation for the reaction of propane with oxygen*

Equation	$C_3H_8(g)$	+	$5O_2(g)$	→	$3CO_2(g)$	+	$4H_2O(g)$
Bonds	H—C—C—C—H (with H's)	+	5× O=O	→	3× O=C=O	+	4× H—O—H
Bonds broken	8(C—H) 2(C—C)		5(O=O)		6(C=O)		8(O—H)
Energy / kJ mol⁻¹	8 × 413 2 × 347 (3304 + 694 + 6488		5 × 498 2490)	−	6 × 805 (4830 + 8542		8 × 464 3712)
$\Delta_r H$			−2054 kJ mol⁻¹				

Limitations

As you are using average bond enthalpies, the actual energy involved in breaking and making individual bonds would be slightly different (the bonds may be in different environments).

Despite this limitation, the calculated enthalpy change of reaction should be in general agreement with the actual enthalpy change of reaction.

Calculations using average bond enthalpies need all species to be gaseous molecules. In the worked example above, you produced $H_2O(g)$ rather than $H_2O(l)$. This means that your calculated $\Delta_r H$ is not a *standard* enthalpy change. You could still work out the standard enthalpy change but you would need to also take into account the enthalpy change for $H_2O(g)$ condensing into $H_2O(l)$.

Bond enthalpies and combustion

Values for the enthalpy change of combustion of alcohols can be calculated from bond enthalpies. $\Delta_c H$ values for three alcohols are shown below together with their equations.

$CH_3OH + 1.5O_2 \rightarrow CO_2 + 2H_2O$ $\qquad \Delta_c H = -658 \text{ kJ mol}^{-1}$

$C_2H_5OH + 3O_2 \rightarrow 2CO_2 + 3H_2O$ $\qquad \Delta_c H = -1276 \text{ kJ mol}^{-1}$

$C_3H_7OH + 4.5O_2 \rightarrow 3CO_2 + 4H_2O$ $\qquad \Delta_c H = -1894 \text{ kJ mol}^{-1}$

The enthalpy changes increase by a constant quantity for each increase in the carbon chain length. The reason is all linked to bond breaking and bond making.

1. Show that the increase in enthalpy change matches the extra bonds broken and made during combustion in progressing from one alcohol?
2. What would be the calculated enthalpy change of combustion of the alcohol with 20 carbon atoms?
3. Give *two* reasons why enthalpy changes calculated from average bond enthalpies are not *standard* enthalpy changes.

Study tip

In a question, if you are only provided with an equation, always draw out all the bonds. It is then much easier to get the number of each type of bond correct. The calculations are not difficult, but many students slip up by short-cutting and getting the wrong number of bonds.

Summary questions

1. State what is meant by the term *average bond enthalpy*. Explain why you would expect the actual C=O bond enthalpy to be different in CO_2 and in $H_2C=O$. *(2 marks)*

2. You are provided with data for the following reaction.
$H_2(g) + I_2(g) \rightarrow 2HI(g) \quad \Delta H = +53 \text{ kJ mol}^{-1}$
$E_a = +183 \text{ kJ mol}^{-1}$
   a Assuming that the activation energy breaks all the bonds in the reactants, calculate how much energy is released during bond formation. *(2 marks)*
   b Explain what can be deduced about the relative strengths of the bonds that are broken and the bonds that are formed. *(2 marks)*

3. Use Table 1 and the following data:
$2SO_2(g) + O_2(g) \rightarrow 2SO_3(g) \quad \Delta H = -192 \text{ kJ mol}^{-1}$
Assume that SO_2 and SO_3 contain only S=O bonds.
   a Calculate the average bond enthalpy for the S=O bond. *(2 marks)*
   b The actual bond enthalpy for the S=O bond in SO_2 is 531 kJ mol^{-1}. Calculate the actual bond enthalpy for the S=O bond in SO_3. *(2 marks)*

9.4 Hess' law and enthalpy cycles
Specification reference: 3.2.1

Learning outcomes
Demonstrate knowledge, understanding, and application of:
- enthalpy cycles
- indirect determination of enthalpy change from enthalpy changes of formation
- indirect determination of enthalpy change from enthalpy changes of combustion
- unfamiliar enthalpy cycles.

Hess' law

The techniques described in Topic 9.2 allow enthalpy changes to be determined *directly* in a single experiment. Unfortunately, the enthalpy changes of many reactions are very difficult to determine directly. **Hess' law** comes to the rescue, allowing enthalpy changes to be determined *indirectly*.

Hess' law states that, if a reaction can take place by two routes, and the starting and finishing conditions are the same, the total enthalpy change is the same for each route.

Hess' law comes from the idea of conservation of energy (see Topic 9.1) and is easy to visualise with a diagram. Figure 1 shows an **enthalpy cycle** with two routes for converting reactants into products.

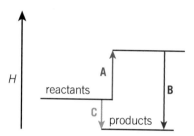

Following the arrows from reactants to products for the two routes:
  Route 1: **A + B**
  Route 2: **C**
By Hess' law, the total enthalpy change is the same for each route.
  A + B = C
If two of A, B, and C are known, the third can be calculated.

▲ **Figure 1** *Enthalpy cycle illustrating Hess' law*

Worked example: An enthalpy cycle

For the enthalpy cycle in Figure 1, $A = +110 \, \text{kJ mol}^{-1}$; $B = -150 \, \text{kJ mol}^{-1}$. Find **C**.

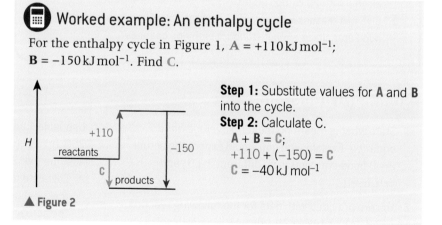

Step 1: Substitute values for **A** and **B** into the cycle.
Step 2: Calculate C.
  A + B = C;
  $+110 + (-150) = C$
  $C = -40 \, \text{kJ mol}^{-1}$

▲ **Figure 2**

This principle can be extended for any number of enthalpy changes. Provided that all enthalpy changes are known except for one, the unknown enthalpy change can always be determined.

ENTHALPY 9

Indirect determination of enthalpy changes

In Topic 9.1, you learnt about several standard enthalpy changes. Chemists have determined the values of many of these enthalpy changes and have listed them in data books. The most useful values are for standard enthalpy changes of formation $\Delta_f H^\ominus$ and combustion $\Delta_c H^\ominus$. The worked examples show how known enthalpy changes can be used with Hess' law to determine enthalpy changes indirectly.

> **Study tip**
>
> Remember, for elements $\Delta_f H^\ominus$ is always zero (Topic 9.1).
>
> Usually you will not be provided with a value of zero for elements.

 Worked example: Enthalpy changes from $\Delta_f H^\ominus$

You can work out the standard enthalpy change of any reaction from the standard enthalpy changes of formation $\Delta_f H^\ominus$ of the reactants and products.

Calculate the standard enthalpy change of reaction for the reaction shown below.

$$Fe_2O_3(s) + 3Ca(s) \rightarrow 2Fe(s) + 3CaO(s)$$

The standard enthalpy changes of formation of the reactants and products are listed in Table 1.

▼ **Table 1** Standard enthalpy changes of formation

Substance	$Fe_2O_3(s)$	$Ca(s)$	$Fe(s)$	$CaO(s)$
$\Delta_f H^\ominus$ / kJ mol^{-1}	−824	0	0	−635

Step 1: Construct the enthalpy cycle between the reactants, the products, and their elements. In the enthalpy cycle, the elements, Fe(s), Ca(s), and O_2(g), form the common link between reactants and products. Using $\Delta_f H^\ominus$, the reactants and products of the original reaction are formed from their elements and the arrow points *upwards*.

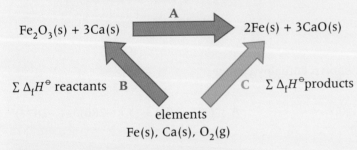

Following the arrows from reactants to products for the two routes:

Route 1: **B + A**

Route 2: **C**

By Hess' law, **B + A = C**

so, **A = C − B**

▲ **Figure 3** Construction of the enthalpy cycle

Step 2: Add $\Delta_f H$ values and calculate the unknown enthalpy change.

$Fe_2O_3(s) + 3Ca(s)$ — A → $2Fe(s) + 3CaO(s)$

(−824) + 0 **B**   **C** 0 + 3 × −635

elements
Fe(s), Ca(s), O_2(g)

A = (3 × −635) − (−824)
  = −1081 kJ mol^{-1}

▲ **Figure 4** Calculation using the enthalpy change

9.4 Hess' law and enthalpy cycles

Worked example: Enthalpy changes from $\Delta_c H^\ominus$

The equation for the formation of butane, $C_4H_{10}(g)$, is shown below.

$$4C(s) + 5H_2(g) \rightarrow C_4H_{10}(g)$$

It would be impossible to measure the enthalpy change of this reaction directly. Carbon and hydrogen form so many compounds that C_4H_{10} would be formed alongside dozens of other compounds.

However, the enthalpy changes of combustion of the reactants C(s) and $H_2(g)$ and of the product $C_4H_{10}(g)$ can all be measured directly. The enthalpy change of reaction can then be found indirectly using the $\Delta_c H^\ominus$ values shown in Table 2.

▼ **Table 2** Standard enthalpy changes of combustion

Substance	C(s)	$H_2(g)$	$C_4H_{10}(g)$
$\Delta_c H^\ominus$ / kJ mol^{-1}	−394	−286	−2877

Step 1 Construct the enthalpy cycle between the reactants, the products, and their common combustion products, $CO_2(g)$ and $H_2O(l)$. Using $\Delta_c H^\ominus$, both the reactants and the products of the original reaction react to form combustion products and the arrows point *downwards*.

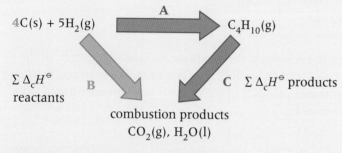

Following the arrows from reactants to products for the two routes:

Route 1: A + C

Route 2: B

By Hess' law, A + C = B

so, A = B − C

▲ **Figure 5** Construction of an enthalpy cycle

Step 2 Add $\Delta_c H$ values and calculate the unknown enthalpy change.

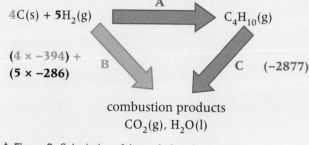

A = (4 × −394) + (5 × −286) − (−2877)
  = −129 kJ mol^{-1}

▲ **Figure 6** Calculation of the enthalpy change

Summary

There are two rules that you can use to help you with your calculations using enthalpy changes of formation and combustion

Using enthalpy changes of formation, $\Delta_f H$,

- $\Delta_r H = \Sigma \Delta_f H$ products $- \Sigma \Delta_f H$ reactants

Using enthalpy changes of combustion, $\Delta_c H$,

- $\Delta_r H = \Sigma \Delta_c H$ reactants $- \Sigma \Delta_c H$ products

9 ENTHALPY

Unfamiliar enthalpy cycle

The energy cycle below and enthalpy change information can be used to calculate an unfamiliar enthalpy change **X**.

▼ **Table 3** Enthalpy information

Reaction	Enthalpy change / kJ mol^{-1}
$C(s) + O_2(g) \rightarrow CO_2(g)$	−393
$H_2(g) + \frac{1}{2}O_2(g) \rightarrow H_2O(l)$	−285
$CaCO_3(s) + 2HCl(aq) \rightarrow CaCl_2(aq) + CO_2(g) + H_2O(l)$	−54
$Ca(s) + 2HCl(aq) \rightarrow CaCl_2(aq) + H_2(g)$	−168

$$Ca(s) + 2HCl(aq) + C(s) + 1\tfrac{1}{2}O_2(g) \xrightarrow{\Delta_f H} CaCO_3(s) + 2HCl(aq)$$

↓ A   ↓ D

$$CaCl_2(s) + H_2(aq) + C(s) + 1\tfrac{1}{2}O_2(g)$$

↓ B

$$CaCl_2(s) + H_2O(l) + C(s) + O_2(g) \xrightarrow{C} CaCl_2(s) + H_2O(l) + CO_2(g)$$

To calculate and identify the unknown enthalpy change, match what has changed at each stage in the cycle with the enthalpy changes provided.

Then work out the two routes to solve the unknown enthalpy change.

1 Use Table 3, to identify the values for enthalpy changes **A–D**.
2 Using the values for **A–D** and the enthalpy cycle, calculate the enthalpy change of formation $\Delta_f H$ of calcium carbonate, $CaCO_3$.

Summary questions

1 Explain why $\Delta_c H^\ominus(C(s))$ and $\Delta_f H^\ominus(CO_2(g))$ have the same value: −394 kJ mol^{-1}. *(1 mark)*

2 You are provided with the following data.

Substance	$NH_3(g)$	$HCl(g)$	$NH_4Cl(s)$	$Ca(OH)_2(s)$	$CaCl_2(s)$	$H_2O(l)$	$CuO(s)$
$\Delta_f H^\ominus$ / kJ mol^{-1}	−46	−93	−314	−986	−796	−286	−157

Calculate $\Delta_r H^\ominus$ for the following reactions.
 a $NH_3(g) + HCl(g) \rightarrow NH_4Cl(s)$ *(1 mark)*
 b $2NH_4Cl(s) + Ca(OH)_2(s) \rightarrow CaCl_2(s) + 2NH_3(s) + 2H_2O(l)$ *(1 mark)*
 c $2NH_3(g) + 3CuO(s) \rightarrow N_2(g) + 3Cu(s) + 3H_2O(l)$ *(1 mark)*

3 You are provided with the following data.

Substance	$C(s)$	$H_2(g)$	$C_6H_{14}(l)$	$C_2H_5OH(l)$
$\Delta_c H^\ominus$ / kJ mol^{-1}	−394	−286	−4163	−1367

Calculate the enthalpy change of formation of   a C_6H_{14} *(1 mark)*   b $C_2H_5OH(l)$. *(1 mark)*

Chapter 9 Practice questions

Practice questions

1. Write equations to represent the following enthalpy changes.

   a. The standard enthalpy change of formation for ethanol, $C_2H_5OH(l)$. *(1 mark)*

   b. The standard enthalpy change of combustion for hexane, $C_6H_{14}(l)$ *(1 mark)*

   c. The standard enthalpy change of neutralisation. *(1 mark)*

   d. The bond enthalpy of H—Br. *(1 mark)*

2. The table below shows enthalpy changes of formation.

Compound	$I_2O_5(s)$	CO(g)	$CO_2(g)$
Δ_fH / kJ mol^{-1}	−158	−110	−394

   a. Define the term *standard enthalpy change of formation*.
   Include the standard conditions in your answer. *(3 marks)*

   b. Calculate the value of Δ_rH, in kJ mol^{-1}, for the reaction in the following equation?
   $$I_2O_5(s) + 5CO(g) \rightarrow I_2(s) + 5CO_2(g)$$
   (3 marks)

3. The equation for the reaction of nitrogen and hydrogen to form ammonia is shown below.
   $$N_2(g) + 3H_2(g) \rightarrow 2NH_3(g)$$
   $\Delta H^\ominus = -97$ kJ mol^{-1}.

   The H—H and N≡N bond enthalpies are +436 and +941 respectively.

   Calculate the N—H bond enthalpy in kJ mol^{-1}.
   Show your working *(3 marks)*

4. 0.766 g of Mg are added 100 cm³ (an excess) of 1.00 mol dm^{-3} HCl. The temperature changed from 22.0 °C to 44.5 °C.

   a. Write the equation for the reaction that takes place. *(1 mark)*

   b. Show that the HCl was in excess. *(2 marks)*

   c. Calculate the energy change in the reaction.
   Assume that the specific heat capacity of the solution is the same as water. *(1 mark)*

   d. Calculate the enthalpy change, in kJ mol^{-1} for the reaction of 1 mol of Mg. *(2 marks)*

5. Enthalpy changes can be calculated from average bond enthalpies. You are supplied with some average bond enthalpies.

bond	C—H	C—C	O—H	C=O	O=O
bond enthalpy / kJ mol^{-1}	+413	+347	+464	+805	+498

   a. Explain why bond enthalpies are always endothermic. *(1 mark)*

   b. The equation for the combustion of pentane is:
   $$C_5H_{12}(g) + 8O_2(g) \rightarrow 5CO_2(g) + 6H_2O(g)$$

   (i) Using the bond enthalpies, calculate the enthalpy change of combustion of pentane *(3 marks)*

   (ii) Explain why this calculated enthalpy change is **not** a standard enthalpy change. *(1 mark)*

   (iii) What are the limitations of using average bond enthalpies for calculating enthalpy changes? *(1 mark)*

6. The enthalpy change of neutralisation can be determined from experimental results.

   a. Define the term enthalpy change of neutralisation. *(1 mark)*

   b. Write the ionic equation for the change that represents enthalpy change of neutralisation. *(1 mark)*

   c. 25.0 cm³ of 2.00 mol dm^{-3} HCl is placed in a plastic cup and its temperature is recorded. 25.0 cm³ of 2.00 mol dm^{-3} KOH is placed in a different plastic cup and its temperature is recorded.

   The initial temperature of both solutions is 19.0 °C. After mixing the maximum temperature of the reaction mixture was 31.9 °C.

   Density of solution = 1.00 g cm^{-3}.

   The specific heat capacity of the solution is the same as water.

   Calculate the enthalpy change of neutralisation. *(4 marks)*

d The same experiment was repeated using 100 cm³ of HCl and NaOH.

What would be the difference, if any in the temperature change and the enthalpy change of neutralisation? Explain your answer. *(2 marks)*

7 Hydrocarbons such as heptane, $C_7H_{16}(l)$, are used as fuels, making use of their combustion reaction with oxygen to form carbon dioxide and water.

a Define the term standard enthalpy change of combustion. Include the standard conditions in your answer. *(3 marks)*

b Write the equation, with state symbols for the equation that represents the enthalpy change of combustion of heptane. *(2 marks)*

c Calculate the enthalpy change of combustion of heptane from the $\Delta_f H$ values in the table below. *(3 marks)*

Substance	$\Delta_f H$ / kJ mol⁻¹
$C_7H_{16}(l)$	−224
$CO_2(g)$	−394
$H_2O(l)$	−286

8 Enthalpy changes of combustion, ΔH_c, are amongst the easiest enthalpy changes to determine directly.

a Define the term *enthalpy change of combustion*. *(2 marks)*

b A student carried out an experiment to determine the enthalpy change of combustion of pentan-1-ol, $CH_3(CH_2)_4OH$.

In the experiment, 1.76 g of pentan-1-ol was burnt. The energy was used to heat 250 cm³ of water from 24.0°C to 78.0°C.

(i) Calculate the energy released, in kJ, during combustion of 1.76 g pentan-1-ol. The specific heat capacity of water = 4.18 J g⁻¹ K⁻¹. Density of water = 1.00 g cm⁻³. *(1 mark)*

(ii) Calculate the amount, in moles, of pentan-1-ol that was burnt. *(2 marks)*

(iii) Calculate the enthalpy change of combustion of pentan-1-ol. Give your answer to three significant figures. *(3 marks)*

c The standard enthalpy change of formation of hexane can be defined as:

The enthalpy change when 1 mol of hexane is formed from its constituent elements in their standard states under standard conditions.

Hexane melts at −95 °C and boils at 69 °C.

(i) What are *standard conditions*? *(1 mark)*

(ii) An incomplete equation is shown below for the chemical change that takes place to produce the standard enthalpy change of formation of hexane.

Add state symbols to the equation to show each species in its standard state.

$6C(.......) + 7H_2(........) \rightarrow C_6H_{14}(.......)$ *(1 mark)*

(iii) It is very difficult to determine the standard enthalpy change of formation of hexane directly. Suggest a reason why. *(1 mark)*

(iv) The standard enthalpy change of formation of hexane can be determined indirectly.

Calculate the standard enthalpy change of formation of hexane using the standard enthalpy changes of combustion below. *(3 marks)*

Substance	$\Delta_c H$ / kJ mol⁻¹
C	−394
H_2	−286
C_6H_{14}	−4163

OCR F322 Jun 09 Q2

10 REACTION RATES AND EQUILIBRIUM

10.1 Reaction rates

Specification reference: 3.2.2, 2.1.3

Learning outcomes

Demonstrate knowledge, understanding, and application of:

→ the effect of concentration on reaction rate
→ the effect of pressure on reaction rate
→ reaction rates from the gradients of graphs.

▲ **Figure 1** *Firework display over Tower Bridge in London. Fireworks are an example of a fast reaction*

▲ **Figure 2** *The Forth Rail Bridge takes many years to paint but fortunately, longer to rust*

How fast is a reaction?

Some chemical reactions are complete within a fraction of a second, whereas others may take centuries. High explosives detonate immediately and fireworks shoot upwards producing amazing displays of colour the instant the fuse burns down (Figure 1), whereas iron rusts relatively slowly – good news if you want to cross the Forth Rail Bridge (Figure 2).

What is meant by rate of reaction?

The **rate of a chemical reaction** measures how fast a reactant is being used up or how fast a product is being formed. The rate of a reaction can be defined as the change in concentration of a reactant or a product in a given time.

$$\text{rate} = \frac{\text{change in concentration}}{\text{time}} \quad \left(\text{units } \frac{\text{mol dm}^{-3}}{\text{s}} = \text{mol dm}^{-3}\text{s}^{-1}\right)$$

- The rate of a reaction is fastest at the start of the reaction, as each reactant is at its highest concentration.
- The rate of reaction slows down as the reaction proceeds, because the reactants are being used up and their concentrations decrease.
- Once one of the reactants has been completely used up, the concentrations stops changing and the rate of reaction is zero.

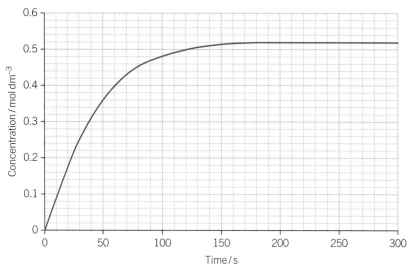

▲ **Figure 3** *Concentration–time graphs can be used to monitor the rate of a chemical reaction*

REACTION RATES AND EQUILIBRIUM 10

Figure 3 shows the formation of a product over the course of a chemical reaction. The slope of the curve is steepest at the start of the reaction, when the rate is greatest. The curve becomes less steep as the reaction proceeds. Eventually the curve becomes a straight line parallel to the x-axis when the reaction is complete and the rate is zero.

Altering the rate of a chemical reaction

You will remember from GCSE that a number of factors can change the rate of a chemical reaction:

- concentration (or pressure when reactants are gases)
- temperature
- use of a catalyst
- surface area of solid reactants.

To understand why the rate changes, you need to think about reactions in terms of the particles involved. The **collision theory** states that two reacting particles must collide for a reaction to occur. Usually only a small proportion of collisions result in a chemical reaction. In most collisions, the molecules collide but then bounce off each other and remain chemically unchanged.

▲ Figure 4 *You can think of the effect of concentration as like a crowded street. The greater the number of people in a crowd, the more chance there is of a collision taking place*

Why are some collisions effective and others ineffective?

An effective collision is one that leads to a chemical reaction (Figure 5). A collision will be effective if two conditions have been met:

- the particles collide with the correct orientation
- the particles have sufficient energy to overcome the activation energy barrier of the reaction.

> **Synoptic link**
>
> You learnt about activation energy is in Topic 9.1, Enthalpy changes. We will look in more detail at activation energy and reaction rate in Topic 10.2, Catalysts.

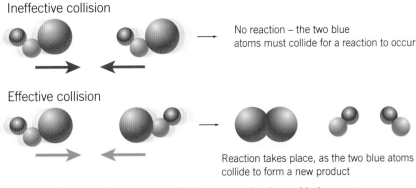

▲ Figure 5 *For a reaction to occur, collisions must take place with the correct orientation and with sufficient energy for a reaction to occur*

How does increasing the concentration affect the rate of reaction?

When the concentration of a reactant is increased, the rate of reaction generally increases. An increase in concentration increases the number of particles in the same volume. The particles are closer together and collide more frequently. In a given period of time there will therefore be more effective collisions (correct orientation and sufficient energy) and an increased rate of reaction (Figure 6).

▲ Figure 6 *Marble chips in different concentrations of acids – the highest concentration on the left and the lowest concentration on the right. The more concentrated the acid, the more frequent the collisions of $H^+(aq)$ ions with the marble chips and the faster hydrogen gas is produced*

10.1 Reaction rates

How does increasing the pressure of a gas affect the rate of reaction?
When a gas is compressed into a smaller volume the pressure of a gas is increased and the rate of reaction increases. The concentration of the gas molecules increases as the same number of gas molecules occupy a smaller volume. The gas molecules are closer together and collide more frequently, leading to more effective collisions in the same time.

Methods for following the progress of a reaction

The progress of a chemical reaction can be followed by:

- monitoring the removal (decrease in concentration) of a reactant
- following the formation (increase in concentration) of a product.

The method chosen will depend on the properties and physical states of the reactants and products in the reaction. In addition to concentration, measurable properties that might change as the reaction proceed include gas volume, mass of reactants or products, and colour.

Reactions that produce gases

If a reaction produces a gas, two methods that can be used to determine the rate of the reaction are:

- monitoring the volume of gas produced at regular time intervals using gas collection
- monitoring the loss of mass of reactants using a balance.

Volume of gas produced and mass loss are both proportional to the change in concentration of a reactant or product. So the change in volume with time or the mass loss with time both give a measure of the rate of reaction.

> **Study tip**
> When asked to identify a method for measuring the rate of a reaction, the equation may give you a clue. If a product has a (g) as its state symbol then the method could involve gas collection.

> **Synoptic link**
> As part of the practical skills required for your course, you need to know how to measure mass, volumes of solutions, and volumes of gases.
>
> This practical application box tells you how to measure volumes of gases.
>
> Measuring mass, volumes of solutions, and volumes of gases are also covered in:
>
> - Topic 3.2, Determination of formulae, for how to measure mass
> - Topic 3.4, Reacting quantities, for how to measure the volumes of gases
> - Topic 4.2, Acid–base titrations, for how to measure volumes of solutions
> - Topic 9.2, Measuring enthalpy changes, for how to measure mass.

> **Study tip**
> When setting up this experiment you should ensure the water level starts close to zero on the scale.

Monitoring the production of a gas using gas collection

The rate of reaction for the decomposition of hydrogen peroxide, H_2O_2, can be measured using the apparatus shown in Figure 7.

The equation for the reaction is shown below.

$$2H_2O_2(aq) \xrightarrow{\text{MnO}_2 \text{ catalyst}} 2H_2O(l) + O_2(g)$$

1. Hydrogen peroxide is added to the conical flask and the bung is replaced.
2. The initial volume of gas in the measuring cylinder is recorded.

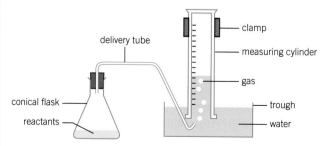

▲ Figure 7 *Apparatus to measure the rate of a reaction in which a gas is produced*

REACTION RATES AND EQUILIBRIUM 10

3. Manganese dioxide, MnO_2, catalyst is then quickly added to the conical flask and the bung is replaced. A stop clock is started.
4. The volume of gas produced in the measuring cylinder is recorded at regular intervals until the reaction is complete.
5. The reaction is complete when no more gas is produced.

Alternatively, a gas syringe can be used instead of a measuring cylinder.

A graph is plotted of total volume of gas produced against time. To calculate the **initial rate** of the reaction, a **tangent** is drawn to the curve at $t = 0$ (Figure 9). The **gradient** of the tangent gives the reaction rate.

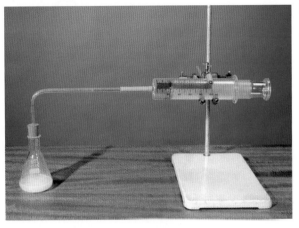

▲ **Figure 8** *Apparatus for collecting a gas with a syringe*

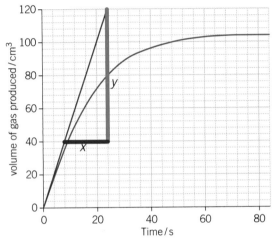

▲ **Figure 9** *Graph of volume of gas produced against time*

> 🖩 **Worked example: Calculating reaction rates from gas produced**
>
> **Step 1:** Plot a graph of volume of gas produced against time (Figure 8).
>
> **Step 2:** Draw a tangent at $t = 0$ (red line in Figure 8). This is the initial rate.
>
> **Step 3:** Calculate the rate from the gradient of the tangent.
>
> $$\text{rate from the gradient} = \frac{y}{x} = \frac{120 - 40}{24 - 8}$$
> $$= \frac{80}{16} = 5.0 \, cm^3 \, s^{-1}$$

Calculate the rate after:

a 24 s

b 40 s

Monitoring the loss of mass of reactants using a balance

The rate of reaction between calcium carbonate and hydrochloric acid can also be determined by monitoring the loss in mass of the reactants over a period of time. The equation for the reaction is shown below.

$$CaCO_3(s) + 2HCl(aq) \rightarrow CaCl_2(aq) + CO_2(g) + H_2O(l)$$

The carbonate and the acid are added to a conical flask on a balance. The mass of the flask and contents is recorded initially and at regular time intervals. The reaction is complete when no more gas is produced so no more mass is then lost. A graph of mass lost against time is plotted.

▲ **Figure 10** *Monitoring mass loss using a balance. The two readings show that mass has been lost as gas is released*

10.1 Reaction rates

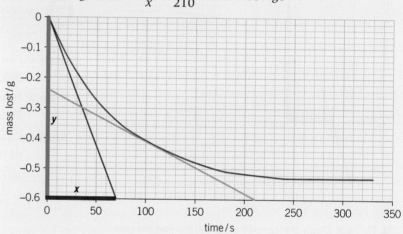

Worked example: Calculating reaction rates from mass loss

Step 1: Plot a graph of mass lost against time (Figure 10).

Step 2: Draw a tangent to the curve at $t = 0$ (red line in Figure 11). This is the initial rate.

Step 3: Calculate the rate from the gradient of the tangent.

$$\text{rate from the gradient} = \frac{y}{x} = \frac{0.6}{70} = 8.6 \times 10^{-3} \, \text{g s}^{-1}$$

Step 4: To calculate the rate at a specific time, the same tangent method is used. In Figure 11, the orange tangent is used to find the rate of reaction after 100 s.

$$\text{gradient} = \frac{y}{x} = \frac{0.36}{210} = 1.7 \times 10^{-3} \, \text{g s}^{-1}$$

▲ **Figure 11** *Graph of mass loss against time*

Summary questions

1. Changing the concentration of a reactant alters the rate of a reaction. State three other factors that can affect the rate of a chemical reaction. Explain how an increase in concentration increases the rate. *(5 marks)*

2. State two possible methods for monitoring the rate of the following reaction:

   $MgCO_3(s) + 2HCl(aq) \rightarrow MgCl_2(aq) + CO_2(g) + H_2O(l)$ *(2 marks)*

3. The reaction $2N_2O_5(g) \rightarrow 4NO_2(g) + O_2(g)$ was carried out. The concentration of the N_2O_5 was recorded every 200 s and the following data obtained.

[$N_2O_5(g)$] / mol dm^{-3}	1.00	0.88	0.78	0.69	0.61	0.54	0.48	0.43	0.38	0.34
Time / s	0	200	400	600	800	1000	1200	1400	1600	1800

   a. Plot a graph of ($N_2O_5(g)$) on the y-axis against time on the x-axis. *(2 marks)*
   b. Calculate the initial rate of reaction and the rate of reaction after 1000 s. *(2 marks)*

10.2 Catalysts

Specification reference: 3.2.2

What does a catalyst do?

A **catalyst** is a substance that changes the rate of a chemical reaction without undergoing any permanent change itself.

- The catalyst is not used up in the chemical reaction.
- The catalyst may react with a reactant to form an **intermediate** or may provide a surface on which the reaction can take place.
- At the end of the reaction the catalyst is regenerated.

A catalyst increases the rate of a chemical reaction by providing an alternative reaction pathway of lower activation energy. See the enthalpy profile diagrams in Figure 1 and Figure 2 for catalysis in exothermic and endothermic reactions (Figure 1).

Learning outcomes

Demonstrate knowledge, understanding, and application of:

→ the role of a catalyst
→ enthalpy profile diagrams
→ homogeneous and heterogeneous catalysts.

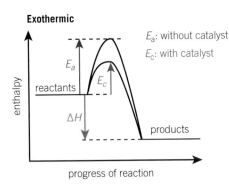

▲ **Figure 1** *Exothermic reaction, with and without a catalyst*

▲ **Figure 2** *Endothermic reaction, with and without a catalyst*

Study tip

When drawing enthalpy profile diagrams:

- Reactants and products should be shown at the correct levels with respect to each other
- ΔH should be shown with an arrow pointing in the correct direction
- The activation energy should be shown with an upward arrow.

Types of catalyst

Homogeneous catalysts

A homogeneous catalyst has the *same* physical state as the reactants. The catalyst reacts with the reactants to form an intermediate. The intermediate then breaks down to give the product and regenerates the catalyst.

Synoptic link

You will learn more about esters and esterification in Topic 26.4, Carboxylic acid derivatives.

Two examples of the many reactions of gases and liquids that use homogeneous catalysis are shown below:

1. Making esters with sulfuric acid as a catalyst
   The equation below shows the preparation of the ester, $CH_3COOC_2H_5$, from ethanoic acid, CH_3COOH, and ethanol, C_2H_5OH. Sulfuric acid, H_2SO_4, is the catalyst.

   $$C_2H_5OH(l) + CH_3COOH(l) \xrightleftharpoons{H_2SO_4(l)} CH_3COOC_2H_5(l) + H_2O(l)$$

The reactants (ethanol and ethanoic acid) and the catalyst (sulfuric acid) are all liquids.

Synoptic link

You learnt about activation energies and endothermic and exothermic reactions in Chapter 9, Enthalpy.

10.2 Catalysts

> **Synoptic link**
> You will learn more about the catalytic breakdown of ozone in Topic 15.2, Organohalogen compounds in the environment.

2 Ozone depletion (Cl• radicals as catalyst)
The equation below shows the depletion of ozone, O_3, in the presence of chlorine radicals, Cl•, which act as a catalyst.

$$2O_3(g) \underset{}{\overset{Cl•(g)}{\rightleftharpoons}} 3O_2(g)$$

The reactant (O_3) and the catalyst (Cl•) are both gases.

Heterogeneous catalysts

A heterogeneous catalyst has a *different* physical state from the reactants. Heterogeneous catalysts are usually solids in contact with gaseous reactants or reactants in solution. Reactant molecules are **adsorbed** (weakly bonded) onto the surface of the catalyst, where the reaction takes place. After reaction, the product molecules leave the surface of the catalyst by **desorption**.

Some of the many common industrial processes that use heterogeneous catalysis are listed in Table 1.

▼ **Table 1** *Industrial processes involving heterogeneous catalysts*

Process	Catalyst	Equation
making ammonia (Haber process)	Fe(s)	$N_2(g) + 3H_2(g) \rightleftharpoons 2NH_3(g)$
reforming	Pt(s) or Rh(s)	$C_6H_{14}(g) \rightarrow C_6H_{12}(g) + H_2(g)$
hydrogenation of alkenes	Ni(s)	$C_2H_4(g) + H_2(g) \rightarrow C_2H_6(g)$
making sulfur trioxide for sulfuric acid (contact process)	$V_2O_5(s)$	$2SO_2(g) + O_2(g) \rightleftharpoons 2SO_3(g)$

▲ **Figure 3** *Fumes from car exhaust pipes*

▲ **Figure 4** *Catalytic converters have a large surface area for heterogeneous catalysis to convert harmful exhaust gases into less harmful gases that can be released into the atmosphere*

Heterogeneous catalysts and atmospheric pollution

Since 1992 all petrol vehicles manufactured for road use in the UK must be fitted with a catalytic converter by law to pass the MOT test.

Catalytic converters contain a catalyst made of platinum, rhodium, and palladium supported on a honeycomb mesh that provides a large surface area on which the reactions can take place (Figure 4). The hot exhaust gases are passed over this heterogeneous catalyst, and harmful gases are converted into less harmful products.

Combustion in a petrol engine forms the toxic gases carbon monoxide and nitrogen monoxide. In the catalytic converter, carbon monoxide is oxidised to carbon dioxide, and nitrogen monoxide is reduced to nitrogen gas. The carbon dioxide and nitrogen products are both non-toxic and can be released into the atmosphere. In addition, any unburnt hydrocarbons are oxidised to water and carbon dioxide.

> Write a balanced equation for the reaction of carbon monoxide and nitrogen monoxide in a catalytic converter.

10 REACTION RATES AND EQUILIBRIUM

Catalysis – sustainability and economic importance

It is estimated that 90% of all chemical materials are produced using a catalyst. Catalysts increase the rate of many industrial chemical reactions by lowering the activation energy. This then reduces the temperature needed for the process and the energy requirements

If a chemical process requires less energy, then less electricity or fossil fuel is used. Making the product faster and using less energy can cut costs and increase profitability. The economic advantages of using a catalyst outweigh any costs associated with developing a catalytic process.

The modern focus on sustainability requires industry to operate processes with high atom economies and fewer pollutants. Using less fossil fuel will cut emissions of carbon dioxide, a gas linked to global warming.

Synoptic link
You learnt about atom economy in Topic 3.4, Reaction quantities.

Autocatalysis

A chemical reaction is said to have undergone autocatalysis if a reaction product acts as a catalyst for that reaction. Typically the reaction starts slowly and then speeds up as the products are formed.

An example of autocatalysis is shown in the equation below

$$2MnO_4^- + 16H^+ + 5C_2O_4^{2-} \rightarrow 2Mn^{2+} + 8H_2O + 10CO_2$$

This reaction is very slow in the absence of a catalyst. However Mn^{3+} ions can act as a catalyst because manganese easily changes between the oxidation states, Mn^{2+} and Mn^{3+}.

In the first step of the autocatalysis, the Mn^{2+} formed reduces MnO_4^- to Mn^{3+}, as shown in the equation below.

$$4Mn^{2+} + MnO_4^- + 8H^+ \rightarrow 5Mn^{3+} + 4H_2O$$

The Mn^{3+} then oxidises the $C_2O_4^{2-}$ to CO_2, reforming Mn^{2+}.

1 What is the catalyst?
2 Write an equation for the reforming of Mn^{2+}.

Summary questions

1 State the difference between a homogeneous catalyst and a heterogeneous catalyst. *(1 mark)*

2 Describe the effect of a catalyst on the activation energy of a chemical reaction and on the enthalpy change of reaction. *(2 marks)*

3 Methanol can be manufactured by the reaction of carbon dioxide with hydrogen as shown in the equation:
$$3H_2(g) + CO_2(g) \rightleftharpoons CH_3OH(g) + H_2O(g) \quad \Delta H = -49 \text{ kJ mol}^{-1}$$
The activation energy of the forward reaction is $+225 \text{ kJ mol}^{-1}$.
  a Draw an enthalpy profile diagram for this reaction. *(2 marks)*
  b Calculate the activation energy of the reverse reaction. *(2 marks)*

151

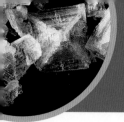

10.3 The Boltzmann distribution

Specification reference: 3.2.2

Learning outcomes

Demonstrate knowledge, understanding, and application of:

→ the Boltzmann distribution
→ the Boltzmann distribution and activation energy
→ the Boltzmann distribution, temperature changes, and catalysts.

Synoptic link

You learnt about activation energy in Topic 9.1, Enthalpy changes.

Study tip

When sketching the Boltzmann curve, make sure it starts at the origin and the curve never crosses the *x*-axis, even at high energy.

The energy of moving particles

You learnt at GCSE that molecules in a gas move at high speed, colliding with each other and with the walls of the container they are held in. These collisions are said to be elastic; the molecules do not slow down as a result of a collision and no energy is lost.

In a gas, a liquid, or a solution, some molecules move slowly with low energy and some molecules move fast with high energy. Most molecules move close to the average speed and have close to the average energy. This spread of molecular energies in gases is known as the Boltzmann distribution (Figure 1). The graph is marked with a line, E_a, that represents the activation energy of a reaction. You can see from the shaded area that only a small proportion of the molecules have more energy than E_a, that is, enough energy to react.

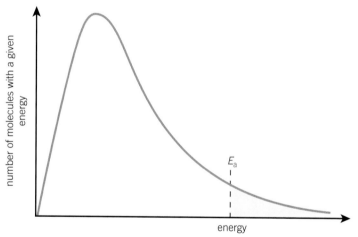

▲ **Figure 1** *The Boltzmann distribution of molecular energies*

There are a number of features of the Boltzmann distribution:

- No molecules have zero energy – the curve starts at the origin.
- The area under the curve is equal to the total number of molecules.
- There is no maximum energy for a molecule – the curve does not meet the *x*-axis at high energy. The curve would need to reach infinite energy to meet the *x*-axis.

The Boltzmann distribution and temperature

The effect of temperature on a Boltzmann distribution curve is shown in Figure 2. As the temperature increases, the average energy of the molecules also increases. A small proportion molecules will still have low energy, but more molecules have higher energy. The graph is now stretched over a greater range of energy values. The peak of the graph is lower on the *y*-axis and further along the *x*-axis – the peak is at a higher energy. The number of molecules is the same, so the area under the curve remains the same.

REACTION RATES AND EQUILIBRIUM 10

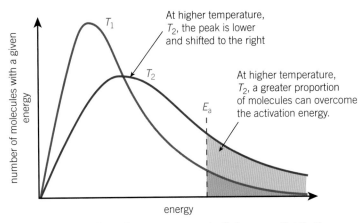

▲ **Figure 2** *The effect of temperature on the Boltzmann distribution*

At higher temperature:
- More molecules have an energy greater than or equal to the activation energy.
- Therefore a greater proportion of collisions will lead to a reaction, increasing the rate of reaction.
- Collisions will also be more frequent as the molecules are moving faster, but the increased energy of the molecules is much more important than the increased frequency of collisions.

The Boltzmann distribution and catalysts

In Topic 10.2 you learnt that a catalyst lowers the activation energy of a reaction. The effect of a catalyst on activation energy is shown on a Boltzmann distribution curve in Figure 3.

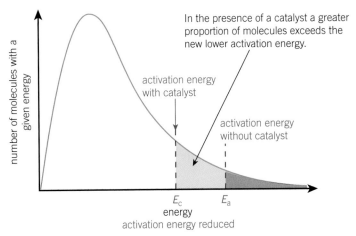

▲ **Figure 3** *The effect of a catalyst on the number of molecules with enough energy to react.*

A catalyst provides an alternative reaction route with a lower activation energy (E_c on the graph). Compared to E_a, a greater proportion of molecules now have an energy equal to, or greater than the lower activation energy, E_c. On collision, more molecules will react to form products. The result is an increase in the rate of reaction.

Study tip

You may be asked to explain how increasing the temperature or using a catalyst increases the rate of a chemical reaction. You will need to include ideas about the Boltzmann distribution.

Summary questions

1. Explain what is meant by the term activation energy. *(1 mark)*

2. Describe and explain how the rate of reaction is affected by a decrease in temperature *(2 marks)*

3. a. Sketch a Boltzmann distribution curve for a volume of gas at temperature T_1. Add a second curve for the distribution at a higher temperature, T_2. *(2 marks)*
   b. Explain how raising the temperature increases the rate of reaction, using your answer to (a) above. *(2 marks)*
   c. Using the Boltzmann distribution, explain how the presence of a catalyst increases the rate of reaction. *(2 marks)*

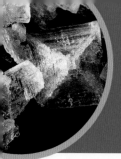

10.4 Dynamic equilibrium and le Chatelier's principle

Specification reference: 3.2.3

Learning outcomes

Demonstrate knowledge, understanding, and application of:

→ dynamic equilibrium
→ le Chatelier's principle
→ the effect of temperature, concentration, and pressure on the position of equilibrium
→ catalysts and equilibrium.

Introducing reversible reactions

When ignited, hydrogen reacts with oxygen in the air to form water. At the end of the reaction all the hydrogen has been used up. The reaction has gone to completion. The equation contains an arrow pointing from reactants to products.

$$2H_2(g) + O_2(g) \rightarrow 2H_2O(l)$$
$$\text{reactants} \quad\quad\quad \text{products}$$

In this topic, you will look at some **reversible reactions**, reactions that take place in both 'forward' and 'reverse' directions. Many of these reactions are important industrial processes, for example, the Haber process for manufacturing ammonia:

$$N_2(g) + 3H_2(g) \rightleftharpoons 2NH_3(g)$$

The $\rightleftharpoons$ symbol indicates that the reversible reaction is in equilibrium.

Dynamic equilibrium

In an equilibrium system:

- the rate of the forward reaction is equal to the rate of the reverse reaction
- the concentrations of reactants and products do not change.

Equilibrium systems are **dynamic**. At equilibrium both the forward and reverse reactions are taking place. As fast as the reactants are becoming products, the products are reacting to become reactants. Therefore in an equilibrium system the concentrations of the reactants and products remain unchanged even though the forward and reverse reactions are still taking place.

For a reaction to remain in equilibrium, the system must be closed. A **closed system** is isolated from its surroundings, so the temperature, pressure, and concentrations of reactants and products are unaffected by outside influences.

Synoptic link

See Topic 9.1, Enthalpy changes, for details of the chemical system and the surroundings.

le Chatelier's principle

The **position of equilibrium** indicates the extent of the reaction. In a reversible reaction, if the temperature, pressure (for reactions involving gases), or concentration of the reactants or products is changed, then the position of equilibrium may change.

le Chatelier's principle states that when a system in equilibrium is subjected to an external change the system readjusts itself to minimise the effect of that change.

REACTION RATES AND EQUILIBRIUM

10

Figure 1 illustrates a system in which the equilibrium has been disrupted by adding more reactant molecules. The position of equilibrium shifts to the right of the equation. More products are made than reactants until a new equilibrium is established.

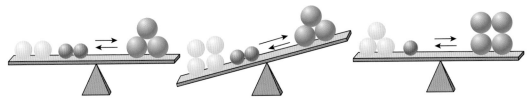

▲ **Figure 1** *When an equilibrium system is subjected to a change the position of equilibrium moves in such a way as to minimise the change*

The effect of concentration changes on equilibrium

Changing the concentration of a reactant or a product in an equilibrium system will change the rate of the forward or reverse reactions. The position of equilibrium will then change. Figure 2 shows the effect of changing the concentration of reactants or products.

When an equilibrium system adjusts as a result of a change:

- if there are more *products formed*, the position of the equilibrium has shifted to the *right*
- if there are more *reactants formed*, the position of the equilibrium has shifted to the *left*.

If you choose equilibria where the reactants and products have different colours, simple experiments can illustrate how the position of equilibrium changes with an external change.

▲ **Figure 2** *The effect of concentration on the position of the equilibrium*

Investigating changes to the position of equilibrium with concentration

The equilibrium between aqueous chromate ions, CrO_4^{2-}, and dichromate ions, $Cr_2O_7^{2-}$, is sensitive to changes in acid concentration. Solutions of chromate and dichromate ions have different colours so it is easy to see any shift in the position of equilibrium (Figure 3).

$$2CrO_4^{2-}(aq) + 2H^+(aq) \rightleftharpoons Cr_2O_7^{2-}(aq) + H_2O(l)$$
yellow orange

▲ **Figure 3** *The chromate/dichromate equilibrium*

155

10.4 Dynamic equilibrium and le Chatelier's principle

Looking to Figure 3, you would expect that the position of equilibrium could be changed by altering the concentrations of the reactants or products. You can carry out a simple experiment to show this.

1 Add a solution of yellow potassium chromate, K_2CrO_4, to a beaker.
2 Add dilute sulfuric acid, H_2SO_4, dropwise until there is no further change. The solution turns an orange colour.
3 Add aqueous sodium hydroxide, NaOH(aq), until there is no further change. The solution changes back to a yellow colour.

You can repeat steps 2 and 3 many times and the colour will change each time from yellow to orange and back to yellow again. So how does it work?

When you add dilute sulfuric acid, H_2SO_4, you are increasing the concentration of H^+(aq) ions. This increases the rate of the forward reaction and so causes the position of equilibrium to shift to minimise the change in H^+(aq) concentration.

1 This shift *decreases* the concentration of the added *reactant*, H^+(aq).
2 The position of equilibrium shifts to the right of the equation, making more products
3 A new position of equilibrium is established towards the products.
   - The solution turns orange as $Cr_2O_7^{2-}$ forms.

When you add aqueous sodium hydroxide, NaOH(aq), the added OH^-(aq) ions react with H^+(aq) ions, decreasing the concentration of H^+(aq) ions.

$$H^+(aq) + OH^-(aq) \rightarrow H_2O(l)$$

The decreased concentration of the reactant, H^+(aq) decreases the rate of the forward reaction and so causes the position of equilibrium to shift to minimise the change in concentration.

1 The shift *increases* the concentration of the *reactant* that has been removed, H^+(aq).
2 The position of equilibrium shifts to the left, making more of the H^+(aq) reactant.
3 A new position of equilibrium is established.
   - The solution turns yellow as CrO_4^{2-}(aq) forms.

Figure 4 summarises these changes.

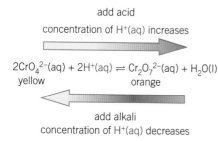

▲ Figure 4 *The effect of changing the concentration of H^+(aq) on the CrO_4^{2-}(aq)/$Cr_2O_7^{2-}$(aq) equilibrium*

Investigating changes to the position of equilibrium with temperature

Changing the temperature of a system in equilibrium will result in the position of equilibrium changing.

The direction in which the equilibrium shifts depends on the sign of ΔH.

- Forward and reverse directions have the same value for the enthalpy change – but the signs are opposite.
- An *increase* in temperature shifts the equilibrium position in the endothermic direction (ΔH is positive)
- An *decrease* in temperature shifts the equilibrium position in the exothermic direction (ΔH is negative)

Synoptic link

You met ΔH in Topic 9.1, Enthalpy changes.

REACTION RATES AND EQUILIBRIUM

10

Cobalt chloride, $CoCl_2$, dissolves in water to form a pink solution. The dissolving process actually produces an equilibrium between two complexes of cobalt that are different colours:

ΔH $[Co(H_2O)_6]^{2+}(aq) + 4Cl^-(aq) \rightleftharpoons CoCl_4^{2-}(aq) + 6H_2O(l)$ ΔH
is negative   pink   blue   is positive

> **Study tip**
>
> When discussing changes in temperature, always state whether the forward reaction is exothermic or endothermic.

▲ **Figure 5** *The $[Co(H_2O)_6]^{2+}(aq)/CoCl_4^{2-}(aq)$ equilibrium*

This equilibrium is sensitive to changes in temperature and the different colours makes it easy to follow any change in equilibrium position. You can carry out a simple experiment to show this.

1. Dissolve cobalt chloride in water in a boiling tube. Add a small quantity of hydrochloric acid. Place the boiling tube in some iced water. The solution is a pink colour.
2. Set up a boiling water bath and transfer the boiling tube into the boiling water. The solution turns a blue colour.
3. Transfer the boiling tube back to the iced water. The solution changes back to a pink colour.

You can repeat steps 2 and 3 many times and the colour will change each time from pink to blue and back to pink again. So how does it work?

In the boiling water, you are increasing the heat energy of the system. This causes the position of equilibrium to shift to minimise the change.

- As the forward reaction is endothermic (ΔH is positive), the position of equilibrium shifts to the right in the endothermic direction, to take heat energy in and minimise the increase in temperature.
- The solution turns a blue colour.

Decreasing the temperature shifts the position of equilibrium in the opposite direction, in the direction that gives out energy, to the reverse exothermic side (ΔH is negative) on the left.

Figure 6 summarises these changes.

> **Study tip**
>
> A small amount of hydrochoric acid can be added to provide more $Cl^-(aq)$ ions. This shifts the equilibrium slightly towards the right and helps to achieve the colour changes. (See effect of concentration above)

Increase temperature
Shift in endothermic direction
→

ΔH $[Co(H_2O)_6]^{2+}(aq) + 4Cl^-(aq) \rightleftharpoons CoCl_4^{2-}(aq) + 6H_2O(l)$ ΔH
is negative   pink   blue   is positive

←
decrease temperature
Shift in exothermic direction

▲ **Figure 6** *The effect of changing temperature on the $[Co(H_2O)_6]^{2+}(aq)/CoCl_4^{2-}(aq)$ equilibrium*

10.4 Dynamic equilibrium and le Chatelier's principle

The effect of temperature on the position of equilibrium for exothermic and endothermic reactions is summarised in Table 1.

▼ **Table 1** *The effect of temperature on the position of equilibrium*

Forward reaction	Increase temperature	Decrease temperature
exothermic (ΔH is negative)	Position of equilibrium shifts to the left. More reactants are made.	Position of equilibrium shifts to the right. More products are made.
endothermic (ΔH is positive)	Position of equilibrium shifts to the right. More products are made.	Position of equilibrium shifts to the left. More reactants are made.

> **Study tip**
> When discussing changes in temperature, always state whether the forward reaction is exothermic or endothermic. An exothermic change has a negative ΔH value, and an endothermic change has a positive ΔH value.

The effect of pressure changes on equilibrium

Changing the pressure of a system containing gases in equilibrium may result in the position of equilibrium changing, but only if there are more gaseous molecules on one side of the equation than the other.

The gases nitrogen dioxide, $NO_2(g)$, and dinitrogen tetroxide, $N_2O_4(g)$, have different colours. The gases form the equilibrium below.

$$2NO_2(g) \rightleftharpoons N_2O_4(g)$$
brown   colourless
2 mol   1 mol

The pressure of a gas is proportional to its concentration. In the same container, two moles of $NO_2(g)$ would have twice the concentration and twice the pressure as the same container holding one mole of $N_2O_4(g)$.

Increasing the pressure of the system will shift the position of equilibrium to the side with the fewer molecules, reducing the pressure of the system.

- As there are fewer gaseous moles on the right-hand side of the equilibrium, the position of equilibrium shifts to the right reducing the number of gaseous moles to minimise the increase in pressure.
- More colourless $N_2O_4(g)$ is formed and the brown colour fades.

Decreasing the pressure shifts the position of equilibrium in the opposite direction, to the side with more gaseous moles on the left and making the brown colour deeper (Figure 7).

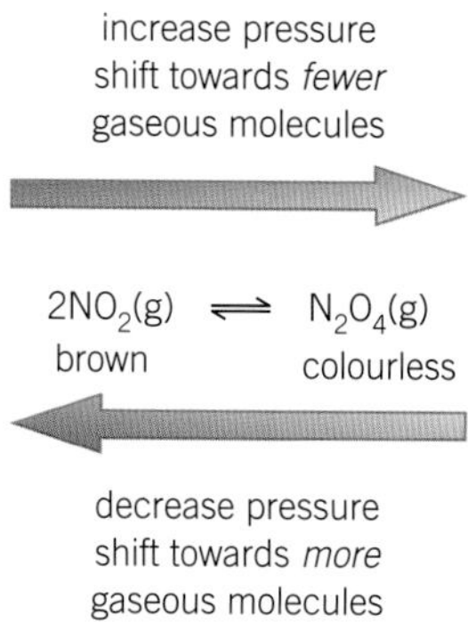

> **Study tip**
> When discussing changes in pressure, always state whether there are more moles of gas on the left or right of the equation.

The effect of a catalyst on equilibrium

A catalyst does not change the position of equilibrium; it merely speeds up the rates of the forward and reverse reactions equally. A catalyst will, however, increase the rate at which an equilibrium is established.

increase pressure
shift towards *fewer*
gaseous molecules

→

$2NO_2(g) \rightleftharpoons N_2O_4(g)$
brown   colourless

←

decrease pressure
shift towards *more*
gaseous molecules

▲ **Figure 7** *Effect and increasing and decreasing the pressure on the equilibrium* $2NO_2(g) \rightleftharpoons N_2O_4(g)$

10 REACTION RATES AND EQUILIBRIUM

Making ammonia with the Haber process

The equation for the Haber process is shown below.

$$N_2(g) + 3H_2(g) \rightleftharpoons 2NH_3(g) \quad \Delta H = -92 \text{ kJ mol}^{-1}$$

le Chatelier's principle can be used to predict the best conditions of temperature and pressure to force the equilibrium to the right in order to produce the maximum yield of ammonia.

- A low temperature will push the equilibrium to the right.
- A high pressure will push the equilibrium to the right.

Why not use these conditions?

- A low temperature would produce a high yield of product, but would do so very slowly. If the temperature is too low, the rate may be so slow that equilibrium may not even be established.
- A high pressure not only increases the yield but also forces the molecules closer together, increasing the concentration and increasing the rate of reaction. However, a very high pressure requires a very strong container and a large quantity of energy, increasing the cost of the process. Safety is also a concern, as a failure in the steelwork or seals of the plant could lead to hot gases, including toxic ammonia, leaking under pressure, endangering the workforce and the surrounding area.

A typical ammonia plant operates under compromise conditions using a high enough temperature to give a reasonable rate without shifting the equilibrium position too far away from ammonia and back to the reactants. This ensures that a good yield of ammonia is achieved quickly, cheaply, and safely. The actual conditions vary depending on the process; however, temperatures of 350–500 °C and pressures of 100–200 atm are usual. An iron catalyst is used to speed up the reaction so that lower temperatures can be used and operating costs are reduced. Only about 15% of the nitrogen and hydrogen is converted to ammonia.

However, unreacted nitrogen and hydrogen are recycled repeatedly, so nearly all of the nitrogen and hydrogen used is eventually converted into ammonia.

▲ **Figure 8** *Air-separating factory for producing industrial gases such as nitrogen for the Haber process*

Explain the two predictions made by le Chatelier's principle for producing the maximum yield of ammonia in the Haber process

Summary questions

1 State the result of an increase in pressure on the following equilibria.
   a $N_2O_4(g) \rightleftharpoons 2NO_2(g)$ *(1 mark)*
   b $CO(g) + 2H_2(g) \rightleftharpoons CH_3OH(g)$ *(1 mark)*
   c $H_2(g) + Br_2(g) \rightleftharpoons 2HBr(g)$ *(1 mark)*

2 State the result of an increase in temperature on the following equilibria.
   a $N_2(g) + O_2(g) \rightleftharpoons 2NO(g) \quad \Delta H^\ominus = +180 \text{ kJ mol}^{-1}$ *(1 mark)*
   b $2SO_2(g) + O_2(g) \rightleftharpoons 2SO_3(g) \quad \Delta H^\ominus = -197 \text{ kJ mol}^{-1}$ *(1 mark)*

3 State how you would alter the temperature *and* pressure to increase the yield of the products in the equilibria below.
   a $CO(g) + 2H_2(g) \rightleftharpoons CH_3OH(g) \quad \Delta H^\ominus = -92 \text{ kJ mol}^{-1}$ *(2 marks)*
   b $PCl_5(g) \rightleftharpoons PCl_3(g) + Cl_2(g) \quad \Delta H^\ominus = +124 \text{ kJ mol}^{-1}$ *(2 marks)*

10.5 The equilibrium constant K_c – part 1

Specification reference: 3.2.3

Learning outcomes

Demonstrate knowledge, understanding, and application of:

→ expressions for K_c
→ calculation of K_c from equilibrium concentrations
→ estimation of the position of equilibrium from the magnitude of K_c.

The position of equilibrium

In Topic 10.4, you learnt that a dynamic equilibrium exists when the rate of the forward reaction is equal to the rate of the reverse reaction. You also learnt about how changing the concentration, temperature, and pressure can change the position of equilibrium.

- In this topic you will learn how to calculate **equilibrium constants**. An equilibrium constant provides the actual position of equilibrium.
- The magnitude of an equilibrium constant indicates whether there are more reactants or more products in an equilibrium system.

The equilibrium law

The exact position of equilibrium is calculated using the equilibrium law. For any reaction at equilibrium it is possible to write an expression for the equilibrium constant K_c in terms of equilibrium concentrations.

Consider a general reversible reaction:

$$aA + bB \rightleftharpoons cC + dD$$

The equilibrium law defines the equilibrium constant K_c in terms of concentrations.

$$K_c = \frac{[C]^c [D]^d}{[A]^a [B]^b} \qquad \frac{[\text{products}]}{[\text{reactants}]}$$

- Square brackets, [ ], are shorthand for 'concentration of'.
- a, b, c, d are the balancing numbers in the overall equation.
- [A], [B], [C], [D] are the *equilibrium* concentrations of the reactants and products of this equilibrium.

For example, the expression for K_c for the equilibrium reaction $N_2(g) + O_2(g) \rightleftharpoons 2NO(g)$ is:

$$K_c = \frac{[NO(g)]^2}{[N_2(g)] [O_2(g)]}$$

Synoptic link

You will learn more about equilibrium constants, including where the units come from, in Chapter 19, Equilibrium.

Study tip

When writing equilibrium constants, remember that the concentrations of the products are always divided by the concentrations of the reactants.

Study tip

Don't forget that the balancing numbers in the equation are accounted for in the expression for K_c, so there's no need to multiply any concentrations by the balancing number. Here, for example, the concentration of hydrogen iodide is squared – you don't need to multiply it by two as well.

Calculating K_c from equilibrium concentrations

To calculate an equilibrium constant you need to know the equilibrium concentrations of the reactants and products.

 Worked example 1: Writing and calculating K_c

Calculate the value of K_c for the following equilibrium:
$$H_2(g) + I_2(g) \rightleftharpoons 2HI(g).$$
Give your answer to three significant figures.

160

The equilibrium concentrations are:

- $[H_2(g)] = 8.00 \times 10^{-4}\,\text{mol dm}^{-3}$
- $[I_2(g)] = 1.20 \times 10^{-4}\,\text{mol dm}^{-3}$
- $[HI(g)] = 2.28 \times 10^{-3}\,\text{mol dm}^{-3}$

Step 1: Write the expression for K_c.

$$K_c = \frac{[HI(g)]^2}{[H_2(g)][I_2(g)]}$$

Step 2: Calculate K_c.

$$K_c = \frac{(2.28 \times 10^{-3})^2}{(8.00 \times 10^{-4}) \times (1.20 \times 10^{-4})} = 54.2$$

 Worked example 2: Writing and calculating K_c

In the manufacture of ammonia at 400 °C, the equilibrium concentrations are $N_2(g)$: 18.6 mol dm^{-3}, $H_2(g)$: 0.900 mol dm^{-3}, $NH_3(g)$: 1.50 mol dm^{-3}.

$$N_2(g) + 3H_2(g) \rightleftharpoons 2NH_3(g)$$

Calculate the value for K_c, to three significant figures.

Step 1: Write the expression for K_c.

$$K_c = \frac{[NH_3(g)]^2}{[N_2(g)][H_2(g)]^3}$$

Step 2: Calculate K_c.

$$K_c = \frac{(1.50)^2}{(18.6)(0.900)^3} = 0.166\ (\text{dm}^6\,\text{mol}^{-2})$$

What does the value of K_c tell us?

In the two examples above, the calculated K_c values were 54.2 and 0.166. The magnitude of K_c indicates the relative proportions of reactants and products in the equilibrium system.

As a rough guide,

- a K_c value of 1 indicates a position of equilibrium that is halfway between reactants and products.
- a K_c value > 1 indicates a position of equilibrium that is towards the products, for example, 54.2 in worked example 1.
- a K_c value < 1 indicates a position of equilibrium that is towards the reactants, for example, 0.166 in worked example 2.

So the larger the value of K_c, the further the position of equilibrium lies to the right-hand side and the greater the concentrations of the products compared to the reactants.

Summary questions

1. The reversible reaction below is allowed to reach equilibrium:

   $$CH_3COOH(l) + C_2H_5OH(l) \rightleftharpoons CH_3COOC_2H_5(l) + H_2O(l)$$

   At equilibrium, the concentrations of $CH_3COOC_2H_5(l)$ and $H_2O(l)$ are 0.25 mol dm^{-3}. The concentration of $CH_3COOH(l)$ is 0.10 mol dm^{-3} and of $C_2H_5OH(l)$ is 0.40 mol dm^{-3}.

   a Write the expression for K_c for this equilibrium.
   (1 mark)

   b Calculate a value for K_c.
   (1 mark)

2. The reversible reaction below is allowed to reach equilibrium at constant temperature.

   $$H_2(g) + I_2(g) \rightleftharpoons 2HI(g)$$

   The equilibrium concentrations are 0.40 mol dm^{-3} $H_2(g)$, 0.30 mol dm^{-3} $I_2(g)$, and 0.14 mol dm^{-3} $HI(g)$. Write the expression for K_c and calculate a value for K_c.
   (2 marks)

3. For the equilibrium system

   $$2SO_2(g) + O_2(g) \rightleftharpoons 2SO_3(g)$$

   the equilibrium concentrations are:

   $[SO_2(g)] = 0.23$ mol dm^{-3},
   $[O_2(g)] = 1.37$ mol dm^{-3},
   $[SO_3(g)] = 0.92$ mol dm^{-3}.

   Write the expression for K_c and calculate the value of K_c at this temperature.
   (2 marks)

Chapter 10 Practice questions

Practice questions

1 The diagram below shows an energy profile diagram for an endothermic reaction.

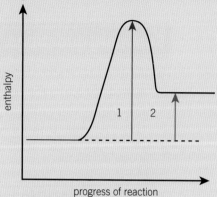

progress of reaction

a What physical quantities are shown by the labels 1 and 2? *(2 marks)*

b Explain the effect of the presence of a catalyst on the two quantities. *(2 marks)*

2 a Reaction rates can be increased or decreased by changing conditions of temperature and pressure.

(i) Explain how increasing the temperature increases the rate of reaction.

Include a labelled sketch of the Boltzmann distribution. Label the axes. *(4 marks)*

Your answer needs to be clear and well organised using the correct terminology.

(ii) Describe and explain the effect of decreasing the pressure on the rate of a reaction. *(2 marks)*

OCR F322 June 2014 Q6 (a)

3 The uses of catalysts have great economic and environmental importance. For example, catalysts are used in ammonia production and in catalytic converters.

a Nitrogen and hydrogen react together in the production of ammonia, NH_3.

$N_2(g) + 3H_2(g) \rightleftharpoons 2NH_3(g)$
$\Delta H = -92$ kJ mol^{-1}

The activation energy for the forward reaction, E_a, is +250 kJ mol^{-1}.

(i) Complete the enthalpy profile diagram for this reaction between nitrogen and hydrogen.

Include the
- products
- enthalpy change of reaction, ΔH
- activation energy for the forward reaction, E_a. *(3 marks)*

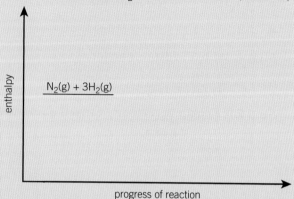

(ii) What is the value of the enthalpy change of formation of ammonia? *(1 mark)*

(iii) The reaction between nitrogen and hydrogen can be catalysed.

Suggest a possible value for the activation energy of the catalysed forward reaction. *(1 mark)*

(iv) What is the value of the activation energy for the uncatalysed reverse reaction (the decomposition of ammonia into nitrogen and hydrogen)? *(1 mark)*

OCR June 2012 Q4 (a)

4 State and explain the effect, if any, of the following changes to the position of the equilibrium system below.

$2SO_2(g) + O_2(g) \rightleftharpoons 2SO_3(g)$ $\Delta H = -197$ kJ mol^{-1}

a Increasing the concentration of SO_2. *(1 mark)*

b Increasing the pressure. *(1 mark)*

c Increasing the temperature *(1 mark)*

d Adding a suitable catalyst. *(1 mark)*

5 Hydrogen has many industrial uses including making margarine and ammonia.

Hydrogen can be made by the reaction between methane and steam.

REACTION RATES AND EQUILIBRIUM

$CH_4(g) + H_2O(g) \rightleftharpoons CO(g) + 3H_2(g)$
$\Delta H = +210 \text{ kJ mol}^{-1}$

a The pressure of the equilibrium mixture is increased.
Explain what happens to the position of the equilibrium. *(2 marks)*

b The temperature of the equilibrium mixture is increased.
Explain what happens to the position of the equilibrium. *(2 marks)*

c The reaction is actually carried out in the presence of a nickel catalyst at a pressure of 30 atmospheres.
  (i) Suggest why the manufacturer uses a pressure of 30 atmospheres. *(1 mark)*
  (ii) The nickel catalyst increases the rate. Use a labelled diagram of the Boltzmann distribution of molecular energies to explain why. *(3 marks)*

OCR F322 Jun 2012 Q3 (a) (b) (c)

6 In the chemical industry, methanol, CH_3OH, is synthesised by reacting together carbon monoxide and hydrogen in the presence of copper, zinc oxide and alumina which act as a catalyst. This is a reversible reaction.
$CO(g) + 2H_2(g) \rightleftharpoons CH_3OH(g)$
$\Delta H = -91 \text{ kJ mol}^{-1}$

a High pressures and low temperatures would give the maximum equilibrium yield of methanol.
Explain why. *(2 marks)*

b Explain why the actual conditions used in the chemical industry might be different from those in (a) above. *(2 marks)*

F322 June 2009 Q4

7 This question looks at two systems in equilibrium.

a The equilibrium system below is set up.
$N_2O_4(g) \rightleftharpoons 2NO_2(g)$
$K_c = 5.0 \times 10^{-3} \text{ mol dm}^{-3}$
  (i) Write the expression for K_c. *(1 mark)*
  (ii) At equilibrium, the concentration of N_2O_4 is 0.50 mol dm^{-3}. Calculate the equilibrium concentration of $NO_2(g)$ in mol dm^{-3}? *(1 mark)*

b Nitrogen gas and hydrogen gas produce ammonia gas as shown below.
$N_2(g) + 3H_2(g) \rightleftharpoons 2NH_3(g)$
  (i) Write the expression for K_c for this equilibrium. *(1 mark)*
  (ii) At 200 °C, the equilibrium concentrations are
  $N_2(g)$: 9.35 mol dm^{-3};
  $H_2(g)$: 0.32 mol dm^{-3};
  $NH_3(g)$ 1.54 mol dm^{-3}.
  Calculate the numerical value of K_c. *(2 marks)*

8 Hydrogen and iodine are mixed together in a sealed container at constant temperature and the mixture is allowed to reach equilibrium.
$H_2(g) + I_2(g) \rightleftharpoons 2HI(g) \quad \Delta H = -9 \text{ kJ mol}^{-1}$
At a particular temperature the equilibrium constant K_c is 35.0. At this temperature:
the concentration of
$H_2(g) = 2.10 \times 10^{-4} \text{ mol dm}^{-3}$
the concentration of
$HI(g) = 1.507 \times 10^{-2} \text{ mol dm}^{-3}$.

a (i) Write an expression for K_c. *(1 mark)*
  (ii) Calculate the equilibrium concentration of $I_2(g)$. *(2 mark)*

b A closed system is required for dynamic equilibrium to be established.
State **two** other features of a *dynamic equilibrium*. *(1 mark)*

c The equilibrium mixture is heated whilst keeping the volume constant.
Predict how the composition of the equilibrium mixture changes.
Explain your answer. *(2 mark)*

d The equilibrium mixture is compressed at constant temperature.
Predict the effect on the composition of the equilibrium mixture changes.
Explain your answer. *(2 mark)*

163

Module 3 summary

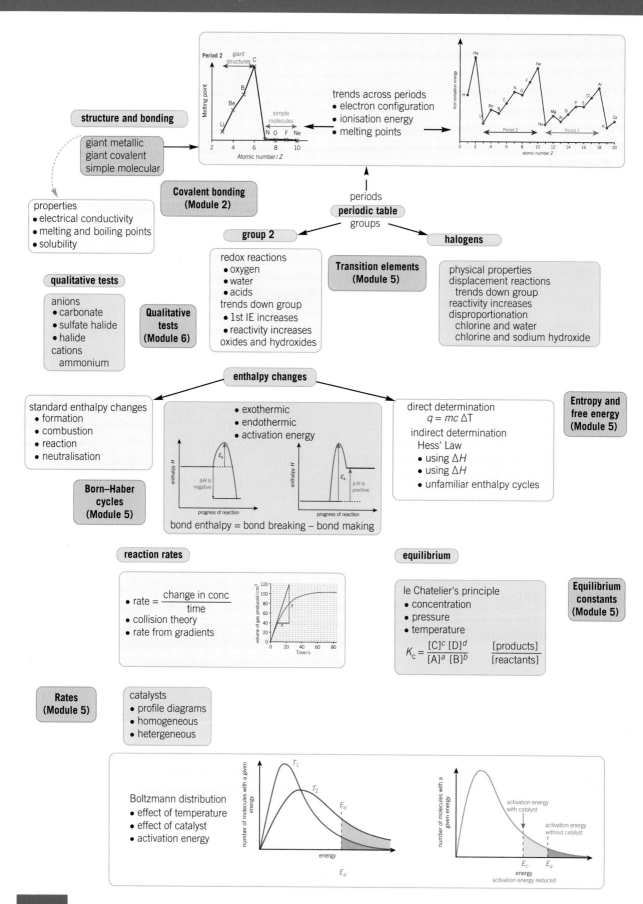

Module 3 Periodic table and energy

Hydrazine

Hydrazine, N_2H_4, has been used for many years as a rocket fuel.

In the presence of suitable catalysts and at high pressure, it decomposes in a series of rapid, exothermic reactions that produce a large volume of gaseous products. These are forced out of the combustion chamber to provide thrust to the rocket. One of these reactions is shown in Equation 1.

$N_2H_4(g) \rightleftharpoons N_2(g) + 2H_2(g)$ **Equation 1**

H\N—N/H /H \H   →   N≡N + 2 H—H

More modern rockets use derivatives of hydrazine, such as methylhydrazine. This, in combination with an oxidiser – dinitrogen tetroxide, N_2O_4, was used to propel the Rosetta probe which succeeded in landing on a comet in 2014

1. Use the bond enthalpy data below to deduce a value for the enthalpy change when one mole of hydrazine decomposes.

Bond	N—H	N—N	N≡N	H—H
Average bond enthalpy / kJ mol^{-1}	+391	+158	+945	+436

2. Suggest and explain three reasons why the decomposition of hydrazine occurs very rapidly in the combustion chamber of a rocket
3. Write an expression for the equilibrium constant K_c for the reaction represented by Equation 1.
4. Describe the effect on the position of the equilibrium represented by Equation 1 when:
   a. the temperature is increased
   b. the pressure is increased
   c. a catalyst is introduced into the system.
5. Is it likely that the reaction represented by Equation 1 will reach equilibrium when it is used to power a rocket? Give your reasoning.

Extension

1. Research some of the unusual redox reactions involving fluorine, and use ideas about the structure of a fluorine atom to explain why it is such a powerful oxidising agent.
2. Prepare a summary of the trends in properties that can be observed in the periodic table. You should consider both trends across periods and down groups and include an explanation of each trend.
3. Carbon dioxide and silicon dioxide have similar formulae but form very different structures. Carbon dioxide molecules have two C═O double bonds. Silicon dioxide has a giant structure with each silicon atom forming four Si—O single bonds. Use the relevant bond energy data to suggest a reason for this difference.

Bond	C—O	C═O	Si—O	Si═O
Average bond enthalpy / kJ mol^{-1}	358	805	466	638

165

Module 3 Practice questions

1. For the 3rd ionisation energy of Mg, which sub-shell loses the electron?
   A 1s;   B 2s;   C 2p;   D 3s   *(1 mark)*
   AS Paper 1 style question

2. Which equation represents the second ionisation energy of calcium?
   A $Ca(s) \rightarrow Ca^{2+}(g) + 2e^-$
   B $Ca(g) \rightarrow Ca^{2+}(g) + 2e^-$
   C $Ca^+(s) \rightarrow Ca^{2+}(g) + e^-$
   D $Ca^+(g) \rightarrow Ca^{2+}(g) + e^-$ *(1 mark)*
   AS Paper 1 style question

3. Which sequence shows the melting point **decreasing** from left to right?
   A Al, Si, P
   B Si, P, S
   C P, S, Cl
   D S, Cl, Ar   *(1 mark)*
   AS Paper 1 style question

4. Which element in Period 3 is **not** described correctly?

	Element	Bonding	Structure
A	chlorine	covalent	simple
B	silicon	covalent	simple
C	sodium	metallic	giant
D	sulfur	covalent	simple

   (1 mark)
   AS Paper 1 style question

5. Which row correctly describes the trend **down** Group 2?

	First ionisation energy	Alkalinity of hydroxide
A	increases	decreases
B	decreases	increases
C	increases	increases
D	decreases	decreases

   (1 mark)
   AS Paper 1 style question

6. Which statement about elements in the halogens is correct?
   A F_2 has the least tendency to be reduced
   B Cl_2 will oxidise I^-
   C Br_2 will oxidise Cl^-
   D I_2 will reduce F^- *(1 mark)*
   AS Paper 1 style question

7. Chlorine reacts with hot NaOH(aq) as follows.
   $Cl_2 + 6NaOH(aq) \rightarrow 5NaCl(aq) + NaClO_3(aq) + 3H_2O(l)$
   Which row shows the correct oxidation numbers of chlorine?

	Cl_2	NaCl	$NaClO_3$
A	0	+1	−5
B	−1	+1	−3
C	0	−1	+5
D	+2	−1	+3

   (1 mark)
   AS Paper 1 style question

8. Which reaction is **not** a disproportionation?
   A $Cl_2 + H_2O \rightarrow Cl^- + ClO^- + 2H^+$
   B $Ca + 2H_2O \rightarrow Ca(OH)_2 + H_2$
   C $3S + 6NaOH \rightarrow 2Na_2S + Na_2SO_3 + 3H_2O$
   D $Cu_2O + H_2SO_4 \rightarrow Cu + CuSO_4 + H_2O$ *(1 mark)*
   AS Paper 1 style question

9. During an experiment 11.50 g of ethanol is completely burnt in air. During the combustion, 341.5 kJ of heat energy is released.
   What is the enthalpy change of combustion of ethanol?
   A $-341.5 \,kJ\,mol^{-1}$ B $-1366 \,kJ\,mol^{-1}$
   C $-85.38 \,kJ\,mol^{-1}$ D $-34.15 \,kJ\,mol^{-1}$ *(1 mark)*
   AS Paper 1 style question

10. Which equation represents the standard enthalpy of formation for ethanol, C_2H_5OH?
    A $2C(s) + 3H_2(g) + ½O_2(g) \rightarrow C_2H_5OH(l)$
    B $C(s) + 1½H_2(g) + ¼O_2(g) \rightarrow ½C_2H_5OH(l)$
    C $4C(s) + 6H_2(g) + O_2(g) \rightarrow 2C_2H_5OH(l)$
    D $2C(s) + 6H(g) + O(g) \rightarrow C_2H_5OH(l)$ *(1 mark)*
    AS Paper 1 style question

11. Which reactions is the enthalpy change equal to the bond enthalpy of H–Br?
    A $HBr(g) \rightarrow H(g) + Br(g)$
    B $HBr(g) \rightarrow H^+(g) + Br^-(g)$

C $HBr(g) \rightarrow \frac{1}{2}H_2(g) + \frac{1}{2}Br_2(l)$

D $HBr(g) \rightarrow \frac{1}{2}H_2(g) + \frac{1}{2}Br_2(g)$ *(1 mark)*

AS Paper 1 style question

12 Which statement explains why a catalyst is used in a chemical reaction?

A Catalysts increase the rate of reaction by increasing the activation energy.

B Catalysts increase the yield of products formed in chemical reactions.

C Catalysts increase the rate of reaction by providing an alternative path of lower energy.

D Catalysts increase the purity of products, meaning less separation of products is required. *(1 mark)*

AS Paper 1 style question

13 Which statement is **not** correct about the equilibrium system?

$H_2(g) + I_2(g) \rightleftharpoons 2HI(g)$

A Rate of forward reaction is equal to the rate of reverse reaction.

B Increasing the concentration of H_2 moves the position of equilibrium left.

C Increasing the pressure does not affect the position of equilibrium.

D Adding a catalyst increases the rate of both the forward and reverse reactions equally. *(1 mark)*

AS Paper 1 style question

14 At equilibrium, the concentration of N_2O_4 was 0.5 mol dm^{-3}.

$N_2O_4(g) \rightleftharpoons 2NO_2(g)$ $K_c = 5.0 \times 10^{-3}$ mol dm^{-3}

What is the equilibrium concentration of $NO_2(g)$ in mol dm^{-3}?

A 0.5; B 0.05; C 0.005; D 0.0025

(1 mark)

AS Paper 1 style question

15 For the reaction: $2HI(g) \rightleftharpoons H_2(g) + I_2(g)$ at a certain temperature, the equilibrium concentrations, in mol dm^{-3}, are

$[H_2(g)] = 0.20$, $[I_2(g)] = 0.20$, and $[HI(g)] = 2.0$.

The value of K_c in mol dm^{-3} is:

A 0.01   B 0.02

C 50   D 100   *(1 mark)*

AS Paper 1 style question

16 Pentane is used as a fuel, making use of its exothermic combustion reaction.

a Explain the term *standard enthalpy change of combustion*. Include the standard conditions in your answer. *(3 marks)*

b The enthalpy change of formation of pentane cannot be measured directly but can be calculated indirectly using enthalpy changes of combustion.

(i) Use the information below to calculate the enthalpy change of formation of pentane. *(3 marks)*

Substance	$\Delta_cH^\ominus$ / kJ mol^{-1}
$C_5H_{12}(l)$	−3509
$C(s)$	−394
$H_2(g)$	−286

$5C(s) + 6H_2(g) \rightarrow C_5H_{12}(l)$

(ii) Suggest **one** reason, other than a high activation energy or slow rate, why the enthalpy change of formation of pentane cannot be obtained directly. *(1 mark)*

AS Paper 1 style question

17 Some properties and reactions of the Group 2 element barium and its compounds are shown below.

a Barium conducts electricity as a result of its structure and bonding.

Draw a diagram to show the bonding in barium and explain why barium conducts electricity. *(3 marks)*

b Identify substances **D**–**G** for the series of reactions starting from barium.

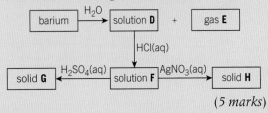

(5 marks)

AS Paper 1 style question

18 A student carries out an experiment to determine the enthalpy change of reaction, Δ_rH, for the equation below:

$Zn(s) + 2AgNO_3(aq) \rightarrow 2Ag(s) + Zn(NO_3)_2(aq)$

directly. The student follows the procedure below.

1. Add 25.0 cm³ of 0.200 mol dm⁻³ silver nitrate to a polystyrene cup.
2. Take the temperature of the solution.
3. Add an excess of powdered zinc to this solution. Stir the solution and record the temperature when it no longer changes. The results are shown below.

Initial temperature = 20.5 °C

Final temperature = 24.0 °C

a Calculate the amount, in mol, of silver nitrate used in the experiment. *(1 mark)*

b Calculate the minimum mass of zinc that the student needed to add in this procedure. *(2 marks)*

c Calculate the energy change.
Assume that the specific heat capacity of the solution is the same as water. *(1 mark)*

d Calculate the enthalpy change of reaction, in kJ mol⁻¹, for the reaction of 1 mol of Zn, as shown in the equation. *(2 marks)*

e Explain, with reasons, how the enthalpy change from this procedure would compare with accurate enthalpy determinations recorded in data books. *(2 marks)*

AS Paper 1 style question

19 This question looks at periodic trends in first ionisation energy and melting point.

The diagram below shows the trend in first ionisation energies for the elements Na to K.

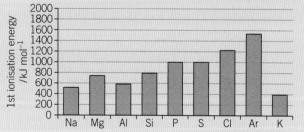

(i) Define the term *first ionisation energy*. *(3 marks)*

(ii) Explain why there is a general increase in first ionisation energy from Na to Ar. *(3 marks)*

(iii) Explain why the first ionisation energy of oxygen is less than that of nitrogen. *(2 marks)*

(iv) Suggest a value for the first ionisation energy of calcium. *(1 mark)*

b The diagram below shows the trend in melting points for the elements Na to Ar.

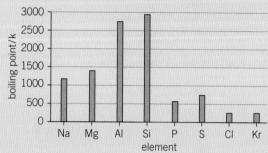

Explain in terms of structure and bonding, why phosphorus has a much lower boiling point than silicon. *(3 marks)*

AS Paper 2 style question

20 This question looks at some equilibrium reactions.

a State two characteristics of a *dynamic equilibrium*. *(2 marks)*

b A dynamic equilibrium is set up between dichromate(VI) and chromate(VI) ions.

$$Cr_2O_7^{2-}(aq) + H_2O(l) \rightleftharpoons 2CrO_4^{2-}(aq) + 2H^+(aq)$$
orange   yellow

The colour of the equilibrium changes on addition first of aqueous acid and then of aqueous alkali.

Predict the colour changes and explain this observation in terms of le Chatelier's principle. *(3 marks)*

c A dynamic equilibrium is set up between nitrogen dioxide, NO_2, and dinitrogen tetroxide, N_2O_4.

$$2NO_2(g) \rightleftharpoons N_2O_4(g) \quad \Delta H = -57 \, kJ \, mol^{-1}$$
brown   colourless

(i) Predict the effect on the colour of the equilibrium mixture and the position of equilibrium of the following changes in conditions.
Explain your answers in terms of le Chatelier's principle.

- Increasing the pressure
- Increasing the temperature.
(*4 marks*)

(ii) The enthalpy change of formation, $\Delta_f H$, for $N_2O_4(g)$ is $+9\,kJ\,mol^{-1}$.

Calculate the enthalpy change of formation of $NO_2(g)$. (*1 mark*)

AS Paper 2 style question

21 $M(NO_3)_2$ and $SrCl_2 \cdot 6H_2O$ are both salts.

When heated, the metal nitrate, $M(NO_3)_2$, decomposes as below.

$2M(NO_3)_2(s) \rightarrow 2MO(s) + 4NO_2(g) + O_2(g)$

a Using oxidation numbers, show that this is a redox reaction. (*3 marks*)

b A student carries out an experiment to try to identify metal **M**.

The student heats 1.15 g of $M(NO_3)_2$ and collects the gas evolved until no further gas was produced. The student plotted a graph of volume of gas against time, as shown below.

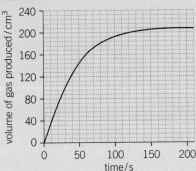

(i) Using the graph, determine the initial rate of reaction, in $cm^3\,s^{-1}$.

Show your working. (*1 mark*)

(ii) Use the graph to find the total amount, in mol, of gas molecules produced. (*2 marks*)

(iii) Determine the molar mass of $M(NO_3)_2$ and the identity of metal **M**.

Show your working. (*6 marks*)

AS Paper 2 style question

22 The following results were obtained from a series of experiments.

Experiment 1 2.50 g of calcium carbonate was added to an excess of hydrochloric acid in a polystyrene cup. 418 J of heat were produced.

Experiment 2 1.40 g of calcium oxide was reacted with 50.0 cm³ of hydrochloric acid (an excess) in a polystyrene cup, producing a temperature rise of 10.0 °C.

Assumptions: Density of solution = 1.00 g cm⁻³.

The specific heat capacity of the solution is the same as water.

a (i) Write equations, including state symbols for the reactions in **Experiment 1** and **Experiment 2**. (*2 marks*)

(ii) Determine the enthalpy change of reaction for
- the reaction of 1 mole of $CaCO_3$ with HCl
- the reaction of 1 mole of CaO with HCl (*7 marks*)

b The equation for the decomposition of calcium carbonate is shown below:

$CaCO_3(s) \rightarrow CaO(s) + CO_2(g)$

Using your calculated enthalpy values from **(a)(ii)**, calculate this enthalpy change (*3 marks*)

AS Paper 2 style question

23 Chlorine and its compounds have many uses.

a Chlorine bleach, used to kill bacteria, is prepared by reacting chlorine with aqueous sodium hydroxide.

$Cl_2(g) + 2NaOH(aq) \rightarrow NaClO(aq) + NaCl(aq) + H_2O(l)$

(i) What is the oxidation number of chlorine in Cl_2, NaClO and NaCl. (*2 marks*)

(ii) The ClO⁻ ion is the bleaching agent. In sunlight, ClO⁻ ions decompose, releasing chlorine gas. Construct an equation for this reaction. (*1 mark*)

b Chlorine can be used to extract bromine from sea water. In this process, chlorine oxidises bromide ions.

(i) Write an ionic equation for this reaction. (*1 mark*)

(ii) What would you see during this extraction? Explain your answer. (*1 mark*)

MODULE 4
Core organic chemistry

Chapters in this module
11 Basic concepts of organic chemistry
12 Alkanes
13 Alkenes
14 Alcohols
15 Haloalkanes
16 Organic synthesis
17 Spectroscopy

Introduction

Organic chemistry is the chemistry of carbon compounds, an incredible range of compounds that outnumber those of all the other elements put together. Organic chemicals mould our modern-day lives, finding use in fuels, plastics, pharmaceuticals, clothing, – the list is almost endless. The module will provide you with the foundations for studying of organic chemistry.

Basic **concepts of organic chemistry** introduces the various types of formula and structures used in organic chemistry. This section is fundamental to the study of all organic chemistry.

Alkanes looks at the important reactions of combustion, exploited in hydrocarbon fuels such as natural gas, petrol and diesel, and with halogens.

Alkenes contains a double bond which gives rise to addition reactions including the formation of the important polymers required for life in the plastic age.

Alcohols looks at the properties and reactions of alcohols. The important oxidation reactions introduce new functional groups: aldehydes, ketones, and carboxylic acids

Haloalkanes considers the importance of the carbon-halogen bond in reactions like hydrolysis. Environmental issues surrounding the disposal of haloalkanes is also studied.

Organic synthesis allows you to acquire organic practical skills. You will also develop synthetic routes for organic compounds using all the reaction from this module.

Spectroscopy studies the important techniques of infrared spectroscopy and mass spectrometry. These are valuable tools for identifying organic compounds.

Knowledge and understanding checklist

From your Key Stage 4 study you should be able to answer the following questions. Work through each point, using your Key Stage 4 notes and the support available on Kerboodle.

- ☐ Recall that crude oil is a main source of hydrocarbons and is a feedstock for the petrochemical industry.
- ☐ Explain how modern life is crucially dependent upon hydrocarbons and recognise that crude oil is a finite resource.
- ☐ Describe and explain the separation of crude oil by fractional distillation.
- ☐ Describe the fractions as largely a mixture of compounds of formula C_nH_{2n+2} which are members of the alkane homologous series.
- ☐ Describe the production of materials that are more useful by cracking.

Maths skills checklist

In this module, you will need to use the following maths skills. You can find support for these skills on Kerboodle and through MyMaths.

- ☐ **Using angles and shapes in regular 2-D and 3-D structures**, for predicting the shapes of and bond angles in molecules.
- ☐ **Translating information between graphical, numerical, and algebraic forms**, for interpreting and analysing spectra.
- ☐ **Visualising and representing 2-D and 3-D forms including 2-D representations of 3-D objects**, for drawing different types of isomer.
- ☐ **Understand the symmetry of 2-D and 3-D shapes**, for describing the types of stereoisomerism shown by molecules.

MyMaths.co.uk
Bringing Maths Alive

11 BASIC CONCEPTS OF ORGANIC CHEMISTRY

11.1 Organic chemistry

Specification reference: 4.1.1

Learning outcomes

Demonstrate knowledge, understanding, and application of:

→ saturated and unsaturated hydrocarbons
→ homologous series
→ functional groups.

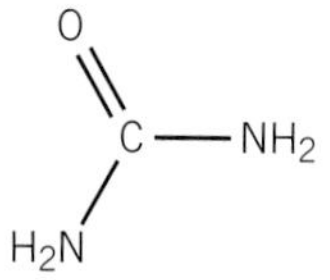

▲ **Figure 1** *The chemical structure of urea, a compound found in living organisms. It is a common fertiliser that was the first organic material synthesised in the laboratory*

Synoptic link

You have met the idea of a covalent bond as a shared pair of electrons in Topic 5.3, Covalent bonding.

▲ **Figure 2** *Organic molecules are responsible for many of the vivid colours found in dyes and pigments*

What is organic chemistry?

Organic chemistry was originally defined by the Swedish chemist Jöns Jacob Berzelius as the chemistry of compounds derived from living systems. Today about sixteen million organic compounds are known, which is surprising since organic chemistry is only about 200 years old. Before 1828, scientists did not believe it was possible to make organic compounds from inorganic compounds. Many experiments were carried out in attempts to synthesise organic compounds but all ended in failure.

In 1828, the German chemist Friedrich Wöhler accidentally synthesised urea (Figure 1) whilst attempting to prepare ammonium cyanate from silver cyanide and ammonium chloride. This was the first organic synthesis carried out from inorganic compounds dispelling the myth that only living organisms could produce organic compounds.

Modern organic chemistry studies the structure, properties, composition, reactions, and preparation of carbon-containing compounds. Organic compounds are vital in every area of modern life: pharmaceuticals, detergents, dyes and pigments, cosmetics, plastics, and agricultural chemicals are all organic compounds – they all contain the element carbon. Today, the vast majority of the organic materials used are produced from fractions of crude oil as fuels in domestic central heating systems, for electrical generation, and to power many forms of transport.

Why is carbon so special?

Carbon is in Group 14(4) of the periodic table, with four electrons in its outer shell. Each carbon atom can form four covalent bonds to other atoms. These can be single, double, and even triple bonds. Carbon atoms can bond to other carbon atoms to form long chains.

Hydrocarbons

A **hydrocarbon** is a compound containing carbon and hydrogen only. The hydrocarbons methane, but-2-ene, and propyne are shown in Figure 3.

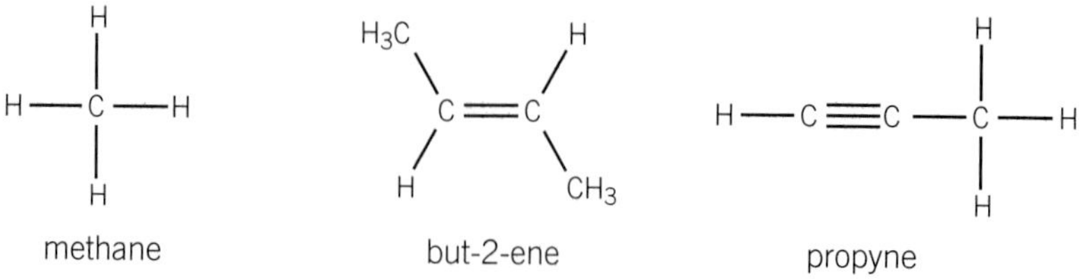

▲ **Figure 3** *The structures of the hydrocarbons methane, but-2-ene, and propyne*

172

BASIC CONCEPTS OF ORGANIC CHEMISTRY 11

Saturated and unsaturated

Hydrocarbons can be **saturated** or **unsaturated**.

- A saturated hydrocarbon has single bonds only, as shown in the structure of methane.
- An unsaturated hydrocarbon contains carbon-to-carbon multiple bonds, as shown in the structures but-2-ene and propyne.

Homologous series

Carbon compounds are so numerous that it is convenient to organise them into families of compounds with similar chemical structures and properties. A **homologous series** is a family of compounds with similar chemical properties whose successive members differ by the addition of a $-CH_2-$ group.

The simplest homologous series is the **alkanes**. Alkanes contain single carbon-to-carbon bonds. The first three members of the homologous series of the alkanes are shown in Figure 4.

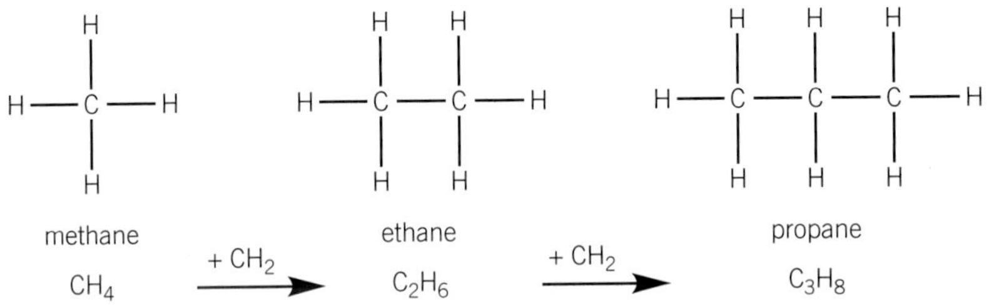

▲ **Figure 4** *The first three members of the alkanes homologous series – methane, ethane, and propane*

Study tip

Remember that the bond angles around carbon in an alkane is 109.5° not 90° as suggested by displayed formula.

Functional groups

A **functional group** is the part of the organic molecule that is largely responsible for the molecule's chemical properties. In addition to hydrogen, carbon can bond to other elements, including oxygen, nitrogen, and halogens. This results in the formation of molecules containing different functional groups, such as alcohol and amine groups (Figure 5).

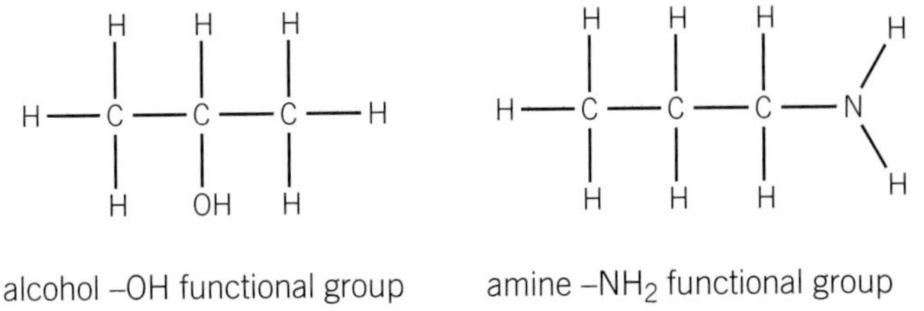

alcohol –OH functional group   amine $-NH_2$ functional group

▲ **Figure 5** *Carbon can form bonds to oxygen in alcohols and to nitrogen in amines*

Summary questions

1. Explain what is meant by the term functional group. *(1 mark)*

2. Define the term hydrocarbon. *(1 mark)*

3. C_2H_5OH, C_3H_7OH, and C_4H_9OH are all members of the same homologous series. Deduce the formula of the next member of the homologous series. *(1 mark)*

4. Give three reasons why carbon forms a large number of compounds. *(3 marks)*

11.2 Nomenclature of organic compounds

Specification reference: 4.1.1

Learning outcomes

Demonstrate knowledge, understanding, and application of:

→ the terms alkyl, aliphatic, alicyclic, and aromatic

→ the general formula of a homologous series

→ IUPAC rules of nomenclature for organic compounds.

Naming hydrocarbons

As there are so many organic compounds a universal system of naming organic molecules is needed to keep track of them all. IUPAC is an organisation that was founded in 1919 by chemists from industry and education who recognised the need for standardisation of the names of compounds. This means that chemists across the globe can communicate clearly.

Hydrocarbons can be classified as:

- **aliphatic** – carbon atoms are joined to each other in unbranched (straight) or branched chains, or non-aromatic rings
- **alicyclic** – carbon atoms are joined to each other in ring (cyclic) structures, with or without branches
- **aromatic** – some or all of the carbon atoms are found in a benzene ring.

There are three **homologous series** of aliphatic **hydrocarbons** that you should be aware of.

- **alkanes** – containing single carbon-to-carbon bonds
- **alkenes** – containing at least one double carbon-to-carbon bond
- **alkynes** – containing at least one triple carbon-to-carbon bond

Stem, prefix, and suffix

- The stem of the name indicates the number of carbon atoms in the *longest continuous chain* in the molecule.
- A prefix can be added before the stem, often to indicate the presence of side chains or a functional groups.
- A suffix is added after the stem to indicate functional groups.

Naming aliphatic alkanes

Follow these steps when naming alkanes:

1. All alkanes have the suffix -ane.
2. Identify the longest continuous chain of carbon atoms (the 'parent' chain) and name it from Table 1.
3. Identify any side chains attached to the parent chain. These side chains are known as **alkyl groups**. An alkyl group has a hydrogen atom removed from an alkane parent chain. The name of the alkyl group (from Table 1) is added as a prefix to the name of the parent chain.
4. Add numbers before any alkyl groups to show the position of the alkyl groups on the parent chain.

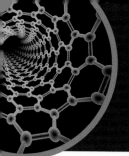

▲ **Figure 1** *An example of an aliphatic, alicyclic, and aromatic hydrocarbon*

Synoptic link

The chemistry of benzene and other aromatic compounds will be covered in Chapter 25, Aromatic chemistry.

11 BASIC CONCEPTS OF ORGANIC CHEMISTRY

▼ **Table 1** *The names of the first ten alkanes and their alkyl groups*

Number of carbon atoms	Stem	Suffix	Parent alkane		Alkyl group	
1	meth-	-ane	methane	CH_4	methyl	CH_3
2	eth-	-ane	ethane	C_2H_6	ethyl	C_2H_5
3	prop-	-ane	propane	C_3H_8	propyl	C_3H_7
4	but-	-ane	butane	C_4H_{10}	butyl	C_4H_9
5	pent-	-ane	pentane	C_5H_{12}	pentyl	C_5H_{11}
6	hex-	-ane	hexane	C_6H_{14}	hexyl	C_6H_{13}
7	hept-	-ane	heptane	C_7H_{16}	heptyl	C_7H_{15}
8	oct-	-ane	octane	C_8H_{18}	octyl	C_8H_{17}
9	non-	-ane	nonane	C_9H_{20}	nonyl	C_9H_{19}
10	dec-	-ane	decane	$C_{10}H_{22}$	decyl	$C_{10}H_{21}$

Study tip
You should be able to name the first ten alkanes and alkenes.

 Worked example: Naming branched alkanes

How do you work out the IUPAC name of the compound in Figure 2?

Step 1: Identify which suffix to use.

The compound is an alkane so the suffix is -ane.

Step 2: Identify the longest continuous chain of carbon atoms

The longest continuous chain contains four carbon atoms so the stem name is but-.

Step 3: Identify any side chains and which number carbon the side chains are on.

There are two side chains – each is a methyl group.

The side chain can be considered to be on carbon 2 or carbon 3, depending on which end the counting starts from (Figure 2). When naming compounds, always use the lowest combination of numbers possible, so use 2- instead of 3-.

Step 4: Combine the suffix, stem, and side chains to name the compound.

When two or more side chains are the same, the prefixes di-, tri-, or tetra- are used, corresponding to two, three, or four of the same side chain respectively.

The compound is 2,2-dimethylbutane.

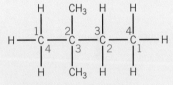

▲ **Figure 2** *2,2-dimethylbutane*

Sometimes it is difficult to identify the longest chain. When there are two or more possible chains of the same length in a molecule, the chain with most branches is considered the longest chain.

Figure 3 shows a molecule that seems to have two different chains containing six carbon atoms, coloured pink and blue. The pink chain has three branches but the blue chain only has two. So, the pink chain is the parent chain.

175

11.2 Nomenclature of organic compounds

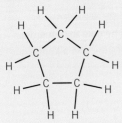

▲ **Figure 3** *3-ethyl-2,4-dimethylhexane*

The longest chain contains six carbons so the stem is hex-.

There are three side chains, methyl groups on carbon 2 and carbon 4 and an ethyl group on carbon 3. The side chains are ordered alphabetically so ethyl comes before methyl

The compound is called 3-ethyl-2,4-dimethylhexane.

Naming alicyclic alkanes

For cyclic alkanes, the same rules apply as when naming aliphatic alkanes. The prefix cyclo- in front of the stem is used to show that the carbon atoms are arranged in a ring structure.

▲ **Figure 4** *Cyclopentane*

Worked example: Naming alicyclic alkanes

How do you work out the IUPAC name of the compound in Figure 4?

Step 1: Identify the longest continuous chain of carbon atoms.

There are five carbon atoms in the ring structure; the stem is pent-.

Step 2: Add the prefix cyclo-.

The compound is called cyclopentane.

Study tip

As you get used to naming molecules, you won't need to go follow a step-by-step procedure.

Naming alkenes

Alkenes are named using the same rules as for alkanes, except the suffix is -ene. The position of the C=C bond in the chain must be stated for alkenes that have four or more carbon atoms in the longest chain.

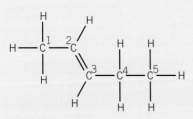

▲ **Figure 5** *Pent-2-ene*

Worked example: Naming alkenes

How do you work out the IUPAC name of the compound in Figure 5?

Step 1: Identify which suffix to use.

The compound is an alkene and the suffix is -ene.

Step 2: Identify the longest continuous chain of carbon atoms.

There are five carbon atoms in the chain so the stem is pent-.

Step 3: Identify where the double bond is.

The double bond is between carbons 2 and 3. Only the smaller number is needed to indicate the position of the double bond, so we use 2- only.

Step 4: Combine the suffix, stem, and position of the double bond to name the compound.

The compound is called pent-2-ene.

11 BASIC CONCEPTS OF ORGANIC CHEMISTRY

Naming compounds containing functional groups

The same basic principles apply as for naming alkanes.

1. Identify the longest unbranched chain of carbon atoms. The stem is now the name of the corresponding alkane.
2. Identify any functional groups and any alkyl side chains, and select the appropriate prefixes or suffixes for them.
3. Number any alkyl groups and functional groups to indicate their position on the longest unbranched chain.

Table 2 lists some common functional groups.

▼ **Table 2** *Common functional groups*

Type of compound	Functional group		Prefix	Suffix
alkene		C=C		-ene
alcohol		–OH	hydroxy-	-ol
haloalkane		–Cl –Br –I	chloro- bromo- iodo-	
aldehyde	(C=O with H)	–CHO		-al
ketone	(C–C(=O)–C)	–C(CO)C–		-one
carboxylic acid	(C=O with OH)	–COOH		-oic acid
ester	(C=O with O–C)	–COOC–		-oate
acyl chloride	(C=O with Cl)	–COCl		-oyl chloride
amine		–NH$_2$	amino-	-amine
nitrile		–CN		-nitrile

> **Study tip**
>
> Remove the final letter -e from the alkane stem when the suffix starts with a vowel– propane-1-ol becomes propan-1-ol and hexane-2-one becomes hexan-2-one.
>
> Keep the final letter -e of the alkane stem when the suffix starts with a consonant– ethanenitrile.

 Worked example: Naming alcohols

What is the IUPAC name or the compound in figure 6?

Step 1: Identify the functional group and suffix.

Alcohol functional group present, the suffix is –ol.

Step 2: Identify the longest chain of carbon atoms.

The longest chain contains four carbon atoms, the stem is butan-.

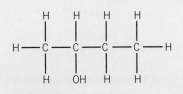

▲ **Figure 6**

11.2 Nomenclature of organic compounds

Step 3: Identify which carbon atom the functional group is on.
The –OH group is on carbon 2
Step 4: Combine the suffix, and stem to name the compound.
The name of the compound is butan-2-ol.

Worked example: Naming aldehydes

What is the IUPAC name of the compound in Figure 7?
Step 1: Identify the functional group and suffix.
Aldehyde functional group present, the suffix is –al.
Step 2: Identify the longest chain of carbon atoms.

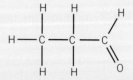

▲ **Figure 7** *Propanal*

The longest chain contains 3 carbon atoms, the stem is propan-.
Step 3: Combine the suffix and stem to name the compound. Aldehydes do not need numbers to show the position of the carbonyl group. It is always on position 1.
The name of the compound is propanal.

Worked example: Naming multiple functional groups

What is the IUPAC name of the compound in Figure 8?
Step 1: Identify the longest chain of carbons.
The longest chain contains 4 carbon atoms, the stem is butan-

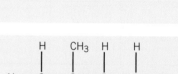

▲ **Figure 8** *2-chloro-2-methylbutane*

Step 2: Identify the functional groups present, which carbon atom they are on, and the prefixes.
Halogen present on carbon 2, the prefix is 2-chloro
Methyl group on carbon 2, the prefix is 2-methyl
Step 3: Combine the stem and the prefixes to name the compound. Remember side chains are named in alphabetical order.
The name of the compound is 2-chloro-2-methylbutane.

Summary questions

1 State the functional group in each of the following molecules.

   a (1 mark)   b (1 mark)   c (1 mark)   d (1 mark)

2 Name each of the molecules in **1** (4 marks)

3 Draw the structures of each of the following molecules:
   a 2,2-dimethylpentane   (1 mark)   b 2-chloro-3-methylpent-1-ene   (1 mark)
   c 4-hydroxypentan-2-one   (1 mark)   d 3-chloropentanoic acid   (1 mark)
   e 1-chloro-2-methylcyclohexane (1 mark)   f 2-chloro-3-methylbutanal   (1 mark)

11.3 Representing the formulae of organic compounds

Specification reference: 4.1.1

Chemical formulae

Different types of formula are used in organic chemistry. For example, in the previous topic some molecules were drawn as displayed formulae, with every atom and every bond shown, to give a clear picture of how the atoms are bonded together in a molecule. You already know about **molecular** and **empirical formulae**. In this topic you will learn how to show the **structural, displayed,** and **skeletal formulae** of molecules.

Molecular formula

The molecular formula shows the number and type of atoms of each element present in a molecule. The molecular formula does not show how the atoms are joined together and different molecules can have the same molecular formula.

A molecule of ethanol is shown in Figure 1. It contains two carbon atoms, six hydrogen atoms, and one oxygen atom. The molecular formula of ethanol is C_2H_6O.

Empirical formula

The empirical formula is the simplest whole-number ratio of the atoms of each element present in a compound.

For example, the alkenes all have the same empirical formula CH_2 and so there will always be twice as many hydrogen atoms as carbon atoms in an alkene. Glucose has the molecular formula $C_6H_{12}O_6$ and therefore the empirical formula CH_2O.

General formula

The **general formula** is the simplest algebraic formula for any member of a homologous series. You can use the general formula to generate the molecular formula for any member of that homologous series. Table 1 shows the general formulae of some homologous series.

Learning outcomes

Demonstrate knowledge, understanding, and application of:

→ general formula
→ displayed formula
→ structural formula
→ skeletal formula.

Synoptic link

You learnt how to calculate empirical and molecular formulae in Topic 3.2, Determination of formulae.

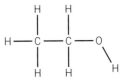

▲ **Figure 1** *The displayed formula of ethanol*

▼ **Table 1** *The general formulae for some homologous series*

Homologous series	General formula
alkanes	C_nH_{2n+2}
alkenes	C_nH_{2n}
alcohols	$C_nH_{2n+1}OH$
carboxylic acids	$C_nH_{2n}O_2$
ketones	$C_nH_{2n}O$

 Worked example: General and molecular formulae

Find the molecular formula of the carboxylic acid that contains six carbon atoms.

Step 1: The general formula of carboxylic acids is $C_nH_{2n}O_2$ (Table 1)

Step 2: There are six carbon atoms so $n = 6$

Step 3: Molecular formula = $C_6H_{12}O_2$

11.3 Representing the formulae of organic compounds

> **Study tip**
>
> When using the general formula for an alcohol you have to be careful—the general formula is $C_nH_{2n+1}OH$, but the molecular formula is $C_nH_{2n+2}O$.

Displayed formula

A displayed formula shows the relative positioning of all of the atoms in a molecule and the bonds between them.

1-Bromobutane has the molecular formula C_4H_9Br. Its displayed formula is shown in Figure 2. Butanoic acid has the molecular formula $C_4H_8O_2$. Its displayed formula is shown in Figure 3.

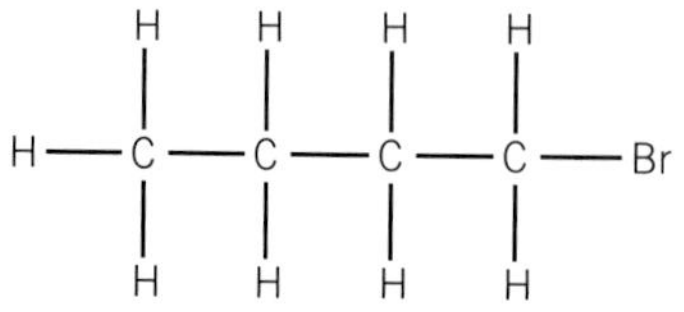

▲ **Figure 2** *The displayed formula of 1-bromobutane*

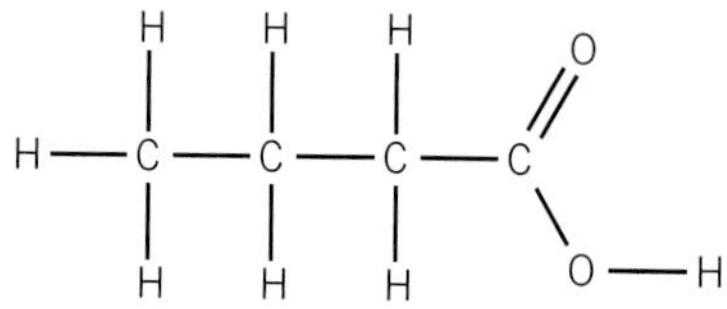

▲ **Figure 3** *The displayed formula of butanoic acid*

Structural formula

The structural formula uses the smallest amount of detail necessary to show the arrangement of the atoms in a molecule. It shows clearly which groups are bonded together. For example, the structural formula of butane, is $CH_3CH_2CH_2CH_3$ or $CH_3(CH_2)_2CH_3$.

Figure 4 shows three molecules that have the molecular formula C_5H_{12} but have different structural formulae.

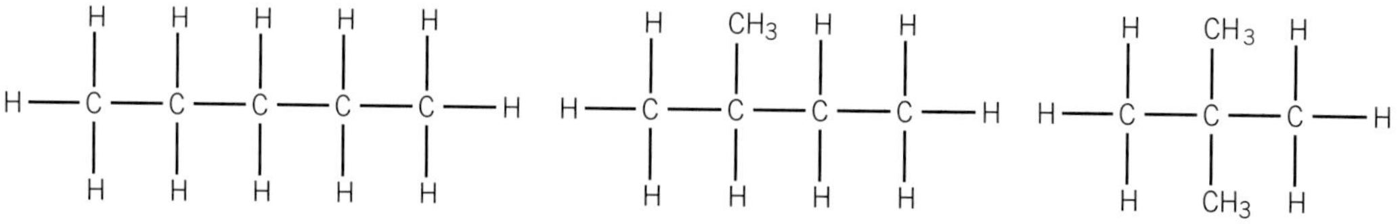

$CH_3CH_2CH_2CH_2CH_3$ $(CH_3)_2CHCH_2CH_3$ $C(CH_3)_4$

▲ **Figure 4** *The displayed structures (top) and structural formulae (bottom) of three different molecules with molecular formula C_5H_{12}*

Skeletal formula

A skeletal formula is a simplified organic formula. You remove:

- all of the carbon and hydrogen labels from carbon chains
- any bonds to hydrogen atoms.

This leaves just a carbon skeleton and any functional groups.

In skeletal formulae:

- a line represents a single bond
- an intersection of two lines represents a carbon atom
- the end of a line represents a $-CH_3$ group.

11 BASIC CONCEPTS OF ORGANIC CHEMISTRY

The displayed and skeletal formulae of 2-methylhexane are shown in Figure 6.

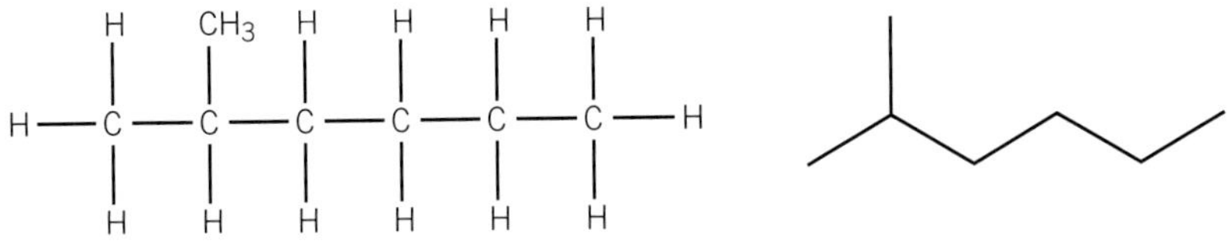

▲ **Figure 6** *The displayed (left) and skeletal (right) formulae of 2-methylhexane*

Cyclic and aromatic compounds

One of the most common uses of skeletal formulae is in representing alicyclic and aromatic compounds, such as cyclohexane, cyclohexene, and benzene (Figures 7 and 8). The skeletal formula of benzene, C_6H_6, can be represented in two different ways.

Molecules containing functional groups

When functional groups are present in a molecule, they must be included in the skeletal formula.

In Figure 9 the skeletal formula for 2-chlorobut-1-ene is shown on the left. The double bond between carbon atoms 1 and 2 is represented by two parallel lines. The chlorine atom is shown on carbon 2, and the carbon and hydrogen atoms in the carbon skeleton are unlabelled.

In Figure 9 the skeletal formula on the right is 2-methylpentan-3-one. Notice how the ketone functional group and the methyl group are represented in a skeletal formula.

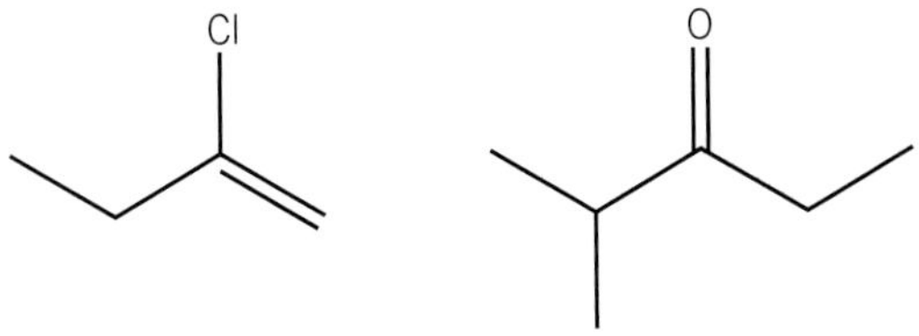

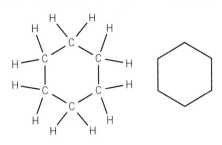

▲ **Figure 7** *The displayed and skeletal formulae of cyclohexane (top) and cyclohexene (bottom)*

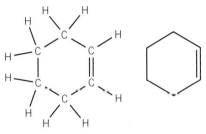

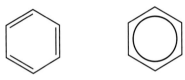

▲ **Figure 8** *Skeletal formulae showing the two representations of benzene*

> **Synoptic link**
>
> You will learn the reasons behind these two representations of benzene when you study the physical properties and chemical reactions of aromatic compounds in Chapter 25, Aromatic chemistry.

▲ **Figure 9** *The skeletal formulae of 2-chlorobut-1-ene (left) and 2-methylpentan-3-one (right)*

Summary questions

1. Deduce the empirical and molecular formulae of the following molecules:
   a. hexanoic acid *(1 mark)*
   b. benzene *(1 mark)*
   c. $HOCH_2CH_2OH$ *(1 mark)*
   d. 2,3-dichlorobutane *(1 mark)*

2. Draw the two possible displayed formulae for the molecules with the molecular formula C_4H_{10}. Write out structural formulae for these two molecules. *(4 marks)*

3. Draw the skeletal formulae of the following molecules:
   a. the molecule with molecular formula C_7H_{16} that contains only two branches, methyl groups on carbon 2 and 3 *(1 mark)*
   b. a branched-chain alcohol with molecular formula $C_4H_{10}O$ where the alcohol group is not on an end carbon *(1 mark)*
   c. a molecule with empirical formula C_4H_5 containing one side chain and a benzene ring *(1 mark)*

11.4 Isomerism

Specification reference: 4.1.1

Learning outcomes

Demonstrate knowledge, understanding, and application of:

→ structural isomerism

→ determination of possible structural formulae from a molecular formula.

The same molecular formula can often apply to different compounds. Different compounds with the same molecular formula are known as isomers. There are several types of isomers. This topic looks at structural isomerism.

Structural isomerism

Structural isomers are compounds with the same molecular formula but different structural formulae.

Many examples of structural isomerism occur in the alkane homologous series. Different structures of C_5H_{12} were shown in Figure 4 in Topic 11.3. Similarly, the molecular formula C_4H_{10} does not unambiguously identify the structure of the molecule. There are two possible structural isomers that can be drawn for C_4H_{10} (Figure 1).

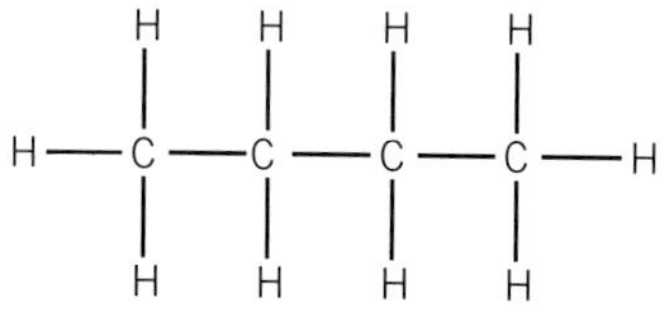

$CH_3CH_2CH_2CH_3$ 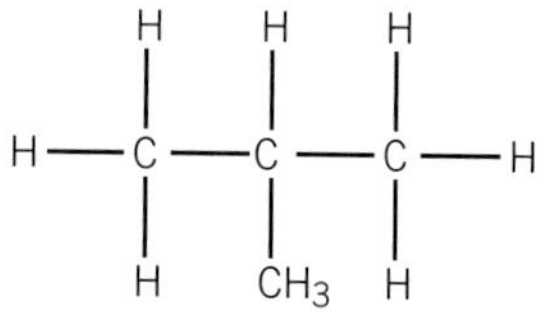 $CH_3CH(CH_3)CH_3$

▲ **Figure 1** *Butane and 2-methylpropane are structural isomers and have the same molecular formula, C_4H_{10}*

Isomers with the same functional group

In compounds containing a functional group, the functional group can be at different positions along the carbon chain. The two possible structural isomers of C_3H_7Cl are shown in Figure 2.

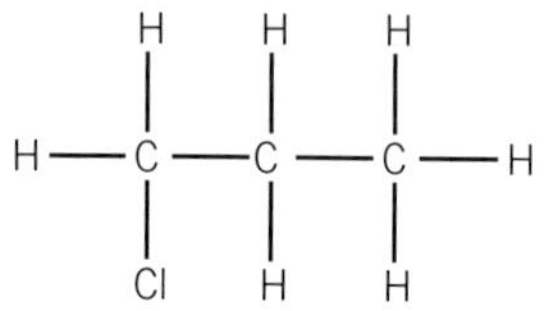

1-chloropropane

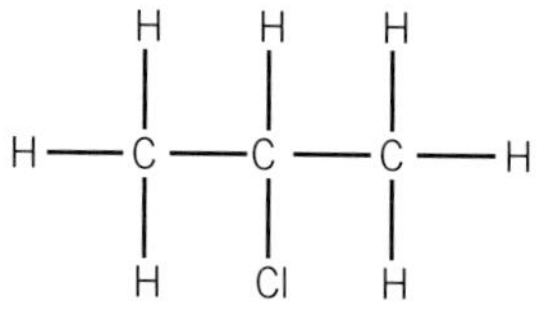

2-chloropropane

▲ **Figure 2** *The two structural isomers of C_3H_7Cl*

🖩 Worked example: Branching alcohols

Draw the skeletal formulae of the four structural isomers that are alcohols and have the molecular formula $C_4H_{10}O$.

Remember that structural isomers can be formed by changing both the position of the functional group and the branching of the carbon chain.

Step 1: Draw out the different ways that four carbons can be connected in aliphatic chains.

Step 2: Add the –OH group in different positions to create all the possible isomers that are alcohols (Figure 3).

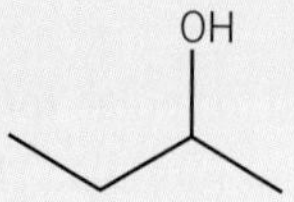

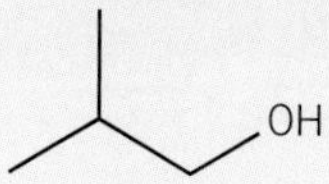

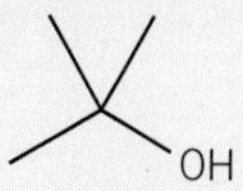

◀ **Figure 3** *The structural isomers of molecular formula $C_4H_{10}O$ that are alcohols*

Synoptic link

The four isomers of $C_4H_{10}O$ will be important later in your study of organic chemistry when you come to look at the classification of the alcohols (Topic 14.1, Properties of alcohols) and their oxidation reactions (Topic 14.2, Reactions of alcohols).

11 BASIC CONCEPTS OF ORGANIC CHEMISTRY

Isomers with different functional groups

Sometimes two molecules containing different functional groups have the same molecular formula. Aldehydes and ketones with the same number of carbon atoms have the same molecular formulae. The molecular formula C_3H_6O could represent either the aldehyde, propanal, or the ketone, propanone (Figure 4).

propanal, CH₃CH₂CHO   propanone, CH₃COCH₃

▲ **Figure 4** *Propanal and propanone are structural isomers of C_3H_6O*

Detecting isomerism by smell

Two of the structural isomers of $C_{10}H_{18}O$ are linalool and geraniol. Both isomers are unsaturated alcohols. They have similar chemical properties and are both used in perfumes. Linalool has the characteristic smell of lavender, whereas geraniol smells of roses (Figure 5).

The systematic names for linalool and geraniol are shown below.

linalool   3,7-dimethylocta-1,6-dien-3-ol

geraniol   3,7-dimethylocta-2,6-dien-1-ol

You can see how adapted your scent buds in your nose are to detecting small differences in chemical compounds.

▲ **Figure 5** *Linalool has the characteristic smell of lavender, whereas geraniol smells of roses*

Draw out the skeletal formula of these two isomeric alcohols.

Summary questions

1. Write the structural formulae for the structural isomers of following:
   a C_3H_7OH *(1 mark)*   b C_4H_9Cl *(1 mark)*

2. Draw the three branched structural isomers with the molecular formula C_5H_{10}. *(3 marks)*

3. Draw the two possible structural isomers of a molecule with molecular formula $C_4H_8O_2$ that contains a carboxylic acid group. Name both of the isomers. *(4 marks)*

4. A compound with a molar mass of 72.0 g mol⁻¹ is composed of 66.63% C, 11.18% H, and 22.19% O. This information leads to three possible structural isomers that contain a carbon to oxygen double bond. Draw the skeletal formula of these three isomers. *(5 marks)*

11.5 Introduction to reaction mechanisms

Specification reference: 4.1.1

Learning outcomes

Demonstrate knowledge, understanding, and application of:

→ homolytic and heterolytic bond fission

→ curly arrows and reaction mechanisms

→ addition, substitution, and elimination reactions.

Synoptic link

You will study homolytic fission reactions and radicals in more depth in Chapter 12.

Types of bond fission

A covalent bond is defined as a shared pair of electrons between two atoms. When a chemical reaction takes place, bonds in the reactants break and new bonds are formed in the products. Covalent bonds can be broken by either **homolytic fission** or **heterolytic fission**.

Homolytic fission

When a covalent bond breaks by homolytic fission, each of the bonded atoms takes *one* of the shared pair of electrons from the bond.

- Each atom now has a single unpaired electron.
- An atom or groups of atoms with an unpaired electron is called a **radical**.

The homolytic fission of the carbon–carbon bond in ethane is shown in the equation below:

$$H_3C–CH_3 \rightarrow H_3C\bullet + \bullet CH_3$$
$$\text{radicals}$$

Heterolytic fission

When a covalent bond breaks by heterolytic fission, one of the bonded atoms takes *both* of the electrons from the bond.

- The atom that takes both electrons becomes a negative ion.
- The atom that does not take the electrons becomes a positive ion.

The heterolytic fission of the carbon–chlorine bond in chloromethane, CH_3Cl, is shown in the equation below.

$$H_3C–Cl \rightarrow H_3C^+ + Cl^-$$
$$\text{ions}$$

Organic reaction mechanisms

An equation tells you about the reactants, products, and stoichiometry of a reaction, but it does not provide any information about *how* the reaction takes place. This process is known as the **reaction mechanism**.

Curly arrows

In a reaction mechanism, curly arrows are used to show the movement of electron pairs when bonds are being broken or made. The heterolytic fission of a carbon–chlorine bond in a haloalkane is shown in Figure 1. Both the bonded electrons go to the chlorine atom to form a positive ion and a negative ion.

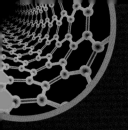

▲ Figure 1 *Curly arrow showing the movement of a pair of electrons in heterolytic bond fission*

11 BASIC CONCEPTS OF ORGANIC CHEMISTRY

Curly arrows and homolytic fission

You now know that the movement of a pair of electrons is shown using an arrow with a full arrowhead. However, in some text books and on the Internet, you may see fish-hook arrows – arrows with half of an arrow head. A fish-hook arrow represents the movement of a single, unpaired electron in mechanisms involving radicals. In Figure 2, a single covalent bond in bromine is broken homolytically. Each atom takes one electron.

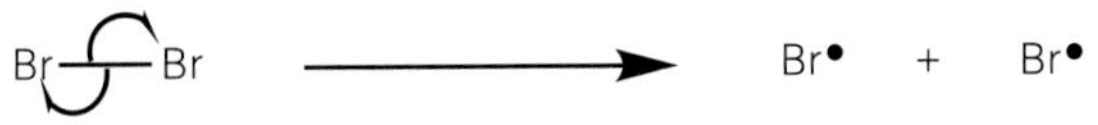

Br—Br ⟶ Br• + Br•

▲ **Figure 2** *In homolytic fission, fish-hook arrows can be used to represent the movement of an unpaired electron*

Draw the reaction mechanism for the homolytic fission of a carbon–hydrogen bond in methane.

Types of reaction

Addition

In an addition reaction, two reactants join together to form one product.

In the addition reaction shown in Figure 3, a molecule is added to the unsaturated alkene, breaking the double bond, to form a single saturated compound. In this reaction, but-2-ene and water add together to form a single product, butan-2-ol.

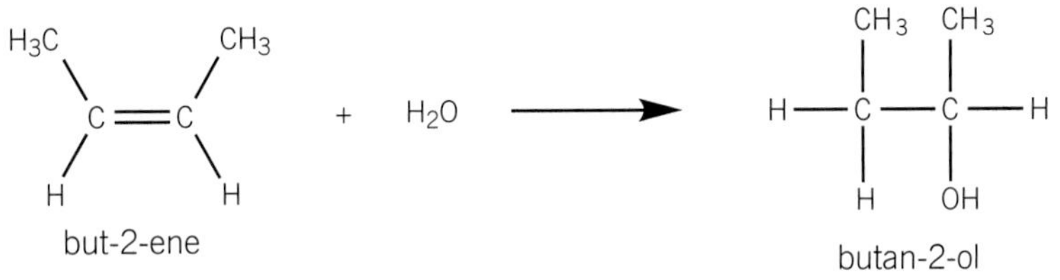

but-2-ene + H_2O ⟶ butan-2-ol

▲ **Figure 3** *An addition reaction of but-2-ene*

Substitution

In a substitution reaction, an atom or group of atoms is replaced by a different atom or group of atoms. Figure 4 shows a substitution reaction of the haloalkane 1-bromopropane.

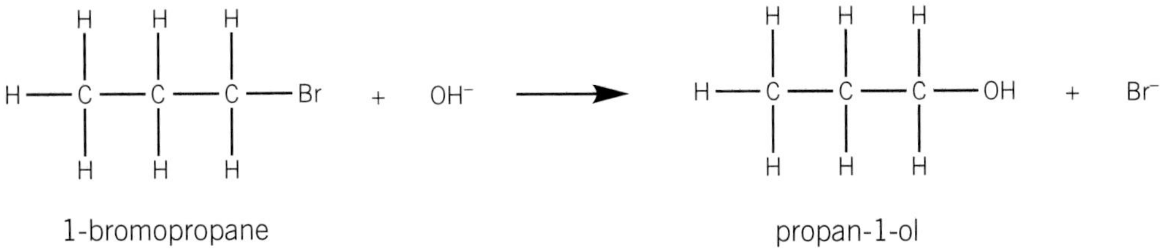

1-bromopropane + OH^- ⟶ propan-1-ol + Br^-

▲ **Figure 4** *A substitution reaction of a haloalkane*

> **Synoptic link**
>
> You will meet important addition reactions when you study the chemistry of the alkenes in Chapter 13, Alkene, and the chemistry of the aldehydes and ketones in topic 26.1, Carbonyl compounds.

> **Synoptic link**
>
> Substitution reactions occur in the chemistry of the haloalkane functional group and in the reactions of benzene. You will meet these reactions later in your course, in Chapters 15, Haloalkanes, and chapter 25, Aromatic chemistry.

11.5 Introduction to reaction mechanisms

Synoptic link
You will meet this elimination reaction when you study the chemistry of alcohols in Topic 14.2, Reactions of alcohols.

Elimination
An elimination reaction involves the removal of a small molecule from a larger one. In an elimination reaction, one reactant molecule forms two products. An example of an elimination reaction is shown in Figure 5.

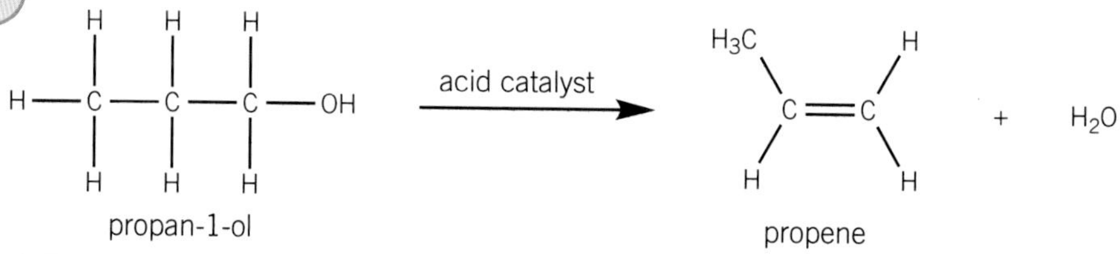

▲ **Figure 5** *The elimination of a water molecule from an alcohol*

Summary questions

1. Define the term heterolytic fission. *(1 mark)*

2. Define the terms radical and homolytic fission. Illustrate your answer by showing the homolytic fission of the chlorine–chlorine bond in a chlorine molecule. *(3 marks)*

3. Classify each of the following reactions as addition, substitution, or elimination.

   a. CH$_3$CH$_2$CH$_2$I + OH$^-$ → CH$_3$CH$_2$CH$_2$OH + I$^-$ *(1 mark)*

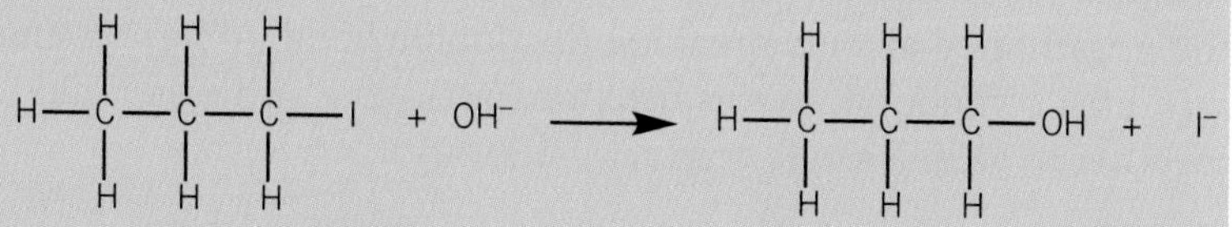

   b. CH$_2$=CHCH$_3$ + HBr → CH$_3$CHBrCH$_3$ *(1 mark)*

   c. CH$_3$CH(OH)CH$_3$ —H$_2$SO$_4$→ CH$_3$CH=CH$_2$ + H$_2$O *(1 mark)*

   d. CH$_3$COCH$_3$ + HCN → CH$_3$C(OH)(CN)CH$_3$ *(1 mark)*

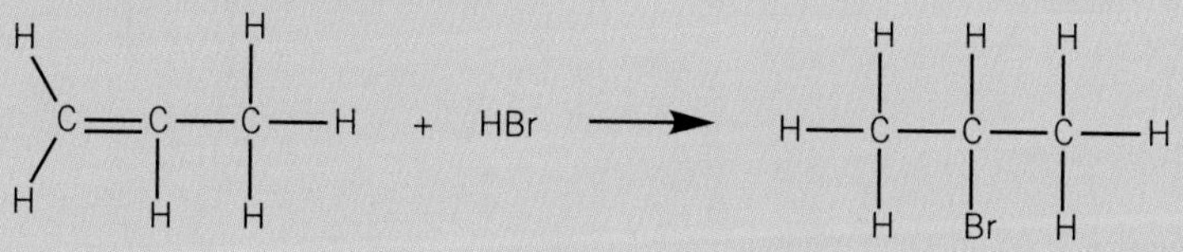

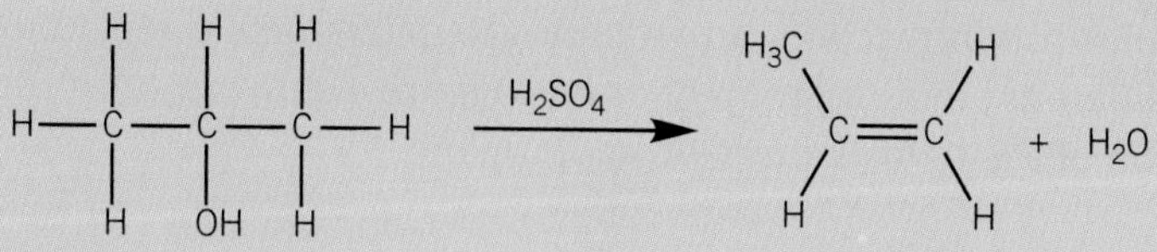

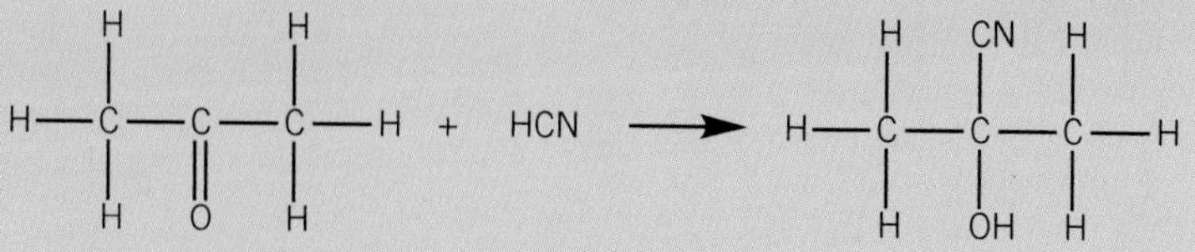

BASIC CONCEPTS OF ORGANIC CHEMISTRY

Practice questions

1. Draw the displayed formula of:
   a. 2-bromobutane (1 mark)
   b. 2,3-dimethylhexane (1 mark)
   c. octane-2,5-diol (1 mark)
   d. 2,2-dichloro-3-methylhexane (1 mark)
   e. 2-methylbut-2-ene (1 mark)

2. Name each of the following molecules. (4 marks)

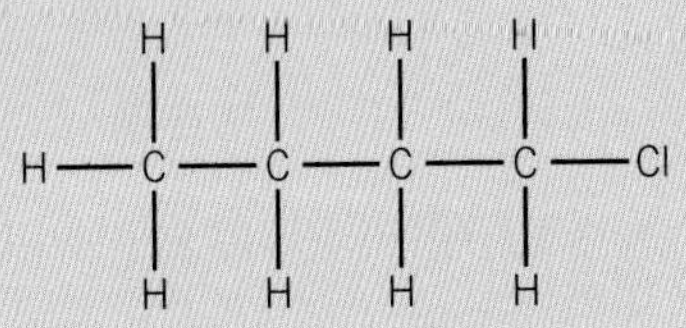

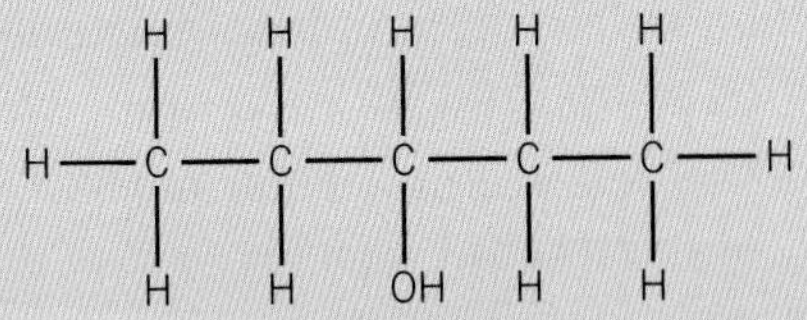

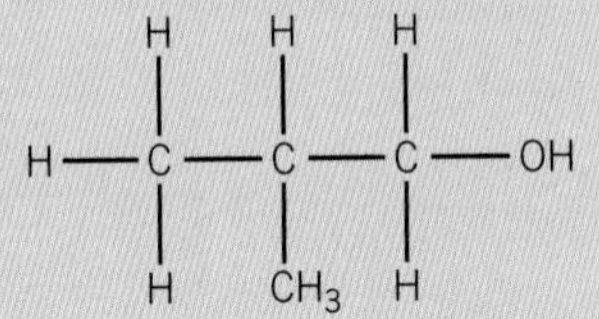

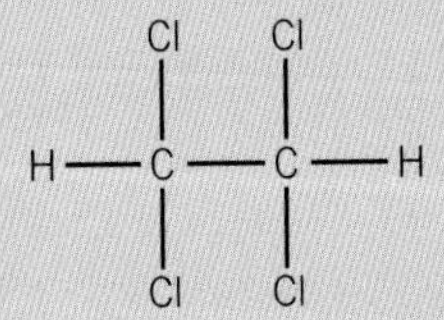

3. Define each of the following terms and give an example to illustrate your answer.
   a. Hydrocarbon (1 mark)
   b. Functional group (1 mark)
   c. Homologous series (2 marks)

4. Draw all the structural isomers of formula C_6H_{14}. (5 marks)

5. Name the following straight-chain molecules.
   a. $CH_3CH_2CH_2CH_3$
   b. $CH_3CH_2CH_2OH$
   c. $CH_3CHOHCH_2CH_3$
   d. $CH_3CH_2CH=CH_2$
   e. $CH_3CH_2CH=CHCH_3$
   f. $CH_3CH_2CHClCH_2OH$ (6 marks)

6. State the molecular formula of the following.
   a. methane
   b. hex-1-ene
   c. ethane
   d. heptan-1-ol (4 marks)

7. Describe the difference between homolytic and heterolytic fission, using suitable examples. (4 marks)

8. What is the molecular formula of the following?
   a. nonane;
   b. octan-4-ol
   c. 3-methylhexan-2-ol (3 marks)

9. This question is about the compounds A–F below:

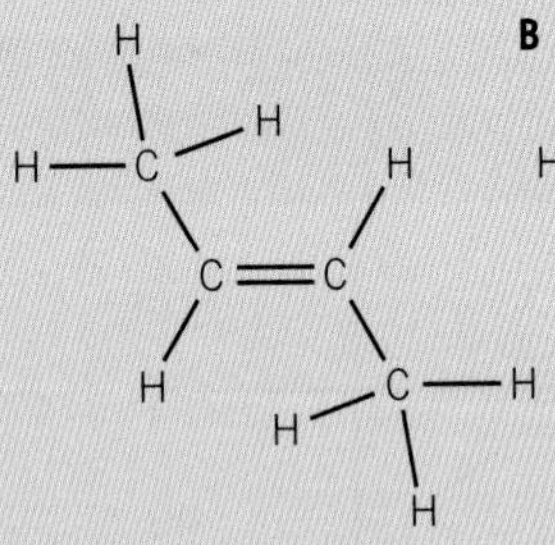

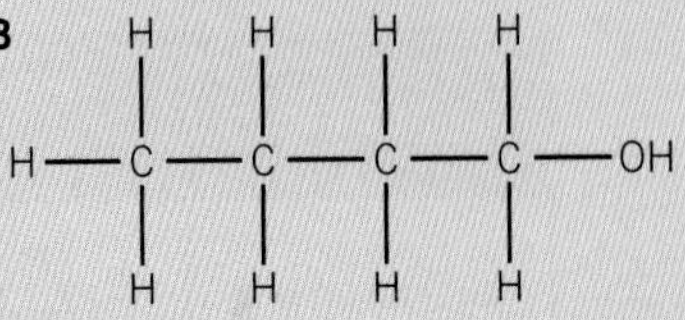

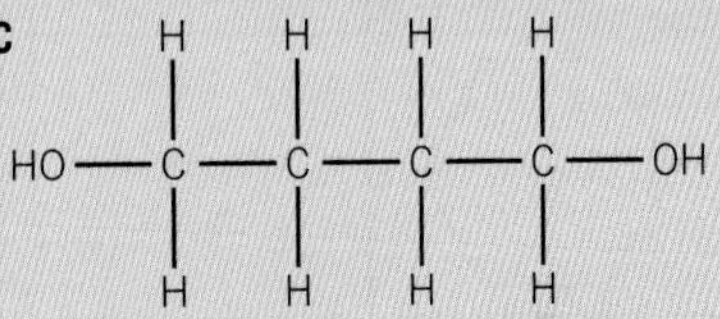

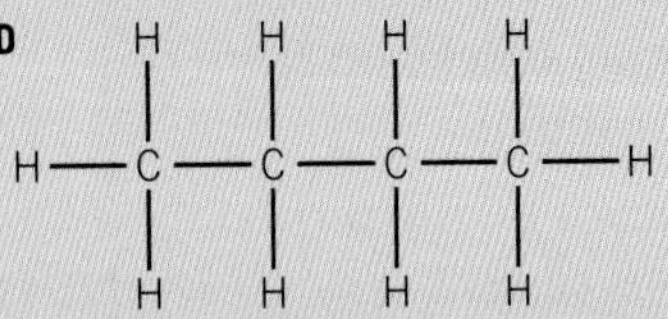

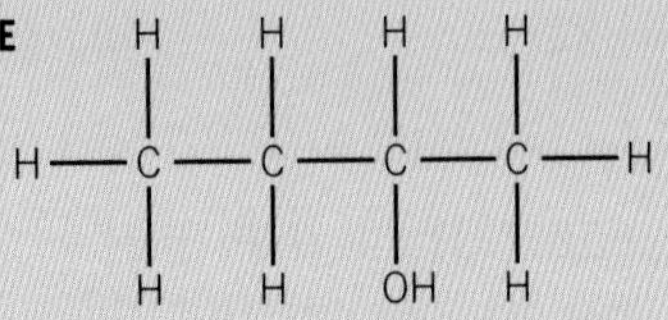

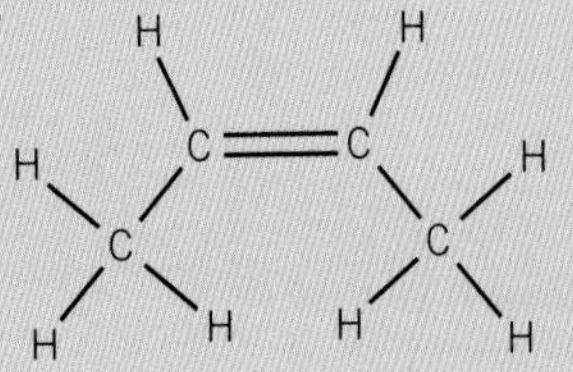

Chapter 11 Practice questions

a Answer the following questions by referring to the compounds **A–F**.

   (i)   What is the molecular formula of compound **D**? *(1 mark)*

   (ii) What is the empirical formula of compound **C**? *(1 mark)*

   (iii) What is the name of compound **E**? *(1 mark)*

   (iv) Which two compounds are structural isomers of each other? *(1 mark)*

   (v)   Draw the skeletal formula for compounds **C** and **F**. *(1 mark)*

10 This question is about cyclic organic compounds. The table shows some information about cycloalkanes.

Cycloalkane	Skeletal formula	Boiling point/°C
cyclopropane	△	−33
cyclopentane	⬠	49
cyclohexane	⬡	81

a These cycloalkanes are members of the same homologous series and have the same general formula.

   (i) What is meant by the term homologous series? *(2 marks)*

   (ii) State the general formula for these cycloalkanes. *(1 mark)*

   OCR F322 June 2014 Q1(a)(i)(ii)

11 The chlorinated alkene, $C_3H_4Cl_2$, has five structural isomers that are alkenes and two other structural isomers that are not alkenes.

a (i) What is meant by the term *structural isomers*? *(1 mark)*

   (ii) Two of the structural isomers of $C_3H_4Cl_2$ are drawn below.

isomer 1

isomer 2

   (ii) Draw the other three structural isomers of $C_3H_4Cl_2$ that are alkenes. *(3 marks)*

   (iii) Name isomer 1 *(1 mark)*

   (iv) Draw one structural isomer of C_3H_4Cl that is not an alkene. *(1 mark)*

   2812 June 2009 Q2

12 Compound **X** is a hydrocarbon containing 85.7% C by mass.

a (i) Calculate the empirical formula of compound **X**. *(2 marks)*

   (ii) The relative molecular mass of compound X is 56.
        Find the molecular formula. *(1 mark)*

13 Decane has the formula, $C_{10}H_{22}$.

State what is meant by each of the following terms. Use decane to illustrate your answers.

a empirical formula;

b structural formula

c displayed formula

d skeletal formula *(8 marks)*

14 Compound **G** is a chloroalkane with the percentage composition by mass: C, 24.7%; H, 2.1%; Cl, 73.2%.

a (i) Calculate the empirical formula of compound **G**. Show your working. *(2 marks)*

   (ii) The relative molecular mass of compound **G** is 145.5.

Show that the molecular formula is $C_3H_3Cl_3$. *(1 mark)*

b Compound **G** is one of six possible structural isomers of $C_3H_3Cl_3$ that are chloroalkenes.

Two of these isomers are shown below as isomer 1 and isomer 2.

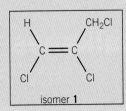
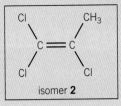

(i) Draw skeletal formulae of three other structural isomers of $C_3H_3Cl_3$ that are chloroalkenes. *(3 marks)*

(ii) Name isomer **1**. *(1 mark)*

15 a With the aid of examples, explain each of the following terms used in organic chemistry.
 (i) Electrophile;
 (ii) Nucleophile;
 (iii) Radical *(3 marks)*

b Write balanced equations to illustrate
 (i) addition;
 (ii) substitution;
 (iii) elimination;
 (iv) homolytic fission *(4 marks)*

16 An organic compound of bromine, **X**, has a molecular mass of 136.9 and the following percentage composition by mass: C, 35.0%; H, 6.60%; Br, 58.4%

a Calculate the empirical formula of **X** *(2 marks)*

b Show that the molecular formula of **X** is the same as the empirical formula of **X**. *(1 mark)*

c Draw all possible isomers of **X**, and name each one of them. *(4 marks)*

17 Butan-1-ol C_4H_9OH has the structural formula $CH_3CH_2CH_2CH_2OH$. Give the following formulae for butan-1-ol.
a empirical formula
b general formula
c displayed formula
d skeletal formula *(4 marks)*

18 A saturated hydrocarbon contains 82.8% carbon and has a relative molecular mass of 58.

Calculate the empirical and molecular formula of the hydrocarbon. *(3 marks)*

19 The skeletal formulae of some saturated hydrocarbons are shown below, labelled **A–F**.

a Some of the compounds **A–F** are alkanes.
 (i) What is the general formula of the alkanes? *(1 mark)*
 (ii) Which of the formulae, **A–F**, has a formula that does **not** fit the general formula of the alkanes? *(1 mark)*
 (iii) What is the molecular formula of an alkane with 16 carbon atoms?? *(1 mark)*

b Some of the structures are structural isomers of one another
 (i) What is meant by the term *structural isomer*? *(1 mark)*
 (ii) Which compounds are structural isomers of one another? *(1 mark)*

20 The structure below is a naturally occurring alcohols.

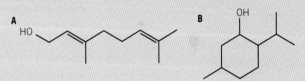

a What is meant by the term *unsaturated*? *(1 mark)*

b What is the molecular formula of these alcohols? *(2 marks)*

12 ALKANES
12.1 Properties of the alkanes
Specification reference: 4.1.2

Learning outcomes
Demonstrate knowledge, understanding, and application of:
→ alkanes as saturated hydrocarbons
→ bonding in alkanes
→ shapes and bond angles of alkanes
→ variations in the boiling points of alkanes.

▼ **Table 1** *Common uses of the first twenty members of the alkane homologous series*

gas, used in domestic fuel CH_4, C_3H_8, C_4H_{10}
petrol, used in cars C_5H_{12}—C_9H_{20}
kerosene, used in aircraft $C_{10}H_{22}$—$C_{16}H_{34}$
diesel, used in cars and lorries $C_{12}H_{26}$—$C_{20}H_{42}$

Synoptic link
You were introduced to the names and formula of the first 10 alkanes in Topic 11.2, Nomenclature of organic compounds. Make sure you learn these as alkanes form the basis for naming all organic compounds.

Synoptic link
Look back at Topic 6.1, Shapes of molecules and ions, to remind yourself how the shapes of molecules depends on electron pair repulsion and what a tetrahedral geometry looks like.

What are alkanes?
Alkanes are the main components of natural gas and crude oil. They are amongst the most stable organic compounds, and their lack of reactivity has allowed crude oil deposits to remain in the Earth for many millions of years.

Alkanes are mainly used as fuels, exploiting their reaction with oxygen to generate heat.

The alkanes in Table 1 show the general formula C_nH_{2n+2} – doubling the carbon and adding two gives the hydrogen number.

Table 1 lists some of the alkanes used in everyday life.

Properties of alkanes
The bonding in alkanes
Alkanes are saturated hydrocarbons, containing only carbon and hydrogen atoms joined together by single covalent bonds.

Each carbon atom in an alkane is joined to four other atoms by single covalent bonds. These are a type of covalent bond called a **sigma bond** (σ-bond).

A covalent bond is defined as a shared pair of electrons. A σ-bond is the result of the overlap of two orbitals, one from each bonding atom. Each overlapping orbital contains one electron, so the σ-bond has two electrons that are shared between the bonding atoms. A σ-bond is positioned on a line directly between bonding atoms.

Each carbon atom in an alkane has four sigma bonds, either C—C or C—H. Figure 1 shows the electron density of a σ-bond between two carbon atoms.

▲ **Figure 1** *σ-bond between two carbon atoms.*

The shape of alkanes
Each carbon atom is surrounded by four electron pairs in four σ-bonds. Repulsion between these electron pairs results in a 3D tetrahedral arrangement around each carbon atom. Each bond angle is approximately 109.5°.

Figure 2 shows the 3D shapes of the first four alkanes. Each formula matches the general formula of C_nH_{2n+2}. The σ-bonds act as axes around which the atoms can rotate freely, so these shapes are not rigid, for example, butane is shown in a zigzag in Figure 2, but it can also rotate into a U shape.

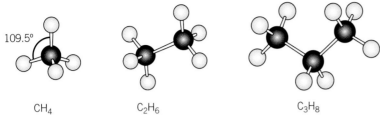

▲ **Figure 2** *Shapes of methane, ethane, propane, and butane*

Variations in the boiling points of alkanes

Crude oil contains hundreds of different alkanes. Oil refineries separate the crude oil into fractions by **fractional distillation** in a distillation tower (Figure 3). Each fraction contains a range of alkanes. Separation like this is possible because the boiling points of the alkanes are different, increasing as their chain length increases. Table 2 shows the boiling points of the first ten alkanes.

> **Study tip**
>
> When drawing 3D shapes, you will need to use solid wedges and dashed wedges. This is how ethane is drawn using wedges.

▼ **Table 2** *The boiling points of the first ten members of the alkane homologous series*

Alkane	CH_4	C_2H_6	C_3H_8	C_4H_{10}	C_5H_{12}	C_6H_{14}	C_7H_{16}	C_8H_{18}	C_9H_{20}	$C_{10}H_{22}$
Boiling point /°C	−164	−89	−42	−1	36	69	98	126	151	174

Why does the boiling point increase? The answer lies with the weak intermolecular forces called London forces. These forces hold molecules together in solids and liquids but, once broken, the molecules move apart from each other and the alkane becomes a gas. The greater the intermolecular forces, the higher the boiling point.

> **Synoptic link**
>
> You learned about London forces and the factors affecting their size in Topic 6.3, Intermolecular forces.

Effect of chain length on boiling point

London forces act between molecules that are in close surface contact. As the chain length increases, the molecules have a larger surface area, so more surface contact is possible between molecules. The London forces between the molecules will be greater and so more energy is required to overcome the forces.

Figure 4 shows how the surface contact increases with increasing chain length, illustrated with skeletal formulae.

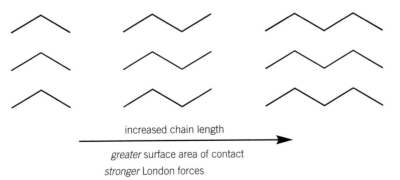

increased chain length →

greater surface area of contact
stronger London forces

▲ **Figure 4** *The effect of increasing chain length on the boiling points of alkanes*

▲ **Figure 3** *A distillation tower in which crude oil is separated into fractions. The oil is heated to about 400 °C. Hydrocarbon gases rise upwards and condense at different levels depending on their boiling points*

12.1 Properties of the alkanes

Effect of branching on boiling point

Isomers of alkanes have the same molecular mass. If you compare the boiling points of branched isomers with straight-chain isomers, you find that the branched isomers have lower boiling points.

Table 3 shows the boiling points for the isomers of pentane, C_5H_{12}.

▼ **Table 3** *The skeletal formulae and boiling points of the structural isomers of C_5H_{12}*

Name	pentane	2-methylbutane	2,2-dimethylpropane
Skeletal formula	∧∧∧	⊥∧	✕
Boiling point / °C	36	28	9

2,2-dimethylpropane is a gas at room temperature, 2-methylbutane would only be a gas on a warm day, and at normal UK temperatures pentane is a liquid.

The reason for this difference lies again with London forces. There are fewer surface points of contact between molecules of the branched alkanes, giving fewer London forces. Another factor lies with the shape of the molecules. The branches get in the way and prevent the branched molecules getting as close together as straight-chain molecules, decreasing the intermolecular forces further. It is easy to picture this difference by looking at the skeletal formulae in Figure 6.

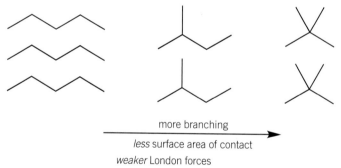

more branching →
less surface area of contact
weaker London forces

▲ **Figure 5** *The effect of branching on the boiling points of alkanes*

Summary questions

1. Kerosene is obtained from crude oil. Name the process used to obtain kerosene from crude oil and explain how the process works. *(2 marks)*

2. Explain why the straight-chain isomer of $C_{10}H_{22}$ has a higher boiling point than any of its branched-chain structural isomers. *(2 marks)*

3. Explain why the boiling points increase down the alkane homologous series. *(2 marks)*

4. Describe the bonding in ethane. State and explain the H—C—H bond angle in a molecule of ethane. *(4 marks)*

12.2 Chemical reactions of the alkanes

Specification reference: 4.1.2

Reactivity of alkanes

Alkanes do not react with most common reagents. The reasons for their lack of reactivity are:

- C—C and C—H σ-bonds are strong
- C—C bonds are non-polar
- the electronegativity of carbon and hydrogen is so similar that the C—H bond can be considered to be non-polar.

Combustion of alkanes

Despite their low reactivity, all alkanes react with a plentiful supply of oxygen to produce carbon dioxide and water. This reaction is called combustion. All combustion processes give out heat, and alkanes are used as fuels because they are readily available, easy to transport, and burn in a plentiful supply oxygen without releasing toxic products.

Methane

Methane is a greenhouse gas with a global warming potential of 22 (meaning that it has 22 times the warming effect of carbon dioxide in the atmosphere). Vast quantities of methane are stored in the ice-saturated arctic tundra of Siberia and northern Canada. Global warming is very likely to thaw the tundra, releasing this methane into the atmosphere and greatly increasing global warming.

The methane locked into the frozen earth of the tundra was formed by the decomposition of organic material in the absence of oxygen. The same chemistry lies behind the production of biogas, methane made as a fuel from organic waste or plant material.

> If methane is such a powerful greenhouse gas, why could the production of biogas help to slow down global warming?

Learning outcomes

Demonstrate knowledge, understanding, and application of:

→ the low reactivity of the σ-bonds in alkanes
→ the combustion of alkanes
→ radical substitution in alkanes.

Synoptic link

You may find looking back at Topic 6.2, Electronegativity and polarity useful.

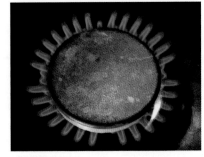

▲ Figure 1 Methane is the main component of the natural gas used for cooking, heating, and the generation of electricity. Natural gas typically contains about 80% methane, with varying proportions of ethane, propane, and butane

Complete combustion of alkanes

In a plentiful supply of oxygen, alkanes burn completely to produce carbon dioxide and water. The reactions below show the complete combustion of four alkanes

methane, as natural gas: $CH_4(g) + 2O_2(g) \rightarrow CO_2(g) + 2H_2O(l)$

ethane: $C_2H_6(g) + 3\frac{1}{2}O_2(g) \rightarrow 2CO_2(g) + 3H_2O(l)$

propane: $C_3H_8(g) + 5O_2(g) \rightarrow 3CO_2(g) + 4H_2O(l)$

hexane, in petrol: $C_6H_{14}(l) + 9\frac{1}{2}O_2(g) \rightarrow 6CO_2(g) + 7H_2O(l)$

▲ Figure 2 The Arctic Tundra in Greenland. Global warming is threatening to melt the permafrost releasing trapped carbon dioxide and methane

12.2 Chemical reactions of the alkanes

> **Study tip**
>
> You don't need to learn an equation for each alkane – the equations are easy to balance. The C in the alkane formula gives the number of CO_2 molecules formed. And halving the number of H gives the number of H_2O molecules. Then you just have to balance the oxygen atoms in O_2.

If you like algebra, the equation below can be used for balancing the complete combustion of any alkane.

$$C_xH_y + (x + \tfrac{y}{4})O_2 \rightarrow xCO_2 + \tfrac{y}{2}H_2O$$

Incomplete combustion of alkanes

In a limited supply of oxygen, there is not enough oxygen for complete combustion. If you look at the equations above, you can see that as you descend the homologous series each alkane molecule (each extra –CH_2–) needs an extra $1\tfrac{1}{2}$ O_2 molecules for complete combustion.

When oxygen is limited during combustion, the hydrogen atoms in the alkane are always oxidised to water, but combustion of the carbon may be incomplete, forming the toxic gas carbon monoxide (CO) or even carbon itself as soot. The equations below compare different degrees of combustion for heptane, a component in petrol.

complete combustion: CO_2 formed $C_7H_{16}(l) + 11O_2(g) \rightarrow 7CO_2(g) + 8H_2O(l)$
incomplete combustion: CO formed $C_7H_{16}(l) + 7\tfrac{1}{2}O_2(g) \rightarrow 7CO(g) + 8H_2O(l)$
incomplete combustion: C formed $C_7H_{16}(l) + 4O_2(g) \rightarrow 7C(s) + 8H_2O(l)$

Look at the different amounts of oxygen that are needed for each equation. It is hardly surprising that it can be difficult to get all of the carbon to combust when combustion is attempted in a closed space, such as in a car engine, faulty heating systems, or inadequate ventilation in living areas.

The silent killer

Carbon monoxide is a colourless, odourless, and highly toxic gas. It combines irreversibly with haemoglobin in red blood cells to form a cherry-pink compound called carboxyhaemoglobin, which prevents the haemoglobin from transporting oxygen around the body. Severe CO poisoning will often turn a victim's lips this bright colour.

The danger lies in the lack of any odour, so a person can be poisoned without noticing any danger.

Write equations for the incomplete combustion of methane.

▲ **Figure 3** *An audible carbon monoxide alarm is a good way to ensure you are immediately alerted to dangerous levels of carbon monoxide in your home*

> **Synoptic link**
>
> To estimate how much CO_2 is produced during combustion, we must refer to the earlier topics on the mole and quantities in equations (Chapter 3).

 Worked example: How much CO_2 is produced during combustion?

Carbon dioxide production is linked to global warming, but how much difference does a car journey make? Let's think about a short trip, a 10-mile journey to a town centre and back. For short journeys in cities, petrol consumption may be 30 miles per gallon (about 4.5 litres). So 10 miles would use about 1.5 litres of petrol, which will weigh about 1 kg (density of petrol ~ $0.67\,\text{g cm}^{-3}$).

Petrol contains a range of alkanes, roughly $n = 5\text{--}9$, so let's take C_7H_{16} (heptane) as typical ($M_r = 100$). Assume complete combustion and room temperature and pressure (RTP).

Step 1: Calculate the amount in moles of C_7H_{16} in 1 kg of heptane.

$1\,\text{kg} = 1000\,\text{g}$; $M_r(C_7H_{16}) = 100$

amount of C_7H_{16} $n(C_7H_{16}) = \dfrac{\text{mass } m}{\text{molar mass } M} = \dfrac{1000}{100} = 10\,\text{mol}$

Step 2: Write the balanced equation for the complete combustion of heptane.

$C_7H_{16}(l) + 11O_2(g) \rightarrow 7CO_2(g) + 8H_2O(l)$

so, 10 mol → 70 mol

Step 3: Rearrange the equation to calculate the volume of CO_2 produced in the combustion of 1 kg of heptane.

amount $n = \dfrac{\text{volume } V}{\text{molar gas volume } V_m}$

$V = n \times V_m$

$= 70 \times 24 = 1680\,\text{dm}^3$ at room temperature and pressure

That is a lot of CO_2 for a short trip.

Reactions of alkanes with halogens

In the presence of sunlight, alkanes react with halogens. The high-energy ultraviolet (UV) radiation present in sunlight provides the initial energy for a reaction to take place. For example, methane reacts with bromine as shown below.

$CH_4(g) + Br_2(l) \rightarrow CH_3Br(g) + HBr(g)$
methane   bromomethane

This is a substitution reaction, as a hydrogen atom in the alkane has been substituted by a halogen atom.

Study tip

You need only look at reactions of alkanes with chlorine and bromine. Fluorine, the most reactive halogen, is just too reactive and would likely cause an explosion. Iodine is not reactive enough and barely reacts with alkanes.

Mechanism for bromination of alkanes

A chemical reaction can often be represented by a series of steps (the reaction mechanism) showing how electrons are thought to move during the reaction. The mechanism for the bromination of methane is an example of **radical substitution**.

The mechanism takes place in three stages, called **initiation**, **propagation**, and **termination**.

Step 1: Initiation

In the initiation stage, the reaction is started when the covalent bond in a bromine molecule is broken by homolytic fission. Remember that a covalent bond is a shared pair of electrons. Each bromine atom

Synoptic link

See Topic 11.5, Introduction to reaction mechanisms, for an introduction into mechanisms and homolytic fission.

12.2 Chemical reactions of the alkanes

takes one electron from the pair, forming two highly reactive bromine radicals. The energy for this bond fission is provided by UV radiation.

$$Br-Br \rightarrow Br\bullet + \bullet Br$$

Remember that a radical is a very reactive species with an unpaired electron. The radical is shown with a single dot to represent the electron.

Step 2: Propagation

In the propagation stage, the reaction propagates through two propagation steps, a **chain reaction**.

Propagation step 1 $CH_4 + Br\bullet \rightarrow \bullet CH_3 + HBr$

Propagation step 2 $\bullet CH_3 + Br_2 \rightarrow CH_3Br + Br\bullet$

- In the first propagation step, a bromine radical, Br•, reacts with a C—H bond in the methane, forming a methyl radical, •CH_3, and a molecule of hydrogen bromide, HBr.
- In the second propagation step, each methyl radical reacts with another bromine molecule, forming the organic product bromomethane, CH_3Br, together with a new bromine radical (Br•).

The new bromine radical then reacts with another CH_4 molecule as in the first propagation step, and the two steps can continue to cycle through in a chain reaction.

In theory the propagation steps could continue until all the reactants have been used up. In practice propagation is terminated whenever two radicals collide. It has been estimated that up to a million propagation cycles take place before a termination step stops the reaction.

Step 3: Termination

In the termination stage, two radicals collide, forming a molecule with all electrons paired. There are a number of possible termination steps with different radicals in the reaction mixture.

$$Br\bullet + \bullet Br \rightarrow Br_2$$
$$\bullet CH_3 + \bullet CH_3 \rightarrow C_2H_6$$
$$\bullet CH_3 + \bullet Br \rightarrow CH_3Br$$

When two radicals collide and react, both radicals are removed from the reaction mixture, stopping the reaction.

Limitations of radical substitution in organic synthesis

Although radical substitution gives us a way of making haloalkanes, this reaction has problems that limit its importance for synthesis of just one organic compound.

Further substitution

In the mechanism above, bromomethane, CH_3Br, was formed in the second propagation step. Another bromine radical can collide with a bromomethane molecule, substituting a further hydrogen atom to

> **Study tip**
>
> For radical substitution, a description of the mechanism should include the equations. However, other mechanisms use diagrams.
>
> Mechanisms are essential for understanding how organic reactions take place.

> **Study tip**
>
> Make sure that the equations are balanced. Notice that there is always one radical on each side of a propagation step. Always show a dot to indicate the unpaired electron on a radical.
>
> Also note that the first step forms HBr – it is a common mistake to show •H forming instead of •CH_3.

form dibromomoethane, CH_2Br_2. Further substitution can continue until all hydrogen atoms have been substituted. The result is a mixture of CH_3Br, CH_2Br_2, $CHBr_3$, and CBr_4.

$$CH_4 \xrightarrow[-HBr]{+Br_2} CH_3Br \xrightarrow[-HBr]{+Br_2} CH_2Br_2 \xrightarrow[-HBr]{+Br_2} CHBr_3 \xrightarrow[-HBr]{+Br_2} CBr_4$$

Substitution at different positions in a carbon chain

For methane, all four hydrogen atoms are bonded to the same carbon atom, so only one monobromo compound, CH_3Br, is possible. With ethane, similarly, only one monosubstituted product, C_2H_5Br, is possible.

If the carbon chain is longer, we will get a mixture of monosubstituted isomers by substitution at different positions in the carbon chain. For example, pentane could form three monosubstituted isomers (Figure 4).

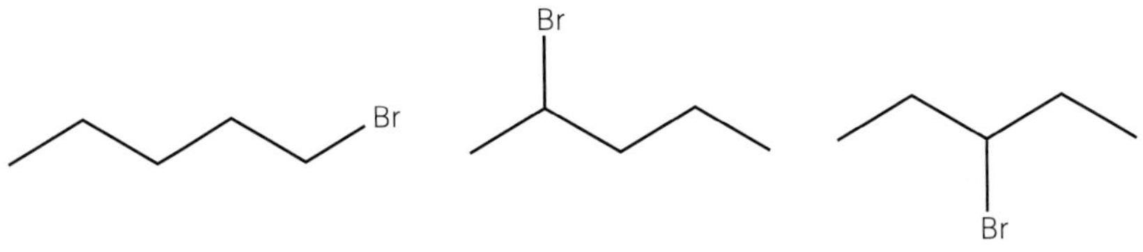

▲ **Figure 4** *Isomers of bromopentane: different possible monosubstituted bromo compounds from substitution of pentane with bromine.*

With further substitution, there are even more possibilities, as shown in Figure 3 for ethane, which can form two different dibromo isomers.

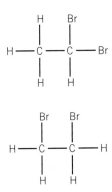

▲ **Figure 3** *Isomers of dibromoethane: different possible disubstituted bromo compounds from substitution of ethane*

Summary questions

1. Write equations for the complete and incomplete combustion of dodecane, the alkane with 12 carbon atoms. *(2 marks)*

2. Explain why the alkanes do not react with common laboratory reagents. *(2 marks)*

3. Write the equation for the incomplete combustion of 3-methylpentane. *(2 marks)*

4. Write an equation for the reaction between propane and bromine to form a monosubstituted product. *(1 mark)*

5. Show the mechanism for the reaction between ethane and chlorine to form a monosubstituted product. *(5 marks)*

6. For propane, write equations for the two propagation steps that convert $C_3H_6Cl_2$ into $C_3H_5Cl_3$. *(2 marks)*

Practice questions

1. What do you understand by the term *saturated hydrocarbon*. *(1 mark)*

2. Name the following compounds
   a. $CH_3CH_2CH_2CH_2CH_2CH_2CH_2CH_3$ *(1 mark)*
   b. $CH_3CH_2CH(CH_3)CH_3$ *(1 mark)*
   c. $CH_3C(CH_3)_2CH_2CH_3$ *(1 mark)*

3. In each of the following pairs, predict which of the following alkanes has the higher boiling point. In each case explain your answer.
   a. hexane and 3-methylpentane. *(2 marks)*
   b. 2-methyloctane and 2,2-dimethylheptane *(2 marks)*
   c. hexane and pentane *(2 marks)*

4. Bromine will react in a similar way as chlorine with methane.
   a. Write an equation for the reaction of bromine with methane. *(1 mark)*
   b. The first step in the mechanism is the breaking of the Br—Br bond.
      State the conditions required for this process to occur.
      Write an equation for this step and state the type of bond fission taking place. *(3 marks)*
   c. Pentane reacts with chlorine to form a mixture of monochlorinated alkanes.
      How many monochlorinated isomers can be formed in this reaction? *(1 mark)*

5. Write equations for the combustion of nonane in
   a. a plentiful supply of oxygen *(1 mark)*
   b. a limited supply of oxygen *(1 mark)*

6. 1.45 g of butane combusts in a plentiful supply of oxygen to form carbon dioxide and water.
   a. Write a balanced equation for the complete combustion of butane (C_4H_{10}) *(1 mark)*
   b. Calculate the amount, in moles, in 1.45 g of butane. *(1 mark)*
   c. Calculate the volume of oxygen required for complete combustion of 1.45 g of butane at RTP. *(2 marks)*
   d. Calculate the volume of carbon dioxide produced at RTP. *(2 marks)*

7. Explain why the boiling points of the alkanes increase with the number of carbon atoms in the chain. *(3 marks)*

8. Some of the hydrocarbons in kerosene have the formula $C_{10}H_{22}$.
   (i) What is the name of the straight chain hydrocarbon with the formula $C_{10}H_{22}$? *(1 mark)*
   (ii) Draw the skeletal formula of one branched chain isomer with the formula $C_{10}H_{22}$. *(1 mark)*
   (iii) Explain why the straight chain isomer of $C_{10}H_{22}$ has a higher boiling point than any of its branched chain structural isomers. *(2 marks)*

   OCR F322 Jan 2010 Q1

9. In the presence of ultraviolet radiation, cyclohexane reacts with bromine.
   a. A mixture of cyclic organic compounds is formed, including $C_6H_{11}Br$.
      (i) Copy and complete the table below to show the mechanism of the reaction between bromine and cyclohexane to form $C_6H_{11}Br$.
         Include all possible equations in your answer.

Step	Equation(s)
Initiation	
Propagation	
Termination	

         (5 marks)

      (ii) The initiation step involves homolytic fission.
         Explain why the initiation step is an example of homolytic fission. *(1 mark)*

   b. The reaction between cyclohexane and bromine also forms $C_6H_{10}Br_2$.
      (i) Write an equation, using molecular formulae, for the reaction of cyclohexane and bromine in the presence of ultraviolet radiation to form $C_6H_{10}Br_2$. *(1 mark)*

(ii) Name one of the structural isomers of $C_6H_{10}Br_2$ formed in the reaction between cyclohexane and bromine.
(1 mark)

OCR F322 June 14 Q1

10 a Methane reacts with IBr to form many products. Two of these products are iodomethane and hydrogen bromide.

(i) Suggest the essential condition needed for this reaction. *(1 mark)*

(ii) The mechanism of the reaction involves three steps, one of which is called termination. Describe the mechanism of the reaction that forms iodomethane and hydrogen bromide.

Include in your answer:

- the name of the mechanism
- the names for the other two steps of the mechanism
- equations for these two steps of the mechanism
- the type of bond fission
- one equation for a termination step.

Your answer should link the named steps to the relevant equations. *(7 marks)*

OCR F322 June 2013 Q4(b)

11 The reaction between butane and chlorine is an example of radical substitution. Initially, chlorobutane, C_4H_9Cl, is formed, which then reacts with more chlorine to form $C_4H_8Cl_2$.

a The first step of the reaction of C_4H_9Cl with chlorine is the homolytic fission of a chlorine molecule.

What is meant by the term homolytic fission? *(1 mark)*

b Complete the missing species in the propagation steps below.

Cl• + C_4H_9Cl → .... + ...

...... + ... → $C_4H_8Cl_2$ + ...
(2 marks)

c Butane, C_4H_{10}, undergoes incomplete combustion when there is a shortage of oxygen. Write an equation for the incomplete combustion of butane.
(1 mark)

OCR F322 Jan 2013 Q1(a) (b)(c) (d) (f)

12 Petrol and camping gas are examples of fuels that contain hydrocarbons.

a Petrol is a mixture of alkanes containing 6 to 10 carbon atoms per molecule. Some of these alkanes are structural isomers of one another.

(i) Explain the term *structural isomers*.
(1 mark)

(ii) State the molecular formula of an alkane present in petrol. *(1 mark)*

b The major hydrocarbon in camping gas is butane. Some camping gas was reacted with chlorine to form a mixture of structural isomers.

(i) What conditions are required for this reaction to take place? *(1 mark)*

(ii) Two structural isomers, **A** and **B**, were separated from this mixture. These isomers had a molar mass of 92.5 g mol^{-1}.

Deduce the molecular formula of these two isomers. *(1 mark)*

(iii) Draw the displayed formulae of **A** and **B** and name each compound.
(2 marks)

13 Octane, C_8H_{18}, is one of the alkanes present in petrol.

a What is the general formula for an alkane? *(1 mark)*

b Carbon monoxide, CO, is formed during the incomplete combustion of octane.

(i) Write an equation for the incomplete combustion of octane, forming carbon monoxide and water. *(1 mark)*

(ii) Why does incomplete combustion sometimes take place? *(1 mark)*

c 22.8 g of octane undergoes complete combustion. Calculate the volume of oxygen, at RTP, required for complete combustion and incomplete combustion.
(3 marks)

13 ALKENES

13.1 The properties of the alkenes

Specification reference: 4.1.3

Learning outcomes

Demonstrate knowledge, understanding, and application of:

- → the general formula of alkenes
- → the nature and shape of the C=C bond.

Synoptic link

You met sigma bonds (σ-bonds) in Topic 12.1, Properties of alkanes.

Structure and bonding in the alkenes

Structure of the alkenes

Alkenes are **unsaturated hydrocarbons**; alkene molecules contain at least one carbon to carbon double bond in their structure. Aliphatic alkenes that contain one double bond have the general formula C_nH_{2n}. The first three members of the alkene homologous series are shown in Figure 1.

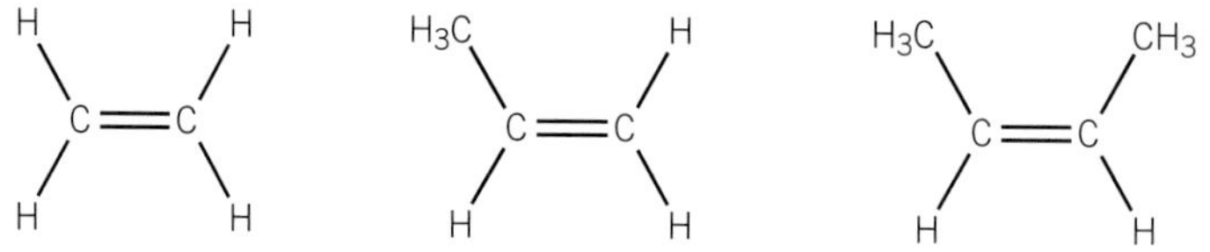

ethene propene but-2-ene

▲ **Figure 1** *The first three members of the alkene homologous series*

Alkenes can be branched, contain more than one double bond, or be cyclic (Figure 2). Whilst branched alkenes obey the general formula C_nH_{2n}, cyclic alkenes and alkenes with more than one double bond do not.

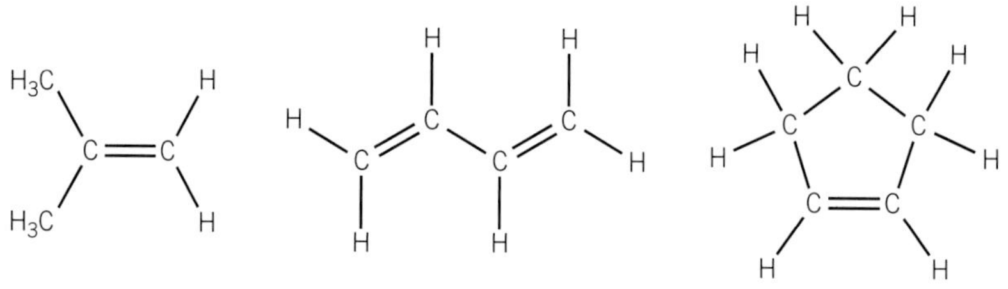

2-methylpropene buta-1,3-diene cyclopentene

▲ **Figure 2** *Examples of branched, polyunsaturated, and cyclic alkenes*

The nature of the double bond

Each carbon atom has four electrons in its outer shell and can use these electrons to form bonds.

For each carbon atom of the double bond, three of the four electrons are used in three σ-bonds, one to the other carbon atom of the double bond and the other two electrons to two other atoms (carbon or hydrogen).

This leaves one electron on each carbon atom of the double bond not involved in σ-bonds. This electron is in a p-orbital. A **π-bond** is formed by the sideways overlap of two p-orbitals, one from each carbon atom of the double bond. Each carbon atom contributes one electron to the electron pair in the π-bond. The π-electron density is concentrated above and below the line joining the nuclei of the bonding atoms.

The π-bond locks the two carbon atoms in position and prevents them from rotating around the double bond. This makes the geometry of the alkenes different from that of the alkanes, where rotation is possible around every atom.

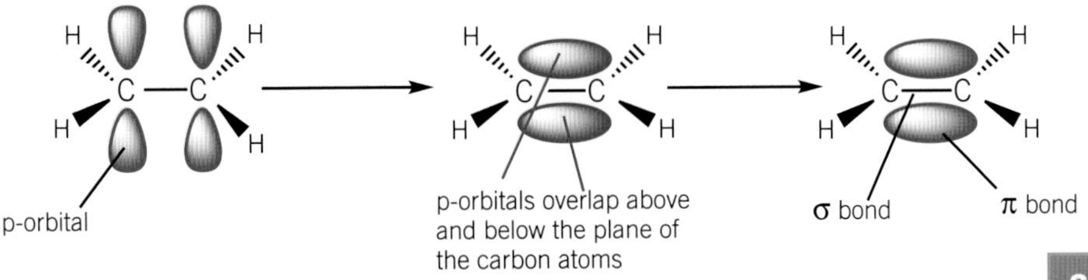

Figure 3 *The formation of the C=C bond in ethene*

The shape around a double bond

The shape around each of the carbon atoms in the double bond is trigonal planar (Figure 4), because:

- there are three regions of electron density around each of the carbon atoms
- the three regions repel each other as far apart as possible, so the bond angle around each carbon atom is 120°
- all of the atoms are in the same plane.

> **Synoptic link**
>
> You learnt how to use electron-pair repulsion theory to predict the shape and bond angles around carbon atoms in organic molecules in Topic 6.1, Shapes of molecules and ions.

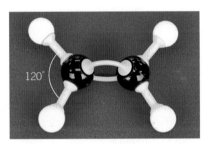

Figure 4 *A model of ethene showing the 120° bond angle around each of the carbon atoms*

> Worked example: But-2-en-1-ol
>
> You can apply the electron-pair repulsion theory to but-2-en-1-ol, which contains both alkene and alcohol functional groups, to predict the shape and bond angles around different atoms.
>
> Name the shape and state the bond angles shown around each of the atoms a, b, and c in Figure 5.
>
> **a** This carbon atom has three bonded regions and forms a trigonal planar shape with bond angles of 120°.
>
> **b** This carbon atom has four bonded electron pairs, which repel each other as far as they can, to give a tetrahedral shape with bond angles of 109.5°.
>
> **c** This oxygen atom has two bonded pairs and two lone pairs. The lone pairs repel more than the bonded pairs and the bond angle will be 104.5°. The two bonds therefore form a non-linear arrangement around the oxygen atom.
>
>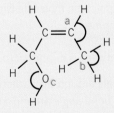
>
> ▲ **Figure 5** *The structure of but-2-en-1-ol*

13.1 The properties of the alkenes

Alkenes in the natural world

Alkenes are responsible for making flamingos pink and for the smell of oranges and lemons.

The colour of a flamingo's plumage is due to its diet. Flamingos in the wild typically eat algae and crustaceans that contain pigments called carotenoids. Enzymes in the liver convert the carotenoids into pink and orange pigments deposited in the feathers. Beta-carotene, shown in Figure 7, is a related chemical that makes carrots orange, and is important because it is a building block that our bodies use to manufacture vitamin A.

▲ **Figure 6** *Flamingos in search of crustaceans to eat*

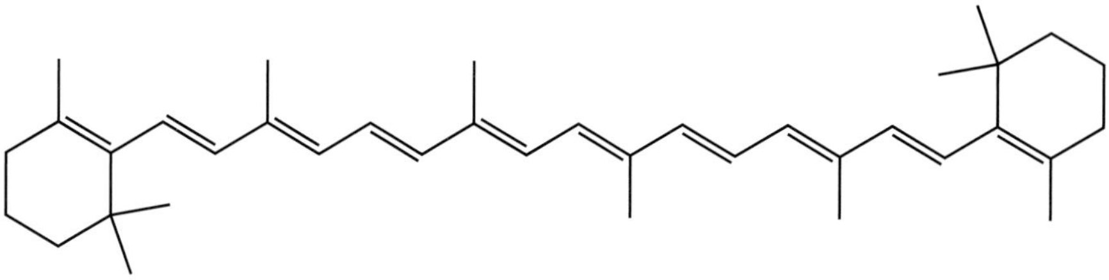

▲ **Figure 7** *Carotene and related alkenes are responsible for the colour of flamingo plumage*

Limonene (Figure 8) is a cyclic alkene with the molecular formula $C_{10}H_{16}$. It is found naturally in the rinds of citrus fruits and is largely responsible for the smell and flavour of oranges and lemons. Limonene is also used in perfumes and household cleaning products because of its fragrance.

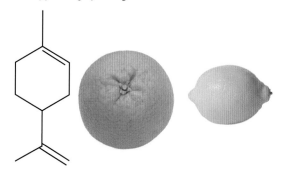
▲ **Figure 8** *Limonene is responsible for the smell of oranges and lemons*

1. Find two other alkenes that have a role in the natural world.
2. Draw the skeletal formulae of these alkenes, state where they are found, and describe their role in nature.

Summary questions

1. State what is meant by the term unsaturated. *(1 mark)*

2. State the molecular formula of the alkene containing nine carbon atoms. The compound is a straight-chain molecule with the double bond on the middle carbon. State its name. *(2 marks)*

3. Explain the difference between a σ- and a π-bond. *(3 marks)*

4. Explain why the bond angle around the carbon atom in an alkene is different to the bond angle around a carbon atom in an alkane. *(3 marks)*

5. Draw and name all of the structural isomers of the alkene C_4H_8. *(3 marks)*

13.2 Stereoisomerism
Specification reference: 4.1.3

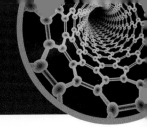

Stereoisomers

Stereoisomers have the same structural formula but a different arrangement of the atoms in space. You will study two types of stereoisomerism, **E/Z isomerism** and **optical isomerism**. *E/Z* isomerism only occurs in compounds with a C=C double bond, whereas **optical isomerism** can occur in a much wider range of compounds, including alkanes with no functional groups.

> **Learning outcomes**
> Demonstrate knowledge, understanding, and application of:
> → stereoisomerism
> → the Cahn–Ingold–Prelog priority rules.

Solving the mystery of maleic and fumaric acid

Maleic acid can be isolated from unripe fruit (especially tomatoes and apples) and fumaric acid is isolated from a wild flower. Both acids have the structural formula HOOCCH=CHCOOH and contain the same functional groups, yet they behave differently when heated. This difference puzzled chemists for some time. In 1874, Jacobus Henricus van't Hoff, a Dutch chemist, proposed that free rotation around a double bond was not possible. He concluded that two molecules with the same structural formula could exist in two different forms (Figure 1). He called these two forms of the same structure *cis* and *trans* (Latin for 'same side' and 'across'), terms that are still in use today. He was awarded the first Nobel Prize in chemistry in 1901 in recognition of his contribution to the field of stereochemistry.

▲ **Figure 2** *Fumaric acid is isolated from the wild flower Fumaria officinalis*

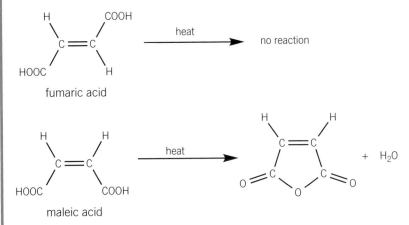

▲ **Figure 1** *Fumaric acid (top) and maleic acid respond differently to heat*

▲ **Figure 3** *Maleic acid is isolated from unripe fruit, especially tomatoes and apples*

1 What is the empirical and molecular formula for fumaric acid and maleic acid?
2 Why do maleic acid (the *cis* form) and fumaric acid (the *trans* form) behave differently when heated?

E/Z isomerism

Stereoisomerism around double bonds arises because rotation about the double bond is restricted and the groups attached to each carbon atom are therefore fixed relative to each other. The reason for the rigidity is the position of the π-bond's electron density above and below the plane of the σ-bond (see Topic 13.1).

> **Synoptic link**
> You will study optical isomerism in Topic 27.2, Amino acids, amides, and chirality.

13.2 Stereoisomerism

> **Study tip**
>
> You may be asked whether a structure has *E/Z* isomerism. Many students remember the first condition, the double bond, but not the second, different groups. You need *both* conditions.

If a molecule satisfies *both* of the following conditions it will have *E/Z* isomerism:

- a C=C double bond
- different groups attached to each carbon atom of the double bond.

Both but-1-ene and but-2-ene (Figure 4) contain a C=C bond, but only but-2-ene can form *E/Z* isomers. If you examine the two structures you can see why.

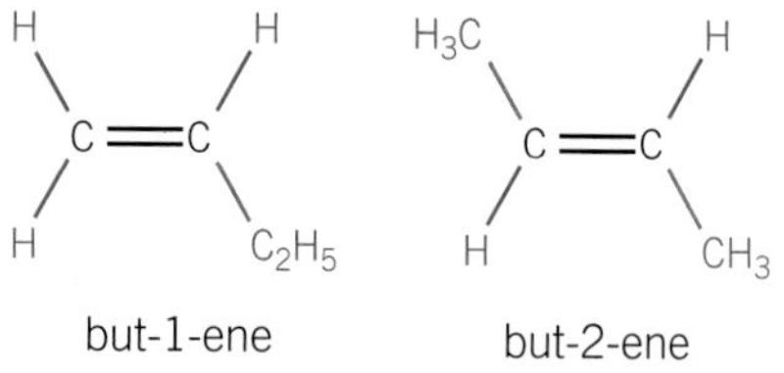

In but-1-ene, the carbon atom on the left-hand side of the molecule is attached to two hydrogen atoms. This means that its structure does not satisfy the second condition. In contrast, but-2-ene has a **methyl group** and a **hydrogen atom** on each of the carbon atoms of the double bond, so it has *E* and *Z* isomers (Figure 5).

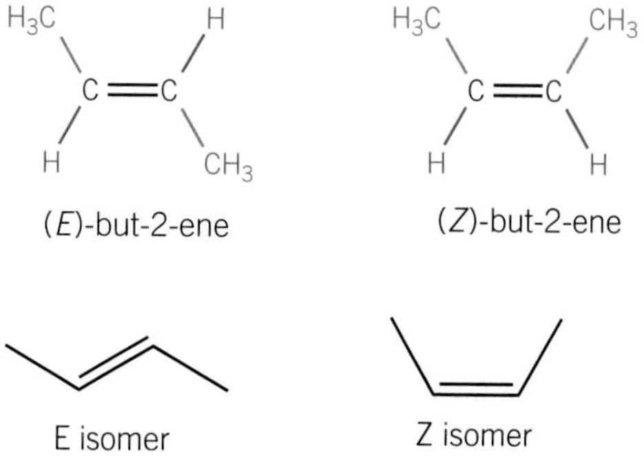

▲ **Figure 4** *But-1-ene does not have E/Z isomers but but-2-ene does*

▲ **Figure 5** *The E and Z isomers of but-2-ene, displayed about the double bond, and as skeletal formulae*

Later in this topic you will learn how to identify which of the isomers is *E* and which is *Z*.

Cis–trans isomerism

Cis–trans isomerism is the name commonly used to describe a special case of *E/Z* isomerism. Molecules must have a C=C double bond and each carbon in the double bond must be attached to two different groups, as for all *E/Z* isomers. However, in *cis–trans* isomers one of the attached groups on each carbon atom of the double bond *must* be the same.

But-2-ene is an example of a molecule with *E/Z* isomers that can also be described with *cis–trans* nomenclature. The *cis–trans* isomers of but-2-ene are shown in Figure 6.

The *cis* isomer has the hydrogen atoms and methyl groups on each carbon in the double bond on the *same* side of the molecule, whereas the *trans* isomer has the hydrogen atoms and methyl groups diagonally *opposite* each other.

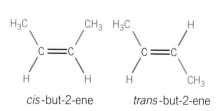

▲ **Figure 6** *The cis–trans isomers of but-2-ene*

Where there is a hydrogen on each of the double-bonded carbon atoms in *cis–trans* isomerism:

- the *cis* isomer is the *Z* isomer
- the *trans* isomer is the *E* isomer.

Cahn–Ingold–Prelog nomenclature

The *cis*–*trans* system of naming can only be used when one of the substituent groups on each carbon atom in the double bond is the same. However, many other isomeric compounds contain C=C bonds and these must be named. For example, the stereoisomer of 2-penten-3-ol shown in Figure 7 cannot be classified as either *cis* or *trans*. In this case the *E*/*Z* system of naming has to be used.

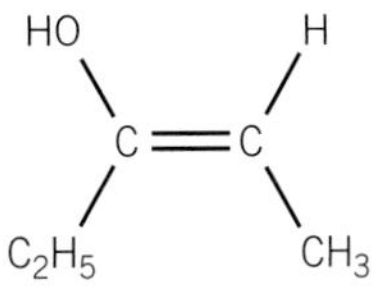

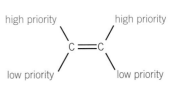

▲ **Figure 7** *The structure of one stereo isomer of 2-penten-3-ol*

Up until 1951, the whole system of naming the isomers of compounds containing C=C bonds was confusing. Robert Sidney Cahn, Christopher Kelk Ingold, and Vladimir Prelog came up with a naming system that is now widely used:

Using the Cahn–Ingold–Prelog rules

In this system the atoms attached to each carbon atom in a double bond are given a priority based upon their atomic number.

- If the groups of higher priority are on the same side of the double bond, the compound is the *Z* isomer.
- If the groups of higher priority are diagonally placed across the double bond, the compound is the *E* isomer.

Assigning priority – step 1
Examine the atoms attached directly to the carbon atoms of the double bond and decide which of the two atoms has the highest priority. The higher the atomic number, the higher the priority.

▼ **Table 1** *The relative priorities of some of the common atoms found in organic molecules*

Atomic number	1	6	7	8	17	35	Atomic number
Lower priority	H	C	N	O	Cl	Br	Higher priority

Increasing priority →

▲ **Figure 8** *How to name E and Z isomers by allocating priority to groups*

 Worked example: Group priorities in 2-bromo-1-chloropropene

Consider the molecule 2-bromo-1-chloropropene, shown in Figure 9.

Step 1: The left-hand carbon atom is bonded to a chlorine atom and a hydrogen atom. The chlorine has the higher atomic number and the higher priority.

▲ **Figure 9** *2-Bromo-1-chloropropene*

13.2 Stereoisomerism

Step 2: The right-hand C atom is bonded to another C atom and a Br atom. The Br atom has the higher atomic number and the higher priority.

Step 3: The two higher-priority groups are diagonally opposite each other.

Step 4: The molecule is called (E)-2-bromo-1-chloropropene.

Assigning priority – step 2

If the two atoms attached to a carbon atom in the double bond are the same, then you will need to find the first point of difference. The group which has the higher atomic number at the first point of difference is given the higher priority.

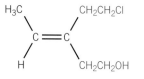

▲ **Figure 10** *Allocating group priorities*

 Worked example: Group priorities at point of difference

Consider the molecule shown in Figure 10.

Step 1: On the left-hand C atom, the C atom has priority over the H atom.

Step 2: On the right-hand side, the atoms immediately attached to the C atom in the double bond are both C atoms.

Step 3: Continue down these two branches (marked in blue in Figure 10) to the first point of difference in the two side chains. This is at the Cl and O atoms (marked in red).

Step 4: Cl has a higher atomic number, so the chain containing the CH_2CH_2Cl has the higher priority.

Step 5: The molecule is the Z isomer.

Summary questions

1. Define the term stereoisomers. *(1 mark)*

2. State the two features a molecule must have to show E/Z isomerism. *(2 marks)*

3. Identify whether the following molecules show E/Z isomerism.
   a. pent-1-ene *(1 mark)*
   b. 2-methylpent-2-ene *(1 mark)*
   c. 2,3-dichlorobut-2-ene *(1 mark)*
   d. 2,3-dichlorobut-2-en-1,4-diol *(1 mark)*
   e. hex-3-ene *(1 mark)*

4. For each of the molecules that show E/Z isomerism in question 3, draw and label the two possible stereoisomers. *(3 marks)*

13.3 Reactions of alkenes

Specification reference: 4.1.3

The reactivity of alkenes

Alkenes are much more reactive than alkanes because of the presence of the π-bond. You learnt in Topic 13.1 that the C=C double bond is made up of a σ-bond and a π-bond, and that the π-electron density is concentrated above and below the plane of the σ-bond (Figure 1). Being on the outside of the double bond, the π-electrons are more exposed than the electrons in the σ-bond. A π-bond readily breaks and alkenes undergo addition reactions relatively easily.

By considering the exposed nature of the π-electrons and the bond enthalpy data below, it is easy to see why the π-bond breaks and the σ-bond remains intact when alkenes react.

- A C—C single bond (σ-bond) in ethene has a bond enthalpy of 347 kJ mol^{-1}.
- The C=C double bond (σ-bond + π-bond) has a bond enthalpy of 612 kJ mol^{-1}.
- The bond enthalpy of the π-bond in an alkene can be calculated as 612 − 347 = 265 kJ mol^{-1}.
- The π-bond is weaker than the σ-bond and is therefore broken more readily.

Learning outcomes

Demonstrate knowledge, understanding, and application of:
→ the reactivity of alkenes
→ the addition reactions of alkenes.

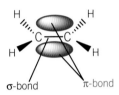

▲ **Figure 1** *The double bond in an alkene is made of a σ-bond and a π-bond; the π-electrons are on the outside of the double bond*

Addition reactions of the alkenes

Alkenes undergo many addition reactions For example, with:

- hydrogen in the presence of a nickel catalyst
- halogens
- hydrogen halides
- steam in the presence of an acid catalyst.

Each of these reactions involves the addition of a small molecule across the double bond, causing the π-bond to break and for new bonds to form.

Synoptic link

You have learnt about addition reactions in Topic 11.5, Introduction to reaction mechanisms.

Hydrogenation of alkenes

When an alkene, such as propene, is mixed with hydrogen and passed over a nickel catalyst at 423 K, an addition reaction takes place to form an alkane (Figure 2). This addition reaction, in which hydrogen is added across a double bond, is known as hydrogenation.

Study tip

An addition reaction is one in which two molecules react together to make one product. The product is saturated.

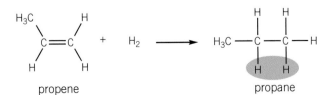

▲ **Figure 2** *Addition of hydrogen across the double bond of an alkene*

13.3 Reactions of alkenes

All C=C bonds react with hydrogen in this way. For example, buta-1,3-diene reacts with excess hydrogen in the presence of a nickel catalyst to form butane (Figure 3). Notice that both double bonds are hydrogenated, requiring two molecules of hydrogen per molecule of buta-1,3-diene.

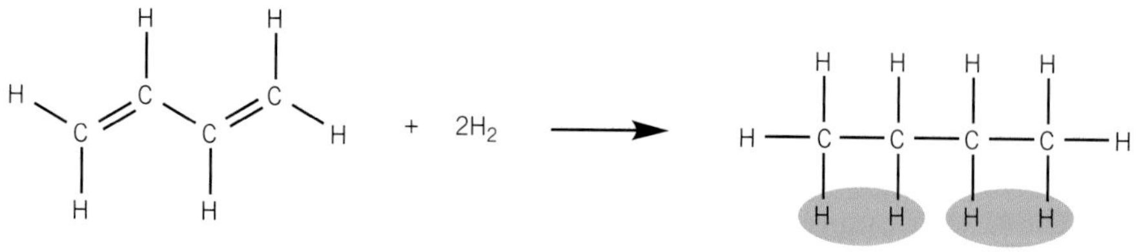

▲ **Figure 3** *Addition of hydrogen across the double bonds in buta-1,3-diene*

Focus on margarine

Margarine (Figure 4) was created in the early 1800s as an inexpensive substitute for butter. Early margarines were made from animal fat. However, in the 1900s, chemists discovered how to harden liquid oils, and vegetable oil replaced animal fat in margarine production.

Vegetable oil (Figure 5) contains molecules with long unsaturated hydrocarbon chains with the C=C bonds in the *cis* orientation. Hydrogen gas is bubbled through the oil in the presence of a catalyst (usually nickel). Many of the unsaturated double bonds are hydrogenated to form saturated carbon chains. The hydrogenated products have a lower melting point, so are more solid. The more complete the hydrogenation process, the firmer the finished margarine product.

The hydrogenation process does actually also form some *trans* double bonds. As the catalyst works, the π-bond is first broken but it then can reform in the *trans* orientation. As with saturated fats, *trans* fats have a lower melting point than *cis* fats. There are health concerns arising from *trans* fats and manufacturers are trying to reduce their formation. Labels on margarines sometimes state 'low in trans fats'.

▲ **Figure 4** *Margarine*

Olive oil is based on a fatty acid called oleic acid which has 18 carbon atoms. Its systematic name is *E*–octadec-9-enoic acid. In the name, the carbon atom in the COOH carboxylic acid group is carbon number 1. The *Z* stereoisomer of oleic acid is called elaidic acid and its hydrogenation product is called stearic acid.

▲ **Figure 5** *Vegetable oils can be converted into solids by hydrogenation of their double bonds*

1. Draw skeletal formulae for oleic acid, elaidic acid, and stearic acid.
2. Suggest why the melting point of oleic acid is much lower than either elaidic acid or stearic acid.

ALKENES 13

Halogenation of alkenes

Alkenes undergo a rapid addition reaction with the halogens chlorine or bromine at room temperature. The reaction for the bromination of propene is shown in Figure 6.

▲ **Figure 6** *Addition of bromine across the double bond of an alkene*

Testing for unsaturation

The reaction of alkenes with bromine can be used to identify if there is a C=C bond present and the organic compound is unsaturated. When bromine water (an orange solution) is added dropwise to a sample of an alkene, bromine adds across the double bond. The orange colour disappears (Figure 6), indicating the presence of a C=C bond. If the same test is carried out with a saturated compound, there is no addition reaction and no colour change. Any compound containing a C=C bond will decolourise bromine water.

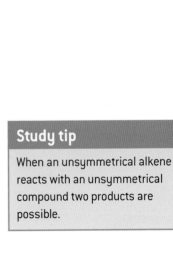

▲ **Figure 7** *Test tubes showing an alkene and orange bromine water (left). The test tube is shaken to mix the two layers, allowing the reaction to take place that decolorises the bromine water (right)*

Addition reactions of alkenes with hydrogen halides

Alkenes react with gaseous hydrogen halides at room temperature to form haloalkanes. If the alkene is a gas, like ethene, then the reaction takes place when the two gases are mixed. If the alkene is a liquid, then the hydrogen halide is bubbled through it. Alkenes also react with concentrated hydrochloric or concentrated hydrobromic acid, which are solutions of the hydrogen halides in water.

Figure 7 shows the reaction of propene, an unsymmetrical alkene, with hydrogen chloride, also an unsymmetrical molecule. There are two possible products. You will find out more about this addition in Topic 13.4.

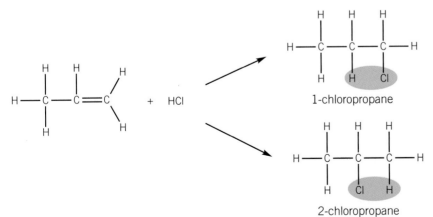

▲ **Figure 8** *When propene reacts with hydrogen chloride, two possible products are formed*

> **Study tip**
>
> When an unsymmetrical alkene reacts with an unsymmetrical compound two products are possible.

13.3 Reactions of alkenes

> **Study tip**
> Concentrated sulfuric acid can also be used as the acid catalyst.

Hydration reactions of alkenes

Alcohols are formed when alkenes react with steam, $H_2O(g)$, in the presence of a phosphoric acid catalyst, H_3PO_4 (Figure 9). Steam adds across the double bond. This addition reaction is used widely in industry to produce ethanol from ethene. As with the addition with hydrogen halides, there are two possible products.

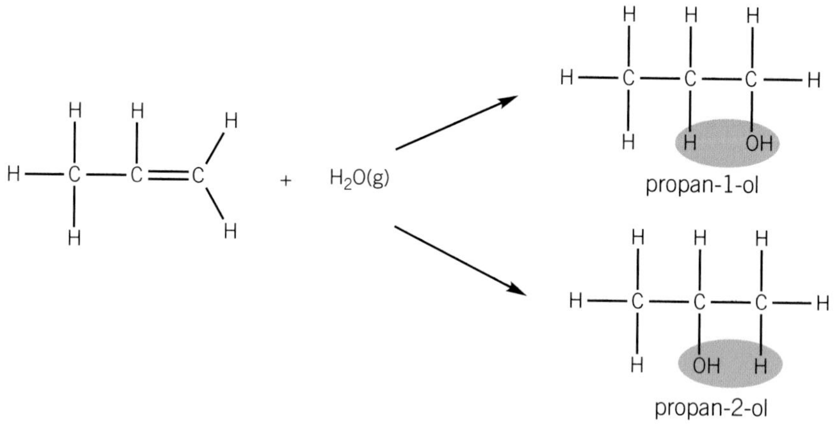

▲ **Figure 9** *The reaction of propene with steam in the presence of phosphoric acid results in a mixture of products*

Summary questions

1. Write an equation, using displayed formula, for the reaction between pent-2-ene and hydrogen in the presence of a nickel catalyst. *(2 marks)*

2. Draw the full displayed structure of the product formed when 2-methylpent-2-ene decolourises bromine. *(2 marks)*

3. Draw the structures of three alkenes that can undergo hydrogenation to produce 2-methylbutane. *(2 marks)*

4. Pent-2-ene can react with hydrogen bromide to form two haloalkanes that are structural isomers.
   a. Draw the full displayed formula of each of the two haloalkanes formed. Name each isomer. *(4 marks)*
   b. Write a balanced equation for the formation of one of these products. *(1 mark)*

5. Write equations for the following reactions.
   a. 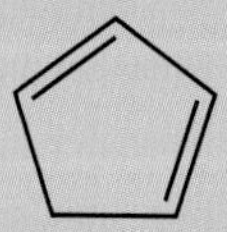 cyclopenta-1,3-diene and excess bromine *(1 mark)*
   b. 2-methylbuta-1,3-diene and excess hydrogen. *(1 mark)*

6. Draw skeletal formulae of all the structural isomers that can be formed when buta-1,3-diene reacts with hydrogen bromide. *(3 marks)*

13.4 Electrophilic addition in alkenes

Specification reference: 4.1.3

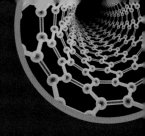

Electrophilic addition reactions

In Topic 11.5, you found out that a reaction mechanism is a series of steps that shows *how* a reaction takes place. Alkenes usually take part in addition reactions to form saturated compounds. The mechanism for this reaction is called **electrophilic addition**.

- The double bond in an alkene represents a region of high electron density because of the presence of the π-electrons.
- The high electron density of the π-electrons attracts **electrophiles**.
- An electrophile is an atom or group of atoms that is attracted to an electron-rich centre and accepts an electron pair. An electrophile is usually a positive ion or a molecule containing an atom with a partial positive (δ+) charge.

Learning outcomes

Demonstrate knowledge, understanding, and application of:

→ electrophiles
→ mechanism of electrophilic addition in alkenes
→ Markownikoff's rule.

Study tip

An electrophile is defined simply as an electron pair acceptor.

The reaction between but-2-ene and hydrogen bromide

Hydrogen bromide adds to but-2-ene to form a single addition product (Figure 1).

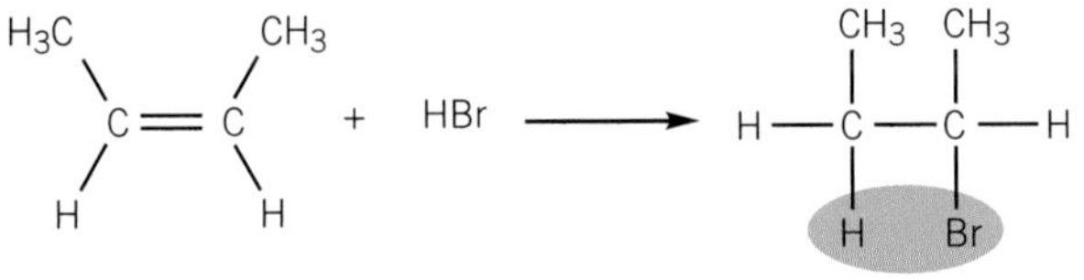

▲ **Figure 1** *The addition reaction between but-2-ene and hydrogen bromide*

Study tip

Curly arrows are used in the reaction mechanism to show the movement of electron pairs.

The mechanism for the reaction

The mechanism for this electrophilic addition is shown in Figure 2.

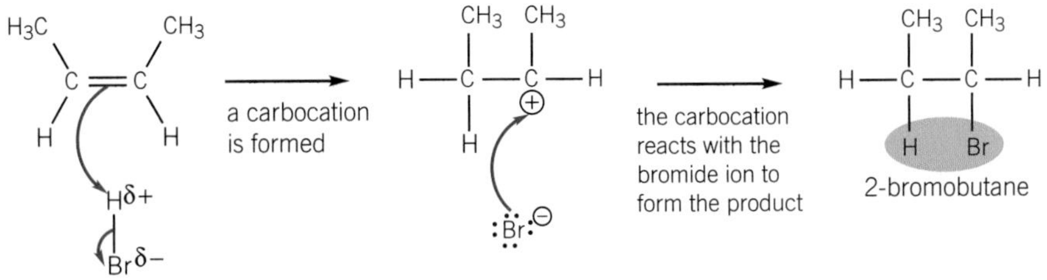

▲ **Figure 2** *The mechanism for the electrophilic addition reaction of but-2-ene with hydrogen bromide*

Study tip

Your curly arrows *must* be drawn precisely.

- Start the arrow at the bond or lone pair.
- Draw the head of the arrow at the atom to which the electron pair transfers.

Incorrect or unclear curly arrows is an easy mistake to make.

1. Bromine is more electronegative than hydrogen, so hydrogen bromide is polar and contains the dipole $H^{\delta+}$—$Br^{\delta-}$.
2. The electron pair in the π-bond is attracted to the partially positive hydrogen atom, causing the double bond to break.
3. A bond forms between the hydrogen atom of the H—Br molecule and a carbon atom that was part of the double bond.
4. The H—Br bond breaks by heterolytic fission, with the electron pair going to the bromine atom.

Study tip

When you are asked to describe a mechanism, it is easier to draw the mechanism than to write about it, even if you are provided with lines.

211

13.4 Electrophilic addition in alkenes

> **Study tip**
>
> Remember, in this reaction:
> - electrophile: HBr
> - mechanism: electrophilic addition
> - bond fission: heterolytic

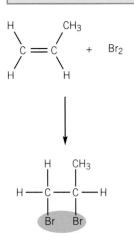

▲ **Figure 3** *The addition reaction between propene and bromine*

> **Study tip**
>
> Practise drawing reaction mechanisms as they are an important skill for organic chemistry.

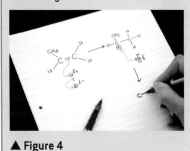

▲ **Figure 4**

> **Study tip**
>
> Remember that, in this mechanism, the Br—Br bond breaks by heterolytic fission.

5. A bromide ion (Br⁻) and a **carbocation** are formed. A carbocation contains a positively charged carbon atom.

6. In the final step the Br⁻ ion reacts with the carbocation to form the addition product.

The reaction between propene and bromine

Hydrogen bromide is a polar molecule, so it is easy to see how it acts as an electrophile in its reactions with the alkenes (Figure 2). However, alkenes can also react with non-polar molecules, such as bromine, Br_2, by electrophilic addition.

Bromine adds to propene to form a single addition product. The equation for the reaction is shown in Figure 3.

The mechanism for the reaction

Bromine is a non-polar molecule. When bromine approaches an alkene, the π-electrons interact with the electrons in the Br—Br bond.

1. This interaction causes polarisation of the Br—Br bond, with one end of the molecule becoming $Br^{\delta+}$ and the other end of the molecule becoming $Br^{\delta-}$. This is known as an induced dipole (Figure 5).

[Diagram showing polarisation of Br—Br bond as bromine approaches the π-bond. Labels: "bonding pair of electrons equally distributed between the two atoms" and "as bromine approaches π-bond. Electrons in bromine bond are repelled. Polarity has been induced in the bromine."]

▲ **Figure 5** *The polarisation of the Br—Br bond in bromine. As bromine approaches the π-bond, the electrons in bromine are repelled, inducing a dipole*

2. The electron pair in the π-bond is attracted to the $Br^{\delta+}$ end of the molecule, causing the double bond to break.

3. A bond has now been formed between one of the carbon atoms from the double bond and a bromine atom.

4. The Br—Br bond breaks by heterolytic fission, with the electron pair going to the $Br^{\delta-}$ end of the molecule.

5. A bromide ion (Br⁻) and a carbocation are formed.

6. In the final stage of the reaction mechanism the Br⁻ ion reacts with the carbocation to form the addition product of the reaction (Figure 6).

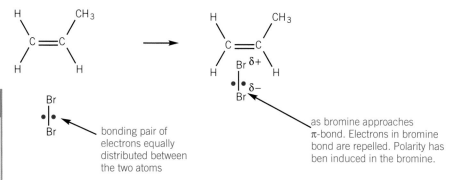

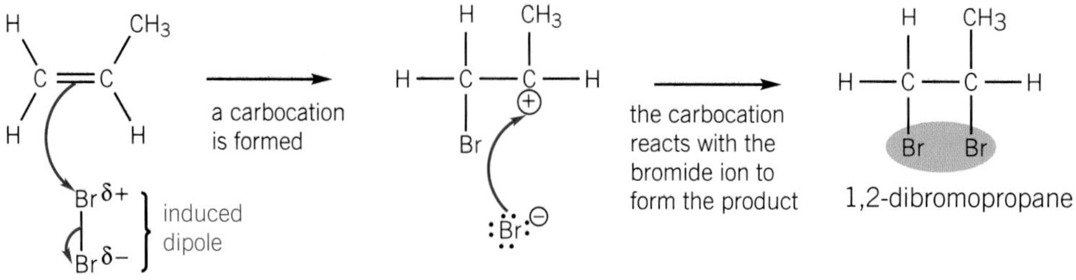

▲ **Figure 6** *The mechanism for the electrophilic addition reaction of propene with bromine*

ALKENES 13

Markownikoff's rule

You learnt in Topic 13.3 that two isomeric products are formed when propene, an unsymmetrical alkene, reacts with hydrogen bromide, also an unsymmetrical molecule. Now you will learn how to predict which of the two possible isomers is the major product.

In 1870 Vladimir Vasilyevich Markownikoff, a Russian chemist, stated that when a hydrogen halide reacts with an unsymmetrical alkene the hydrogen of the hydrogen halide attaches itself to the carbon atom of the alkene with the greater number of hydrogen atoms and the smaller number of carbon atoms.

Electrophilic addition occurs in two steps. In the first step a carbocation is formed. In the reaction between propene and a hydrogen halide, two carbocations are possible – a **primary** and a **secondary** carbocation (Figure 6). In the primary carbocation the positive charge is on a carbon atom at the end of a chain. In the secondary carbocation the positive charge is on a carbon atom with two carbon chains attached.

> **Study tip**
> A carbocation has a positive charge on a carbon atom

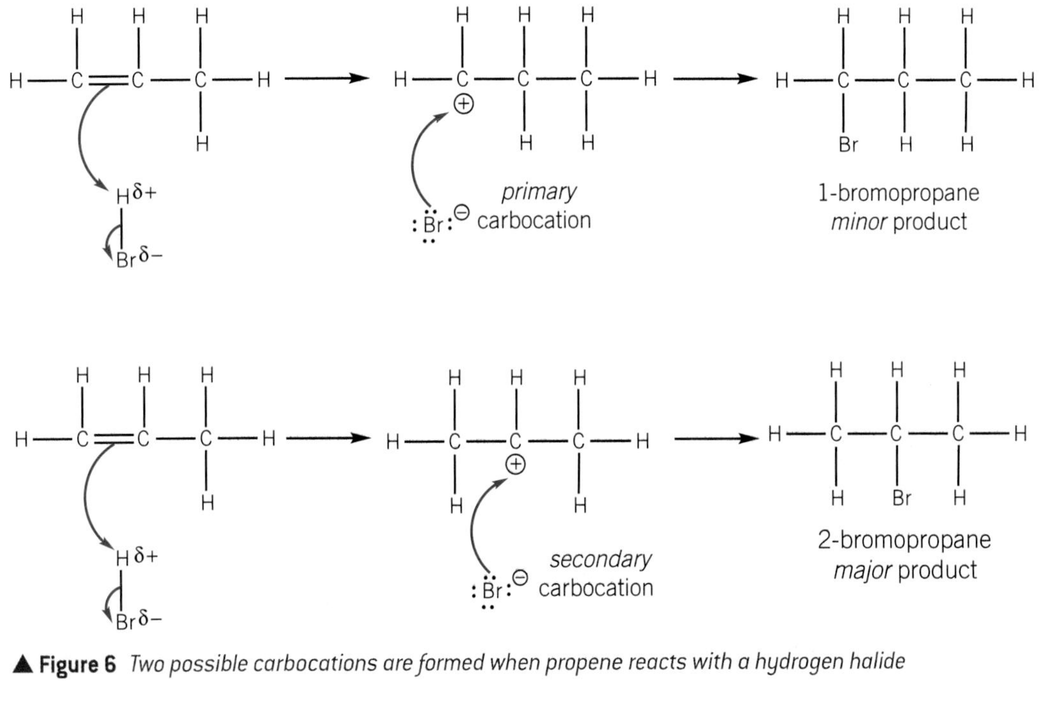

When propene reacts with hydrogen bromide, the major product is 2-bromopropane, formed from the secondary carbocation. The yield of 1-bromopropane is much smaller.

Figure 6 shows two mechanisms for the reaction between propene and hydrogen bromide. The top mechanism shows a primary carbocation as the intermediate. The lower mechanism shows a secondary carbocation as the intermediate. To understand the different yields of the two isomers, you need to consider the stability of carbocations.

▲ **Figure 6** *Two possible carbocations are formed when propene reacts with a hydrogen halide*

13.4 Electrophilic addition in alkenes

Carbocation stability

Carbocations are classified by the number of alkyl groups attached to the positively charged carbon atom (Figure 7). An alkyl group is normally represented by the symbol –R. Tertiary carbocations (with three R groups) are the most stable, and primary carbocations are the least stable.

Carbocation stability is linked to the electron-donating ability of alkyl groups. Each alkyl group donates and pushes electrons towards the positive charge of the carbocation. The positive charge is spread over the alkyl groups. The more alkyl groups attached to the positively-charged carbon atom, the more the charge is spread out, making the ion more stable. Therefore tertiary carbocations are more stable than secondary carbocations, which are more stable than primary carbocations.

Addition of a hydrogen halide to an unsymmetrical alkene forms the major product via the most stable carbocation.

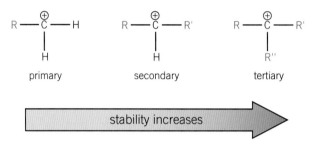

▲ **Figure 7** *The relative stabilities of primary, secondary, and tertiary carbocations*

Summary questions

1. Define the term electrophile. *(1 mark)*

2. Show the mechanism for the reaction between bromine and ethene. Explain how bromine acts as an electrophile. *(3 marks)*

3. Draw the displayed formula of the major product of the reaction of the following molecules with hydrogen bromide. Name each product.
   a $CH_3CH_2CH\!=\!CH_2$ *(2 marks)*
   b $CH_2\!=\!CHCH_2CH_2CH_2CH_3$ *(2 marks)*
   c $(CH_3)_2C\!=\!CHCH_3$ *(2 marks)*
   d (cyclohexene with CH₃ substituent)

   (2 marks)

4. Draw the mechanism for the reaction of the compound in **3d** with hydrogen chloride. *(3 marks)*

13.5 Polymerisation in alkenes

Specification reference: 4.1.3

Polymers

The Mayan civilisation in Central America is believed to be among the first to find an application for **polymers**: in the sixteenth century they were playing ball games using rubber collected from local trees. The first synthetic polymer, known as Bakelite, was made in 1907 by the Belgian Leo Bakeland in the USA, and was used as an electrical insulator. It was not until the late 1920s that now-common addition polymers such as poly(vinylchloride) (PVC) were synthesised on a large scale.

Addition polymerisation

Polymers are extremely large molecules formed from many thousands of repeat units of smaller molecules known as monomers. Unsaturated alkene molecules undergo **addition polymerisation** to produce long saturated chains containing no double bonds. Many different polymers can be formed, each with its own specific properties depending on the **monomer** used. Industrial polymerisation is carried out at high temperature and high pressure using catalysts.

- Addition polymers have high molecular masses.
- Synthetic polymers are usually named after the monomer that reacts to form their giant molecules, prefixed by 'poly'.

A general equation can be written for any addition polymerisation reaction (Figure 2). The equation must be balanced using the symbol n.

> **Learning outcomes**
> Demonstrate knowledge, understanding, and application of:
> → Addition polymerisation
> → Processing waste polymers
> → Biodegradable and photodegradable polymers.

▲ Figure 1 *Raw rubber is obtained from a milky latex exuded by the rubber tree in response to injury*

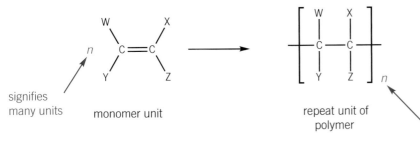

▲ Figure 2 *A general equation for addition polymerisation*

- A repeat unit is the specific arrangement of atoms in the polymer molecule that repeats over and over again.
- The repeat unit is always written in square brackets.
- After the bracket you place a letter n to show that there is a large number of repeats.

> **Study tip**
> Make sure polymerisation equations are balanced. You will need the letter n before the monomer and outside the bracket after the repeat unit in the polymer.

Poly(ethene)

Addition polymers are usually made from one type of monomer unit. For example the polymer poly(ethene) is made by heating a large number of ethene monomers at high pressure (Figure 3).

13.5 Polymerisation in alkenes

▼ **Figure 4** *Multicoloured drinking straws made from polyethene*

Study tip

You may come across many other addition polymers during your A Level chemistry course, but the concepts will always be very similar. You need to be able to draw one or more repeat units of a polymer, given the monomer.

Use the examples in Table 1 to test your ability to draw the repeat units of a polymer.

▲ **Figure 3** *Polymerisation of ethene monomers at high temperature and pressure to prepare poly(ethene)*

Poly(ethene) is one of the most commonly used polymers – you will come across it in supermarket bags, shampoo bottles, and children's toys.

The history of polyethene

Poly(ethene) was made by accident by Eric Fawcett and Reginald Gibson in 1933. They were carrying out high-pressure experiments with ethene at a chemical plant in Cheshire. A test vessel leaked, allowing a trace of oxygen to contaminate a fresh sample of ethene. The following day they found a white, waxy residue had been produced.

Nowadays poly(ethene) is manufactured in large quantities using carefully controlled processes. Depending on the method used, different densities of poly(ethene) can be made that have different structures, properties, and uses. For example high density poly(ethene) (HDPE) has linear chains. This gives the plastic some strength, making it ideal for use in children's toys, detergent bottles, and water pipes. Low density poly(ethene) (LDPE) has branched chains. This results in a plastic that has little strength but is flexible, making it ideal for use in plastic films and plastic bags.

1. Why does HDPE have a greater density and greater strength than LDPE?
2. The molecular mass of HDPE is typically in the range 200 000–500 000. Estimate the number of monomer units in this range.

Poly(chloroethene)

Poly(chloroethene), also known as poly(vinyl chloride) or PVC, can be prepared to make a polymer that is flexible or rigid. The equation for the formation of PVC is shown in Figure 5, and its common uses are shown in Figure 6.

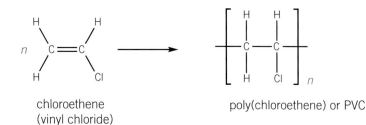

chloroethene (vinyl chloride)   poly(chloroethene) or PVC

▲ **Figure 5** *The polymerisation of chloroethene*

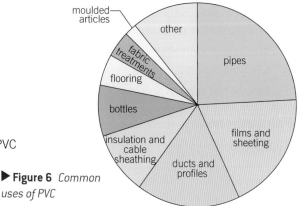

▶ **Figure 6** *Common uses of PVC*

ALKENES 13

Other polymers and monomers

▼ **Table 1** *Some common addition polymers, their monomers, and their uses*

Monomer	Polymer	Uses
propene 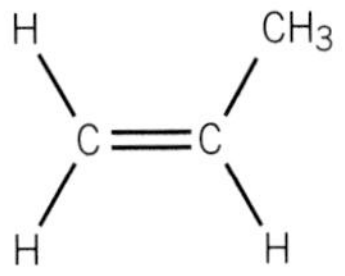 — structure showing $CH_2=CHCH_3$ with H, H on one carbon and H, CH_3 on the other	poly(propene) — repeat unit with H, H on one C and H, CH_3 on the other, bracketed with subscript n 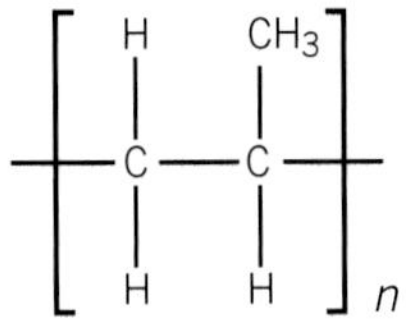	 Used to make children's toys, packing crates, guttering, uPVC windows, and fibre for ropes.
phenylethene / styrene 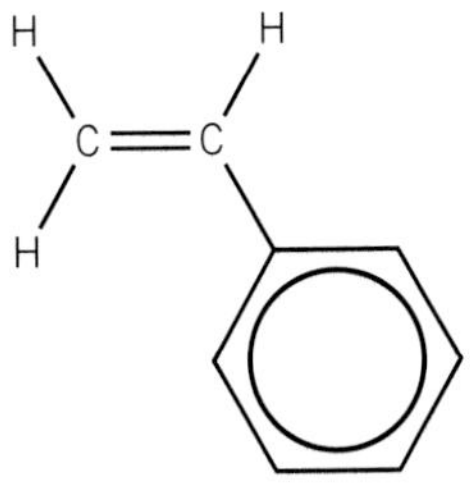 — $CH_2=CH$–phenyl	poly(phenylethene) or poly(styrene) 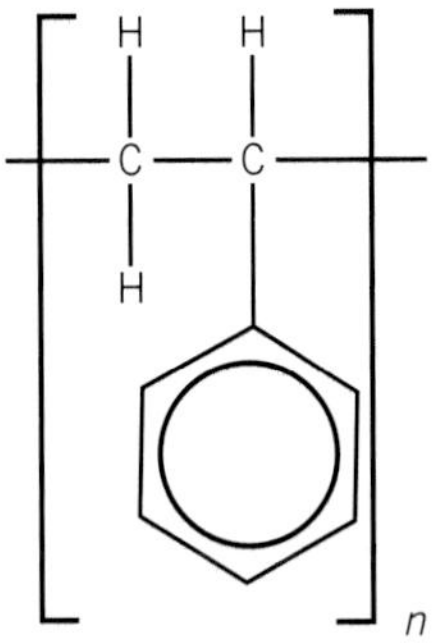 — repeat unit with phenyl group, subscript n	 Used for packaging material and also in food trays and cups due to its thermal insulating properties.
tetrafluoroethene 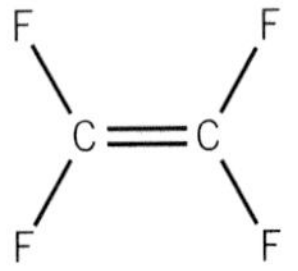 — $CF_2=CF_2$	poly(tetrafluoroethene) (PTFE, Teflon) 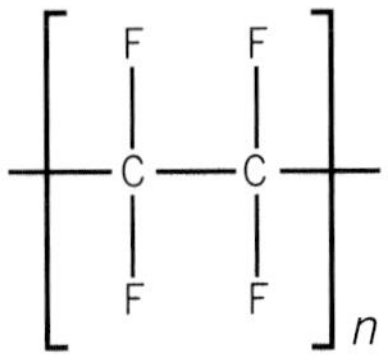 — repeat unit with F, F on each C, subscript n	 Used as coating for non-stick pans, permeable membrane for clothing and shoes, and cable insulation.

Identifying monomers from polymer chains

Once you have identified the repeat unit in a polymer, you should be able to identify the original monomer from which the polymer was made. For example, the structure of poly(methyl methacrylate), a glass substitute used as perspex, is shown in Figure 7 with its repeat unit shaded. The monomer can be identified by copying the shaded repeat unit and changing the single C—C bond in the chain into a double bond.

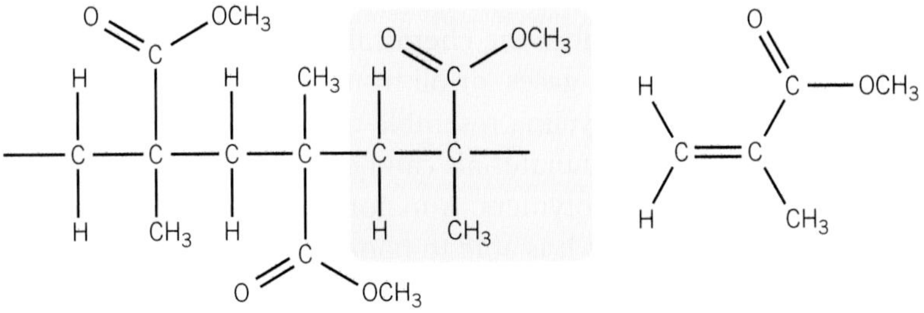

polymer   monomer

▲ **Figure 7** *The structure of the polymer poly(methyl methacrylate) with its corresponding monomer*

13.5 Polymerisation in alkenes

Environmental concerns

Disposing of waste polymers

Polymers are readily available, cheap to purchase, and more convenient for our throwaway society than the alternatives, such as glass bottles, metal dustbins, paper bags, and cardboard packaging.

The lack of reactivity that makes polymers suitable for storing food and chemicals safely also presents chemists with a challenge in their disposal. Many alkene-based polymers are non-biodegradable. The growing amount of polymer waste has serious environmental effects, for example, killing marine life.

> **Study tip**
>
> It does not matter how complicated the polymer looks. Look at the chain and see where it repeats. Then you will be able to identify the monomer from its polymer.

▲ **Figure 8** *Waste polymers can wrap around marine life, suffocating them*

Recycling

Recycling polymers reduces their environmental impact by conserving finite fossil fuels as well as decreasing the amount of waste going to landfill. Discarded polymers have to be sorted by type. The recycling process is undermined if polymers are mixed as this renders the product unusable. Once sorted, the polymers are chopped into flakes, washed, dried, and melted. The recycled polymer is cut into pellets and used by manufacturers to make new products.

PVC recycling

The disposal and recycling of PVC is hazardous due to the high chlorine content and the range of additives present in the polymer. Dumping PVC in landfill is not sustainable and, when burnt, PVC releases hydrogen chloride, a corrosive gas, and other pollutants like toxic dioxins.

Previously, recycling involved grinding PVC and reusing it to manufacture new products. New technology uses solvents to dissolve the polymer. High-grade PVC is then recovered by precipitation from the solvent, and the solvent is used again.

▲ **Figure 9** *Plastic bottles are sorted based on their resin identification codes*

Using waste polymers as fuel

Some polymers are difficult to recycle. As they are derived from petroleum or natural gas, they have a high stored energy value. Waste polymers can be incinerated to produce heat, generating steam to drive a turbine producing electricity. In Sheffield an energy recovery facility incinerates non-recyclable material including waste polymers and uses the heat to generate electricity for the National Grid and to heat buildings across the city centre.

Feedstock recycling

Feedstock recycling describes the chemical and thermal processes that can reclaim monomers, gases, or oil from waste polymers. The products from feedstock recycling resemble those produced from crude oil in refineries. These materials can be used as raw materials for the production of new polymers. A major advantage of feedstock recycling is that it is able to handle unsorted and unwashed polymers.

▲ **Figure 10** *Incinerator in Sheffield. Domestic refuse is burnt and the energy used to heat buildings in Sheffield city centre*

ALKENES 13

Biodegradable and photodegradable polymers

Bioplastics produced from plant starch, cellulose, plant oils, and proteins offer a renewable and sustainable alternative to oil-based products. The use of bioplastics not only protects our environment but also conserves valuable oil reserves.

Biodegradable polymers

Biodegradable polymers are broken down by microorganisms into water, carbon dioxide, and biological compounds. These polymers are usually made from starch or cellulose, or contain additives that alter the structure of traditional polymers so that microorganisms can break them down.

Compostable polymers degrade and leave no visible or toxic residues. Compostable polymers based on poly(lactic acid) (Figure 11) are becoming more common as an alternative to alkene-based polymers.

Supermarket bags made from plant starch can be used as bin liners for food waste so that the waste and bag can be composted together. Compostable plates, cups, and food trays made from sugar cane fibre are replacing expanded polystyrene. As the technology advances, bioplastics are likely to be more extensively used in packaging, electronics, and more fuel-efficient and recyclable vehicles.

Photodegradable polymers

Where the use of plant-based polymers is not possible, photodegradable oil-based polymers are being developed. These polymers contain bonds that are weakened by absorbing light to start the degradation. Alternatively, light-absorbing additives are used.

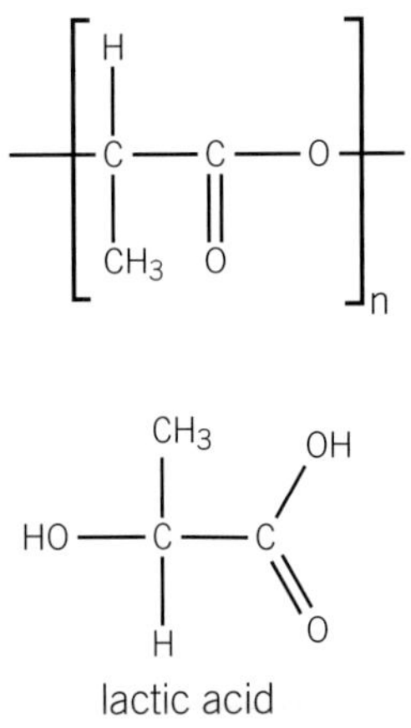

▲ **Figure 11** *A section of poly(lactic acid), showing the repeat unit and its monomer (it polymerises by a different mechanism from the polyalkene addition polymers studied in this topic)*

▲ **Figure 12** *Sugar cane is now used as a starting material in the production of bioplastics*

Summary questions

1 Define the terms:
   a monomer *(1 mark)*
   b repeat unit *(1 mark)*

2 State one atmospheric pollutant caused by the burning of PVC. *(1 mark)*

3 But-2-ene can undergo addition polymerisation. Draw the repeat unit for poly(but-2-ene). *(2 marks)*

4 A section of Orlon, an addition polymer, is shown below.

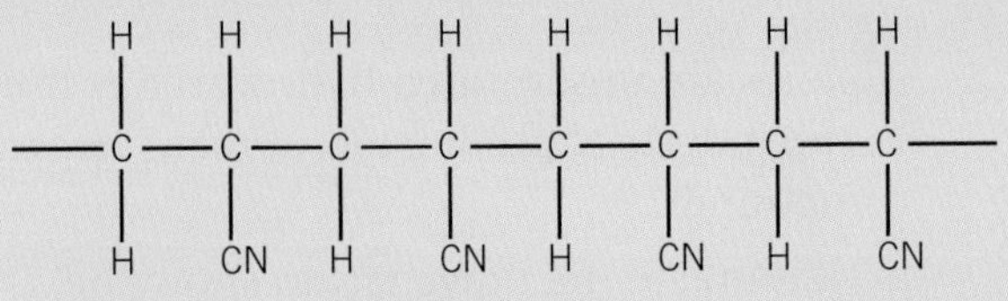

   [structure showing —C(H)(H)—C(H)(CN)—C(H)(H)—C(H)(CN)—C(H)(H)—C(H)(CN)—C(H)(H)—C(H)(CN)—]

   a Identify the repeat unit.
   b Draw the skeletal formula of the monomer that forms Orlon. *(2 marks)*

5 Scientists have developed a way to form a copolymer of two or more alkenes alternating to make an addition polymer. One such polymer is made from phenylethene, $C_6H_5CH{=}CH_2$, and propanenitrile, $CH_3CH{=}CHCN$. Draw a section of this polymer showing two repeat units. *(2 marks)*

Chapter 13 Practice questions

Practice questions

1 Alkenes are unsaturated hydrocarbons. The structures of but-1-ene and methylpropene are shown below.

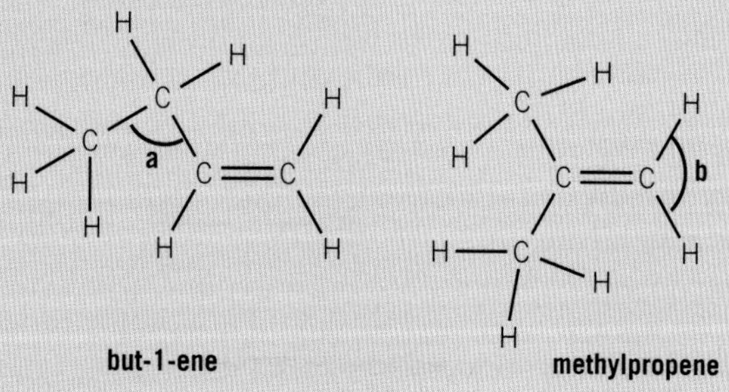

but-1-ene   methylpropene

a What is meant by the terms *unsaturated* and *hydrocarbon*? (2 marks)

b Suggest values for the bond angle **a** in but-1-ene and bond angle **b** in methylpropene. (2 marks)

c Explain, with the aid of a sketch, how p-orbitals are involved in the formation of the C=C double bond. (2 marks)

2 Alkenes undergo addition reactions to form saturated compounds.

a Define the term *electrophile*. (1 mark)

b The reaction between bromine and methylpropene is an electrophilic addition.
Describe, with the aid of curly arrows, the mechanism for this reaction. Show the intermediate and the product along with any relevant dipoles and lone pairs of electrons.

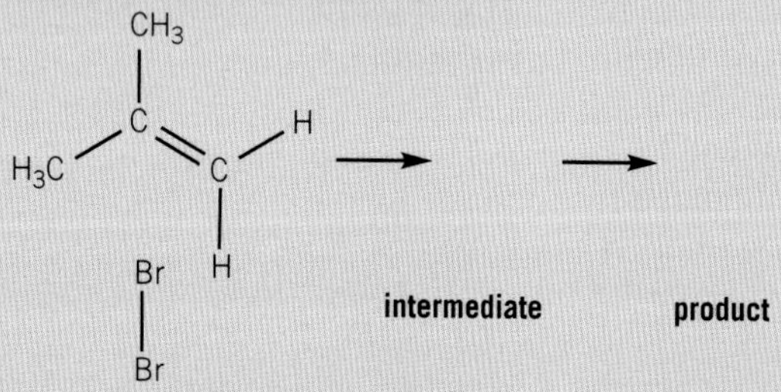

(4 marks)

3 Iodine monobromide, IBr, has a permanent dipole. Alkenes react with IBr in a similar way to the reactions of alkenes with HBr.

a Propene reacts with IBr to make two possible organic products. One of these products is 2-bromo-1-iodopropane.

(i) Using the curly arrow model, complete the mechanism to make 2-bromo-1-iodopropane.

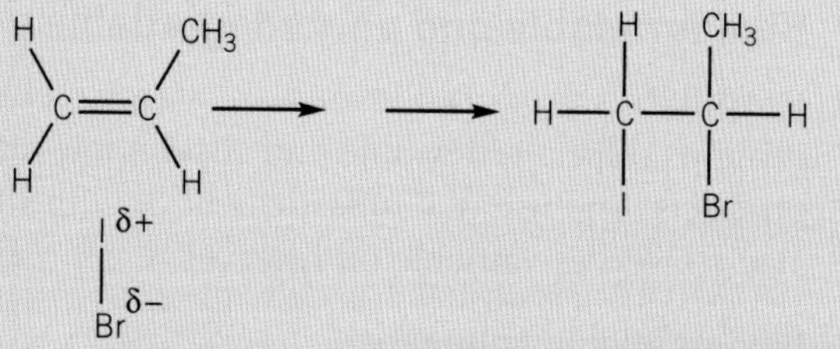

(3 marks)

(ii) What is the name of this mechanism? (1 mark)

(iii) Draw the structure of the other possible organic product of the reaction of propene with IBr. (1 mark)

OCR F322 June 2013 Q4(a)

4 Ethene can be converted into petrochemicals.

a Describe how ethene can be converted into 1,2-dibromoethane, bromoethane and ethanol.
Name and describe the mechanism for the conversion of ethene into 1,2-dibromoethane using the 'curly arrow' model. Include any relevant dipoles. (9 marks)

b Draw and explain the shape of an ethene molecule. State the H—C—H bond angle in ethene. (3 marks)

c Addition polymers are made by the polymerisation of alkenes. *E*-Pent-2-ene can be made into an addition polymer.

(i) Draw the structure of *E*-pent-2-ene. (1 mark)

(ii) Draw the structure of poly(pent-2-ene). Include two repeat units. (1 mark)

OCR F322 June 2012 Q1(d)(e)

5 Alkenes are unsaturated hydrocarbons used in the industrial production of many organic compounds.

a Complete the flowchart to show the organic product formed in each addition reaction of methylpropene. (4 marks)

13 ALKENES

[Reaction scheme showing alkene (H)(H)C=C(CH₃)(CH₃) with H₂/Ni, Br₂, and HBr reactions, producing a mixture of isomers]

b Curly arrows are used in reaction mechanisms to show the movement of electron pairs during chemical reactions.

Use curly arrows to outline the mechanism for the addition reaction of methylpropene with chlorine. The structure of methylpropene has been drawn for you. Include relevant dipoles in your answer.

[Structure of methylpropene: (H)(H)C=C(CH₃)(CH₃)]

(3 marks)

OCR F322 June 2009 Q3

6 In this question, you are asked to suggest structures for several organic compounds.

a Compounds **F**, **G** and **H** are **unbranched** alkenes that are isomers, each with a relative molecular mass of 70.0. Compounds **F** and **G** are E/Z stereoisomers. Compound **H** is a structural isomer of compounds **F** and **G**.

- Explain what is meant by the terms structural isomer and stereoisomer.
- Explain why some alkenes have E/Z isomerism.
- Analyse this information to suggest possible structures for compounds **F**, **G** and **H**.

In your answer you should make clear how each structure fits with the information given above. *(11 marks)*

OCR F322 June 2009 Q7(a)

7 Reduction of a branched alkene, **A**, with hydrogen formed a compound **B**. On analysis, **B** had the composition by mass of **C**, 82.8%; **H**, 17.2% ($M_r = 58$). Reaction of **A** with steam in the presence of a catalyst produced a mixture of two isomers **C** and **D**.

Identify possible structures for compounds **A–D**. Your answer should include displayed formulae for each structure and equations for each reaction. *(9 marks)*

8 Two alkenes, **Q** and **R**, are E and Z isomers respectively. Using hydrogen in the presence of a nickel catalyst, **Q** and **R** can each be converted into the alkane, butane.

Draw the structure of **Q** and of **R**. *(3 marks)*

9 Compound **C** can be polymerised to form compound **E**.

$$H_3C—CH=CH_2$$
C

a State the type of polymerisation. *(1 mark)*
b Name compound **C**. *(1 mark)*
c Draw a section of compound **E**. Show two repeat units. *(1 mark)*

10 Draw the four isomers of C_4H_8 containing a C=C bond and name them. State whether the compounds are structural isomers or stereoisomers of each other. *(6 marks)*

11 Use the Cahn-Ingold-Prelog rules to identify each of the following as E or Z isomers. Draw the structure of the other stereoisomer for each molecule. *(4 marks)*

a (Cl)(H)C=C(H)(C₂H₅)

b (Cl)(H₃C)C=C(Br)(CH₂CH₃)

c (H₃C)(C₂H₅)C=C(H)(CH₂CH₂Cl)

d (H)(H₃C)C=C(CH₂CH₂OH)(CH₂CH₂Cl)

12 3-bromo-4-chloropent-2-ene can undergo addition polymerisation.

Write an equation to illustrate this reaction showing clearly the repeat unit of the polymer. *(2 marks)*

14 ALCOHOLS
14.1 Properties of alcohols
Specification reference: 4.2.1

Learning outcomes
Demonstrate knowledge, understanding, and application of:
- polarity, solubility, and volatility of alcohols
- classification of alcohols.

The alcohol homologous series
Alcohols contain the –OH functional group, known as the hydroxyl group. The hydroxyl group is responsible for both the physical and chemical properties of the alcohols.

The simplest alcohol, methanol, CH_3OH, is used as a high-performance fuel because of its efficient combustion. Methanol is also an important chemical feedstock – the starting material in many industrial syntheses. It can be converted into polymers, paints, solvents, insulation, adhesives, and many other useful products.

The second member of the alcohol homologous series, ethanol, is used primarily in alcoholic drinks and as a fuel, and also finds use as a solvent and a feedstock.

Naming alcohols
You will recall that the suffix -ol is added to the stem name of the longest carbon chain. The position of the alcohol functional group in the chain is indicated using a number.

▲ Figure 1 *Ethanol is used in alcoholic drinks*

 Worked example: Naming an alcohol

What is the IUPAC name of this alcohol?

Step 1: There are four carbons in the longest chain, so the stem is butane.

Step 2: There are two –OH functional groups, so the suffix is -diol.

Step 3: The –OH groups are on carbons 2 and 3, so the infix is -2,3-.

Step 4: The suffix -diol does not start with a vowel. The alkane chain name is not shortened.

Step 5: There is a methyl group (–CH_3) on carbon 2. The prefix 2-methyl is added to the name.

Step 6: The compound is 2-methylbutane-2,3-diol.

Synoptic link
Alcohols are named using the rules introduced in Topic 11.2, Nomenclature of organic compounds.

Physical properties

When you compare the physical properties of alcohols with alkanes of the same number of carbon atoms (e.g., methanol and methane), there are some interesting differences. Alcohols are less volatile, have higher melting points, and greater water solubility than the corresponding alkanes. The differences become much smaller as the length of the carbon chain increases.

These differences can be explained by considering the polarity of the bonds in both the alkanes and alcohols, and the effect that these bonds have on the strength of the intermolecular forces.

- The alkanes have non-polar bonds because the electronegativity of hydrogen and carbon are very similar.
- The alkane molecules are therefore non-polar.
- The intermolecular forces between non-polar molecules are very weak London forces.
- Alcohols have a polar O—H bond because of the difference in electronegativity of the oxygen and hydrogen atoms.
- Alcohol molecules are therefore polar.
- The intermolecular forces will be very weak London forces but there will also be much stronger hydrogen bonds between the polar O—H groups.

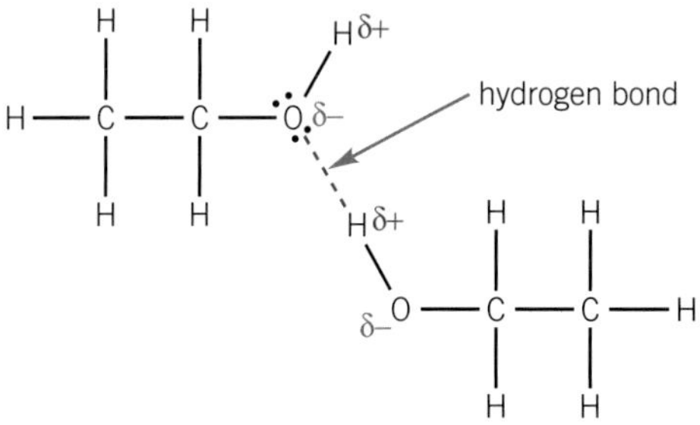

Synoptic link

If you are not sure about electronegativity, polarity, and intermolecular forces you should look back at Topics 6.3, Intermolecular forces, and 12.1, The properties of alkanes.

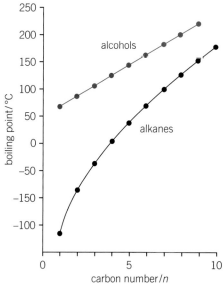

▲ **Figure 3** *The relationship between carbon number and boiling point for the alcohols and the alkanes. As chain length increases, the contribution of the –OH group decreases and the alcohols resemble the alkanes more closely*

▲ **Figure 2** *Hydrogen bonding between two ethanol molecules*

Volatility and boiling points

The difference in the boiling points of the alkanes and alcohols can be seen in Figure 3. In the liquid state, intermolecular hydrogen bonds hold the alcohol molecules together. These bonds must be broken in order to change the liquid alcohol into a gas. This requires more energy than overcoming the weaker London forces in alkanes, so alcohols have a lower volatility than the alkanes with the same number of carbon atoms.

Study tip

Compounds with low boiling points are volatile – they are easily converted from a liquid to a gas. The *higher* the boiling point, the *lower* the volatility.

Solubility in water

A compound that can form hydrogen bonds with water is far more water-soluble than a compound that cannot. Alkanes are non-polar molecules and cannot form hydrogen bonds with water. Alcohols such as methanol and ethanol are completely soluble in water, as hydrogen bonds form between the polar –OH group of the alcohol and the water molecules (Figure 4).

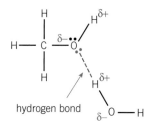

▲ **Figure 4** *Hydrogen bonds form between methanol and water*

14.1 Properties of alcohols

As the hydrocarbon chain increases in size, the influence of the –OH group becomes relatively smaller, and the solubility of longer-chain alcohols becomes more like that of hydrocarbons – solubility decreases.

Ethylene glycol to the rescue

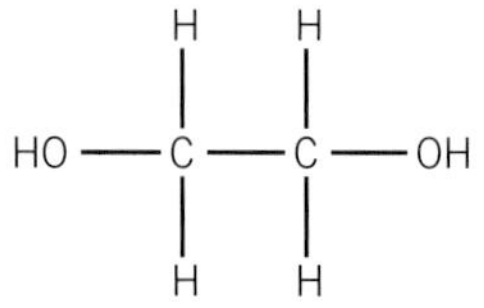

Scraping ice from the windscreen of a car is one of the least pleasant tasks of the early-morning motorist. Spraying a solution of ethane-1,2-diol, also known as ethylene glycol (Figure 6), onto the cold windscreen can melt away the ice in seconds. Water freezes at 0 °C and ethane-1,2-diol at −13 °C, but when the two are mixed the freezing point of the mixture can be as low as −40 °C.

▲ **Figure 6** *Ethane-1,2-diol has two alcohol functional groups in its structure*

▶ **Figure 5** *Deicing fluid, a mixture of a chemical called ethylene glycol and water, is generally heated and sprayed under pressure to remove ice and snow from aircraft*

How does this molecule work its magic? Use a clear diagram in your answer to show the role of intermolecular forces.

Synoptic link

See Topic 13.4, Electrophilic addition in alkenes, for the classification of carbocations as primary, secondary, or tertiary.

Classifying alcohols

Alcohols can be classified as primary, secondary, or tertiary. This classification depends on the number of hydrogen atoms and alkyl groups attached to the carbon atom that contains the alcohol functional group.

Primary alcohols

Methanol and ethanol, the two simplest alcohols in the alcohols homologous series, are both **primary alcohols**.

In a primary alcohol the –OH group is attached to a carbon atom that is attached to *two* hydrogen atoms and *one* alkyl group. (Methanol, with three hydrogen atoms and no carbon atoms attached, is an exception that is still classified as a primary alcohol.)

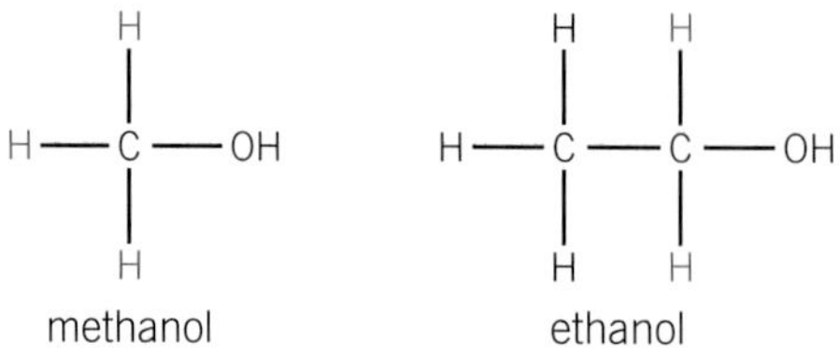

▲ **Figure 7** *The structures of the primary alcohols methanol and ethanol*

Secondary alcohols

In a **secondary alcohol** the –OH group is attached to a carbon atom that is attached to *one* hydrogen atom and *two* alkyl groups. Propan-2-ol and pentan-3-ol are both examples of secondary alcohols.

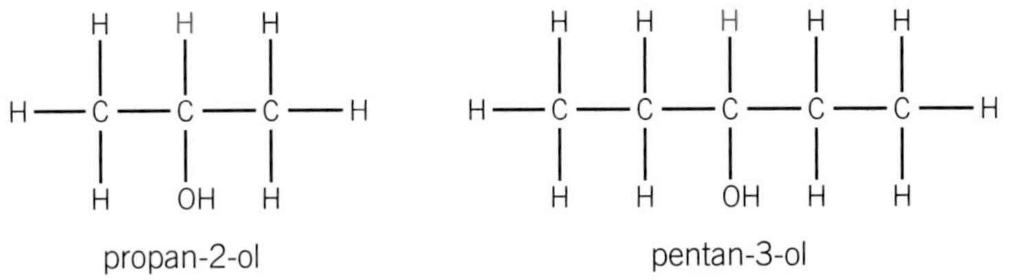

propan-2-ol   pentan-3-ol

▲ **Figure 8** *The structure of two secondary alcohols*

Tertiary alcohols

In a **tertiary alcohol** the –OH group is attached to a carbon atom that is attached to *no* hydrogen atoms and *three* alkyl groups. 2-Methylpropan-2-ol and 2-methylbutan-2-ol are both examples of tertiary alcohols.

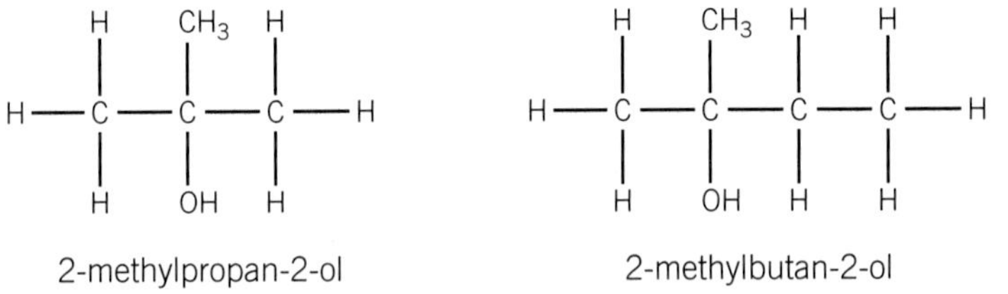

2-methylpropan-2-ol   2-methylbutan-2-ol

▲ **Figure 9** *The structure of two tertiary alcohols*

You must be able to recognise the three different classes of alcohol in order to predict how the alcohol will react with oxidising agents. You will cover oxidation of alcohols in the next topic.

Summary questions

1. **a** Name the following alcohols. *(3 marks)*

   i   ii 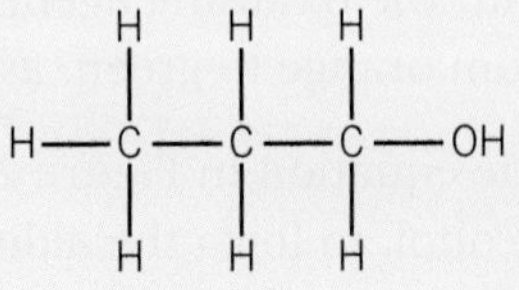   iii

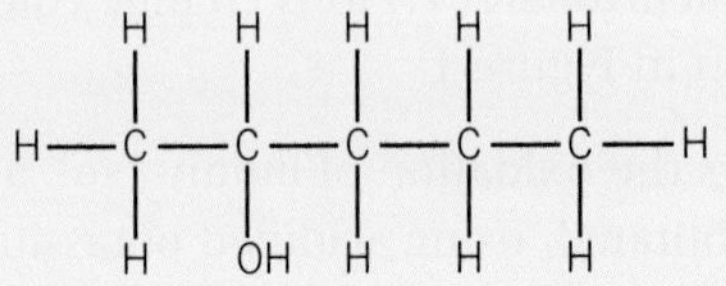

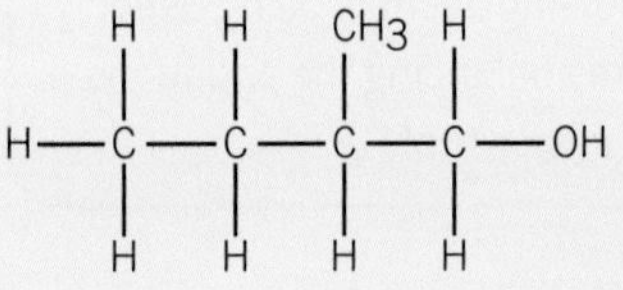

   b Classify the three structures in part (a) as primary, secondary, or tertiary. *(3 marks)*

2. Four of the isomers with the molecular formula $C_4H_{10}O$ are alcohols. Draw skeletal formulae of these four alcohols and classify them as primary, secondary, or tertiary. *(8 marks)*

3. State and explain the trend in the boiling point of the alcohols as the length of the carbon chain increases. *(3 marks)*

4. Show, using a diagram, how hydrogen bonds act between molecules of propan-1-ol and water. *(2 marks)*

5. Propane-1,2,3-triol is more soluble in water than propan-1-ol. Explain why. *(4 marks)*

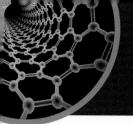

14.2 Reactions of alcohols

Specification reference: 4.2.1

Learning outcomes

Demonstrate knowledge, understanding, and application of:

→ combustion of alcohols
→ oxidation of alcohols
→ elimination from alcohols
→ substitution of alcohols by halide ions.

Combustion of alcohols

Alcohols burn completely in a plentiful supply of oxygen to produce carbon dioxide and water. The equation shows the complete combustion of ethanol.

$$C_2H_5OH(l) + 3O_2(g) \rightarrow 2CO_2(g) + 3H_2O(l)$$

The reaction is exothermic, releasing a large quantity of energy in the form of heat. As the number of carbon atoms in the alcohol chain increases the quantity of heat released per mole also increases.

Oxidation of alcohols

Primary and secondary alcohols can be oxidised by an oxidising agent. The usual oxidising mixture is a solution of potassium dichromate(VI), $K_2Cr_2O_7$, acidified with dilute sulfuric acid, H_2SO_4. If the alcohol is oxidised, the orange solution containing dichromate(VI) ions is reduced to a green solution containing chromium(III) ions (Figure 1).

$$Cr_2O_7^{2-} \rightarrow Cr^{3+}$$
dichromate(VI) ions → chromium(III) ions

Oxidation of primary alcohols

Primary alcohols can be oxidised to either aldehydes or carboxylic acids. The product of the oxidation depends on the reaction conditions used because aldehydes are themselves also oxidised to carboxylic acids.

Preparation of aldehydes

On gentle heating of primary alcohols with acidified potassium dichromate, an aldehyde is formed. To ensure that the aldehyde is prepared rather than the carboxylic acid, the aldehyde is distilled out of the reaction mixture as it forms. This prevents any further reaction with the oxidising agent. The dichromate(VI) ions change colour from orange to green, as shown in Figure 1.

The equation in Figure 2 shows the oxidation of butan-1-ol, a primary alcohol, to form the aldehyde butanal, using acidified potassium dichromate(VI). Notice that [O] is used to indicate the oxidising agent.

▲ **Figure 1** *Oxidation of the primary alcohol butan-1-ol by acidified potassium dichromate(VI) solution, a strong oxidising agent that is reduced to a green solution containing chromium(III) ions*

Study tip

Use [O] in brackets to show that an oxidising agent is being used. It makes balancing the equations much easier, too!

butan-1-ol + [O] →(K₂Cr₂O₇/H₂SO₄, distil)→ butanal + H₂O

oxidising agent

▲ **Figure 2** *The oxidation of butan-1-ol to with acidified potassium dichromate(VI) to form butanal*

Preparation of carboxylic acids

If a primary alcohol is heated strongly under reflux, with an excess of acidified potassium dichromate(VI), a carboxylic acid is formed. Use of an excess of the acidified potassium dichromate(VI) ensures that all of the alcohol is oxidised. Heating under reflux ensures that any aldehyde formed initially in the reacton also undergoes oxidation to the carboxylic acid.

The complete oxidation of butan-1-ol to butanoic acid is shown in Figure 3.

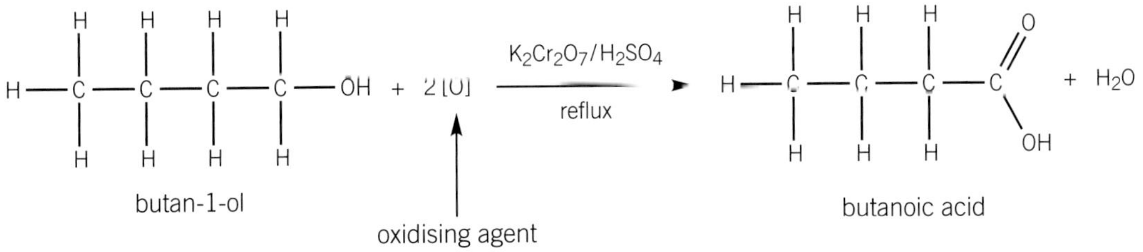

▲ **Figure 3** *The oxidation of butan-1-ol with excess acidified potassium dichromate(VI) to form butanoic acid*

You can see that the conditions of the oxidation of a primary alcohol, such as whether a reagent in excess the conditions and the technique used, influence the product formed.

- When preparing the aldehyde, use distillation to remove the aldehyde from the reaction mixture.
- When preparing the carboxylic acid, heat the alcohol under reflux.

Study tip
Remember the order of oxidation for primary alcohols.

primary alcohol → aldehyde
→ carboxylic acid

Oxidation of secondary alcohols

Secondary alcohols are oxidised to ketones. It is not possible to further oxidise ketones using acidified dichromate(VI) ions.

To ensure the reaction goes to completion, the secondary alcohol is heated under reflux with the oxidising mixture. The dichromate(VI) ions once again change colour from orange to green. The oxidation of propan-2-ol to propanone is shown in Figure 4.

Synoptic link
Distillation and heat under reflux are practical techniques that you will need to use as part of your practical course. You will meet both in Topic 16.1, Practical techniques in organic chemistry.

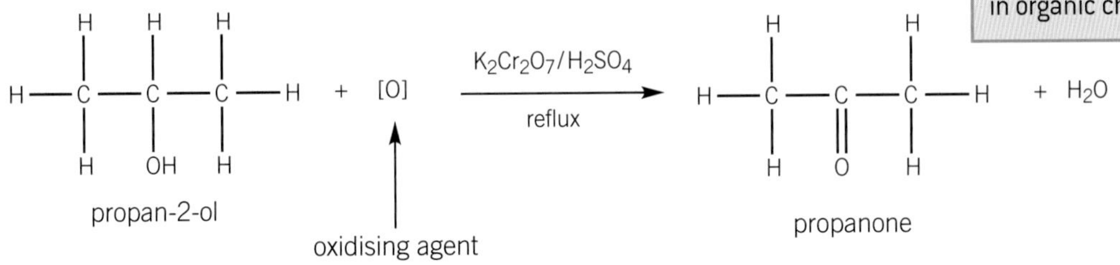

▲ **Figure 4** *The oxidation of propan-2-ol with excess acidified potassium dichromate(VI) to form propanone*

Oxidation of tertiary alcohols

Tertiary alcohols do not undergo oxidation reactions. The acidified dichromate(VI) remains orange when added to a tertiary alcohol.

Study tip
On oxidation:

primary alcohol → aldehyde
→ carboxylic acid

secondary alcohol → ketone

tertiary alcohol → *no reaction*

14.2 Reactions of alcohols

Synoptic link
You will recall from Topic 11.5, Introduction to reaction mechanisms, that an elimination reaction involves the removal of a small molecule from a molecule, forming two products.

Summary questions

1. Alcohols can be oxidised using acidified potassium dichromate. show the full displayed formula of the organic products formed when acidified potassium dichromate is heated under reflux with:
   a. butan-2-ol *(1 mark)*
   b. hexan-1-ol *(1 mark)*
   c. ethane-1,2-diol *(1 mark)*
   d. butane-1,2-diol *(1 mark)*

2. Write the overall equation for the reaction between butan-1-ol, sodium iodide and sulfuric acid. Name the organic product and state the type of reaction taking place. *(3 marks)*

3. Draw the full displayed formula of the organic product of the reactions between:
   a. propan-2-ol and concentrated sulfuric acid *(1 mark)*
   b. 2-methylpropan-2-ol and a mixture of sodium chloride and concentrated sulfuric acid. *(1 mark)*

Dehydration of alcohols

Dehydration is any reaction in which a water molecule is removed from the starting material. An alcohol is heated under reflux in the presence of an acid catalyst such as concentrated sulfuric acid, H_2SO_4, or concentrated phosphoric acid, H_3PO_4. The product of the reaction is an alkene.

Dehydration of an alcohol is an example of an elimination reaction.

The equation for the dehydration of the alcohol cyclohexanol is shown in Figure 7. The products of the reaction are cyclohexene and water.

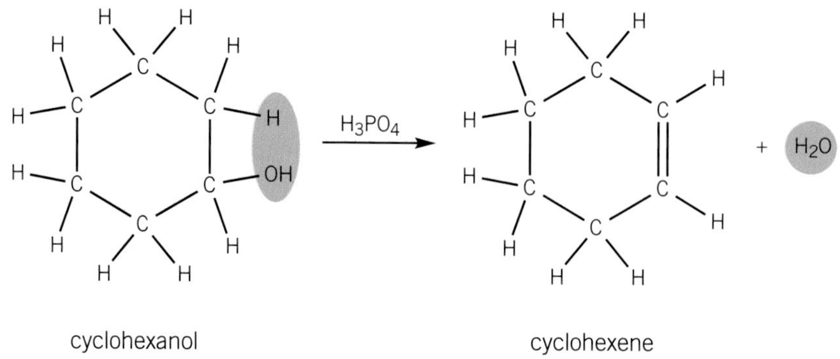

▲ **Figure 7** *The dehydration of cyclohexanol to form cyclohexene*

Substitution reactions of alcohols

Alcohols react with hydrogen halides to form haloalkanes. When preparing a haloalkane, the alcohol is heated under reflux with sulfuric acid and a sodium halide the hydrogen bromide is formed *in situ* (in place).

$$NaBr(s) + H_2SO_4(aq) \rightarrow NaHSO_4(aq) + HBr(aq)$$

The HBr formed reacts with the alcohol to produce the haloalkane. The reaction of propan-2-ol with HBr is shown in Figure 8.

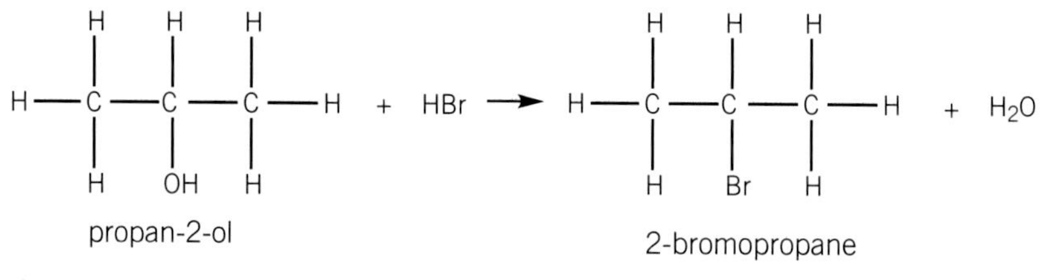

▲ **Figure 8** *Substitution of the alcohol group –OH in propan-2-ol*

The overall equation for the reaction of propan-2-ol with H_2SO_4 and NaBr is shown below.

$$CH_3CHOHCH_3 + NaBr + H_2SO_4 \rightarrow CH_3CHBrCH_3 + NaHSO_4 + H_2O$$

ALCOHOLS 14

Practice questions

1. There are four structural isomers that are alcohols with the molecular formula $C_4H_{10}O$.
   a. Draw the displayed formula of these alcohols and classify them as primary, secondary or tertiary. *(4 marks)*
   b. One of these alcohols does not react with acidified dichromate.
      Identify this alcohol *(1 mark)*

2. Identify, by name and structure, the alcohol needed to be reacted with acidified potassium dichromate solution to make each of the following oxidation products.
   a. hexanal *(2 marks)*
   b. 2-methylbutanoic acid *(2 marks)*
   c. 3-methylheptan-2-one *(2 marks)*

3. Alcohols are soluble in water due to their ability to hydrogen bond with water molecules; Draw a diagram to show the hydrogen bonding between water and a molecule of propan-1-ol. *(2 marks)*

4. Describe the oxidation reactions of propan-1-ol when using a suitable oxidising agent.
   Indicate how the use of different reaction conditions can control which organic product forms.
   Include reagents, observations and equations in your answer. In your equations, use structural formulae and use [O] to represent the oxidising agent. *(6 marks)*

   OCR F322 June 2014 Q7 (d)

5. The skeletal formulae of six alcohols, **C**, **D**, **E**, **F**, **G**, and **H**, are shown below.

   a. (i) Which two alcohols are structural isomers of one another? *(1 mark)*
      (ii) Which alcohol is a tertiary alcohol? *(1 mark)*
      (iii) Which alcohol can be oxidised to a carboxylic acid using acidified $K_2Cr_2O_7$? *(1 mark)*
   b. (i) What is the molecular formula of alcohol **G**? *(1 mark)*
      (ii) What is the name of alcohol **C**? *(1 mark)*
   c. The alcohols are members of a homologous series.
      Explain the term homologous series. *(2 marks)*

      OCR F322 June 2013 Q2 (a–c)

6. Pentan-2-ol is a secondary alcohol.
   a. Pentan-2-ol can be converted into three alkenes, **A**, **B**, and **C**, by the elimination of water.
      - Two of the alkenes, **A** and **B**, are stereoisomers.
      - The third alkene, **C**, is a structural isomer of both **A** and **B**.

      This elimination often uses a catalyst.
      (i) What is a suitable catalyst for this reaction? *(1 mark)*
      (ii) Construct an equation, using molecular formulae, for the elimination of water from pentan-2-ol. *(1 mark)*
      (iii) Explain what is meant by the terms structural isomers and stereoisomers. *(4 marks)*
      (iv) Draw the structures of stereoisomers **A** and **B** and the structure of compound **C**. *(3 marks)*
      (v) Stereoisomers **A** and **B** show E/Z isomerism. State two features of these molecules that enable them to show E/Z isomerism. *(2 marks)*
   b. Pentan-2-ol can be oxidised by heating under reflux with acidified aqueous potassium dichromate(VI).
      Write the equation for this oxidation.
      Use skeletal formulae.
      Use [O] to represent the oxidising agent. *(2 marks)*
   c. Pentan-1-ol can also be oxidised but it gives two different products.
      Show the structures of the two organic products formed. State the reagents and different conditions for each oxidation. *(3 marks)*

      OCR F322 Jan 2013 Q3

229

15 HALOALKANES

15.1 The chemistry of the haloalkanes

Specification reference: 4.2.2

Learning outcomes

Demonstrate knowledge, understanding, and application of:

→ hydrolysis of haloalkanes
→ nucleophiles
→ nucleophilic substitution
→ rates of hydrolysis of primary haloalkanes.

Naming the haloalkanes

Haloalkanes are compounds containing the elements carbon, hydrogen, and at least one halogen. When naming haloalkanes, a prefix is added to the name of the longest chain to indicate the identity of the halogen (Table 1). When two or more halogens are present in a structure they are listed in alphabetical order.

In this chapter you will mainly learn about aliphatic haloalkanes, in which the halogen is joined to a straight or branched carbon chain. These compounds, like alcohols, can be classified as primary, secondary, and tertiary.

▼ **Table 1** *Prefixes used to represent the different halogens in haloalkanes*

Halogen	Prefix
F	fluoro-
Cl	chloro-
Br	bromo-
I	iodo-

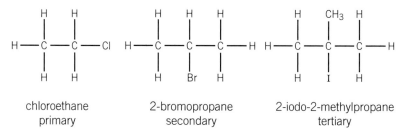

▲ **Figure 1** *The structure of some primary, secondary and tertiary haloalkanes*

Reactivity of the haloalkanes

Haloalkanes have a carbon–halogen bond in their structure. Halogen atoms are more electronegative than carbon atoms. The electron pair in the carbon–halogen bond is therefore closer to the halogen atom than the carbon atom. The carbon–halogen bond is polar (Figure 2).

In haloalkanes the carbon atom has a slightly positive charge and can attract species containing a lone pair of electrons. Species that donate a lone pair of electrons are known as **nucleophiles**. A nucleophile is an atom or group of atoms that is attracted to an electron deficient carbon atom, where it donates a pair of electrons to form a new covalent bond.

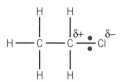

▲ **Figure 2** *Polarity in a carbon–chlorine bond. The bonding pair of electrons is closer to the chlorine atom than the carbon atom*

Nucleophiles include:

- hydroxide ions, :OH$^-$
- water molecules, H$_2$O:
- ammonia molecules, :NH$_3$.

Study tip

You can define a nucleophile as an electron pair donor.

Synoptic link

You should recall polarity and electronegativity from Topic 6.2, Electronegativity and polarity.

HALOALKANES 15

When a haloalkane reacts with a nucleophile, the nucleophile replaces the halogen in a substitution reaction. A new compound is produced containing a different functional group. The reaction mechanism is **nucleophilic substitution**.

Nucleophilic substitution in the haloalkanes

Primary haloalkanes undergo nucleophilic substitution reactions with a variety of different nucleophiles to produce a wide range of different compounds. Substitution is a reaction in which one atom or group of atoms is replaced by another atom or group of atoms.

Hydrolysis

Hydrolysis is a chemical reaction involving water or an aqueous solution of a hydroxide that causes the breaking of a bond in a molecule. This results in the molecule being split into two products.

In the hydrolysis of a haloalkane, the halogen atom is replaced by an –OH group. This is an example of a nucleophilic substitution reaction which takes place as follows:

1. The nucleophile, OH^-, approaches the carbon atom attached to the halogen on the opposite side of the molecule from the halogen atom (Figure 3).
2. This direction of attack by the OH^- ion minimises repulsion between the nucleophile and the $\delta-$ halogen atom.
3. A lone pair of electrons on the hydroxide ion is attracted and donated to the $\delta+$ carbon atom.
4. A new bond is formed between the oxygen atom of the hydroxide ion and the carbon atom.
5. The carbon–halogen bond breaks by heterolytic fission.
6. The new organic product is an alcohol. A halide ion is also formed.

The mechanism for the reaction is shown in Figure 3.

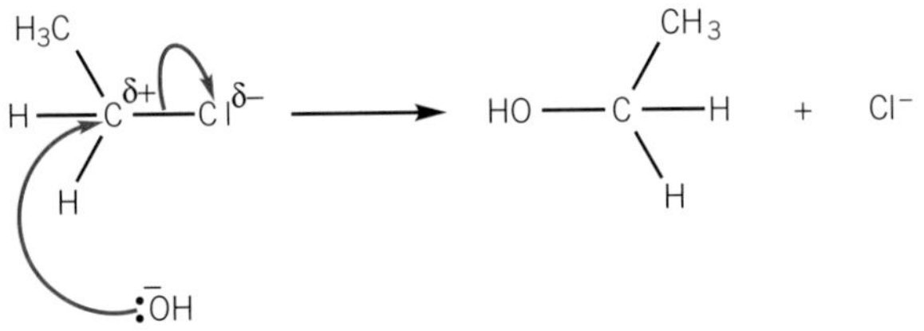

▲ **Figure 3** *Nucleophilic substitution mechanism for the hydrolysis of chloroethane*

> **Study tip**
> It is important to show the curly arrows starting and ending in the correct places when you draw a mechanism. Curly arrows show the movement of electron pairs. Curly arrows start from either a lone pair or a bonding pair of electrons and point to the atom to which the electron pair moves.

Haloalkanes can be converted to alcohols using aqueous sodium hydroxide. The reaction is very slow at room temperature so the mixture is heated under reflux to obtain a good yield of product. The equation for the hydrolysis of 1-bromobutane is shown below.

$$CH_3CH_2CH_2CH_2Br + NaOH \rightarrow CH_3CH_2CH_2CH_2OH + NaBr$$
1-bromobutane → butan-1-ol

15.1 The chemistry of the haloalkanes

Hydrolysis and carbon–halogen bond strength

In hydrolysis, the carbon–halogen bond is broken and the –OH group replaces the halogen in the haloalkane. The rate of hydrolysis depends upon the strength of the carbon–halogen bond in the haloalkane. The bond enthalpies for the carbon–halogen bonds are shown in Figure 4.

> **Synoptic link**
>
> You will recall from Topic 5.3, Covalent bonding, that the larger the value of the average bond enthalpy, the stronger the covalent bond.

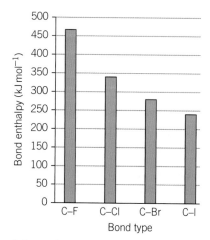

▲ Figure 4 *Graph showing the bond enthalpies for the carbon–halogen bond*

You can see from the graph that the C—F bond is the strongest carbon–halogen bond and the C—I bond the weakest. Less energy is required to break the C—I bond than other carbon–halogen bonds. From these bond enthalpies we can predict that:

- iodoalkanes react faster than bromoalkanes
- bromoalkanes react faster than chloroalkanes
- fluoroalkanes are unreactive as a large quantity of energy is required to break the C—F bond.

You can test out some of these predictions by carrying out an experiment to hydrolyse some haloalkanes.

Measuring the rate of hydrolysis of primary haloalkanes

In this experiment you compare the rate of hydrolysis of three haloalkanes: 1-chlorobutane, 1-bromobutane, and 1-iodobutane.

The general equation for hydrolysis of these haloalkanes with water is shown in the equation below. X represents any of the halogens.

$$CH_3CH_2CH_2CH_2X + H_2O \rightarrow CH_3CH_2CH_2CH_2OH + H^+ + X^-$$

The rate of each reaction can be followed by carrying out the reaction in the presence of aqueous silver nitrate. As the reaction takes place halide ions, $X^-(aq)$, are produced which react with $Ag^+(aq)$ ions to form a precipitate of the silver halide.

$$Ag^+(aq) + X^-(aq) \rightarrow AgX(s)$$
Precipitate of silver halide

> **Synoptic link**
>
> See Topic 8.3, Qualitative analysis, for details of the formation of the silver halide precipitates using aqueous silver nitrate.

The nucleophile in the reaction is water, which is present in the aqueous silver nitrate. Haloalkanes are insoluble in water, and the reaction is carried out in the presence of an ethanol solvent. Ethanol allows water and the haloalkane to mix and produce a single solution rather than two layers.

Hydrolysis of haloalkanes

1. Set up three test tubes as follows:
   - Tube 1: Add 1 cm³ of ethanol and two drops of 1-chlorobutane
   - Tube 2: Add 1 cm³ of ethanol and two drops of 1-bromobutane
   - Tube 3: Add 1 cm³ of ethanol and two drops of 1-iodobutane

2. Stand the test tubes in a water bath at 60 °C.

3. Place a test tube containing 0.1 mol dm⁻³ silver nitrate in the water bath and allow all tubes to reach a constant temperature.

4. Add 1 cm³ of the silver nitrate quickly to each of the test tubes. Immediately start a stop-clock.

5. Observe the test tubes for five minutes and record the time taken for the precipitate to form.

You should observe the results shown in Table 2 and Figure 5.

▲ **Figure 5** *The precipitates formed when the haloalkanes 1-chlorobutane, 1-bromobutane, and 1-iodobutane react with aqueous silver nitrate in ethanol*

▼ **Table 2** *Observations from the hydrolysis of the haloalkanes*

Haloalkane	Observation
1-chlorobutane	A white precipitate forms very slowly.
1-bromobutane	A cream precipitate forms slower than with 1-iodobutane but faster than with 1-chlorobutane.
1-iodobutane	A yellow precipitate forms rapidly.

Why is it important that the test tubes are placed in a water bath at 60 °C at the same time.

These observations are explained by considering the bond enthalpies of the carbon–halogen bonds. The compound with the slowest rate of reaction is the one that has the strongest carbon–halogen bond:

- 1-Chlorobutane reacts slowest and the C—Cl bond is the strongest
- 1-Iodobutane reacts fastest and the C—I bond is the weakest.

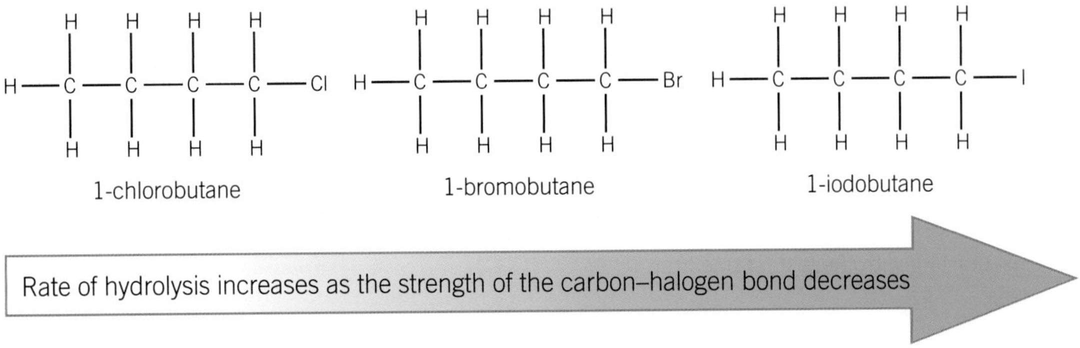

1-chlorobutane   1-bromobutane   1-iodobutane

Rate of hydrolysis increases as the strength of the carbon–halogen bond decreases →

▲ **Figure 6** *The pattern in the rates of hydrolysis of the haloalkanes*

15.1 The chemistry of the haloalkanes

Hydrolysis of primary, secondary, and tertiary haloalkanes

You have just learnt that the rate of hydrolysis depends upon the strength of the carbon–halogen bond, but this does not explain why three structural isomers of C_4H_9Br hydrolyse at different rates (Table 3).

▼ **Table 3** *Relative rate of hydrolysis for some bromoalkanes*

Formula of bromoalkane	Relative rate of reaction
$CH_3CH_2CH_2CH_2Br$	slowest
$CH_3CH_2CHBrCH_3$	
$(CH_3)_3CBr$	fastest

Clearly the strength of the carbon–halogen bond is not the only factor that influences the rate of hydrolysis. If you examine the structures of the three haloalkanes, you will notice that one haloalkane is primary, one is secondary and one is tertiary (Figure 7).

The data shows that the tertiary haloalkane is hydrolysed fastest, whilst hydrolysis of the primary haloalkane is the slowest. The main reason lies with the reaction mechanism. A primary haloalkane will react by a one-step mechanism (look back at Figure 3), whereas a tertiary haloalkane reacts by a two-step mechanism.

- In the first step, the carbon–halogen bond of the tertiary haloalkane breaks by heterolytic fission, forming a tertiary carbocation and a halide ion.
- In the second step, a hydroxide ion attacks the carbocation to form the organic product.

The increased rate and different can be explained by the increased stability of the tertiary carbocation compared to that of the primary carbocation.

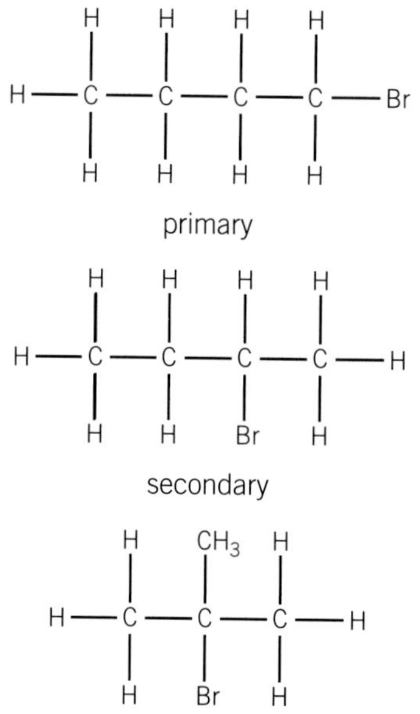

▲ **Figure 7** *The structures of the primary, secondary, and tertiary bromoalkanes*

1. Predict, using curly arrows, the mechanism for the reaction of $(CH_3)_3CBr$ with OH^- ions.
2. Referring back to the reasons behind Markownikoff's rule in Topic 13.4, explain why a tertiary carbocation is more stable than a primary carbocation.

Summary questions

1. Explain why a carbon–halogen bond is polar. *(2 marks)*

2. Show the mechanism for the reaction between 1-chloropropane and sodium hydroxide. *(2 marks)*

3. Place the following haloalkanes in the order of their rates of hydrolysis, starting with the fastest one first. Explain your answer.
   A $CH_3CH_2CH_2Cl$
   B $CH_3CH_2CH_2I$
   C $CH_3CH_2CH_2Br$

4. Ethanol reacts with sodium metal to form an ethoxide ion, $CH_3CH_2O^-$. The ethoxide ion acts as a nucleophile.
   a. Predict the structure of the organic product formed when the ethoxide ion reacts with 1-bromopropane.
   b. Show the mechanism of this nucleophilic substitution reaction. *(3 marks)*

15.2 Organohalogen compounds in the environment

Specification reference: 4.2.2

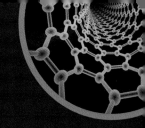

Uses of organohalogen compounds

Organohalogen compounds are molecules that contain at least one halogen atom joined to a carbon chain. They have found many practical uses, some of which are listed in Table 1.

Organohalogen compounds are also used in many pesticides.

Organohalogen compounds are rarely found in nature and, as they are not broken down naturally in the environment, they have become the focus of some concern.

The ozone layer

The ozone layer is found at the outer edge of the stratosphere, at a height that varies from about 10 to 40 km above the Earth's surface. Only a tiny fraction of the gases making up the ozone layer is ozone, but this is enough to absorb most of the biologically damaging ultraviolet radiation (called UV-B) from the Sun's rays, allowing only a small amount to reach the Earth's surface. UV-B radiation is the radiation most commonly linked to sunburn and much research has been carried out on its harmful effects. It is feared that continued depletion of the ozone layer will allow more UV-B radiation to reach the Earth's surface. This is very bad news for living organisms, leading to increased genetic damage and a greater risk of skin cancer in humans.

In the stratosphere, ozone is continually being formed and broken down by the action of ultraviolet (UV) radiation. Initially very high energy UV breaks oxygen molecules into oxygen radicals, O:

$$O_2 \rightarrow 2O$$

A steady state is then set up involving O_2 and the oxygen radicals in which ozone forms and then breaks down. In this steady state, the rate of formation of ozone is the same as the rate at which it is broken down:

$$O_2 + O \rightleftharpoons O_3$$

Human activity, especially in the production and use of chlorofluorocarbons (CFCs), has upset this delicate equilibrium.

CFCs and the ozone layer

Until recently, CFCs and HCFCs were the most common compounds used as refrigerants, in air-conditioning units, and as aerosol propellants. CFCs are very stable because of the strength of the carbon–halogen bonds within their molecules. In 1973 two chemists, Frank Sherwood Rowland and Mario Molina, began to look at the

Learning outcomes

Demonstrate knowledge, understanding, and application of:

→ uses of organohalogen compounds
→ the ozone layer
→ breakdown of the ozone layer by radicals.

▼ **Table 1** Uses of organohalogen compounds

Use	Examples of compounds
general solvents	$CHCl_3$
dry cleaning solvents	$C_2H_2Cl_2$, C_2HCl_3
making polymers	C_2H_3Cl, C_2F_4
flame retardants	CF_3Br
refrigerants	F_2CCl_2, $HCClF_2$, $HCCl_2F$

Study tip

Though O is not drawn with the typical 'radical dot', it is still a radical.

15.2 Organohalogen compounds in the environment

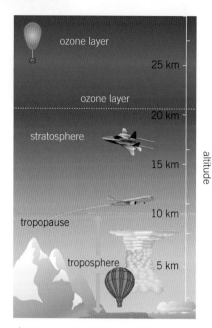

▲ **Figure 1** *Earth's atmosphere is divided into five layers. We exist in the lowest level, the troposphere. The stratosphere is the next layer up, and includes the ozone layer*

impact of CFCs on the Earth's atmosphere. They concluded that CFCs remain stable until they reach the stratosphere (Figure 1). Here the CFCs begin to break down, forming chlorine radicals, which are thought to catalyse the breakdown of the ozone layer. In 1995, Sherwood and Molina were awarded a Nobel Prize for their work in discovering this ozone depletion.

How do CFCs deplete the ozone layer?

The stability of CFCs, due to the strength of their carbon–halogen bonds, means that CFCs have a long residence time in the troposphere. It may take them many years to reach the stratosphere. Once in the stratosphere UV radiation provides sufficient energy to break a carbon–halogen bond in CFCs by homolytic fission to form radicals. The C—Cl bond has the lowest bond enthalpy and so is the bond that breaks.

As radiation initiates the breakdown, this process is called photodissociation. The photodissociation of CF_2Cl_2 is shown below.

$$CF_2Cl_2 \rightarrow CF_2Cl\bullet + Cl\bullet$$

The chlorine radical formed, $Cl\bullet$, is a very reactive intermediate. It can react with an ozone molecule, breaking down the ozone into oxygen. This breakdown occurs by a two-step process:

propagation step 1 $Cl\bullet + O_3 \rightarrow ClO\bullet + O_2$

propagation step 2 $ClO\bullet + O \rightarrow Cl\bullet + O_2$

The overall equation is $O_3 + O \rightarrow 2O_2$

Propagation step 2 regenerates a chlorine radical, which can attack and remove another molecule of ozone in propagation step 1. The two propagation steps repeat in a cycle over and over again in a chain reaction. It has been estimated that a single CFC molecule can promote the breakdown of 100 000 molecules of ozone.

Are CFCs responsible for all ozone-depleting reactions?

The simple answer is no. Other radicals also catalyse the breakdown of ozone. Nitrogen oxide radicals are formed naturally during lightning strikes, and also as a result of aircraft travel in the stratosphere.

Nitrogen oxide radicals cause the breakdown of ozone by a mechanism similar to that involving chlorine radicals.

propagation step 1 $NO\bullet + O_3 \rightarrow NO_2\bullet + O_2$

propagation step 2 $NO_2\bullet + O \rightarrow NO\bullet + O_2$

The overall equation is the same as with chlorine radicals, showing that the radicals act as a catalyst for the process.

$$O_3 + O \rightarrow 2O_2$$

> **Synoptic link**
>
> This photodissociation is the initiation step in a chain reaction. See Topic 12.2, Chemical reactions of alkanes, to review the reaction mechanism of radical substitution, which is similar.

▲ **Figure 2** *Lightning is a major source of NO radicals, which are an important contributor to ozone depletion*

HALOALKANES 15

The end of the road for organohalogen compounds?

In 1987, the Montreal Protocol was signed, which introduced steps for the complete removal of CFCs in all but a limited number of products where no suitable alternative can be found. Research into alternatives for refrigeration and air conditioning units have led to the development of coolants that use hydrocarbons, ammonia, or even carbon dioxide. For sprays, some companies have produced pump-action spray dispensers to replace products in aerosol form. Where aerosols are still being used, the propellant is most likely to be a hydrocarbon such as butane.

Brominated flame retardants (BFRs) are organobromine compounds commonly used by the electronics industry in components such as circuit boards, outer coverings, and cables to reduce the flammability of the product. Organobromine compounds are currently under close scrutiny as scientists suspect that they are toxins that may interfere with the effective function of the human endocrine system.

Both Apple and Dell claim to have reduced or eliminated environmentally damaging flame retardants from their products and to have stopped using the organohalogen polymer, PVC.

▲ **Figure 3** *Organohalogen compounds found use in aerosols*

The following organohalogen compounds have all found a use over the last 50 years. Draw the structures of each compound.
- 1,1,2-trichloro-1,2,2-trifluoroethane (a CFC)
- 1,2,5,6,9,10-hexabromocyclododecane (a BFR with 12 carbon atoms)
- PVC

Summary questions

1. List three sources of CFCs in the atmosphere. *(1 mark)*

2. Name the layer of the atmosphere above the troposphere. *(1 mark)*

3. Explain why ozone depletion should be of major concern to humans. *(2 marks)*

4. Write equations to show how the HCFC $HCClF_2$ breaks down ozone in the upper atmosphere. *(2 marks)*

5. OH• radicals are known to deplete the ozone layer. Complete the two propagation steps given below and write an overall equation for the process.

   Step 1   ___ + O_3 → ___ + O_2

   Step 2   ___ + O_3 → ___ + O_2 *(3 marks)*

Chapter 15 Practice questions

Practice questions

1 Name each of the following haloalkanes

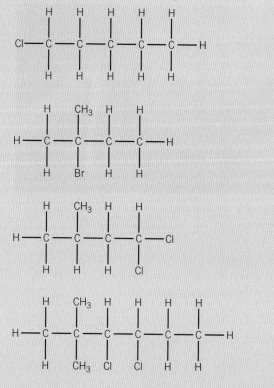

(4 marks)

2 Nucleophilic substitution takes place when a bromoalkane is hydrolysed by heating with an alkali in aqueous solution.

   a Explain the term nucleophilic substitution. (2 marks)
   b Write an equation for the reaction of 2-bromopropane with a hydroxide ion in aqueous solution. (1 mark)
   c Outline the mechanism for this reaction. (3 marks)
   d State and explain what would happen to the rate of the hydrolysis reaction if 2-bromopropane was replaced with 2-iodopropane. (1 mark)

3 Haloalkanes are useful synthetic reagents for the preparation of many important chemicals. Some reactions of 1-chlorobutane is shown below:

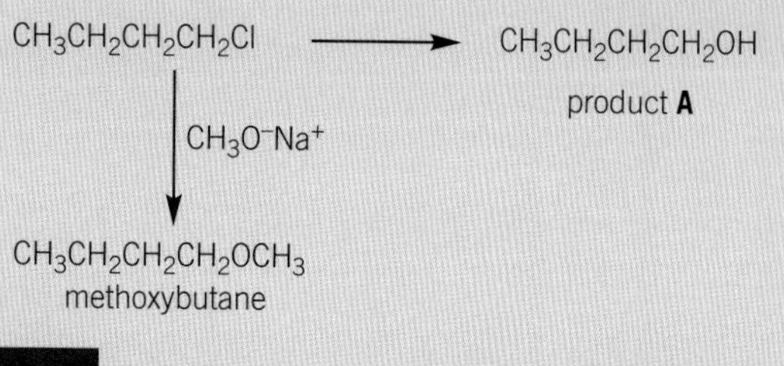

   $CH_3CH_2CH_2CH_2Cl \longrightarrow CH_3CH_2CH_2CH_2OH$ product **A**

   $\downarrow CH_3O^-Na^+$

   $CH_3CH_2CH_2CH_2OCH_3$
   methoxybutane

   a Write an equation for the preparation of product **A** from 1-chlorobutane and name product **A**. (1 mark)
   b The formation of product **A** is an example of a hydrolysis reaction.
      Name the type of mechanism for the hydrolysis of haloalkanes. (1 mark)
   c State and explain the effect on the rate of hydrolysis of replacing 1-chlorobutane with 1-bromobutane. (1 mark)
   d Methoxybutane can be made from 1-chlorobutane and sodium methoxide. The methoxide ion acts as a nucleophile.
      Suggest a mechanism for this reaction. (3 marks)

4 Haloalkane, **H**, has a relative molecular mass of 127 and has the following composition by mass:

   C, 37.8 %; H, 6.3%; Cl, 55.9%.

   a Deduce the empirical formula and molecular formula of **H**. (3 marks)
   b Compound **H** reacts with an excess of aqueous sodium hydroxide under reflux to form compound **J**.
      What is the molecular formula of **J**. (1 mark)
   c Write an equation for the formation of **J** from **H**. (1 mark)

5 In the stratosphere, nitrogen monoxide, NO, is linked with ozone depletion.
   Complete the equations below that describe how NO contributes to ozone depletion.
   Step 1: $NO + O_3 \longrightarrow$ .... + ...
   Step 2: $NO_2 +$ .... $\longrightarrow NO +$ ....
   Overall: .... + ... $\longrightarrow 2O_2$
   (3 marks)
   OCR F322 June 2012 Q4(d)

6 A student carried out an investigation to compare the rates of hydrolysis of 1-iodopropane and 1-bromopropane. The student heated hot aqueous sodium hydroxide with each haloalkane and found that 1-iodopropane was hydrolysed faster. The equation for the reaction with 1-iodopropane is shown below.

   $CH_3CH_2CH_2I + OH^- \longrightarrow CH_3CH_2CH_2OH + I^-$

HALOALKANES 15

a (i) Outline the mechanism for this hydrolysis of 1-iodopropane.

Show curly arrows and relevant dipoles. *(3 marks)*

(ii) State the name of this type of mechanism. *(1 mark)*

b Explain why 1-iodopropane is hydrolysed faster than 1-bromopropane. *(1 mark)*

OCR F322 June 2009 Q6

7 Haloalkanes are polar molecules and react with nucleophiles.

a The displayed formula of chloromethane is shown below. Label the dipole on the C—Cl bond.

```
      H
      |
  H — C — Cl
      |
      H
```
(1 mark)

b Chloromethane is hydrolysed by a solution of sodium hydroxide as shown in the equation below.

$CH_3Cl + OH^- \rightarrow CH_3OH + Cl^-$

(i) State the solvent in which the sodium hydroxide is dissolved. *(1 mark)*

(ii) State and explain the role of the hydroxide ion, OH^-, in this reaction. *(1 mark)*

(iii) Show, with the aid of curly arrows, the mechanism of this hydrolysis. *(3 marks)*

(iv) What would happen to the rate of hydrolysis if chloromethane were replaced by iodomethane? Explain your answer. *(1 mark)*

8 This question is about halogenated hydrocarbons.

a Haloalkanes undergo nucleophilic substitution reactions with ammonia to form amines. Amines contain the $-NH_2$ functional group. For example, 1-bromopropane reacts with ammonia to form propylamine, $CH_3CH_2CH_2NH_2$.

$CH_3CH_2CH_2Br + 2NH_3 \rightarrow CH_3CH_2CH_2NH_2 + NH_4Br$

(i) Iodoethane is reacted with ammonia. Write an equation for this reaction. *(1 mark)*

(ii) The first step in the mechanism of the reaction between $CH_3CH_2CH_2Br$ and NH_3 is shown below.

```
        H                          H
        |                          |
CH3CH2—C—Br    →    CH3CH2—C—+NH3  + ........
        |                          |
        H                          H
```

Complete the mechanism. Include relevant dipoles, lone pairs, curly arrows and the missing product. *(3 marks)*

b A student investigates the rate of hydrolysis of six haloalkanes. The student mixes 5 cm³ of ethanol with five drops of haloalkane. This mixture is warmed to 50 °C in a water bath. The student adds 5 cm³ of aqueous silver nitrate, also heated to 50 °C, to the haloalkane. The time taken for a precipitate to form is recorded in the results table. The student repeats the whole experiment at 60 °C instead of 50 °C.

Haloalkane	Time taken for a precipitate to form / s	
	50 °C	60 °C
$CH_3CH_2CH_2CH_2Cl$	243	121
$CH_3CH_2CH_2CH_2Br$	121	63
$CH_3CH_2CH_2CH_2I$	40	19
$CH_3CH_2CHBrCH_3$	89	42
$(CH_3)_2CHCH_2Br$	110	55
$(CH_3)_3CBr$	44	21

Describe and explain the factors that affect the rate of hydrolysis of haloalkanes. Include ideas about

- the halogen in the haloalkanes
- the groups attached to the carbon of the carbon–halogen bond (the type of haloalkane)
- the temperature of the hydrolysis.

In your answer you should link the evidence with your explanation. *(7 marks)*

OCR F322 Jan 11 Q5

239

16 ORGANIC SYNTHESIS
16.1 Practical techniques in organic chemistry

Specification reference: 4.2.3

Learning outcomes

Demonstrate knowledge, understanding, and application of:

→ Quickfit apparatus for distillation and heating under reflux

→ techniques for preparation and purification of an organic liquid.

Study tip

As well as being able to carry out these practical techniques, you should be able to describe them.

Preparing an organic liquid

As part of your practical endorsement you will need to be able to prepare a sample of an organic liquid.

Quickfit apparatus

To carry out an organic preparation you will need to use Quickfit apparatus. A basic set of Quickfit apparatus is shown in Figure 1 and contains:

1. round-bottom or pear-shaped flask
2. receiver
3. screw-tap adaptor
4. condenser
5. still head.

Heating under reflux

Many organic reactions occur slowly at room temperature. It is common for organic reactions to be heated to overcome the activation energy and increase the rate of reaction. Heating under **reflux** is a common procedure used to prepare an organic liquid without boiling off the solvent, reactants, or products. To heat under reflux, you will need the following pieces of apparatus:

- round-bottom or pear-shaped flask
- condenser
- rubber tubing
- stand and clamp
- heat source (usually a Bunsen burner, tripod, and gauze, or a heating mantle).

Figure 2 shows the experimental setup for a reflux experiment using a Bunsen burner to heat the reaction mixture. This allows a reaction under reflux to be carried out at a fixed temperature. A water bath can be used if the reaction can be carried out below 100 °C.

For heating flammable liquids, a heating mantle can be used so that there is no naked flame present. This provides an added level of safety should any of the apparatus leak or crack.

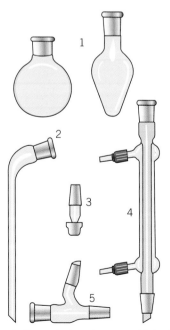

▲ **Figure 1** *Standard set of Quickfit apparatus*

ORGANIC SYNTHESIS 16

In Figure 2, the flask is clamped by its neck. Before fitting the condenser you need to add the reaction mixture and anti-bumping granules to the flask. Anti-bumping granules (Figure 3) are added to the liquid before the flask is heated so that the contents will boil smoothly. If the granules are not used, large bubbles form at the bottom of the liquid and make the glassware vibrate or jump violently.

Finally, apply a thin layer of grease to the ground-glass joint on the condenser (Figure 4). Place the condenser carefully into the flask and gently rotate the condenser back and forth to provide a good seal and ensure that the apparatus comes apart easily at the end of the experiment.

During reflux the condenser is kept in the upright position. Condensers should be clamped only loosely as the glass outer jacket is very fragile and is easily broken. Never put a stopper in the top of the condenser. Otherwise you would have a closed system and pressure would build up inside as the heated air expanded. This could result in the apparatus exploding!

Rubber tubing is used to connect the inlet of the condenser to the tap and the outlet to the sink. Water always enters the condenser at the bottom and leaves at the top to ensure that the outer jacket is full.

Heating under reflux enables a liquid to be continually boiled whilst the reaction takes place. This prevents volatile components from escaping and the flask from boiling dry. The vapour from the mixture rises up the inner tube of the condenser until it meets the outer jacket containing cold water. The vapour then condenses and drips back into the flask. The process is similar to putting a lid on a saucepan when cooking.

Distillation

Chemical reactions may not go to completion or may produce by-products as well as the desired product. Once reflux is complete, the crude liquid or solid product present in the flask needs to be purified to remove any by-products and remaining reactants. **Distillation** is a common method used to separate a pure liquid from its impurities.

To carry out a distillation, you will need the following apparatus:

- round-bottom or pear-shaped flask
- condenser
- rubber tubing
- heat source
- stand and clamp
- screw-cap adaptor
- receiver adaptor
- still head
- thermometer.

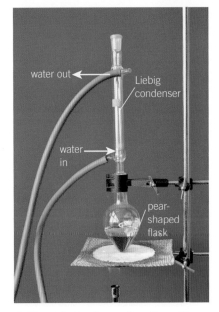

▲ **Figure 2** *Heating under reflux with a Bunsen burner*

▲ **Figure 3** *Anti-bumping granules are used when heating liquids to ensure that the liquid boils smoothly*

▲ **Figure 4** *The ground-glass joints are greased lightly so that the apparatus comes apart easily after the experiment*

16.1 Practical techniques in organic chemistry

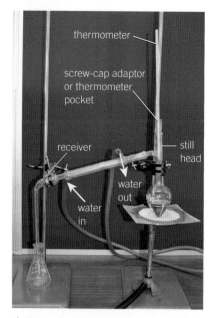

▲ Figure 5 *Apparatus set-up for distillation*

The apparatus is set up for distillation as in Figure 5.

The flask is clamped by its neck and the still head is connected to the flask. The still-head adaptor is T shaped and has two ground-glass joints, one to fit the screw-cap adaptor and one to fit the condenser. You should grease the joints so that the apparatus comes apart easily after the experiment.

A second clamp is placed round the receiver adaptor at the point at which it is attached to the condenser. This removes the need to clamp the condenser, as it will be supported sufficiently at both ends.

As with reflux, rubber tubing is used to connect the inlet of the condenser to the tap and the outlet to the sink. Water always enters the condenser at the lowest point. For distillation, this is the closest point to the receiver adaptor.

A flask is used to collect the distillate so that the distillation apparatus is not completely airtight.

Once the apparatus is set up, the flask is heated and the mixture in the flask will start to boil. The different liquids in the mixture will have different boiling points. The liquid with the lowest boiling point is the most volatile and will boil first.

The vapour moves out of the flask up into the other parts of the apparatus, leaving behind the less volatile components of the mixture. When the vapours reach the cold condenser, they condense and become a liquid. This liquid then drips into the collecting flask.

Purifying organic products

When preparing samples of organic liquids water may be obtained along with the product. If this has happened you will see two liquid layers inside your collection flask, one the organic layer and one the aqueous or water layer. It is easy to separate these two layers although it is important that you know which is the organic layer so that you don't throw away your product by mistake. An easy way to identify the organic layer is to add some water to your mixture. The layer that gets bigger is the aqueous layer.

Once the organic layer, has been identified, the two layers are separated using a separating funnel.

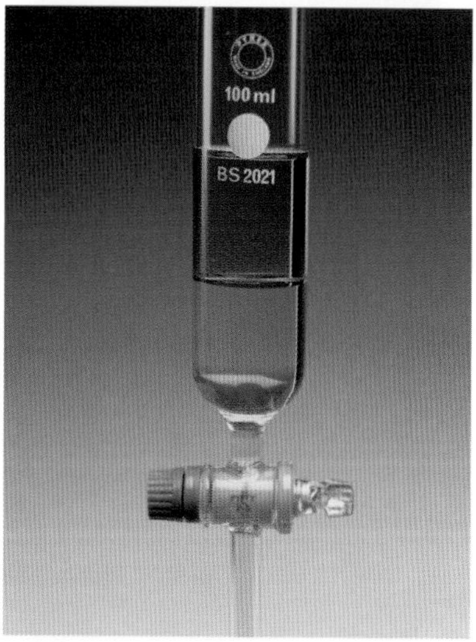

1 Ensure that the tap of the separating funnel is closed.
2 Pour the mixture of liquids into the separating funnel, place a stopper in the top of the funnel, and invert to mix the contents.
3 Allow the layers to settle.
4 Add some water to see which layer increase in volume – this is the aqueous layer.
5 Place a conical flask under the separating funnel, remove the stopper and open the tap until the whole of the lower layer has left the funnel.

▲ Figure 6 *Separating funnels are used to separate an organic liquid layer from an aqueous layer. The bottom layer can be run off, leaving the upper layer behind in the funnel*

ORGANIC SYNTHESIS   16

6 Place a second conical flask under the separating funnel to collect the other layer.

7 You will now have one conical flask containing the organic layer and another containing the aqueous layer. Label the flasks so that you don't muddle them.

In preparation using acids, your impure product may contain acid impurities. These can be removed by adding aqueous sodium carbonate and shaking the mixture in the separating funnel.

Any acid present will react with sodium carbonate releasing carbon dioxide gas. The tap needs to be slowly opened, holding the stoppered separating funnel upside down, to release any gas pressure that may build up.

Finally the aqueous sodium carbonate layer is removed and the organic layer washed with water before running both layers off into two separate flasks.

Drying the organic product

There may be some water left in the organic product. Traces of water are removed by adding a drying agent to the organic liquid. A drying agent is an anhydrous inorganic salt that readily takes up water to become hydrated. The most common drying agents are shown in Table 1.

▼ **Table 1** *Common inorganic salts used to dry organic liquids*

Name of compound	Formula	Use
anhydrous calcium chloride	$CaCl_2$	Drying hydrocarbons
anhydrous calcium sulfate	$CaSO_4$	General drying
anhydrous magnesium sulfate	$MgSO_4$	General drying

The procedure for drying an organic liquid is given below.

1 Add the organic liquid to a conical flask.

2 Using a spatula, add some of the drying agent to the liquid and gently swirl the contents to mix together.

3 Place a stopper on the flask to prevent your product from evaporating away. Leave for about ten minutes.

4 If the solid has all stuck together in a lump, there is still some water present. Add more drying agent until some solid is dispersed in the solution as a fine powder.

5 Decant the liquid from the solid into another flask. If the liquid is dry it should be clear.

Redistillation

Sometimes organic liquids have boiling points that are relatively close together, so your prepared sample may still contain some organic impurities. The distillation apparatus is cleaned and dried and set up again so that a second distillation can be carried out. This time, only collect the product with the boiling point of the compound you are trying to make. The narrower the boiling range, the purer the product. You now will have separated your product from any impurities.

Summary questions

1 Name two drying agents that can be used to remove water from an organic product. *(2 marks)*

2 Outline the steps that you would need to carry out to separate the organic liquid 1-bromobutane from a mixture of 1-bromobutane, water, and sulfuric acid. *(3 marks)*

3 In many organic reactions, the materials are heated under reflux to increase the yield of products. Explain why the reactants are not heated under reflux when preparing a sample of ethanal from ethanol. *(2 marks)*

4 To prepare a sample of 1-bromobutane, butan-1-ol is heated under reflux with a mixture of sodium bromide and sulfuric acid. After distillation, the sample contains two layers. The densities of the components in these layers are: water, $1.00\,g\,cm^{-3}$; butan-1-ol, $0.81\,g\,cm^{-3}$; 1-bromobutane, $1.30\,g\,cm^{-3}$.

  a Draw the displayed formula of 1-bromobutane. *(1 mark)*

  b Construct an equation for the reaction which takes place. *(1 mark)*

  c Explain which of the layers will contain your product. *(1 mark)*

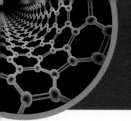

16.2 Synthetic routes

Specification reference: 4.2.3

Learning outcomes

Demonstrate knowledge, understanding, and application of:

→ properties and reactions of functional groups

→ constructing two-stage synthetic routes.

Organic synthesis

Organic synthesis is the preparation of complex molecules from simple starting materials. With organic synthesis, chemists can make entirely new structures that can be investigated for their uses. The manufacture of modern medicines targeted at diseases owes much to the work of organic chemists. Many chemists are employed by large pharmaceutical companies to synthesise and test new medicines, for example, those designed to target the growth of cancer cells in the human body.

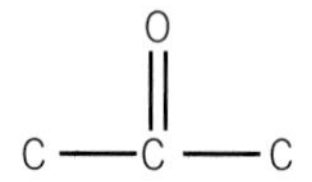

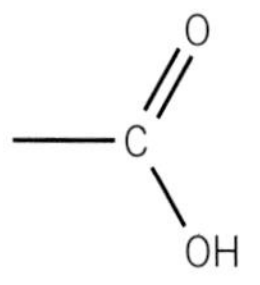

Identifying functional groups in organic molecules

Before you attempt organic synthesis, you need to know the different functional groups that you have met so far in your course (Table 1). You should be able to identify these functional groups in unfamiliar molecules containing several functional groups.

▼ **Table 1** *Functional groups for AS Chemistry*

Type of compound	Functional group
alkene	C=C
alcohol	−OH
haloalkane	−Cl −Br −I
aldehyde	−CHO
ketone	C−CO−C
carboxylic acid	−COOH
ester	−COO−C
amine	−NH$_2$
acyl chloride	−COCl
nitrile	−CN

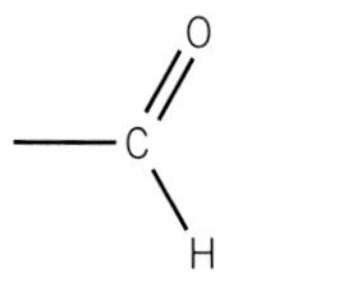

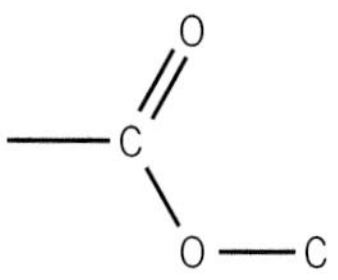

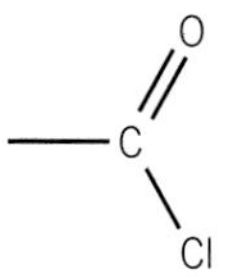

Worked example: Identifying functional groups 1

Identify the functional groups in the molecule shown in Figure 1.

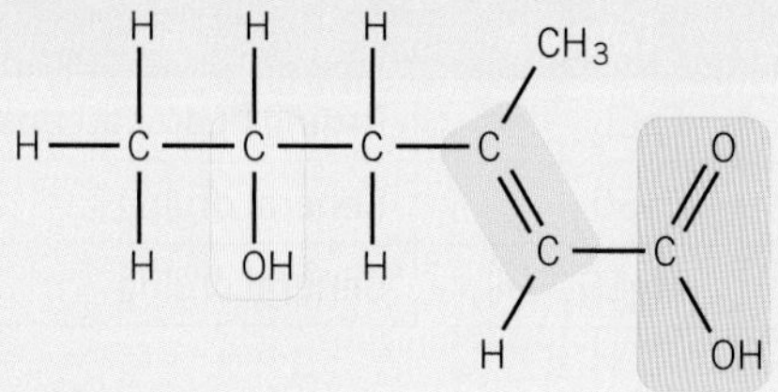

▲ **Figure 1**

The molecule contains three functional groups. From left to right:

- −OH is an alcohol
- −C=C− is an alkene
- −COOH is a carboxylic acid.

Worked example: Identifying functional groups 2

Identify the functional groups in the molecule shown in figure 2.

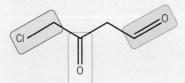

▲ **Figure 2**

There are three functional groups. From left to right, the functional groups are haloalkane, ketone, and aldehyde.

ORGANIC SYNTHESIS 16

Synthetic routes

The flowchart in Figure 3 links together all of the functional groups that you have studied so far in your course. The flowchart will be useful for predicting reactions and in solving synthesis problems.

> **Study tip**
>
> You need to be able to suggest reagents, conditions, and equations for converting one compound into another.

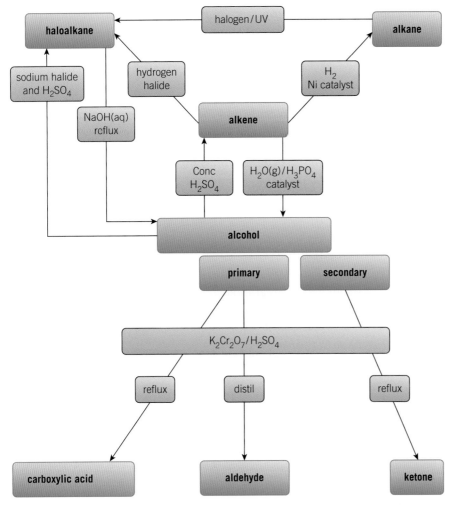

▲ **Figure 3** *Connections between the functional groups covered in AS Level chemistry*

Predicting properties and reactions

Prenol, $(CH_3)_2CCHCH_2OH$, is a natural compound found in many fruits and widely used in organic synthesis. By drawing out its structure, you can identify its functional groups (Figure 4).

After identifying the functional groups, you can now predict properties and reactions of prenol.

Solubility	Reactions of alkene	Reactions of primary alcohol
soluble in water: the −OH group will form hydrogen bonds with water	→ alkane → haloalkane → alcohol	→ alkene → haloalkane → aldehyde → carboxylic acid

▲ **Figure 4** *Functional groups in prenol*

245

16.2 Synthetic routes

Target molecules
The term **target molecule** is used to describe the compound that the chemist is attempting to prepare by organic synthesis.

In a simple synthesis, the target molecule can be obtained by reacting a readily available starting material with a readily available reagent in a one-step reaction. However, in most cases synthesis is not that straightforward. Some organic syntheses require many steps to change functional groups or to add carbon atoms to the chain length in order to obtain the target molecule.

Two stage synthesis
To convert a starting molecule into the target molecule, you need to:
- identify the functional groups in your starting and target molecules
- identify the intermediate that links the starting and target molecules
- state the reagents and conditions for each step.

> **Study tip**
> At AS you will only be required to solve synthesis questions with a maximum of two steps. At A Level, multi-step synthesis may be required.

Worked example: Synthesis of propanal
State the reagents and conditions for the two-stage synthesis of propanal from 1-chloropropane. Give equations for each stage.

Step 1: Identify the functional groups in the starting and target molecule.

CH₃CH₂CH₂Cl →(stage 1)→ INTERMEDIATE →(stage 2)→ CH₃CH₂CHO

STARTING MOLECULE — haloalkane

TARGET MOLECULE — aldehyde

Step 2: Identify the intermediate compound that links the target and starting molecules together.

- An aldehyde (TARGET) can be made from the oxidation of a **primary alcohol**.
- A **primary alcohol** can be made by the hydrolysis of a haloalkane (STARTING MOLECULE).

INTERMEDIATE — primary alcohol (CH₃CH₂CH₂OH)

Step 3: Identify the reagents and conditions.
- Stage 1: Reagents: NaOH(aq); Conditions: Reflux

CH₃CH₂CH₂Cl + NaOH → CH₃CH₂CH₂OH + NaCl

- Stage 2: Reagents: Potassium dichromate(VI) and sulfuric acid; Conditions: Distil

CH₃CH₂CH₂OH + [O] → CH₃CH₂CHO + H₂O

ORGANIC SYNTHESIS 16

 Worked example: Synthesis of butanone

State the reagents and conditions for the two-stage synthesis of butanone from but-2-ene. Show skeletal formulae throughout and Give equations for each stage.

Step 1: Identify the functional groups in your starting and target molecule.

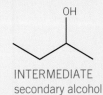

Step 2: Identify the intermediate compound that links the target and starting molecules together.

- A ketone (TARGET) can be made from the oxidation of a **secondary alcohol**.
- A **secondary alcohol** can be made by the hydration of an alkene (STARTING MOLECULE).

Step 3: Identify the reagents and conditions.

- Stage 1: Reagents: steam, $H_2O(g)$; Conditions: Acid catalyst, e.g. H_3PO_4 or H_2SO_4

- Stage 2: Reagents: Potassium dichromate(VI) and sulfuric acid; Conditions: Reflux

 Worked example: Predicting reactions

In this worked example, you are provided with a synthetic route. This time, the starting molecule has two functional groups.

$$\text{CH}_2\text{=CHCHBrCH}_3 \xrightarrow{\text{step 1}} \text{CH}_3\text{CH}_2\text{CHBrCH}_3 \xrightarrow{\text{step 2}} \text{CH}_3\text{CH}_2\text{CH(OH)CH}_3 \xrightarrow{\text{step 3}} \text{CH}_3\text{CH}_2\text{COCH}_3$$

Predict the reagents and conditions for each of each step.

Step 1: Requires H_2/Ni catalyst to convert the alkene into an alkane.

Step 2: Requires NaOH(aq) and reflux to convert the haloalkane into an alcohol.

Step 3: Requires $K_2Cr_2O_7$/H_2SO_4 and reflux convert the secondary alcohol into a ketone.

Study tip

All the information used for these three reaction steps has been taken from Figure 5. It is important that you learn all these reactions and conditions

16.2 Synthetic routes

Does nature really provide a cure for everything?

Some people believe that the natural resources of our planet provide cures for every disease or illness, and only time and the limits of our knowledge prevent us from finding a cure for every illness. You may already know that the bark of the willow tree had been used for centuries for the treatment of fever and pain. Salicylic acid was found to be the active component in willow bark, from which aspirin was developed. As far back as 400 BC willow leaf tea was used to relieve the pain of childbirth.

Just 50 years ago, paclitaxel was obtained from the bark of the Pacific yew tree (Figure 5). Taxol is currently used as an anti-cancer drug, specifically for the treatment of ovarian, breast, and colon cancer. Demand for Taxol is much higher than the small amounts that can be obtained from yew bark. To meet this demand, chemists synthesise Taxol from more readily available substances.

About 45 years ago, ibuprofen was introduced in the UK as an anti-inflamatory drug. Its structure has some similarities to the structure of aspirin. Ibuprofen is also marketed under several other names such as Nurofen.

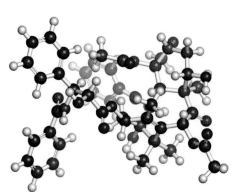

▲ **Figure 5** *The Pacific yew tree (left), Taxus brevifolia, the source of the anti-cancer drug Taxol (right)*

The skeletal formulae of the painkiller drugs aspirin and ibuprofen are shown below. Both structures contain a benzene ring and functional groups on their side chains.

aspirin   ibuprofen

1 What functional group is present in both molecules?
2 What is the molecular formula of each molecule?
3 Predict a two-stage synthesis of ibuprofen from the compound shown in Figure 6

> **Synoptic link**
>
> A good knowledge of the functional groups and their reactions is important when studying organic synthesis.

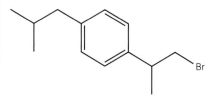
▲ **Figure 6**

ORGANIC SYNTHESIS 16

Ozonolysis

Ozonolysis is a technique used in organic chemistry to break open a C=C double bond. The reaction is useful in organic synthesis for the preparation of aldehydes and ketones. During the ozonolysis of but-2-ene two molecules of ethanal are formed as shown below.

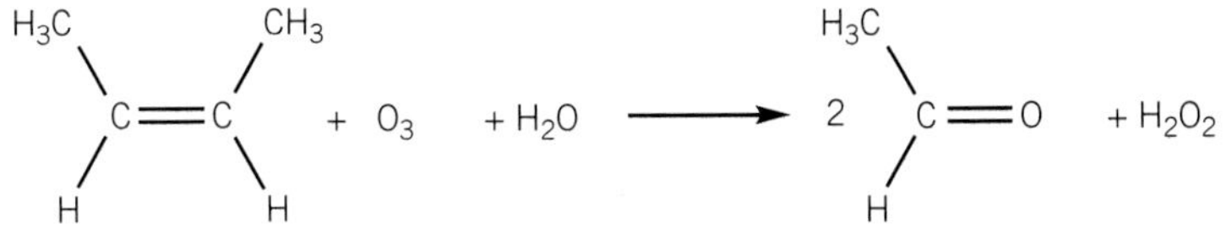

$$H_3C-CH=CH-CH_3 + O_3 + H_2O \longrightarrow 2\ CH_3CHO + H_2O_2$$

Propanoic acid can be made in a two-step process starting from pent-3-ene.

Step 1: Ozonolysis

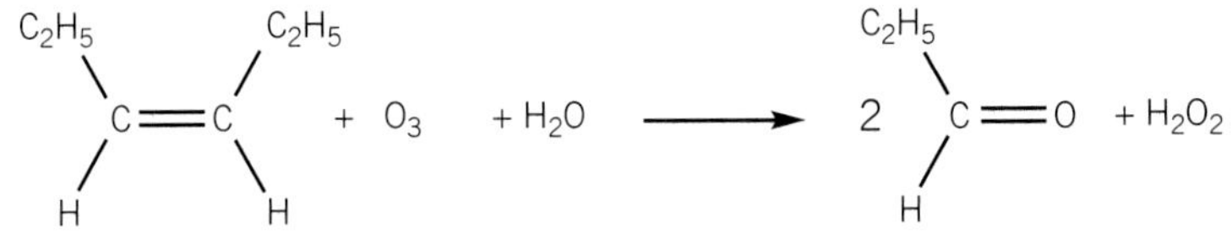

$$C_2H_5-CH=CH-C_2H_5 + O_3 + H_2O \longrightarrow 2\ C_2H_5CHO + H_2O_2$$

Step 2: Reflux with acidified potassium dichromate(VI) – oxidation

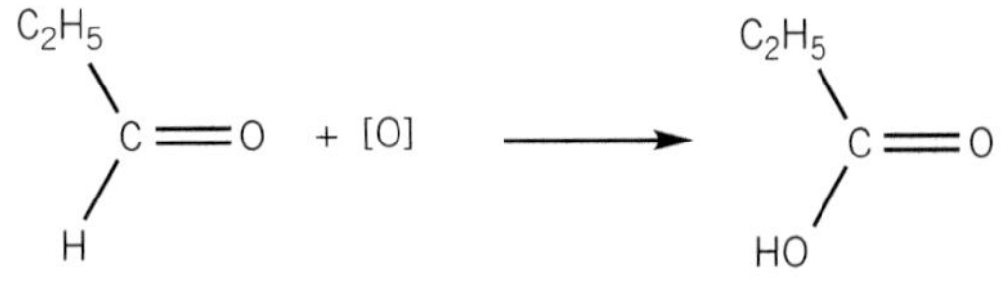

$$C_2H_5CHO + [O] \longrightarrow C_2H_5COOH$$

Suggest a two-stage synthesis of hexanedioic acid from cyclohexene. Use skeletal formula to illustrate any equations.

Summary questions

1. Identify the functional groups in the following molecules.

    a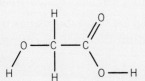

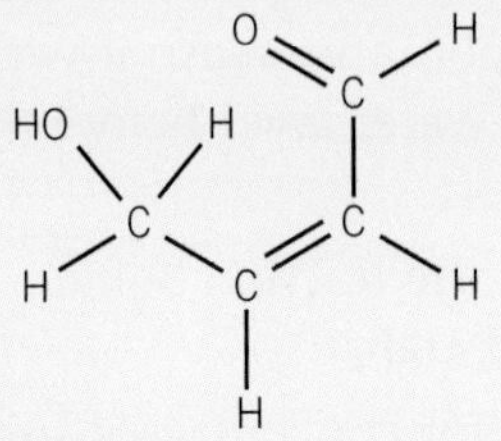

    (3 marks)

    b

    (2 marks)

2. State the reagents and conditions required to form 2-chloropropane from propan-2-ol. *(2 marks)*

3. Suggest two different synthetic routes that could be used to convert but-2-ene into butan-2-ol. *(4 marks)*

4. For each of the reactions shown in the reaction scheme below, state the reagents and conditions, and give the chemical equation for each of the steps. *(6 marks)*

    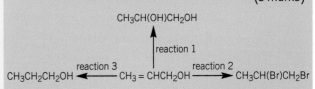

Chapter 16 Practice questions

Practice questions

1 Alcohols can be converted into chloroalkanes by reaction with hydrochloric acid, HCl.

2-Chloro-2-methylpropane can be prepared by shaking together 4.7 cm³ (3.70 g) of 2-methylpropan-2-ol with 20 cm³ of concentrated HCl. After 10 minutes two separate layers begin to form.

Use the data in the table below to answer the questions that follow.

Compound	relative molecular mass	density / g cm^{-3}	boiling point / °C
2-methylpropan-2-ol	74.0	0.78	83
2-chloro-2-methylpropane	92.5	0.84	51
water	18.0	1.00	100

One of the layers is aqueous and the other contains the organic product.

  a Suggest whether the upper or lower layer is likely to contain the organic product. Explain your reasoning. *(1 mark)*

  b The organic layer was shaken with a dilute solution of sodium hydrogencarbonate, NaHCO$_3$. A gas was evolved.

    Identify this gas. *(1 mark)*

  c Suggest the chemical that could have reacted with the NaHCO$_3$ to form the gas. *(1 mark)*

  d The resulting impure organic liquid was dried with anhydrous calcium chloride and then distilled. 2.22 g of pure 2-chloro-2-methylpropane was produced.

    (i) At what temperature would you expect the pure organic product to distil? *(1 mark)*

    (ii) Calculate the amount, in mol, of 2-methylpropan-2-ol used in the experiment. *(1 mark)*

    (iii) Calculate the amount, in mol, of pure 2-chloro-2-methylpropane that was produced. *(1 mark)*

    (iv) Calculate the percentage yield of 2-chloro-2-methylpropane in this experiment. *(1 mark)*

OCR 2812 Jan 01 Q5(b)–(d) adapted

2 A teacher provided the following instructions for her students so that they could prepare a sample of 1-bromobutane C$_4$H$_9$Br, (boiling point 102 °C).

Reagents
- Concentrated sulfuric acid
- Sodium bromide
- Butan-1-ol, C$_4$H$_9$OH

Procedure

1 Transfer 5 g of crushed sodium bromide to a 100 cm³ round-bottomed flask: add 10 cm³ of distilled water and 5 cm³ of butan-1-ol (boiling point 117 °C). Swirl the flask gently to dissolve the sodium bromide.

2 Carefully add 6 cm³ of concentrated sulfuric acid to the mixture in the flask. The sulfuric acid will react with the sodium bromide to form hydrogen bromide. If the contents of the flask become too hot then cool the flask in cold water.

3 Fit a water-cooled condenser and boil the contents of the flask for 30 minutes.

4 Allow the apparatus to cool and arrange it for distillation.

5 Heat the flask gently and allow the mixture to distil over into a small conical flask. Collect the fraction boiling between 95 °C and 125 °C.

6 Add sodium hydrogencarbonate solution to the distillate and shake the mixture until no more gas is given off. Two layers of liquid will form in the conical flask.

7 Separate the two layers of liquid and collect the impure 1-bromobutane. Wash it with water. Separate the water layer before drying the 1-bromobutane.

8 Re-distil and collect the fraction boiling between 101 °C and 103 °C.

  a Suggest hazards for concentrated sulfuric acid and butan-1-ol. *(2 marks)*

  b In stage 5, suggest what impurities might be present within the fraction collected. *(3 marks)*

  c In stage 6, a gas is given off.

    Name the gas and explain why it is formed. *(2 marks)*

  d In stage 7, the two layers of liquid have to be separated from each other. The 1-bromobutane is in the lower layer.

ORGANIC SYNTHESIS 16

Suggest how the two layers might be separated. *(2 marks)*

e In Stage 7, how could the 1-brombutane be dried? *(1 mark)*

OCR 2813/04 Jun 06 adapted

3 A student prepared an organic compound and distilled the crude product using the apparatus below. The crude distillate was obtained over the range 90 °C to 130 °C. The boiling point of the pure organic compound is 116 °C.

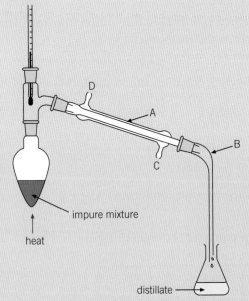

a Identify the apparatus labelled **A** and **B**. *(2 marks)*

b State and explain whether water should enter apparatus **A** at position **C** or **D**. *(1 mark)*

c The distillate contained an aqueous and an organic layer. The density of the layers is shown below.

Layer	aqueous layer	organic layer
Density / g cm^{-3}	1.05	0.72

Explain how you would obtain a dry, pure sample of the organic product, starting from the two layers *(4 marks)*

4 With the aid of equations and conditions, state how each of the following can be made from butan-1-ol.

a but-1-ene
b butanoic acid
c 1-bromobutane *(6 marks)*

5 Identify the functional groups in each of the following molecule **A** and **B**.

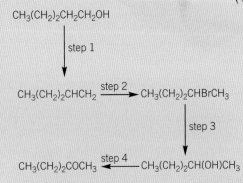

(5 marks)

6 Propene can be converted into propan-2-ol by two different synthetic routes. One route can be carried out as a one-step process whilst the other route requires two steps.

a State reagents, conditions and provide an equation for a one-step process. *(3 marks)*

b Draw a reaction scheme, including reagents and conditions for a two-step synthetic route. *(3 marks)*

7 State the reagents, conditions and provide equations for each of the following two-step conversions.

a propene to propanone *(4 marks)*
b propan-2-ol to propane *(4 marks)*

8 An organic synthetic route starting with pentan-1-ol is shown below.

a Identify the reagents and conditions for each of the steps in the reaction scheme. *(4 marks)*

$CH_3(CH_2)_2CH_2CH_2OH$

↓ step 1

$CH_3(CH_2)_2CHCH_2$ —step 2→ $CH_3(CH_2)_2CHBrCH_3$

↓ step 3

$CH_3(CH_2)_2COCH_3$ ←step 4— $CH_3(CH_2)_2CH(OH)CH_3$

b State the type of reaction taking place in step 1. *(1 mark)*

c Show the mechanism for step 2. *(3 marks)*

d What would you observe in step 4. *(1 mark)*

e How would you convert the product from step 1 into an alkane? *(2 marks)*

17 SPECTROSCOPY
17.1 Mass spectrometry
Specification reference: 4.2.4

Learning outcomes
Demonstrate knowledge, understanding, and application of:
→ mass spectra for identification
→ fragmentation peaks.

Synoptic link
You have already met mass spectrometry in Topic 2.2, Relative mass.

Mass spectra can be used to identify the molecular mass of an organic compound and to gain further information about its structure.

Molecular ions and fragment ions

When an organic compound is placed in the mass spectrometer, it loses an electron and forms a positive ion, the **molecular ion**. The **mass spectrometer** detects the mass-to-charge ratio (m/z) of the molecular ion which gives the molecular mass of the compound.

For propan-1-ol, the following equation shows the formation of the molecular ion.

$$CH_3CH_2CH_2OH \rightarrow CH_3CH_2CH_2OH^+ + e^-$$
$$\text{molecular ion}$$
$$m/z = 60$$

The molecular ion M^+ is the positive ion formed when a molecule loses an electron.

The mass spectrum of propan-1-ol is shown in Figure 2.

▲ **Figure 1** *A researcher examines the mass spectrum of a protein sample*

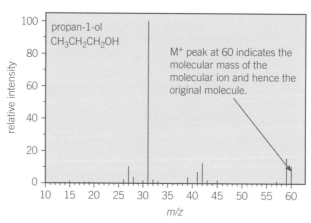

▲ **Figure 2** *The mass spectrum of propan-1-ol*

Molecular mass from a mass spectrum

To find the molecular mass, the molecular ion peak (M^+ peak) has to be located. The molecular ion peak is the clear peak at the highest m/z value on the right-hand side of the mass spectrum. In Figure 2, the molecular ion peak is located at $m/z = 60$. This shows that the molecular mass of propan-1-ol is 60.

You will usually also see a very small peak one unit after the M^+ peak. This is referred to as the $M+1$ peak. The $M+1$ peak exists because 1.1% of carbon is present as the carbon-13 isotope. For example, propan-1-ol has a molecular mass of 60, but a small proportion of the alcohol molecules will contain an atom of ^{13}C and thus have a molecular mass of 61. This gives the small M+1 peak.

Fragmentation

In the mass spectrometer some molecular ions break down into smaller pieces known as fragments in a process called **fragmentation**. The other peaks in a mass spectrum are caused by **fragment ions**, formed from the breakdown of the molecular ion.

The simplest fragmentation breaks a molecular ion into two species – a positively charged fragment ion and a radical. Any positive ions formed will be detected by the mass spectrometer, but the uncharged radicals are not detected. In the mass spectrum for propan-1-ol (Figure 2), the largest peak has an *m/z* value of 31. The equation below shows the possible identity of this fragment ion and its formation from the molecular ion.

$$CH_3CH_2CH_2OH^+ \rightarrow CH_2OH^+ + CH_3CH_2\bullet$$
$$\text{molecular ion} \quad\quad \text{fragment ion} \quad\quad \text{radical}$$
$$m/z = 31$$

You should be able to see how the molecular ion could break to form fragment ion and radical.

Using fragmentation peaks to identify an organic molecule

The mass spectrum of each compound is unique, as molecules will all fragment in slightly different ways depending on their structures. Mass spectra can therefore be used to help identify molecules. So even though two molecules may have the same molecular mass and the same molecular ion peak, the fragment ions found in the spectrum may be different.

Figure 3 shows the different mass spectra of hexane and 2-methylpentane – two isomers of C_6H_{14}. Both spectra have the molecular ion peak at 86, which is the molecular mass of both hexane and 2-methylpentane, but the fragmentation patterns are different.

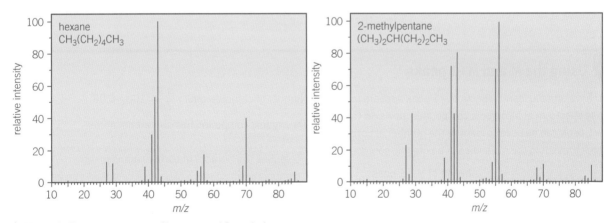

▲ **Figure 3** *The mass spectra of hexane and 2-methylpentane*

In Figure 3, a number of peaks in the spectrum have *m/z* values lower than the molecular ion peak. These fragment peaks provide valuable information about the structure of the organic compound.

17.1 Mass spectrometry

Table 1 *Some common fragment ions and their corresponding m/z values*

Fragment ion	m/z
CH_3^+	15
$C_2H_5^+$	29
$C_3H_7^+$	43
$C_4H_9^+$	57

Table 1 shows some common fragment ions and their *m/z* values.

The displayed formula and mass spectrum of ethanol is shown in Figure 4.

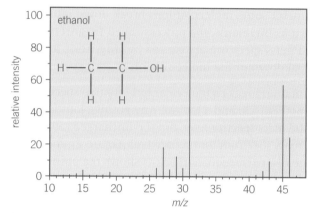

▲ **Figure 4** *The mass spectrum of ethanol*

The main features of the spectrum are:

- molecular ion peak, M^+ at $m/z = 46$
- small $M+1$ peak at $m/z = 47$
- a number of fragment ion peaks.

You should be able to link fragments together to obtain an ethanol molecule:

- peak at $m/z = 15$ for CH_3^+
- peak at $m/z = 29$ for $CH_3CH_2^+$ ($C_2H_5^+$)
- peak at $m/z = 31$ for CH_2OH^+
- peak at $m/z = 45$ for $CH_3CH_2O^+$.

It can also be useful to look at the differences between *m/z* for the molecular ion and fragments ion (Figure 4) although the peak for CH_3^+ at $m/z = 15$ is very small, there is a peak at $m/z = 31$, which is 15 less than the value for M^+.

➕ Using the M and M+1 peaks

You have already learnt why an $M+1$ peak appears in the mass spectrum of all organic compounds; however the $M+1$ peak can be used to identify the number of carbon atoms present in the molecules of an organic compound.

$$\text{Number of carbon atoms} = \frac{\text{height of } M+1 \text{ peak}}{\text{height of } M \text{ peak}} \times 100$$

A spectrum of an unknown compound has an M peak of height 74 mm and an $M+1$ peak of height 4.5 mm. How many carbon atoms are there in the organic compound?

$$\text{Number of carbon atoms} = \frac{4.5}{74} \times 100 = 6.08$$

The organic compound has six carbon atoms.

Calculate the number of carbon atoms in the compound which has a M peak of height 64 mm and a $M+1$ peak at 5.8 mm.

17 SPECTROSCOPY

Drug testing in sport

During the London Olympics, a team of 150 scientists analysed more than 6000 samples in an attempt to ensure that the games were not tainted by disqualifications due to the use of prohibited drugs. The London anti-doping laboratory operated by GlaxoSmithKline and King's College London was operational for 24 hours a day throughout both the Olympics and Paralympics. Every medal winner was automatically tested, and many random tests were carried out both in Olympic venues and within the accommodation in the Olympic Village. The state-of-the-art laboratories employed a number of analytical techniques including mass spectrometry to test urine samples. This technique, along with many others, is used to ensure that every athlete is 'clean'. In the Athens Olympics of 2004, twenty-four doping violations were uncovered.

Clenbuterol is believed to increase the development of muscle. It is considered a performance-enhancing drug and is banned in athletic competitions. Athletes who test positive for clenbuterol often claim that they must have eaten contaminated meat, as clenbuterol has been used to increase yields in livestock production. However, clenbuterol is banned for use in cattle in both the USA and much of Europe.

▲ **Figure 5** *Paralympic sprinters, London 2012: a place on the podium would mean a compulsory drugs test after the race*

▲ **Figure 6** *Clenbuterol is considered a performance-enhancing drug and is banned in most athletic competitions*

What is the m/z value of the molecular ion, M^+, of clenbuterol?

Summary questions

1. Write equations to show how the following molecules are converted into their molecular ions, and predict the m/z value for their molecular ions.
   a. nonane *(2 marks)*
   b. butan-1-ol *(2 marks)*

2. Explain why there sometimes a peak in the spectrum after the molecular ion peak. *(2 marks)*

3. Use the spectra provided to identify the molecular ion peak and hence the molecular mass of compound A and compound B. *(2 marks)*

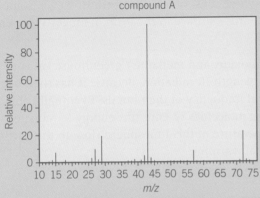

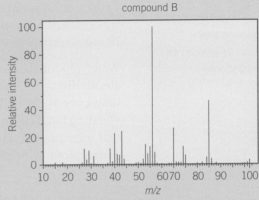

4. Identify the fragments of $m/z = 29$ and $m/z = 31$ formed from the molecular ion of butan-1-ol. *(2 marks)*

5. The mass spectrum of the ketone, hexan-2-one has a large fragment ion at $m/z = 43$.
   a. Suggest two possible structures for this fragment ion. *(2 marks)*
   b. Write an equation for the formation of each structure from the molecular ion. *(4 marks)*

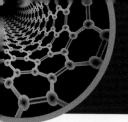

17.2 Infrared spectroscopy

Specification reference: 4.2.4

Learning outcomes

Demonstrate knowledge, understanding, and application of:

→ absorption of infrared radiation by atmospheric gases
→ infrared spectroscopy
→ combining spectroscopic techniques.

Infrared radiation and covalent bonds

Atoms in molecules are joined by covalent bonds. These bonds possess energy and vibrate naturally about a central point, the amount of vibration increasing with increasing temperature. The atoms in molecules are therefore in constant motion. The bonds can absorb infrared (IR) radiation, which makes them bend or stretch more.

One type of vibration, a stretch, is a rhythmic movement along the line between the atoms so that the distance between the two atomic centres increases and decreases (Figure 1). If you take A level physics, the vibration is similar to the simple harmonic motion of a spring.

The second type of vibration, a bend, results in a change in bond angle, as shown in Figure 2.

The amount that a bond stretches or bends depends on:

- the mass of the atoms in the bond – heavier atoms vibrate more slowly than lighter atoms
- the strength of the bond – stronger bonds vibrate faster than weaker bonds.

Any particular bond can only absorb radiation that has the same frequency as the natural frequency of the bond. The frequency values are very large, so chemists use a more convenient scale called **wavenumber**, which is proportional to frequency. The vibrations of most bonds are observed in the IR wavenumber range of $200\,\text{cm}^{-1}$ to $4000\,\text{cm}^{-1}$.

▲ **Figure 1** *A bond in a molecule can stretch so that the distance between the atomic centres changes*

bond angle decreases

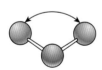

bond angle increases

▲ **Figure 2** *Bonds in a molecule can bend, causing the bond angle to change*

Infrared radiation and atmospheric gases

Much of the Sun's visible and IR radiation is relatively unaffected by atmospheric gases. This radiation passes through the atmosphere to the Earth's surface, where most of it is absorbed. However, some is re-emitted from the Earth's surface in the form of longer-wavelength IR radiation.

Water vapour, carbon dioxide, and methane ('greenhouse gases') absorb this longer-wavelength IR radiation, because it has the same frequency as the natural frequency of their bonds. Eventually, the vibrating bonds in these molecules re-emit this energy as radiation that increases the temperature of the atmosphere close to the Earth's surface, leading to global warming.

SPECTROSCOPY 17

Water vapour, carbon dioxide, and methane are the three most abundant greenhouse gases in the atmosphere. You have probably heard about the need to reduce emissions of carbon dioxide. Today many incentives are available to householders for reducing pollution and for converting to renewable sources of energy that do not emit carbon dioxide.

Infrared spectroscopy and organic molecules

Organic chemists use **infrared spectroscopy** as a means of identifying the functional groups present in organic molecules.

1. The sample under investigation is placed inside an IR spectrometer.
2. A beam of IR radiation in the range 200–4000 cm^{-1} is passed through the sample.
3. The molecule absorbs some of the IR frequencies, and the emerging beam of radiation is analysed to identify the frequencies that have been absorbed by the sample.
4. The IR spectrometer is usually connected to a computer that plots a graph of transmittance against wavenumber.

A typical IR spectrum is shown in Figure 4. The dips in the graph are still called 'peaks'. Each peak is observed at a wavenumber that can be related to a particular bond in the molecule.

▲ **Figure 3** *The increasingly familiar sight of solar photovoltaic panels on a house. The cells in a solar panel contain a semi-conducting material that converts sunlight directly into electricity. Solar power is a renewable, inexhaustible source of energy and does not contribute to global warming*

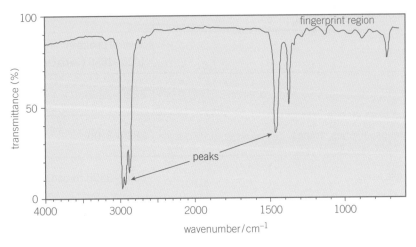

▲ **Figure 4** *An infrared spectrum of a molecule with a number of peaks that can be interpreted to identify the functional groups in the molecule*

Below 1500 cm^{-1}, there are a number of peaks in what is known as the **fingerprint region** of the spectrum. The fingerprint contains unique peaks which can be used to identify the particular molecule under investigation, either using computer software or by physically comparing the spectrum to booklets of published spectra.

Figure 5 shows the fingerprint region in the IR spectrum of the naturally occurring compound, vanillin, responsible for vanilla flavouring.

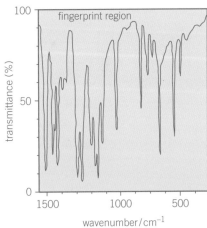

▲ **Figure 5** *Fingerprint region for vanillin*

17.2 Infrared spectroscopy

The infrared spectra of common functional groups

It is difficult to predict with certainty the identity of a functional group from a peak in the fingerprint region but, outside that region, peaks are clearer. Chemists have studied the IR spectra of thousands of organic molecules and have developed tables of data to help identify a particular functional group in an unknown molecule. The OCR data table for IR spectroscopy is shown in Table 1.

▼ **Table 1** *Characteristic IR absorptions for bonds within functional groups*

Bond	Location	Wavenumber / cm^{-1}
C—C	alkanes, alkyl chains	750–1100
C—X	haloalkanes (X = Cl, Br, I)	500–800
C—F	fluoroalkanes	1000–1350
C—O	alcohols, esters, carboxylic acids	1000–1300
C=C	alkenes	1620–1680
C=O	aldehydes, ketones, carboxylic acids, esters, amides, acyl chlorides, and acid anhydrides	1630–1820
aromatic C=C	arenes	Several peaks in range 1450–1650 (variable)
C≡N	nitriles	2220–2260
C—H	alkyl groups, alkenes, arenes	2850–3100
O—H	carboxylic acids	2500–3300 (broad)
N—H	amines, amides	3300–3500
O—H	alcohols, phenols	3200–3600

You should be able to identify the following functional groups in compounds from an IR spectrum.

- O—H group in alcohols
- C=O group in aldehydes, ketones, and carboxylic acids
- COOH group in carboxylic acids.

You should also be aware that all organic compounds produce a characteristic peak between 2850 and 3100 cm^{-1} from the presence of C—H bonds. This is often confused with the O—H peak in alcohols, so you will need to take care.

Study tip

When interpreting IR spectra it is important to identify both the bond and the functional group to which it belongs. For example, the O—H bond in an alcohol.

SPECTROSCOPY 17

Infrared spectrum of an alcohol

The IR spectrum of an alcohol has an absorbance peak within the range 3200–3600 cm^{-1} caused by the O—H bond in an alcohol. There is also a peak between 1000–1300 cm^{-1} caused by the C—O bond, though this peak is often difficult to assign because of the many other peaks in the fingerprint region. The IR spectrum of propan-2-ol, CH$_3$CHOHCH$_3$, is shown in Figure 5.

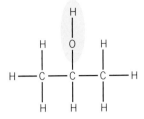

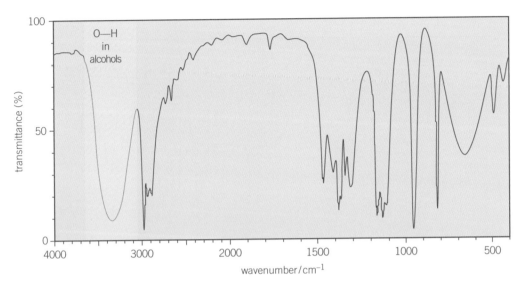

▲ **Figure 5** *The infrared spectrum of propan-2-ol*

Infrared spectrum of an aldehyde or ketone

The IR spectrum of an aldehyde or ketone has a key absorbance peak within the range 1630–1820 cm^{-1} caused by the C═O bond. This peak typically absorbs close to 1700 cm^{-1}. The IR spectrum of pentan-3-one is shown in Figure 6.

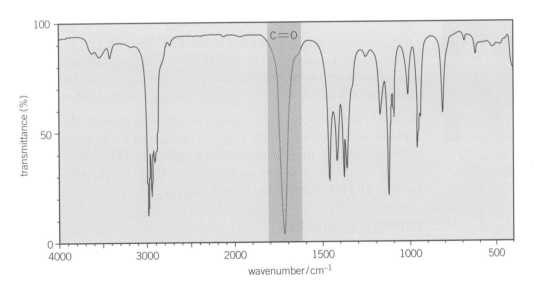

▲ **Figure 6** *The infrared spectrum of pentan-3-one showing the key absorption peak for the C═O group in red between 1630 and 1820 cm^{-1}*

17.2 Infrared spectroscopy

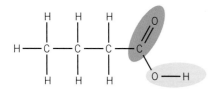

Infrared spectrum of a carboxylic acid

The IR spectrum of carboxylic acid has a key absorbance peak within the range 1630–1820 cm^{-1} caused by the C=O bond, and a broad peak at 2500–3330 cm^{-1} caused by the O—H group in the carboxylic acid. As with aldehydes and ketones, the C=O peak typically absorbs close to 1700 cm^{-1}. There is also a peak at 1000–1300 cm^{-1} that represents the C—O bond. The C—O peak is not always reliable, as it is in the fingerprint region. The IR spectrum of butanoic acid is shown in Figure 7.

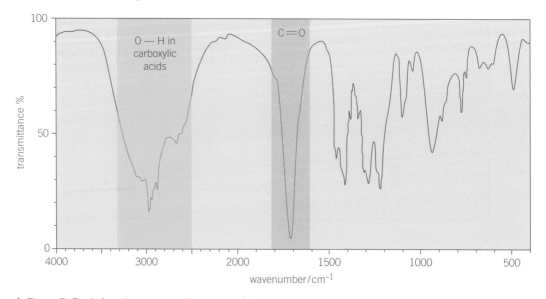

▲ **Figure 7** *The infrared spectrum of butanoic acid, showing the key absorption peak for the C=O group between 1630 and 1820 cm^{-1} in red and the peak for the O—H group in a carboxylic acid between 2500 and 3300 cm^{-1} in green*

A C=O peak at around 1700 cm^{-1} together with a very broad O—H absorption within the range 2500–3000 cm^{-1} indicates for the presence of a carboxylic acid in an organic molecule.

Applications of infrared spectroscopy

Many pollutants can be identified by their IR spectral fingerprints. Remote sensors analyse the IR spectra of vehicle emissions to detect and measure carbon monoxide, carbon dioxide, and hydrocarbons in busy town centres or by motorways to monitor localised pollution.

IR-based breathalysers pass a beam of IR radiation through the captured breath in the sample chamber and detect the IR absorbance of the compounds in the breath. The characteristic bonds present in ethanol are detected. The more IR radiation absorbed, the higher the reading, and the more ethanol in the breath.

▲ **Figure 8** *Infrared spectroscopy is also used in roadside breathalysers to detect ethanol*

Putting it all together

In this chapter, you have found out how organic chemists make use of infrared spectroscopy and mass spectrometry for obtaining different information about an organic compound.

In practice, organic chemists interpret and information from a variety of sources when determining the structure of an organic molecule.

SPECTROSCOPY 17

A typical sequence for identification would include all of the following.

- **Elemental analysis** – use of percentage composition data to determine the empirical formula
- **Mass spectrometry** – use of the molecular ion peak from a mass spectrum to determine the molecular mass; use of fragment ions to identify sections of a molecule.
- **Infrared spectroscopy** – use of absorption peaks from an infrared spectrum to identify bonds and functional groups present in the molecule.

Once you have both the empirical formula and the molecular mass of a compound, you can determine the molecular formula of your unknown compound. By then using evidence from the infrared spectrum, it may be possible to identify an unknown compound.

You will find out more about combining different techniques in the A level course.

Synoptic link
See Topic 3.2 Determination of formulae for calculations of empirical and molecular formulae

Study tip
You may be provided with percentage compositions by mass, mass spectra, and infrared spectra. You might then have to combine evidence from these sources of information to determine an unknown structure.

Summary questions

1. Name two types of vibration caused by infrared radiation in molecules. *(2 marks)*

2. State the approximate wavenumbers for the infrared absorption peaks for the functional groups –OH (alcohols), –C=O (aldehydes and ketones), and –COOH (carboxylic acids) in the following molecules:
   a. propan-2-ol *(1 mark)*
   b. octanoic acid *(1 mark)*
   c. 2-hydroxybutanal *(1 mark)*
   d. 2-hydroxypropanoic acid *(1 mark)*

3. Identify the bonds responsible for the labelled absorptions in the infrared spectra below. *(9 marks)*

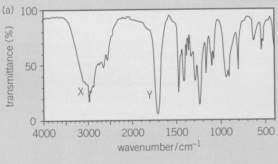
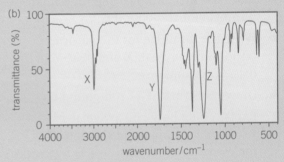
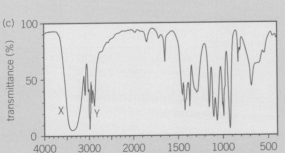
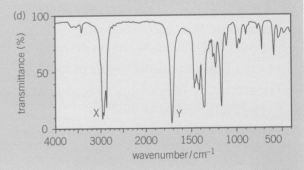

Chapter 17 Practice questions

Practice questions

1. Compound **G** is a branched-chain organic compound that does not have *E* and *Z* isomers. Elemental analysis of compound **G** gave the following percentage composition by mass:

   C, 55.8%; H, 7.0%; O, 37.2%.

   The mass spectrum and infrared spectrum of compound **G** are shown below.

   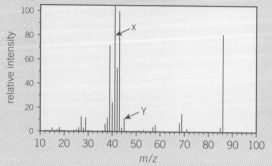

   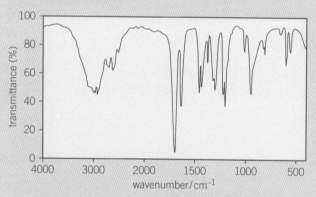

   - Calculate the empirical and molecular formulae for compound **G**.
   - Write the formulae for the particles responsible for peak **X** and peak **Y** in the mass spectrum.
   - Draw the structure of compound **G**.

     Explain fully how you arrive at a structure for compound **G** using all the evidence provided. *(7 marks)*

   OCR F322 June 2014 Q8 (b)

2. The solvent, **M**, is an organic compound used in paints. The solvent **M** was analysed.

   M has a relative molecular mass of 72.0.

   The percentage composition by mass of **M** is C, 66.7%; H, 11.1%; O, 22.2%.

   The infrared spectrum of **M** is shown below.

   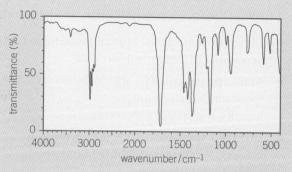

   The analysis produces several possible organic structures.

   Suggest, with reasons, **three** possible structures for **M**. *(6 marks)*

   OCR F322 June 2013 Q6(d) adapted

3. Compound **X** is a saturated compound that contains carbon, hydrogen and oxygen only.

   A scientist analyses a 1.00 g sample of compound **X** and finds it contains 0.133 g of hydrogen and 0.600 g of carbon.

   The scientist also analyses compound **X** using mass spectrometry and infrared spectroscopy.

   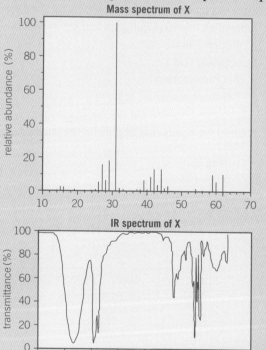

   Using all the information, show the structures of compound **X**.

   In your answer you should link the evidence with your explanation. *(7 marks)*

   OCR F322 Jan 2013 Q 8

SPECTROSCOPY 17

4 An analytical chemist was provided with a compound **J** which has an unbranched carbon skeleton. After analysis, the chemist obtained the following results.

Type of analysis	Evidence
infrared spectroscopy	broad absorption at 3350 cm^{-1}
percentage composition by mass	C, 70.59%; H, 13.72%; O, 15.69%
mass spectrometry	molecular ion peak at m/z = 102.0

Use this information to suggest all the possible structures for the unbranched compound **J**. In your answer you should make clear how your explanation is linked to the evidence. *(8 marks)*

OCR F322 June 2009 Q7(b)

5 Alcohol **E** is one of the following alcohols.

Butan-2-ol, ethane-1,2-diol, 2-methylpentan-3-ol, 2-methylpentan-3-ol, propan-1-ol, propan-2-ol

A student oxidises alcohol **E** by heating under reflux with excess acidified potassium dichromate(VI). An organic product **F** is isolated. The mass spectrum of the alcohol **E** and the IR spectrum of **F** are shown below.

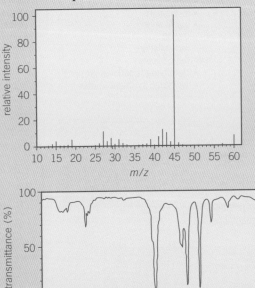

- Name **or** draw the structures of the alcohol **E** and the organic product **F**.
- Write an equation for the reaction of alcohol **E** with acidified potassium dichromate(VI).

Use [O] to represent the oxidising agent, acidified potassium dichromate(VI).

In your answer, you should make clear how each structure fits with the information given above.

(7 marks)

OCR F322 Jan 10 Q6(g)

6 Infrared spectroscopy and mass spectrometry are used in the search for organic molecules in outer space. Compound **A** has been analysed by infrared spectroscopy.

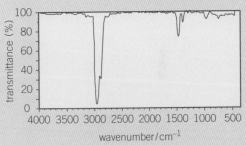

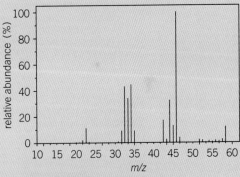

a A research chemist concludes that compound **A** is a hydrocarbon. What evidence is there to support this conclusion? *(2 marks)*

b How does the mass spectrum confirm that compound **A** has a molecular formula of C_4H_{10}? *(1 mark)*

c Identify the fragment ions that give rise to the peaks m/z 15, m/z 29 and m/z 43 in the mass spectrum. *(3 marks)*

d Identify molecule **A**. Explain your answer *(1 mark)*

OCR F322 June 10 Q4 (b)

Module 4 Summary

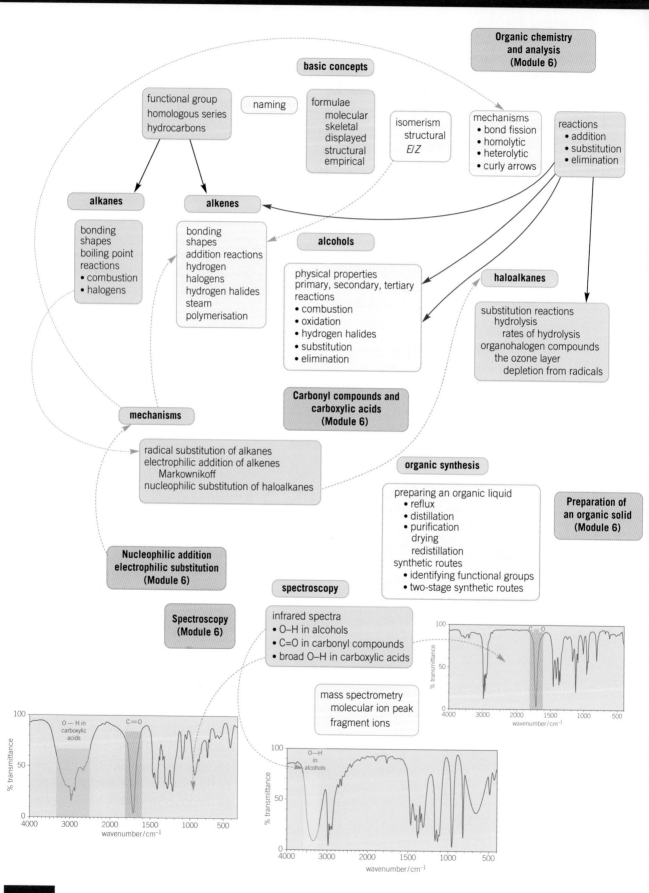

Module 4 Core organic chemistry and analysis

Acrolein and crotonaldehyde

The molecule acrolein, which has the systematic name propenal, is a unpleasant-smelling liquid with toxic properties. You may have unintentionally formed acrolein when you have burnt food, because the glycerol formed from the breakdown of fats in food can decompose to acrolein at temperatures above 280 °C

Acrolein takes part in a range of chemical reactions. Because of its reactivity, it has been useful as a starting material for the synthesis of more complex molecules, including perfumes, plastics, and adhesives.

A similar molecule, crotonaldehyde (butenal), has many of the same characteristics and is used in the manufacture of food preservatives and pesticides

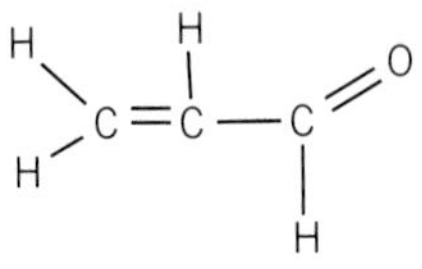

1. Name the two functional groups present in acrolein and crotonaldehyde.
2. Crotonaldehyde can exist as a pair of E/Z isomers, whereas acrolein cannot. Explain this difference and illustrate the meaning of E/Z isomers by drawing out the two isomers of crotonaldehdye.
3. Both of these molecules can be formed from alcohols. What reagents and conditions are required? Draw the structure of the alcohol that can be used to form crotonaldehyde.
4. Acrolein can form an addition polymer in the presence of oxygen and water. Draw out the structure of this polymer, showing the repeat unit clearly.
5. Acrolein reacts with bromine in an addition reaction. Outline the two stage mechanism.
6. Predict two peaks that would be present in the infrared spectrum of crotonaldehyde.

acrolein

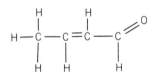

crotonaldehyde

Extension

1. Cyclopropane is the smallest member of the cycloalkane homologous series. The structure of cyclopropane makes it unusually reactive and it will take part in an electrophilic addition reaction with HBr to form 1-bromopropane. Research the reason for this unusual reactivity and draw out a possible mechanism for the reaction.
2. Summarise the key details of the following reactions. In your summary, give an example of each reaction and describe why the molecules in your example are able to react with one another.
   a radical substitution
   b electrophilic addition
   c nucleophilic substitution
3. Research some examples of photodegradable polymers. Use your research to identify structural features that seem to cause the polymer to be photodegradable.

265

Module 4 Practice questions

1. What is the systematic name for $CH_3CH_2CH_2C(CH_3)_2CH(CH_3)_2$?
   - A 3,3,5-trimethylhexane
   - B 2,3,3-trimethylhexane
   - C 4,4,5-trimethylhexane
   - D 1,1,2,2-tetramethylpentane *(1 mark)*

   AS Paper 1 style question

2. Which statement describes a nucleophile?
   - A donates an electron pair
   - B accepts an electron pair
   - C must be negatively charged
   - D has a single unpaired electron *(1 mark)*

   AS Paper 1 style question

3. Which name classifies a hydrocarbon as containing a benzene ring?
   - A cyclic
   - B hexagonal
   - C aliphatic
   - D aromatic *(1 mark)*

   AS Paper 1 style question

4. An alkane contains 30 hydrogen atoms per molecule. Its empirical formula is:
   - A C_6H_{15}
   - B C_7H_{15}
   - C $C_{14}H_{30}$
   - D $C_{15}H_{30}$ *(1 mark)*

   AS Paper 1 style question

5. Butane reacts with bromine in the presence of UV. How many dibrominated structural isomers could be formed?
   - A 2   B 3   C 5   D 6 *(1 mark)*

   AS Paper 1 style question

6. Alkenes normally react by:
   - A electrophilic substitution
   - B nucleophilic substitution
   - C electrophilic addition
   - D nucleophilic addition *(1 mark)*

   AS Paper 1 style question

7. Which compound is a likely product from addition of Br_2 to but-1-ene?
   - A 1,2-dibromobutane
   - B 1,3-dibromobutane
   - C 1,4-dibromobutane
   - D 2,3-dibromobutane *(1 mark)*

   AS Paper 1 style question

8. Which molecule has E/Z isomers?
   - A $CH_3CH_2CH_2CH=CH_2$
   - B $(CH_3)_2C=CHCH_3$
   - C $(CH_3)_2CHCH=CH_2$
   - D $CH_3CH_2CH=CHCH_3$ *(1 mark)*

   AS Paper 1 style question

9. Which compound **cannot** be oxidised under mild reaction conditions?
   - A 3-methylhexane-2,3-diol
   - B 4-ethylhexan-3-ol
   - C 3-ethylpentan-3-ol
   - D 2,3-dimethylhexan-1-ol *(1 mark)*

   AS Paper 1 style question

10. Which alcohol forms a mixture of alkenes when dehydrated?
    - A 2-methylpropan-1-ol
    - B propan-2-ol
    - C butan-1-ol
    - D butan-2-ol *(1 mark)*

    AS Paper 1 style question

11. Which property is **not** correct for ozone?
    - A Ozone in the upper atmosphere absorbs UV radiation.
    - B Ozone is broken down by chlorine radicals
    - C One of the steps in the breakdown of ozone is $Cl\bullet + O_3 \rightarrow ClO_2 + O\bullet$
    - D Ozone can be broken down by NO radicals formed during lightning strikes. *(1 mark)*

12. The boiling points of three liquids, **A**, **B** and **C**, are shown in the table below.

Name of liquid	Boiling point / °C
A	65
B	141
C	80

    Liquid **A** reacts with liquid **B** to produce liquid **C** and water. Distillation of the reaction mixture produces four pure liquids, which

266

are collected as they form in separate beakers. Which beaker would contain water?

A Beaker 1
B Beaker 2
C Beaker 3
D Beaker 4   (1 mark)

AS Paper 1 style question

13 Which piece of apparatus would you not require to carry out heating by reflux?

A round-bottomed or pear-shaped flask
B condenser
C rubber tubing
D receiver adaptor   (1 mark)

AS Paper 1 style question

14 Which information from an infrared spectrum suggests the presence of a carboxylic acid in an organic compound?

A A narrow peak at $1740\,cm^{-1}$ and a narrow peak at $1000\,cm^{-1}$
B A broad peak between 3200–$3550\,cm^{-1}$
C A narrow peak at $1700\,cm^{-1}$ and a broad peak at 3400–$3550\,cm^{-1}$
D A narrow peak at $1700\,cm^{-1}$ and a broad peak at 2500–$3100\,cm^{-1}$ (1 mark)

AS Paper 1 style question

15 Which peak would **not** be visible in the mass spectrum of ethanol?

A A molecular ion peak at 46
B A fragment ion peak at 32
C A M+1 peak at 47
D A fragment ion peak at 17   (1 mark)

AS Paper 1 style question

16 An alcohol **A** has the molecular formula $C_4H_{10}O$.

a Compound **A** could be one of four structural isomers.
   (i) What is meant by the term *structural isomers*?   (1 mark)
   (ii) Draw the structures of the four isomers.   (4 marks)

b Alcohol **A** is heated under reflux with a solution containing acidified dichromate(VI) ions to form an organic product **B** with the molecular formula C_4H_8O. The identity of compound **A** can now be deduced

Write an equation for this oxidation. Your equation should show the structures of alcohol **A** and compound **B**.   (3 marks)

c The mass spectrum for alcohol **A** has peaks at $m/z = 74$ and 45.

Suggest the molecular formulae for the species responsible for these two peaks.
                                               (2 marks)

AS Paper 1 style question

17 a Complete the flowchart below which illustrates some of the common reactions of 1-chloro-2-methylpropene.   (6 marks)

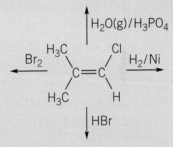

b 1-chloro-2-methylpropene undergoes an electrophilic addition reaction with bromine.
   (i) Explain how bromine acts as an electrophile   (1 mark)
   (ii) Show the reaction mechanism for the reaction between 1-chloro-2-methylpropene and bromine.
                                               (3 marks)
   (iii) 1-chloro-2-methylpropene can undergo polymerisation.
Draw a section of the polymer formed from this reaction showing two repeat units.
                                               (1 mark)

AS Paper 1 style question

18 Draw the stereoisomers of the following molecules. Use the Cahn-Ingold-Prelog rules to classify each of the isomers as either *E* or *Z*.

a Pent-2-ene
b Hex-3-ene   (4 marks)

AS Paper 1 style question

Module 4 Practice questions

19 Compound **A** is an organic acid. Its formula can be represented as RCOOH. R represents an alkyl group C_nH_{2n+1}.

A student carries out an experiment to try to identify the alkyl group R and compound **A**. The procedure included preparation of a standard solution and a titration.

  a The student weighs out 2.077 g of compound **A** and prepares 250 cm³ of solution for use in the titration.

  Describe how the student would prepare this solution using the weighed sample of compound **A**. *(3 marks)*

  b The student fills a burette with their prepared solution of compound **A**.

  The student adds 25.0 cm³ of 0.118 mol dm⁻³ KOH to a conical flask.

  The student added 31.25 cm³ of the solution of **A** from the burette to reach the end point.

  The equation in the titration is:

  RCOOH(aq) + KOH(aq) →

  RCOOK(aq) + H₂O(l)

  (i) Calculate the amount, in mol of compound **A** used in the titration.
  Show your working.

  (ii) Calculate the molar mass of compound **A** and suggest the identity of the alkyl group R.
  Show your working. *(5 marks)*

  c Draw out possible structure(s) for compound **A**. *(2 marks)*

  AS Paper 1 style question

20 Cyclohexene, C_6H_{10} can be prepared from cyclohexanol, $C_6H_{11}OH$, as described below.

17.0 g of cyclohexanol and 4 g of phosphoric acid are heated in the apparatus shown below.

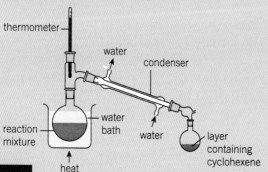

The equation for the reaction is shown below.
$C_6H_{12}O \rightarrow C_6H_{10} + H_2O$

  a Draw skeletal formulae for cyclohexanol and cyclohexene. *(2 marks)*

  b State the role of the phosphoric acid in the reaction. *(1 mark)*

  c The distilled mixture contains an aqueous and an organic layer.

  The student attempts to purify the distilled mixture to obtain a pure sample of cyclohexene.

  Outline the steps that the student would need to carry out. *(3 marks)*

  d The student obtains 3.69 g of purified cyclohexene.

  Calculate the percentage yield of cyclohexene.

  Give you answer to an appropriate number of significant figures. *(3 marks)*

  e The student carries out a simple test on their sample of cyclohexene to show that it is unsaturated.

  Describe this test.

  Include the expected observation and the structure of any organic product. *(2 marks)*

  f Explain how IR spectroscopy could be used to show that the purified product contained no unreacted cyclohexanol. *(2 marks)*

  AS Paper 1 style question

21 This question looks at properties and reactions of the alkane and alkene homologous series.

  a Explain what is meant by the term homologous series. *(2 marks)*

  b Explain, with examples, how the boiling points of alkanes are affected by chain length and branching. *(4 marks)*

  c The structures of but-1-ene and but-2-ene differ only in the position of the double bond but they have significant differences in their properties.

  (i) But-2-ene has *E*/*Z* stereoisomers but but-1-ene does not.

268

Explain this difference.

Your answer should include the following

- the meaning and origin of *E/Z* stereoisomerism
- the structures of *E* and *Z* but-2-ene.
- reasons why but-1-ene does not show *E/Z* stereoisomerism.

(4 marks)

(ii) But-1-ene reacts with hydrogen bromide to form two structural isomers, one in a much greater quantity than the other. However but-2-ene forms only one structural isomer.

Explain this observation. Include a mechanism for the reaction with but-2-ene in your answer. *(6 marks)*

AS Paper 1 style question

22 1-iodobutane, $CH_3(CH_2)_2CH_2I$, is a structural isomer of C_4H_9I.

a Draw structures to show the other structural isomers of C_4H_9I. Name each structure. *(3 marks)*

b $CH_3(CH_2)_2CH_2I$ can be converted into butan-1-ol by alkaline hydrolysis.

(i) Write an equation for this reaction *(1 mark)*

(ii) Outline the mechanism for this reaction. *(3 marks)*

c Butan-1-ol is much more soluble in water than 1-iodobutane. Explain why? *(3 marks)*

d Butan-1-ol fully combusts in an excess of air.

(i) Write an equation for this combustion. *(1 mark)*

(ii) What volume of air is needed to fully combust 4.07 g of butan-1-ol? *(3 marks)*

AS Paper 1 style question

23 Crude oil contains many hydrocarbons.

The table shows information about some of these hydrocarbons.

Hydrocarbon	Molecular formula	Boiling point / °C
hexane	C_6H_{14}	69
3-methylpentane	C_6H_{14}	63
2,2-dimethylbutane	C_6H_{14}	50

a What is the empirical formula of hexane? *(1 mark)*

b Explain why hexane is both *saturated* and a *hydrocarbon*. *(2 marks)*

c Draw the skeletal formula for 2,2-dimethylbutane. *(1 mark)*

d Describe and explain the trend shown by the boiling points of the hydrocarbons in the table. *(3 marks)*

24 1-Bromobutane, $CH_3CH_2CH_2CH_2Br$, is one of four structural isomers of C_4H_9Br.

a Draw the structures of the other three structural isomers of C_4H_9Br. Name each isomer. *(3 marks)*

b 1-Bromobutane can be prepared by reacting butane with bromine. The mechanism takes place in three stages, initiation, propagation and termination.

(i) Write an overall equation for this reaction. *(1 mark)*

(ii) What conditions are required for this reaction? *(1 mark)*

(iii) Name this mechanism. *(1 mark)*

(iv) Write equations for the initiation and propagation stages in the mechanism for this reaction. *(3 marks)*

(v) One possible termination equation is the reverse of the initiation stage.

Write equations for **two** other possible termination equations of this reaction. *(2 marks)*

MODULE 5
Physical chemistry and transition elements

Chapters in this module
18 Rates of reaction
19 Equilibrium
20 Acids, bases, and pH
21 Buffers and neutralisation
22 Enthalpy and entropy
23 Redox and electrode potentials
24 Transition elements

Introduction

In this module many of the physical chemistry topics that you studied during the first year of your course will be further developed and extended. You will learn how to apply mathematical skills to a quantitative study of equilibrium, rates, and energy, and to predict the extent, speed, and feasibility of a chemical reaction. Your knowledge of inorganic chemistry will be expanded by a detailed study of the transition elements.

The study of chemical equilibrium and reaction rates enables chemists to understand the quantitative effect of temperature, pressure, and concentration on reactions, including the control of industrial processes. You will also learn how an understanding of reaction rates allows you to predict the very steps that make up reaction mechanisms.

Acids, bases, and pH are important in many chemical and biological processes. The role of buffers in maintaining the pH of blood is used to show the essential role of acids and bases to our health and well-being.

Chemical reactions cause us to age, allow sugar to dissolve in water, and explain the rusting of iron – these reactions are spontaneous. The study of entropy, free energy, and electrode potentials allow you to predict whether reactions are feasible, and how conditions could be changed so that they may become feasible. You will also study the importance of redox reactions in electrochemical cells, including fuel cells and those used to power our portable devices.

Important practical quantitative techniques studied include the determination of reaction rates, measuring pH, and carrying out redox titrations.

Transition elements looks at the chemical and physical properties of the central region of the periodic table. You will learn about the colours, shapes, and reactions of transition metal ions and learn about the different types of isomerism shown by transition metal complexes.

Knowledge and understanding checklist

From your previous studies you should be able to answer the following questions. Work through each point, using your notes and the support available on Kerboodle.

- [ ] Calculate reaction rates from measurement of gradients for concentration–time graphs.
- [ ] Know how to investigate reaction rates by gas collection or mass loss over time.
- [ ] Write K_c expressions from an equation and calculate K_c from equilibrium concentrations.
- [ ] Apply le Chatelier's principle for changes in concentration, pressure, and temperature.
- [ ] Write neutralisation equations of acids with bases.
- [ ] Calculate an enthalpy change from enthalpy changes of formation.
- [ ] Apply oxidation number rules and describe redox in terms of electrons and oxidation number.
- [ ] Recognise tetrahedral and octahedral shapes.
- [ ] Know how to prepare a standard solution, to carry out a titration, and solve straightforward titration calculations.
- [ ] Identify anions and ammonium ions by qualitative analysis.

Maths skills checklist

In this module, you will need to use the following maths skills. You can find support for these skills on Kerboodle and through MyMaths.

- [ ] **Changing the subject of, substituting numbers into, and solving algebraic equations.** You will need this when carrying out rates, equilibrium, pH, energy, and titration calculations.
- [ ] **Plotting two variables from experimental or other data.** You will need this when you plot graphs using collected or supplied data from rates experiments.
- [ ] **Drawing and using the gradient of a tangent to a curve as a measure of rate of reaction.** You will need this when you calculate the reaction rate from a concentration–time graph.
- [ ] **Finding logarithms and their inverse.** You will need $\log_{10}$ and 10^x when calculating pH and H^+ concentrations, and $\ln$ and e^x when calculating rate constants and using the Arrhenius equation.
- [ ] **Finding arithmetic means.** You will need this when calculating mean titres.
- [ ] **Using angles and shapes in regular 2-D and 3-D structures.** You will need this when predicting the shapes of and bond angles in complex ions.
- [ ] **Visualising and representing 2-D and 3-D forms including 2-D representations of 3-D objects.** You will need this when drawing different types of isomer.
- [ ] **Understand the symmetry of 2-D and 3-D shapes.** You will need this when describing the types of stereoisomerism shown by complex ions.

MyMaths.co.uk
Bringing Maths Alive

18 RATES OF REACTIONS
18.1 Orders, rate equations, and rate constants

Specification reference: 5.1.1

Learning outcomes

Demonstrate knowledge, understanding, and application of:

→ the terms rate of reaction, order, overall order, rate constant

→ the rate equation and calculation of the rate constant k

→ deducing orders from experimental results.

Synoptic link

In Topic 10.1, Reaction rates, you learnt about some different methods for monitoring the rate of a reaction and how the rate can be determined graphically. In this chapter you will find out how reaction rates are linked to the concentration of reactants.

Study tip

The symbol $\propto$ means proportional to.

Synoptic link

In Topic 10.5, The equilibrium constant – part 1, you first met the idea of using square brackets for the concentration of a substance.

Rate of reaction

Every reaction has a reaction rate. Reaction rates are measured by observing the changes in the quantities of reactants or products over time.

In experiments, different quantities can be measured to calculate the rate of reaction. The units are then those of the quantity measured per unit of time.

$$\text{rate} = \frac{\text{quantity reacted or produced}}{\text{time}}$$

For consistency, chemists measure rates of reaction as the change of concentration with a change in time:

$$\text{rate} = \frac{\text{change in concentration}}{\text{change in time}}$$

Concentration is measured in $mol\,dm^{-3}$ and time can be any convenient measurement, depending on the overall rate of the reaction. If time has been measured in seconds, rate has the units of $\frac{mol\,dm^{-3}}{s}$, which should be written $mol\,dm^{-3}\,s^{-1}$.

Because concentration is such a common term, chemists use a square bracket as shorthand for concentration. For a reactant **A**:

- [**A**] is shorthand for 'concentration of **A**'
- [**A**] has the usual units of concentration, $mol\,dm^{-3}$.

Order of reaction

Changing the concentration often changes the rate of a reaction. The rate of reaction is proportional to the concentration of a particular reactant raised to a power. For example, for reactant [**A**] and power n, the rate is given by:

$$\text{rate} \propto [\mathbf{A}]^n$$

For each reactant, the power is the **order of reaction** for that reactant. In a reaction, different reactants can have different orders and each may affect the rate in different ways. Common orders are **zero order** (0), **first order** (1), and **second order** (2).

Zero order

When the concentration of a reactant has no effect on the rate, the reaction is zero order with respect to the reactant:

zero order: $\quad \text{rate} \propto [\mathbf{A}]^0$

In a zero order reaction:
- any number raised to the power zero is 1
- concentration does not influence the rate.

First order
A reaction is first order with respect to a reactant when the rate depends on its concentration raised to the power of one:

first order: rate $\propto [A]^1$

In a first order reaction:
- if the concentration of **A** is doubled (×2), the reaction rate increases by a factor of $2^1 = 2$
- if the concentration of **A** is tripled (×3), the reaction rate increases by a factor of $3^1 = 3$.

Second order
A reaction is second order with respect to a reactant when the rate depends on its concentration raised to the power of two:

second order: rate $\propto [A]^2$

In a second order reaction:
- if the concentration of **A** is doubled (×2), the reaction rate increases by a factor of $2^2 = 4$
- if the concentration of **A** is tripled (×3), the reaction rate increases by a factor of $3^2 = 9$.

The rate equation and the rate constant
The **rate equation** gives the mathematical relationship between the concentrations of the reactants and the reaction rate. For two reactants, **A** and **B**, the rate equation is shown below.

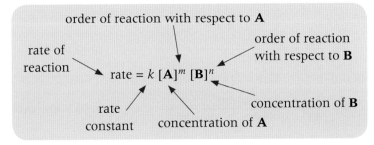

The **rate constant** k is the proportionality constant. It is the number that mathematically converts between the rate of reaction and concentration and orders.

Overall order
The overall order of reaction gives the overall effect of the concentrations of all reactants on the rate of reaction. It is calculated as follows:

overall order = sum of orders with respect to each reactant.

So, for the rate equation:

rate = $k[A]^m[B]^n$ overall order = $m + n$

Study tip
You can test this maths on your calculator.

Try raising any number to the power zero and you will always get 1 as the answer.

Study tip
In a first order reaction, any change in concentration gives the same change to the rate.

Study tip
In a second order reaction, any change in concentration changes the rate by the square of the change.

Study tip
A proportionality constant is just like an exchange rate.

Here the rate constant is the 'exchange rate' that converts concentration into rate (and vice versa).

18 18.1 Orders, rate equations, and rate constants

 Worked example: The rate equation and overall order for a reaction

For a reaction **A** + **B** + **C** → products, the orders are **A**: 0, **B**: 1, and **C**: 2.

Step 1: Write the overall order.

overall order = 1 + 2 = 3

Step 2: Write the rate equation.

rate = $k[\mathbf{A}]^0[\mathbf{B}]^1[\mathbf{C}]^2$

As $[\mathbf{A}]^0 = 1$, zero order reactants are usually omitted.

As $[\mathbf{B}]^1 = [\mathbf{B}]$, 1st order powers are usually omitted.

The rate equation is simplified to:

rate = $k[\mathbf{B}][\mathbf{C}]^2$

Units of the rate constant k

The units of the rate constant depend upon the number of concentration terms in the rate equation. The units of k can be determined by:

1. rearranging the equation to make k the subject
2. substitute units into the expression for k
3. cancel common units and show the final units on a single line.

Three examples are shown below but the principle is the same for any rate equation.

overall order: 0   rate = $k[A]^0 = k$

k = rate = mol dm^{-3} s^{-1} units = mol dm^{-3} s^{-1}

overall order: 1   rate = $k[A]^1$

$k = \dfrac{\text{rate}}{[A]^1} = \dfrac{\cancel{\text{mol dm}^{-3}}\,\text{s}^{-1}}{\cancel{\text{mol dm}^{-3}}}$ units = s^{-1}

overall order: 2   rate = $k[A]^2$

$k = \dfrac{\text{rate}}{[A]^2} = \dfrac{\cancel{\text{mol dm}^{-3}}\,\text{s}^{-1}}{(\cancel{\text{mol dm}^{-3}})\,(\text{mol dm}^{-3})}$ units = dm^3 mol^{-1} s^{-1}

 Worked example: Units of the rate constant k

What are the units of the rate constant k in the rate equation below?

rate = $k[A]^2[B]$

Step 1: Rearrange the rate equation to make the k the subject.

$k = \dfrac{\text{rate}}{[A]^2[B]}$

> **Study tip**
>
> Remember that the units of rate are mol dm^{-3} s^{-1} and the units of concentration are mol dm^{-3}.

> **Study tip**
>
> Notice that $[A]^2 = [A] \times [A]$ so there are two concentration units multiplied together:
>
> (mol dm^{-3})^2
>
> = (mol dm^{-3}) (mol dm^{-3})

> **Synoptic link**
>
> In Topic 19.1, The equilibrium constant K_c – part 2, you will meet the same idea of substituting and cancelling units.

Step 2: Substitute units and cancel the common units.

$$\frac{\text{mol dm}^{-3}\,\text{s}^{-1}}{(\text{mol dm}^{-3})(\text{mol dm}^{-3})(\text{mol dm}^{-3})} = \text{mol}^{-2}\,\text{dm}^{6}\,\text{s}^{-1}$$

The convention for writing units is to put positive indices before negative indices. So units are $\text{dm}^{6}\,\text{mol}^{-1}\,\text{s}^{-1}$.

Orders from experimental results

Orders of reaction must be determined experimentally by monitoring how a physical quantity changes over time. Orders cannot be found directly from the chemical equation. The example below shows how the rate equation and rate constant can be determined from experimental results. You will find out how these experimental results could be obtained in Topic 18.3.

When comparing the effect of different concentrations of reactants on reaction rates, it is important that the rate is always measured after the same time, ideally as close to the start of the experiment as possible. The example below shows the **initial rate**.

The initial rate is the instantaneous rate at the beginning of an experiment when $t = 0$.

Worked example: Rate constant from experimental results

Nitrogen dioxide reacts with ozone as shown.

$$2NO_2(g) + O_3(g) \rightarrow N_2O_5(g) + O_2(g)$$

The rate of the reaction is investigated and the following experimental results are obtained.

Experiment	$[NO_2(g)]$ / mol dm^{-3}	$[O_3(g)]$ / mol dm^{-3}	Initial rate / mol dm^{-3} s^{-1}
1	1.00×10^{-3}	2.50×10^{-3}	3.20×10^{-8}
2	2.00×10^{-3}	2.50×10^{-3}	6.40×10^{-8}
3	2.00×10^{-3}	5.00×10^{-3}	1.28×10^{-7}

Step 1: Determine the orders, overall order, and rate equation.

Comparing Experiment 1 and 2:

- $[NO_2(g)]$ doubles and $[O_3(g)]$ stays the same
- the rate also doubles
- the reaction is 1st order with respect to $NO_2(g)$.

Comparing Experiment 2 and 3:

- $[O_3(g)]$ doubles and $[NO_2(g)]$ stays the same
- the rate also doubles
- the reaction is 1st order with respect to $O_3(g)$.
  overall order = sum of individual orders = 1 + 1 = 2.

Study tip

Zero order – no matter what you do to the concentration, the rate does not change

1st order – whatever you do to the concentration, the rate changes by the same factor.

2nd order – whatever you do to the concentration, the rate changes by the same factor squared.

18.1 Orders, rate equations, and rate constants

> **Study tip**
>
> The results from Experiment 1 were used to calculate k but the results from any of the three experiments could have been used. For practice, try calculating k for Experiments 2 and 3 – you should get exactly the same numerical value for k.

Step 2: Write the rate equation.

$$\text{rate} = k[NO_2(g)][O_3(g)]$$

Step 3: Calculate the rate constant, including units.

Rearranging the rate equation: $k = \dfrac{\text{rate}}{[NO_2(g)][O_3(g)]}$

Substituting values from Experiment 1:

$$k = \dfrac{3.20 \times 10^{-8}}{1.00 \times 10^{-3} \times 2.50 \times 10^{-3}} = 1.28 \times 10^{-2}$$

Substituting units into k expression and cancel common units:

$$\dfrac{\cancel{mol\,dm^{-3}}\,s^{-1}}{(\cancel{mol\,dm^{-3}})(mol\,dm^{-3})} = dm^3\,mol^{-1}\,s^{-1}$$

$$k = 1.28 \times 10^{-2}\,dm^3\,mol^{-1}\,s^{-1}$$

Summary questions

1. In a reaction between reactants **A**, **B**, and **C**, the rate equation is:
   rate = $k[B]^2[C]$
   a State the orders with respect to **A**, **B**, and **C**. *(3 marks)*
   b Determine the overall order for the reaction. *(1 mark)*

2. A reaction is zero order with respect to **A**, first order with respect to **B**, and second order with respect to **C**.
   a What is the effect on the rate of doubling the concentration of   i **A**   ii **B**   iii **C**? *(3 marks)*
   b What is the effect on the rate of doubling all three concentrations at the same time? *(1 mark)*

3. Three reactants **A**, **B**, and **C** are reacted together. Doubling the concentration of **A** increases the rate by a factor of 2. Changing the concentration of **B** has no effect on the rate. Halving the concentration of **C** decreases the rate to half its original value.
   a State the orders with respect to **A**, **B**, and **C**. *(3 marks)*
   b Determine the overall order. *(1 mark)*
   c Determine the rate equation for the reaction. *(2 marks)*
   d Calculate the units of k. *(1 mark)*

4. The rate of reaction during an experiment is investigated and the following results are obtained.

Experiment	[A] / mol dm^{-3}	[B] / mol dm^{-3}	Initial rate / mol dm^{-3} s^{-1}
1	0.010	0.200	5.40×10^{-4}
2	0.020	0.200	2.16×10^{-3}
3	0.020	0.300	3.24×10^{-3}

   Determine the rate equation and calculate k, including units. *(6 marks)*

18.2 Concentration–time graphs
Specification reference: 5.1.1

Continuous monitoring of rate

Concentration–time graphs can be plotted from continuous measurements taken during the course of a reaction. This is called **continuous monitoring**. You have already seen two methods for continuous monitoring of reactions that produce a gas as one of the products:

- monitoring by gas collection
- monitoring by mass loss.

Not all reactions produce gases, so another property is needed that can be measured with time. A useful property is a colour change, which can be estimated by eye or monitored using a colorimeter.

Monitoring rate with a colorimeter

In a **colorimeter**, the wavelength of the light passing through a coloured solution is controlled using a filter. The amount of light absorbed by a solution is measured (Figure 1).

Learning outcomes

Demonstrate knowledge, understanding, and application of:

→ continuous monitoring of reaction rates
→ concentration–time graphs
→ half-life and first order reactions
→ determination of the rate constant k from half-lives.

Synoptic link

Revisit Topic 10.1, Reaction rates, to revise continuous monitoring using gases.

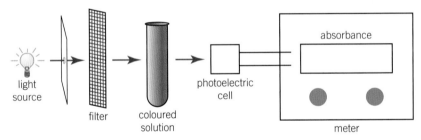

▲ **Figure 1** *A colorimeter measures the intensity of light passing through a sample. The filter is chosen so that it is the complementary colour to the colour being absorbed in the reaction. Absorbance is recorded, which is directly linked to the concentration of the solution*

▲ **Figure 2** *A colorimeter and the filter settings*

Analysing by colorimetry

The equation for the reaction of propanone and iodine, in the presence of an acid catalyst, is shown below.

$$CH_3COCH_3(aq) + I_2(aq) \rightarrow CH_3COCH_2I(aq) + H^+(aq) + I^-(aq)$$
orange/brown

As this reaction proceeds, iodine is used up and its orange/brown colour fades. The absorbance of the colour is measured precisely by the colorimeter.

The reaction can be monitored and the results analysed using the method outlined below.

1. Prepare standard solutions of known concentration of the coloured chemical, iodine, $I_2(aq)$, in this reaction.

18.2 Concentration–time graphs

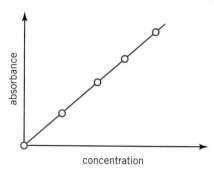

▲ Figure 3 A calibration curve can be used to determine concentrations from absorbance measurements from a colorimeter

2. Select a filter with the complementary colour of the coloured chemical. For iodine, this would be a green/blue filter but the colorimeter will usually tell you which setting to use.
3. Zero the colorimeter with water.
4. Measure the absorbance readings of the standard solutions of iodine.
5. Plot a calibration curve (Figure 3) of absorbance against iodine concentration. You now have a way of converting an absorbance reading into a concentration of iodine.
6. Carry out the reaction between propanone and iodine. Take absorbance readings of the reacting mixture at measured time intervals.
7. Use the calibration curve to measure the concentration of iodine at each absorbance reading.
8. Finally plot a second graph of concentration of iodine against time. From the **concentration–time graph**, you can determine the order of reaction with respect to the coloured chemical (in this case iodine).

> Look back at the equation for this reaction. Suggest another method that could be used for continuous monitoring of the rate of this reaction with time.

Synoptic link

You first encountered concentration–time graphs in Topic 10.1, Reaction rates.

Concentration–time graphs

Orders from shapes

The gradient of a concentration–time graph is the rate of the reaction. The order with respect to a reactant can also be deduced from the shape of a concentration–time graph for zero and first order reactions. The order with respect to a reactant can only be obtained if all other reactant concentrations remain effectively constant. Figure 4 shows the variation of concentration with time for a zero order reaction and a first order reaction.

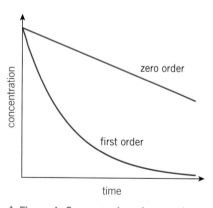

▲ Figure 4 Concentration–time graphs for zero and first order reactants

Zero order
A zero order reaction produces a straight line with a negative gradient. The reaction rate does not change at all during the course of the reaction. The value of the gradient is equal to the rate constant k. The straight-line graph makes a zero order relationship easy to identify.

First order
A first order reaction produces a downward curve with a decreasing gradient over time. As the gradient decreases with time, the reaction gradually slows down. In a first order concentration–time graph, the time for the concentration of the reactant to halve is constant. This time is called the **half-life** and the rate constant of a first order reaction can be determined using its value.

Study tip

You will not be expected to analyse a concentration–time graph for a second order reaction.

Second order
The graph for a second order is also a downward curve, steeper at the start, but tailing off more slowly.

RATES OF REACTIONS 18

Half-life and first order reactions

Half-life

Half-life $t_{1/2}$ is the time taken for half of a reactant to be used up. First order reactions have a constant half-life with the concentration halving every half life. This pattern is called **exponential decay**. A first order relationship can be confirmed from a concentration–time graph by measuring successive half-lives. If they are the same, the reaction is first order with respect to the reactant.

> **Study tip**
>
> Half life is the time taken for the concentration of a reactant to decrease to half its original value.

Dinitrogen pentoxide, decomposes to nitrogen dioxide and oxygen.

$$2N_2O_5(g) \rightarrow 4NO_2(g) + O_2(g)$$

Figure 5 shows a concentration–time graph for this decomposition. A tangent has been drawn when the concentration is $0.050 \, mol \, dm^{-3}$. This enables the rate to be calculated at this concentration.

- The graph shows three successive half-lives with a constant value of 100 s.
- The constant half-life means that this reaction is first order with respect to N_2O_5.
- $[N_2O_5]_0$ is the initial concentration of N_2O_5 at time = 0 s.

The rate equation for this reaction is then:

$$\text{rate} = k[N_2O_5(g)]$$

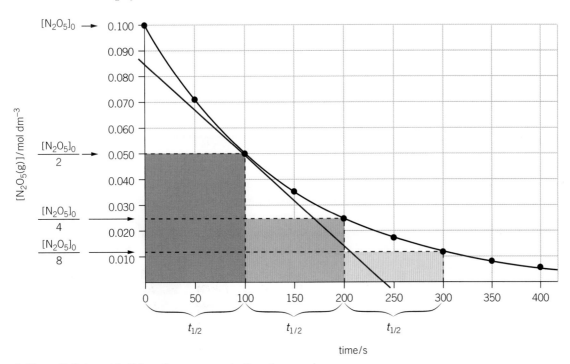

▲ Figure 5 *Constant half-lives from a concentration–time graph*

18.2 Concentration–time graphs

Determination of k for a first order reaction

There are two methods for determining the rate constant from a concentration–time graph for a first order reaction.

Calculating the rate constant from the rate

A tangent to the curve on the concentration–time graph is drawn at a particular concentration. The gradient of the tangent is calculated giving the rate of reaction.

The rate constant is calculated by rearranging the rate equation and substituting the value of rate (the gradient of the tangent) and the concentration at the position where the tangent has been drawn.

$$\text{rate} = k[\text{N}_2\text{O}_5(g)] \qquad \text{so, } k = \frac{\text{rate}}{[\text{N}_2\text{O}_5(g)]}$$

In Figure 4, a tangent has been drawn at $[\text{N}_2\text{O}_5(g)] = 0.050\,\text{mol}\,\text{dm}^{-3}$.

$$\text{rate} = \text{gradient} = \frac{0.084}{240} = 3.5 \times 10^{-4}\,\text{mol}\,\text{dm}^{-3}$$

$$\text{so, } k = \frac{\text{rate}}{[\text{N}_2\text{O}_5(g)]} = \frac{3.5 \times 10^{-4}}{0.050} = 7.0 \times 10^{-3}\,\text{s}^{-1}$$

Calculating the rate constant from the half-life

A much easier way of calculating the value of the rate constant is to make use of the exponential relationship for a constant half-life:

$$k = \frac{\ln 2}{t_{1/2}}$$

This method is also much more accurate than drawing a tangent – it is always difficult to judge how steep the tangent should be drawn. From Figure 2:

$$k = \frac{\ln 2}{t_{1/2}} = \frac{\ln 2}{100} = \frac{0.693}{100} = 6.93 \times 10^{-3}\,\text{s}^{-1}$$

> **Synoptic link**
>
> Refer back to Topic 10.1, Reaction rates, to see how the gradient of a tangent gives the rate of reaction.

> **Study tip**
>
> Remember that you can find the units of a rate constant by substituting units into the rate expression and cancelling the common units.
>
> $$k = \frac{\cancel{\text{mol}\,\text{dm}^{-3}}\,\text{s}^{-1}}{\cancel{\text{mol}\,\text{dm}^{-3}}}$$
>
> units $= \text{s}^{-1}$

> **Study tip**
>
> You will find the ln button on a scientific calculator.
>
>
>
> On this calculator, it is next to the log button but it may be in a different position amongst the scientific functions of a calculator.

Exponential decay in medicine

Drug breakdown in the body is affected by many different factors, but many drugs break down by exponential decay. The average half-lives of two medicines used to combat asthma are:

- salbutamol: 1.6 hours
- salmeterol: 5.5 hours.

Doctors will advise that these drugs should be used differently – salbutamol for asthma attacks and salmeterol late at night before sleeping.

▲ **Figure 6** *Salbutamol is used in inhalers to combat asthma*

1. What approximate proportion of each drug will remain in the body 11 hours after use?
2. Why should the doctor advise patients to use these drugs differently?

RATES OF REACTIONS 18

Summary questions

1. Describe how you can find the rate constant from a concentration–time graph for a zero order reaction. *(2 marks)*

2. The half-life of two first order reactions are:
   - Reaction 1 $t_{1/2} = 46.0$ s
   - Reaction 2 $t_{1/2} = 165$ s
   - Use the equation $k = \dfrac{\ln 2}{t_{1/2}}$ to calculate the rate constant of each reaction. *(2 marks)*

3. The reaction $CH_3COCH_3(aq) + I_2(aq) \rightarrow CH_3COCH_2I(aq) + H^+(aq) + I^-(aq)$ was carried out. The concentration of I_2 was recorded every 10 seconds and the results in Table 1 obtained.

   a. Plot a graph of $[I_2(aq)]$ on the y-axis against time on the x-axis. *(4 marks)*

   b. What is the order with respect to I_2? Explain your answer. *(1 mark)*

▼ Table 1

$[I_2(aq)]$ / mol dm^{-3}	Time / s
0.004 10	0
0.003 85	10
0.003 60	20
0.003 30	30
0.003 10	40
0.002 85	50
0.002 60	60

4. When heated strongly, nitrous oxide, N_2O, decomposes into its elements.

   $2N_2O(g) \rightarrow 2N_2(g) + O_2(g)$

   This reaction is first order with respect to N_2O. The graph below shows how nitrous oxide decomposes.

   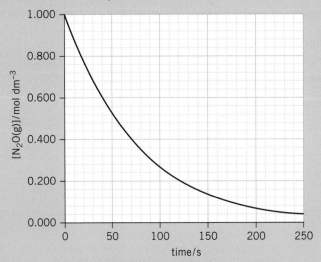

   a. Use half-life to show that the reaction is first order with respect to N_2O and deduce the rate equation. *(3 marks)*

   b. Determine the rate when $t = 70$ s. *(2 marks)*

   c. Calculate the rate constant using
       i the rate at 70 s   ii the ln2 method. *(4 marks)*

18.3 Rate–concentration graphs and initial rates

Specification reference: 5.1.1

Learning outcomes

Demonstrate knowledge, understanding, and application of:

→ deducing orders from the shapes of rate–concentration graphs
→ determination of rate constant for a first order reaction from the gradient
→ initial rate investigations and clock reactions.

Rate–concentration graphs

Rate–concentration graphs can be plotted from measurements of the rate of reaction at different concentrations. Rate–concentration graphs are very important as they offer a route into the direct link between rate and concentration in the rate equation.

Orders from shapes

Figure 1 shows the variation of rate with concentration for zero order, first order, and second order reactants. The order with respect to a reactant can be deduced from the shapes of these graphs.

Zero order

A zero order reactant produces a horizontal straight-line with zero gradient (Figure 1).

From the rate equation:

$$\text{rate} = k[\mathbf{A}]^0 \quad \text{so, rate} = k$$

- The intercept on the y-axis gives the rate constant k.
- The reaction rate does not change with increasing concentration.

First order

A first order reactant produces a straight-line graph through the origin.

From the rate equation:

$$\text{rate} = k[\mathbf{A}]^1 \quad \text{so, rate} = k[\mathbf{A}]$$

- Rate is directly proportional to concentration for a first order relationship.
- The rate constant can be determined by measuring the gradient of the straight line of this graph.

Second order

A second order reactant produces an upward curve with increasing gradient.

From the rate equation:

$$\text{rate} = k[\mathbf{A}]^2$$

- As this rate–concentration graph is a curve, the rate constant cannot be obtained directly from this graph.
- By plotting a second graph of the rate against the concentration squared, the result is a straight line through the origin. The gradient of this straight line graph is equal to the rate constant k.

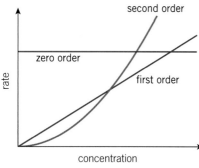

▲ Figure 1 *Rate–concentration graphs for zero, first, and second order reactants*

RATES OF REACTIONS 18

Worked example: Rate constant from a rate–concentration graph

Dinitrogen pentoxide, N_2O_5, decomposes to form nitrogen dioxide, NO_2, and oxygen, O_2.

$$2N_2O_5(g) \rightarrow 4NO_2(g) + O_2(g)$$

Figure 2 shows a rate–concentration graph for this decomposition.

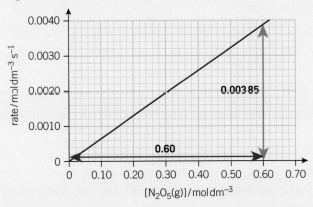

▲ **Figure 2** *Calculating gradient of rate–concentration graph*

Step 1: Determine the order and the rate equation.

The graph is a straight line through the origin – first order with respect to N_2O_5. The rate equation is:

$$\text{rate} = k[N_2O_5(g)]$$

Step 2: Determine the rate constant from the gradient.

The reaction is first order and the gradient gives the rate constant.

$$\text{gradient} = \frac{\text{change in rate}}{\text{change in concentration}} = \frac{0.00385}{0.60}$$

So the rate constant $k = 6.4 \times 10^{-3}\,s^{-1}$

Log–Log graphs

A graph of log(rate) against log(concentration) is a convenient way of finding both the order and the rate constant.

The rate equation for a reactant **A** is:

$$\text{rate} = k[\mathbf{A}]^n$$

The logarithmic version of this relationship fits the $y = mx + c$ equation for a straight-line graph:

$$\underbrace{\log(\text{rate})}_{y} = \underbrace{n}_{m} \underbrace{\log[\mathbf{A}]}_{x} + \underbrace{\log k}_{c}$$

The gradient gives the order n and the intercept gives $\log k$, from which k can be calculated.

Analyse the log–log graph in Figure 3 to work out the order with respect to **A** and the value of the rate constant.

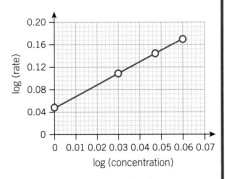

▲ **Figure 3** *Graph of log(rate) against log(concentration)*

283

18.3 Rate–concentration graphs and initial rates

The initial rates method

The initial rate is the instantaneous rate at the start of a reaction when the time $t = 0$. The initial rate can be found by measuring the gradient of a tangent drawn at $t = 0$ on a concentration–time graph.

A **clock reaction** is a more convenient way of obtaining the initial rate of a reaction by taking a single measurement. The time t from the start of an experiment is measured for a visual change to be observed, often a colour or a precipitate.

Provided that there is no significant change in rate during this time, it can be assumed that the average rate of reaction over this time will be the same as the initial rate.

The initial rate is then proportional to $\frac{1}{t}$.

The clock reaction is repeated several times with different concentrations, and values of $\frac{1}{t}$ are calculated for each experimental run.

> **Synoptic link**
>
> Look back to Topic 10.1, Reaction rates, to revise measuring and calculating the initial rate of reaction using concentration–time graphs.

Iodine clocks

A common type of clock reaction relies on the formation of iodine. As aqueous iodine is coloured orange-brown, the time from the start of the reaction and the appearance of the iodine colour can be measured. Starch is usually added since it forms a complex with iodine which is a an intense dark blue-black colour.

Iodine clock procedure

Separate experiments are carried out using different concentrations of one of the reactants and all other concentrations are kept constant. The colour change is delayed by including a small amount of another chemical (e.g., aqueous sodium thiosulfate, $Na_2S_2O_3(aq)$) which actually removes of iodine as it forms. As soon as this chemical is all used up, the blue-black colour appears.

- In each experiment, the solution is colourless at the start and the time t is measured for the blue-black colour of the starch–iodine to appear.
- The initial rate is proportional to $\frac{1}{t}$.
- A graph of $\frac{1}{t}$ (which is proportional to rate) against concentration is then plotted and the shape matched to the shapes in Figure 1.

Further series of experiments are then carried out in which the concentration of one of the other reactants is changed.

From the results, the order with respect to each reactant is determined and a rate equation is written. The rate constant can then be calculated.

▲ **Figure 4** *An iodine clock reaction. The left conical flask shows the colourless solution at the start of the reaction. The right conical flask shows the blue-black colour of the starch–iodine complex. You need to time to the first appearance of the blue-black colour which is initially very pale*

> The equation between thiosulfate ions, S_2O_3, and iodine, I_2, in shown below.
>
> $$2S_2O_3^{2-}(aq) + I_2(aq) \rightarrow S_4O_6^{2-}(aq) + 2I^-(aq)$$
>
> 1. Use the equation to explain why thiosulfate ions delay the colour.
> 2. Why is it important to use the same volume of thiosulfate ions each time?

RATES OF REACTIONS 18

 Worked example: Rate equation from a clock reaction

Hydrogen peroxide, $H_2O_2(aq)$, reacts with iodide ions in acid solution to form iodine.

$$H_2O_2(aq) + 2I^-(aq) + 2H^+(aq) \rightarrow I_2(aq) + 2H_2O(l)$$

Five experiments are carried out using different concentrations of hydrogen peroxide with the same volumes and concentrations of iodide ions and acid. In each experiment, the time taken for the blue-black colour of the starch–iodine complex to form is measured. The results are shown in Table 1.

▼ **Table 1** *Time taken for the starch–iodine complex to form at different $H_2O_2(aq)$ concentrations*

$[H_2O_2(aq)]$ / mol dm^{-3} s^{-1}	Time / s	$\frac{1}{t}$ / s^{-1}
0.050	23	0.043
0.040	29	0.034
0.030	39	0.026
0.020	68	0.015
0.010	129	0.007

A graph of $\frac{1}{t}$ ($\propto$ rate) against $[H_2O_2(aq)]$ is plotted (Figure 5).

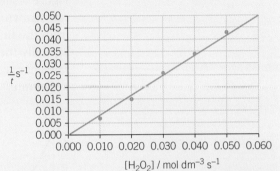

▲ **Figure 5** *Rate-concentration graph for $H_2O_2(aq)$*

Two further series of experiments are then carried out – one where $[I^-(aq)]$ is varied and one where $[H^+(aq)]$ is varied.

Rate concentration graphs are plotted for each series of experiments (Figure 6 and Figure 7).

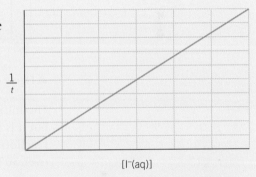

▲ **Figure 6** *Rate-concentration graph for $I^-(aq)$*

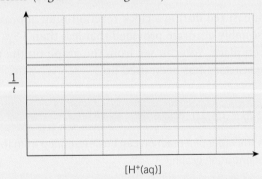

▲ **Figure 7** *Rate-concentration graph for $H^+(aq)$*

Analyse the results to determine the rate equation.

Step 1: Work out the orders from the shapes of the graphs.

    Figure 5: straight line through the origin   1st order for $H_2O_2(aq)$

    Figure 6: straight line through the origin   1st order for $I^-(aq)$

    Figure 7: horizontal straight-line   zero order for $H^+(aq)$

Step 2: Write the rate equation by combining the orders.

$$\text{rate} = k[H_2O_2(aq)][I^-(aq)]$$

> **Study tip**
>
> The reaction is zero order with respect to H^+. You can write the rate equation including H^+:
>
> $$\text{rate} = k[H_2O_2(aq)][I^-(aq)][H^+(aq)]^0$$
>
> However, any number to the power zero is equal to 1 and so zero order species are usually omitted to give the final rate equation as:
>
> $$\text{rate} = k[H_2O_2(aq)][I^-(aq)]$$

18.3 Rate–concentration graphs and initial rates

Clock reactions – how accurate?

In a clock reaction you are measuring the average rate during the first part of the reaction. Over this time, you can assume that the average rate of reaction is constant and is the same as the initial rate. In a clock reaction, you are measuring an average rate of a change in reactant over time. The shorter the period of time over which an average rate is measured, the less the rate changes over that time period.

Figure 7 shows two concentration–time graphs, with the measured rate shown by the dashed lines. On the left, the measurement of t is from the first steepest part of the curve. The rate hardly changes and is virtually the same as the initial rate. On the right, you can see that the rate changes over the time measured and there is a difference from the initial rate.

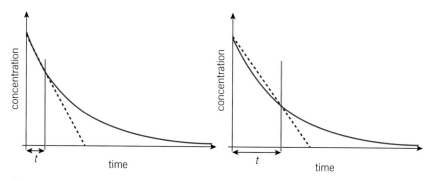

▲ **Figure 8** *The accuracy of a clock reaction is dependent in the time measured*

The initial rate measured during a clock reaction is an approximation but it is still reasonably accurate provided that less than 15% of the reaction has taken place. You can see from the graph on the right that over 50% of the reactant has been used up.

Summary questions

1. Describe how the order of a reaction can be determined from the shape of a rate–concentration graph. *(3 marks)*

2. Describe how a rate constant can be determined from a rate–concentration graph. *(3 marks)*

3. The table below shows the results of a clock reaction involving species **A**.

[A] / mol dm^{-3} s^{-1}	Time / s
0.050	33
0.040	43
0.030	58
0.020	80
0.010	185

   a. Determine the initial rates for each concentration as $\frac{1}{t}$. *(2 marks)*
   b. Plot a rate–concentration graph. *(4 marks)*
   c. Determine the order of reaction with respect to **A**. *(1 mark)*

18.4 Rate-determining step

Specification reference: 5.1.1

The rate-determining step

Multi-step reactions

An overall chemical equation compares the reactants and products. The balancing numbers give the **stoichiometry**, the relative amounts of the species in the reaction. For example:

$$H_2O_2(aq) + 2I^-(aq) + 2H^+(aq) \rightarrow I_2(aq) + 2H_2O(l)$$

  1 mol   2 mol   2 mol

A reaction can only take place when particles collide. For the reaction above to take place in a single step, one molecule of H_2O_2, two I^- ions, and two H^+ ions would have to collide together simultaneously, which is an extremely unlikely event.

Such reactions are much more likely to take place in a series of steps and it is unlikely that more than two particles will collide together at the same time. The series of steps that make up an overall reaction is called the **reaction mechanism**.

The rate-determining step

The steps in a multi-step reaction will take place at different rates. The slowest step in the sequence is called the **rate-determining step**.

Predicting reaction mechanisms

Chemists use their knowledge and understanding of chemical principles to propose possible mechanisms for reactions. But how do you know whether a reaction mechanism is likely to be correct?

- The rate equation only includes reacting species involved in the rate-determining step.
- The orders in the rate equation match the number of species involved in the rate-determining step.

So the rate-determining step provides important evidence in supporting or rejecting a proposed reaction mechanism.

The hydrolysis of haloalkanes

Haloalkanes are hydrolysed by hot aqueous alkali:

$$RBr + OH^- \rightarrow ROH + Br^-$$

The hydrolysis reactions of haloalkanes can be investigated experimentally to determine the overall order of reaction, the rate equation, and a possible mechanism for the reaction.

Learning outcomes

Demonstrate knowledge, understanding, and application of:

→ the rate-determining step
→ predicting a rate equation that is consistent with the rate-determining step
→ predicting possible steps in a reaction mechanism from the rate equation and the balanced equation.

Synoptic link

See Topic 3.4, Reacting quantities, for more details of stoichiometry

Study tip

Imagine a production line in a factory. Three steps are involved in making a product. The first step takes 30 minutes to complete. The other two steps take two minutes. The first step is the slowest step and takes up the vast majority of the time required to make the product. The first step is the rate-determining step.

Synoptic link

In Topic 15.1, The chemistry of haloalkanes, you first studied the hydrolysis of haloalkanes.

18.4 Rate-determining step

 Worked example: The hydrolysis of tertiary haloalkanes

The tertiary haloalkane, $(CH_3)_3CBr$, is hydrolysed by aqueous alkali.

$$(CH_3)_3CBr + OH^- \rightarrow (CH_3)_3COH + Br^-$$

The rate equation has been determined experimentally as:

rate = $k[(CH_3)_3CBr]$

Use this information to propose a reaction mechanism for this reaction.

Stage 1: Summarise the information that you know

The rate equation shows that the slow rate-determining step involves only one molecule of $(CH_3)_3CBr$. OH^- has *no effect* on the reaction rate and must be involved in a fast step.

Stage 2: Compare the overall equation with the information from Stage 1. Any sequence of steps must add up to give the overall equation.

Step 1 $(CH_3)_3CBr$ $\rightarrow$ slow

Step 2 $+ OH^- \rightarrow$ fast

Overall $(CH_3)_3CBr + OH^- \rightarrow (CH_3)_3COH + Br^-$

Stage 3: Identify the intermediate and work out the two-step mechanism:

- Step 1 must form two species, one the other product of the reaction, Br^-.
- The other species formed must be $(CH_3)_3C^+$.
- $(CH_3)_3C^+$ doesn't feature in the overall equation – it must be formed in Step 1 and used up in Step 2 and is an **intermediate**.

The completed two-step mechanism is shown below.

Step 1 $(CH_3)_3CBr \rightarrow (CH_3)_3C^+ + Br^-$ slow (rate-determining step)

Step 2 $(CH_3)_3C^+ + OH^- \rightarrow (CH_3)_3COH$ fast

Overall $(CH_3)_3CBr + OH^- \rightarrow (CH_3)_3COH + Br^-$

- The slow Step 1 matches the rate equation
- The sum of the steps gives the overall equation.

Summary questions

1. Nitrogen and hydrogen react as follows:
   $N_2(g) + 3H_2(g) \rightarrow 2NH_3(g)$.
   Explain why it is likely that this reaction proceeds by more than one step. *(1 mark)*

2. A proposed two-step mechanism for a reaction is shown below.
   $H_2O_2(aq) + Br^-(aq) \rightarrow H_2O(l) + BrO^-(aq)$ slow
   $H_2O_2(aq) + BrO^-(aq) \rightarrow H_2O(l) + Br^-(aq) + O_2(g)$ fast
   a Determine the rate equation for this reaction. *(1 mark)*
   b Determine the overall equation. *(1 mark)*
   c Identify the intermediate. *(1 mark)*
   d State the role of Br^-. *(1 mark)*

3. The overall equation for the decomposition of N_2O is:
   $2N_2O(g) \rightarrow 2N_2(g) + O_2(g)$
   The rate equation is: rate = $k[N_2O(g)]$
   Suggest a possible two-step mechanism for the reaction. *(2 marks)*

18.5 Rate constants and temperature

Specification reference: 5.1.1

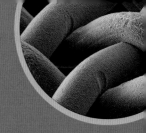

The effect of temperature on rate constants

You have already seen how temperature changes can affect the rate of a reaction in terms of the Boltzmann distribution.

As temperature increases, the rate increases and the value of the rate constant k will also increase. For many reactions each 10 °C rise in temperature doubles the rate constant and doubles the rate of the reaction. Figure 1 shows the typical variation of a rate constant with temperature.

What factors affect the rate constant?

When temperature increases, two factors contribute to the increased rate and rate constant.

- Increasing the temperature shifts the Boltzmann distribution to the right, increasing the proportion of particles that exceed the activation energy E_a.
- As the temperature increases, particles move faster and collide more frequently.

To react, particles must also collide with the correct orientation.

With increasing temperature, the increased frequency of collisions is comparatively small compared with the increase in the proportion of molecules that exceed E_a from the shift in the Boltzmann distribution. So the change in rate is mainly determined by E_a.

The Arrhenius equation

You have seen that rate constants can be determined experimentally. By carrying out the same experiment at different temperatures, rate constants can be calculated at different temperatures. The graph in Figure 1 shows the variation of the rate constant with temperature.

The Arrhenius equation, shown below, is an exponential relationship between the rate constant k and temperature T.

Learning outcomes

Demonstrate knowledge, understanding, and application of:

→ the effect of temperature change on reaction rate and rate constant

→ the Arrhenius equation

→ determination of E_a and A graphically from the Arrhenius equation.

Synoptic link

Look back at Topic 10.3, The Boltzmann distribution, to revise the basics of the effect of temperature on reaction rates.

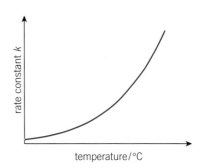

▲ Figure 1 *Variation of rate constant k with temperature T*

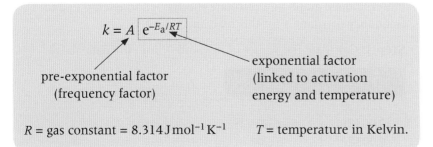

$$k = A\, e^{-E_a/RT}$$

pre-exponential factor (frequency factor)

exponential factor (linked to activation energy and temperature)

R = gas constant = 8.314 J mol⁻¹ K⁻¹ T = temperature in Kelvin.

289

18 18.5 Rate constants and temperature

The exponential factor $e^{-E_a/RT}$ represents the proportion of molecules that exceed E_a and that have sufficient energy for a reaction to take place.

The pre-exponential term (frequency factor) A takes into account the frequency of collisions with the correct orientation. This term does increase slightly with temperature as the frequency of collisions increases but it is essentially constant over a small temperature range. The frequency factor essentially gives the rate if there were no activation energy.

Logarithmic form of the Arrhenius equation

The Arrhenius equation can also be expressed as a logarithmic relationship:

$$\ln k = -\frac{E_a}{RT} + \ln A$$

This form of the Arrhenius equation is very useful as it enables both E_a and A to be determined graphically. A plot of $\ln k$ against $\frac{1}{T}$ gives a straight line graph of the type $y = mx + c$ (Figure 2).

$$\ln k = \underbrace{\left(-\frac{E_a}{R}\right)}_{m} \underbrace{\left(\frac{1}{T}\right)}_{x} + \underbrace{\ln A}_{c}$$

So a plot of $\ln k$ against $\frac{1}{T}$ gives a downward straight line with:

- gradient m of $-\frac{E_a}{R}$
- intercept c of $\ln A$ on the y axis.

Depending on whether $\ln k$ is positive or negative, you may find that your graph comes either side of the origin (it may even straddle the x-axis). However, the principle is the same. The gradient is $-\frac{E_a}{R}$ and intercept is $\ln A$.

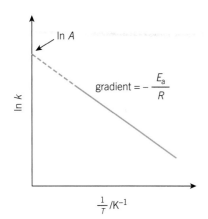

▲ **Figure 2** *Arrhenius plot with ln k positive*

> **Study tip**
>
> As you have seen, there are two forms of the Arrhenius equation. You will mainly be using the logarithmic form of the equation to determine E_a and A graphically, but it is important that you recognise the exponential form of the Arrhenius equation, $k = A\,e^{-E_a/RT}$.

Worked example: Using the Arrhenius equation to calculate E_a and A

Hydrogen peroxide decomposes to form water and oxygen.

$$2H_2O_2(aq) \rightarrow 2H_2O(l) + O_2(g)$$

The rate constant k varies with temperature as shown in Table 1.

▼ **Table 1**

T / K	k / s^{-1}
295	0.000 493
298	0.000 656
305	0.001 400
310	0.002 360
320	0.006 120

Plot a suitable graph and calculate the activation energy E_a and frequency constant A for this reaction.

Step 1: Calculate the values of $\ln k$ and $\frac{1}{T}$ from the data above:

$\frac{1}{T}$ / K^{-1}	ln k
0.003 39	−7.615
0.003 36	−7.329
0.003 28	−6.571
0.003 26	−6.049
0.003 13	−5.096

Step 2: Plot a graph of $\ln k$ against $\frac{1}{T}$.

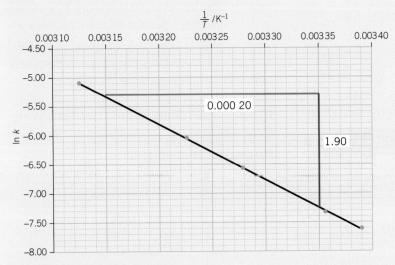

▲ **Figure 3** *Graph of ln k against $\frac{1}{T}$*

Step 3: Calculate the activation energy E_a from the gradient of the straight line

$$\text{Gradient} = -\frac{E_a}{R} = \frac{-1.90}{0.00020} = -9500$$

$$E_a = 8.314 \times 9500 = 78983 \, \text{J mol}^{-1} = 79 \, \text{kJ mol}^{-1}$$

Step 4: Calculate $\ln A$ from the intercept.

Figure 4 shows extrapolation of the straight-line graph from Figure 3 back to the y-axis at $\frac{1}{T} = 0 \, \text{K}^{-1}$.

The intercept gives the value for $\ln A$.

$\ln A = 25.40$.

To work out A on your calculator, you need to use the exponential function. The exponential function is the inverse of $\ln$, so $A = e^{\ln A}$

$$A = e^{25.40} = 1.07 \times 10^{11}$$

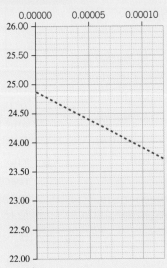

▲ **Figure 4** *Graph of ln k against $\frac{1}{T}$ showing intercept on y axis*

RATES OF REACTIONS

Study tip

The e (or EXP) function is normally accessed as the inverse of the ln button. You will need to use the [shift] or [2nd] button to access it.

Summary questions

1. Describe how the activation energy and frequency factor can be found graphically. *(2 marks)*

2. In the Arrhenius equation, state the effect on k of an increase in:
   a A *(1 mark)*
   b E_a *(1 mark)*
   c T *(1 mark)*

3. The variation of the rate constant with temperature for the first-order reaction
   $2N_2O_5(g) \rightarrow 2N_2O_4(g) + O_2(g)$
   is given below.

   Determine graphically the activation energy and frequency factor for the reaction. *(8 marks)*

T/K	k/s^{-1}
298	1.74×10^{-5}
308	6.61×10^{-5}
318	2.51×10^{-4}
328	7.59×10^{-4}
338	2.40×10^{-3}

Chapter 18 Practice questions

Practice questions

1. The rate equation for a reaction is:

   rate = $k[A]^2[B]$

   When [A] and [B] are both 0.010 mol dm^{-3}, rate = 1.2×10^{-4} mol dm^{-3} s^{-1}.

   a. Calculate k, including units. *(2 marks)*
   b. The experiment was repeated three times at the same temperature. Copy and complete the missing values in the table.

[A] / mol dm^{-3}	[B] / mol dm^{-3}	Rate / mol dm^{-3} s^{-1}
0.0100		2.40×10^{-4}
	0.0100	1.08×10^{-3}
0.0400	0.0400	
	0.0100	6.00×10^{-3}

   (4 marks)

2. Nitrogen pentoxide, N$_2$O$_5$(g), decomposes as below.

   $N_2O_5(g) \rightarrow 2NO_2(g) + \frac{1}{2}O_2(g)$

   The rate equation is:

   rate = $k[N_2O_5(g)]$

   When [N$_2$O$_5$(g)] = 0.20 mol dm^{-3}, the initial rate of disappearance of N$_2$O$_5$ is 2.8×10^{-3} mol dm^{-3} s^{-1}.

   a. Calculate k, including units. *(2 marks)*
   b. (i) What is meant by half-life? *(1 mark)*
      (ii) What is [N$_2$O$_5$(g)] and the rate after three half lives? *(2 marks)*
      (iii) Calculate the half-life. *(1 mark)*
   c. What is the initial rate of **formation** of NO$_2$ gas? *(1 mark)*

3. HBr reacts with O$_2$. The four-step mechanism below has been proposed for the reaction.

   HBr + O$_2$ → HBrO$_2$
   HBrO$_2$ + HBr → 2HBrO
   HBrO + HBr → H$_2$O + Br$_2$
   HBrO + HBr → H$_2$O + Br$_2$

   a. What is the overall equation for the reaction? *(2 marks)*
   b. For this reaction, the rate constant k is determined at different temperatures, T. From the results, a graph of ln k against $\frac{1}{T}$ is plotted.
      (i) What is the shape of the graph? *(1 mark)*
      (ii) How would you determine values for the activation energy E_a and the pre-exponential factor (frequency factor) A from the graph? *(2 marks)*

4. A student investigates the reaction between iodine, I$_2$, and propanone, (CH$_3$)$_2$CO, in the presence of aqueous hydrochloric acid, HCl(aq). The results of the investigation are shown below.

   Rate–concentration graph

   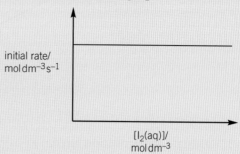

   Results of initial rates experiments

Experiment	[(CH$_3$)$_2$CO(aq)] / mol dm^{-3}	[HCl(aq)] / mol dm^{-3}	Initial rate / mol dm^{-3} s^{-1}
1	1.50×10^{-3}	2.00×10^{-2}	2.10×10^{-9}
2	3.00×10^{-3}	2.00×10^{-2}	4.20×10^{-9}
3	3.00×10^{-3}	5.00×10^{-2}	1.05×10^{-8}

   a. Determine the orders with respect to I$_2$, (CH$_3$)$_2$CO, and HCl, the rate equation, and the rate constant for the reaction. Explain all of your reasoning. *(9 marks)*
   b. The student then investigates the reaction of hydrogen, H$_2$, and iodine monochloride, ICl. The equation for this reaction is shown below.

   H$_2$(g) + 2ICl(g) → 2HCl(g) + I$_2$(g)

   The rate equation for this reaction is shown below.

   rate = $k[H_2(g)][ICl(g)]$

   Predict a possible two-step mechanism for this reaction. The first step should be the rate-determining step. *(2 marks)*

   F325 Jan 12 Q1

5. Iodide ions, I$^-$, react with S$_2$O$_8^{2-}$ ions as shown in the equation below.

   $2I^-(aq) + S_2O_8^{2-}(aq) \rightarrow I_2(aq) + 2SO_4^{2-}(aq)$

A student investigates the rate of this reaction using the initial rates method.

The student measures the time taken for a certain amount of iodine to be produced.

a   Outline a series of experiments that the student could have carried out using the initial rates method.

How could the results be used to show that the reaction is first-order with respect to both I^- and $S_2O_8^{2-}$? In your answer you should make clear how the results are related to the initial rates. *(4 marks)*

b   In one of the experiments, the student reacts together:
   - 8.0×10^{-2} mol dm^{-3} I^-(aq)
   - 4.0×10^{-3} mol dm^{-3} $S_2O_8^{2-}$(aq).

The initial rate is 1.2×10^{-3} mol dm^{-3} s^{-1}. The reaction is first-order with respect to I^- and first-order with respect to $S_2O_8^{2-}$.

Calculate the rate constant k for this reaction. State the units, if any. *(3 marks)*

OCR F325 Jan 13 Q4(a) (b)

6   In aqueous solution, benzenediazonium chloride, $C_6H_5N_2Cl$, decomposes above 10 °C.

$C_6H_5N_2Cl(aq) + H_2O(l) \rightarrow C_6H_5OH(aq) + N_2(g) + HCl(aq)$

A student investigates the rate of this reaction using an excess of water at 50 °C. The student takes measurements at intervals during the reaction and then plots his experimental results to give the graph shown below.

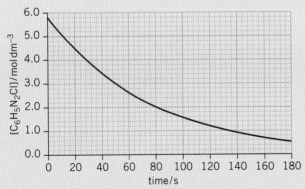

a   The student uses a half-life method to suggest the order of reaction with respect to $C_6H_5N_2Cl$.

   (i)   What is meant by the half-life of a reaction? *(1 mark)*

   (ii) Confirm the order of reaction with respect to $C_6H_5N_2Cl$. Show your working on the graph. *(2 marks)*

   (iii) What would be the effect, if any, on the half-life of this reaction of doubling the initial concentration of $C_6H_5N_2Cl$? *(1 mark)*

b   The student predicts that the rate equation is:
   rate = $k[C_6H_5N_2Cl]$.

   (i)   Using the graph and this rate equation, determine the rate of reaction after 40 s. Show your working on the graph. *(2 marks)*

   (ii) Calculate the rate constant k for this reaction and give its units. *(2 marks)*

c   The order of this reaction with respect to H_2O is effectively zero. Explain why. *(1 mark)*

OCR F325 Jun 13 Q2

7   A student carries out an initial rates investigation on the reaction below.

$5I^-(aq) + IO_3^-(aq) + 6H^+(aq) \rightarrow 3I_2(aq) + 3H_2O(l)$

From the results, the student determines the rate equation for this reaction:

rate = $k[I^-(aq)]^2 [IO_3^-(aq)] [H^+(aq)]^2$

a   (i)   What is the overall order of reaction? *(1 mark)*

   (ii) A proposed mechanism for this reaction takes place in several steps. Suggest two reasons why it is unlikely that this reaction could take place in one step. *(2 marks)*

b   On the rate–concentration graphs below, sketch lines to show the relationship between initial rate and concentration for IO_3^-(aq) and H^+(aq).

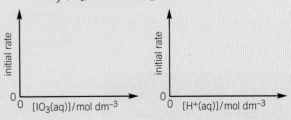

(2 marks)

F325 Jun 14 Q4 (a) (b)

19 EQUILIBRIUM

19.1 The equilibrium constant K_c – part 2

Specification reference: 5.1.2

Learning outcomes

Demonstrate knowledge, understanding, and application of:

→ expressions for K_c for homogeneous and heterogeneous equilibria

→ the techniques and procedures used to determine quantities present at equilibrium

→ calculation of quantities present at equilibrium

→ calculations of K_c including determination of units.

Synoptic link

Look back at Topic 10.5, The equilibrium constant K_c – part 1, to revisit the introductory work on equilibrium constants.

Study tip

The convention for writing units is to put positive indices *before* negative indices.

The equilibrium constant K_c

You have already seen how to write the equilibrium constant K_c for an equilibrium system in terms of the equilibrium concentrations of the species present at equilibrium.

For example, the expression for K_c for the equilibrium reaction $N_2(g) + 3H_2(g) \rightleftharpoons 2NH_3(g)$ is:

$$K_c = \frac{[NH_3(g)]^2}{[N_2(g)][H_2(g)]^3}$$

You also learnt how to calculate the numerical value for K_c from the equilibrium concentrations. You were able to then compare values of K_c – the larger the value, the further the position of equilibrium towards the products. This topic looks at how K_c can be calculated from experimental results.

Units of K_c

The units of K_c depend upon the number of concentration terms on the top and bottom of the equilibrium constant term. To work out the units:

- substitute units into the expression for K_c
- cancel common units and show the final units on a single line.

 Worked example: Calculating units for K_c

Calculate units for K_c in each of the following reactions.

Example 1

$$H_2(g) + I_2(g) \rightleftharpoons 2HI(g) \qquad K_c = \frac{[HI(g)]^2}{[H_2(g)][I_2(g)]}$$

$$= \frac{\cancel{(mol\,dm^{-3})^2}}{\cancel{(mol\,dm^{-3})}\cancel{(mol\,dm^{-3})}}$$

No units as all concentration terms cancel.

Example 2

$N_2O_4(g) \rightleftharpoons 2NO_2(g)$

$$K_c = \frac{[NO_2(g)]^2}{[N_2O_4(g)]} = \frac{(mol\,dm^{-3})^{\cancel{2}}}{\cancel{(mol\,dm^{-3})}} \qquad \text{units} = mol\,dm^{-3}$$

Example 3

$2SO_2(g) + O_2(g) \rightleftharpoons 2SO_3(g)$

$$K_c = \frac{[SO_3(g)]^2}{[SO_2(g)]^2[O_2(g)]} \Rightarrow \frac{\cancel{(mol\,dm^{-3})^2}}{\cancel{(mol\,dm^{-3})^2}(mol\,dm^{-3})}$$

units = $dm^3\,mol^{-1}$

Homogeneous and heterogeneous equilibria

Equilibria are divided into two main types – homogeneous and heterogeneous.

Homogeneous equilibria

A **homogeneous equilibrium** contains equilibrium species that *all* have the same state or phase. In all the examples above and the one below, the equilibria are **homogeneous** since all the equilibrium species have the same state and are gases:

$N_2(g) + 3H_2(g) \rightleftharpoons 2NH_3(g)$ homogeneous – all species are gases

Heterogeneous equilibria

A **heterogeneous equilibrium** contains equilibrium species that have different states or phases. For example, in the equilibrium below carbon is a solid but the other three species are all gases.

$C(s) + H_2O(g) \rightleftharpoons CO(g) + H_2(g)$ heterogeneous – mixture of states

In homogeneous equilibria, the K_c expression contains concentrations of all species. However, in heterogeneous equilibria, the concentration of solids and liquids are essentially constant. So any species that are solids and liquids are omitted from the K_c expression. They are automatically incorporated within the overall equilibrium constant. K_c only includes species that are (g) or (aq).

So for the equilibrium above, C(s) is constant and is omitted:

$$K_c = \frac{[CO(g)][H_2(g)]}{[H_2O(g)]} \quad \text{units} = \text{mol dm}^{-3}$$

Calculating equilibrium quantities and calculating K_c

You have seen how to calculate K_c from equilibrium concentrations. When working from experimental data, you will need to calculate the equilibrium concentrations from your experimental results.

> **Worked example: Calculation of K_c from equilibrium amounts**
>
> Nitrogen monoxide, NO, and oxygen, O_2, react together in a reversible reaction.
>
> $2NO(g) + O_2(g) \rightleftharpoons 2NO_2(g)$
>
> 1.60 mol NO(g) and 1.40 mol O_2(g) are mixed together in a container with a volume 4.0 dm³. At equilibrium, 1.20 mol NO_2(g) has formed. Calculate K_c.

19

19.1 The equilibrium constant K_c – part 2

Study tip

In Step 1, the numbers in bold are supplied in the question.

NO_2 has increased in amount by +1.20 mol

Using the balancing numbers in the equation, the changes in the amounts of other species can be calculated.

Finally the initial amounts and changes are compared to give the equilibrium amounts.

With practice, it is easier than it looks.

Synoptic link

This reaction is called esterification. You will study this reaction in Topic 26.4, Carboxylic acid derivatives.

Step 1: Using the equation, work out equilibrium amounts of all equilibrium species.

equation:	$2NO(g)$	+	$O_2(g)$	⇌	$2NO_2(g)$
reacting amounts	2 mol		1 mol	→	2 mol
initial/mol	**1.60**		**1.40**		0
change/mol	−1.20		−0.60		**+1.20**
equilibrium/mol	0.40		0.80		1.20

Step 2: Find the equilibrium concentrations in $mol\,dm^{-3}$.

Total volume is $4.0\,dm^3$, so the equilibrium amounts are divided by 4 to give concentrations in $mol\,dm^{-3}$.

$[NO(g)]$ $\quad \dfrac{0.40}{4} \quad = \quad 0.10\,mol\,dm^{-3}$

$[O_2(g)]$ $\quad \dfrac{0.80}{4} \quad = \quad 0.20\,mol\,dm^{-3}$

$[NO_2(g)]$ $\quad \dfrac{1.20}{4} \quad \quad 0.30\,mol\,dm^{-3}$

Step 3: Write the expression for K_c, substitute values, and calculate K_c.

$$K_c = \dfrac{[NO_2(g)]^2}{[NO(g)]^2[O_2(g)]} = \dfrac{0.30^2}{0.10^2 \times 0.20} = 45\,dm^3\,mol^{-1}$$

▲ **Figure 1** *Flask left to reach equilibrium*

Determining K_c from experimental results

This application describes an experimental procedure that would allow you to experimentally determine a value for K_c in the laboratory.

The reaction of a carboxylic acid with an alcohol shown below is a reversible reaction.

$$CH_3COOH + C_2H_5OH \rightleftharpoons CH_3COOC_2H_5 + H_2O.$$

1. In a conical flask, mix together 0.100 mol CH_3COOH and 0.100 mol C_2H_5OH. Add 0.0500 mol of HCl(aq) as an acid catalyst to the flask.
   The total volume of the mixture in the flask is $20.0\,cm^3$.
   The amount of water in the aqueous acid catalyst is 0.500 mol.
2. Add 0.0500 mol of HCl(aq) to a second conical flask as a control.
3. Stopper both flasks and leave for a week to reach equilibrium (Figure 1).
4. Carry out a titration on the equilibrium mixture using a standard solution of sodium hydroxide.
5. Repeat the titration with the control to determine the amount of acid catalyst that had been added.

Results

By analysing the two titrations:

- amount of HCl(aq) in control = 0.0500 mol
- amount of acid (HCl and CH_3COOH) in equilibrium mixture = 0.115 mol

EQUILIBRIUM 19

 Worked example: Calculation of K_c

The calculation of K_c from the results is shown below. The method is similar to the first worked example.

Step 1: Determine the equilibrium amount of CH_3COOH.

From the results of the two titrations:

equilibrium amount of $CH_3COOH = 0.115 - 0.0500$
$= 0.065$ mol

Step 2: Use the equilibrium equation to determine the equilibrium amounts of each component.

equation:	CH_3COOH	$+ C_2H_5OH$	$\rightleftharpoons CH_3COOC_2H_5$	$+ H_2O$
reacting amounts	1 mol	1 mol	→ 1 mol	1 mol
initial/mol	0.100	0.100	0	0.500
change/mol	−0.035	−0.035	+0.035	+0.035
equilibrium/mol	**0.065**	0.065	0.035	0.535

Step 3: Find the equilibrium concentrations, in $mol\,dm^{-3}$.

The volume of the solution is $20\,cm^3 = 0.0200\,dm^3$ so equilibrium amounts are divided by 0.0200 to give the concentrations in $mol\,dm^{-3}$

$[CH_3COOH] = \dfrac{0.065}{0.0200} = 3.25\,mol\,dm^{-3}$

$[C_2H_5OH] = \dfrac{0.065}{0.0200} = 3.25\,mol\,dm^{-3}$

$[CH_3COOC_2H_5] = \dfrac{0.035}{0.0200} = 1.75\,mol\,dm^{-3}$

$[H_2O] = \dfrac{0.535}{0.020} = 26.75\,mol\,dm^{-3}$

Step 4: Write the expression for K_c, substitute values, and calculate K_c.

$K_c = \dfrac{[CH_3COOC_2H_5]\,[H_2O]}{[CH_3COOH]\,[C_2H_5OH]} = \dfrac{1.75 \times 26.75}{3.25 \times 3.25} = 4.43$

(no units, all units cancel)

---

1. In Stage 3 of the method above, the flask is left for a week to reach equilibrium. Suggest how the method could be modified to be sure that equilibrium had been reached.
2. A student set up another experiment using the same method but with different initial amounts: $0.150\,mol\,CH_3COOH$, $0.200\,mol\,C_2H_5OH$. The same amount of acid catalyst was used (containing $0.500\,mol\,H_2O$). The total volume is $25\,cm^3$.
Analysis of the mixture showed that $0.0750\,mol$ of CH_3COOH were present at equilibrium. Use these results to calculate K_c for the reaction (including units).

Study tip

At equilibrium the conical flask contains a mixture of four liquids in a single layer. The four equilibrium species all have the same state or phase. Therefore this is a homogeneous equilibrium and all four species are included in the K_c expression.

Summary questions

1. For each of the following equilibria, determine the expression for K_c, including units.
   a $H_2(g) + Br_2(g) \rightleftharpoons 2HBr(g)$
   (2 marks)
   b $CO(g) + 2H_2(g) \rightleftharpoons CH_3OH(g)$
   (2 marks)
   c $MgCO_3(s) \rightleftharpoons MgO(s) + CO_2(g)$
   (2 marks)

2. H_2 reacts with I_2 in a reversible reaction:
   $H_2(g) + I_2(g) \rightleftharpoons 2HI(g)$
   $0.450\,mol$ of H_2 and $0.300\,mol$ of I_2 were mixed together in a container of volume $1\,dm^3$. At equilibrium, $0.060\,mol$ of I_2 was present.
   a Calculate the equilibrium concentrations of I_2 and HI.
   (2 marks)
   b Calculate K_c. *(2 marks)*

3. N_2 reacts with H_2 in the manufacture of ammonia, NH_3:
   $N_2(g) + 3H_2(g) \rightleftharpoons 2NH_3(g)$
   $3.28\,mol$ of N_2 and $6.64\,mol$ H_2 were mixed together in a container of volume $2\,dm^3$. At equilibrium, $1.76\,mol\,NH_3$ had formed.
   a Calculate the equilibrium amounts in mol of N_2 and H_2.
   (2 marks)
   b Calculate K_c. *(2 marks)*

19.2 The equilibrium constant K_p

Specification reference: 5.1.2

Learning outcomes

Demonstrate knowledge, understanding, and application of:

→ the terms mole fraction and partial pressure
→ expressions for K_p for homogeneous and heterogeneous equilibria
→ calculations of K_p including determination of units
→ application of K_c and K_p to other equilibrium constants.

Synoptic Link

In Topic 2.3, Formulae and equations, you learnt that 1 mole of gas molecules occupies 24.0 dm³ at RTP.

You also learnt about the ideal gas equation which allows gas volumes to be calculated at any temperature and pressure.

Mole fraction and partial pressure

Equilibria involving gases are usually expressed in terms of K_p, the equilibrium constant in terms of partial pressures. For gases, it is easier to measure pressure than concentration. Concentration and pressure are proportional to one another and K_p has a direct relationship to K_c. To understand how to calculate K_p, you will first need to understand mole fractions and partial pressure.

Mole fraction

Under the same conditions of temperature and pressure, the same volume of different gases contains the same number of moles of gas molecules.

The **mole fraction** of a gas is the same as its proportion by volume to the total volume of gases in a gas mixture.

For a gas **A** in a gas mixture:

$$\text{mole fraction } x(\mathbf{A}) = \frac{\text{number of moles of } \mathbf{A}}{\text{total number of moles in gas mixture}}$$

As with all fractions, the sum of the mole fractions in a gas mixture must equal one.

Worked example: Calculating mole fractions

Air has approximate molar proportions of 78% $N_2(g)$, 21% $O_2(g)$, and 1% other gases. Calculate the mole fractions of the gases in air.

Step 1: Mole fraction of N_2 $x(N_2) = \frac{78}{100} = 0.78$

Step 2: Mole fraction of O_2 $x(O_2) = \frac{21}{100} = 0.21$

Step 3: Mole fraction of other gases $x = \frac{1}{100} = 0.01$

Sum of mole fractions equals 1 $0.78 + 0.21 + 0.01 = 1$

Partial pressure

In a gas mixture, the **partial pressure** p of a gas is the contribution that the gas makes towards the total pressure P. The sum of the partial pressures of each gas equals the total pressure.

For gas **A** in a gas mixture:

$$\text{partial pressure } p(\mathbf{A}) = \text{mole fraction of } \mathbf{A} \times \text{total pressure } P$$

$$p(\mathbf{A}) = x(\mathbf{A}) \times P$$

EQUILIBRIUM
19

 Worked example: Calculating partial pressures

Use the mole fractions of the gases in air to calculate the partial pressures of gases in air at a pressure of 100 kPa.

Step 1: Calculate partial pressure of nitrogen.
$$p(N_2) = x(N_2) \times P = 0.78 \times 100 = 78 \text{ kPa}$$

Step 2: Calculate partial pressure of oxygen.
$$p(O_2) = x(O_2) \times P = 0.21 \times 100 = 21 \text{ kPa}$$

Step 3: Calculate partial pressure of other gases.
$$p = x \times P = 0.01 \times 100 = 1 \text{ kPa}$$

The sum of the partial pressures equals the total pressure
78 + 21 + 1 = 100 kPa

Study tip
Always check your calculations:
- sum of mole fractions = 1
- sum of partial pressures = total pressure

 Dissolving oxygen in the blood

You breathe as a way of getting oxygen into your blood and getting carbon dioxide out. But how does it work?

The amount of oxygen that can dissolve in the blood is proportional to the partial pressure of oxygen breathed into the lungs. At equilibrium in the lungs, each cubic decimetre of the blood can dissolve about 3 cm^3 of gaseous oxygen. An adult human only has about 5 dm^3 of blood and so 15 cm^3 of oxygen does not seem much. Luckily, as quickly as the oxygen dissolves, it combines with haemoglobin in the blood. This allows more than 200 cm^3 of oxygen to effectively dissolve in each dm^3 of blood.

Normal air at 1 atmosphere pressure has a pressure of 101 kPa. The 21% of oxygen in the air has a partial pressure of just over 21 kPa.

On Earth, as altitude increases, the atmospheric pressure decreases (Table 1).

▼ **Table 1** Altitude and pressure

Altitude / m	sea level	2000	4000	6000	8000	10 000
Pressure / kPa	101	81	63	49	38	29

So how do people survive at altitude? The body compensates by producing more haemoglobin, but at very high altitudes the partial pressure of oxygen may be so small that very little may dissolve.

High altitude mountaineers compensate by using oxygen tanks and breathing pure oxygen.

1. Why does less oxygen dissolve in the blood with increasing altitude?
2. Mount Everest has an altitude of 8848 m above sea level.
   a. Estimate the partial pressure of oxygen at the summit of Mount Everest.
   b. Without breathing apparatus, estimate the volume of oxygen that can be dissolved per dm^3 of blood at the top of Mount Everest?
   c. Breathing pure oxygen at altitude allows mountaineers to breathe normally. However, breathing pure oxygen for a length of time at lower altitudes is unhealthy. Suggest why there is a difference.

▲ **Figure 1** High-altitude mountaineer with breathing apparatus containing oxygen

19.2 The equilibrium constant K_p

Study tip

You should *never* use square brackets [ ] when writing expressions for K_p. Expressions for K_p involve partial pressures, not concentration. This is a common error made by students.

▲ **Figure 2** *Ammonia production. Part of an industrial plant used to make ammonia from nitrogen and hydrogen gases. Around 80% of the ammonia produced is used to produce fertilisers*

Study tip

$x(A) = \dfrac{\text{number of moles of } A}{\text{total number of moles}}$

Study tip

Check that the sum of the partial pressures matches the total pressure:

Partial pressures:
30 + 90 + 80 = 200 atm

Total pressure = 200 atm

Synoptic link

There are many different equilibrium constants but the principles that you have learnt for K_c and K_p can be applied to them all. In Topic 20.1, Brønsted–Lowry acids and bases, you will meet K_a and K_w, variations on K_c with special labels to match their context.

The equilibrium constant K_p

K_p is written in a similar way to K_c but with partial pressures replacing concentration terms.

For the equilibrium:

$$H_2(g) + I_2(g) \rightleftharpoons 2HI(g) \qquad K_p = \dfrac{p(HI)^2}{p(H_2) \times p(I_2)}$$

p is the equilibrium partial pressure.

Suitable units for partial pressures are kilopascals (kPa), pascals (Pa), or atmospheres (atm) – but the same unit must be used for all gases. As with concentration terms for K_c, the power of a partial pressure in the K_p expression is the *balancing number* in the chemical equation.

K_p only includes gases because only gases have partial pressures. Any other species *must* be ignored.

Worked example: Calculating K_p for a homogeneous equilibrium

Nitrogen, hydrogen, and ammonia coexist in a homogeneous equilibrium, with all species being gases:

$$N_2(g) + 3H_2(g) \rightleftharpoons 2NH_3(g)$$

This equilibrium reaction is very important, since it is used in the industrial production of ammonia (Figure 2).

An equilibrium mixture at 400 °C contains 18 mol of $N_2(g)$, 54 mol of $H_2(g)$, and 48 mol $NH_3(g)$. The total equilibrium pressure is 200 atm. Use this information to calculate K_p.

Step 1: Find the mole fractions of nitrogen, hydrogen, and ammonia

Total number of gas moles = 18 + 54 + 48 = **120** mol

$$x(N_2) = \dfrac{18}{120} \qquad x(H_2) = \dfrac{54}{120} \qquad x(NH_3) = \dfrac{48}{120}$$

Step 2: Find the partial pressures.

$$p(N_2) = \dfrac{18}{120} \times 200 = 30 \text{ atm}$$

$$p(H_2) = \dfrac{54}{120} \times 200 = 90 \text{ atm}$$

$$p(NH_3) = \dfrac{48}{120} \times 200 = 80 \text{ atm}$$

Step 3: Calculate K_p.

For the equilibrium:

$$N_2(g) + 3H_2(g) \rightleftharpoons 2NH_3(g),$$

$$K_p = \dfrac{p(NH_3)^2}{p(N_2) \times p(H_2)^3} \qquad \text{units: } K_p = \dfrac{(\text{atm})^2}{(\text{atm})(\text{atm})^3} = \text{atm}^{-2}$$

$$K_p = \dfrac{80^2}{30 \times 90^3} = 2.9 \times 10^{-4} \text{ atm}^{-2}$$

EQUILIBRIUM 19

 Worked example: Calculating K_p for a heterogeneous equilibrium

When calcium carbonate, $CaCO_3$, is heated in a closed system, a heterogeneous equilibrium is set up. It is a heterogeneous equilibrium because there is a mixture of solids and a gas.

$$CaCO_3(s) \rightleftharpoons CaO(s) + CO_2(g)$$
$$\text{solid} \qquad \text{solid} \qquad \text{gas}$$

In the equilibrium mixture, CO_2 has a partial pressure of 2.5×10^{-2} atm at 600 °C. Calculate K_p.

Step 1: Write down the expression for K_p.

Only gaseous species are included in the K_p expression.

$$K_p = p(CO_2)$$

Step 2: Calculate K_p.

$$K_p = 2.5 \times 10^{-2} \text{ atm}$$

▲ **Figure 3** *A lime kiln in the Derbyshire Peak District. Limestone contains calcium carbonate, $CaCO_3$. The limestone is heated to a high temperature and carbon dioxide is allowed to escape. This prevents an equilibrium being set up so that the maximum yield of lime (calcium oxide, CaO) is formed*

Summary questions

1. **a** A gas mixture at a total pressure of 320 kPa contains three moles of $N_2(g)$ and one mole of $O_2(g)$. Calculate the mole fraction and partial pressure of each gas. *(2 marks)*

   b A gas mixture contains 10 cm³ of $O_2(g)$, 25 cm³ of $N_2(g)$, and 15 cm³ of $H_2(g)$. The total pressure of the mixture is 500 kPa. Calculate the mole fraction and partial pressure of each gas. *(3 marks)*

2. For the following equilibria, write an expression for K_p and calculate the value for K_p, including units.

   a $2HI(g) \rightleftharpoons H_2(g) + I_2(g)$
   partial pressures: $p(HI)$ 56 kPa, $p(H_2)$ 22 kPa, $p(I_2)$ 22 kPa *(3 marks)*

   b $2NO_2(g) \rightleftharpoons 2NO(g) + O_2(g)$
   partial pressures: $p(NO_2)$ 45 kPa, $p(NO)$ 60 kPa, $p(O_2)$ 30 kPa *(3 marks)*

   c $2SO_2(g) + O_2(g) \rightleftharpoons 2SO_3(g)$
   partial pressures: $p(SO_2)$ 74 kPa, $p(O_2)$ 23 kPa, $p(SO_3)$ 142 kPa *(3 marks)*

3. In the equilibrium:

   $$2SO_2(g) + O_2(g) \rightleftharpoons 2SO_3(g)$$

   2.0 mol of $SO_2(g)$ was mixed with 1.0 mol $O_2(g)$ and the mixture was allowed to reach equilibrium at constant temperature. At equilibrium, 0.5 mol of the $SO_2(g)$ had reacted.

   a Calculate the amounts, in mol, of SO_2, O_2, and SO_3 in the equilibrium mixture. *(3 marks)*

   b The total equilibrium pressure was 900 kPa. Calculate the mole fractions and partial pressures of SO_2, O_2, and SO_3 in the equilibrium mixture. *(6 marks)*

   c Calculate K_p and state its units. *(3 marks)*

19.3 Controlling the position of equilibrium

Specification reference: 5.1.2

Learning outcomes

Demonstrate knowledge, understanding, and application of:

→ the effect of temperature on equilibrium constants

→ the effect of concentration, pressure, and catalysts on equilibrium constants

→ equilibrium constants and controlling the position of equilibrium.

Synoptic link

Revisit Topic 10.4, Dynamic equilibrium and le Chatelier's principle, if you need to revise le Chatelier's principle before continuing.

Synoptic link

You first encountered the relationship between the value of K and the extent of an equilibrium reaction in Topic 10.5, The equilibrium constant K_c – part 1.

Why does the equilibrium position shift?

In Topic 10.4, you saw how le Chatelier's principle can be used to predict how the position of equilibrium shifts in response to a change.

You learnt three rules:

1. If the concentration of a species is increased, the equilibrium position shifts in the direction that reduces the concentration.
2. If the pressure is increased, the equilibrium position shifts towards the side with fewer gaseous moles.
3. If the temperature is increased, the equilibrium position shifts in the endothermic direction.

So why does le Chatelier's principle work? The answer lies with equilibrium constants.

Do equilibrium constants change?

The magnitude of an equilibrium constant K indicates the extent of a chemical equilibrium.

- $K = 1$ indicates an equilibrium halfway between reactants and products.
- $K = 100$ indicates an equilibrium well in favour of the products.
- $K = 1 \times 10^{-2}$ indicates an equilibrium well in favour of the reactants.

The value of K gives the exact position of equilibrium. When the reactants or products of a reversible reaction are mixed together, the reaction proceeds until the concentrations of the equilibrium species give the value of K when placed in the equilibrium constant expression.

At a set temperature, K is constant and *does not* change despite any modifications to concentration, pressure, or the presence of a catalyst.

But K *does* change if the temperature is changed – a temperature change is the only condition that will causes K to change its value.

The effect of temperature on equilibrium constants K

The value of K is changed by altering the temperature. Whether the value of K gets larger or smaller depends on whether the forward reaction is exothermic or endothermic.

Exothermic reactions

If the forward reaction is exothermic:

- the equilibrium constant *decreases* with increasing temperature
- raising the temperature decreases the equilibrium yield of products.

EQUILIBRIUM 19

The effect of increasing temperature on K_p for the $SO_2/O_2/SO_3$ equilibrium is shown below.

$$2SO_2(g) + O_2(g) \rightleftharpoons 2SO_3(g) \qquad \Delta H^\ominus = -197 \text{ kJ mol}^{-1}$$
$$\text{exothermic}$$

equilibrium position shifts to the *left*

Temperature / K	K_p / atm^{-1}	Exothermic change
500	2.5×10^{10}	• K_p *decreases* with increasing temperature
700	3.0×10^{4}	
1100	1.3×10^{-1}	• equilibrium position shifts to *left*

> **Study tip**
> Exothermic reactions give out energy.

Explaining the equilibrium shift

At equilibrium, $K_p = \dfrac{p(SO_3)^2}{p(SO_2)^2 \times p(O_2)}$

If the temperature increases from 500 K to 700 K:

- K_p decreases from 2.5×10^{10} to 3.0×10^4 atm^{-1}
- the system is no longer in equilibrium
- the ratio $\dfrac{p(SO_3)^2}{p(SO_2)^2 \times p(O_2)}$ is now greater than K_p

The equilibrium partial pressures that gave a K_p value of 2.5×10^{10} at 500 K must change to give the new K_p value of 3.0×10^4 at 700 K.

- The partial pressure of the product SO_3 must *decrease*.
- The partial pressures of the reactants SO_2 and O_2 must *increase*.
- The position of equilibrium shifts towards the reactants (to the left).
- A new equilibrium will be reached where $\dfrac{p(SO_3)^2}{p(SO_2)^2 \times p(O_2)}$ is equal to the new K_p value of 3.0×10^4 atm^{-1}.

Endothermic reactions

If the forward reaction is endothermic:

- the equilibrium constant *increases* with increasing temperature
- raising the temperature increases the equilibrium yield of products.

The effect of temperature on K_p for the $N_2/O_2/NO$ equilibrium is shown below.

$$N_2(g) + O_2(g) \rightleftharpoons 2NO(g) \qquad \Delta H^\ominus = +180 \text{ kJ mol}^{-1}$$
$$\text{endothermic}$$

equilibrium position shifts to the *right*

Temperature / K	K_p / atm^{-1}	Endothermic change
700	5×10^{-13}	• K_p *increases* with increasing temperature
1100	4×10^{-8}	
1500	1×10^{-5}	• Equilibrium position shifts to *right*

> **Study tip**
>
> | Effect of increasing temperature on equilibrium ||
> |---|---|
> | exothermic | endothermic |
> | K decreases | K increases |
> | equilibrium shifts to *left* | equilibrium shifts to *right* |

> **Study tip**
> Endothermic reactions take in energy.

19.3 Controlling the position of equilibrium

Explaining the equilibrium shift in endothermic reactions

At equilibrium: $K_p = \dfrac{p(NO)^2}{p(N_2) \times p(O_2)}$

If the temperature increases from 700 K to 1100 K:

- K_p increases from 5×10^{-13} to 4×10^{-8} atm^{-1}
- the system is no longer in equilibrium
- the ratio $\dfrac{p(NO)^2}{p(N_2) \times p(O_2)}$ is now less than K_p

The equilibrium partial pressures that gave a K_p value of 5×10^{-13} at 700 K must change to give the new K_p value of 4×10^{-8} at 1100 K.

- the partial pressure of the product NO must *increase*
- the partial pressures of the reactants N_2 and O_2 must *decrease*
- The position of equilibrium shifts towards the products (to the right).

A new equilibrium is established where $\dfrac{p(NO)^2}{p(N_2) \times p(O_2)}$ is equal to the new K_p value of 4×10^{-8} atm^{-1}.

What about K_c?

The explanations for K_p work also for K_c. If you were supplied with K_c information, just use concentrations instead of partial pressures.

How do changes in concentration and pressure affect equilibrium constants?

The value of an equilibrium constant K is *unaffected* by changes in concentration or pressure. This may seem strange as you know from le Chatelier's principle that the equilibrium position can be shifted by changes in concentration or pressure. The equilibrium shift actually results from the very fact that the equilibrium constant does *not* change.

Equilibrium constants and concentration changes

Consider the following equilibrium:

$$N_2O_4(g) \rightleftharpoons 2NO_2(g)$$

At a constant temperature, the equilibrium concentrations are:

- $[NO_2(g)] = 0.400 \text{ mol dm}^{-3}$
- $[N_2O_4(g)] = 0.010 \text{ mol dm}^{-3}$

$$K_c = \dfrac{[NO_2(g)]^2}{[N_2O_4(g)]} = \dfrac{0.400^2}{0.010} = 16.0 \text{ mol dm}^{-3}$$

If $[N_2O_4(g)]$ is increased to $0.020 \text{ mol dm}^{-3}$

- the ratio $\dfrac{[NO_2(g)]^2}{[N_2O_4(g)]}$ is now *less than* K_c and is equal to

$$\dfrac{0.400^2}{0.020} = 8.0 \text{ mol dm}^{-3}$$

- the system is now no longer in equilibrium.

> **Study tip**
>
> It is important to realise that changes in concentration do *not* change K. The *ratio* of reactants to products changes. The ratio then changes to get back to K – this is why the equilibrium position shifts.

The concentrations must change to return the ratio back to the K_c value of $16.0 \, \text{mol dm}^{-3}$.

- the concentration of the product NO_2 must *increase*
- the concentration of the reactant N_2O_4 must *decrease*
- A new equilibrium is established where $\dfrac{[NO_2(g)]^2}{[N_2O_4(g)]}$ is restored to its K_c value of $16.0 \, \text{mol dm}^{-3}$.

$$N_2O_4(g) \rightleftharpoons 2NO_2(g)$$

equilibrium position shifts to the *right*

By le Chatelier's principle, you would also predict a shift in the equilibrium position from left to right to decrease $[N_2O_4(g)]$. le Chatelier's principle only works because K_c controls the relative concentrations of reactants and products present at equilibrium.

The same principle could have been shown using K_c for the equilibrium constant instead of K_p

Equilibrium constants and pressure changes

For the same equilibrium as above, a doubling of pressure will result in a doubling of the partial pressures and concentration of *both* NO_2 and N_2O_4.

Consider the following equilibrium:

$$N_2O_4(g) \rightleftharpoons 2NO_2(g)$$

At a constant temperature, the equilibrium partial pressures are:

- $p(NO_2) = 9.6 \text{ atm}$
- $p(N_2O_4) = 0.24 \text{ atm}$

$$K_p = \frac{p(NO_2)^2}{p(N_2O_4)} = \frac{9.6^2}{0.24} = 384 \text{ atm}$$

If the total pressure is doubled, $p(N_2O_4)$ is increased to 19.2 atm and $p(NO_2)$ is increased to 0.48 atm and

- the ratio $\dfrac{p(NO_2)^2}{p(N_2O_4)}$ is now *greater than* K_p and equal to $\dfrac{19.6^2}{0.48} = 768 \text{ atm}$
- the system is now no longer in equilibrium.

The partial pressures must change to return the ratio back to the K_p value of 384 atm.

- the partial pressure of the product NO_2 must *decrease*
- the partial pressure of the reactant N_2O_4 must *increase*
- A new equilibrium position will be reached where $\dfrac{p(NO_2)^2}{p(N_2O_4)}$ is restored to its K_p value of 384 atm.

$$N_2O_4(g) \rightleftharpoons 2NO_2(g)$$

equilibrium position shifts to the *left*

Using le Chatelier's principle, you predict a shift in the equilibrium position from right to left to the side with fewer gaseous moles. This shift has been directed by the value of K_p being restored. The same principle could have been shown using K_c for the equilibrium constant instead of K_p.

19.3 Controlling the position of equilibrium

Table 1 *Summary of the effect of increasing pressure*

Gaseous moles	Ratio of products/reactants	Effect of increasing pressure	Equilibrium shift
fewer moles of gaseous products	ratio < K	products increase, reactants decrease	right
more moles of gaseous products	ratio > K	products decrease, reactants increase	left
same number of moles of gaseous reactants and products	ratio = K	no change	no effect

How does a catalyst affect equilibrium constants?

Equilibrium constants are unaffected by the presence of a catalyst.

- Catalysts affect the *rate* of a chemical reaction but *not* the position of equilibrium.
- Catalysts speed up *both* the forward and reverse reactions in the equilibrium by the same factor.

Equilibrium is reached quicker but the equilibrium position is *not* changed.

Summary questions

1. Predict whether the two reactions with the K_c values below are exothermic or endothermic. Explain your answer.

Temperature / K	Numerical value of K_c	
	Reaction A	Reaction B
200	5.51×10^{-8}	4.39×10^4
400	1.46	4.03
600	3.62×10^2	3.00×10^{-2}

   (2 marks)

2. The reaction $A(aq) + 2B(aq) \rightleftharpoons C(aq)$ is endothermic and at room temperature both forward and reverse reactions are relatively slow. **A** and **B** are both colourless, whilst **C** is a deep blue colour.
   a. State an expression for the equilibrium constant K_c for this reaction. *(1 mark)*
   b. Explain the effect on the value of K_c of each of the following:
      i Increasing temperature
      ii Adding more **A** at constant temperature
      (3 marks)
   c. Explain what would you see if more **C** were added at room temperature? *(3 marks)*

3. Study the equilibrium: $H_2O(g) + CO(g) \rightleftharpoons H_2(g) + CO_2(g)$ $\Delta H = -41.2 \text{ kJ mol}^{-1}$
   a. State an expression for K_p for this reaction. *(1 mark)*
   b. When 1.00 mol of steam and 1.00 mol of carbon monoxide were allowed to reach equilibrium, 0.60 mol of hydrogen was formed. Calculate a value for K_p. *(4 marks)*
   c. Explain, with a reason, the effect on the percentage of hydrogen present at equilibrium of the following changes. Also state the effect on K_p.
      i increasing the pressure
      ii increasing the temperature. *(4 marks)*

4. For the equilibrium $N_2(g) + 3H_2(g) \rightleftharpoons 2NH_3(g)$, explain, in terms of K_c, the effect of increasing pressure on the equilibrium yield of $NH_3(g)$. *(3 marks)*

19 EQUILIBRIUM

Practice questions

1. A 0.300 mol sample of HBr(g) is added to a 1.00 dm³ container maintained at constant temperature. The equilibrium is:

   $$2HBr(g) \rightleftharpoons H_2(g) + Br_2(g)$$

   At equilibrium, there are 0.0520 mol of both $H_2(g)$ and $Br_2(g)$.

   a What is the equilibrium amount of HBr(g)? *(1 mark)*
   b Write the expression for K_c. *(1 mark)*
   c Calculate the value of K_c. *(1 mark)*

2. 6.00 mol SO_2 and 4.50 mol O_2 are mixed together and allowed to reach equilibrium in a sealed container of volume 3.00 dm³. The equilibrium is:

   $$2SO_2(g) + O_2(g) \rightleftharpoons 2SO_3(g)$$

   At equilibrium, 2.40 mol of O_2 remained.
   Calculate K_c. Show your working and give units *(5 marks)*

3. A mixture of equal numbers of moles of CO(g) and $H_2O(g)$ are left to reach equilibrium in a sealed container at constant temperature. The equilibrium equation is shown below.

   $$CO(g) + H_2O(g) \rightleftharpoons CO_2(g) + H_2(g)$$

   At equilibrium, the partial pressure of CO_2 is 15.0 kPa. The total pressure is 150 kPa.
   Calculate K_p, including units. *(4 marks)*

4. The amount, in mol, of an equilibrium mixture at 200 kPa total pressure is shown below.

   $$CH_4(g) + H_2O(g) \rightleftharpoons CO(g) + 3H_2(g)$$
   $$0.800 \quad\quad 0.800 \quad\quad 1.20 \quad\quad 3.60$$

   a Write the expression for K_p *(1 mark)*
   b Calculate K_p, including units. Give your answer to three significant figures. *(5 marks)*

5. In an industrial process involving a reversible gas phase reaction, the following yields of products are obtained under different operating conditions.

Temperature / K	Pressure / atm	Yield / %
300	100	35
300	300	60
500	100	5
500	300	12

   a Use this information to answer the following.

   (i) Explain whether the reaction exothermic or endothermic. *(1 mark)*

   (ii) Explain whether there are more or fewer moles of gaseous products than reactants. *(1 mark)*

   b What would be the effect (if any) on the value of K_c of each of the following?

   (i) Changing the conditions from 300 K and 100 atm to 500 K and 100 atm. *(1 mark)*

   (ii) Changing the conditions from 300 K and 100 atm to 300 K and 300 atm. *(1 mark)*

   (iii) Introducing a catalyst at 300 K and 100 atm. *(1 mark)*

6. Methanol can be prepared industrially by reacting together carbon monoxide and hydrogen. This is a reversible reaction:

   $$CO(g) + 2H_2(g) \rightleftharpoons CH_3OH(g) \quad \Delta H = -94 \text{ kJ mol}^{-1}$$

   A chemist mixes together 0.114 mol CO(g) and 0.152 mol $H_2(g)$ in a container. The container is pressurised and then sealed. The total volume is 200 cm³.

   The mixture is heated to 500 K and left to reach equilibrium. The volume of the sealed container is kept at 200 cm³.

   The chemist analyses the equilibrium mixture and finds that 0.052 mol CH_3OH has formed.

   a Calculate the value of K_c, including units, for the equilibrium at 500 K. Give your answer to **three** significant figures. *(6 marks)*

   b The chemist repeats the experiment using the same initial amounts of CO and H_2. The same procedure is used but the mixture is heated in the 200 cm³ sealed container to a higher temperature than 500 K. As the gas volume is kept at 200 cm³, the increased temperature also increases the pressure.

   Explain why it is difficult to predict how the yield of CH_3OH would change.
   Explain what happens to the value of K_c. *(4 marks)*

   OCR F325 Jan 13 Q5

Chapter 19 Practice questions

7 In the Contact Process for the production of sulfuric acid, highly purified sulfur dioxide and oxygen react together to form sulfur trioxide. The process is carried out at about 700 K and at a pressure of 120 kPa. The equilibrium is shown below.

$$2SO_2(g) + O_2(g) \rightleftharpoons 2SO_3(g) \quad \Delta H = -190 \text{ kJ mol}^{-1}$$

The partial pressures of SO_2 and SO_3 at equilibrium are 33 kPa and 39 kPa, respectively.

- **a** What would be the effect on the yield of SO_3 of increasing the temperature? Explain your answer. *(2 marks)*
- **b** Determine the partial pressure and the mole fraction of oxygen in the equilibrium mixture. *(2 marks)*
- **c** (i) Write an expression for K_p for this equilibrium. *(1 mark)*
  (ii) Calculate K_p, including units. *(3 marks)*

8 Ammonia is one of our most important chemicals, produced in enormous quantities because of its role in the production of fertilisers. Much of this ammonia is manufactured from nitrogen and hydrogen gases using the Haber process. The equilibrium is shown below.

$$N_2(g) + 3H_2(g) \rightleftharpoons 2NH_3(g) \quad \Delta H = -92 \text{ kJ mol}^{-1}$$

- **a** (i) Write an expression for K_c for this equilibrium. *(1 mark)*
  (ii) Deduce the units of K_c for this equilibrium. *(1 mark)*
- **b** A research chemist was investigating methods to improve the synthesis of ammonia from nitrogen and hydrogen at 500 °C.

  The chemist mixed together nitrogen and hydrogen and pressurised the gases so that their total gas volume was 6.0 dm³. The mixture was allowed to reach equilibrium at constant temperature and without changing the total gas volume. The equilibrium mixture contained 7.2 mol N_2 and 12.0 mol H_2.

  At 500 °C, the numerical value of K_c for this equilibrium is 8.00×10^{-2}.

  Calculate the amount, in mol, of ammonia present in the equilibrium mixture at 500 °C. *(4 marks)*

- **c** The research chemist doubled the pressure of the equilibrium mixture whilst keeping all other conditions the same. As expected the equilibrium yield of ammonia increased.
  (i) Explain in terms of le Chatelier's principle why the equilibrium yield of ammonia increased. *(2 marks)*
  (ii) Explain in terms of K_c why the equilibrium yield of ammonia increased. *(3 marks)*

F325 Jun 10 Q5(a)–(c) adapted

9 Phosgene, $COCl_2$, can be manufactured by the reaction of carbon monoxide and chlorine in the presence of a catalyst. An equilibrium system exists between CO, Cl_2 and $COCl_2$.

$$CO(g) + Cl_2(g) \rightleftharpoons COCl_2(g) \quad \Delta H = -108 \text{ kJ mol}^{-1}$$

- **a** Explain the conditions of temperature and pressure needed to obtain a high equilibrium yield of phosgene. *(2 marks)*
- **b** A chemist mixes together 0.300 mol CO with 0.350 mol Cl_2 in a sealed container of volume 2.00 dm³.

  The mixture is heated and allowed to reach equilibrium. At equilibrium, the mixture contains 0.272 mol $COCl_2$.

  Calculate K_c including units. Show your working. *(5 marks)*

- **c** The gas mixture was compressed whilst keeping the temperature constant. The system was left to reach equilibrium. The position of equilibrium shifted to the right.
  (i) Explain the effect, if any, on K_c. *(1 mark)*
  (ii) Explain in terms of K_c why the equilibrium moved to the right. *(3 marks)*

(iii) State and explain the rates of the forward and reverse reactions
- when the gas mixture was first compressed
- when the system reached equilibrium (*3 marks*)

d The temperature was increased whilst keeping the pressure constant. The system was left to reach equilibrium. The value of K_c decreased.

(i) Explain what happens to the position of equilibrium. (*2 marks*)

(ii) Deduce the sign of the enthalpy change for the forward reaction in the equilibrium. Explain your reasoning. (*1 mark*)

20 ACIDS, BASES, AND pH
20.1 Brønsted–Lowry acids and bases

Specification reference: 5.1.3

Learning outcomes

Demonstrate knowledge, understanding, and application of:

→ Brønsted–Lowry acids and bases
→ conjugate acid–base pairs
→ monobasic, dibasic, and tribasic acids
→ the role of H^+ in acid reactions.

Development of models to explain acid–base behaviour

Acids have been known for thousands of years. As scientific knowledge has increased, the definitions of an acid and a base have been modified over time. The flowchart in Figure 1 highlights some of the changing models of acid and base behaviour. Hydrogen was recognised as the important chemical component in acid compounds in the 1830s and this idea has been further refined since.

Brønsted–Lowry acids and bases

You have already seen that acids and alkalis release H^+ and OH^- ions in water. This is the Arrhenius model of acids and bases:

- acids dissociate and release H^+ ions in aqueous solution
- alkalis dissociate and release OH^- ions in aqueous solution
- H^+ ions are neutralised by OH^- ions to form water:
  $$H^+(aq) + OH^-(aq) \rightarrow H_2O(l)$$

An alkali is a soluble base.

The Brønsted–Lowry model for acids and bases extends this model to emphasise the role of proton transfer between species.

- A **Brønsted–Lowry acid** is a proton donor.
- A **Brønsted–Lowry base** is a proton acceptor.

Conjugate acid–base pairs

The equation below shows the dissociation of hydrochloric acid, HCl(aq).

$$HCl(aq) \rightleftharpoons H^+(aq) + Cl^-(aq)$$

Even though an equilibrium sign has been shown, HCl is a strong acid and the equilibrium position is well over to the right-hand side. A single arrow can be used to indicate that the forward reaction effectively goes to completion.

HCl(aq) and Cl^-(aq) are called a **conjugate acid–base pair**. A conjugate acid–base pair contains two species that can be interconverted by transfer of a proton.

- In the forward direction, HCl *releases* a proton to form its *conjugate base*, Cl^-.
- In the reverse direction, Cl^- *accepts* a proton to forms its *conjugate acid*, HCl.

Figure 1 *Some models that have been used to describe acid and base behaviour*

ACIDS, BASES, AND pH 20

Figure 2 shows several examples of conjugate acid–base pairs.

	acid				base
nitric acid	$HNO_3(aq)$	$\rightleftharpoons$	$H^+(aq)$	+	$NO_3^-(aq)$
sulfuric acid	$H_2SO_4(aq)$	$\rightleftharpoons$	$H^+(aq)$	+	$HSO_4^-(aq)$
ethanoic acid	$CH_3COOH(aq)$	$\rightleftharpoons$	$H^+(aq)$	+	$CH_3COO^-(aq)$

▲ Figure 2 Conjugate acid–base pairs

Synoptic link

Look back at Topic 4.1, Acids, bases, and neutralisation, to revise the Arrhenius dissociation model, weak acids, and strong acids. You will also cover dissociation in detail later in this chapter.

Neutralisation can be shown by a simple equation.

$$H^+(aq) + OH^-(aq) \rightarrow H_2O(l)$$

In hydrochloric acid, H^+ ions have been supplied by dissociation:

$$HCl(aq) \rightleftharpoons H^+(aq) + Cl^-(aq)$$

Combining the two equations gives the acid–base equilibrium shown in Figure 3. There are now two conjugate acid–base pairs which are labelled acid 1–base 1 and acid 2–base 2.

$$HCl(aq) + OH^-(aq) \rightleftharpoons H_2O(l) + Cl^-(aq)$$
acid 1   base 2   acid 2   base 1

▲ Figure 3 Acid–base pairs: acid 1–base 1 and acid 2–base 2

In the forward direction:
- HCl is an acid as it donates H^+
- OH^- is a base as it accepts H^+.

In the reverse direction:
- H_2O is an acid as it donates H^+
- Cl^- is a base as it accepts H^+.

Study tip

You need to be able to identify acid–base pairs in equations for acid–base equilibria.

In their formulae, the acid and base in a conjugate acid–base pair differ by H^+.

Always show clearly how the acid–base pairs are linked together either by labelling with 1 and 2 or by showing arrow links as in Figure 3.

The hydronium ion, $H_3O^+(aq)$

In aqueous solution, dissociation requires a proton to be transferred from an acid to a base. The dissociation does not take place unless water is present. Figure 4 shows the acid–base equilibrium set up in hydrochloric acid. Water accepts a proton and is behaving as a base.

$$HCl(aq) + H_2O(l) \rightleftharpoons H_3O^+(aq) + Cl^-(aq)$$
acid 1   base 2   acid 2   base 1

▲ Figure 4 Acid–base equilibrium of HCl in aqueous solution

H_2O has accepted a proton to form its conjugate acid, the hydronium ion, $H_3O^+(aq)$. The hydronium ion is very important, as it is the active acid ingredient in any aqueous acid. You can use H_3O^+ in equations but you will see that it is far more common to use the simpler $H^+(aq)$.

20.1 Brønsted–Lowry acids and bases

> **Study tip**
> The simplified form $H^+(aq)$ is often used instead of $H_3O^+(aq)$. Just remember that $H^+(aq)$ really represents the hydronium ion, $H_3O^+(aq)$.

For example, neutralisation of an acid by an alkali can be written using $H_3O^+(aq)$ or $H^+(aq)$:

$$H_3O^+(aq) + OH^-(aq) \rightarrow 2H_2O(l)$$
$$H^+(aq) + OH^-(aq) \rightarrow H_2O(l)$$

Monobasic, dibasic, and tribasic acids

The terms **monobasic**, **dibasic**, and **tribasic acids** refer to the total number of hydrogen ions in the acid that can be replaced per molecule in an acid–base reaction. This would typically be replacement of protons by metal ions or ammonium ions to form a salt.

> **Study tip**
> You may come across the terms monoprotic, diprotic, and triprotic, which are also used for monobasic, dibasic, and tribasic.

The number of hydrogen atoms in the formula gives a clue for the type of acid. Table 1 shows some common monobasic, dibasic, and tribasic acids.

▼ **Table 1** *Mono-, di-, and tribasic acids. The replaceable H is shown in bold*

Acid	Name	Type
HCl	hydrochloric acid	monobasic
CH_3COO**H**	ethanoic acid	monobasic
H$_2CO_3$	carbonic acid	dibasic
H$_3BO_3$	boric acid	tribasic

> **Study tip**
> Organic acids do not replace any hydrogen atoms from the carbon chain.

Worked example: Writing equations for complete neutralisations

Write equations for the complete neutralisation of nitric acid, sulfuric acid, and phosphoric acid with an excess of NaOH(aq).

Step 1 Decide on whether an acid is monobasic, dibasic, or tribasic by counting the number of hydrogen atoms in the formula, not including organic carbon chains.

HNO_3 has **one** hydrogen atom, therefore it is **mono**basic.

H_2SO_4 has **two** hydrogen atoms, therefore it is **di**basic.

H_3PO_4 has **three** hydrogen atoms, therefore it is **tri**basic.

Step 2 Write the equation using as many NaOH units needed to replace the hydrogen atoms in the acid.

HNO_3 has *one* replaceable H^+ ion

$$HNO_3(aq) + NaOH(aq) \rightarrow NaNO_3(aq) + H_2O(l)$$

one H^+ replaced

H_2SO_4 has *two* replaceable H^+ ions

$$H_2SO_4(aq) + 2NaOH(aq) \rightarrow Na_2SO_4(aq) + 2H_2O(l)$$

two H^+ replaced

ACIDS, BASES, AND pH

20

H_3PO_4 has *three* replaceable H^+ ions

$$H_3PO_4(aq) + 3NaOH(aq) \rightarrow Na_3PO_4(aq) + 3H_2O(l)$$

three H^+ replaced

These equations can be used to calculate the volume of $0.100 \, mol \, dm^{-3}$ NaOH(aq) that is required to completely neutralise $25.0 \, cm^3$ of $0.100 \, mol \, dm^{-3}$ of each of the three acids above:

The monobasic acid, HNO_3, would react with $25.0 \, cm^3$ NaOH(aq).

The dibasic acid, H_2SO_4, would react with $50.0 \, cm^3$ NaOH(aq).

The tribasic acid, H_3PO_4, would react with $75.0 \, cm^3$ NaOH(aq).

▲ **Figure 4** *Phosphoric acid, H_3PO_4, with a pH of 2–3, is found in rust remover as well as cola drinks. Try testing the drink with pH indicator paper*

The role of H^+ in the reactions of acids

Acids react with metals, carbonates, metal oxides, and alkalis to form salts. The active species from the acid is $H^+(aq)$ and ionic equations can be written for each type of reaction to emphasise the role of H^+ ions.

When writing ionic equations with $H^+(aq)$, the acid does not matter – it is the same reaction. For example, the equations below are for reaction of magnesium with dilute hydrochloric and sulfuric acids:

Synoptic link

Look back at Chapter 4, Acid and redox, to revise these reactions.

Reaction with HCl(aq)

full equation: $\quad 2HCl(aq) + Mg(s) \rightarrow MgCl_2(aq) + H_2(g)$

cancel spectator ions: $\quad 2H^+(aq) + \cancel{2Cl^-(aq)} + Mg(s) \rightarrow Mg^{2+}(aq) + \cancel{2Cl^-(aq)} + H_2$

ionic equation: $\quad 2H^+(aq) + Mg(s) \rightarrow Mg^{2+}(aq) + H_2(g)$

Reaction with H_2SO_4(aq)

full equation: $\quad H_2SO_4(aq) + Mg(s) \rightarrow MgSO_4(aq) + H_2(g)$

cancel spectator ions: $\quad 2H^+(aq) + \cancel{SO_4^{2-}(aq)} + Mg(s) \rightarrow Mg^{2+}(aq) + \cancel{SO_4^{2-}(aq)} + H_2$

ionic equation: $\quad 2H^+(aq) + Mg(s) \rightarrow Mg^{2+}(aq) + H_2(g)$

The ionic equation for the reaction of magnesium with dilute hydrochloric acid and dilute sulfuric acid is the same.

Spectator ions are ions that do not change during the reaction. In the equations, spectator ions can simply be cancelled out. You then know that *any* acid will react with magnesium to form $Mg^{2+}(aq)$ ions and hydrogen gas. All the ionic equations in the next section use $H^+(aq)$ only for the acid.

Study tip

Spectator ions are the spectators of chemical reactions. They do not themselves react during the reaction.

Redox reactions between acids and metals

Dilute acids undergo redox reactions with some metals to produce salts and hydrogen gas.

$$acid + metal \rightarrow salt + hydrogen$$

For zinc, the ionic equation for a reaction with *any* dilute acid is:

$$2H^+(aq) + Zn(s) \rightarrow Zn^{2+}(aq) + H_2(g)$$

Study tip

A few acids, such as nitric acid, are also strong oxidising agents. Except when it is very dilute, nitric acid reacts differently with most metals.

20.1 Brønsted–Lowry acids and bases

Reactions between acids and bases

Neutralisation of acids with carbonates

Carbonates are bases that neutralise acids to form a salt, water, and carbon dioxide.

$$\text{acid} + \text{carbonate} \rightarrow \text{salt} + \text{water} + \text{carbon dioxide}$$

Solid copper(II) carbonate reacts with *any* acid as follows:

$$2H^+(aq) + CuCO_3(s) \rightarrow Cu^{2+}(aq) + H_2O(l) + CO_2(g)$$

In aqueous solutions of carbonates spectator ions cancel and the final equation is simpler, for example, with aqueous sodium carbonate:

$$2H^+(aq) + \cancel{2Na^+(aq)} + CO_3^{2-}(aq) \rightarrow \cancel{2Na^+(aq)} + H_2O(l) + CO_2(g)$$

$$2H^+(aq) + CO_3^{2-}(aq) \rightarrow H_2O(l) + CO_2(g)$$

> **Study tip**
>
> For a solid, you still write the ionic equation with the full formula of the carbonate because the carbonate changes state during the reaction.

Neutralisation of acids with metal oxides

An acid is neutralised by a solid metal oxide or hydroxide to form a salt and water only.

$$\text{acid} + \text{base} \rightarrow \text{salt} + \text{water}$$

For magnesium oxide the reaction with *any* acid is:

$$2H^+(aq) + MgO(s) \rightarrow Mg^{2+}(aq) + H_2O(l)$$

Neutralisation of acids with alkalis

With alkalis, the acid and base are in solution. The overall reaction is the same as for metal oxides:

$$\text{acid} + \text{alkali} \rightarrow \text{salt} + \text{water}$$

The ionic equation, shown below, is much simpler than the full equation – neutralisation of $H^+(aq)$ ions by $OH^-(aq)$ ions to form neutral water, $H_2O(l)$:

$$H^+(aq) + OH^-(aq) \rightarrow H_2O(l)$$

> **Study tip**
>
> Full and ionic equations for acids are much more difficult to write than they look. Practise writing equations from scratch for carbonates and bases.

Summary questions

1. **a** Define the terms:
      i Brønsted–Lowry acid   **ii** Brønsted–Lowry base.   *(2 marks)*
   b State the conjugate base of:
      i HNO_3 **ii** HBr   **iii** CH_3CH_2COOH *(3 marks)*

2. Label the two acid–base pairs in the following acid–base equilibria.
   a $CH_3COOH(aq) + H_2O(aq) \rightleftharpoons H_3O^+(aq) + CH_3COO^-(aq)$ *(1 mark)*
   b $NH_3(aq) + H_2O(l) \rightleftharpoons NH_4^+(aq) + OH^-(aq)$ *(1 mark)*
   c $HCO_3^-(aq) + HCl(aq) \rightleftharpoons H_2CO_3(aq) + Cl^-(aq)$ *(1 mark)*

3. Write the full and ionic equations for the following reactions.
   a aluminium metal and sulfuric acid   *(2 marks)*
   b solid iron(II) oxide with nitric acid   *(2 marks)*
   c aqueous calcium hydroxide and hydrochloric acid   *(2 marks)*
   d aqueous sodium carbonate and phosphoric acid   *(2 marks)*

20.2 The pH scale and strong acids

Specification reference: 5.1.3

The pH scale

Søren Sørensen, the founder of pH

Many of the things used on a daily basis, like tap water, food, drinks, cosmetics, and medicines, are routinely tested for pH. Just over 100 years ago, the Danish chemist Søren Sørensen introduced pH as a simple numerical scale for measuring hydrogen ion concentrations.

Sørensen found that hydrogen ion concentrations in solutions had a very large range of values with negative powers of 10. Sørensen used the negative logarithm of these hydrogen ion concentrations to produce a more manageable scale. Using Sørensen's scale, hydrogen ion concentrations of 10^{-1} to 10^{-14} are converted to pH values of 1 to 14 (Table 1). The strength of Sørensen's pH scale is its simplicity and ease of use. For any solution at 25 °C:

- pH less than 7 shows increasing acidity
- pH greater than 7 shows increasing alkalinity
- pH 7 is neutral.

Before Sørensen's pH scale, the degree of acidity and basicity in solutions was measured as colours using indicators – Sørensen added numbers to the colours. Sørensen's method relied on measuring hydrogen ion using an electrochemical cell. Nowadays, pH meters are used for measuring pH accurately (Figure 1) and are based on the same principle. pH indicator paper or universal indicator solution is still used as an easier and cheaper, though less accurate, alternative.

pH – a logarithmic scale

The pH scale in Table 1 shows the relationship between pH and the concentration of $H^+(aq)$.

- A *low* value of $[H^+(aq)]$ matches a *high* value of pH.
- A *high* value of $[H^+(aq)]$ matches a *low* value of pH.

The mathematical relationship between pH and $[H^+(aq)]$ is:

$$pH = -\log[H^+(aq)]$$

The reverse of this relationship is:

$$[H^+(aq)] = 10^{-pH}$$

As a logarithmic scale, a change of one pH number is equal to a 10 times difference in $[H^+(aq)]$. So there is a large difference in hydrogen ion concentration between solutions with a pH of 2 and a pH of 3.

Learning outcomes

Demonstrate knowledge, understanding, and application of:

→ the pH scale

→ converting between pH and $[H^+(aq)]$

→ calculating the pH of a strong monobasic acid.

▼ **Table 1** *pH and hydrogen ion concentrations at 25 °C*

	pH	$[H^+(aq)]$ /mol dm^{-3}
acid	−1	10^1
	0	$10^0 = 1$
	1	10^{-1}
	2	10^{-2}
	3	10^{-3}
	4	10^{-4}
	5	10^{-5}
	6	10^{-6}
neutral	7	10^{-7}
	8	10^{-8}
	9	10^{-9}
	10	10^{-10}
	11	10^{-11}
	12	10^{-12}
	13	10^{-13}
	14	10^{-14}
alkali	15	10^{-15}

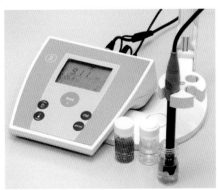

◀ **Figure 1** *A pH meter showing a weakly alkaline pH of 9.11 for the solution in the container. The probe of the pH meter, dipped in the solution, contains an electrode that measures the electrical potential of the hydrogen ions. This is directly related to the pH*

20.2 The pH scale and strong acids

- A pH of 1 has 10 times the concentration of H⁺ ions as a solution with a pH of 2.
- To dilute a solution from a pH 1 to a pH of 4 (just 3 pH units) would require dilution by 10 × 10 × 10 = 1000 times.
- A solution with a pH of 1 contains 10^{13} times more H⁺(aq) ions as a solution with a pH of 14.
- Comparing 1 with 14 is a lot more manageable than comparing 1×10^{-1} with 1×10^{-14}.

Converting between pH and [H⁺(aq)]

You need to be able to convert pH into [H⁺(aq)] and vice versa.

> **Study tip**
> You need to to learn the two equations linking pH with [H⁺(aq)] as you will use both many times.

Worked example: Converting from [H⁺(aq)] to pH

What is the pH of a solution with a H⁺(aq) concentration of 2.45×10^{-3} mol dm⁻³?

$$pH = -\log [H^+(aq)]$$
$$pH = -\log(2.45 \times 10^{-3})$$

You should get 2.61 (to two decimal places).

> **Study tip**
> Check your answers for both worked examples. Are they sensible? The [H⁺(aq)] should have a negative power of 10 within one of the pH values.

Worked example: Converting from pH to [H⁺(aq)]

What is the [H⁺(aq)] of a solution with a pH of 8.75?

$$[H^+(aq)] = 10^{-pH}$$
$$[H^+(aq)] = 10^{-8.75}$$

You should get 1.78×10^{-9} (to two decimal places).

▲ **Figure 1** *The log button and the 10^x function above the log button*

> **Study tip**
> Practise with any pH of a [H⁺(aq)] value. Then try to go back again using pH = −log[H⁺(aq)].
>
> Repeat until you are comfortable converting between pH and [H⁺(aq)] and get used to the buttons on your calculator.

Calculating the pH of strong acids

In aqueous solution, a strong monobasic acid, HA, completely dissociates:

$$HA(aq) \rightarrow H^+(aq) + A^-(aq)$$

1 mol   1 mol

So, for a strong acid, [H⁺(aq)] is equal to the concentration of the acid, [HA(aq)]:

$$[H^+(aq)] = [HA(aq)]$$

The pH of a strong acid can be calculated directly from the *concentration* of the acid.

Worked example: Calculating pH from acid concentration

A sample of nitric acid has a concentration of $1.35 \times 10^{-2}\,\text{mol}\,\text{dm}^{-3}$. What is the pH?

Step 1: Convert $[HA(aq)]$ into $[H^+(aq)]$.

HNO_3 is a strong acid and completely dissociates.

$[H^+(aq)] = [HNO_3(aq)] = 1.35 \times 10^{-2}\,\text{mol}\,\text{dm}^{-3}$

Step 2: Use your calculator to find the pH.

$pH = -\log[H^+(aq)] = -\log(1.35 \times 10^{-2}) = 1.87$

Synoptic link

You first met strong and weak acids in Topic 4.1, Acids, bases, and neutralisation.

Study tip

If the acid is strong, you can use the simple method here. For weak acids, there is another important step. See Topic 20.4, The pH of weak acids, for details.

Worked example: Calculating acid concentration from pH

A sample of hydrochloric acid has a pH of 3.79. What is the concentration of the hydrochloric acid?

Step 1: Use your calculator to find $[H^+(aq)]$.

$[H^+(aq)] = 10^{-pH} = 10^{-3.79} = 1.62 \times 10^{-4}\,\text{mol}\,\text{dm}^{-3}$

Step 2: Convert $[H^+(aq)]$ into $[HA(aq)]$.

HCl is a strong acid and completely dissociates.

$[HCl(aq)] = [H^+(aq)] = 1.62 \times 10^{-4}\,\text{mol}\,\text{dm}^{-3}$

Study tip

The whole number before a decimal place is the logarithmic way for showing powers of 10.

Significant figures for pH values starts after the decimal place. So a pH of 2.66 is to two significant figures only. So don't round pH values to one decimal place as this is only one significant figure. As a general rule, show pH values to two decimal places unless you are asked for a different number of places. This reflects the accuracy of most pH meters.

pH changes on dilution

A pH of 1 has 10 times the concentration of $H^+(aq)$ ions as a solution with a pH of 2. The next worked example shows that diluting by half changes the pH by just 0.30.

pH on dilution

$50\,\text{cm}^3$ of $0.100\,\text{mol}\,\text{dm}^{-3}$ hydrochloric acid is diluted to $100\,\text{cm}^3$ with water. What is the change in pH?

Step 1: Find the concentration of the diluted hydrochloric acid, $[HCl(aq)]$.

On diluting to $100\,\text{cm}^3$, the concentration has been halved to $0.0500\,\text{mol}\,\text{dm}^{-3}$.

20.2 The pH scale and strong acids

Step 2: Find the pH values before and after dilution.

HCl is a strong acid and completely dissociates:
$$[H^+(aq)] = [HCl(aq)]$$

Before dilution, $[H^+(aq)] = 0.100$
$$pH = -\log(0.100) = 1.00$$

After dilution, $[H^+(aq)] = 0.0500$
$$pH = -\log(0.0500) = 1.30$$

Changing the concentration by a factor of 2 changes the pH by 0.30.

Summary questions

1. **a** Calculate how many times more hydrogen ions there are in a 1 dm³ solution of pH 3 than a 1 dm³ solution of pH 9. *(1 mark)*
   b A solution has 100 000 000 000 times more hydrogen ions than a solution of pH 13, both solutions having the same volume. Calculate the pH of the first solution. *(1 mark)*
   c Calculate the pH for the following concentrations of H⁺ ions.
      i 0.01 mol dm^{-3} **ii** $1 \times 10^{-13} \text{ mol dm}^{-3}$ **iii** $0.0001 \text{ mol dm}^{-3}$
      (3 marks)
   d Calculate $[H^+(aq)]$ for solutions with the following pH values.
      i pH 3 **ii** pH 6 **iii** pH 12 *(3 marks)*

2. Calculate the pH of solutions with the following concentrations of HCl. Give answers to two decimal places.
   a $2.50 \times 10^{-3} \text{ mol dm}^{-3}$ *(1 mark)*
   b $8.10 \times 10^{-6} \text{ mol dm}^{-3}$ *(1 mark)*
   c $3.72 \times 10^{-8} \text{ mol dm}^{-3}$ *(1 mark)*
   d $4.42 \times 10^{-10} \text{ mol dm}^{-3}$ *(1 mark)*
   e $1.23 \times 10^{-12} \text{ mol dm}^{-3}$ *(1 mark)*
   f 2.31 mol dm^{-3} *(1 mark)*

3. Calculate the concentrations of nitric acid with the following pH values.
   a pH 2.83 *(1 mark)*
   b pH 7.91 *(1 mark)*
   c pH 12.27 *(1 mark)*
   d pH 9.69 *(1 mark)*
   e pH 4.71 *(1 mark)*
   f pH −1.08 *(1 mark)*

4. **a** Calculate the pH of the following diluted solutions of HCl(aq).
      i 25 cm³ of $0.0500 \text{ mol dm}^{-3}$ is diluted to 75 cm³ with water.
      ii 5.0 cm³ of $0.0250 \text{ mol dm}^{-3}$ is diluted to 100 cm³ with water *(4 marks)*
   b You are required to prepare 250 cm³ of $0.100 \text{ mol dm}^{-3}$ HCl by diluting $0.600 \text{ mol dm}^{-3}$ HCl with water. Calculate the volume of $0.600 \text{ mol dm}^{-3}$ HCl that must be diluted with water. *(4 marks)*

20.3 The acid dissociation constant K_a

Specification reference: 5.1.3

Strong and weak acids

Strong and weak are terms used to describe the extent of dissociation of an acid. A strong acid, such as hydrochloric acid, HCl, completely dissociates in aqueous solution.

$$HCl(aq) \rightarrow H^+(aq) + Cl^-(aq)$$

A weak acid, such as ethanoic acid, CH_3COOH, **partially dissociates**.

$$CH_3COOH(aq) \rightleftharpoons H^+(aq) + CH_3COO^-(aq)$$

In this topic you will learn about the extent of dissociation of weak acids in terms of an equilibrium constant.

The acid dissociation constant K_a

The acid dissociation constant K_a is one of several special equilibrium constants used for acid–base equilibria. All the constants are just versions of the equilibrium constant K_c in terms of concentrations in $mol\,dm^{-3}$, with special labels to show their use.

The dissociation of any weak acid, HA, can be shown as a general form:

$$HA(aq) \rightleftharpoons H^+(aq) + A^-(aq).$$

The acid dissociation constant, K_a is calculated as:

$$K_a = \frac{[H^+(aq)][A^-(aq)]}{[HA(aq)]} \quad \text{units:} \quad \frac{(mol\,dm^{-3}) \times (mol\,dm^{-3})}{(mol\,dm^{-3})} = mol\,dm^{-3}$$

Using this principle, you can write the K_a expression for any weak acid. For ethanoic acid:

$$K_a = \frac{[H^+(aq)][CH_3COO^-(aq)]}{[CH_3COOH(aq)]}$$

As with all equilibrium constants, K_a changes with temperature and recorded K_a values are usually standardised at 25 °C. The larger the numerical value of K_a, the further the equilibrium is to the right. If you look at the K_a expression, the concentration of the dissociated ions are on top and so the larger the K_a value, the greater the dissociation and the greater the acid strength.

K_a and pK_a

The K_a values highlight a potential problem that you have already seen with $[H^+(aq)]$ – it is difficult to compare numbers with negative indices. As with the pH scale, the problem has been solved by converting the K_a value into a negative logarithm called pK_a.

$$pK_a = -\log K_a$$

The reverse of this relationship is:

$$K_a = 10^{-pK_a}$$

> **Learning outcomes**
>
> Demonstrate knowledge, understanding, and application of:
> → strong and weak acids
> → the acid dissociation constant K_a
> → converting between K_a and pK_a.

> **Synoptic link**
>
> For more details of strong and weak acids, see Topic 4.1, Acids, bases, and neutralisation.

> **Synoptic link**
>
> For more details of equilibrium constants, see Chapter 19, Equilibrium.

> **Study tip**
>
> K_a is simply K_c applied to the dissociation of an acid.
>
> As the terms are always the same as the general form of $\frac{[H^+(aq)][A^-(aq)]}{[HA(aq)]}$, the units are always $mol\,dm^{-3}$.

20.3 The acid dissociation constant K_a

pK_a values are much more manageable than K_a and it is much easier to compare relative acidic strengths using pK_a values than K_a values.

> **Worked example: Converting from K_a to pK_a**
>
> What is the pK_a value of a weak acid with a K_a value of $1.48 \times 10^{-4}\,\text{mol}\,\text{dm}^{-3}$?
>
> $pK_a = -\log K_a$
>
> $pK_a = -\log(1.48 \times 10^{-4})$
>
> You should get 3.83 (to two decimal places).

> **Worked example: Converting from pK_a to K_a**
>
> What is the K_a value of a weak acid with a pK_a of 4.82?
>
> $K_a = 10^{-pK_a}$
>
> $[H^+(aq)] = 10^{-4.82}$
>
> You should get 1.51×10^{-5} (to two decimal places).

Table 1 compares the K_a and pK_a values of four weak acids.

▼ **Table 1** K_a and pK_a values for four acids

Acid		K_a / mol dm^{-3}	pK_a	Relative acid strength
nitrous acid	HNO_2	4.10×10^{-4}	3.39	strongest acid
methanoic acid	HCOOH	1.77×10^{-4}	3.75	
ethanoic acid	CH_3COOH	1.76×10^{-5}	4.75	
chloric(I) acid	HClO	3.00×10^{-8}	7.53	weakest acid

Study tip

As with pH and $[H^+(aq)]$, K_a should have a negative power within one of the pK_a values.

The *stronger* the acid, the *larger* the K_a value and the *smaller* the pK_a value.

The *weaker* the acid, the *smaller* the K_a value and the *larger* the pK_a value.

pK_a values are used extensively for comparing the strengths of weak acids particularly in biological systems.

 Weak acids in wine

Sulfurous acid

Wines often contain traces of sulfurous acid, H_2SO_3, added as a preservative. Sulfurous acid is a dibasic acid and its dissociation is shown below.

$H_2SO_3(aq) \rightleftharpoons H^+(aq) + HSO_3^-(aq)$ $pK_a = 1.92$

$HSO_3^-(aq) \rightleftharpoons H^+(aq) + SO_3^{2-}(aq)$ $pK_a = 7.18$

For the first dissociation, H_2SO_3 acts as a weak acid. From the pK_a values, the weak acid, HSO_3^-, in the second dissociation is a far weaker acid than H_2SO_3.

This behaviour is typical for dibasic and tribasic acids.

20 ACIDS, BASES, AND pH

Malic and tartaric acid

The main acids in wine grapes are two weak dibasic acids – tartaric acid and malic acid. Figure 1 shows the structures of malic acid and tartaric acids.

▲ **Figure 1** *The structures of malic acid and tartaric acid, found in wine grapes*

Both acids contain two carboxyl groups, COOH. When the acids ionise, hydrogen ions dissociate from the COOH groups, one at a time. Figure 2 and 3 show the stepwise dissociation of H^+ from the two COOH groups in malic acid.

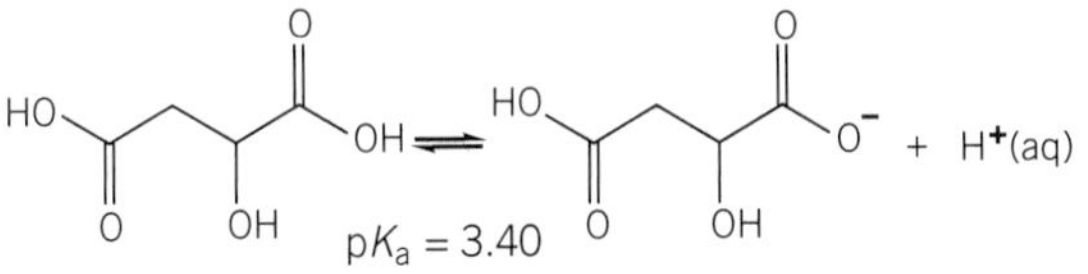

▲ **Figure 2** *Dissociation of H^+ from the first COOH group* ($pK_a = 3.40$)

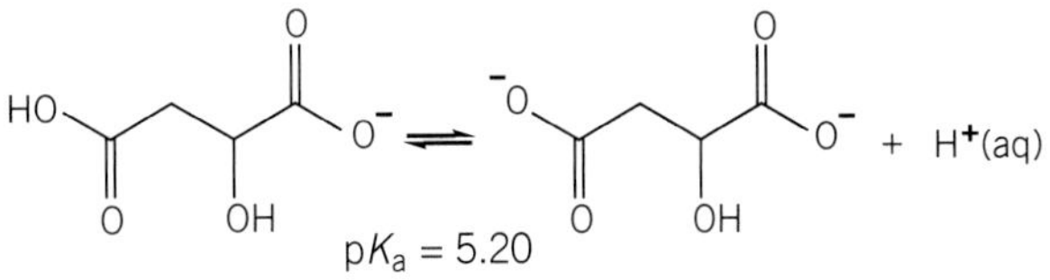

▲ **Figure 3** *Dissociation of H^+ from the second COOH group* ($pK_a = 5.20$)

The alcohol groups in malic acid do *not* dissociate as the –OH group is much less acidic than the COOH group.

> Citric acid is a tribasic acid with the structure: 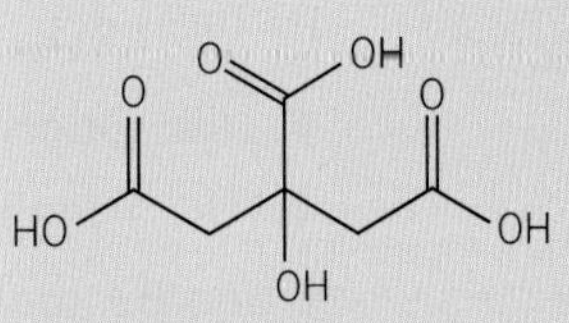
>
> ▲ **Figure 4** *Citric acid*
>
> 1 Write equations for the stepwise dissociation of citric acid using skeletal formulae.
> 2 The formula of citric acid can be written as $C_3H_4OH(COOH)_3$. Using this formula as your starting point, write K_a expressions for each of the three dissociations.

Summary questions

1 For each of the following acid–base equilibria, write down the expression for K_a. State the units of K_a for each reaction.
   a $HNO_2(aq) \rightleftharpoons H^+(aq) + NO_2^-(aq)$ *(1 mark)*
   b $C_6H_5COOH(aq) \rightleftharpoons H^+(aq) + C_6H_5COO^-(aq)$ *(1 mark)*
   c $H_2S(aq) \rightleftharpoons H^+(aq) + HS^-(aq)$ *(1 mark)*

2 a Calculate the pK_a of solutions with the following K_a values. Give your answers to two decimal places.
   i 9.5×10^{-8} mol dm^{-3}
   ii 7.1×10^{-4} mol dm^{-3}
   iii 6.3×10^{-8} mol dm^{-3} *(3 marks)*
   b Calculate K_a of solutions with the following pK_a values. Give answers in standard form and to one decimal place.
   i pK_a 1.92
   ii pK_a 7.70
   iii pK_a 4.77 *(3 marks)*

3 Phosphoric acid, H_3PO_4, is a tribasic acid.
   a Write expressions for K_a for the three successive dissociations of phosphoric acid. *(3 marks)*
   b Predict how the relative values of K_a and pK_a would change for each dissociation. *(3 marks)*

20.4 The pH of weak acids

Specification reference: 5.1.3

Learning outcomes

Demonstrate knowledge, understanding, and application of:

→ calculations of pH and K_a for weak acids

→ the limitations of pH and K_a calculations for weak acids.

The pH of weak acids

For a monobasic acid, HA, in aqueous solution:

- a strong acid HA completely dissociates: $[H^+(aq)] = [HA(aq)]$
- a weak acid HA partially dissociates: $[H^+(aq)] \neq [HA(aq)]$

For a weak acid there is an equilibrium:

$$HA(aq) \rightleftharpoons H^+(aq) + A^-(aq)$$

$[H^+(aq)]$ depends upon:

- the concentration of the acid, $[HA(aq)]$
- the acid dissociation constant K_a.

When HA molecules dissociate, $H^+(aq)$ and $A^-(aq)$ ions are formed in equal quantities. The start and equilibrium concentrations of HA(aq), $H^+(aq)$, and $A^-(aq)$ are shown in Figure 1.

	HA(aq) $\rightleftharpoons$	$H^+(aq)$ +	$A^-(aq)$
start	$[HA(aq)]_{start}$	0	0
equilibrium (eqm)	$[HA(aq)]_{eqm}$ $[HA(aq)]_{start} - [H^+(aq)]_{eqm}$	$[H^+(aq)]_{eqm}$	$[A^-(aq)]_{eqm}$

▲ **Figure 1** *Dissociation of HA at equilibrium*

K_a can be calculated using equilibrium concentrations:

$$K_a = \frac{[H^+(aq)]_{eqm}[A^-(aq)]_{eqm}}{[HA(aq)]_{eqm}} = \frac{[H^+(aq)]_{eqm}[A^-(aq)]_{eqm}}{[HA(aq)]_{start} - [H^+(aq)]_{eqm}}$$

Approximations

The equation above can be greatly simplified by making two approximations.

Approximation 1

From Figure 1, HA dissociates to produce equilibrium concentrations of $H^+(aq)$ and $A^-(aq)$ that are equal. There will also be a very small concentration of $H^+(aq)$ from the dissociation of water but this will be extremely small and can be neglected compared with the H^+ concentration from the acid.

$$[H^+(aq)]_{eqm} = [A^-(aq)]_{eqm}$$

Approximation 2

From Figure 1, the equilibrium concentration of HA is smaller than the undissociated concentration.

$$[HA(aq)]_{eqm} = [HA(aq)]_{start} - [H^+(aq)]_{eqm}$$

As the dissociation of weak acids is small, you can assume that $[HA(aq)]_{start} \gg [H^+(aq)]$ and you can neglect any decrease in the concentration of HA from dissociation.

$$[HA(aq)]_{eqm} = [HA(aq)]_{start}$$

Study tip

For a weak acid with a K_a value of 1×10^{-3}, only 1 molecule in every 1000 dissociates.

So at equilibrium, there are still 999 molecules of the undissociated weak acid present.

999 ~ 1000, only a 0.1% difference.

For a value of 1×10^{-4}, the approximation is even better:

9999 ~ 10000

ACIDS, BASES, AND pH

Simplifying the K_a expression
By applying the two approximations, the K_a expression can be greatly simplified.

$$K_a = \frac{[H^+(aq)]_{eqm}[A^-(aq)]_{eqm}}{[HA(aq)]_{start} - [H^+(aq)]_{eqm}} \longrightarrow \frac{[H^+(aq)]_{eqm}^2}{[HA(aq)]_{start}}$$

Therefore $K_a = \dfrac{[H^+(aq)]^2}{[HA(aq)]}$

Provided that two of [H$^+$(aq)] (or pH), K_a and [HA(aq)] are known, you can always calculate the third quantity.

Calculating pH
To calculate pH using the K_a expression above, you first need to make [H$^+$(aq)] the subject of the equation:

$$[H^+(aq)]^2 = K_a \times [HA(aq)] \qquad [H^+(aq)] = \sqrt{K_a \times [HA(aq)]}$$

The worked example below shows how the pH of a weak acid can be calculated from the acid concentration and K_a value.

> **Study tip**
> The key to solving the pH of a weak acid is the relationship:
> $[H^+(aq)] = \sqrt{K_a \times [HA(aq)]}$
> It is well worth memorising this equation.

🖩 Worked example: Calculating the pH of a weak acid

Calculate the pH of 0.0245 mol dm^{-3} ethanoic acid, CH$_3$COOH, at 25 °C, where $K_a = 1.7 \times 10^{-5}$ mol dm^{-3}.

Step 1: Calculate [H$^+$(aq)] from K_a and [HA(aq)].

CH$_3$COOH is a weak acid and partially dissociates:
CH$_3$COOH(aq) $\rightleftharpoons$ H$^+$(aq) + CH$_3$COO$^-$(aq)

$$K_a = \frac{[H^+(aq)][CH_3COO^-(aq)]}{[CH_3COOH(aq)]} \sim \frac{[H^+(aq)]^2}{[CH_3COOH(aq)]}$$

$[H^+(aq)]^2 = K_a \times [CH_3COOH(aq)]$

$[H^+(aq)] = \sqrt{K_a \times [CH_3COO(aq)]} = \sqrt{(0.0245 \times 1.7 \times 10^{-5})}$

$= 6.45 \times 10^{-4}$ mol dm^{-3}

Step 2: Use your calculator to find the pH.

pH $= -\log[H^+(aq)] = -\log(6.45 \times 10^{-4}) = 3.19$

> **Study tip**
> Take care when using the calculator for the square root. You must use brackets around the values for $K_a[HA(aq)]$ or calculate $K_a[HA(aq)]$ first, and then take the square root of the answer afterwards.

Determination of K_a
Experimentally the K_a for a weak acid can be determined by
- preparing a standard solution of the weak acid of known concentration
- measuring the pH of the standard solution using a pH meter.

The worked example below shows how the K_a value of a weak acid can be calculated from the acid concentration and a pH measurement.

20.4 The pH of weak acids

 Worked example: Calculating K_a for a weak acid

The pH of $0.065\,\text{mol}\,\text{dm}^{-3}$ propanoic acid, CH_3CH_2COOH, is 3.04. Calculate K_a.

Step 1: Use your calculator to find $[H^+(aq)]$.

$$[H^+(aq)] = 10^{-pH} = 10^{-3.04} = 9.12 \times 10^{-4}\,\text{mol}\,\text{dm}^{-3}$$

Step 2: Calculate K_a from $[H^+(aq)]$ and $[HA(aq)]$.

$$K_a = \frac{[H^+(aq)]\,[CH_3CH_2COO^-(aq)]}{[CH_3CH_2COOH(aq)]} \sim \frac{[H^+(aq)]^2}{[CH_3CH_2COOH(aq)]}$$

$$K_a = \frac{(9.12 \times 10^{-4})^2}{0.065} = 1.28 \times 10^{-5}\,\text{mol}\,\text{dm}^{-3}$$

Approximations in calculations involving weak acids

Both the worked examples make use of two approximations. Are these justified? Are there any situations where the approximations break down?

Approximation 1

The first approximation assumes that the dissociation of water is negligible.

$$[H^+(aq)]_{eqm} \sim [A^-(aq)]_{eqm}$$

At 25 °C, $[H^+(aq)]$ from dissociation of water = $10^{-7}\,\text{mol}\,\text{dm}^{-3}$. If the pH > 6, then $[H^+(aq)]$ from the dissociation of water will be significant compared with dissociation of the weak acid. This approximation breaks down for very weak acids or very dilute solutions.

Approximation 2

The second approximation assumes that the concentration of acid is much greater than the H^+ concentration at equilibrium:

$$[HA(aq)]_{start} \gg [H^+(aq)]_{eqm}.$$

$[HA(aq)]_{eqm} = [HA(aq)]_{start} - [H^+(aq)]_{eqm}$ approximates to $[HA]_{eqm} = [HA]_{start}$

This approximation will hold for weak acids with small K_a values. It breaks down when $[H^+(aq)]$ becomes significant and there is a real difference between $[HA(aq)]_{eqm}$ and $[HA(aq)]_{start} - [H^+(aq)]_{eqm}$

This approximation is not justified for *stronger* weak acids with $K_a > 10^{-2}\,\text{mol}\,\text{dm}^{-3}$ and for very dilute solutions.

> **Synoptic link**
>
> See Topic 20.5, pH and strong bases, for details of the dissociation of water.

 Checking the approximations

You can carry out the weak acid pH calculation without making the $[HA(aq)]_{start} - [H^+(aq)]_{eqm}$ approximation, but you will then need to solve a quadratic equation. A far easier method is to use one of the many quadratic equation solver websites that you can find online.

Problem: Calculate the pH of a $0.100\,\text{mol}\,\text{dm}^{-3}$ solution of chlorous acid $HClO_2$ (K_a: $1.2 \times 10^{-2}\,\text{mol}\,\text{dm}^{-3}$).

$$K_a = \frac{[H^+(aq)][ClO_2^-(aq)]}{[HClO_2(aq)]_{eqm}} = \frac{[H^+(aq)]^2}{[HClO_2(aq)]_{start} - [H^+(aq)]}$$

Rearranging the equation gives:

$$[H^+(aq)]^2 = K_a \times [HClO_2(aq)]_{start} - K_a \times [H^+(aq)]$$
$$= 1.2 \times 10^{-2} \times 0.100 - 1.2 \times 10^{-2} \times [H^+(aq)]$$
$$[H^+(aq)]^2 + 1.2 \times 10^{-2} \times [H^+(aq)] - 1.2 \times 10^{-3} = 0$$
$$a = 1 \quad b = 1.2 \times 10^{-2} \quad c = -1.2 \times 10^{-3}$$

Solving the quadratic equation (using a quadratic solver website), gives:

$[H^+(aq)] = 0.0292$ mol dm^{-3} pH = $-\log(0.0292) = 1.53$

Using the much simpler and quicker approximation method:

$$[H^+(aq)] = \sqrt{K_a \times [HClO_2(aq)]} = \sqrt{0.100 \times 1.2 \times 10^{-2}}$$
$$= 0.0346 \text{ mol dm}^{-3}$$
$$pH = -\log(0.0346) = 1.46$$

Comparing the pH values of 1.53 and 1.46, the approximation does not seem to be justified. There is an 18% difference between the [H$^+$(aq)] values from both methods, which is significant.

> 1 Calculate the pH of a 0.100 mol dm^{-3} solution of nitrous acid HNO$_2$ (K_a: 4.0×10^{-4} mol dm^{-3}) by both methods and show whether the [HA(aq)]$_{start}$ − [H$^+$(aq)]$_{eqm}$ approximation is justified.
> 2 Suggest when the approximation is valid.

Study tip

Don't worry. You do not need to know how to solve quadratic equations as part of your chemistry course.

Summary questions

1. **a** Write K_a expressions for the following weak acids:
   i HCN   **ii** C$_6$H$_5$COOH   *(2 marks)*
   b Two approximations are often used to simplify pH calculation of weak acids.
   i State the two approximations.
   ii State the simplified square root expression for calculating [H$^+$(aq)] for a weak acid after using these two approximations. *(3 marks)*

2. **a** Calculate the pH of the following weak acids. Give your answers to two decimal places.
   i 0.075 mol dm^{-3} HCOOH ($K_a = 1.77 \times 10^{-4}$ mol dm^{-3})
   ii 3.42×10^{-2} mol dm^{-3} C$_6$H$_5$COOH ($K_a = 6.46 \times 10^{-5}$ mol dm^{-3}) *(4 marks)*
   b Calculate the K_a and pK_a values for the following acid solutions.
   i a 0.125 mol dm^{-3} solution with a pH of 2.32
   ii a 1.75×10^{-3} mol dm^{-3} solution with a pH of 3.84 *(4 marks)*

3. Sulfuric acid is a dibasic acid. For its first dissociation, sulfuric acid behaves as a strong acid. The second dissociation has a K_a value of 1.20×10^{-2} mol dm^{-3}. The pH of 0.0100 mol dm^{-3} sulfuric acid is 1.84.
   a Use the information above to suggest why the pH is not simply $-\log(2 \times 0.0100) = 1.70$ *(1 mark)*
   b Describe the steps that would be needed to calculate the actual pH of the sulfuric acid. *(3 marks)*
   c Explain why [H$^+$(aq)] >> [A$^-$(aq)] after the second dissociation of sulfuric acid (the dissociation of water being insignificant at this pH).
   Note: Calculating the pH as 1.84 is extension work and requires some algebra for [H$^+$(aq)] and [A$^-$(aq)] and use of a quadratic equation. *(2 marks)*

20.5 pH and strong bases

Specification reference: 5.1.3

Learning outcomes

Demonstrate knowledge, understanding, and application of:

→ calculating the pH of strong bases

→ the ionic product of water K_w.

The ionisation of water

Water ionises very slightly, acting as both an acid and as a base, setting up the acid–base equilibrium as shown in Figure 1.

$$H_2O(l) + H_2O(l) \rightleftharpoons H_3O^+(aq) + OH^-(aq)$$
acid 1   base 2   acid 2   base 1

▲ **Figure 1** *Acid–base equilibrium of water*

Or more simply:

$$H_2O(l) \rightleftharpoons H^+(aq) + OH^-(aq)$$

Treating water as a weak acid:

$$K_a = \frac{[H^+(aq)]\,[OH^-(aq)]}{[H_2O(l)]}$$

The dissociation of water is very small:

- 1 dm³ (1000 g) of water is mainly undissociated $H_2O(l)$
- so $[H_2O(l)] = \dfrac{1000}{18.0} = 55.6\,\text{mol}\,\text{dm}^{-3}$, a constant.

Rearranging the K_a expression above gives:

$$K_a \times [H_2O(l)] = [H^+(aq)]\,[OH^-(aq)]$$

K_w is called the **ionic product of water** – the ions in water (H^+ and OH^-) multiplied together.

$$K_w = [H^+(aq)]\,[OH^-(aq)]$$

As with all equilibrium constants, K_w varies with temperature. The value of K_w at 298 K (25 °C) is $1.00 \times 10^{-14}\,\text{mol}^2\,\text{dm}^{-6}$.

Study tip
$K_a \times [H_2O(l)]$ is a constant $= K_w$

Study tip
$[H^+(aq)]\,[OH^-(aq)]$ is always equal to K_w.

The importance of K_w

The significance of K_w having a value of $1.00 \times 10^{-14}\,\text{mol}^2\,\text{dm}^{-6}$ at 25 °C is huge. The value sets up the neutral point in the pH scale. The examples below all apply to 298 K (25 °C). K_w controls the concentration of $H^+(aq)$ and $OH^-(aq)$ ions in aqueous solutions.

The pH of pure water at 25 °C

On dissociation, water is neutral – it produces the same number of $H^+(aq)$ and $OH^-(aq)$ ions.

So $[H^+(aq)] = [OH^-(aq)]$

$K_w = [H^+(aq)]\,[OH^-(aq)] = [H^+(aq)]^2 = 1.00 \times 10^{-14}\,\text{mol}^2\,\text{dm}^{-6}$ at 25 °C

$[H^+(aq)] = \sqrt{1.00 \times 10^{-14}} = 1.00 \times 10^{-7}\,\text{mol}\,\text{dm}^{-3}$

pH $= -\log[H^+(aq)] = -\log(1.00 \times 10^{-7}) = 7$

ACIDS, BASES, AND pH

You know this fact already, but the very reason that neutral water has a pH of 7 is the K_w value of 1.00×10^{-14} mol² dm⁻⁶.

What about solutions of acids and alkalis?

The ionic product of water K_w is essentially an equilibrium constant that controls the concentrations of H⁺(aq) and OH⁻(aq) in aqueous solutions.

In any aqueous solution, there will always be both H⁺(aq) and OH⁻(aq) ions present such that [H⁺(aq)] [OH⁻(aq)] = K_w.

- A solution is acidic when [H⁺(aq)] > [OH⁻(aq)]
- A solution is neutral when [H⁺(aq)] = [OH⁻(aq)]
- A solution is alkaline when [H⁺(aq)] < [OH⁻(aq)]

So a solution that is acidic still contains OH⁻(aq) ions, it is just that there are more H⁺(aq) ions (and vice versa for an alkaline solution). The value of K_w, 1.00×10^{-14} mol² dm⁻⁶ at 298 K (25 °C), controls the concentrations of H⁺(aq) and OH⁻(aq).

For pH values that are whole numbers, it is easy to work out the H⁺(aq) and OH⁻(aq) concentrations as the indices for [H⁺(aq)] and [OH⁻(aq)] add up to −14.

In an acid solution with a pH of 3:

- [H⁺(aq)] = 10^{-3} mol dm⁻³
- [OH⁻(aq)] = 10^{-11} mol dm⁻³

If the pH value is not a whole number, you will need to do a little more work as in the worked example below.

▲ **Figure 2** *A pH meter measuring the pH of rainwater. Pure (distilled and deionised) water is considered neutral at pH 7. There are several reasons why water may deviate from pH 7, including the presence of dissolved salts and gases. Rainwater tends to have a pH somewhere in the range 5–6 (the sample here has a pH of 5.48)*

> **Study tip**
>
> For [H⁺(aq)] = 10^{-3} mol dm⁻³
> [OH⁻(aq)] = 10^{-11} mol dm⁻³
> (−3) + (−11) = −14

 Worked example: Calculating [H⁺(aq)] and [OH⁻(aq)] in aqueous solutions 1

What are the concentrations of H⁺(aq) and OH⁻(aq) ions in a solution with a pH of 3.25 at 25 °C?

Step 1: Use your calculator to find [H⁺(aq)].

$$[H^+(aq)] = 10^{-pH} = 10^{-3.25} = 5.62 \times 10^{-4} \text{ mol dm}^{-3}$$

Step 2: Calculate [OH⁻(aq)] from K_w and [H⁺(aq)].

$$K_w = [H^+(aq)][OH^-(aq)] = 1.00 \times 10^{-14}$$

$$[OH^-(aq)] = \frac{K_w}{[H^+(aq)]} = \frac{1.00 \times 10^{-14}}{5.62 \times 10^{-4}}$$

$$= 1.78 \times 10^{-11} \text{ mol dm}^{-3}$$

> **Study tip**
>
> K_w is the key to working out [H⁺(aq)] and [OH⁻(aq)].

Worked example: Calculating [H⁺(aq)] and [OH⁻(aq)] in aqueous solutions 2

What is the pH of a solution with [OH⁻(aq)] = 2.00×10^{-2} mol dm⁻³ at 25 °C?

Step 1: Calculate [H⁺(aq)] from K_w and [OH⁻(aq)].

$$K_w = [H^+(aq)][OH^-(aq)] = 1.00 \times 10^{-14}$$

$$[H^+(aq)] = \frac{K_w}{[OH^-(aq)]} = \frac{1.00 \times 10^{-14}}{2.00 \times 10^{-2}}$$

$$= 5.00 \times 10^{-13} \text{ mol dm}^{-3}$$

Step 2: Use your calculator to find pH.

$$pH = -\log[H^+(aq)] = -\log(5.00 \times 10^{-13}) = 12.30$$

The pH of solutions of strong bases

An alkali is a soluble base that releases OH⁻ ions in aqueous solution. A strong base is an alkali that completely dissociates in solution. For example, sodium hydroxide, NaOH, is a strong alkali and completely dissociates:

NaOH(aq) → Na⁺(aq) + OH⁻(aq)
 1 mol   1 mol

NaOH is a monoacidic base as each mole of NaOH releases one mole of OH⁻(aq) ions.

The pH of a strong base can be calculated from:

- the concentration of the base
- the ionic product of water K_w.

Synoptic link

For bases and alkalis, see Topic 4.1, Acids, bases, and neutralisation.

Worked example: Calculating the pH of a solution of a strong base

A solution of NaOH has a concentration of 0.0750 mol dm⁻³. What is the pH at 25 °C?

Step 1: Convert [NaOH(aq)] into [OH⁻(aq)].

NaOH is a strong monoacidic alkali and completely dissociates.

[OH⁻(aq)] = [NaOH(aq)] = 0.0750 mol dm⁻³

Step 2: Use K_w and [OH⁻(aq)] to find [H⁺(aq)].

$$K_w = [H^+(aq)][OH^-(aq)] = 1.00 \times 10^{-14} \text{ mol}^2 \text{ dm}^{-6}$$

$$[H^+(aq)] = \frac{K_w}{[OH^-(aq)]} = \frac{1.00 \times 10^{-14}}{0.0750} = 1.33 \times 10^{-13} \text{ mol dm}^{-3}$$

Step 3: Use your calculator to find pH.

$$pH = -\log[H^+(aq)] = -\log(1.33 \times 10^{-13}) = 12.88$$

Study tip

Each mole of a *dibasic* base, such as Ca(OH)₂, releases two moles of OH⁻ ions.

Ca(OH)₂(aq) → Ca²⁺(aq) + 2OH⁻(aq)
1 mol   2 mol

So in a pH calculation:

[OH⁻(aq)] = 2 × [Ca(OH)₂(aq)]

20 ACIDS, BASES, AND pH

What about weak bases?

As with strong bases, a weak base is also an alkali. Ammonia gas is an example of a weak base – it dissolves in water releasing OH^- ions from water molecules. An equilibrium is set up with the equilibrium positioned well to the left.

$$NH_3(aq) + H_2O(l) \rightleftharpoons NH_4^+(aq) + OH^-(aq)$$

So one mole of NH_3 releases far less than one mole of $OH^-(aq)$ ions.

The pH of weak bases can be calculated using a similar method to that used for weak acids.

> **Study tip**
> Calculation of the pH of weak bases goes beyond the scope of this Chemistry course.

pOH

pOH is a similar scale to pH, used as a convenient way of showing OH^- concentrations, $[OH^-(aq)]$:

$$pOH = -\log[OH^-(aq)] \quad [OH^-(aq)] = 10^{-pOH}$$

The pOH scale has a similar range to pH, centred at the neutral point where $[H^+(aq)]$ and $[OH^-(aq)]$ are both equal to 10^{-7} mol dm^{-3}. The pH and pOH scales are shown in Table 1.

▼ **Table 1** pH and pOH scales

pH	0	1	2	3	4	5	6	7	8	9	10	11	12	13	14
pOH	14	13	12	11	10	9	8	7	6	5	4	3	2	1	0

At 25 °C, $K_w = 10^{-14}$ mol^2 dm^{-6}, and pH + pOH = 14

The pOH scale gives an alternative method for calculating the pH of a strong alkali.

What is the pH of a solution with $[OH^-(aq)] = 4.50 \times 10^{-3}$ mol dm^{-3} at 25 °C?

Step 1: Calculate pOH from $[OH^-(aq)]$.
$pOH = -\log[OH^-(aq)] = -\log(4.50 \times 10^{-3})$
$= 2.34$

Step 2: Calculate pH using pH + pOH = 14
$pH = 14 - pOH$
$pH = 14 - 2.34 = 11.66$

Use pOH to solve the following

1. Calculate the pH of the following solutions:
   a 0.125 mol dm^{-3} NaOH   b 3.25×10^{-3} KOH
2. Calculate the $[OH^-(aq)]$ in solutions with
   a pH = 10.76   b pH = 12.38

Summary questions

1. **a** State the expression for the ionic product of water. *(1 mark)*
   b Calculate the $[OH^-(aq)]$ of a solution with a $[H^+(aq)]$ of 10^{-4} mol dm^{-3}. *(1 mark)*
   c Explain why the units of K_w are mol^2 dm^{-6}. *(1 mark)*

2. Calculate the pH of the following solutions of strong bases at 25 °C:
   a 0.0120 mol dm^{-3} KOH(aq) *(1 mark)*
   b 1.35×10^{-3} mol dm^{-3} NaOH(aq) *(1 mark)*
   c 8.75×10^{-2} mol dm^{-3} Ca(OH)$_2$(aq) *(1 mark)*

3. Calculate the concentration of the following strong bases:
   a a solution of NaOH with a pH of 12.65 *(1 mark)*
   b a solution of KOH with a pH of 11.57 *(1 mark)*
   c a solution of Ca(OH)$_2$ with a pH of 12.83 *(1 mark)*

4. K_w changes with temperature.
   At 60 °C, $K_w = 9.31 \times 10^{-14}$ mol^2 dm^{-6}.
   At 10 °C, $K_w = 2.93 \times 10^{-15}$ mol^2 dm^{-6}.
   a Calculate the pH of water at
      i 60 °C   **ii** 10 °C *(2 marks)*
   b Discuss whether water is neutral at these two temperatures. *(2 marks)*
   c Deduce whether the ionisation of water is exothermic or endothermic. *(1 mark)*

Chapter 20 Practice questions

Practice questions

1. Three solutions **A**, **B**, and **C**, all have the same concentration of $0.0150 \text{ mol dm}^{-3}$.
   - Solution **A** is a HCl(aq).
   - Compound **B** is NaOH(aq).
   - Solution **C** is CH_3CH_2COOH(aq).

   a Calculate the pH of:
      (i) solution **A** *(1 mark)*
      (ii) solution **B** *(2 marks)*

   b Solution **A** is diluted with an equal volume of water. What is the new pH of the diluted solution of **A**? *(2 marks)*

   c Solution **C** has a pH of 2.60. Calculate the value of K_a of CH_3CH_2COOH? *(1 mark)*

2. Hydrogencarbonate ions, HCO_3^-, can behave as a weak Brønsted–Lowry acid and as a weak Brønsted–Lowry base.

   a What is meant by the terms Brønsted–Lowry acid and base? *(1 mark)*

   b Write the formula of the conjugate acid and conjugate base of HCO_3^-. *(2 marks)*

   c NaOH(aq) is added to $NaHCO_3$(aq). An acid–base equilibrium is set up between OH^-(aq) and HCO_3^-(aq).
      (i) Complete the equation for the equilibrium that would be set up and label the conjugate acid–base pairs.
      $OH^- + HCO_3^- \rightleftharpoons \ldots\ldots + \ldots\ldots$
      $\ldots\ldots \quad \ldots\ldots \quad \ldots\ldots \quad \ldots\ldots$ *(2 marks)*
      (ii) What would happen to HCO_3^- if HCl(aq) were added to HCO_3^- instead of NaOH(aq)? *(1 marks)*

3. Boric acid, H_3BO_3, is a tribasic weak acid.

   a What is meant by the term *tribasic acid*? *(1 mark)*

   b Write equilibria for the three stages of dissociation of boric acid. *(3 marks)*

   c Write the K_a expression for the second dissociation stage of boric acid. *(1 mark)*

   d Write an equation for the reaction of boric acid with an excess of NaOH(aq). *(1 mark)*

4. A $1.50 \times 10^{-2} \text{ mol dm}^{-3}$ solution of methanoic acid, HCOOH, has $[H^+] = 1.55 \times 10^{-3} \text{ mol dm}^{-3}$.

   a (i) Use an equation for the dissociation of methanoic acid to show what is meant by a *weak acid*. *(1 mark)*
      (ii) How does this information indicate that methanoic acid is a weak acid? *(1 mark)*
      (iii) What is the formula of the conjugate base of methanoic acid. *(1 mark)*

   b (i) Calculate the pH of this solution and give one reason why the pH scale is a more convenient measurement for measuring acid concentrations than $[H^+]$. *(2 marks)*
      (ii) Write the K_a expression for methanoic acid. *(1 mark)*
      (iii) Calculate the values of K_a and pK_a for methanoic acid. *(3 marks)*
      (iv) Calculate the percentage of HCOOH molecules that have dissociated in this aqueous solution of methanoic acid. *(1 mark)*

   c Aqueous methanoic acid takes part in usual acid reactions. Write full equations for the following reactions.
      (i) methanoic acid with $CaCO_3$ *(1 mark)*
      (ii) methanoic acid with Al *(1 mark)*

5. Sour milk contains lactic acid, $CH_3CH(OH)COOH$ ($K_a = 1.38 \times 10^{-4} \text{ mol dm}^{-3}$)

   a (i) Write the K_a expression for lactic acid. *(1 mark)*
      (ii) Calculate the pH of a $0.0145 \text{ mol dm}^{-3}$ solution of lactic acid. Give your answer to **two** decimal places. *(3 marks)*

   b An excess of aqueous sodium hydroxide with a concentration of $0.125 \text{ mol dm}^{-3}$ is added to the solution of lactic acid.
      (i) Calculate the pH of the aqueous sodium hydroxide. *(2 marks)*
      (ii) Write an overall equation for the reaction. *(1 mark)*
      (iii) Write an ionic equation for the reaction. *(1 mark)*

6. At 50 °C, the pH of water is 6.63

   a Write the expression for K_w. *(1 mark)*

   b What name is given to K_w? *(1 mark)*

c (i) Calculate the concentration of H^+(aq) ions in water at 50 °C. *(1 mark)*

(ii) Deduce the concentration of OH^-(aq) ions in the water at 50 °C. *(1 mark)*

(iii) Calculate K_w at 50 °C. *(2 marks)*

d Predict, with a reason, the sign of ΔH for the dissociation of water. *(1 mark)*

OCR F325 Jun 11 Q4(a)–(d)

7 Sodium hydroxide, NaOH, is a strong alkali and vitamin C is a weak Brønsted-Lowry acid.

a What is meant by:
(i) a strong alkali *(1 mark)*
(ii) a weak Brønsted-Lowry acid? *(2 marks)*

b A student needs to prepare 50.0 cm³ of a solution of NaOH with a pH 12.75 at 25 °C.
(i) Calculate the concentration, in mol dm⁻³, of this solution of NaOH. *(2 marks)*
(ii) Calculate the mass of NaOH that the student needs to use. *(2 marks)*

c Vitamin C, $C_6H_8O_6$, has a K_a value of 6.76×10^{-5} mol dm⁻³. The equilibrium for the dissociation of vitamin C in water is shown below:

$C_6H_8O_6$(aq) $\rightleftharpoons$ H^+(aq) + $C_6H_7O_6^-$(aq)

0.500 g of Vitamin C was dissolved in water to form a solution with a volume of 125 cm³.
(i) Write an expression for the acid dissociation constant of vitamin C.
(ii) Calculate the pH of the solution formed. *(6 marks)*

8 a Nitric acid, HNO_3, is a strong Brønsted–Lowry acid. Nitrous acid, HNO_2, is a weak Brønsted–Lowry acid with a K_a value of 4.43×10^{-4} mol dm⁻³.
(i) What is the difference between a strong acid and a weak acid? *(1 mark)*
(ii) What is the expression for the acid dissociation constant K_a of nitrous acid, HNO_2? *(1 mark)*
(iii) Calculate the pH of 0.375 mol dm⁻³ nitrous acid, HNO_2. Give your answer to two decimal places. *(2 marks)*
(iv) A student suggests that an acid–base equilibrium is set up when nitric acid is mixed with nitrous acid.

Complete the equation for the equilibrium that would be set up and label the conjugate acid–base pairs.

$HNO_3 + HNO_2 \rightleftharpoons$ .... + ...

.... *(2 marks)*

b Calcium hydroxide, $Ca(OH)_2$, is a strong Brønsted–Lowry base.
(i) Explain what is meant by the term Brønsted–Lowry base. *(1 mark)*
(ii) Calculate the pH of 0.0400 mol dm⁻³ $Ca(OH)_2$. Give your answer to two decimal places. *(3 marks)*

c Aqueous calcium hydroxide is added to nitrous acid, HNO_2.

Write the overall equation and the ionic equation for the reaction that takes place. *(2 marks)*

OCR F325 Jan 12 Q4(a)–(c)

9 Compounds **A** and **HA** are two carboxylic acids.

a Compound **A** is a straight chain organic acid. A chemist analysed a sample of acid **A** as below.

The chemist first prepared a 250 cm³ solution of **A** by dissolving 10.8 g of **A** in water.

In a titration, 25.00 cm³ 0.500 mol dm⁻³ NaOH were neutralised by exactly 21.40 cm³ of solution **A**.

Calculate the pH of the NaOH(aq) used in the titration.

Use the results to calculate the molar mass of acid **A** and suggest its identity. *(7 marks)*

2816/01 Jan 2003 Q4(b)

b A carboxylic acid **HA** is a food additive used as a preservative in cakes.

The K_a value of **HA** is 1.51×10^{-5} mol dm⁻³.

A student analyses a sample of **HA** using the procedure below.

A student dissolves 0.7369 g of **HA** in water and makes the solution up to 1.00 dm³. The student measures the pH of the resulting solution as 3.52.

Determine the molar mass of **HA** and suggest a possible formula for **HA**.

Show all your working. *(6 marks)*

F325 Jan 2013 7(c)(i)

21 BUFFERS AND NEUTRALISATION

21.1 Buffer solutions

Specification reference: 5.1.3

Learning outcomes

Demonstrate knowledge, understanding, and application of:

→ buffer solutions
→ preparing buffer solutions
→ the role of acid–base pairs in buffer solutions
→ calculating the pH of buffer solutions.

Synoptic link

You first learnt about conjugate acid–base pairs in Topic 20.1, Brønsted–Lowry acids and bases.

Synoptic link

For more information about dissolving ionic compounds, see Topic 5.2 Ionic bonding and structure, Topic 22.2, Enthalpy changes in solution, and Topic 22.3, Factors affecting lattice enthalpy and hydration.

What is a buffer solution?

A **buffer solution** is a system that minimises pH changes when small amounts of an acid or a base are added. Buffer solutions contain two components to remove added acid or alkali – a weak acid (as component 1) and its conjugate base (as component 2).

- The weak acid, HA, removes added alkali.
- The conjugate base, A^-, removes added acid.

When alkalis and acids are added to a buffer, the two components in the buffer solution react and will eventually be used up. As soon as one component has all reacted, the solution loses its buffering ability towards added acid or alkalis. As the buffer works, the pH does change but only by a small amount – you should not assume that the pH stays completely constant.

Preparing weak acid buffer solutions

A buffer solution based on a weak acid needs a weak acid and its conjugate base. Two methods for preparing a buffer based on ethanoic acid are outlined below.

Preparation from a weak acid and its salt

A buffer solution can be prepared by mixing a solution of ethanoic acid, CH_3COOH, with a solution of one of its salts, for example, sodium ethanoate, CH_3COONa.

When ethanoic acid is added to water, the acid partially dissociates and the amount of ethanoate ions in solution is very small. Ethanoic acid is the source of the weak acid component of the buffer solution.

$$CH_3COOH(aq) \rightleftharpoons H^+(aq) + CH_3COO^-(aq)$$
component 1

Salts of weak acids are ionic compounds and provide a convenient source of the conjugate base. When added to water, the salt completely dissolves. Dissociation into ions is complete and so the salt is the source of the conjugate base component of the buffer solution.

$$CH_3COONa(s) + aq \rightarrow CH_3COO^-(aq) + Na^+(aq)$$
component 2

Preparation by partial neutralisation of the weak acid

A buffer solution can also be prepared by adding an aqueous solution of an alkali, such as $NaOH(aq)$, to an excess of the weak acid. The weak acid is partially neutralised by the alkali, forming the conjugate

BUFFERS AND NEUTRALISATION

base. Some of the weak acid is left over unreacted. The resulting solution contains a mixture of the salt of the weak acid and any unreacted weak acid.

Two reservoirs to remove added acid and alkali

In the ethanoic acid equilibrium, the equilibrium position lies well towards ethanoic acid. When CH_3COO^- ions are added to CH_3COOH, the equilibrium position shifts even further to the left, reducing the already small concentration of $H^+(aq)$ ions, and leaving a solution containing mainly the two components, CH_3COOH and CH_3COO^-:

$$CH_3COOH(aq) \rightleftharpoons H^+(aq) + CH_3COO^-(aq)$$
component 1   component 2

CH_3COOH and CH_3COO^- act as two reservoirs that are able to act independently to remove added acid and alkali. This is achieved by shifting the buffer's equilibrium system either to the right or left.

Action of the buffer solution

The conjugate acid–base pair, $HA(aq)/A^-(aq)$, in an acid buffer solution controls the pH. The equilibrium is shown below.

$$HA(aq) \rightleftharpoons H^+(aq) + A^-(aq)$$

The control of pH can be explained in terms of shifts in the equilibrium position using le Chatelier's principle.

> **Synoptic link**
>
> For details of equilibrium position and le Chatelier's principle, see Topic 10.4, Dynamic equilibrium and le Chatelier's principle.

Conjugate base removes added acid

On addition of an acid, $H^+(aq)$:

1. $[H^+(aq)]$ increases.
2. $H^+(aq)$ ions react with the conjugate base, $A^-(aq)$
3. The equilibrium position shifts to the left, removing most of the $H^+(aq)$ ions

$$HA(aq) \rightleftharpoons H^+(aq) + A^-(aq)$$
$\longleftarrow$
added acid

Weak acid removes added alkali

On addition of an alkali, $OH^-(aq)$:

1. $[OH^-(aq)]$ increases.
2. The small concentration of $H^+(aq)$ ions reacts with the $OH^-(aq)$ ions:
$$H^+(aq) + OH^-(aq) \rightarrow H_2O(l)$$
3. HA dissociates, shifting the equilibrium position to the right to restore most of $H^+(aq)$ ions.

added alkali $\longrightarrow$
$$HA(aq) \rightleftharpoons H^+(aq) + A^-(aq)$$

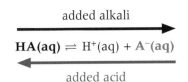

▲ **Figure 1** *Shifting the buffer equilibrium. Additions of acid and alkali shift the equilibrium position in opposite directions*

The role of the weak acid, HA, and its conjugate base, A^-, in controlling pH is summarised in Figure 1.

pH and buffer solutions

Choosing the components for a buffer solution

Different weak acids result in buffer solutions that operate over different pH ranges. How do you know which weak acid to use?

21.1 Buffer solutions

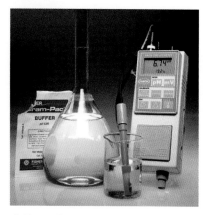

▲ Figure 2 *Measuring the pH of a buffer solution with a pH probe and a pH meter*

A buffer is most effective at removing either added acid or alkali when there are equal concentrations of the weak acid and its conjugate base.

When [HA(aq)] = [A⁻(aq)]:

- the pH of the buffer solution is the same as the pK_a value of HA
- the operating pH is typically over about two pH units, centred at the pH of the pK_a value.

The ratio of the concentrations of the weak acid and its conjugate base can then be adjusted to fine-tune the pH of the buffer solution.

Table 1 shows the pK_a values of several weak acids together with their typical operating pH ranges as buffers.

▼ Table 1 *Buffer pH ranges for weak acids*

Acid	Name	pK_a	Typical operating pH range
CH_3COOH	ethanoic acid	4.76	3.76–5.76
$HOCH_2COOH$	glycolic acid	3.83	2.83–4.83
$HCOOH$	methanoic acid	3.75	2.75–4.75
$CH_3CHOHCOOH$	lactic acid	3.08	2.08–4.08

Calculating the pH of a buffer solution

The equilibrium and K_a expression for a weak acid is shown below:

$$HA(aq) \rightleftharpoons H^+(aq) + A^-(aq) \qquad K_a = \frac{[H^+(aq)][A^-(aq)]}{[HA(aq)]}$$

When you calculate the pH of a weak acid, you make an approximation that $[H^+(aq)] = [A^-(aq)]$. For a buffer solution, this is no longer true as $A^-(aq)$ has been added as one of the components of the buffer.

To work out the pH of a buffer solution, rearrange the K_a expression as shown below.

$$[H^+(aq)] = K_a \times \frac{[HA(aq)]}{[A^-(aq)]}$$

- K_a — acid dissociation constant
- $\frac{[HA(aq)]}{[A^-(aq)]}$ — ratio of weak acid and its conjugate base

Provided that K_a and the concentrations of HA and A^- are known, $[H^+(aq)]$ and the pH can be calculated.

When the concentrations of HA and A^- are the same:

- $[HA(aq)] = [A^-(aq)]$ and $K_a = \frac{[H^+(aq)][\cancel{A^-(aq)}]}{[\cancel{HA(aq)}]}$
- $K_a = [H^+(aq)]$ and $pK_a = pH$

> **Study tip**
>
> The pH of a buffer solution depends upon:
> - the pK_a (or K_a) value of the weak acid
> - the ratio of the concentrations of the weak acid, HA, and its conjugate base, A^-.

> **Study tip**
>
> Take care. Weak acid and buffer pH calculations use the same K_a expression but the methods are different.
>
> For weak acids: $[H^+(aq)] = [A^-(aq)]$
> For buffers: $[H^+(aq)] \neq [A^-(aq)]$

21 BUFFERS AND NEUTRALISATION

 Worked example: Calculating the pH of a buffer solution

Calculate the pH when the buffer solution contains 0.100 mol dm^{-3} CH$_3$COOH and 0.300 mol dm^{-3} CH$_3$COONa.

K_a(CH$_3$COOH) = 1.74×10^{-5} mol dm^{-3}

Step 1: Calculate [H$^+$(aq)] from K_a, [HA(aq)], and [A$^-$(aq)].

$$[H^+(aq)] = K_a \times \frac{[CH_3COOH(aq)]}{[CH_3COO^-(aq)]}$$

$$= 1.74 \times 10^{-5} \times \frac{0.100}{0.300}$$

$$= 5.80 \times 10^{-6} \text{ mol dm}^{-3}$$

Step 2: Use your calculator to find the pH.

pH = $-\log[H^+(aq)] = -\log(5.80 \times 10^{-6}) = 5.24$

> **Study tip**
>
> If you are given pK_a and the concentrations of HA and A$^-$ are the same, the pH is equal to the pK_a value.

 Worked example: Calculating the pH of a buffer solution made from a weak acid and its salt

150 cm^3 of 1.00 mol dm^{-3} HCOOH is mixed with 100 cm^3 0.750 mol dm^{-3} HCOONa. Calculate the pH of the buffer solution formed.

K_a(HCOOH) = 1.78×10^{-4} mol dm^{-3}

Step 1: Calculate the amounts, in mol, of HCOOH and HCOO$^-$ in the buffer solution.

$$n(HCOOH) = \frac{1.00 \times 150}{1000} = 0.150 \text{ mol}$$

$$n(HCOO^-) = \frac{0.750 \times 100}{1000} = 0.0750 \text{ mol}$$

Step 2: Calculate the concentrations of HCOOH and HCOO$^-$ in the buffer solution.

Total volume of buffer solution = 250 cm^3

$$[HCOOH(aq)] = \frac{1000 \times 0.150}{250} = 0.600 \text{ mol dm}^{-3}$$

$$[HCOO^-(aq)] = \frac{1000 \times 0.0750}{250} = 0.300 \text{ mol dm}^{-3}$$

Step 3: Calculate [H$^+$(aq)] from K_a, [HA(aq)] and [A$^-$(aq)].

$$[H^+(aq)] = K_a \times \frac{[HCOOH(aq)]}{[HCOO^-(aq)]}$$

$$= 1.78 \times 10^{-4} \times \frac{0.600}{0.300}$$

$$= 3.56 \times 10^{-4} \text{ mol dm}^{-3}$$

Step 4: Use your calculator to find the pH.

pH = $-\log[H^+(aq)] = -\log(3.56 \times 10^{-4}) = 3.45$

> **Study tip**
>
> Remember:
> $$n = \frac{c \times V \text{ (in cm}^3\text{)}}{1000}$$

21.1 Buffer solutions

> **Study tip**
> This is the hardest worked example but is also mirrors the commonest method of preparation of a buffer solution, by partial neutralisation of a weak acid.

 Worked Example: Calculating the pH of a buffer solution made by partial neutralisation

100 cm^3 of 0.750 mol dm^{-3} NaOH(aq) is added to 150 cm^3 of 1.50 mol dm^3 HCOOH. Calculate the pH of the buffer solution formed. K_a(HCOOH) = 1.78 × 10^{-4} mol dm^{-3}.

Step 1: Calculate the amount, in mol, of HCOO$^-$ in the buffer solution.

The partial neutralisation is:
$$HCOOH(aq) + NaOH(aq) \rightarrow HCOONa + H_2O$$

n(NaOH) added = $\dfrac{0.750 \times 100}{1000}$ = 0.0750 mol

n(HCOO$^-$) formed = 0.0750 mol

Step 2: Calculate the amount, in mol, of HCOOH in the buffer solution.

n(HCOOH) used = $\dfrac{1.50 \times 150}{1000}$ = 0.225 mol

n(HCOOH) remaining = n(HCOOH) used − n(NaOH) added
= 0.225 − 0.0750 = 0.150 mol

Step 3: These are the same amounts as obtained from **Step 1** of the previous worked example.

The remainder of the calculation is identical to give the pH of 3.45.

Summary questions

1. A CH$_3$(CH$_2$)$_2$COOH / CH$_3$(CH$_2$)$_2$COO$^-$ buffer contains equal concentrations of CH$_3$(CH$_2$)$_2$COOH and CH$_3$(CH$_2$)$_2$COONa. (K_a for CH$_3$(CH$_2$)$_2$COOH = 1.51 × 10^{-5} mol dm^{-3} at 25 °C)
   a. Calculate the pH of the buffer solution. *(1 mark)*
   b. State the equilibrium equation for the buffer. *(1 mark)*
   c. Explain in terms of equilibrium how the buffer solution removes added alkali. *(2 marks)*

2. Three buffer solutions are made from benzoic acid, C$_6$H$_5$COOH, and sodium benzoate, C$_6$H$_5$COONa, with the following compositions. Calculate the pH of each buffer solution shown below.
   (K_a for C$_6$H$_5$COOH = 6.3 × 10^{-5} mol dm^{-3} at 25 °C)
   a. Buffer A: 0.10 mol dm^{-3} C$_6$H$_5$COOH and 0.10 mol dm^{-3} C$_6$H$_5$COONa *(2 marks)*
   b. Buffer B: 1.25 mol dm^{-3} C$_6$H$_5$COOH and 0.25 mol dm^{-3} C$_6$H$_5$COONa *(2 marks)*
   c. Buffer C: 0.20 mol dm^{-3} C$_6$H$_5$COOH and 0.80 mol dm^{-3} C$_6$H$_5$COONa *(2 marks)*

3. K_a of CH$_3$COOH = 1.74 × 10^{-5} mol dm^{-3} at 25 °C
   a. 80 cm^3 of 0.200 mol dm^{-3} CH$_3$COOH is mixed with 20 cm^3 0.500 mol dm^{-3} CH$_3$COONa. Calculate the pH of the buffer solution. *(2 marks)*
   b. A CH$_3$COOH / CH$_3$COO$^-$ buffer is prepared by adding 150 cm^3 0.250 mol dm^{-3} NaOH to 100 cm^3 0.500 mol dm^{-3} CH$_3$COOH. Calculate the pH of the buffer solution. *(2 marks)*

21.2 Buffer solutions in the body
Specification reference: 5.1.3

Controlling blood pH
Buffer solutions are widespread in living systems. The well-being of the human body relies on precise pH control with different parts of the body requiring specific pH values for effective functioning. Enzymes are particularly sensitive and each has an optimum pH. The role of pH control in the body falls to buffer solutions, for example, in the plasma of the blood.

Blood plasma needs to be maintained at a pH between 7.35 and 7.45. The pH is controlled by a mixture of buffers, with the carbonic acid–hydrogencarbonate (H_2CO_3/HCO_3^-) buffer system being the most important. Normal healthy blood should have a pH of 7.40.

What happens if the pH slips outside this range?
If the pH falls below 7.35, people can develop a condition called acidosis, which can cause fatigue, shortness of breath, and in extreme cases, shock or death. If the pH rises above 7.45, the condition is called alkalosis, which can cause muscle spasms, light-headedness, and nausea.

Remember that pH is a sensitive scale – a difference of just 0.30 pH units is a two-fold difference in $H^+(aq)$ concentration. So although 7.40 ± 0.30 (i.e. 7.10 – 7.70), does not sound very different from the healthy blood pH of 7.40, the difference is very large in terms of acidity or alkalinity.

The carbonic acid–hydrogencarbonate buffer system
A general buffer system based on HA/A⁻ controls pH. The carbonic acid–hydrogencarbonate operates in a similar way:

On addition of an acid, $H^+(aq)$:

1 $[H^+(aq)]$ increases.
2 $H^+(aq)$ ions react with the conjugate base, $HCO_3^-(aq)$.
3 The equilibrium position shifts to the left, removing most of the $H^+(aq)$ ions.

$$\xleftarrow{\text{added acid}}$$
$$H_2CO_3(aq) \rightleftharpoons H^+(aq) + HCO_3^-(aq)$$

On addition of an alkali, $OH^-(aq)$:

1 $[OH^-(aq)]$ increases.
2 The small concentration of $H^+(aq)$ ions reacts with the $OH^-(aq)$ ions:

$$H^+(aq) + OH^-(aq) \rightarrow H_2O(l)$$

Learning outcome
Demonstrate knowledge, understanding, and application of:
→ buffer solutions in the control of blood pH.

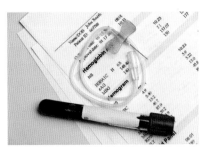

▲ Figure 1 *A sample tube of human blood for analysis. Healthy blood has a pH of 7.40*

Synoptic link
You first encountered the pH scale in Topic 20.2, The pH scale and strong acids.

21.2 Buffer solutions in the body

$$H_2CO_3(aq) \underset{\text{added acid}}{\overset{\text{added alkali}}{\rightleftharpoons}} H^+(aq) + HCO_3^-(aq)$$

▲ **Figure 2** *The carbonic acid–hydrogencarbonate ion equilibrium*

3 H_2CO_3 dissociates, shifting the equilibrium position to the right to restore most of $H^+(aq)$ ions:

$$H_2CO_3(aq) \xrightarrow{\text{added alkali}} H^+(aq) + HCO_3^-(aq)$$

> **Study tip**
>
> If you take A level biology, you will learn far more about how carbon dioxide is removed from the body to control carbonic acid in blood plasma.

Worked example: Calculating the concentration ratio of HCO_3^-/H_2CO_3 in healthy blood

The pK_a for the carbonic acid–hydrogencarbonate equilibrium is 6.1 at body temperature. What is the ratio of HCO_3^-/H_2CO_3 in healthy blood at a pH of 7.40?

Step 1: Express HCO_3^-/H_2CO_3 in terms of K_a and $[H^+(aq)]$.

$$H_2CO_3(aq) \rightleftharpoons H^+(aq) + HCO_3^-(aq) \qquad K_a = \frac{[H^+(aq)][HCO_3^-(aq)]}{[H_2CO_3(aq)]}$$

$$\frac{[HCO_3^-(aq)]}{[H_2CO_3(aq)]} = \frac{K_a}{[H^+(aq)]}$$

Step 2: Convert pH into $[H^+(aq)]$ and pK_a into K_a.

$$[H^+(aq)] = 10^{-pH} = 10^{-7.40} = 3.98 \times 10^{-8} \text{ mol dm}^{-3}$$

$$K_a = 10^{-pK_a} = 10^{-6.1} = 7.9 \times 10^{-7} \text{ mol dm}^{-3}$$

Step 3: Calculate the HCO_3^-/H_2CO_3 ratio.

$$\frac{[HCO_3^-(aq)]}{[H_2CO_3(aq)]} = \frac{K_a}{[H^+(aq)]} = \frac{7.9 \times 10^{-7}}{3.98 \times 10^{-8}} = \frac{20}{1}$$

The body produces far more acidic materials than alkaline, which the conjugate base HCO_3^- converts to H_2CO_3. The body prevents H_2CO_3 building up by converting it to carbon dioxide gas, which is then exhaled by the lungs.

The Henderson–Hasselbalch equation

You have already learnt a method for calculating pH using K_a, $[HA(aq)]$, and $[A^-(aq)]$. The Henderson–Hasselbalch equation is an alternative way of calculating the pH of a buffer solution that allows you to really see how pK_a and the base/acid ratio control the pH:

$$pH = pK_a + \log \frac{[A^-(aq)]}{[HA(aq)]}$$

In the Henderson–Hasselbalch equation, it is easy to appreciate the effect of the base/acid concentration ratio. When the acid and base concentrations are the same:

$$\frac{[A^-(aq)]}{[HA(aq)]} = 1 \quad \text{and} \quad \log 1 = 0$$

$pH = pK_a$, a relationship you have already seen.

BUFFERS AND NEUTRALISATION 21

Although the equation appears more complex than the method you have learnt, the calculation is often easier. For a buffer solution containing 0.100 mol dm^{-3} CH$_3$COOH (pK_a = 4.76) and 0.300 mol dm^{-3} CH$_3$COONa, the pH can be solved in a single step:

$$pH = pK_a + \log \frac{[CH_3COO^-(aq)]}{[CH_3COOH(aq)]} = 4.76 + \log \frac{0.300}{0.100}$$

$$= 4.76 + \log 3 = 4.76 + 0.48$$
$$= 5.24$$

Use the Henderson–Hasselbalch equation to calculate the pH of the following buffer solutions.

a 0.250 mol dm^{-3} HCOOH and 0.750 mol dm^{-3} HCOONa
   (pK_a HCOOH = 3.75)
b 0.800 mol dm^{-3} HOCH$_2$COOH and 0.400 mol dm^{-3} HOCH$_2$COONa
   (pK_a HOCH$_2$COOH = 3.83)
c 0.475 mol dm^{-3} CH$_3$CHOHCOOH and 0.625 mol dm^{-3} CH$_3$CHOHCOONa
   (pK_a CH$_3$CHOHCOOH = 3.08)

Study tip
If you prefer, you can use this method for calculating the pH of a buffer solution. But be very careful that you show the ratio the correct way round.

Summary questions

1 The carbonic acid–hydrogencarbonate buffer system controls blood pH.
   a State the equilibrium in this buffer system. *(1 mark)*
   b Explain in terms of equilibrium how this buffer system removes excess acid from the blood. *(2 marks)*

2 Normal healthy blood has a pH of 7.40. A sample of blood in a patient suffering from acute acidosis has a pH of 6.92.
   a Calculate the [H$^+$(aq)] concentration at these two pH values in the blood. *(2 marks)*
   b Calculate how many times greater the [H$^+$(aq)] in blood is at pH 6.92 compared to blood at pH 7.40. *(2 marks)*

3 a Calculate the concentration ratio of CH$_3$COOH / CH$_3$COO$^-$ in a buffer solution with a pH of 4.11.
      K_a(CH$_3$COOH) = 1.74 × 10^{-5} mol dm^{-3} *(3 marks)*
   b Healthy blood at a pH of 7.40 has a hydrogencarbonate : carbonic acid ratio of 20 : 1. A patient is admitted to hospital. The patient's blood pH is measured as 7.20.
      i   Calculate K_a for carbonic acid.
      ii Calculate the hydrogencarbonate : carbonic acid ratio in the patient's blood. *(6 marks)*

21.3 Neutralisation

Specification reference: 5.1.3

Learning outcomes

Demonstrate knowledge, understanding, and application of:

→ pH titration curves
→ choosing an indicator
→ using a pH meter.

Synoptic link

Look back to Topic 4.2, Acid–base titrations.

▲ Figure 1 *A student using a digital pH meter to measure the pH of a solution in a conical flask during an acid–base titration*

pH titration curves

Acid–base titrations use indicators to monitor neutralisation reactions accurately. The results can then be analysed to find out some unknown information about the acid or base. This topic looks closely at the pH changes that take place during a titration. A pH meter can be used for monitoring these changes as outlined below.

 Using a pH meter

A pH meter consists of an electrode, that is dipped into a solution, and connected to a meter that displays the pH reading. A pH meter typically records pH values to two decimal places. Indicator paper is usually matched from colour charts to the nearest whole number, so a pH meter is able to give more accurate measurements of pH during a titration. The procedure below monitors the pH as an aqueous base is added to an acid solution.

1. Using a pipette, add a measured volume of acid to a conical flask.
2. Place the electrode of the pH meter in the flask.
3. Add the aqueous base to the burette and add to the acid in the conical flask, 1 cm^3 at a time.
4. After each addition, swirl the contents. Record the pH and the total volume of the aqueous base added.
5. Repeat steps 3 and 4 until the pH starts to change more rapidly. Then add the aqueous base dropwise for each reading until the pH changes less rapidly.
6. Now add the aqueous base 1 cm^3 at a time again until an excess has been added and the pH has been basic, with little change, for several additions.

A graph of pH against total volume of aqueous base added is then plotted. Figure 2 shows the curve for a titration between 0.1 mol dm^{-3} solutions of a strong monobasic acid and a strong monobasic base.

For an alternative automatic method, you could attach the pH meter to a datalogger and use a magnetic stirrer in the flask. The aqueous base would then be added from the burette to the flask slowly, and the pH titration curve could be plotted automatically using the datalogger or appropriate software on a computer.

> A pH meter records a pH as 4.65 but the best match with indicator paper is a pH of 5. Calculate the [H$^+$(aq)] concentration at each pH reading and estimate the percentage error in pH and [H$^+$(aq)] made by measuring the pH with indicator paper.

The pH titration curve

The key features of a typical **pH titration curve** are shown labelled on Figure 2.

BUFFERS AND NEUTRALISATION

When the base is first added, the acid is in great excess and the pH increases very slightly. As the vertical section is approached, the pH starts to increase more quickly as the acid is used up more quickly.

Eventually, the pH increases rapidly during addition of a very small volume of base, producing the vertical section. Only drops of solution will be needed for the whole vertical section.

After the vertical section, the pH will rise very slightly as the base is now in great excess.

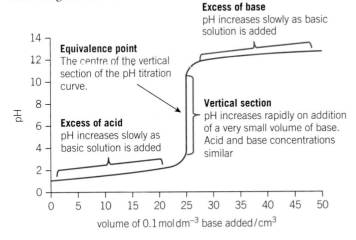

▲ **Figure 2** *pH titration curve for addition of a base to an acid*

The **equivalence point** of the titration is the volume of one solution that *exactly* reacts with the volume of the other solution. The solutions have then exactly reacted with one another and the amounts used matching the stoichiometry of the reaction. The equivalence point is the centre of the vertical section of the pH titration curve.

The titration curve may be a different shape for combinations of acid and base with different strengths. These curves and shapes are discussed later in this topic and you will see how the shape helps in choosing a suitable indicator for a titration.

Adding an acid to a base

A pH titration curve can also be plotted with the acid added from the burette to the base in the flask. The shape is essentially the same, just the other way around, going from high pH to low pH (Figure 3).

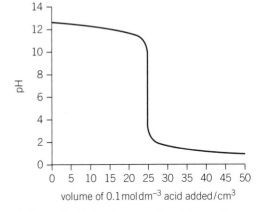

▲ **Figure 3** *pH titration curve for addition of an acid to a base*

Acid–base indicators

The end point

An acid–base indicator is a weak acid, HA, that has a distinctively different colour from its conjugate base, A⁻, for example, for the common indicator methyl orange (Figure 4):

- the weak acid, HA, is red
- the conjugate base A⁻ is yellow.

At the **end point** of a titration, the indicator contains equal concentrations of HA and A⁻ and the colour will be in between the

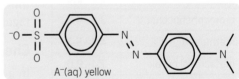

▲ **Figure 4** *HA and A⁻ forms of methyl orange*

21.3 Neutralisation

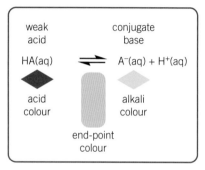

▲ Figure 5 Colours for methyl orange indicator

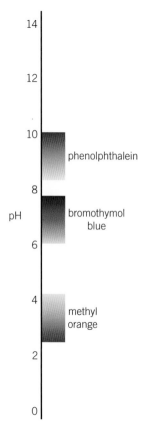

▲ Figure 6 pH colour ranges for some common indicators

▼ Table 1 pH colours and approximate ranges

Indicator	pH range
thymol blue	1.2–2.8
bromophenol blue	3.6–4.6
methyl red	4.4–6.2
metacresol purple	7.4–9.0
indigo carmine	11.4–13.0

two extreme colours. For methyl orange, the colour at its end point is orange (Figure 5).

Explaining indicator colour changes

An indicator is a weak acid. The equilibrium position is shifted towards the weak acid in acidic conditions or towards the conjugate base in basic conditions, changing the colour as it does so.

In a titration in which a strong base is added to a strong acid, methyl orange is initially red as the presence of H^+ ions forces the equilibrium position well to the left.

On addition of a basic solution containing $OH^-(aq)$ ions:

- $OH^-(aq)$ ions react with $H^+(aq)$ in the indicator:
$$H^+(aq) + OH^-(aq) \rightarrow H_2O(l)$$
- The weak acid, HA, dissociates, shifting the equilibrium position to the right.
- The colour changes, first to orange at the end point and finally to yellow as the equilibrium position is shifted to the right.

$$HA(aq) \xrightarrow{\text{equilibrium}} A^-(aq) + H^+(aq)$$

If methyl orange is added initially to a basic solution and acid is added:

- $H^+(aq)$ ions react with the conjugate base, $A^-(aq)$.
- The equilibrium position shifts to the left.
- The colour changes, first to orange at the end point and finally to red when the equilibrium position has shifted to the left.

$$HA(aq) \xleftarrow{\text{equilibrium}} A^-(aq) + H^+(aq)$$

This is very similar to the mode of action of a buffer solution. All acid–base indicators work in a similar way.

How sensitive is the end point?

Different indicators have different K_a values and change colour over different pH ranges.

At the end point:

- $[HA(aq)] = [A^-(aq)]$ and $K_a = \dfrac{[H^+(aq)]\cancel{[A^-(aq)]}}{\cancel{[HA(aq)]}}$
- $K_a = [H^+(aq)]$ and $pK_a = pH$

The pH of the end point is the same as the pK_a value of HA.

The sensitivity of an indicator depends upon the indicator itself and eyesight. Most indicators change colour over a range of about two pH units. The pH ranges and colours for three indicators are shown in Figure 6. Table 1 shows pH ranges of more indicators.

Choosing the indicator

In a titration, you must use an indicator that has a colour change which coincides with the vertical section of the pH titration curve. Ideally the end point and equivalence point would coincide. However, this may not be possible and the end point may give a volume that is slightly different from the equivalence point. But any difference will be very small in the order of one or two drops (i.e., 0.05–0.10 cm^3).

Figures 7–10 show pH titration curves for combinations of 0.1 mol dm^{-3} strong and weak acids with 0.1 mol dm^{-3} strong and weak bases.

The pH ranges for methyl orange and phenolphthalein indicators have been added to show which indicator is suitable for each titration.

No indicator is suitable for a weak acid–weak base titration as there is no vertical section and, even at its steepest, the pH requires several cm^3 to pass through a typical pH indicator range of 2 pH units.

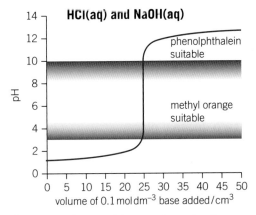

▲ Figure 7 Strong acid–strong base titration

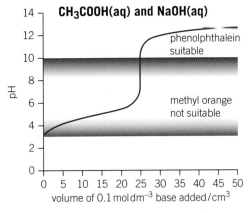

▲ Figure 8 Weak acid–strong base titration

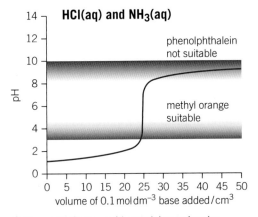

▲ Figure 9 Strong acid–weak base titration

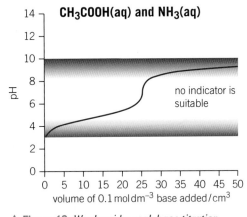

▲ Figure 10 Weak acid–weak base titration

Summary questions

1. Explain the difference between the terms equivalence point and end point? *(2 marks)*

2. For the indicator chlorophenol red, the HA form is yellow and the A$^-$ form is red.
   a. Write an equation to link these two forms of the indicator. *(1 mark)*
   b. Predict the colour of chlorophenol red:
      i in a strong acid   ii in a strong alkali
      iii at the end-point of a titration. *(3 marks)*

3. This question refers to the titration curve in Figure 8 for 0.100 mol dm^{-3} CH$_3$COOH with 0.100 mol dm^{-3} NaOH.

   Predict how the curve would differ if 0.050 mol dm^{-3} CH$_3$COOH had been used instead. *(2 marks)*

Chapter 21 Practice questions

Practice questions

1. A weak acid HA has a pK_a value of 4.32.
   a. A buffer solution is prepared by mixing equal volumes of 0.100 mol dm^{-3} HA and 0.100 mol dm^{-3} A$^-$. What is the pH? *(1 mark)*
   b. Calculate the ratio of HA : A$^-$ needed to prepare a buffer solution with a pH of 3.50. *(2 marks)*
   c. 25 cm^3 of 0.200 mol dm^{-3} HA is added to 75 cm^3 of 0.100 mol dm^{-3} A$^-$.
      Calculate the pH of the resulting buffer solution. *(3 marks)*
   d. A chemist needs to prepare a buffer solution with a pH of exactly 4 using HNO$_2$ and NaNO$_2$.
      K_a for HNO$_2$ = 4.00 × 10^{-4} mol dm^{-3}
      Calculate the volumes of 0.250 mol dm^{-3} HNO$_2$ and 0.250 mol dm^{-3} needed to make 250 cm^3 of this buffer solution. *(4 marks)*

2. The titration curve below was produced by adding 0.100 mol dm^{-3} CH$_3$COOH to NaOH(aq).

   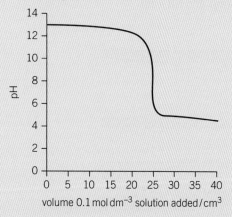

   a. What evidence from the curve shows that the NaOH(aq) used had a concentration of 0.10 mol dm^{-3}. *(2 marks)*
   b. What volume of CH$_3$COOH was needed to reach the end point? *(1 mark)*
   c. What pH range should the indicator have for this titration? *(1 mark)*
   d. The titration was repeated using 0.10 mol dm^{-3} H$_2$SO$_4$ instead of CH$_3$COOH. State and explain **two** differences to the titration curve. *(2 marks)*

3. A chemist prepares a buffer solution containing CH$_3$COOH and CH$_3$COO$^-$ ions by mixing together the following:
   200 cm^3 of 3.20 mol dm^{-3} CH$_3$COOH (K_a = 1.74 × 10^{-5} mol dm^{-3}) and 800 cm^3 of 0.500 mol dm^{-3} NaOH. The volume of the buffer solution is 1.00 dm^3.
   a. Write an equation for the reaction that takes place during the preparation. *(1 mark)*
   b. Explain why a buffer solution is formed when these two solutions are mixed together. *(1 mark)*
      (i) Calculate the amount, in mol, of CH$_3$COOH and CH$_3$COO$^-$ present in the buffer solution. *(2 marks)*
      (ii) Calculate the pH of this buffer solution. Give your answer to **two** decimal places. *(2 marks)*

4. Blood is an example of a buffered solution. Healthy human blood is slightly basic and has a pH of 7.40. If the pH falls, a condition known as acidosis can occur which can be fatal if the pH drops below 6.80.
   a. Explain the term *buffer solution*. *(1 mark)*
   b. (i) Calculate the hydrogen ion concentration, [H$^+$(aq)], in healthy human blood. *(1 mark)*
      (ii) How many times greater is the hydrogen ion concentration in blood at pH 6.8 compared to blood at pH 7.4? *(2 marks)*
   c. Mild acidosis may occur during strenuous exercise from the build up of lactic acid, CH$_3$CH(OH)COOH.
      (K_a of 8.40 × 10^{-4} mol dm^{-3})
      (i) Write an equation for the dissociation of lactic acid. *(1 mark)*
      (ii) Write the expression for K_a of lactic acid. *(1 mark)*
      (iii) Calculate the pH of a 0.100 mol dm^{-3} solution of lactic acid. Include any approximations that you make in your calculation. *(4 marks)*

5. The most important buffer system in blood is a mixture of carbonic acid, H$_2$CO$_3$, and hydrogencarbonate ions, HCO$_3^-$.
   a. Write the equilibrium present in this buffer system. *(1 mark)*

b Explain how this buffer system acts as a buffer in the control of blood pH. *(4 marks)*

c Healthy blood at a pH of 7.40 has a $HCO_3^-:H_2CO_3$ ratio of 20:1.

A patient is admitted to hospital. The patient's blood pH is measured as 7.20. Calculate the $HCO_3^-:H_2CO_3$ ratio in the patient's blood. *(4 marks)*

6 Glycolic acid, $HOCH_2COOH$, and thioglycolic acid, $HSCH_2COOH$, are weak acids.

a Glycolic acid reacts with bases, such as aqueous sodium hydroxide, NaOH(aq), to form salts.

A student pipetted 25.0 cm³ of 0.125 mol dm⁻³ glycolic acid into a conical flask. The student added NaOH(aq) from a burette. A pH meter and data logger were used to measure continuously the pH of the contents of the conical flask.

The pH curve obtained is shown below.

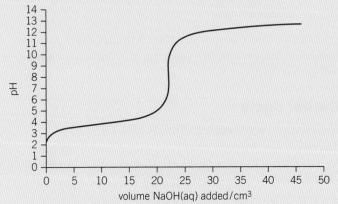

1 mol of glycolic acid reacts with 1 mol of sodium hydroxide.

(i) Write the equation for the reaction that takes place in the titration. *(1 mark)*

(ii) Determine the concentration, in mol dm⁻³, of the NaOH. *(2 marks)*

(iii) The student decided to carry out this titration using an acid–base indicator.

What important factor does the student need to consider when deciding on the most suitable indicator to use? *(1 mark)*

b The 0.125 mol dm⁻³ glycolic acid had a pH of 2.37.

(i) What is the expression for the acid dissociation constant K_a of glycolic acid? *(1 mark)*

(ii) Calculate K_a for glycolic acid. *(3 marks)*

(iii) Calculate the percentage dissociation of the glycolic acid. *(1 mark)*

c A buffer of glycolic acid and ammonium glycolate is used in a facial cleanser.

Explain, using equations:

• how a solution containing glycolic acid and glycolate ions can act as a buffer

• how this buffer could be prepared from ammonia and glycolic acid.

In your answer you should explain how the equilibrium system allows the buffer solution to control the pH. *(7 marks)*

d Ammonium thioglycolate, $HSCH_2COONH_4$, is the ammonium salt of thioglycolic acid, $HSCH_2COOH$.

When ammonium thioglycolate is dissolved in water, an acid–base equilibrium is set up. The equilibrium lies well to the left-hand side.

$HSCH_2COO^-(aq) + NH_4^+(aq) \rightleftharpoons$
$\qquad HSCH_2COOH(aq) + NH_3(aq)$

• Label one conjugate acid–base pair as '**Acid 1**' and '**Base 1**'.

• Label the other conjugate acid–base pair as '**Acid 2**' and '**Base 2**'. *(2 marks)*

OCR F325 Jan 2011 Q3 (a)–(d)

7 Benzoic acid, C_6H_5COOH, is a weak acid, used for preserving fruit juices. The acid dissociation constant K_a of benzoic acid is 6.30×10^{-5} mol dm⁻³ at 25 °C.

a Write the equation for the dissociation of benzoic acid when dissolved in water. *(1 mark)*

b Write the expression for the acid dissociation constant K_a of benzoic acid. *(1 mark)*

c The solubility of benzoic acid in water is 3.40 g dm⁻³.

Calculate the pH of a saturated solution of benzoic acid in water at 25 °C. *(5 marks)*

2816/01 June 2009 Q3 (a)–(c)

22 ENTHALPY AND ENTROPY
22.1 Lattice enthalpy

Specification reference: 5.2.1

Learning outcomes

Demonstrate knowledge, understanding, and application of:
→ lattice enthalpy
→ Born–Haber cycles
→ calculation of lattice enthalpy.

Lattice enthalpy

Solid ionic compounds tend to be very stable – the stability arises from the strength of the ionic bonds, electrostatic attractions between oppositely-charged ions in the ionic lattice structure. This creates a substantial energy barrier that must be overcome to break down the lattice, reflected in the high melting points of many ionic compounds. **Lattice enthalpy** is a measure of the strength of ionic bonding in a giant ionic lattice.

Lattice enthalpy is the enthalpy change that accompanies the formation of one mole of an ionic compound from its gaseous ions under standard conditions. For example:

$$K^+(g) + Cl^-(g) \rightarrow KCl(s) \qquad \Delta_{LE}H^\ominus = -711 \text{ kJ mol}^{-1}$$

gaseous ions → solid ionic compound

Lattice enthalpy involves ionic bond formation from separate gaseous ions. It is an *exothermic* change and the value for the enthalpy change will always be negative (Figure 1).

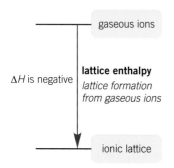

▲ **Figure 1** *Lattice enthalpy is exothermic*

Synoptic link

In Topic 5.2, Ionic bonding and structure, you learnt that the solid structure of an ionic compound is a giant ionic lattice.

The Born–Haber cycle

Lattice enthalpy cannot be measured directly and must be calculated indirectly using known energy changes in an energy cycle. You have seen how to calculate enthalpy changes indirectly. The indirect determination of lattice enthalpy requires a special type of energy cycle called a **Born–Haber cycle**. Figure 2 shows two routes for changing elements in their standard states into an ionic lattice.

Synoptic link

See Topic 9.4, Hess' law and enthalpy cycles, for details of the indirect determination of enthalpy changes.

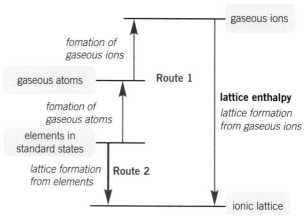

▲ **Figure 2** *Energy cycle linking elements to ionic lattice and gaseous ions*

22 ENTHALPY AND ENTROPY

Route 1
Route 1 requires three different processes.

Formation of gaseous atoms:
- Changing the elements in their standard states into gaseous atoms.
- This change is endothermic as it involves bond breaking.

Formation of gaseous ions:
- Changing the gaseous atoms into positive and negative gaseous ions.
- Overall, this change is endothermic.

Lattice formation:
- Changing the gaseous ions into the solid ionic lattice.
- This is the lattice enthalpy and is exothermic.

Route 2
Route 2 converts the elements in their standard states directly to the ionic lattice. There is just one enthalpy change, the enthalpy change of formation $\Delta_f H$, and this is exothermic.

Key enthalpy changes

The key enthalpy changes involved in a full Born–Haber cycle are listed below. Some, such as ionisation energies and enthalpy change of formation, will be familiar. The example equations all apply to enthalpy changes in the Born–Haber cycle for sodium chloride.

The **standard enthalpy change of formation** $\Delta_f H^\ominus$ is the enthalpy change that takes place when one mole of a compound is formed from its elements under standard conditions, with all reactants and products in their standard states.

$$\underbrace{Na(s) + \tfrac{1}{2}Cl_2(g)}_{\text{elements in standard states}} \rightarrow \underbrace{NaCl(s)}_{\text{one mole of compound}} \quad \Delta_f H^\ominus = -711\,\text{kJ mol}^{-1}$$

The compound will always be an ionic compound in its solid lattice.

The **standard enthalpy change of atomisation** $\Delta_{at} H^\ominus$ is the enthalpy change that takes place for the formation of one mole of gaseous atoms from the element in its standard state under standard conditions.

$$Na(s) \rightarrow Na(g) \quad \Delta_{at} H^\ominus = +107\,\text{kJ mol}^{-1}$$

$$\underbrace{\tfrac{1}{2}Cl_2(g)}_{\text{element in standard state}} \rightarrow \underbrace{Cl(g)}_{\text{1 mole of gaseous atoms}} \quad \Delta_{at} H^\ominus = +121\,\text{kJ mol}^{-1}$$

$\Delta_{at} H^\ominus$ is always an *endothermic* process because bonds are broken to form gaseous atoms. When the element is a gas in its standard state, $\Delta_{at} H^\ominus$ is related to the bond enthalpy of the bond being broken.

For example, for $Cl_2(g) \rightarrow 2Cl(g)$, a Cl—Cl bond is being broken, which has a bond enthalpy of $+242\,\text{kJ mol}^{-1}$. So for $\tfrac{1}{2}Cl_2(g)$, the enthalpy change is $+121\,\text{kJ mol}^{-1}$.

> **Synoptic link**
>
> To revise ionisation energies, look back at Topic 7.2, Ionisation energies. To revise enthalpy change of formation, see Topic 9.1, Enthalpy changes.

> **Study tip**
>
> Take care. $\Delta_{at} H^\ominus$ is for the formation of one mole of gaseous atoms. Only $\tfrac{1}{2}$ mol of Cl_2 is needed to form 1 mol of Cl.

22.1 Lattice enthalpy

The **first ionisation energy** $\Delta_{IE}H^\ominus$ is the enthalpy change required to remove one electron from each atom in one mole of gaseous atoms to form one mole of gaseous 1+ ions. For example:

$$\underset{\text{gaseous atoms}}{Na(g)} \rightarrow \underset{\text{gaseous 1+ ions}}{Na^+(g)} + e^- \qquad \Delta_{IE}H^\ominus = +496 \, kJ \, mol^{-1}$$

Ionisation energies are *endothermic* because energy is required to overcome the attraction between a negative electron and the positive nucleus.

Electron affinity is the opposite of ionisation energy.

- Electron affinity measures the energy to *gain* electrons.
- Ionisation energy measures the energy to *lose* electrons.

The **first electron affinity** $\Delta_{EA}H^\ominus$ is the enthalpy change that takes place when one electron is added to each atom in one mole of gaseous atoms to form one mole of gaseous 1− ions. For example:

$$\underset{\text{gaseous atoms}}{Cl(g)} + e^- \rightarrow \underset{\text{gaseous 1− ions}}{Cl^-(g)} \qquad \Delta_{EA}H^\ominus = -346 \, kJ \, mol^{-1}$$

First electron affinities are *exothermic* because the electron being added is attracted in towards the nucleus.

Determination of lattice enthalpies

By constructing a full Born–Haber cycle you can calculate lattice enthalpies.

> **Worked example: Calculating the lattice enthalpy of sodium chloride**
>
> You are provided with the following enthalpy changes.
>
> Construct a Born–Haber cycle and calculate the lattice enthalpy of sodium chloride, NaCl.
>
> $Na(s) + \frac{1}{2}Cl_2(g) \rightarrow NaCl(s)$ $\Delta_f H^\ominus = -411 \, kJ \, mol^{-1}$
>
> $Na(s) \rightarrow Na(g)$ $\Delta_{at}H^\ominus = +108 \, kJ \, mol^{-1}$
>
> $\frac{1}{2}Cl_2(g) \rightarrow Cl(g)$ $\Delta_{at}H^\ominus = +121 \, kJ \, mol^{-1}$
>
> $Na(g) \rightarrow Na^+(g) + e^-$ $\Delta_{IE}H^\ominus = +496 \, kJ \, mol^{-1}$
>
> $Cl(g) + e^- \rightarrow Cl^-(g)$ $\Delta_{EA}H^\ominus = -346 \, kJ \, mol^{-1}$
>
> **Step 1:** Construct a Born–Haber cycle. Use Figure 2 as a guide.
>
> - The formation of gaseous atoms requires two separate energy changes, $\Delta_{at}H^\ominus(Na)$ and $\Delta_{at}H^\ominus(Cl)$.
> - The formation of gaseous ions requires two separate energy changes, $\Delta_{IE}H^\ominus(Na)$ and $\Delta_{EA}H^\ominus(Cl)$.

> **Study tip**
>
> You need to learn the definitions for lattice enthalpy, first ionisation energy, and enthalpy change of formation only. For the other enthalpy changes, you are not required to recall the definition but you must understand what the energy changes are and use them in constructing energy cycles and in calculations.

> **Study tip**
>
> Enthalpy change of atomisation is endothermic.

> **Study tip**
>
> Ionisation energies are endothermic. First electron affinity is exothermic.

ENTHALPY AND ENTROPY 22

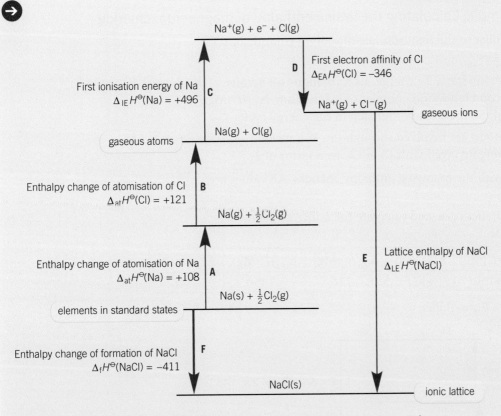

▲ **Figure 3** Born–Haber cycle for NaCl(s). All enthalpy changes have units of kJ mol⁻¹

Between each horizontal energy level:

- only one species has changed, matching the enthalpy change that has taken place
- all the species are balanced.

For example, for change **A**, Na(s) has been atomised to Na(g) and the enthalpy change is the atomisation of Na. $\frac{1}{2}Cl_2(g)$ has not changed but it is still included to account for all the species that are present.

Step 2: Calculate the lattice enthalpy of sodium chloride.

Route 1: **A + B + C + D + E, $\Delta_{LE}H^\ominus$(NaCl)**

Route 2: **F**

Using Hess' law: **A + B + C + D + $\Delta_{LE}H^\ominus$(NaCl) = F**

$\Delta_{at}H^\ominus(Na) + \Delta_{at}H^\ominus(Cl) + \Delta_{IE}H^\ominus(Na) + \Delta_{EA}H^\ominus(Cl) + \Delta_{LE}H^\ominus(NaCl) = \Delta_f H^\ominus(NaCl)$

$108 + 121 + 496 + (-346) + \Delta_{LE}H^\ominus(NaCl) = -411$

$\Delta_{LE}H^\ominus(NaCl) = -411 - \{108 + 121 + 496 + (-346)\} = -790 \text{ kJ mol}^{-1}$

22.1 Lattice enthalpy

 Worked example: Calculating the lattice enthalpy of magnesium chloride

This example is similar to the first example of sodium chloride but with some differences:

- The magnesium ion has a 2+ charge. Its formation will involve the first *and* second ionisation energies of magnesium, $\Delta_{IE1}H^\ominus(Mg)$ and $\Delta_{IE2}H^\ominus(Mg)$. So there is another step in the energy cycle.

- Two chlorine atoms are involved throughout. You need to multiply anything involved with Cl or Cl$^-$ by a factor of 2.

You are provided with the following enthalpy changes. All values are in units of kJ mol^{-1}.

Construct a Born–Haber cycle and calculate the lattice enthalpy of magnesium chloride.

$\Delta_f H^\ominus(MgCl_2)$	$\Delta_{at} H^\ominus(Mg)$	$\Delta_{at} H^\ominus(Cl)$	$\Delta_{IE1} H^\ominus(Mg)$	$\Delta_{IE2} H^\ominus(Mg)$	$\Delta_{EA} H^\ominus(Cl)$
−642	+150	+121	+736	+1450	−346

Step 1: Construct a Born–Haber.

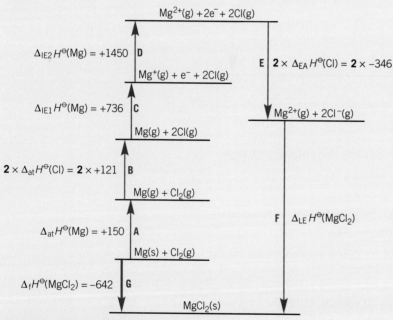

▲ **Figure 4** *Born–Haber cycle for MgCl$_2$(s). All enthalpy changes have units of kJ mol^{-1}*

Step 2: Calculate the lattice enthalpy of magnesium chloride.

Route 1:   **A + B + C + D + E + F**, $\Delta_{LE}H^\ominus$
Route 2:   **G**
Using Hess's law: **A + B + C + D + E** + $\Delta_{LE}H^\ominus$ = **G**

$\Delta_{at}H^\ominus(Mg) + 2 \times \Delta_{at}H^\ominus(Cl) + \Delta_{IE1}H^\ominus(Mg) + \Delta_{IE2}H^\ominus(Mg) + 2 \times \Delta_{EA}H^\ominus(Cl) + \Delta_{LE}H^\ominus(MgCl_2) = \Delta_f H^\ominus(MgCl_2)$

$+150 + 2 \times 121 + 736 + 1450 + (2 \times -346) + \Delta_{LE}H^\ominus(MgCl_2) = -642$

$\Delta_{LE}H^\ominus(MgCl_2) = -642 - \{+150 + 2 \times 121 + 736 + 1450 + (2 \times -346)\} = -2528 \text{ kJ mol}^{-1}$

22 ENTHALPY AND ENTROPY

Successive electron affinities

The example for magnesium chloride included two ionisation energies to account for the formation of $Mg^{2+}(g)$. When an anion has a greater charge than 1−, such as O^{2-}, successive electron affinities are required. These are dealt with in a similar way to successive ionisation energies.

The first two electron affinities (EA) for oxygen are shown below.

First EA $\quad$ $O(g) + e^- \rightarrow O^-(g) \quad \Delta_{EA1}H^\ominus = -141\,kJ\,mol^{-1}$
$\qquad$ gaseous $\qquad\qquad$ gaseous
$\qquad$ atoms $\qquad\qquad\quad$ 1− ions

Second EA $\quad$ $O^-(g) + e^- \rightarrow O^{2-}(g) \quad \Delta_{EA2}H^\ominus = +790\,kJ\,mol^{-1}$
$\qquad\quad$ gaseous $\qquad\qquad$ gaseous
$\qquad\quad$ 1− ions $\qquad\qquad\,\,$ 2− ions

Second electron affinities are *endothermic*. A second electron is being gained by a negative ion, which repels the electron away. So energy must be put in to force the negatively-charged electron onto the negative ion. A Born–Haber cycle including $O^{2-}(g)$ would contain the two energy changes shown in Figure 5.

> **Synoptic link**
>
> For details of successive ionisation energies, see Topic 7.2, Ionisation energies.

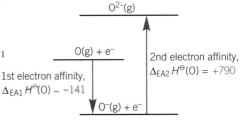

▲ **Figure 5** *Energy changes for formation of $O^{2-}(g)$. All enthalpy changes have units of $kJ\,mol^{-1}$. The first electron affinity is exothermic, but the second electron affinity is endothermic*

Summary questions

1 a State the name of the enthalpy changes for the following:
  i $\,\,$ $Ca(g) \rightarrow Ca^+(g) + e^-$
  ii $\,\,$ $S^-(g) + e^- \rightarrow S^{2-}(g)$
  iii $\,\,$ $Ca(s) + Cl_2(g) \rightarrow CaCl_2(s)$ $\qquad$ (3 marks)

b Write equations for the changes that accompany the following:
  i $\,\,$ lattice enthalpy of magnesium oxide
  ii $\,\,$ first electron affinity of fluorine
  iii $\,\,$ enthalpy change of atomisation of oxygen. $\qquad$ (3 marks)

2 You are provided with the information below:

Enthalpy change	Equation	$\Delta H^\ominus / kJ\,mol^{-1}$
A	$Na(s) + \frac{1}{2}Br_2(l) \rightarrow NaBr(s)$	−361
B	$Br(g) + e^- \rightarrow Br^-(g)$	−325
C	$Na(s) \rightarrow Na(g)$	+108
D	$Na(g) \rightarrow Na^+(g) + e^-$	+496
E	$\frac{1}{2}Br_2(l) \rightarrow Br(g)$	+112

a State the name of each enthalpy change **A** to **E**. (5 marks)

b Construct a Born–Haber cycle for sodium bromide. (5 marks)

c Calculate the lattice enthalpy of sodium bromide. (2 marks)

3 a Draw a Born–Haber cycle for the formation of sodium oxide, Na_2O, showing all species with state symbols and labelling all energy gaps. (5 marks)

b Use your Born–Haber cycle and the data below to calculate the lattice enthalpy of sodium oxide.
$\Delta_{at}H^\ominus(Na) = +108\,kJ\,mol^{-1}$
$\Delta_{at}H^\ominus(O) = +249\,kJ\,mol^{-1}$
$\Delta_{IE}H^\ominus(Na) = +496\,kJ\,mol^{-1}$
$\Delta_f H^\ominus(Na_2O) = -414\,kJ\,mol^{-1}$
$\Delta_{EA1}H^\ominus(O) = -141\,kJ\,mol^{-1}$
$\Delta_{EA2}H^\ominus(O) = +790\,kJ\,mol^{-1}$
(2 marks)

22.2 Enthalpy changes in solution

Specification reference: 5.2.1

Learning outcomes

Demonstrate knowledge, understanding, and application of:

→ enthalpy changes in solution

→ enthalpy cycles related to hydration.

Synoptic link

In Topic 5.2, Ionic bonding and structure, you learnt that ionic compounds tend to dissolve in polar solvents.

Dissolving ionic compounds

Water molecules are able to break up the giant ionic lattice structure and overcome the strong electrostatic attractions between oppositely-charged ions. This is what happens when salt dissolves in water.

Enthalpy change of solution

The overall energy change associated with the dissolving process is called the enthalpy change of solution.

The **standard enthalpy change of solution** $\Delta_{sol}H^\ominus$ is the enthalpy change that takes place when one mole of a solute dissolves in a solvent. If the solvent is water, the ions from the ionic lattice finish up surrounded with water molecules as aqueous ions.

The equation below represents the enthalpy change of solution of sodium chloride in water. In the equation, aq represents an excess of water.

$$Na^+Cl^-(s) + aq \rightarrow \underbrace{Na^+(aq) + Cl^-(aq)}_{\text{aqueous ions}} \quad \Delta_{sol}H^\ominus = +4\,kJ\,mol^{-1}$$

solid ionic lattice

The enthalpy change of solution can be exothermic or endothermic. Figure 1 shows part of the sodium chloride lattice and compares the ions in the ionic lattice and in solution. Remember that the attraction exists in three dimensions, so there will also be ions and water molecules above and below the plane of the paper.

NaCl(s):   Na$^+$ and Cl$^-$ ions attracted together in a giant ionic lattice

NaCl(aq): Na$^+$ and Cl$^-$ ions are separate, but now surrounded by water molecules.

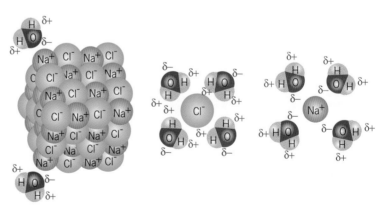

▲ Figure 1 *Comparison of positive and negative ions in ionic lattice and aqueous ions*

In the aqueous ions, the δ+ and δ− partial charges in the water molecules are attracted towards the positive and negative ions.

- The δ− oxygen atom is attracted to the positive sodium ion.
- The δ+ hydrogen atoms is attracted to the negative chloride ion.

22 ENTHALPY AND ENTROPY

 Experimental determination of the enthalpy change of solution

The enthalpy change of solution can be easily determined directly in the laboratory using the procedure below for potassium chloride, KCl. The thermometer used is graduated in 1 °C units so that the temperature can be estimated to the nearest 0.5 °C.

- The same procedure can be used for any ionic compound.
- Take care in the calculation to get the sign correct. Remember that $\Delta_{sol}H$ can be exothermic or endothermic.
- In the example, the mass of potassium chloride has been chosen as a direct multiple of the molar mass. This is good and common practice as it makes the calculation easier.

Procedure

1 Weigh out a sample of KCl(s).
2 Using a measuring cylinder, pour 25.0 cm³ of distilled water into the plastic cup in the beaker. Measure the temperature of the water, to the nearest 0.5 °C.
3 Quickly tip all of the KCl(s) into the water in the plastic cup. Stir the mixture with the thermometer until all of the KCl has dissolved and the temperature no longer changes.
Record this value to the nearest 0.5 °C.

Results

mass of potassium chloride	=	3.73 g
initial temperature of water	=	22.0 °C
final temperature of solution	=	15.5 °C
temperature change ΔT of solution	=	6.5 °C

Calculation

Step 1: Calculate the energy change q in the solution in kJ.

density of water = 1.00 g cm⁻³
so 25.0 cm³ of water has a mass of 25.0 g

mass of solution = mass of water + mass KCl
= 25.0 + 3.73 = 28.73 g

For the solution, specific heat capacity c is close to that of water = (4.18 J g⁻¹ K⁻¹)

$q = mc\Delta T = 28.73 \times 4.18 \times 6.5$
$= 780.5941$ J $= 0.780\,594\,1$ kJ

Step 2: Calculate the amount, in moles, of KCl that dissolved.

$n(KCl) = \dfrac{m}{M} = \dfrac{3.73}{74.6} = 0.0500$ mol

Step 3: Calculate $\Delta_{sol}H$ in kJ mol⁻¹.

$\Delta_{sol}H$ is for the reaction: KCl(s) → K⁺(aq) + Cl⁻(aq)
    1 mol → aqueous ions

In the experiment, 0.0500 mol KCl(s) gains 0.780 594 1 kJ of energy from the solution.

1 mol KCl would gain $\dfrac{0.7805941}{0.0500}$
= 15.6 kJ of energy from the solution

$\Delta_{sol}H = +15.6$ kJ mol⁻¹

Calculate the enthalpy change of solution from the following results.

a 6.40 g of NH_4NO_3 is dissolved in 25.0 g of water. The temperature decreased by 17.5 °C.
b 3.00 g of NaOH is dissolved in 50.0 g of water. The temperature increased by 14.0 °C.

What mass is used in enthalpy calculations?

In the application box, 3.73 g of solid was added and dissolved in 25.0 g of water.

When using the equation $q = mc\Delta T$ to calculate the energy change, m is the mass that is changing temperature (where the thermometer is). The thermometer is in the solution that is changing temperature and so the mass should be the mass of the solution and not the mass of water.

Whenever you carry out an enthalpy experiment, you should consider what is changing temperature before using $q = mc\Delta T$.

Synoptic link

See Topic 9.2, Measuring enthalpy changes, for examples of other experiments to determine enthalpy changes directly.

If you were to have used 25.0 g instead of 28.73 g in the calculation, you would have obtained an final enthalpy change of +13.6 kJ mol^{-1} rather than +15.6 kJ mol^{-1}, a 13% error.

The dissolving process

When a solid ionic compound dissolves in water, two processes take place:

- the ionic lattice breaks up
- water molecules are attracted to, and surround, the ions.

Two types of energy change are involved.

1 The ionic lattice is broken up forming separate gaseous ions.

This is the opposite energy change from lattice energy, which forms the ionic lattice from gaseous ions.

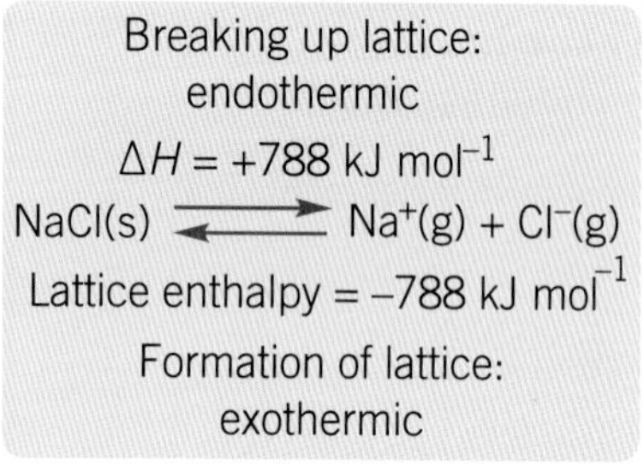

2 The separate gaseous ions interact with polar water molecules to form hydrated aqueous ions. The energy change involved is called the **enthalpy change of hydration.**

The enthalpy change of hydration $\Delta_{hyd}H$ is the enthalpy change that accompanies the dissolving of gaseous ions in water to form one mole of aqueous ions. The equations below show the two enthalpy changes of hydration that are involved during the dissolving of NaCl(s). In the equations, aq represents an excess of water.

$$Na^+(g) + aq \rightarrow Na^+(aq) \qquad \Delta_{hyd}H^\ominus = -406 \text{ kJ mol}^{-1}$$
gaseous ions   aqueous ions

$$Cl^-(g) + aq \rightarrow Cl^-(aq) \qquad \Delta_{hyd}H^\ominus = -378 \text{ kJ mol}^{-1}$$
gaseous ions   aqueous ions

Breaking up lattice: endothermic
$\Delta H = +788 \text{ kJ mol}^{-1}$
NaCl(s) ⇌ Na$^+$(g) + Cl$^-$(g)
Lattice enthalpy = -788 kJ mol^{-1}
Formation of lattice: exothermic

▲ **Figure 2** *Energy transfers involved in the formation of a lattice and the breaking up of a lattice*

Figure 2 shows the energy cycle that links enthalpy change of solution, lattice enthalpy, and hydration enthalpies.

The enthalpy change of solution $\Delta_{sol}H$ can be exothermic or endothermic, depending on the relative sizes of the lattice enthalpy and the enthalpy changes of hydration. The enthalpy change of solution in Figure 3 is endothermic.

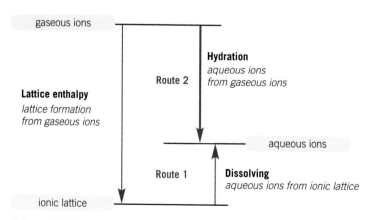

▲ **Figure 3** *Enthalpy changes involved in the dissolving of an ionic compound in water*

22 ENTHALPY AND ENTROPY

Determination of unknown enthalpy changes

You can use the energy changes from Figure 3 as a basis for constructing an energy cycle to calculate an unknown enthalpy change.

Worked example: Calculating the lattice enthalpy of sodium chloride

Construct an energy cycle and calculate the lattice enthalpy of sodium chloride, NaCl. You are provided with the following enthalpy changes.

$Na^+(g) + aq \rightarrow Na^+(aq)$ $\Delta_{hyd}H^\ominus = -406 \text{ kJ mol}^{-1}$

$Cl^-(g) + aq \rightarrow Cl^-(aq)$ $\Delta_{hyd}H^\ominus = -378 \text{ kJ mol}^{-1}$

$Na^+Cl^-(s) + aq \rightarrow Na^+(aq) + Cl^-(aq)$ $\Delta_{sol}H^\ominus = +4 \text{ kJ mol}^{-1}$

Step 1: Construct the energy cycle.

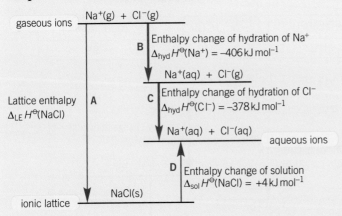

▲ **Figure 5** Energy cycle for NaCl(s). All enthalpy changes have units of kJ mol^{-1}

Between each horizontal energy level:

- only one species has changed, matching the enthalpy change that has taken place
- all the species are balanced.

Step 2: Calculate the lattice enthalpy.

Route 1:   A + D

Route 2:   B + C

Using Hess's law: A + D = B + C

$\Delta_{LE}H^\ominus(NaCl) + \Delta_{sol}H^\ominus(NaCl) = \Delta_{hyd}H^\ominus(Na^+) + \Delta_{hyd}H^\ominus(Cl^-)$

$\Delta_{LE}H^\ominus(NaCl) + (+4) = (-406) + (-378)$

$\Delta_{LE}H^\ominus(NaCl) = (-406) + (-378) - (+4) = -788 \text{ kJ mol}^{-1}$

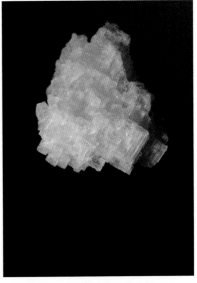

▲ **Figure 4** Sodium chloride crystals

Study tip

Remember that between each horizontal energy level only one species changes and all the species are balanced.

Worked example: Calculating the enthalpy change of hydration of Ca^{2+} ions

This approach is similar to the first example of sodium chloride but there is two important differences.

1. Two Br^- ions are involved. As such, you need to multiply enthalpy changes involving $2Br^-$ by a factor of 2.
2. The enthalpy change of solution of calcium bromide is exothermic. This affects the positioning of the energy levels in the energy cycle and the direction of change **D**.

Compare the energy cycles in Figure 5 and Figure 6 to see the difference.

You are provided with the following enthalpy changes. All values are in units of $kJ\,mol^{-1}$. Construct an energy cycle and calculate the enthalpy change of hydration of Ca^{2+} ions.

$\Delta_{LE}H^\ominus(CaBr_2)$	$\Delta_{hyd}H^\ominus(Br^-)$	$\Delta_{sol}H^\ominus(CaBr_2)$
−2176	−348	−99

Step 1: Construct the energy cycle.

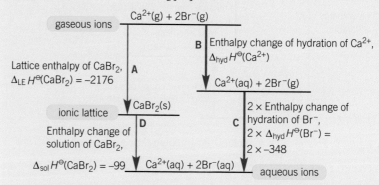

▲ **Figure 6** Energy cycle for $CaCl_2(s)$, all enthalpy changes have units of $kJ\,mol^{-1}$

Between each horizontal energy level:
- only one species has changed, matching the enthalpy change that has taken place
- all the species are balanced.

Step 2: Calculate the enthalpy change of hydration.

Route 1:   **A + D**

Route 2:   **B + C**

Using Hess's law: **A + D = B + C**

$\Delta_{LE}H^\ominus(CaBr_2) + \Delta_{sol}H^\ominus(CaBr_2) = \Delta_{hyd}H^\ominus(Ca^{2+}) + 2 \times \Delta_{hyd}H^\ominus(Br^-)$

$(-2176) + (-99) = \Delta_{hyd}H^\ominus(Ca^{2+}) + (2 \times -348)$

$\Delta_{hyd}H^\ominus(Ca^{2+}) = (-2176) + (-99) - (2 \times -348) = -1579\,kJ\,mol^{-1}$

22 ENTHALPY AND ENTROPY

Cold packs – where does the heat go?

Cold packs are used for sporting injuries to reduce swelling to muscles and sprains. They are made of a pouch containing two compartments — one containing a chemical, the other water. The cold pack is activated by squeezing the pack to break the seal between the two compartments. The pack is then shaken vigorously. The chemical dissolves in the water, cooling the pack and the area around the injury.

Various chemicals can be used, the most common being the ionic compound ammonium nitrate, NH_4NO_3.

▲ **Figure 7** *Applying a cool ice pack to an injured elbow*

In terms of bond breaking and bond making, how do you think the cold pack works?

Summary questions

1. Define the terms:
   i enthalpy change of solution
   ii enthalpy change of hydration. *(2 marks)*

2. You are provided with the following enthalpy changes:
   $$K^+(g) + F^-(g) \rightarrow KF(s) \quad \Delta H = -801 \text{ kJ mol}^{-1}$$
   $$K^+(g) + aq \rightarrow K^+(aq) \quad \Delta H = -320 \text{ kJ mol}^{-1}$$
   $$F^-(g) + aq \rightarrow F^-(aq) \quad \Delta H = -524 \text{ kJ mol}^{-1}$$
   a Name each enthalpy change. *(3 marks)*
   b Write an equation, including state symbols, that represents the enthalpy change of solution of KF. *(1 mark)*
   c Construct an energy cycle for KF that links the enthalpy changes in a and b ($\Delta_{sol}H$ (KF) is exothermic). *(3 marks)*
   d Calculate the enthalpy change of solution of KF. *(1 mark)*

3. 5.18 g of solid lithium fluoride were dissolved in 50.0 cm³ of water in a polystyrene cup of negligible heat capacity. The temperature of the solution increased by 29.0 °C. Calculate the enthalpy change of solution of lithium fluoride, showing all your working.

   (Assume that the density of water = 1.00 g cm⁻³ and the specific heat capacity of solution = 4.18 J g⁻¹ K⁻¹.) *(3 marks)*

4. You are provided with the information below.
   $\Delta_{hyd}H(Ca^{2+}) = -1579 \text{ kJ mol}^{-1}$
   $\Delta_{hyd}H(F^-) = -524 \text{ kJ mol}^{-1}$
   $\Delta_{sol}H(CaF_2) = -16 \text{ kJ mol}^{-1}$
   a Construct an energy cycle that would allow you to find the lattice enthalpy of calcium fluoride using these enthalpy changes. Show all species with state symbols and label all energy gaps. *(5 marks)*
   b Calculate the lattice enthalpy of calcium fluoride. *(2 marks)*

22.3 Factors affecting lattice enthalpy and hydration

Specification reference: 5.2.1

Learning outcomes

Demonstrate knowledge, understanding, and application of:

→ factors affecting lattice enthalpy and hydration enthalpy.

Synoptic link

In Topic 5.2, Ionic bonding and structure, you learnt about the giant ionic lattice structures and their properties.

Ionic compounds

Ionic compounds tend to have the following general properties:

- high melting and boiling points
- soluble in polar solvents
- conduct electricity when molten or in aqueous solution.

Amongst the many ionic compounds, there is actually a wide range in melting points and solubilities.

Some ionic compounds can be melted with the heat of a Bunsen burner, whilst others have such high melting points that they can be used to coat the inside of furnaces. There are even some ionic compounds that are liquids at room temperature – these are called ionic liquids.

Although many ionic compounds are soluble in polar solvents such as water, others are insoluble.

The relative magnitude of lattice and hydration enthalpies helps to explain the variety of melting and boiling points and trends in solubility seen in ionic solids.

Factors affecting lattice enthalpy

The values of lattice enthalpies depend upon ionic size and ionic charge.

Effect of ionic size

The effect of increasing ionic size can be seen by comparing the lattice enthalpies of Group 1 chlorides (Table 1).

▼ Table 1 *Effect of ionic size on lattice enthalpy*

Cation size	Compound	Lattice enthalpy / kJ mol^{-1}	Melting point / °C	Effect of ionic size
Na$^+$	NaCl	−786	801	• ionic radius increases • attraction between ions decreases • lattice energy less negative • melting point decreases
K$^+$	KCl	−715	771	
Rb$^+$	RbCl	−689	718	

358

Effect of ionic charge

Table 2 summarises the effect on increasing ionic charge by comparing sodium oxide with calcium oxide, chosen for the similar sizes of the cations Na⁺ and Ca²⁺.

▼ **Table 2** *Effect of ionic charge on lattice enthalpy*

Compound	Na₂O	CaO
Cation charge	Na⁺	Ca²⁺
Lattice enthalpy / kJ mol⁻¹	−2455	−3414
Melting point / °C	1132	2900
Effect of cation charge	• Ionic charge increases • Attraction between ions increases • Lattice energy becomes more negative • Melting point increases	

Study tip

Although there are more ions in one mole of Na₂O than one mole of CaO, this does not outweigh the increased attraction from the 2+ charge in Ca²⁺ which is much more significant

Table 3 shows the change in ionic size and ionic charge across Period 3 in the periodic table. The effects of both ionic size and charge affect the attraction of these ions for oppositely charged ions. Across Period 3, the ionic size of the cations Na⁺, Mg²⁺, and Al³⁺ decreases as there are more protons attracting the same number of electrons.

▼ **Table 3** *Effect of ionic charge and size across Period 3*

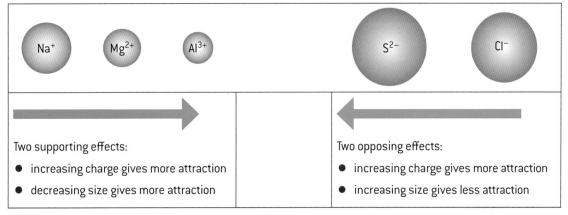

Two supporting effects:	Two opposing effects:
• increasing charge gives more attraction • decreasing size gives more attraction	• increasing charge gives more attraction • increasing size gives less attraction

Predicting melting points

The magnitude of lattice energy gives a good indication of the melting point of an ionic compound, as shown in Table 1 and Table 2. Some metal oxides, such as MgO, Al₂O₃, and ZrO, have very exothermic lattice enthalpies and very high melting points. These stable metal oxides find a use as a protective coating for the inside of furnaces and refractories. Lattice enthalpy is a very good indicator for the size of the melting point but other factors, such as the packing of ions in an ionic lattice, may need to be considered as well.

▲ **Figure 1** *Rubies and sapphires are mainly aluminium oxide, Al₂O₃, with traces of elements giving the different colours. The very exothermic lattice enthalpy of Al₂O₃ ensures that the jewels are long lasting*

22.3 Factors affecting lattice enthalpy and hydration

Factors affecting hydration

Hydration enthalpies are affected by ionic size and ionic charge in a similar way as lattice enthalpies. Tables 4 and 5 summarise the effects of increasing ionic size and charge on hydration enthalpies.

▼ Table 4 *Effect of ionic size on hydration enthalpy down Group 1*

Cation size	Hydration enthalpy / kJ mol^{-1}	Effect of ionic size
Na^+	−406	• ionic radius increases • attraction between ion and water molecules decreases • hydration energy less negative
K^+	−320	
Rb^+	−296	

▼ Table 5 *Effect of ionic charge on hydration enthalpy*

Ionic charge	Na^+	Ca^{2+}
Hydration enthalpy / kJ mol^{-1}	−406	−1579
Effect of ionic charge	• ionic charge increases • attraction with water molecules increases • hydration energy becomes more negative	

Predicting solubility

To dissolve an ionic compound in water, the attraction between the ions in the ionic lattice must be overcome. This requires a quantity of energy equal to the lattice enthalpy. Water molecules are attracted to the positive and negative ions, surrounding them and releasing energy equal to hydration enthalpy.

If the sum of the hydration enthalpies is larger than the magnitude of the lattice enthalpy, the overall enthalpy change (the enthalpy change of solution) will be exothermic and the compound *should* dissolve.

However, many compounds with endothermic enthalpy changes of solution are soluble so this does not provide the total picture. The reasons for solubility also depends on temperature and another variable called entropy, which is discussed in the Topic 22.4 and 22.5.

> **Synoptic link**
>
> In Topic 22.2, Enthalpy changes in solution, you learnt about the energy processes involved in dissolving an ionic compound.

> **Synoptic link**
>
> In Topic 22.5, Free energy, you will find out why some endothermic processes are able to take place on their own at room temperature.

22 ENTHALPY AND ENTROPY

➕ Ionic liquids

You have learnt that ionic compounds generally have high melting and boiling points, but ionic liquids have a low enough melting point that they are liquids below 100 °C. So what is an ionic liquid?

Most ionic compounds have regular-shaped ions of reasonably similar size so they can pack together tightly to form a solid ionic lattice.

In an ionic liquid, the ions can have irregular shapes and can be very different sizes. This means that they are unable to pack together into a lattice and can be liquids at room temperature.

Ionic liquids conduct electricity, do not vaporise easily, and can dissolve an amazing range of substances. The chemical structure of an ionic liquid typically contains an organic ion as shown by the ionic liquid [EMIM]Cl in Figure 2. Ionic liquids can be made from biomass, and are seen as potential green solvents in the future to replace oil-derived solvents.

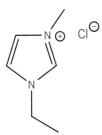

▲ **Figure 2** *Structures of the ionic liquid, [EMIM]Cl*

a Why doesn't the ionic liquid [EMIM]Cl form an ionic lattice?
b What is the empirical formula of the ionic liquid [EMIM]Cl?
c Carry out an Internet search to find some potential uses for ionic liquids and their synthesis from biomass.

Summary questions

1. State and explain the factors that affect the values of hydration enthalpies. *(4 marks)*

2. Explain the differences between the following lattice enthalpies:
   a $\Delta_{LE}H(NaCl)$ and $\Delta_{LE}H(KCl)$ *(2 marks)*
   b $\Delta_{LE}H(NaCl)$ and $\Delta_{LE}H(MgCl_2)$ *(2 marks)*

3. Explain why it is difficult to predict which compound in the following pairs of compounds has the more exothermic lattice enthalpy:
   a $\Delta_{LE}H(NaBr)$ and $\Delta_{LE}H(KCl)$ *(2 marks)*
   b $\Delta_{LE}H(MgO)$ and $\Delta_{LE}H(MgF_2)$ *(2 marks)*

22.4 Entropy

Specification reference: 5.2.2

Learning outcomes

Demonstrate knowledge, understanding, and application of:

→ entropy changes
→ calculation of entropy changes.

What is entropy?

Entropy S can be used to explain things that occur naturally:

- a gas spreading through a room
- heat from a fire spreading through a room
- ice melting in a hot room.

In all these examples, energy is being dispersed and becoming more spread out. There is always a natural tendency for energy to spread out rather than be concentrated in one place.

The greater the entropy, the greater the dispersal of energy and the greater the disorder.

The term entropy is used for the dispersal of energy within the chemicals making up the chemical system.

The units of entropy are $J\,K^{-1}\,mol^{-1}$. The greater the entropy value, the greater that energy is spread out per Kelvin per mole.

In general:

- solids have the smallest entropies
- liquids have greater entropies
- gases have the greatest entropies.

These are very general rules though. Each state has substances that can have very different entropies.

Predicting entropy changes

At 0 K, there would be no energy and all substances would have an entropy value of zero. Above 0 K, energy becomes dispersed amongst the particles and all substances have positive entropy.

Systems that are more chaotic have a higher entropy value.

- If a system changes to become more random, energy can be spread out more – there will be an entropy change ΔS which will be positive.
- If a system changes to become less random, energy becomes more concentrated – the entropy change ΔS will be negative.

In an equation for a physical or chemical change, you can predict whether entropy increases or decreases by comparing the physical states and amount of gas molecules on either side of an equation.

Changes of state

Entropy increases during changes in state that give a more random arrangement of particles:

$$\text{solid} \rightarrow \text{liquid} \rightarrow \text{gas}$$

ENTHALPY AND ENTROPY 22

So when any substance changes state from solid to liquid to gas, its entropy increases:

- Melting and boiling increase the randomness of particles.
- Energy is spread out more and ΔS is positive.

Figure 1 shows the values for the entropy S of the three states of H_2O and shows that the entropy change ΔS is positive as the molecules become more disordered in changing from solid to liquid to gas.

▼ Table 1 *Entropy increases as H_2O changes to more random states*

State		Entropy S / J K^{-1} mol^{-1}	Entropy change ΔS / J K^{-1} mol^{-1}
increasing disorder ↓	$H_2O(s)$	+41	70 − 41 = +29
	$H_2O(l)$	+70	189 − 70 = +119
	$H_2O(g)$	+189	

Change in the number of gaseous molecules

Reactions that produce gases result in an increase in entropy, for example, calcium carbonate reacting with hydrochloric acid:

$$CaCO_3(s) + 2HCl(aq) \rightarrow CaCl_2(aq) + CO_2(g) + H_2O(l)$$

- Production of a gas increases the disorder of particles.
- Energy is spread out more and ΔS is positive.

You can predict the sign of the entropy change for reactions where the reactants and products have different numbers of gas molecules. For example, the reaction of nitrogen and hydrogen gases to produce ammonia results in a *decrease* in the number of gas molecules:

$$N_2(g) + 3H_2(g) \rightarrow 2NH_3(g)$$

4 moles of gas → 2 moles of gas

- There is a decrease in the randomness of particles.
- The energy is spread out less and ΔS is negative.

▲ Figure 1 *Reaction producing a gas and increasing entropy*

Standard entropies

Every substance has a **standard entropy** $S^{\ominus}$, which can be found in data books.

The standard entropy $S^{\ominus}$ of a substance is the entropy of one mole of a substance, under standard conditions (100 kPa and 298 K).

- Standard entropies have units of J K^{-1} mol^{-1}.
- Standard entropies are always positive.

363

22.4 Entropy

Synoptic link

This calculation is similar to finding the enthalpy change of reaction using enthalpy changes of formation. See Topic 9.4, Hess' law and enthalpy cycles.

Calculating entropy changes

Standard entropies can be used to calculate the entropy change of a reaction $\Delta S^\ominus$.

$$\Delta S^\ominus = \Sigma S^\ominus(\text{products}) - \Sigma S^\ominus(\text{reactants}) \quad (\Sigma = \text{sum of})$$

Worked example: Calculating an entropy change of reaction

Calculate the entropy change of reaction for the reaction below:

$$2NO(g) + O_2(g) \rightarrow 2NO_2(g)$$

You are provided with the standard entropies in Table 2.

▼ Table 2 *Standard entropies*

Substance	$S^\ominus$ / J K^{-1} mol^{-1}
NO(g)	+211
O$_2$(g)	+205
NO$_2$(g)	+240

Step 1: Link the standard entropies with the equation.

$$\Delta S^\ominus = \Sigma S^\ominus(\text{products}) - \Sigma S^\ominus(\text{reactants})$$
$$= (2 \times S^\ominus(NO_2)) - \{(2 \times S^\ominus(NO)) + S^\ominus(O_2)\}$$

Step 2: Calculate $\Delta S^\ominus$.

$$\Delta S^\ominus = (2 \times 240) - \{(2 \times 211) + 205\} = -147 \, \text{J K}^{-1} \text{mol}^{-1}$$

Study tip

The final entropy change has a negative value. This is expected as 3 moles of gas molecules are forming 2 moles of gas molecules

Summary questions

1. State and explain whether the following changes would be accompanied by an increase or decrease in entropy:
   a $H_2O(g) \rightarrow H_2O(l)$ *(1 mark)*
   b $Mg(s) + 2HCl(aq) \rightarrow MgCl_2(aq) + H_2(g)$ *(1 mark)*
   c $N_2(g) + 3H_2(g) \rightarrow 2NH_3(g)$ *(1 mark)*

2. Use the data in Table 3 to calculate $\Delta S^\ominus$ for the following reactions.
   a $2SO_2(g) + O_2(g) \rightarrow 2SO_3(l)$ *(1 mark)*
   b $2H_2S(g) + 3O_2(g) \rightarrow 2SO_2(g) + 2H_2O(l)$ *(1 mark)*

▼ Table 3 *Standard entropies*

	SO$_2$(g)	O$_2$(g)	SO$_3$(l)	H$_2$S(g)	H$_2$O(l)	CO$_2$(g)
$S^\ominus$ / J K^{-1} mol^{-1}	+248	+205	+96	+206	+70	+214

3. Calculate the standard entropy of $C_2H_6(g)$ from Table 3 and the information below. *(2 marks)*
   $2C_2H_6(g) + 7O_2(g) \rightarrow 4CO_2(g) + 6H_2O(l)$ $\Delta S^\ominus = -619 \, \text{J K}^{-1} \text{mol}^{-1}$

22.5 Free energy
Specification reference: 5.2.2

Why do reactions happen?
Feasibility
Some reactions happen whereas others do not. What factors control whether a reaction is able to occur? The answer comes from energy – a reaction can happen if the products have a lower overall energy than the reactants. The term **feasibility** is used to describe whether a reaction is able to happen and is energetically feasible. The word spontaneous may also be used for energetically feasible.

Free energy
The overall change in energy during a chemical reaction is called the **free energy change** ΔG and is made up of two types of energy.

1. The enthalpy change ΔH. This is the heat transfer between the chemical system and the surroundings.
2. The entropy change at the temperature of the reaction $T\Delta S$. This is the dispersal of energy within the chemical system itself.

The Gibbs' equation
The relationship between the two types of energy, ΔH and $T\Delta S$, is given by the Gibbs' equation.

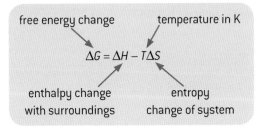

$$\Delta G = \Delta H - T\Delta S$$

where ΔG is the free energy change, T is the temperature in K, ΔH is the enthalpy change with surroundings, and ΔS is the entropy change of system.

> **Learning outcomes**
> Demonstrate knowledge, understanding, and application of:
> → free energy
> → the Gibbs' equation
> → using free energy to predict feasibility.

> **Study tip**
> Make sure that you learn the Gibbs' equation.

Condition for feasibility
The feasibility of a reaction depends on the balance between ΔH and $T\Delta S$ in the Gibbs' equation. For a reaction to be feasible, there must be a decrease in free energy:

- $\Delta G < 0$

The value for ΔH is usually much larger than for ΔS and often dominates the Gibbs' equation. Typically ΔH values are shown in units of kJ mol^{-1} whereas the much smaller ΔS values are shown in units of J K^{-1} mol^{-1}. For example, the ΔH and ΔS values for the reaction between carbon and oxygen to form carbon dioxide are shown below.

$$C(s) + O_2(g) \rightarrow CO_2(g) \quad \Delta H = -393.5 \text{ kJ mol}^{-1}$$
$$\Delta S = 0.9 \text{ J K}^{-1}\text{mol}^{-1}$$

22.5 Free energy

> **Study tip**
>
> To convert 25 °C to Kelvin, add 273:
> 25 °C = 25 + 273
>   = 298 K

In the Gibbs' equation, the value and units of ΔS must be changed to kJ K^{-1} mol^{-1} by dividing by 1000. This then matches the kJ in the ΔH.

For the reaction of carbon and oxygen to form carbon dioxide, at 25 °C:

$$\Delta G = -393.5 - 298 \times \frac{0.9}{1000} = -393.8 \text{ kJ mol}^{-1}$$

For many reactions at room temperature, ΔH has a much larger magnitude that $T\Delta S$ and so ΔG is largely dependent on ΔH. As temperature increases, the $T\Delta S$ term clearly becomes more important.

The worked example below shows how feasibility can change as temperature increases and $T\Delta S$ becoming more significant.

> **Study tip**
>
> Ensure that the units are consistent.
>
> ΔH: kJ mol^{-1} ΔS: J K^{-1} mol^{-1}
>
> Convert ΔS into kJ K^{-1} mol^{-1}
>
> To convert J K^{-1} mol^{-1} to kJ K^{-1} mol^{-1}, divide by 1000.
>
> Remember that $T\Delta S$ uses temperature in K, not °C.

Worked example: Feasibility at different temperatures

The ΔH and ΔS values for the decomposition of calcium carbonate are shown below.

$$CaCO_3(s) \rightarrow CaO(s) + CO_2(g) \quad \begin{array}{l} \Delta H = +178 \text{ kJ mol}^{-1}; \\ \Delta S = 161 \text{ J K}^{-1} \text{ mol}^{-1} \end{array}$$

Determine the feasibility of this reaction at 25 °C and at 1000 °C.

Step 1: Make units consistent.

Convert T to K: $T = 273 + 25 = 298$ K

Convert ΔS to kJ K^{-1} mol^{-1}: $\Delta S = \dfrac{161}{1000}$
                                           $= 0.161$ kJ K^{-1} mol^{-1}

Step 2: Use Gibbs' equation to calculate ΔG at 25 °C.

$$\Delta G = +178 - 298 \times 0.161$$
$$= +130 \text{ kJ mol}^{-1}$$

As $\Delta G > 0$, the reaction is *not* feasible at 25 °C

Step 3: Use Gibbs' equation to calculate ΔG at 1000 °C.

$T = 273 + 1000 = 1273$ K

$\Delta G = \Delta H - T\Delta S$

$\Delta G = +178 - 1273 \times 0.161$
       $= -27.0$ kJ mol^{-1}

As $\Delta G < 0$, the reaction *is* feasible at 1000 °C

The reaction can proceed at the high temperature of 1000 °C. This fits with the process for making calcium oxide (lime) by heating calcium carbonate (limestone) in a furnace or lime kiln.

- At 25 °C, ΔG is positive.
  The reaction is *not* feasible and does *not* take place.

- Above 833 °C (inside the lime kiln), ΔG is negative.
  The reaction is feasible and does take place.

▲ **Figure 1** *A limestone quarry. At high temperatures, limestone is processed into cement which contains calcium oxide*

> **Study tip**
>
> For this reaction at 883 °C, $\Delta G = 0$.
>
> You should be able to prove this using the Gibbs' equation.
>
> At temperatures above 883 °C, $\Delta G < 0$ and the reaction is feasible. The worked example on the next page shows how to calculate the minimum temperature for a reaction to be feasible.

22 ENTHALPY AND ENTROPY

The balance between ΔH and TΔS

The feasibility of a reaction depends upon the balance between ΔH and $T\Delta S$.

Table 1 summarises shows how feasibility can vary with different signs for ΔH and ΔS.

▼ **Table 1** *Feasibility differences at high and low temperatures*

ΔH	ΔS	T	ΔG = ΔH − TΔS	Feasible?
negative	positive	low or high	negative	yes
positive	negative	low or high	positive	no
negative	negative	low	negative	yes
negative	negative	high	positive	no
positive	positive	low	positive	no
positive	positive	high	negative	yes

> **Study tip**
>
> Look carefully at each row in Table 1
>
> Feasibility is supported when:
> - ΔH is negative
> - ΔS is positive.
>
> But T increases the significant of ΔS because $T\Delta S$ is in the Gibbs' equation.

The worked example below shows how the minimum temperature for feasibility can be calculated.

 Worked example: Minimum temperature for feasibility

The equation for the reduction of aluminium oxide with carbon is shown below.

$$Al_2O_3(s) + 3C(s) \rightarrow 2Al(s) + 3CO(g)$$

▼ **Table 2** *Entropy and enthalpy data for the reduction of aluminium oxide*

	Al_2O_3(s)	C(s)	Al(s)	CO(g)
$S^\ominus$ / J K^{-1} mol^{-1}	+51	+6	+28	+198
$\Delta_f H^\ominus$ / kJ mol^{-1}	−1676	0	0	−111

Show whether the reaction is feasible at 25 °C. Calculate the minimum temperature, in °C, for feasibility and whether this is a realistic method for extracting aluminium from its ore.

The key to this problem is the Gibbs' equation. You will first need to calculate ΔH and ΔS for the reaction.

Step 1: Calculate $\Delta S^\ominus$.

$\Delta S^\ominus = \Sigma S^\ominus$ (products) − $\Sigma S^\ominus$ (reactants)

$= \{ (2 \times S^\ominus(Al)) + (3 \times S^\ominus(CO)) \} − \{ S^\ominus(Al_2O_3) + (3 \times S^\ominus(C)) \}$

$\Delta S^\ominus = \{ (2 \times 28) + (3 \times 198) \} − \{ 51 + (3 \times 6) \}$
$= +581 \text{ J K}^{-1} \text{ mol}^{-1}$

> **Synoptic link**
>
> Refer back to Topic 22.4, Entropy, for details of how to calculate an entropy change of reaction from standard entropies. Look back to Topic 9.4, Hess' law and enthalpy changes, for calculating an enthalpy change of reaction from standard enthalpies of formation.

22.5 Free energy

Study tip
Although a full energy cycle could have been used, here we are using the convenient equation that links ΔH with $\Delta_f H$ values for reactants and products.

Study tip
You need to remember to convert the ΔS value into units of $kJ\,mol^{-1}\,K^{-1}$ to match the kJ in the ΔH value. You cannot mix units. The calculated ΔG value will have units of $kJ\,mol^{-1}$.

Synoptic link
For details of enthalpy changes during dissolving, see Topic 22.2, Enthalpy changes of solution.

Step 2: Calculate $\Delta H^\ominus$.

$$\Delta H^\ominus = \Sigma H^\ominus\,(\text{products}) - \Sigma H^\ominus\,(\text{reactants})$$
$$= (3 \times \Delta_f H^\ominus(CO) - \Delta_f H^\ominus(Al_2O_3)$$
$$\Delta H^\ominus = (3 \times -111 - (-1676) = +1343\,J\,mol^{-1}$$

Step 3: Use the Gibbs' equation to calculate ΔG at 25 °C.

Convert T to K: $T = 273 + 25 = 298\,K$

Convert ΔS to $kJ\,K^{-1}\,mol^{-1}$

$$\Delta S = \frac{581}{1000} = 0.581\,kJ\,K^{-1}\,mol^{-1}$$
$$\Delta G = \Delta H - T\Delta S$$
$$\Delta G = +1343 - 298 \times 0.581 = +1170\,kJ\,mol^{-1}$$

As $\Delta G > 0$, the reaction is *not* feasible.

Step 4: Calculate the minimum temperature, in °C, for thermal decomposition to take place.

At the minimum temperature for decomposition,
$\Delta G = \Delta H - T\Delta S = 0$

rearrange $\Delta H - T\Delta S = 0$: $T = \dfrac{\Delta H}{\Delta S}$

substitute ΔH and ΔS: $T = \dfrac{1343}{0.581} = 2312\,K$

convert K to °C: $T = 2316 - 273 = 2039\,°C$

The very high temperature makes the reduction of aluminium oxide by carbon impracticable.

How can endothermic reactions take place at room temperature?

Some ionic compounds dissolve in water at room temperature in an endothermic process, cooling down the water. For example, the enthalpy change of solution of potassium chloride, KCl(s), is endothermic ($\Delta H = +16\,kJ\,mol^{-1}$).

To find out why endothermic processes can take place at room temperature, you need to use entropy and free energy. The worked example below shows how an endothermic process can be feasible at room temperature.

 Worked example: An endothermic feasible process

The enthalpy change and standard entropies for KCl(s), K$^+$(aq), and Cl$^-$(aq) are shown below.

$$KCl(s) \rightarrow K^+(aq) + Cl^-(aq) \qquad \Delta H = +16\,kJ\,mol^{-1}$$

$S^\ominus/J\,K^{-1}\,mol^{-1}$ +83   +103   +57

Show that this process is feasible at 25 °C.

ENTHALPY AND ENTROPY 22

Step 1: Calculate ΔS at 25 °C.

$\Delta S = \Sigma S \text{ (products)} - \Sigma S \text{ (reactants)}$

$= \{ (S^{\ominus}(K^+) + (S^{\ominus}(Cl^-) \} - S^{\ominus}(KCl)$

$= \{ 103 + 57 \} - 83 = +77 \text{ J K}^{-1}\text{mol}^{-1} = +0.077 \text{ kJ K}^{-1}\text{mol}^{-1}$

Step 2: Use Gibbs' equation to calculate ΔG at 25 °C.

$T = 273 + 25 = 298 \text{ K}$

$\Delta G = +16 - 298 \times 0.077 = -7 \text{ kJ mol}^{-1}$

As $\Delta G < 0$, the process is feasible.

Limitations of predictions made for feasibility

The free energy change is useful for predicting feasibility, but many reactions have a negative ΔG and do not seem to take place.

For example, the equation for the decomposition of hydrogen peroxide is shown below, together with its ΔG value at 25 °C.

$H_2O_2(l) \rightarrow H_2O(l) + \frac{1}{2}O_2(g) \qquad \Delta G = -117 \text{ kJ mol}^{-1}$

But hydrogen peroxide does not decompose spontaneously at 25 °C. So why doesn't the reaction take place? The answer lies with the very large activation energy resulting in a very slow rate. If hydrogen peroxide is left for long enough, it would decompose. The reaction does take place by addition of a trace of a MnO_2 catalyst, which allows the reaction to take place via an alternative route with a lower activation energy. The reduced energy barrier then allows the reaction to take place.

So, although the sign of ΔG indicates the thermodynamic feasibility, it takes no account of the kinetics or rate of a reaction.

 Diamonds are forever … or are they?

You may have heard the phrase 'diamonds are forever', but this is not quite correct. Diamond and graphite are two different structures of the same element, carbon. Consider the free energy for the conversion of diamond into graphite:

$C(s, \text{diamond}) \rightarrow C(s, \text{graphite})$

$\Delta G^{\ominus} = -3 \text{ kJ mol}^{-1}$

As $\Delta G^{\ominus} < 0$, the reaction for converting diamond into graphite is feasible!

▲ **Figure 2** *Diamond ring*

If the chemistry is correct, why do you not need to worry about a diamond ring turn into graphite?

Summary questions

1. **a** State the Gibbs' equation. *(1 mark)*
   b State the condition for a reaction to be feasible. *(1 mark)*
   c State the limitations of making predictions about feasibility based on ΔG. *(1 mark)*

2. You are provided with the following information:

   $2NaHCO_3(s) \rightarrow Na_2CO_3(s) + H_2O(l) + CO_2(g)$
   $\Delta H = +91 \text{ kJ mol}^{-1}$
   $S^{\ominus}$ values in $\text{J K}^{-1}\text{mol}^{-1}$:
   $NaHCO_3$, +102; $Na_2CO_3(s)$, +135; $H_2O(l)$, +70;
   $CO_2(g)$, +214

   a Calculate ΔS and ΔG at 25 °C. *(3 marks)*
   b Show whether or not this reaction is feasible at 25 °C. *(1 mark)*
   c Calculate the minimum temperature, in °C, for this reaction to take place. *(1 mark)*

3. You are provided with the following information:

   $2NH_4Cl(s) + Ca(OH)_2(s) \rightarrow CaCl_2(s) + 2NH_3(g) + 2H_2O(l)$

	$S^{\ominus}/\text{J K}^{-1}\text{mol}^{-1}$	$\Delta_f H^{\ominus}/\text{kJ mol}^{-1}$
$NH_4Cl(s)$	+95	−314
$Ca(OH)_2(s)$	+83	−986
$CaCl_2(s)$	+105	−796
$NH_3(g)$	+192	−46
$H_2O(l)$	+70	−286

   a Calculate ΔS and ΔH for this reaction. *(2 marks)*
   b Show whether the reaction is feasible at 25 °C. *(2 marks)*
   c Calculate the minimum temperature, in °C, for this reaction to take place. *(1 mark)*

Chapter 22 Practice questions

Practice questions

1 Lattice enthalpy can be used as a measure of ionic bond strength. Lattice enthalpies are determined indirectly using an enthalpy cycle called a Born–Haber cycle.

The table below shows the enthalpy changes that are needed to determine the lattice enthalpy of magnesium chloride, $MgCl_2$.

	Enthalpy change	Energy / kJ mol^{-1}
A	1st electron affinity of chlorine	−349
B	1st ionisation energy of magnesium	+736
C	atomisation of chlorine	+150
D	formation of magnesium chloride	−642
E	atomisation of magnesium	+76
F	2nd ionisation energy of magnesium	+1450
G	lattice enthalpy of magnesium chloride	

a On the cycle below, write the correct letter in each box.

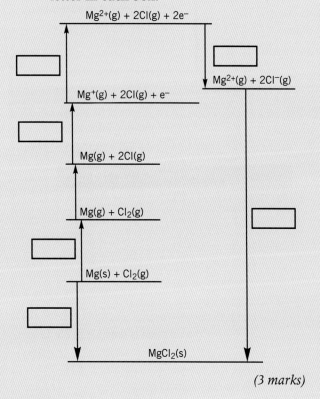

(3 marks)

b Use the Born–Haber cycle to calculate the lattice enthalpy of magnesium chloride. *(2 marks)*

c Deduce the value for the bond enthalpy for Cl—Cl. *(1 mark)*

d Magnesium chloride has stronger ionic bonds than sodium chloride. Explain why. *(3 marks)*

F325 June 2010 Q1

2 The lattice enthalpy of potassium bromide, KBr, can be calculated by a Born–Haber cycle.

a (i) Complete the table below to show the types of energy change which take place in this cycle.

Process	Energy change
$K(s) \rightarrow K(g)$	
$K(g) \rightarrow K^+(g) + e^-$	
$\frac{1}{2} Br_2(l) \rightarrow Br(g)$	
$Br(g) + e^- \rightarrow Br^-(g)$	
$K^+(g) + Br^-(g) \rightarrow KBr(s)$	
$K(s) + \frac{1}{2} Br_2(l) \rightarrow KBr(s)$	

(6 marks)

(ii) Draw a labelled energy diagram of this Born–Haber cycle. *(6 marks)*

b The lattice enthalpy of NaBr is more exothermic than for KBr.

(i) Define the term *lattice enthalpy*. *(2 marks)*

(ii) Explain why the lattice enthalpy of NaBr is more exothermic that for KBr. *(2 marks)*

3 You are provided with the following information.

Ion	ΔH hydration / kJ mol^{-1}
Na$^+$	−405
Mg^{2+}	−1926
Br$^-$	−368

The enthalpy change of solution of $MgBr_2$ is −256 kJ mol^{-1}.

a Define the terms *enthalpy change of solution* and *enthalpy change of hydration*. *(3 marks)*

b Explain the difference between the ΔH hydration values for Na^+ and Mg^{2+}. *(3 marks)*

c (i) The energy cycle below links enthalpy change of solution of $MgBr_2$ with the lattice enthalpy of $MgBr_2$ using hydration enthalpies.

On the two dotted lines, add the species present, including state symbols.

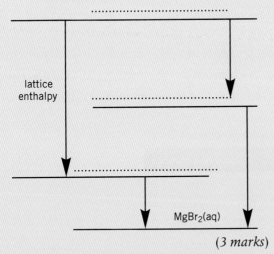

(3 marks)

(ii) Calculate the lattice enthalpy of $MgBr_2$. *(2 marks)*

4 Free energy changes can be used to predict the feasibility of processes.

a Write down the equation that links the free energy change with the enthalpy change and temperature. *(1 mark)*

b You are provided with equations for five processes. For each process, predict the sign of ΔS.

Process	Sign of ΔS
$2CO(g) + O_2(g) \rightarrow 2CO_2(g)$	
$NaCl(s) + aq \rightarrow NaCl(aq)$	
$H_2O(l) \rightarrow H_2O(s)$	
$Mg(s) + H_2SO_4(aq) \rightarrow MgSO_4(aq) + H_2(g)$	
$CuSO_4(s) + 5H_2O(l) \rightarrow CuSO_4 \cdot 5H_2O(s)$	

(2 marks)

c Ammonia can be oxidised as shown in the equation below.

$$4NH_3(g) + 5O_2(g) \rightarrow 4NO(g) + 6H_2O(g)$$

Standard entropies are given in the table below.

Substance	$NH_3(g)$	$O_2(g)$	$NO(g)$	$H_2O(g)$
$S^\ominus / J K^{-1} mol^{-1}$	192	205	211	189

Calculate the standard entropy change, in $J K^{-1} mol^{-1}$, for this oxidation of ammonia. *(2 marks)*

d The exothermic reaction below occurs spontaneously at low temperatures but does **not** occur at very high temperatures.

$$2SO_2(g) + O_2(g) \rightarrow 2SO_3(g)$$

Explain why. *(2 marks)*

e The main ore of iron contains iron(III) oxide, Fe_2O_3. Iron is extracted from this ore by heating with carbon. The equation below shows one of the reactions which take place.

$$Fe_2O_3(s) + 3C(s) \rightarrow 2Fe(s) + 3CO(g)$$

$\Delta S = +543 J K^{-1} mol^{-1}$ and $\Delta H = +493 kJ mol^{-1}$

Calculate the minimum temperature at which this reaction becomes feasible. Show all your working. *(3 marks)*

F325 Jan 2012 Q6

5 The reaction below is the thermal decomposition of lead nitrate, $Pb(NO_3)_2$.

$$2Pb(NO_3)_2(s) \rightarrow 2PbO(s) + 4NO_2(g) + O_2(g)$$

Substance	$Pb(NO_3)_2(s)$	$PbO(s)$	$NO_2(g)$	$O_2(g)$
$\Delta_f H / kJ mol^{-1}$	−452	−217	33	0
$S / J K^{-1} mol^{-1}$	213	69	240	205

a Calculate the standard enthalpy change, the standard entropy change, and the free energy change for this reaction at 25 °C. *(6 marks)*

b State the condition for feasibility and calculate the minimum temperature needed for this reaction to be feasible. *(3 marks)*

23 REDOX AND ELECTRODE POTENTIALS

23.1 Redox reactions

Specification reference: 5.2.3

Learning outcomes

Demonstrate knowledge, understanding, and application of:

→ oxidising and reducing agents
→ redox equations from half-equations
→ redox equations from oxidation numbers
→ interpretation and prediction of reactions involving electron transfer.

Synoptic link

For details of oxidation number rules, oxidation, and reduction, see Topic 4.3, Redox.

Study tip

Remember **OILRIG**:
OIL Oxidation Is Loss of electrons
RIG Reduction Is Gain of electrons

Study tip

You need to be very careful with language here. For example, nitric acid, HNO_3, is a strong oxidising agent in many reactions. But only the nitrogen atom in HNO_3 is reduced using oxidation numbers.

Reducing and oxidising agents

You have seen that reduction and oxidation can be described in terms of transfer of electrons or change in oxidation number:

- Reduction – gain of electrons   OR   decrease in oxidation number.
- Oxidation – loss of electrons   OR   increase in oxidation number.

You also learnt the important rules for assigning oxidation numbers (Table 1).

▼ **Table 1** Important oxidation numbers

Atom	Oxidation number
uncombined element	0
combined hydrogen	+1
combined oxygen	−2
ion of element	ionic charge

Oxidising agents and reducing agents

In a redox reaction, there will always be an **oxidising agent** and a **reducing agent**.

The oxidising agent takes electrons from the species being oxidised. The oxidising agent contains the species that is reduced.

The reducing agent adds electrons to the species being reduced. The reducing agent contains the species that is oxidised.

For example, the equation for the reaction between Ag^+ ions and Cu is shown below.

$$2Ag^+(aq) + Cu(s) \rightarrow Cu^{2+}(aq) + 2Ag(s)$$

$+1$ → 0 Ag^+ reduced
0 → $+2$ Cu oxidised

Ag^+ is the oxidising agent – Ag^+ has oxidised Cu.
Cu is the reducing agent – Cu has reduced Ag^+.

▲ **Figure 1** Redox reaction between Cu and Ag^+ ions – a coil of copper wire is suspended in a solution of silver nitrate. Silver ions are reduced by copper, forming crystalline deposits of silver metal. The copper is oxidised to Cu^{2+} ions indicated by the pale blue solution that gets deeper in colour with time

REDOX AND ELECTRODE POTENTIALS

Construction of redox equations using half-equations

You can write equations for a redox reaction from the half-equations for the reduction and oxidation reactions. The procedure ensures that the number of electrons that are transferred balance.

Worked example: Writing a redox equation from half-equations

The redox reaction between hydrogen peroxide, H_2O_2, and alkaline Cr^{3+} ions has the following half-equations. Write the overall redox equation.

reduction: $\quad H_2O_2 + 2e^- \rightarrow 2OH^-$

oxidation: $\quad Cr^{3+} + 8OH^- \rightarrow CrO_4^{2-} + 4H_2O + 3e^-$

Step 1: Balance the electrons:

The reduction equation is multiplied through by × 3 to give 6e⁻.

The oxidation equation is multiplied through by × 2 to also give 6e⁻.

reduction × 3: $\quad 3H_2O_2 + 6e^- \rightarrow 6OH^-$

oxidation × 2: $\quad 2Cr^{3+} + 16OH^- \rightarrow 2CrO_4^{2-} + 8H_2O + 6e^-$

Step 2: Add and cancel electrons:

$3H_2O_2 + 2Cr^{3+} + 16OH^- + \cancel{6e^-} \rightarrow 6OH^- + 2CrO_4^{2-} + 8H_2O + \cancel{6e^-}$

$3H_2O_2 + 2Cr^{3+} + 16OH^- \rightarrow 6OH^- + 2CrO_4^{2-} + 8H_2O$

Step 3: Finally cancel any species that are on both sides of the equation:

Here, there are $16OH^-$ on the left-hand side and $6OH^-$ on the right-hand side. So you cancel to leave just $10\,OH^-$ on the left-hand side.

$3H_2O_2 + 2Cr^{3+} + \underset{10\,OH^-}{\cancel{16OH^-}} \rightarrow \cancel{6OH^-} + 2CrO_4^{2-} + 8H_2O$

$3H_2O_2 + 2Cr^{3+} + 10\,OH^- \rightarrow 2CrO_4^{2-} + 8H_2O$

Using oxidation numbers to write equations

You can write an equation for a redox reaction from the oxidation number of the species in the redox reaction. The procedure ensures that the changes in oxidation number balance.

Construction of redox equations using oxidation numbers

Oxidation numbers can be used to write and balance a redox equation. By assigning oxidation numbers to all atoms in an equation, it is easy to see which atoms have been reduced and which have been oxidised.

Synoptic link

You learnt in Topic 4.3, Redox, that during a redox reaction, the number of electrons lost during oxidation must equal the number of electrons gained during reduction.

▲ **Figure 2** Oxidation of green Cr^{3+} ions (left) in alkaline conditions by hydrogen peroxide produces yellow chromate(VI) ions, CrO_4^{2-} (right)

Study tip

Look for number patterns. Here, 3 and 2 have the common linked number of 2 × 3 = 6. You can also use fractions and you may find it easier just to multiply the reduction equation by 1.5. This is perfectly fine and just as correct.

Study tip

When you have completed your equation, have a final check to see if there are any species that are on both sides that need cancelling.

Synoptic link

In Topic 4.3, Redox, you learnt that in a redox reaction, the overall increase in oxidation number during oxidation must equal the overall decrease in oxidation number during reduction.

23.1 Redox reactions

To complete the equation, balance the changes in oxidation number as shown in the worked example below.

This example is tricky to balance by trial and error as hydrogen and oxygen appear in more than one product. The oxidation number method takes much of the guesswork out of balancing redox equations.

> **Worked example: Writing a redox equation from oxidation numbers**
>
> Sulfur, S, reacts with concentrated nitric acid, HNO_3, to form sulfuric acid, H_2SO_4, nitrogen dioxide, NO_2, and water, H_2O.
>
> Construct the overall equation for this redox reaction.
>
> **Step 1:** Summarise the information provided.
>
> $$S + HNO_3 \rightarrow H_2SO_4 + NO_2 + H_2O$$
>
> **Step 2:** Assign oxidation numbers to identify the atoms that change their oxidation number.
>
> $S + HNO_3 \rightarrow H_2SO_4 + NO_2 + H_2O$
>
> 0   +6   oxidation number change: +6
>
>   +5   +4   oxidation number change: −1
>
> **Step 3:** Balance *only* the species that contain the elements that have changed oxidation number.
>
> To match the increase of +6 for sulfur, you need a total decrease of −6 from nitrogen.
>
> So HNO_3 and NO_2 also need to be multiplied by 6 to give $6 \times -1 = -6$
>
> $$S + 6HNO_3 \rightarrow H_2SO_4 + 6NO_2 + H_2O$$
>
> **Step 4:** Balance any remaining atoms.
>
> As the oxidation numbers are now balanced, the hydrogen and oxygen atoms are balanced by adding a 2 in front of the H_2O.
>
> $$S + 6HNO_3 \rightarrow H_2SO_4 + 6NO_2 + 2H_2O$$

> **Study tip**
>
> In this equation, there are also atoms of hydrogen and oxygen but their oxidation numbers do not change.

Predicting products of redox reactions

In equations, you might not know all the species involved in the reaction and you might need to predict any missing reactants or products. In aqueous redox reactions, H_2O is often formed. Other likely products are H^+ and OH^- ions, depending on the conditions used.

As a final check, ensure that both sides of the equations are balanced by charge.

The worked example shows how a half-equation can be predicted using oxidation numbers.

> **Study tip**
>
> Always make sensible predictions – if your predicted formula looks strange, you have probably made up something that doesn't exist.

Worked example: Writing a half-equation

Acidified MnO_4^- ions are reduced to Mn^{2+} ions:

$$MnO_4^- + H^+ \rightarrow Mn^{2+}$$

Write the half equation for this reduction.

Step 1: Assign the oxidation numbers and the change in oxidation number.

$$MnO_4^- + H^+ \rightarrow Mn^{2+}$$

+7   +2 oxidation number decreases by 5

Step 2: Balance the electrons.

A decrease in oxidation number of 5 requires $5e^-$ on the left-hand side:

$$MnO_4^- + H^+ + 5e^- \rightarrow Mn^{2+}$$

Step 3: Balance any remaining atoms and predict any further species.

There are four oxygen atoms and H^+ on the left-hand side but no hydrogen or oxygen atoms on the right hand side.

A likely product is H_2O.

Add H_2O to the right hand side and balance the H_2O. Finally balance the H^+.

$$MnO_4^- + 8H^+ + 5e^- \rightarrow Mn^{2+} + 4H_2O$$

Study tip

You need to be methodical and only add any species that look sensible. You are unlikely to add species other than H_2O, H^+, and OH^-.

Summary questions

1. For the following equations, determine what has been oxidised and reduced in terms of oxidation number. Identify the oxidising and reducing agents.
   a $Mg + Zn^{2+} \rightarrow Mg^{2+} + Zn$ (3 marks)
   b $2H_2SO_4 + Cu \rightarrow CuSO_4 + SO_2 + 2H_2O$ (3 marks)
   c $Fe^{2+} + 2H^+ + NO_2^- \rightarrow Fe^{3+} + H_2O + NO$ (3 marks)

2. Construct the overall equation from the following pairs of half-equations.
   a $Fe^{2+} \rightarrow Fe^{3+} + e^-$
      $Cr_2O_7^{2-} + 14H^+ + 6e^- \rightarrow 2Cr^{3+} + 7H_2O$ (2 marks)
   b $Zn \rightarrow Zn^{2+} + 2e^-$
      $VO_2^+ + 4H^+ + 3e^- \rightarrow V^{2+} + 2H_2O$ (3 marks)
   c $H_2O_2 \rightarrow O_2 + 2H^+ + 2e^-$
      $MnO_4^- + 8H^+ + 5e^- \rightarrow Mn^{2+} + 4H_2O$ (3 marks)

3. Use oxidation numbers to balance the following reactions.
   a $MnO_4^- + H^+ + H_2S \rightarrow Mn^{2+} + S + H_2O$ (2 marks)
   b $BrO_3^- + Br^- + H^+ \rightarrow Br_2 + H_2O$ (2 marks)

4. Balance and complete the following half-equations.
   a $MnO_2 + H^+ + e^- \rightarrow Mn^{2+} +$ (2 marks)
   b $NO_3^- + H^+ + e^- \rightarrow NO +$ (2 marks)

5. Use oxidation numbers to write an equation for the following reactions. In each reaction, there is one extra product to predict.
   a Hydrogen iodide, HI, reacts with concentrated sulfuric acid, H_2SO_4, to form iodine, I_2, hydrogen sulfide, H_2S, and one other product. (3 marks)
   b Lead(IV) oxide reacts with concentrated hydrochloric acid to form lead(II) chloride, chlorine gas, and one other product. (3 marks)

23.2 Manganate(VII) redox titrations

Specification reference: 5.2.3

Learning outcomes

Demonstrate knowledge, understanding, and application of:

→ the techniques and procedures used when carrying out redox titrations involving Fe^{2+}/MnO_4^-

→ structured and non-structured calculations for Fe^{2+}/MnO_4^- titrations

→ non-familiar redox systems.

Synoptic link

For full details for preparing standard solutions and carrying out a titration, see Topic 4.2, Acid–base titrations.

Carrying out redox titrations

For redox titrations, the procedures are very similar to acid–base titrations and any differences will be described in examples.

You will be studying two common redox titrations:

- potassium manganate(VII) ($KMnO_4$(aq)) under acidic conditions
- sodium thiosulfate ($Na_2S_2O_3$(aq)) for determination of iodine (I_2(aq)).

Manganate(VII) titrations

In manganate(VII) titrations, MnO_4^-(aq) ions are reduced and so the other chemical used must be a reducing agent that is oxidised.

Procedure

1. A standard solution of potassium manganate(VII), $KMnO_4$, is added to the burette.
2. Using a pipette, add a measured volume of the solution being analysed to the conical flask. An excess of dilute sulfuric acid is also added to provide the H^+(aq) ions required for the reduction of MnO_4^-(aq) ions. You do not need to add an indicator, as the reaction is self-indicating.
3. During the titration, the manganate(VII) solution reacts and is decolourised as it is being added. The end point of the titration is judged by the first permanent pink colour, indicating when there is an excess of MnO_4^- ions present. In titrations, this end point is one of the easiest to judge. You can see these changes in Figure 1.
4. Repeat the titration until you obtain concordant titres (two titres that agree within ±0.10 cm³).

Reading the meniscus

$KMnO_4$(aq) is a deep purple colour and it is very difficult to see the bottom of the meniscus through the intense colour. In manganate(VII) titrations, burette readings are read from the top, rather than the bottom, of the meniscus. The titre is the difference between the two readings, so the titre is the same as reading the bottom of the meniscus, provided that the top is used for both initial and final burette readings.

▲ Figure 1 Carrying out a manganate(VII) titration. On the left, you can see the purple manganate(VII) colour disappearing as the flask is being swirled. On the right, you can see the pale permanent pink colour at the end point

In a manganate(VII) titration, a student obtained a titre of 22.30 cm³. Each burette reading taken had an uncertainty of ±0.05 cm³.

What is the percentage uncertainty in the student's titre?

REDOX AND ELECTRODE POTENTIALS

Examples of manganate(VII) titrations

Manganate(VII) titrations can be used for the analysis of many different reducing agents. Two examples are:
- iron(II) ions, Fe^{2+}(aq)
- ethanedioic acid, $(COOH)_2$(aq).

The same principles can be applied for redox titrations of other reducing agents, provided that they can reduce MnO_4^- to Mn^{2+}.

> **Study tip**
>
> Potassium manganate(VII) is also known as potassium permanganate. Permanganate is the traditional name for the IUPAC name of manganate(VII).

Analysing the percentage purity of an iron(II) compound

The half-equations and full equation for the reaction of acidified manganate(VII) ions with iron(II) ions are shown below.

Reduction: $\quad MnO_4^-(aq) + 8H^+(aq) + 5e^- \rightarrow Mn^{2+}(aq) + 4H_2O(l)$

Oxidation: $\quad Fe^{2+}(aq) \rightarrow Fe^{3+}(aq) + e^-$

Overall: $\quad MnO_4^-(aq) + 8H^+(aq) + 5Fe^{2+}(aq) \rightarrow Mn^{2+}(aq) + 5Fe^{3+}(aq) + 4H_2O(l)$

In this analysis, you will determine the percentage purity of an impure sample of iron(II) sulfate, $FeSO_4 \cdot 7H_2O$.

> **Study tip**
>
> You should be able to combine these two half equations to obtain the overall equation.

Procedure

1. Prepare a 250.0 cm³ solution of the impure $FeSO_4 \cdot 7H_2O$ in a volumetric flask.
2. Using a pipette, measure 25.0 cm³ of this solution into a conical flask. Then add 10 cm³ of 1 mol dm⁻³ H_2SO_4(aq) (an excess).
3. Using a burette, titrate this solution using a standard 0.0200 mol dm⁻³ solution of potassium manganate(VII), $KMnO_4$(aq).
4. Finally analyse your results to determine the percentage purity.

Mass measurements

Mass of weighing bottle + impure $FeSO_4 \cdot 7H_2O$ / g	18.34
Mass of weighing bottle / g	11.37
Mass of impure $FeSO_4 \cdot 7H_2O$ / g	6.97

> **Synoptic link**
>
> The analysis of results from redox titrations is essentially the same method as used for acid–base titrations. See Topic 4.2, Acid–base titrations, for further details.

Titration readings

	Trial	1	2	3
Final burette reading / cm³	24.10	23.70	47.15	23.35
Initial burette reading / cm³	1.00	0.00	23.70	0.00
Titre / cm³	23.10	23.70	23.45	23.35
Mean titre / cm³			23.40	

> **Study tip**
>
> Only the closest titres are chosen for calculating the mean titre.

Analysis

Step 1: Calculate the amount of MnO_4^- that reacted.

$$n(MnO_4^-) = c \times \frac{V}{1000} = 0.0200 \times \frac{23.40}{1000} = 4.68 \times 10^{-4} \text{ mol}$$

> **Study tip**
>
> You need to use the mean titre V and the concentration c of MnO_4^- ions.

23.2 Manganate(VII) redox titrations

> **Study tip**
>
> You need to use the equation and $nMnO_4^-$.

> **Study tip**
>
> The most appropriate number of significant figures (s.f.) to use in this example is 3 s.f. This is the lowest number of significant figures in the quantities used – for the mass of $FeSO_4 \cdot 7H_2O$ as 6.97 g. But only round to three significant figures for your final answer or you will have introduced rounding errors during the stages of your working.

Step 2: Determine the amount of Fe^{2+} that reacted.

$MnO_4^-(aq) + 8H^+(aq) + 5Fe^{2+}(aq) \rightarrow Mn^{2+}(aq) + 5Fe^{3+}(aq) + 4H_2O(l)$

1 mol   5 mol

4.68×10^{-4} mol $5 \times 4.68 \times 10^{-4} = \mathbf{2.34 \times 10^{-3}}$ mol Fe^{2+}

Step 3: Work out the unknown information. There are several stages.

1. Scale up to find the amount of Fe^{2+} in the 250.0 cm³ solution that you prepared.

   $n(Fe^{2+})$ in 25.00 cm³ used in the titration = $\mathbf{2.34 \times 10^{-3}}$ **mol**

   $n(Fe^{2+})$ in 250.0 cm³ solution   = $2.34 \times 10^{-3} \times 10$

                                           = 2.34×10^{-2} mol

2. Find the mass of $FeSO_4 \cdot 7H_2O$ in the impure sample.

   $n(FeSO_4 \cdot 7H_2O) = n(Fe^{2+}) = \dfrac{m}{M}$

   mass m of $FeSO_4 \cdot 7H_2O = n \times M = 2.34 \times 10^{-2} \times 277.9 = 6.50286$ g

3. Find the percentage purity of $FeSO_4 \cdot 7H_2O$ in the impure sample.

   Percentage purity = $\dfrac{\text{mass of } FeSO_4 \cdot 7H_2O}{\text{mass of impure sample}} \times 100$

                    = $\dfrac{6.50286}{6.97} \times 100 = 93.3\%$

Iron supplements containing iron(II) are used to treat iron-deficiency. A 0.304 g iron supplement tablet is crushed and dissolved in 25.0 cm³ of dilute sulfuric acid. The resulting solution is titrated with 0.00750 mol dm⁻³ $KMnO_4$ and 19.20 cm³ are required to reach the end point.

Calculate the percentage by mass of iron in the iron tablet.

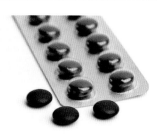

▲ **Figure 2** *Iron supplements are used to treat iron-deficiency (anaemia)*

> **Study tip**
>
> You should be able to see how the two half equations have been combined.
> The common link is 2 × 5 = 10
> The reduction half-equation has been mutiplied by 2.
> The oxidation half-equation has been mutiplied by 5.
> This gives 10e⁻ is each half equation.
> The half-equations are then added and the 10e⁻ are cancelled.

🖩 Worked example: Determination of a formula

Hydrated ethanedioic acid has the formula $(COOH)_2 \cdot xH_2O$. The value of x can be determined by reacting a solution of $(COOH)_2 \cdot xH_2O$ with acidified manganate(VII) ions. The half-equations and full equation are shown below.

Reduction: $MnO_4^-(aq) + 8H^+(aq) + 5e^- \rightarrow Mn^{2+}(aq) + 4H_2O(l)$

Oxidation: $(COOH)_2(aq) \rightarrow 2CO_2(g) + 2H^+(aq) + 2e^-$

Overall:

$2MnO_4^-(aq) + 6H^+(aq) + 5(COOH)_2(aq) \rightarrow 2Mn^{2+}(aq) + 10CO_2(g) + 8H_2O(l)$

Procedure

1. A 0.1203 g sample of $(COOH)_2 \cdot xH_2O$ is dissolved in 25.0 cm³ of 1.0 mol dm⁻³ H_2SO_4 in a conical flask.

2. The contents of the flask are heated to 60 °C.
3. The hot contents of the flask are titrated against 0.0200 mol dm^{-3} MnO$_4^{2-}$ and 19.10 cm^3 are required to reach the end point.

Determine the value of x and the formula of $(COOH)_2 \cdot xH_2O$.

Step 1: Calculate the amount of MnO$_4^-$ that reacted.

$$n(MnO_4^-) = c \times \frac{V}{1000} = 0.0200 \times \frac{19.10}{1000} = 3.82 \times 10^{-4} \text{ mol}$$

Step 2: Determine the amount of $(COOH)_2$ that reacted.

$$2MnO_4^-(aq) + 6H^+(aq) + 5(COOH)_2(aq) \rightarrow 2Mn^{2+}(aq) + 10CO_2(g) + 8H_2O(l)$$

2 mol	5 mol (balancing numbers)
3.82×10^{-4} mol	$2.5 \times 3.82 \times 10^{-4}$ = 9.55×10^{-4} mol

Step 3: Work out the unknown information. There are several stages.

1. Calculate the molar mass of $(COOH)_2 \cdot xH_2O$.

$$n((COOH)_2 \cdot xH_2O) = n((COOH)_2) = \frac{m}{M}$$

Molar mass M of $(COOH)_2 \cdot xH_2O = \frac{m}{M} = \frac{0.1203}{9.55 \times 10^{-4}} = 126.0 \text{ g mol}^{-1}$

2. Determine the value of x and the formula of $(COOH)_2 \cdot xH_2O$.

Molar mass of $(COOH)_2 = (12.0 + 16.0 \times 2 + 1.0) \times 2 = 90.0$

Mass of H_2O in $(COOH)_2 \cdot xH_2O = 126.0 - 90.0 = 36.0$ g

value of $x = \frac{36.0}{18.0} = 2$ Formula = $(COOH)_2 \cdot 2H_2O$

> **Study tip**
> You need to use the mean titre V and the concentration c of MnO$_4^-$ ions.

> **Study tip**
> You need to use the equation and number of moles of MnO$_4^-$.

Non-familiar redox titrations

The principles for redox titrations can be extended to the analysis of many different substances.

- Manganate(VII) titrations can be used to analyse reducing agents that reduce MnO$_4^-$ to Mn^{2+}.
- KMnO$_4$ can be replaced with other oxidising agents, the commonest used being acidified dichromate(VI), H$^+$/Cr$_2$O$_7^{2-}$.

The procedures and calculations are similar to those in the examples although some modifications may be necessary depending on the substance being analysed.

> **Study tip**
> You could be presented with details of a redox titration using different chemicals, but the essential steps in the calculation remain the same.

23.2 Manganate(VII) redox titrations

Summary questions

1. Write the equation for the reaction that takes place in a Fe^{2+}/MnO_4^- titration. *(1 mark)*

2. Chloride ions, Cl^-, are oxidised to chlorine by acidified MnO_4^- ions. Suggest why sulfuric acid is used rather than hydrochloric acid in a manganate(VII) titration. *(1 mark)*

3. In an Fe^{2+}/MnO_4^- titration, what volume of 0.0200 mol dm^{-3} $KMnO_4$ would be required to react exactly with 25.0 cm^3 of 0.114 mol dm^{-3} $FeSO_4$? *(1 mark)*

4. Acidified dichromate(VI) can be used in some redox titrations instead of manganate(VII). The half-equation is shown below.

    $$Cr_2O_7^{2-} + 14H^+ + 6e^- \rightarrow 2Cr^{3+} + 7H_2O$$

    a Write the overall equation for the reaction with Fe^{2+} ions. *(1 mark)*
    b What is x in the following statement?
    1 mol $Cr_2O_7^{2-}$ reacts with x mol Fe^{2+}. *(1 mark)*

5. A metal ore contains Fe(II). 6.46 g of the ore is dissolved in sulfuric acid and the resulting solution is made up to 250.0 cm^3. 25.0 cm^3 of this solution is titrated against 21.40 cm^3 of 0.0200 mol dm^{-3} potassium manganate(VII) solution. Calculate the percentage by mass of iron(II) in the sample of ore. *(1 mark)*

6. A 250.0 cm^3 solution is prepared by dissolving 9.066 g of $Fe(NO_3)_2 \cdot xH_2O$ in water. 25.0 cm^3 samples of this solution are acidified and titrated against 0.0200 mol dm^{-3} $K_2Cr_2O_7$. The mean titre is 26.25 cm^3. Calculate the molar mass, the value of x and the formula of the salt.
    Hint: Refer to question **4** for the equation. *(2 marks)*

23.3 Iodine/thiosulfate redox titrations
Specification reference: 5.2.3

Iodine/thiosulfate titrations

In iodine–thiosulfate titrations, thiosulfate ions, $S_2O_3^{2-}$(aq), are oxidised and iodine, I_2, is reduced.

Oxidation: $\quad 2S_2O_3^{2-}(aq) \rightarrow S_4O_6^{2-}(aq) + 2e^-$

Reduction: $\quad I_2(aq) + 2e^- \rightarrow 2I^-(aq)$

Overall: $\quad 2S_2O_3^{2-}(aq) + I_2(aq) \rightarrow 2I^-(aq) + S_4O_6^{2-}(aq)$

The concentration of aqueous iodine can be determined by titration with a standard solution of sodium thiosulfate. If it were possible to analyse only iodine, this titration would be very limiting. However, by adding another step to the process, it is possible to obtain unknown information about many different oxidising agents.

> **Learning outcomes**
>
> Demonstrate knowledge, understanding, and application of:
>
> → the techniques and procedures used when carrying out redox titrations involving $I_2 / S_2O_3^{2-}$
>
> → structured and non-structured calculations for $I_2 / S_2O_3^{2-}$ titrations.

Analysis of oxidising agents

Iodine/thiosulfate titrations can be used to determine:
- the ClO^- content in household bleach
- the Cu^{2+} content in copper(II) compounds
- the Cu content in copper alloys.

Procedure

1. Add a standard solution of $Na_2S_2O_3$ to the burette.
2. Prepare a solution of the oxidising agent to be analysed. Using a pipette, add this solution to a conical flask. Then add an excess of potassium iodide. The oxidising agent reacts with iodide ions to produce iodine, which turns the solution a yellow-brown colour.
3. Titrate this solution with the $Na_2S_2O_3$(aq). During the titration, the iodine is reduced back to I^- ions and the brown colour fades quite gradually, making it difficult to decide on an end point. This problem is solved by using starch indicator. When the end point is being approached, the iodine colour has faded enough to become a pale straw colour. The fading of the yellow-brown iodine colour as aqueous sodium thiosulfate is added is shown in Figure 1.

▲ **Figure 1** *Stages in an iodine–thiosulfate titration. Aqueous sodium thiosulfate is added until the solution has faded to a pale straw colour*

23.3 Iodine/thiosulfate redox titrations

Using starch for the end point

When the end point is being approached and the iodine colour has faded enough to become a pale straw colour, a small amount of starch indicator is added. A deep blue-black colour forms to assist with the identification of the end point. As more sodium thiosulfate is added, the blue-black colour fades. At the end point, all the iodine will have just reacted and the blue-black colour disappears (Figure 2).

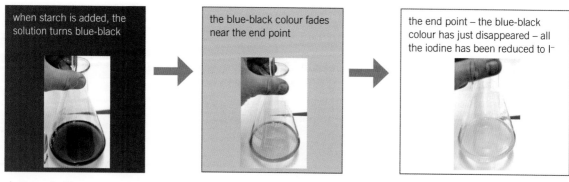

▲ Figure 2 *The starch colour change in an iodine–thiosulfate titration*

Research why it is important to only add the starch indicator towards the end of an iodine–thiosulfate titration.

Examples of iodine–thiosulfate titrations

Two worked examples are provided for the analysis of two different oxidising agents:

- chlorate(I) ions, ClO^-(aq)
- copper(II) ions, Cu^{2+}(aq).

The same principles can be applied for the analysis of other oxidising agents, provided that they are capable of oxidising I^- ions to I_2.

Analysis of household bleach

The active ingredient in household bleach is chlorate(I) ions, ClO^-(aq), commonly known as 'hypochlorite'. Many bottles of bleach show the NaClO content, which supplies the active hypochlorite (ClO^-) ions as labelled on the bottle (Figure 2).

In this analysis, we will calculate the concentration of ClO^- ions in bleach, in mol dm^{-3}.

- ClO^- ions from the bleach are first reacted with I^- and H^+ ions to form I_2:
$$ClO^-(aq) + 2I^-(aq) + 2H^+(aq) \rightarrow Cl^-(aq) + I_2(aq) + H_2O(l)$$

- In the titration, I_2 reacts with $S_2O_3^{2-}$ ions:
$$2S_2O_3^{2-}(aq) + I_2(aq) \rightarrow 2I^-(aq) + S_4O_6^{2-}(aq)$$

1 mol ClO^- mol produces 1 mol I_2 which reacts with 2 mol $S_2O_3^{2-}$.

So 1 mol ClO^- mol is equivalent to 2 mol $S_2O_3^{2-}$.

▲ Figure 3 *Bleach is a solution of sodium chlorate(I), NaClO, commonly named sodium hypochlorite*

REDOX AND ELECTRODE POTENTIALS 23

Procedure

1. Using a pipette, add 10.0 cm³ of the bleach into a 250 cm³ volumetric flask and add water to prepare 250.0 cm³ of solution.
2. Using a pipette, measure 25.0 cm³ of this solution into a conical flask. Then add 10 cm³ of 1 mol dm⁻³ potassium iodide (KI) followed by sufficient 1 mol dm⁻³ HCl(aq) to acidify the solution. The HCl provide H⁺ ions for the reaction.
3. Using a burette, titrate this solution using a standard 0.0500 mol dm⁻³ solution of sodium thiosulfate ($Na_2S_2O_3$(aq)).
4. Repeat the titration to obtain concordant results.
5. Finally analyse your results to determine the concentration of chlorate(I) ions in the bleach.

Results

Mean titre of $Na_2S_2O_3$ = 24.20 cm³

Analysis

Step 1: Calculate the amount of $S_2O_3^{2-}$ that reacted.

$$n(S_2O_3^{2-}) = c \times \frac{V}{1000} = 0.0500 \times \frac{24.20}{1000} = 1.21 \times 10^{-3} \text{ mol}$$

Step 2: Determine the amounts of I_2 and ClO^- that reacted.

1 mol ClO^- is equivalent to 2 mol $S_2O_3^{2-}$

$$n(ClO^-) = \frac{1}{2} \times n(S_2O_3^{2-})$$

$$= \frac{1}{2} \times 1.21 \times 10^{-3} = 6.05 \times 10^{-4} \text{ mol}$$

Step 3: Determine the amount of ClO^- in the 250.0 cm³ solution.

$n(ClO^-)$ in 25.00 cm³ used in the titration = 6.05×10^{-4} mol

$n(ClO^-)$ in 250.0 cm³ solution $= 6.05 \times 10^{-4} \times 10$
$= 6.05 \times 10^{-3}$ mol

Step 4: Determine the concentration of ClO^- in the bleach.

The 250.0 cm³ solution was prepared using 10 cm³ bleach, so:

10 cm³ bleach contains 6.05×10^{-3} mol ClO^- ions

1 dm³ bleach contains $6.05 \times 10^{-3} \times 100 = 0.605$ mol NaClO

concentration of ClO^- ions in the bleach $= 6.05 \times 10^{-3} \times 100$
$= 0.605$ mol dm⁻³

> **Study tip**
>
> You need to use the mean titre V and the concentration c of MnO_4^- ions.

> **Study tip**
>
> See details above to check why 1 mol ClO^- mol is equivalent to 2 mol $S_2O_3^{2-}$.

> **Study tip**
>
> Scale up to find the amount of ClO^- in the 250.0 cm³ solution that you prepared.

---

1. Show that the concentration of the bleach above agrees with the information of the bleach label in Figure 3.
2. 10.0 cm³ of a bleach is diluted to 250.0 cm³ in a volumetric flask. 25.0 cm³ of this solution is added to a conical flask followed by an excess of KI(aq). This solution is titrated with 0.0750 mol dm⁻³ $Na_2S_2O_3$(aq) and 23.20 cm³ are required to reach the end point. Calculate the concentration of ClO^- in the bleach, in mol dm⁻³.

23.3 Iodine/thiosulfate redox titrations

▲ **Figure 4** *The mixture is made up of brown iodine and a white precipitate of copper(I) iodide*

Analysis of copper

Iodine/thiosulfate titrations can be used to determine the copper content of copper(II) salts or alloys. For copper(II) salts, $Cu^{2+}(aq)$ ions are produced simply by dissolving the compound in water. Insoluble copper(II) compounds can be reacted with acids to form $Cu^{2+}(aq)$ ions.

For copper alloys, such as brass or bronze, the alloy is reacted and dissolved in concentrated nitric acid, followed by neutralisation to form $Cu^{2+}(aq)$ ions:

$$Cu(s) \rightarrow Cu^{2+}(aq)$$

In this analysis:

- $Cu^{2+}(aq)$ ions react with $I^-(aq)$ to form a solution of iodine, $I_2(aq)$, and a white precipitate of copper(I) iodide, $CuI(s)$. The mixture appears as a brown colour.

$$2Cu^{2+}(aq) + 4I^-(aq) \rightarrow 2CuI(s) + I_2(aq)$$

Reduction: +2   +1

Oxidation: −1   0

- The iodine in the brown mixture is then titrated with a standard solution of sodium thiosulfate:

$$2S_2O_3^{2-}(aq) + I_2(aq) \rightarrow 2I^-(aq) + S_4O_6^{2-}(aq)$$

2 mol Cu^{2+} produces 1 mol iodine which reacts with 2 mol S_2O_3

So 1 mol Cu^{2+} is equivalent to 1 mol $S_2O_3^{2-}$

> ### 🖩 Worked example: Analysis of brass
>
> Brass, an alloy of copper and zinc, is used in many ornaments and in brass musical instruments. Brass can be analysed using an iodine/thiosulfate titration to find its percentage of copper and zinc.
>
> A sample of brass is analysed using the practical procedure outlined below.
>
> - A 0.500 g sample of brass is reacted with concentrated nitric acid to form a solution containing Cu^{2+} and Zn^{2+} ions. The solution is then neutralised.
>
> - Excess KI(aq) is added. Cu^{2+} ions react with I^- ions, forming I_2.
>
> - The iodine is titrated with $0.200\,\text{mol}\,\text{dm}^{-3}$ sodium thiosulfate and $25.20\,\text{cm}^3$ are required to reach the end point.
>
> Calculate the percentage by mass of copper and zinc in the brass.
>
> **Calculation**
> **Step 1:** Calculate the amount of $S_2O_3^{2-}$ that reacted.
>
> $$n(S_2O_3^{2-}) = c \times \frac{V}{1000} = 0.200 \times \frac{25.20}{1000} = 5.04 \times 10^{-3}\,\text{mol}$$

▲ **Figure 5** *The two main alloys of copper are brass (the saxophone) and bronze (the bell). Brass is made of copper and zinc, whilst bronze is made of copper and tin*

Step 2: Determine the amounts of I_2 and Cu^{2+} that reacted.

1 mol Cu^{2+} is equivalent to 1 mol $S_2O_3^{2-}$

$n(Cu^{2+}) = n(S_2O_3^{2-}) = 5.04 \times 10^{-3}$ mol

Step 3: Work out the unknown information: the percentages of Cu and Zn.

5.04×10^{-3} mol of Cu^{2+} has a mass of $5.04 \times 10^{-3} \times 63.5 = \mathbf{0.320\,g}$

mass of Zn = 0.500 − 0.320 = **0.180 g**

% composition of Cu = $\dfrac{\mathbf{0.320}}{0.500} \times 100 = 64.0\%$

% composition of Zn = $\dfrac{\mathbf{0.180}}{0.500} \times 100 = 36.0\%$

> **Study tip**
>
> See details above to check why 1 mol Cu^{2+} mol is equivalent to 1 mol $S_2O_3^{2-}$.

> **Study tip**
>
> You could be presented with details of an $I_2/S_2O_3^{2-}$ redox titration that analyses different oxidising agents than shown in the worked examples. However, the essential steps in the calculation remain the same.

Summary questions

1. Write the equation for the reaction that takes place in an $I_2/S_2O_3^{2-}$ titration. *(1 mark)*

2. Why is an excess of KI added when analysing oxidising agents in iodine–thiosulfate titrations? *(1 mark)*

3. In an $I_2/S_2O_3^{2-}$ titration, what volume of 0.180 mol dm^{-3} $S_2O_3^{2-}$ would be required to react exactly with the iodine formed from 25.0 cm^3 of 0.144 mol dm^{-3} $CuSO_4$? *(1 mark)*

4. Iodine / thiosulfate titrations can be used to analyse IO_3^- ions. The reaction with I^- is shown below.

   $IO_3^- + 5I^- + 6H^+ \rightarrow 3I_2 + 3H_2O$

   What is x in the following statement?
   1 mol IO_3^- is equivalent to x mol $S_2O_3^{2-}$ *(1 mark)*

5. A student dissolved 1.365 g of $Cu(NO_3)_2 \cdot xH_2O$ in water and made up the solution to 250.0 cm^3. The student added 25.0 cm^3 of the solution to a conical flask and added an excess of KI(aq).

   The resulting mixture was titrated against 0.0235 mol dm^{-3} $Na_2S_2O_3$ and 24.05 cm^3 was required to reach the end point. Determine the molar mass of the copper(II) salt, the value of x, and suggest its formula. *(5 marks)*

23.4 Electrode potentials

Specification reference: 5.2.3

Learning outcomes

Demonstrate knowledge, understanding, and application of:

→ standard electrode potential
→ measuring cell potentials
→ calculating cell potentials from electrode potentials.

Electrochemical cells

In this topic you will be looking at a type of electrochemical cell called a **voltaic cell**, which converts chemical energy into electrical energy. The conversion of chemical energy into electrical energy takes place in the modern cells and batteries that power devices such as mobile phones. So what do you need to make a voltaic cell? As electrical energy results from movement of electrons, you need chemical reactions that transfer electrons from one species to another. These are redox reactions.

Half cells

A **half-cell** contains the chemical species present in a redox half-equation. A voltaic cell can be made by connecting together two different half-cells, which then allows electrons to flow. In the cell, the chemicals in the two half-cells must be kept apart – if allowed to mix, electrons would flow in an uncontrolled way and heat energy would be released rather than electrical energy.

Metal/metal ion half-cells

The simplest half-cell consists of a metal rod dipped into a solution of its aqueous metal ion. This is represented using a vertical line for the phase boundary between the aqueous solution and the metal, for example, $Zn^{2+}(aq)|Zn(s)$ and $Cu^{2+}(aq)|Cu(s)$ half-cells are shown in Figure 1.

At the phase boundary where the metal is in contact with its ions, an equilibrium will be set up. By convention, the equilibrium in a half-cell is written so that the forward reaction shows reduction and the reverse reaction shows oxidation (Figure 2).

In an isolated half-cell, there is no net transfer of electrons either into or out of the metal.

When two half-cells are connected, the direction of electron flow depends upon the relative tendency of each electrode to release electrons.

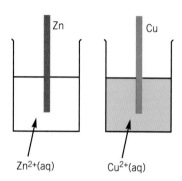

▲ Figure 1 $Zn^{2+}(aq)|Zn(s)$ and $Cu^{2+}(aq)|Cu(s)$ half-cells

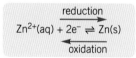

▲ Figure 2 Electrode equilibria

Ion/ion half-cells

An ion/ion half-cell contains ions of the same element in different oxidation states. For example, a half-cell can be made containing a mixture of aqueous iron(II) and iron(III) ions. The redox equilibrium is:

$$Fe^{3+}(aq) + e^- \rightleftharpoons Fe^{2+}(aq)$$

In this type of half-cell there is no metal to transport electrons either into or out of the half-cell, so an inert metal electrode made out of platinum is used. A half-cell based on $Fe^{2+}(aq)$ and $Fe^{3+}(aq)$ ions is shown in Figure 3.

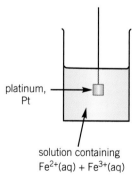

▲ Figure 3 A half-cell based on $Fe^{2+}(aq)$ and $Fe^{3+}(aq)$ – ions of the same element in different oxidation states

23 REDOX AND ELECTRODE POTENTIALS

Electrode potentials

How do you know which electrode has a greater tendency to gain or lose electrons? In a cell with two metal/metal ion half-cells connected, the more reactive metal releases electrons more readily and is oxidised.

In an operating cell:

- the electrode with more reactive metal loses electrons and is oxidised – this is the negative electrode
- the electrode with the less reactive metal gains electrons and is reduced – this is the positive electrode.

Standard electrode potential

The tendency to be reduced and gain electrons is measured as a **standard electrode potential** $E^\ominus$. The standard chosen is a half-cell containing hydrogen gas, $H_2(g)$, and a solution containing $H^+(aq)$ ions. An inert platinum electrode is used to allow electrons into and out of the half-cell. Figure 4 shows a standard hydrogen electrode.

By definition, the standard conditions used are:

- solutions have a concentration of exactly $1\,\text{mol}\,\text{dm}^{-3}$
- the temperature is $298\,\text{K}$ ($25\,°\text{C}$)
- the pressure is $100\,\text{kPa}$ ($1\,\text{bar}$).

The standard electrode potential is the e.m.f. of a half-cell connected to a standard hydrogen half-cell under standard conditions of $298\,\text{K}$, solution concentrations of $1\,\text{mol}\,\text{dm}^{-3}$, and a pressure of $100\,\text{kPa}$.

By definition, the standard electrode potential of a standard hydrogen electrode is exactly $0\,\text{V}$. The sign of a standard electrode potential shows the sign of the half-cell connected to the standard hydrogen electrode and shows the relative tendency to gain electrons compared with the hydrogen half-cell.

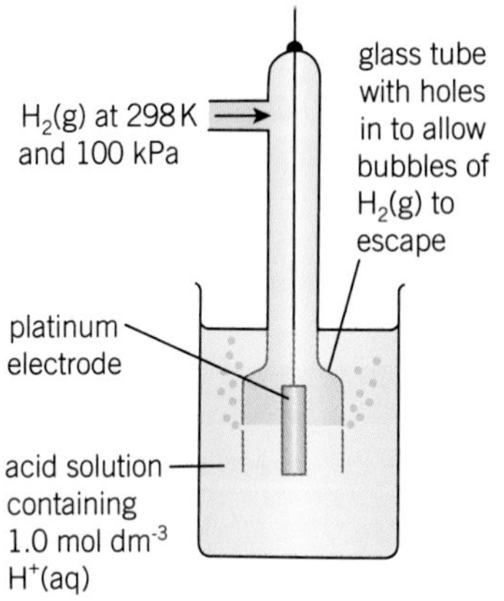

▲ **Figure 4** *Standard hydrogen electrode*

Measuring a standard electrode potential

To measure a standard electrode potential, the half-cell is connected to a standard hydrogen electrode.

- The two electrodes are connected by a wire to allow a controlled flow of electrons.
- The two solutions are connected with a **salt bridge** which allows ions to flow. The salt bridge typically contains a concentrated solution of an electrolyte that does not react with either solution. An example of a salt bridge is a strip of filter paper soaked in aqueous potassium nitrate, $KNO_3\,(aq)$.

Figure 5 shows the experimental set up for measuring the standard electrode potential of a copper half-cell.

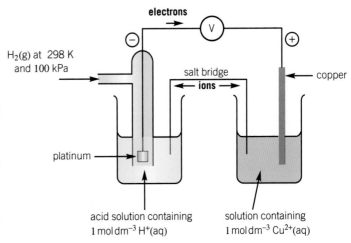

▲ **Figure 5** *Measuring a standard electrode potential*

23.4 Electrode potentials

Standard electrode potentials have been measured for many different redox systems and these are listed in data reference tables. Table 1 shows examples of standard electrode potentials. The equilibrium is always shown so that the forward reaction is reduction. The redox systems in Table 1 have been sorted in order of their standard electrode potentials, with the most negative value at the top.

The *more negative* the $E^\ominus$ value:

- the greater the tendency to *lose* electrons and undergo *oxidation*.
- the less the tendency to *gain* electrons and undergo *reduction*

The *more positive* the $E^\ominus$ value:

- the greater the tendency to *gain* electrons and undergo *reduction*
- the less the tendency to *lose* electrons and undergo *oxidation*.

Metals tend to have negative $E^\ominus$ values and lose electrons. Non-metals tend to have positive $E^\ominus$ values and gain electrons.

In general:

- The *more negative* the $E^\ominus$ value, the greater the *reactivity of a metal* in losing electrons.
- The *more positive* the $E^\ominus$ value, the greater the *reactivity of a non-metal* in gaining electrons.

$Mg^{2+}(aq) + 2e^- \rightleftharpoons Mg(s) \quad E = -2.37\,V$ (reduction / oxidation)

▲ **Figure 6** $Mg^{2+}(aq)|Mg$ has a very negative $E^\ominus$ value and a greater tendency to lose electrons

$Cl_2(g) + 2e^- \rightleftharpoons 2Cl^-(aq) \quad E = +1.36\,V$ (reduction / oxidation)

▲ **Figure 7** $Cl_2(g)|Cl^-(aq)$ has a very positive $E^\ominus$ value and a greater tendency to gain electrons

▼ **Table 1** Standard electrode potentials

Redox system	$E^\ominus$/V
$Mg^{2+}(aq) + 2e^- \rightleftharpoons Mg(s)$	−2.37
$Al^{3+}(aq) + 3e^- \rightleftharpoons Al(s)$	−1.66
$Zn^{2+}(aq) + 2e^- \rightleftharpoons Zn(s)$	−0.76
$Fe^{2+}(aq) + 2e^- \rightleftharpoons Fe(s)$	−0.44
$2H^+(aq) + 2e^- \rightleftharpoons H_2(g)$	0.00
$Cu^{2+}(aq) + 2e^- \rightleftharpoons Cu(s)$	+0.34
$I_2(aq) + 2e^- \rightleftharpoons 2I^-(aq)$	+0.54
$Fe^{3+}(aq) + e^- \rightleftharpoons Fe^{2+}(aq)$	+0.77
$Ag^+(aq) + e^- \rightleftharpoons Ag(s)$	+0.80
$Cl_2(g) + 2e^- \rightleftharpoons 2Cl^-(aq)$	+1.36

Cell potentials

Cells can easily be assembled using any half-cells. The e.m.f. measured is then a cell potential, E_{cell}. The procedure below describes how to set up cells to measure standard cell potentials $E^\ominus_{cell}$.

Measuring standard cell potentials

1. Prepare two standard half-cells.
   - For a metal/metal ion half-cell, the metal ion must have a concentration of 1 mol dm^{-3}.
   - For an ion/ion half-cell, both metal ions present in the solution must have the same concentration. There must be an inert electrode, usually platinum.
   - For a half-cell containing gases (e.g., a hydrogen half-cell), the gas must be at 100 kPa pressure, in contact with a solution with an ionic concentration of 1 mol dm^{-3}. There must be an inert electrode, usually platinum.
   - For all half-cells, the temperature must be at 298 K.
2. Connect the metal electrodes of the half-cells to a voltmeter using wires.
3. Prepare a salt bridge by soaking a strip of filter paper in a saturated aqueous solution of potassium nitrate, KNO_3.
4. Connect the two solutions of the half-cells with a salt bridge.
5. Record the standard cell potential from the voltmeter.

Study tip

In an ion/ion half-cell, both solutions can be 1 mol dm^{-3}, but it may be difficult to dissolve enough solute to get these concentrations. However equal ion concentrations gives the same e.m.f.

Figure 8 shows a standard cell made from $Zn^{2+}(aq)|Zn(s)$ and $Cu^{2+}(aq)|Cu(s)$ half-cells. The digital voltmeter displays the standard cell potential as 1.10 V. The cell potential is the potential difference between the two half-cells.

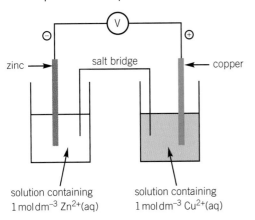

▲ **Figure 8** *A simple cell from $Zn^{2+}(aq)|Zn(s)$ and $Cu^{2+}(aq)|Cu(s)$ half-cells*

In the zinc–copper cell:

- The copper half-cell has the more positive $E^\ominus$ and a greater tendency to undergo reduction and to gain electrons.
- The zinc half-cell has the more negative $E^\ominus$ and a greater tendency to undergo oxidation and to lose electrons.
- Electrons flow along the wire from the more negative zinc half-cell to the less negative copper half-cell.
- The zinc electrode is negative and the copper electrode is positive.

An overall cell equation is written by combining the relevant equilibria from Table 1.

- The copper equilibrium has the more positive $E^\ominus$ and undergoes reduction, gaining electrons. The reduction half equation is the same way round as the equilibrium, with the electrons on the left-hand side.
- The zinc equilibrium has the more negative $E^\ominus$ and undergoes oxidation, losing electrons. So this equilibrium is reversed to give the oxidation half equation with the electrons on the right-hand side.
- The overall cell equation is obtained by combining the reduction and oxidation half-equations.

reduction: $Cu^{2+}(aq) + 2e^- \rightarrow Cu(s)$ positive electrode
oxidation: $Zn(s) \rightarrow Zn^{2+}(aq) + 2e^-$ negative electrode
overall cell reaction: $Zn(s) + Cu^{2+}(aq) \rightarrow Zn^{2+}(aq) + Cu(s)$

> 1 A standard cell is constructed from $Mg^{2+}(aq)|Mg(s)$ and $Fe^{2+}(aq), Fe^{3+}(aq)|Pt(s)$ half-cells. Refer to Table 1.
>   a From which electrode will the electrons flow?
>   b Write half-equations for the oxidation and reduction processes and the overall equation.
> 2 Suggest concentrations of $FeSO_4$ and $Fe_2(SO_4)_3$ for preparing a standard half-cell based on $Fe^{3+}(aq)$ and $Fe^{2+}(aq)$ ions.

Study tip

Remember that electrode potential equilibria are always written with reduction (gaining electrons) in the forward direction and oxidation (losing electrons) in the reverse direction.

Study tip

To get the half-equation for the redox system with the more negative electrode potential, the equilibrium is written the other way round with electrons of the right-hand side.

Calculation of a standard cell potential from standard electrode potentials

Standard electrode potentials essentially quantify the tendency of redox systems to gain or lose electrons. A standard cell potential can be calculated directly from standard electrode potentials. The calculation is just the difference between the $E^\ominus$ values:

$$E^\ominus_{cell} = E^\ominus(\text{positive electrode}) - E^\ominus(\text{negative electrode})$$

For a standard zinc–copper cell,

$$E^\ominus_{cell} = +0.34 - (-0.76) = 1.10 \text{ V}$$

Look back to Figure 8 – you can see that the measured e.m.f. on the voltmeter is the same as the calculated value using standard electrode potentials.

Summary questions

1. In an electrochemical cell, what are the charge carriers:
   a. in the wire *(1 mark)*
   b. in the salt bridge? *(1 mark)*

2. a. Define the term standard electrode potential. *(2 marks)*
   b. State the standard conditions used for electrochemical cells. *(1 mark)*

3. Draw a diagram to show the apparatus needed to measure the standard cell potential for a cell based on Fe^{2+}, Fe and Co^{2+}, Co^{3+} ions. *(3 marks)*

4. Refer to Table 1.
   a. Calculate the standard cell potential for the following cell from the redox systems in Table 1:
      i. $Cu^{2+}(aq)|Cu(s)$ and $Cl_2(g), Cl^-(aq)|Pt(s)$
      ii. $Al^{3+}(aq)|Al(s)$ and $Fe^{2+}(aq)|Fe(s)$
      iii. $Ag^+(aq)|Ag(s)$ and $Zn^{2+}(aq)|Zn(s)$ *(3 marks)*
   b. For each cell given in part a, write half-equations for the oxidation and reduction reactions and the overall cell reaction. *(9 marks)*

5. Four standard cells are set up using different half-cells. Use the results to determine the standard electrode potential of the redox systems in five half-cells used.

Positive half-cell	Negative half-cell	$E^\ominus_{cell}$ / V
$H^+(aq), H_2(g)|Pt(s)$	$Pb^{2+}(aq)|Pb(s)$	0.13
$Ni^{2+}(aq)|Ni(s)$	$Cd^{2+}(aq)|Cd(s)$	0.15
$Co^{3+}(aq), Co^{2+}(aq)|Pt(s)$	$Cd^{2+}(aq)|Cd(s)$	2.21
$Pb^{2+}(aq)|Pb(s)$	$Ni^{2+}(aq)|Ni(s)$	0.12

   (5 marks)

23.5 Predictions from electrode potentials

Specification reference: 5.2.3

Predicting redox reactions

Predictions can be made about the feasibility of any potential redox reactions using standard electrode potentials. Table 1 shows three redox systems sorted with the most negative standard electrode potential at the top.

- The *most negative* system has the greatest tendency to be *oxidised* and *lose* electrons.
- The *most positive* system has the greatest tendency to be *reduced* and *gain* electrons.

An oxidising agent takes electrons away from the species being oxidised. So oxidising agents are reduced and are on the left. A reducing agent adds electrons to the species being reduced. So reducing agents are oxidised and are on the right (Table 1).

You can predict the feasibility of redox reactions from $E^\ominus$ values.

In Table 1, a reaction should take place between an oxidising agent on the left and a reducing agent on the right, provided that the redox system of the oxidising agent has a more positive $E^\ominus$ value than the redox system of the reducing agent.

- The strongest reducing agent is at the top on the right.
- The strongest oxidising agent is at the bottom on the left

Comparing redox system C with A and B

Redox system **C** has a more positive $E^\ominus$ value and will have a greater tendency to be reduced than redox systems **A** and **B**.
So the oxidising agent $Ag^+(aq)$ on the left of **C** should react with reducing agents on the right in redox systems **A** and **B**, that is, with $Cr(s)$ and $Cu(s)$.

- **A** $Cr^{3+}(aq) + 3e^-$ ⟵ $Cr(s)$
- **B** $Cu^{2+}(aq) + 2e^-$ ⟵ $Cu(s)$
- **C** $Ag^+(aq) + e^-$ ⟶ $Ag(s)$

Comparing redox system B with A and C

Redox system **B** has a more positive $E^\ominus$ value and will have a greater tendency to be reduced than redox system **A**.
So the oxidising agent $Cu^{2+}(aq)$ on the left of **B** should react with reducing agents on the right in redox system **A**, that is, only with $Cr(s)$.

- **A** $Cr^{3+}(aq) + 3e^-$ ⟵ $Cr(s)$
- **B** $Cu^{2+}(aq) + 2e^-$ ⟶ $Cu(s)$

Comparing redox system A with B and C

The oxidising agent $Cr^{3+}(aq)$ on the left of **A** will not react as there are no redox systems with a less positive $E^\ominus$ value.

Learning outcomes

Demonstrate knowledge, understanding, and application of:

→ prediction of the feasibility of a reaction using standard cell potentials

→ limitations of feasibility predictions.

▼ **Table 1** *Standard electrode potentials: reducing and oxidising agents*

	Redox system	$E^\ominus / V$
A	$Cr^{3+}(aq) + 3e^- \rightleftharpoons Cr(s)$	−0.77
B	$Cu^{2+}(aq) + 2e^- \rightleftharpoons Cu(s)$	+0.34
C	$Ag^+(aq) + e^- \rightleftharpoons Ag(s)$	+0.80

(oxidation ⟵ reducing agent; oxidising agent ⟶ reduction)

Study tip

The explanation here is in terms of the strength of oxidising agents. This can also be explained in terms of reducing agents. Both approaches are equivalent and give the same answer.

23.5 Predictions from electrode potentials

> **Study tip**
>
> For practice, combine the half equations above to get these overall equations.

Overall equation

To write overall equations, the half-equations must be combined. The reduction half-equation is the same way round as the equilibrium. The oxidation half-equation is obtained by reversing the equilibrium.

Combining the possible redox reactions above gives the following equations:

$$Cr(s) + 3Ag^+(aq) \rightarrow Cr^{3+}(aq) + 3Ag(s)$$
$$Cu(s) + 2Ag^+ \rightarrow Cu^{2+}(aq) + 2Ag$$
$$2Cr(s) + 3Cu^{2+}(aq) \rightarrow 2Cr^{3+}(aq) + 3Cu(s)$$

You can apply this same principle to any pair of redox systems. If they are not ordered then just remember that:

- the redox system with the more positive (less negative) $E^\ominus$ value will react from left to right, and gain electrons
- the redox system with the less positive (more negative) $E^\ominus$ value will react from right to left, and lose electrons.

Limitations of predictions using $E^\ominus$ values

Reaction rate

One limitation of predictions for feasibility based on ΔG lies with reactions that have a very large activation energy, resulting in a very slow rate. This same limitation applied to predictions made based on E values.

So electrode potentials may indicate the thermodynamic feasibility of a reaction, but they give no indication of the rate of a reaction.

> **Synoptic link**
>
> For details of ΔG see Topic 22.5, Free energy.

Concentration

Standard electrode potentials are measured using concentrations of $1 \, mol \, dm^{-3}$. Many reactions take place using concentrated or dilute solutions. If the concentration of a solution is not $1 \, mol \, dm^{-3}$, then the value of the electrode potential will be different from the standard value.

For example, the redox equilibrium and standard electrode potential for the $Zn^{2+}|Zn$ redox system is shown below.

$$Zn^{2+}(aq) + 2e^- \rightleftharpoons Zn(s) \quad E^\ominus = -0.76 \, V$$

If the concentration of $Zn^{2+}(aq)$ is greater than $1 \, mol \, dm^{-3}$, the equilibrium will shift to the right, removing electrons from the system and making the electrode potential less negative.

In concentrations of $Zn^{2+}(aq)$ less than $1 \, mol \, dm^{-3}$, the equilibrium will shift to the left increasing electrons in the system and making the electrode potential more negative.

Any change to the electrode potential will affect the value of the overall cell potential.

Other factors

The actual conditions used for the reaction may be different from the standard conditions used to record $E^\ominus$ values. This will affect the value of the electrode potential.

Standard electrode potentials apply to aqueous equilibria. Many reactions take place that are not aqueous.

REDOX AND ELECTRODE POTENTIALS

23

Standard electrode potentials and ΔG

Feasibility has now been considered in terms of both ΔG and now E. It is perhaps unsurprising that ΔG and E are connected mathematically:

$$\Delta G^\ominus = -nFE^\ominus_{cell}$$

Where n is amount, in moles, of electrons transferred in the balanced equation and F is the Faraday constant, $96\,500$ C mol^{-1}.

When a cell reaction is written, $E^\ominus_{cell}$ is positive, calculated from:

$$E^\ominus_{cell} = E^\ominus \text{(positive electrode)} - E^\ominus \text{(negative electrode)}$$

For a standard cell constructed from $Zn^{2+}(aq)|Zn(s)$ and $Cu^{2+}(aq)|Cu(s)$ half-cells the cell reaction and value of $E^\ominus_{cell}$:

$$Zn(s) + Cu^{2+}(aq) \rightarrow Zn^{2+}(aq) + Cu(s) \qquad E^\ominus_{cell} = 1.10\text{ V}$$

2 mol of electrons are transferred, so:

$$\Delta G^\ominus = -nFE^\ominus_{cell} = -2 \times 96\,500 \times 1.10 = -212\,300\text{ J} = -212\text{ kJ}$$

Providing that $E^\ominus_{cell}$ is positive, ΔG will be negative. The smaller the value of $E^\ominus_{cell}$, the less negative the value of ΔG.

> **Study tip**
>
> The final answer has been rounded to three significant figures to match the least significant measurement of 1.10 V.

Two standard electrode potentials are shown below.

$Fe^{2+}(aq) + 2e^- \rightleftharpoons Fe(s) \qquad E^\ominus = -0.44\text{ V}$

$Cd^{2+}(aq) + 2e^- \rightleftharpoons Cd(s) \qquad E^\ominus = -0.40\text{ V}$

a Calculate $E^\ominus_{cell}$ and write the cell equation that produced this value.

b Calculate $\Delta G^\ominus$ for this cell. Comment on its value.

Summary questions

1. State the limitations of predictions for feasibility based on $E^\ominus$ values. *(2 marks)*

2. You are provided with the following information.

   $U^{3+}(aq) + 3e^- \rightleftharpoons U(s) \qquad E^\ominus = -1.80\text{ V}$
   $V^{3+}(aq) + e^- \rightleftharpoons V^{2+}(aq) \qquad E^\ominus = -0.26\text{ V}$
   $Fe^{3+}(aq) + e^- \rightleftharpoons Fe^{2+}(aq) \qquad E^\ominus = +0.77\text{ V}$
   $Br_2(l) + 2e^- \rightleftharpoons Br^-(aq) \qquad E^\ominus = +1.09\text{ V}$

   a What is the: **i** strongest oxidising agent **ii** strongest reducing agent? *(2 marks)*
   b i Which of the following would be predicted to react with Fe^{3+}?
       $U^{3+}(aq) \quad Br^-(aq) \quad U(s) \quad V^{3+}(aq) \quad V^{2+}(aq)$
   ii Write equations for any predicted reactions from **i**. *(4 marks)*

3. You are provided with the following information.

   $Fe^{3+}(aq) + e^- \rightleftharpoons Fe^{2+}(aq) \qquad E^\ominus = +0.77\text{ V}$
   $Cr_2O_7^{2-}(aq) + 14H^+(aq) + 6e^- \rightleftharpoons 2Cr^{3+}(aq) + 7H_2O(l) \qquad E^\ominus = +1.33\text{ V}$
   $Cl_2(g) + 2e^- \rightleftharpoons 2Cl^-(aq) \qquad E^\ominus = +1.36\text{ V}$

   Predict feasible redox reactions based on these $E^\ominus$ values and write equations for the reactions. *(3 marks)*

23.6 Storage and fuel cells

Specification reference: 5.2.3

Learning outcomes

Demonstrate knowledge, understanding, and application of:

→ modern storage cells
→ fuel cells.

▲ **Figure 1** *A mixture of storage cells. You can see the types on the labels*

Study tip

The MnO_2/Mn_2O_3 system is more positive and undergoes reduction, providing the electrons. The Zn/ZnO system is less positive and the equilibrium equation is reversed. So zinc is oxidised and manganese in MnO_2 is reduced.

Modern storage cells

Modern cells and batteries

Our modern-day cells and batteries are based on the chemistry introduced in Topic 23.4. The key requirement is for the two electrodes to have different electrode potentials. You do not need to learn details of any specific cell but you do need to apply the important principles of electrode potentials, half-equations, and cell reactions.

Cells can be divided into three main types – primary, secondary, and fuel cells. Modern cells and batteries are part of a fast-moving technological field, and much research is being carried out to develop more efficient and effective portable sources of electrical energy.

Primary cells

Primary cells are non-rechargeable and are designed to be used once only. When in use, electrical energy is produced by oxidation and reduction at the electrodes. However, the reactions cannot be reversed. Eventually the chemicals will be used up, voltage will fall, the battery will go flat, and the cell will be discarded or recycled. Primary cells still find use for low-current, long-storage devices such as wall clocks and smoke detectors. Most modern primary cells are alkaline based on zinc and manganese dioxide, Zn/MnO_2, and a potassium hydroxide alkaline electrolyte. The $E^\ominus$ values for the redox systems are shown below.

$ZnO(s) + H_2O(l) + 2e^- \rightleftharpoons Zn(s) + 2OH^-(aq)$ $\qquad E^\ominus = -1.28\,V$

$2MnO_2(s) + H_2O(l) + 2e^- \rightleftharpoons Mn_2O_3(s) + 2OH^-(aq)$ $\qquad E^\ominus = +0.15\,V$

Oxidation: $Zn(s) + 2OH^-(aq) \rightarrow ZnO(s) + H_2O(l) + 2e^-$

Reduction: $2MnO_2(s) + H_2O(l) + 2e^- \rightarrow Mn_2O_3(s) + 2OH^-(aq)$

Cell reaction: $Zn(s) + 2MnO_2(s) \rightarrow ZnO(s) + Mn_2O_3(s)$ $\quad E^\ominus_{cell} = 1.43\,V$

Secondary cells

Secondary cells are rechargeable. Unlike primary cells, the cell reaction producing electrical energy can be reversed during recharging. The chemicals in the cell are then regenerated and the cell can be used again.

Common examples of secondary cells include:

- lead–acid batteries used in car batteries
- nickel–cadmium, NiCd, cells and nickel–metal hydride, NiMH – the cylindrical batteries used in radios, torches, and so on.
- lithium-ion and lithium-ion polymer cells used in our modern appliances – laptops, tablets, cameras, mobile phones – and also being developed for cars.

23 REDOX AND ELECTRODE POTENTIALS

Lithium-ion and lithium-ion polymer cells

Lithium-ion cells are extremely popular.

Lithium is a light metal, which translates into a very high energy density when used in lithium-ion batteries.

The cells can be a regular shape, as in camera cells and also as lithium-ion polymer pouch batteries, with an internal salt bridge made of a micro-porous polymer covered in an electrolytic gel. Because the solid polymer is flexible, flexible batteries can be easily formed into various shapes and sizes, ideal for fitting round other components in a tightly-packed and lightweight laptop, tablet, or mobile phone.

When a lithium-ion cell charges and discharges, electrons move through the connecting wires to power the appliance, whilst Li^+ ions move between the electrodes within the cell (Figure 2). The negative electrode is made of graphite coated with lithium metal. The positive electrode is made out of a metal oxide, typically CoO_2. The simplified reactions at each electrode during use are shown below.

negative electrode: $\quad Li \rightarrow Li^+ + e^-$
positive electrode: $\quad Li^+ + CoO_2 + e^- \rightarrow LiCoO_2$

Figure 2 shows movement of Li^+ ions during discharge and a modern lithium-ion polymer battery.

When fully charged, a lithium-ion cell has a voltage of 4.2 V but this drops with use. The typical operating voltage is about 3.7–3.8 V.

Lithium-ion cells do have some limitations. They can become unstable at high temperatures and, on very rare occasions, they have ignited laptops and mobile phones. Care must be taken in their recycling, as lithium is a very reactive metal.

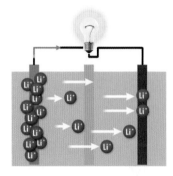

▲ **Figure 2** *The structure of a typical lithium-ion rechargeable battery. The battery consists of a positive electrode (green) and negative electrode (red), with a layer (yellow) separating them. When in use, electrons (green arrow) flow from the negative electrode to the positive electrode and, at the same time, Li^+ ions (blue) travel from the anode to the cathode, to keep the electrical charge balance in the cell. When charging, the process is reversed and lithium ions are transferred back to the negative electrode*

> The standard electrode potential for $Li^+ + e^- \rightleftharpoons Li$ is -3.04 V
>
> **a** Assuming this value, what is the electrode potential of the CoO_2 electrode at the fully charged state of the cell?
>
> **b** Assuming that Co changes oxidation number, what is the change?

Fuel cells

A fuel cell uses the energy from the reaction of a fuel with oxygen to create a voltage.

- The fuel and oxygen flow into the fuel cell and the products flow out. The electrolyte remains in the cell.
- Fuel cells can operate continuously provided that the fuel and oxygen are supplied into the cell.
- Fuel cells do not have to be recharged.

Many different fuels can be used, but hydrogen is the most common. Hydrogen fuel cells produce no carbon dioxide during combustion, with water being the only combustion product. Fuel cells using many other hydrogen-rich fuels, such as methanol, are also being developed.

▲ **Figure 3** *Hydrogenesis fuel-cell ferry. This is the UK's first fuel-cell ferry, shown in Bristol docks, Bristol, UK. The fuel-cell is fuelled by compressed hydrogen gas and oxygen. The only emission is water*

23.6 Storage and fuel cells

Hydrogen fuel cells

A hydrogen cell can have either an alkali or acid electrolyte. The fuel cells, half-equations and $E^\ominus$ values for the redox systems are shown in Table 1 below. The cell voltages are both 1.23 V, despite the alkali and acid cells having different redox systems and half-equations.

▼ Table 1 *Hydrogen fuel cells*

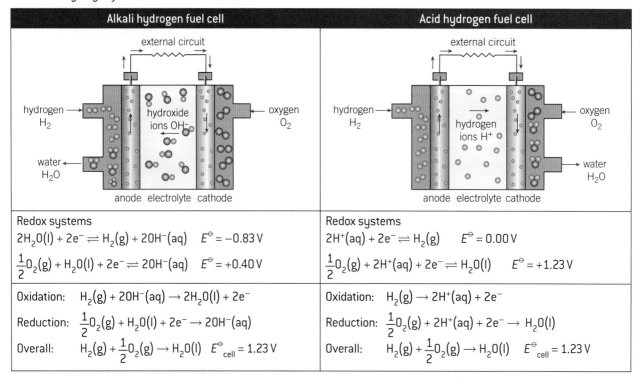

Summary questions

1. State the key difference between a primary cell and a secondary cell. *(1 mark)*
2. State the key characteristics of a fuel cell. *(2 marks)*
3. The redox equilibria for a Ni–Cd cell with $E^\ominus_{cell}$ = 1.30 V are shown below.

   $Cd(OH)_2 + 2e^- \rightleftharpoons Cd + 2OH^-$ $E^\ominus = -0.81$ V

   $2NiO(OH) + 2H_2O + 2e^- \rightleftharpoons 2Ni(OH)_2 + 2OH^-$

   a. Write the equation for the overall cell reaction. (Cd is the negative electrode.) *(1 mark)*
   b. Use oxidation numbers to show the elements that are oxidised and reduced in the working cell. *(2 marks)*
   c. Calculate the standard electrode potential for the half equation involving NiO(OH). *(1 mark)*
   d. Construct the overall equation when the cell is being charged. *(1 mark)*

4. In a methanol fuel cell, the overall reaction is the combustion of methanol.
   The reaction at the positive electrode is
   $O_2 + 4H^+ + 4e^- \rightarrow 2H_2O$
   a. Write an equation for the complete combustion of methanol. *(1 mark)*
   b. Deduce the half-equation for the reaction at the negative electrode in a methanol fuel cell. *(2 marks)*

REDOX AND ELECTRODE POTENTIALS

Practice questions

1. Manganese forms compounds in which manganese has several oxidation numbers.

   a. Manganese(IV) oxide, MnO_2, can be converted into manganate(VI), MnO_4^{2-}, with hot alkaline chlorate(VII) ions, ClO_3^-.

      (i) Unbalanced oxidation and reduction half-equations for this reaction are shown below.
      $$MnO_2 + OH^- \rightarrow MnO_4^{2-} + H_2O$$
      $$H_2O + ClO_3^- \rightarrow OH^- + Cl^-$$
      Balance the equations and add the correct number of electrons to the correct sides of the two half equations. *(2 marks)*

      (ii) Using your half equations, construct the overall equation for this reaction. *(1 mark)*

   b. Manganate(VI), MnO_4^{2-}, disproportionates when it reacts with hot water to form an alkaline solution containing manganese(IV) oxide, MnO_2, and manganate(VII), MnO_4^-. Construct the equation for this disproportionation reaction. *(2 marks)*

   F325 June 2012 Q7

2. Aqueous potassium manganate(VII), $KMnO_4$, in acidic conditions can be used in analysis. A student analyses a sample of sodium sulfite, Na_2SO_3, using the following method.

   The student prepares 250.0 cm³ of a solution of the impure Na_2SO_3. 25.0 cm³ samples are acidified and titrated using 0.0150 mol dm⁻³ $KMnO_4$.

   The equation for the oxidation of Na_2SO_3 is given below.
   $$2MnO_4^-(aq) + 6H^+(aq) + 5SO_3^{2-}(aq) \rightarrow 2Mn^{2+}(aq) + 5SO_4^{2-}(aq) + 3H_2O(l)$$

   Mass readings

Mass of weighing bottle / g	14.47
Mass of weighing bottle + impure Na_2SO_3 / g	12.54

   Titration readings:

	Trial	1	2	3
Final burette reading / cm³	28.70	27.85	37.50	32.75
Initial burette reading / cm³	0.00	0.00	10.00	5.00

   The reduction half equation for acidified manganate(VII) ions is shown below.
   $$MnO_4^-(aq) + 8H^+(aq) + 5e^- \rightarrow Mn^{2+}(aq) + 4H_2O(l)$$

   a. What is the oxidation half-equation for the sulfite ion, $SO_3^{2-}(aq)$, in this titration? *(1 mark)*

   b. What colour change takes place at the end point? *(1 mark)*

   c. Using the results, determine the percentage of sodium sulfite, $Na_2S_2O_3$, in the impure sample. Show all your working. *(7 marks)*

3. The copper content in brass is determined using the procedure below.

   A 2.56 g sample of brass is reacted with an excess of concentrated nitric acid. In this process the copper is converted to Cu^{2+}.

   This solution is neutralised and the resulting solution is made up to 250 cm³.

   A 25.0 cm³ sample of this solution is pipetted into a conical flask and an excess of aqueous potassium iodide is added. The reaction below takes place.
   $$Cu^{2+}(aq) + 2I^-(aq) \rightarrow CuI(s) + \frac{1}{2}I_2(aq)$$

   The iodine formed is titrated with 0.100 mol dm⁻³ thiosulfate ions, $S_2O_3^{2-}(aq)$. 29.8 cm³ is required to reach the end point.
   $$I_2(aq) + 2S_2O_3^{2-}(aq) \rightarrow 2I^-(aq) + S_4O_6^{2-}(aq)$$

   a. When the brass is reacted with nitric acid, a redox reaction takes place.

      The unbalanced equation for this reaction is shown below.
      $$Cu + HNO_3 \rightarrow Cu(NO_3)_2 + NO_2 + H_2O$$

      (i) Using oxidation numbers, show which element has been oxidised and which has been reduced.

      (ii) Balance the equation *(3 marks)*

   b. Explain how you carry out the titration as the end point is approached and how you would detect the end point of this titration. *(2 marks)*

   c. Determine the percentage, by mass, of copper in the brass. Give your answer to **one** decimal place. *(5 marks)*

d Another sample of brass is heated with concentrated sulfuric acid. The copper reacts to produce a blue solution and 90.0 cm³ of a gas **A**, measured at RTP. The mass of the gas **A** collected is 0.240 g.

Suggest a possible identity of gas **A**. Show your working.

Predict the balanced equation for this reaction. *(4 marks)*

4 Redox reactions are used to generate electrical energy from electrochemical cells.

a The table below shows three redox systems, and their standard electrode potentials.

	Redox system	$E^\ominus$/V
A	$Cr^{3+}(aq) + 3e^- \rightleftharpoons Cr(s)$	−0.74
B	$Sn^{4+}(aq) + 2e^- \rightleftharpoons Sn^{2+}(aq)$	+0.15
C	$Ag^+(aq) + e^- \rightleftharpoons Ag(s)$	+0.80

(i) Define the term *standard electrode potential*. *(2 marks)*

(ii) Draw a labelled diagram to show how the standard electrode potential of a Sn^{4+}/Sn^{2+} redox system could be measured. *(3 marks)*

(iii) Using the information in the table, write equations for the reactions that are feasible.

Suggest **two** reasons why these reactions may not actually take place. *(5 marks)*

b From the redox systems in the table, what is the strongest oxidising agent? *(1 mark)*

c A student sets up the standard cell based on two of the redox systems **A–C** in the table which gives the greatest cell potential.

Which redox systems would be used and what is the cell potential? *(1 mark)*

F325 Jan 2011 Q4

5 Modern fuel cells are being developed as an alternative to the direct use of fossil fuels.

The fuel can be hydrogen but many other substances are being considered.

In a methanol fuel cell, the overall reaction is the combustion of methanol.

As with all fuel cells, the reducing agent is supplied at one electrode and the oxidising agent at the other electrode.

a What is meant by the term *oxidising agent*? *(1 mark)*

b Oxygen reacts at one electrode of a methanol fuel cell:

$O_2 + 4H^+ + 4e^- \rightarrow 4H_2O$

(i) Explain with a reason whether oxygen is produced at the positive or negative electrode. *(1 mark)*

(ii) Write an equation for the complete combustion of methanol. *(1 mark)*

(iii) Deduce the half-equation for the reaction that takes place at the other electrode in a methanol fuel cell. *(1 mark)*

c Suggest **one** advantage of using methanol, rather than hydrogen, in a fuel cell for vehicles. Justify your answer. *(1 mark)*

F325 Jan 2011 Q4 adapted

6. Standard electrode potentials for six redox systems are shown in the table below.

	Redox system	$E^\ominus$/V
1	$Mg^{2+}(aq) + 2e^- \rightleftharpoons Mg(s)$	−2.37
2	$Cr^{3+}(aq) + 3e^- \rightleftharpoons Cr(s)$	−0.74
3	$Fe^{3+}(aq) + e^- \rightleftharpoons Fe^{2+}(aq)$	−0.44
4	$Ti^{3+}(aq) + e^- \rightleftharpoons Ti^{2+}(aq)$	−0.37
5	$Ag^+(aq) + e^- \rightleftharpoons Ag(s)$	+0.80
6	$O_2(g) + 4H^+(aq) + 4e^- \rightleftharpoons 2H_2O(l)$	+1.23

a A standard cell can be made based on redox systems **1** and **3**.

(i) Draw a labelled diagram to show how this standard cell can be set up in the laboratory. *(3 marks)*

(ii) State the charge carriers that transfer current through the wire and through the solution. *(1 mark)*

(iii) Write down the cell potential and state the negative electrode. *(1 mark)*

b Select from the table:
  (i) the species which reduces Ti^{3+}(aq) but does **not** reduce Cr^{3+}(aq).
  (ii) the strongest reducing agent? *(2 marks)*

c Write the overall cell reaction for a cell based on redox systems **2** and **6**. *(1 mark)*

7 The standard electrode potentials in the table below can be used to predict redox reactions.

	Redox system	$E^\ominus$/V
1	$Cr^{3+}(aq) + 3e^- \rightleftharpoons Cr(s)$	−0.74
2	$Fe^{3+}(aq) + e^- \rightleftharpoons Fe^{2+}(aq)$	−0.44
3	$Cr_2O_7^{2-}(aq) + 14H^+(aq) + 6e^- \rightleftharpoons 2Cr^{3+}(aq) + 7H_2O(l)$	+1.33

a Define the term standard electrode potential. *(2 marks)*

b Using the information in the table, write equations for the reactions that are feasible. *(3 marks)*

c A student sets up a standard cell using half cells based on redox systems **1** and **3** in the table.
  (i) Draw a labelled diagram to show how the standard cell potential of this cell could be measured. *(3 marks)*
  (ii) State the cell potential and the sign of the electrode in redox system **1** in the cell. *(1 mark)*
  (iii) The half-cell based on redox system **1** is diluted by adding distilled water. Predict what would happen to the cell potential. Explain your reasoning. *(3 marks)*

d Platinum is an extremely unreactive transition element. However, platinum does take part in a redox reaction with *aqua regia*, a mixture of concentrated hydrochloric and nitric acids. Two products of this reaction are hexachloroplatinic acid, H_2PtCl_6, and nitrogen dioxide, NO_2.

  (i) Use oxidation states to show that this is a redox reaction. *(2 marks)*
  (ii) Write an equation for the reaction of platinum metal with *aqua regia*. *(1 mark)*

OCR F325 JUN 2012 Q5c

8 Brass is an alloy which contains copper.

The percentage of copper in brass can be determined using the steps below.

1. 2.80 g of brass is reacted with an excess of concentrated nitric acid, HNO_3.

   The half equations taking place are shown below.

   $Cu(s) \rightarrow Cu^{2+}(aq) + 2e^-$

   $2HNO_3(l) + e^- \rightarrow NO_3^-(aq) + NO_2(g) + H_2O(l)$

2. Excess aqueous sodium carbonate is added to neutralise any acid. The mixture effervesces and a precipitate forms.

3. The precipitate is reacted with ethanoic acid to form a solution which is made up to 250 cm³ with water.

4. A 25.0 cm³ sample of the solution is pipetted into a conical flask and an excess of aqueous potassium iodide is added.

   A precipitate of copper(I) iodide and a solution of iodine, I_2(aq), forms.

5. The resulting mixture is titrated with 0.100 mol dm⁻³ sodium thiosulfate to estimate the iodine present:

   $I_2(aq) + 2S_2O_3^{2-}(aq) \rightarrow 2I^-(aq) + S_4O_6^{2-}(aq)$

6. Steps 4 and 5 are repeated to obtain an average titre of 29.8 cm³.

  a For steps 1, 2, and 4, write overall equations for the reactions taking place. *(4 marks)*
  b Determine the percentage, by mass, of copper in the brass.

  Give your answer to one decimal place. *(5 marks)*

F325 June 11 Q8

24 TRANSITION ELEMENTS
24.1 d-block elements

Specification reference: 5.3.1

Learning outcomes

Demonstrate knowledge, understanding, and application of:

→ d-block elements and transition elements

→ electron configurations of atoms and ions

→ variable oxidation state, the formation of coloured compounds, and catalytic behaviour.

The d-block elements

The **d-block elements** are located between Group 2 and Group 13 of the periodic table. Across the periodic table from scandium to zinc, the 3d sub-shell has the highest energy and electrons are added to 3d orbitals – hence the name d-block elements.

In this topic you will study the first row of the d-block elements, scandium to zinc in Period 4 of the periodic table (Figure 1).

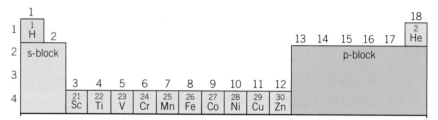

▲ Figure 1 *Part of the periodic table showing the first row of the d-block elements*

The d-block elements are all metallic, displaying the typical physical properties of metals. They have high melting points and boiling points, are shiny in appearance, and conduct both electricity and heat. Their metallic properties give rise to some of their uses.

Elements such as copper, silver, nickel, and zinc have been used in coinage for many years – iron is used extensively in construction and the production of tools, and copper for electrical cables and water pipes. Titanium is known for its great strength and is used in the aerospace industry as well as in medical applications such as joint replacement and in cosmetic dentistry. Figure 2 shows some of the common uses of d-block elements.

▲ Figure 2 *Most of the coinage used in the UK is a mixture of copper, nickel, and zinc, whilst iron is used in construction and tools and copper is used in electrical installations and plumbing*

Electron configuration of d-block elements and their ions

The electron configuration of an atom or ion shows the arrangement of electrons in shells and sub-shells. Electrons occupy orbitals in order of increasing energy (Figure 3).

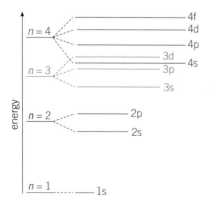

▲ Figure 3 *Energy level diagram showing the overlap of the 3d and 4s sub-shells*

The electron configurations of the elements scandium to zinc are shown in Table 1. Notice how the 3d sub-shell is being filled.

▼ **Table 1** *The electron configurations of the elements from scandium to zinc*

Element	Number of electrons	Electron configuration
scandium	21	$1s^2 2s^2 2p^6 3s^2 3p^6 3d^1 4s^2$
titanium	22	$1s^2 2s^2 2p^6 3s^2 3p^6 3d^2 4s^2$
vanadium	23	$1s^2 2s^2 2p^6 3s^2 3p^6 3d^3 4s^2$
chromium	24	$1s^2 2s^2 2p^6 3s^2 3p^6 3d^5 4s^1$
manganese	25	$1s^2 2s^2 2p^6 3s^2 3p^6 3d^5 4s^2$
iron	26	$1s^2 2s^2 2p^6 3s^2 3p^6 3d^6 4s^2$
cobalt	27	$1s^2 2s^2 2p^6 3s^2 3p^6 3d^7 4s^2$
nickel	28	$1s^2 2s^2 2p^6 3s^2 3p^6 3d^8 4s^2$
copper	29	$1s^2 2s^2 2p^6 3s^2 3p^6 3d^{10} 4s^1$
zinc	30	$1s^2 2s^2 2p^6 3s^2 3p^6 3d^{10} 4s^2$

Synoptic link

See Topic 5.1, Electron structure, for details of filling orbitals.

Study tip

Remember that the 4s sub-shell has a lower energy than the 3d sub-shell. Atoms of d-block elements fill their 4s sub-shell before they fill their 3d sub-shell.

The special case of chromium and copper

In Table 1 the electron configurations of chromium and copper do not follow the expected principle for placing electrons singly in orbitals before pairing. A simplistic explanation is one of *stability*. It is believed that a half-filled d^5 sub-shell and a fully filled d^{10} sub-shell give additional stability to atoms of chromium and copper.

Electron configuration of d-block ions

When the d-block elements, scandium to zinc, form positive ions from their atoms, they lose their 4s electrons before losing any of the 3d electrons. This means that:

- when forming an atom, the 4s orbital *fills* before the 3d orbitals
- when forming an ion, the 4s orbital *empties* before the 3d orbitals.

The electron configurations of an iron atom and its common ions, Fe^{2+} and Fe^{3+}, are shown below.

Fe atom $1s^2 2s^2 2p^6 3s^2 3p^6 3d^6 4s^2$

Fe^{2+} ion $1s^2 2s^2 2p^6 3s^2 3p^6 3d^6$ Two 4s electrons lost.

Fe^{3+} ion $1s^2 2s^2 2p^6 3s^2 3p^6 3d^5$ Two 4s and one 3d electron lost

Study tip

When forming ions of d-block elements, the 4s electrons are removed before any of the 3d electrons are removed.

Transition elements

Transition elements are defined as d-block elements that form at least one ion with a partially filled d-orbital. Although scandium and zinc are d-block elements, they do not match this definition and are not classified as transition elements.

Scandium only forms the ion Sc^{3+} by loss of two 4s electrons and one 3d electron.

- The electron configuration of Sc is $1s^2 2s^2 2p^6 3s^2 3p^6 3d^1 4s^2$.
- Sc^{3+} has an electronic configuration of $1s^2 2s^2 2p^6 3s^2 3p^6$.

Zinc only forms the Zn^{2+} ion by the loss of its two 4s electrons.
- The electron configuration of Zn is $1s^2 2s^2 2p^6 3s^2 3p^6 3d^{10} 4s^2$.
- Zn^{2+} has an electronic configuration of $1s^2 2s^2 2p^6 3s^2 3p^6 3d^{10}$.

Sc^{3+} ions have empty d-orbitals and Zn^{2+} ions have full d-orbitals. So scandium and zinc do not form ions with partially filled d-orbitals and are not therefore classified as transition elements.

Properties of the transition metals and their compounds

The transition elements have a number of characteristic properties that are different from other metals.
- They form compounds in which the transition element has different oxidation states.
- They form coloured compounds.
- The elements and their compounds can act as catalysts.

Variable oxidation states

Transition elements form compounds with more than one oxidation state. For example, iron forms two chlorides – iron(II) chloride, $FeCl_2$, and iron(III) chloride, $FeCl_3$.

The number of oxidation states increases across the transition elements series to manganese, and then decreases. All of the transition elements form compounds with an oxidation number of +2, resulting from the loss of two electrons. Each oxidation state often has a characteristic colour. The common oxidation states of the ions of transition elements are shown in Table 2, together with their typical colours.

▼ Table 2 *Common oxidation states and colours in transition element compounds and ions*

Element	Common oxidation states and their colours					
Sc		+3	By definition not a transition element			
Ti	+2	+3	+4	+5		
V	+2	+3	+4	+5		
Cr	+2	+3	+4	+5	+6	
Mn	+2	+3	+4	+5	+6	+7
Fe	+2	+3	+4	+5	+6	
Co	+2	+3	+4	+5		
Ni	+2	+3	+4			
Cu	+1	+2	+3			
Zn	+2	By definition not a transition element				

> **Study tip**
> You are expected to be able to list these properties and give suitable examples.

▲ Figure 4 *Transition metal compounds are often brightly coloured solids.*

A species containing a transition element in its highest oxidation state is often a strong oxidising agent.

TRANSITION ELEMENTS 24

Worked example: Oxidation numbers of ions containing transition elements

Find the oxidation number of the transition element in each of the ions MnO_4^-, $Cr_2O_7^{2-}$, and VO_3^-.

MnO_4^- $(Mn) + (-2 \times 4) = -1$ Mn has an oxidation number of +7.

$Cr_2O_7^{2-}$ $(Cr \times 2) + (-2 \times 7) = -2$ Each Cr has an oxidation number of +6.
   $2Cr = +12$.

VO_3^- $(V) + (-2 \times 3) = -1$ V has an oxidation number of +5.

Study tip
You will be expected to work out the oxidation states of transition metal ions in compounds using the rules for assigning oxidation states that you first met in Topic 4.3, Redox.

Formation of coloured compounds

Compounds and ions of transition elements are frequently coloured. Potassium dichromate(VI) is bright orange, cobalt(II) chloride is pink-purple, nickel(II) sulfate is green, whilst hydrated copper(II) sulfate is blue. The solid compounds can be dissolved in water to produce coloured solutions.

The colour of a solution is linked to the partially filled d-orbitals of the transition metal ion. The colour of a solution can vary with different oxidation states.

For example, iron forms two common oxidation states, +2 and +3, in its compounds. The oxidation states form different coloured solutions:

Iron(II) Fe^{2+} $1s^2 2s^2 2p^6 3s^2 3p^6 3d^6$ – pale green

Iron(III) Fe^{3+} $1s^2 2s^2 2p^6 3s^2 3p^6 3d^5$ – yellow

Chromium forms two common oxidation states, +3 and +6 in its compounds. The oxidation states form different coloured solutions:

Cr(III) – green

Cr(VI) – yellow or orange

▲ Figure 5 Transition metal compounds dissolve in water to form brighly coloured solutions

▲ Figure 6 The colours of iron in oxidation state +2 (left) and +3 (right)

Transition metals as catalysts

A catalyst is a substance that increases the rate of a chemical reaction without itself changing. A catalyst works by providing an alternative reaction pathway with a lower activation energy.

Transition metals and their compounds are important catalysts used in a variety of industrial processes and in the laboratory.

- The Haber process for the manufacture of ammonia from the reaction between nitrogen and hydrogen. The reaction is catalysed by a finely divided iron catalyst:

    $N_2(g) + 3H_2(g) \rightleftharpoons 2NH_3(g)$

- The Contact process in the production of sulfur trioxide from the oxidation of sulfur dioxide. The reaction is catalysed by vanadium(V) oxide, $V_2O_5(s)$:

    $2SO_2(g) + O_2(g) \rightleftharpoons 2SO_3(g)$

▲ Figure 7 The colours of chromium in oxidation state as $Cr_2O_7^{2-}$, +6 (left) and as Cr^{3+}, +3 (right)

24.1 d-block elements

- The hydrogenation of vegetable fats in the manufacture of margarine uses nickel as the catalyst:

$$CH_2=CH_2 + H_2 \xrightarrow{Ni} CH_3-CH_3$$

- The catalytic decomposition of hydrogen peroxide forming oxygen uses manganese(IV) oxide, $MnO_2(s)$, as the catalyst:

$$2H_2O_2(aq) \rightarrow 2H_2O(l) + O_2(g)$$

These are all examples of **heterogeneous catalysis**, as the catalyst is in a different state to the reactants. However, some chemical reactions are catalysed by **homogeneous catalysts** where the catalyst is in the same physical state as the reactants. For example, the reaction between iodide ions and peroxodisulfate ions, $S_2O_8^{2-}$, is catalysed by $Fe^{2+}(aq)$ ions, with reactants and catalyst all in aqueous solution. The overall reaction is shown below.

$$S_2O_8^{2-}(aq) + 2I^-(aq) \rightarrow 2SO_4^{2-}(aq) + I_2(aq)$$

When the reaction is carried out with a trace of starch, a blue-black colour forms showing the formation of iodine. If this experiment is repeated, with a small amount of $Fe^{2+}(aq)$ added, the blue-black solution forms much more quickly demonstrating the catalytic action of the transition metal ion. The equations below show how $Fe^{2+}(aq)$ ions catalyse the reaction:

$Fe^{2+}(aq)$ reacts $\qquad S_2O_8^{2-}(aq) + Fe^{2+}(aq) \rightarrow 2SO_4^{2-}(aq) + Fe^{3+}(aq)$

$Fe^{2+}(aq)$ regenerated $\quad Fe^{3+}(aq) + 2I^-(aq) \rightarrow I_2(aq) + Fe^{2+}(aq)$

So although Fe^{2+} is used up in the first step, it is regenerated in the second step. Overall Fe^{2+} ions are not consumed.

The reaction of zinc metal with acids is catalysed by the presence of $Cu^{2+}(aq)$ ions, for example:

$$Zn(s) + H_2SO_4(aq) \rightarrow ZnSO_4(aq) + H_2(g)$$

> **Synoptic link**
>
> Catalysis was discussed in detail in Topic 10.2, Catalysts.

> **Study tip**
>
> You should be able to combine the equations from each step to obtain the overall equation.

Summary questions

1. State three properties of the transition elements that are different from other metals. *(3 marks)*

2. Scandium is classified as a d-block metal but not as a transition metal. Explain why. *(2 marks)*

3. What is the oxidation state of:
   a. iron in Fe_2O_3 *(1 mark)*
   b. chromium in K_2CrO_4 *(1 mark)*
   c. vanadium in $VOCl_2$ *(1 mark)*
   d. copper in $CuCl_4^{2-}$ *(1 mark)*

4. Explain why zinc compounds dissolve in water to form colourless solutions. *(2 marks)*

24.2 The formation and shapes of complex ions

Specification reference: 5.3.1

Complex ions

One of the most important properties of the d-block elements is their ability to form **complex ions**. When solid hydrated copper(II) sulfate, $CuSO_4 \cdot 5H_2O$, is dissolved in water, a blue solution forms containing the complex ion $[Cu(H_2O)_6]^{2+}$ (Figure 1). Complex ions are not restricted to d-block elements – other elements such as aluminium can also form complex ions.

Learning outcomes

Demonstrate knowledge, understanding, and application of:

→ the formation of dative covalent bonds between metal ions and ligands in the formation of complex ions

→ examples of complex ions with four and six coordination and their associated shapes.

The formation of complex ions

A complex ion is formed when one or more molecules or negatively charged ions bond to a central metal ion. These molecules or ions are known as **ligands**.

- A ligand is defined as a molecule or ion that donates a pair of electrons to a central metal ion to form a **coordinate bond** or **dative covalent bond**.
- A dative covalent bond or coordinate bond, a special kind of covalent bond, is formed when one of the bonded atoms provides both of the electrons for the shared pair.
- The **coordination number** indicates the number of coordinate bonds attached to the central metal ion.

Representing complex ions

In its formula, the complex ion is enclosed inside square brackets with the overall charge of the complex shown outside of the square brackets. The overall charge on a complex ion is the sum of the charges on the central metal ion and any ligands present.

An example of a complex ion is $[Cr(H_2O)_6]^{3+}$, formed when $CrCl_3 \cdot 6H_2O$ is dissolved in water. Cr^{3+} is the central metal ion – water acts as a ligand with each water molecule donating a lone pair of electrons from the oxygen atom to the central metal ion to form a coordinate bond. The coordination number is 6 as there are six coordinate bonds to the central metal ion (Figure 2).

▲ **Figure 1** *Hydrated copper(II) sulfate, $CuSO_4 \cdot 5H_2O$, dissolves in water to form a blue solution containing $[Cu(H_2O)_6]^{2+}$ complex ions*

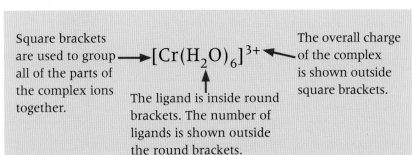

Square brackets are used to group all of the parts of the complex ions together. → $[Cr(H_2O)_6]^{3+}$ ← The overall charge of the complex is shown outside square brackets.

The ligand is inside round brackets. The number of ligands is shown outside the round brackets.

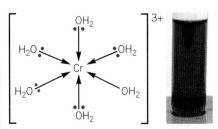

▲ **Figure 2** *The complex ion $[Cr(H_2O)_6]^{3+}$. Each water ligand donates a lone pair to the chromium ion to form a coordinate bond*

24.2 The formation and shapes of complex ions

> **Study tip**
> Ligands usually contain a lone pair of electrons.

Common ligands in transition metal chemistry

A ligand is any molecule or ion that can donate a pair of electrons to a central metal ion. Some ligands like water are neutral and have no charge, whereas other ligands like the hydroxide ion are negatively charged.

Monodentate ligands

A ligand that is able to donate *one* pair of electrons to a central metal ion is called a **monodentate ligand**. Some monodentate ligands are shown in Table 1.

▼ Table 1 *Common monodentate ligands*

Monodentate ligands		
Name	Formula	Charge on ligand
water	$H_2O:$	neutral
ammonia	$:NH_3$	neutral
chloride	$:Cl^-$	−1
cyanide	$:CN^-$	−1
hydroxide	$:OH^-$	−1

Bidentate ligands

Some ligands can donate two lone pairs of electrons to the central metal ion, forming two coordinate bonds. These ligands are called **bidentate** ligands. The most common bidentate ligands are 1,2-diaminoethane (frequently shortened to en) and the ethanedioate (oxalate) ion. Figure 3 shows these ligands and lone pairs that may be able to bond to the central metal ion.

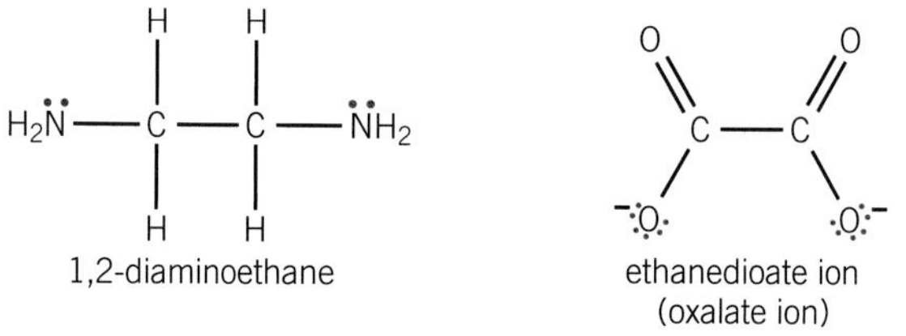

▲ Figure 3 *The structures of the 1,2-diaminoethane and ethanedioate (oxalate) ligands*

In 1,2-diaminoethane, each nitrogen atom donates a pair of electrons to the central metal ion forming a coordinate bond. In the ethanedioate ion each negatively-charged oxygen atom donates a lone pair of electrons to the central metal ion.

An example of a complex ion containing a bidentate ligand is $[Co(NH_2CH_2CH_2NH_2)_3]^{3+}$ (Figure 4).

In this complex:
- the oxidation number of Co is +3
- the coordination number is 6 because there are three 1,2-diaminoethane ligands and each ligand forms two coordinate bonds.

> **Study tip**
> If you are given an unfamiliar ligand, it is important that you show how the ligands bond to the central metal ion. A good place to start is with nitrogen and oxygen atoms as they often have lone pairs of electrons.

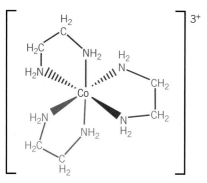

▲ Figure 4 *The complex ion $[Co(NH_2CH_2CH_2NH_2)_3]^{3+}$ is an example of a complex containing a bidentate ligand*

Shapes of complex ions

The shape of a complex ion depends upon its coordination number. The commonest coordination numbers are six and four giving rise to six-coordinate and four-coordinate complexes.

Six-coordinate complexes

Many complex ions have a coordination number of six, giving an octahedral shape.

When manganese sulfate, $MnSO_4$, is dissolved in water, the complex ion $[Mn(H_2O)_6]^{2+}$ is formed. This has an octahedral shape with bond angles around the manganese of 90° (Figure 5). The complex ion $[Co(H_2NCH_2CH_2NH_2)_3]^{3+}$ shown in Figure 4 is also octahedral, with bond angles of 90°, as it has six coordinate bonds.

Four-coordinate complexes

Complexes with a coordination number of four have two common shapes – tetrahedral and square planar.

Tetrahedral complexes

The tetrahedral shape is by far the more common of the two shapes, with bond angles of 109.5° around the central metal ion. $[CoCl_4]^{2-}$ and $[CuCl_4]^{2-}$ are common examples of complexes that have a tetrahedral shape (Figure 6).

> **Synoptic link**
>
> See Topic 6.1, Shapes of molecules and ions, for further details of showing 3D shapes.

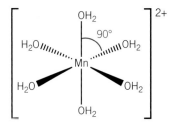

▲ **Figure 5** $[Mn(H_2O)_6]^{2+}$ has an octahedral shape with a bond angle of 90°

▲ **Figure 6** $[CoCl_4]^{2-}$ and $[CuCl_4]^{2-}$ have a tetrahedral shape with a bond angle of 109.5°

Square planar complexes

A square planar shape occurs in complex ions of transition metals with eight d-electrons in the highest energy d-sub-shell. Platinum(II), palladium(II), and gold(III) fall in this category and tend to form square planar complexes. The structure of the $[Pt(NH_3)_4]^{2+}$ complex ion is shown in Figure 7. In this shape, the ligands are arranged at the corners of a square, similar to the octahedral shape but without the ligands above and below the plane.

> **Study tip**
>
> You will be expected to draw 3D shapes of complex ions.

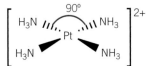

▲ **Figure 7** $[Pt(NH_3)_4]^{2+}$ has a square planar shape with a bond angle of 90°

24.2 The formation and shapes of complex ions

Colours in transition metal chemistry

In general, the colour of a complex ion depends upon the metal at the centre of the complex, its oxidation state, and the ligands that are coordinately bonded to the central metal ion. For a compound to have colour it must absorb visible light. The visible spectrum is shown in Figure 8.

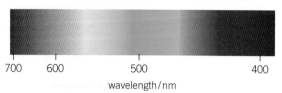

▲ **Figure 8** *The visible spectrum*

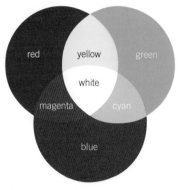

When a compound absorbs visible light, the colour seen is the sum of the remaining colours that are *not* absorbed. If a compound absorbs all wavelengths of visible light, the compound will appear black. If a compound absorbs the wavelengths of every colour except blue, the compound will appear blue.

If magenta is absorbed by a sample then the colour observed is green, on the opposite side of the colour chart (Figure 9). Similarly if the wavelengths of blue and red light are absorbed, the compound will appear green.

Solutions containing $[Cu(H_2O)_6]^{2+}$ ions absorbs red light. The colour on the opposite side of the chart, the complementary colour, is cyan (Figure 9). $[Cu(H_2O)_6]^{2+}$ forms a pale blue solution.

Chromium(VI) solutions absorb green, cyan, blue, and magenta. Chromium(VI) is orange.

What colour are the following complexes?
a $[Co(NH_3)_5H_2O]^{3+}$ which absorbs cyan.
b $[Co(CN)_6]^{3-}$ which absorbs blue.

▲ **Figure 9** *Colour chart. The complementary colour to blue is yellow as it is on the opposite side of the diagram*

Summary questions

1 Define the terms:
   a complex ion (1 mark)
   b coordination number (1 mark)
   c ligand. (1 mark)

2 A complex of chromium contains a chromium(III) ion bonded to four water molecules and two chloride ions.
   a State the formula of the complex ion. (1 mark)
   b Draw the complex ion, state the shape of the complex, and indicate on the diagram appropriate bond angles. (3 marks)

3 $[Fe(H_2O)_6]^{2+}$ reacts with the bidentate ligand 1,10-phenanthroline (phen) to form an orange-red complex $[Fe(phen)_3]^{2+}$. The structure of 1,10-phenanthroline is shown below.

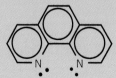

Draw the structure of the complex. (2 marks)

408

24.3 Stereoisomerism in complex ions

Specification reference: 5.3.1

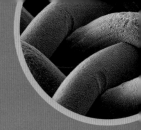

Stereoisomers

Stereoisomers have the same structural formula but a different arrangement of the atoms in space. Complex ions can display two types of stereoisomerism:

- *cis–trans* isomerism
- optical isomerism.

Stereoisomerism is also very important for organic chemistry.

For complex ions, the type of stereoisomerism depends on the number and type of ligands that are attached to the central metal ion, and the shape of the complex.

- Some four-coordinate and six-coordinate complex ions containing two different monodentate ligands show *cis–trans* isomerism.
- Some six-coordinate complex ions containing monodentate and bidentate ligands can show both *cis–trans* and optical isomerism.

Cis–trans isomerism in complex ions

In organic chemistry, *cis–trans* stereoisomerism requires the presence of a C=C double bond which prevents rotation of groups attached to each carbon atom of the C=C bond.

In complex ions, no double bond is involved and the shape of the complex holds groups in different orientations about the central metal ion.

Cis–trans isomerism occurs in some square planar and octahedral complex ions.

Cis–trans isomerism in square planar complexes

The simplest example of *cis–trans* isomerism is found in four coordinate square planar complexes that have no more than two identical ligands attached to the central metal ion. The complex $[Pd(NH_3)_2Cl_2]$, formed from the Pd^{2+} ion, two NH_3 ligands and two Cl^- ligands, shows *cis–trans* isomerism (Figure 1).

In square planar complexes the ligands are arranged in the same plane at the corners of a square with 90° bond angles. In the *cis*-isomer, the two identical groups are adjacent to each other, whereas in the *trans*-isomer the two identical groups are opposite each other. In the *cis*-isomer the coordinate bonds between the identical ligands are 90° apart. In the *trans*-isomer the coordinate bonds between the identical ligands are 180° apart.

Learning outcomes

Demonstrate knowledge, understanding, and application of:

→ *cis–trans* isomerism
→ optical isomerism
→ the role of *cis*-platin in the treatment of cancer.

Synoptic link

In Topic 13.2, Stereoisomerism, you met *cis-trans* and *E/Z* stereoisomerism in alkenes. You will meet optical isomerism in organic chemistry in Topic 27.2, Amino acids, amides, and chirality.

Synoptic link

See Topic 24.2, The formation and shapes of complex ions, for details of square planar complex ions.

24.3 Stereoisomerism in complex ions

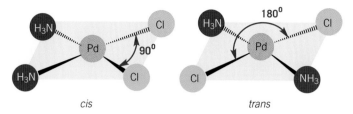

▲ **Figure 1** The cis and trans isomers of $[Pd(NH_3)_2Cl_2]$

You should represent these isomers in three dimensions as shown in Figure 2.

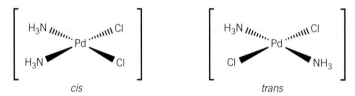

▲ **Figure 2** The cis and trans isomers of $[Pd(NH_3)_2Cl_2]$ shown in three dimensions

Cis–trans isomerism in octahedral complexes

Monodentate ligands

Octahedral complexes containing four of one type of ligand and two of another type of ligand can also exist as *cis* and *trans* isomers. Alfred Werner, a chemist studying in Zurich, successfully explained that there were two isomers of the compound, $[Co(NH_3)_4Cl_2]^+$, one violet in colour, the other green. The *cis* and *trans* isomers of $[Co(NH_3)_4Cl_2]^+$ are shown in Figure 3.

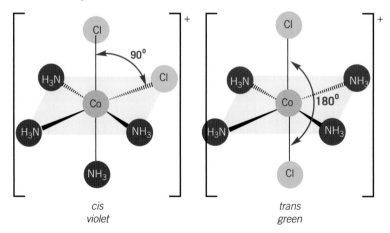

▲ **Figure 3** The cis and trans isomers of $[Co(NH_3)_4Cl_2]^+$

In the *cis*-isomer the two chloride ligands are adjacent to each other with their coordinate bonds separated by 90°, whereas in the *trans*-isomer the two chloride ligands are at opposite corners of the octahedron, with their coordinate bonds 180° apart. You will need to draw the two isomers showing the three-dimensional arrangement of the atoms as shown in Figure 4.

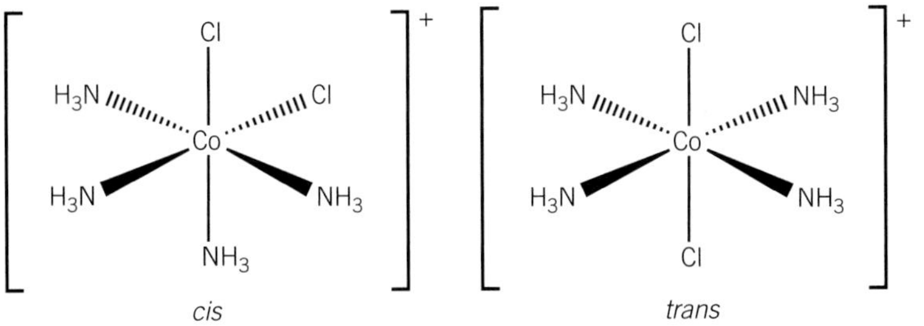

▲ Figure 4 *Three-dimensional diagrams to illustrate the cis and trans isomers of* $[Co(NH_3)_4Cl_2]^+$

Study tip

When drawing isomers you should ensure that the bond to the metal ion comes from the atom in the ligand that has the lone pair. In Figure 4, this is from the chloride ions and the nitrogen atom of the NH_3 ligands.

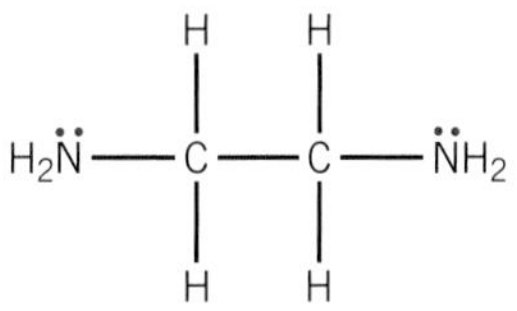

Bidentate ligands

Octahedral complexes containing bidentate ligands can also show *cis–trans* isomerism. Figure 5 shows the structure of the bidentate ligand 1,2-diaminoethane, $NH_2CH_2CH_2NH_2$. Each nitrogen atom in $H_2NCH_2CH_2NH_2$ can form a coordinate bond to the central metal ion.

The complex ion $[Co(NH_2CH_2CH_2NH_2)_2Cl_2]^+$ has *cis* and *trans* isomers. These are shown in Figure 6.

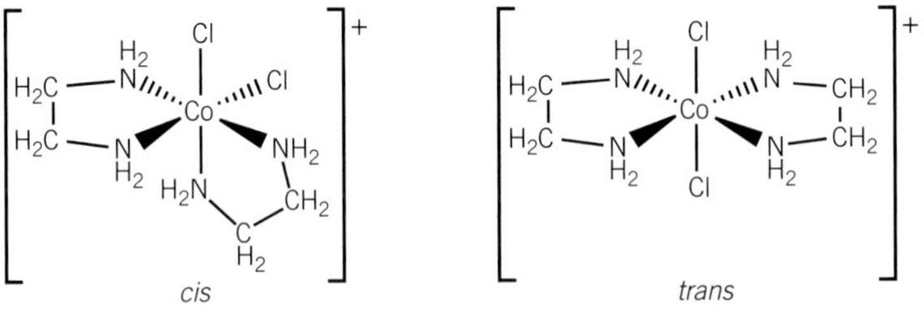

▲ Figure 5 *The ligand 1,2-diaminoethane*

▲ Figure 6 *The cis and trans isomers of* $[Co(NH_2CH_2CH_2NH_2)_2Cl_2]^+$

Study tip

1,2-diaminoethane, $NH_2CH_2CH_2NH_2$, is such a common bidentate ligand that it is often abbreviated to en. So you may see the formula for the complex $[Co(NH_2CH_2CH_2NH_2)_2Cl_2]^+$ shown more simply as $[Co(en)_2Cl_2]^+$. Unless otherwise told, you should show the full formula of $NH_2CH_2CH_2NH_2$.

Optical isomerism in octahedral complexes

Optical isomerism only occurs in octahedral complexes containing two or more bidentate ligands. Optical isomers, called enantiomers, are non-superimposable mirror images of each other, like a left hand and a right hand.

The *cis* isomer with formula $[Co(NH_2CH_2CH_2NH_2)_2Cl_2]^+$ shown in Figure 6 also exists as two optical isomers. *Trans*-isomers cannot form optical isomers as a mirror image is exactly the same and can be superimposed. The two optical isomers of $[Co(NH_2CH_2CH_2NH_2)_2Cl_2]^+$ are shown in Figure 7.

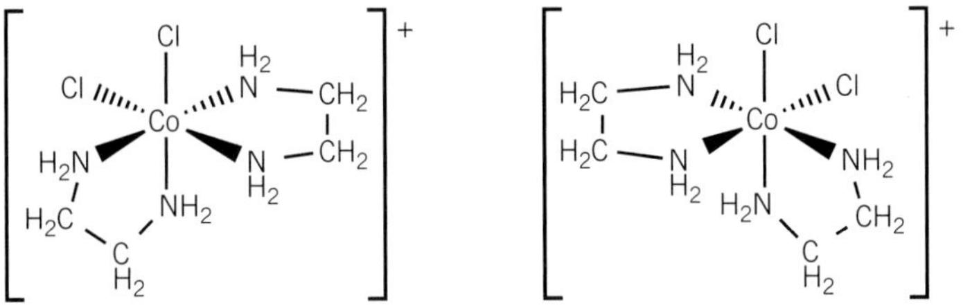

▲ Figure 7 *The two optical isomers of* $[Co(NH_2CH_2CH_2NH_2)_2Cl_2]^+$

24.3 Stereoisomerism in complex ions

Optical isomers can also be seen in complexes containing three bidentate ligands. Figure 8 shows the two optical isomers of $[Ru(NH_2CH_2CH_2NH_2)_3]^{2+}$.

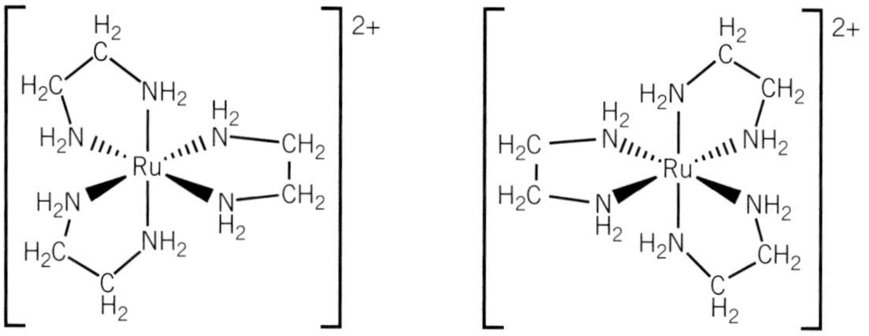

▲ **Figure 8** *The two optical isomers of* $[Ru(NH_2CH_2CH_2NH_2)_3]^{2+}$

The role of *cis–trans* isomerism in medicine

Scientists researching the effect of electric fields on bacteria found that when platinum electrodes were used to apply an electrical current to a colony of *E.coli*, the *E.coli* failed to divide but continued to grow.

The scientists initially thought that the electric current was preventing cell division. It was later discovered that a platinum compound was being formed at an electrode. This compound is *cis*-platin (Figure 9).

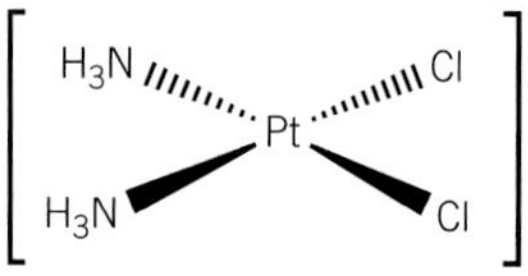

▲ **Figure 9** *The structure of cis-platin*

After carrying out further research and clinical tests it was proven that *cis*-platin did indeed attack tumours and in many cases the tumours were seen to shrink in size. Unfortunately *cis*-platin has many unpleasant side effects and can lead to kidney damage.

Cis-platin is still used extensively in the treatment of cancer, whilst other clinical trials continue into finding other platinum-based drugs that may inhibit tumour growth without the side effects.

How does *cis*-platin work?

Cis-platin works by forming a platinum complex inside of a cell which binds to DNA and prevents the DNA of the cell from replicating. Activation of the cell's own repair mechanism eventually leads to apoptosis, or systematic cell death.

Oxaliplatin (Figure 10) is used for treatment of colorectal cancer.

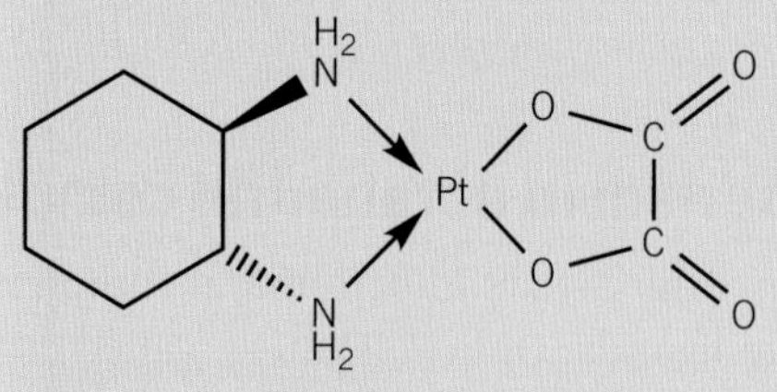

▲ **Figure 10** *Oxaplatin*

a Identify the ligands that make up oxaliplatin by name and structure.
b What is the coordination number and the charge of the platinum ion in oxalaplatin?

Summary questions

1 Draw the structure of the following complex ions:
  a *cis*-$[Co(NH_3)_4(H_2O)_2]^{2+}$ *(1 mark)*
  b *trans*-$[Ru(NH_2CH_2CH_2NH_2)_2Cl_2]$. *(1 mark)*

2 Describe the role of *cis*-platin as a medicine. *(2 marks)*

3 Show the structures of the types of isomerism found in the transition metal complex $[Co(C_2O_4)_2(H_2O)_2]^-$. *(2 marks)*

24.4 Ligand substitution and precipitation

Specification reference: 5.3.1

Ligand substitution reactions

A ligand is a molecule or ion that donates a pair of electrons to a central metal ion to form a coordinate or dative covalent bond. A **ligand substitution** reaction is one in which one ligand in a complex ion is replaced by another ligand.

Reactions of aqueous copper(II) ions

When copper(II) sulfate is dissolved in water, the pale blue complex ion, $[Cu(H_2O)_6]^{2+}$ is formed in aqueous solution.

Ligand substitution with ammonia

When an excess of aqueous ammonia is added to a solution containing $[Cu(H_2O)_6]^{2+}$ the pale blue solution changes colour to form a dark blue solution. The equation below shows this ligand substitution reaction.

$$[Cu(H_2O)_6]^{2+}(aq) + 4NH_3(aq) \rightarrow [Cu(NH_3)_4(H_2O)_2]^{2+}(aq) + 4H_2O(l)$$

pale blue solution → dark blue solution

In the reaction, four ammonia ligands have replaced four of the water ligands. Both $[Cu(H_2O)_6]^{2+}(aq)$ and $[Cu(NH_3)_4(H_2O)_2]^{2+}(aq)$ are octahedral complex ions.

The chemistry of this reaction is a little more complex than it first seems. When carrying out qualitative analysis, you should always add the reagent (ammonia) drop-wise to the solution under test (in this case, the $[Cu(H_2O)_6]^{2+}(aq)$ solution) so that all observations are recorded.

If you are careful you will see two different reactions taking place:

- A pale blue precipitate of $Cu(OH)_2$ is formed in the first stage of the reaction.
- The $Cu(OH)_2$ precipitate then dissolves in excess ammonia to form a dark blue solution.

The colour changes are shown in Figure 1.

$[Cu(H_2O)_6]^{2+}$ $Cu(OH)_2$ $[Cu(NH_3)_4(H_2O)_2]^{2+}$

▲ **Figure 1** *The stepwise addition of ammonia to $[Cu(H_2O)_6]^{2+}$*

Learning outcomes

Demonstrate knowledge, understanding, and application of:

→ ligand substitution reactions of $[Cu(H_2O)_6]^{2+}$ and $[Cr(H_2O)_6]^{3+}$

→ biological importance of iron in haemoglobin

→ precipitation reactions with aqueous sodium hydroxide and aqueous ammonia.

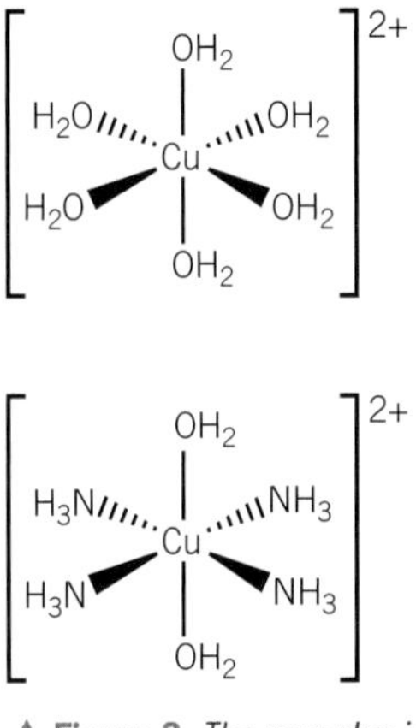

▲ **Figure 2** *The complex ions $[Cu(H_2O)_6]^{2+}$ and $[Cu(NH_3)_4(H_2O)_2]^{2+}$ both have octahedral shapes. In $[Cu(NH_3)_4(H_2O)_2]^{2+}$, the four NH_3 ligands are arranged in the same plane with the two H_2O ligands above and below*

24.4 Ligand substitution and precipitation

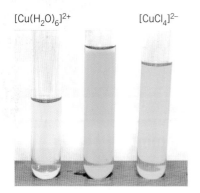

▲ Figure 3 *The addition of concentrated HCl to $[Cu(H_2O)_6]^{2+}$*

Ligand substitution with chloride ions

Concentrated hydrochloric acid, HCl, can be used as a source of chloride ions. When an excess of concentrated hydrochloric acid is added to a solution containing $[Cu(H_2O)_6]^{2+}$, the pale blue solution changes colour to form a yellow solution. This is another example of a ligand substitution reaction as six water ligands have been replaced with four chloride ligands.

If water is added to the yellow solution, a blue solution will be formed, although more dilute and paler in colour than the original blue solution.

If you take care making your observations, you will notice that an intermediate green solution is formed. This is not a new species but is the result of the yellow solution mixing with the blue solution to give a green colour as the reaction proceeds (Figure 3).

The equation below shows that the reaction is in equilibrium.

$$[Cu(H_2O)_6]^{2+}(aq) + 4Cl^-(aq) \rightleftharpoons [CuCl_4]^{2-}(aq) + 6H_2O(l)$$

pale blue solution   yellow solution
octahedral   tetrahedral

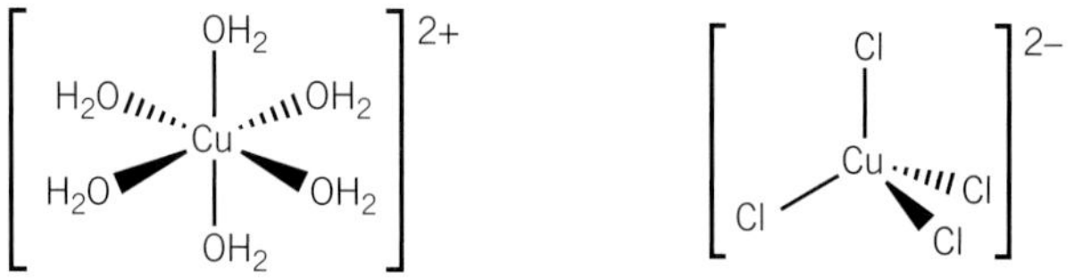

▲ Figure 4 *The shapes of the octahedral $[Cu(H_2O)_6]^{2+}$ and tetrahedral $[CuCl_4]^{2-}$ ions*

In this reaction, you will notice a change in coordination number, a change in colour, and a change in shape. However, the oxidation state of the copper remains as +2. Chloride ligands are larger in size than the water ligands, so fewer chloride ligands can fit round the central Cu^{2+} ion. This explains the change in coordination number (Figure 4).

Reactions of aqueous chromium(III) ions

When chromium(III) potassium sulfate, $KCr(SO_4)_2 \cdot 12(H_2O)$, also known as chrome alum, is dissolved in water, the complex ion $[Cr(H_2O)_6]^{3+}$ is formed. This is a pale purple solution. When chromium(III) sulfate is dissolved in water, a green solution containing chromium(III) is formed. However, this solution is not $[Cr(H_2O)_6]^{3+}$, rather it is the complex ion $[Cr(H_2O)_5SO_4]^+$, where one of the water ligands has been replaced by the sulfate ion, SO_4^{2-}. Both the purple and green solutions, contain chromium(III) ions in oxidation state +3.

▲ Figure 5 *The colours of aqueous solutions of chrome alum (left) and chromium(III) sulfate (right)*

▲ Figure 6 *The colours of the $Cr(OH)_3$ precipitate and $[Cr(NH_3)_6]^{3+}$ ions*

Reaction with ammonia

$[Cr(H_2O)_6]^{3+}$ takes part in a ligand substitution reaction with an excess of aqueous ammonia forming the complex ion $[Cr(NH_3)_6]^{3+}$. When the ammonia is added drop-wise to the chromium(III) solution, the reaction takes place in two distinct steps.

1. Initially a grey-green precipitate of $Cr(OH)_3$ is formed.
2. The $Cr(OH)_3$ precipitate dissolves in excess ammonia to form the complex ion $[Cr(NH_3)_6]^{3+}$.

TRANSITION ELEMENTS

24

The equation for the ligand substitution reaction is shown below:

$$[Cr(H_2O)_6]^{3+}(aq) + 6NH_3(aq) \rightarrow [Cr(NH_3)_6]^{3+}(aq) + 6H_2O(l)$$
   violet   purple

The observations for this reaction are shown in Figure 6.

Ligand substitution and haemoglobin

Blood carries oxygen around the body due to presence of haemoglobin, the iron-containing protein present in all red blood cells. Haemoglobin contains four protein chains held together by weak intermolecular forces. Each protein chain has a haem molecule within its structure. The central metal ion in a haem group is Fe^{2+} which can bind to oxygen gas, O_2 (Figure 7).

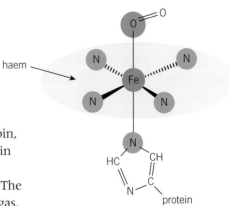

▲ Figure 7 Oxygen binds to the Fe^{2+} ion by the formation of a coordinate bond. For clarity, the 2+ charge has been omitted

As blood passes through the lungs, the haemoglobin bonds to oxygen because of the increased oxygen pressure in the capillaries of the lungs. A compound called oxyhaemoglobin forms, which releases this oxygen to body cells as and when required (Figure 8). In addition, the haemoglobin in red blood cells can bond to carbon dioxide, which is carried back to the lungs. Carbon dioxide is then released from the red blood cells and carbon dioxide is exhaled.

Carbon monoxide can also bind to the Fe^{2+} ion in haemoglobin. The complex formed is known as carboxyhaemoglobin. If carbon monoxide is breathed in, a ligand substitution reaction takes place where the oxygen in haemoglobin is replaced by carbon monoxide. Carbon monoxide binds to haemoglobin more strongly than oxygen, so a small concentration of carbon monoxide in the lungs can prevent a large proportion of the haemoglobin molecules from carrying oxygen. The bond is so strong that this process is irreversible. If the concentration of carboxyhaemoglobin becomes too high, oxygen transport is prevented, leading to death.

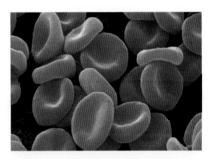

▲ Figure 8 Oxyhaemoglobin in red blood cells is responsible for carrying oxygen round the body

Precipitation reactions

A **precipitation reaction** occurs when two aqueous solutions containing ions react together to form an insoluble ionic solid, called a precipitate.

Transition metal ions in aqueous solution react with aqueous sodium hydroxide and aqueous ammonia to form precipitates. However, some of these precipitates will dissolve in an excess of sodium hydroxide or ammonia to form complex ions in solution.

Precipitation reactions with sodium hydroxide

Figure 9 shows the colours of $Cu^{2+}(aq)$, $Fe^{2+}(aq)$, $Fe^{3+}(aq)$, and $Mn^{2+}(aq)$ ions. Figure 10 shows the precipitates formed when these ions react with aqueous sodium hydroxide. None of these precipitates dissolve in an excess of aqueous sodium hydroxide.

The observations and equations for the reactions between these transition metal ions and aqueous sodium hydroxide are shown in Table 1.

▲ Figure 9 From left to right, solutions of $Cu^{2+}(aq)$, $Fe^{2+}(aq)$, $Fe^{3+}(aq)$, and $Mn^{2+}(aq)$

▲ Figure 10 From left to right the precipitates of $Cu(OH)_2$, $Fe(OH)_2$, $Fe(OH)_3$, and $Mn(OH)_2$

415

24.4 Ligand substitution and precipitation

▼ **Table 1** *The reactions of the aqueous transition metal ions with aqueous sodium hydroxide*

Ion	Observation(s) with NaOH(aq)	Equations(s)
Cu^{2+}	Blue solution reacts to form a blue precipitate of copper(II) hydroxide. The precipitate is insoluble in excess sodium hydroxide.	$Cu^{2+}(aq) + 2OH^-(aq) \rightarrow Cu(OH)_2(s)$ blue precipitate
Fe^{2+}	Pale green solution reacts to form a green precipitate of iron(II) hydroxide. The precipitate is insoluble in excess sodium hydroxide but turns brown at its surface on standing in air as iron(II) is oxidised to iron(III).	$Fe^{2+}(aq) + 2OH^-(aq) \rightarrow Fe(OH)_2(s)$ green precipitate In air: $Fe(OH)_2(s) \rightarrow Fe(OH)_3(s)$ orange-brown precipitate
Fe^{3+}	Pale yellow solution reacts to form an orange-brown precipitate of iron(III) hydroxide. The precipitate is insoluble in excess sodium hydroxide.	$Fe^{3+}(aq) + 3OH^-(aq) \rightarrow Fe(OH)_3(s)$ orange-brown precipitate
Mn^{2+}	Pale pink solution reacts to form a light brown precipitate of manganese(II) hydroxide which darkens on standing in air. The precipitate is insoluble in excess sodium hydroxide.	$Mn^{2+}(aq) + 2OH^-(aq) \rightarrow Mn(OH)_2(s)$ light-brown precipitate

▼ **Table 2** *The reactions of $Cr^{3+}(aq)$ ions with NaOH(aq)*

Ion	Observation(s) with NaOH(aq)	Equations(s)
Cr^{3+}	Violet solution reacts to form a grey-green precipitate of chromium(III) hydroxide. The precipitate is soluble in excess sodium hydroxide forming a dark green solution. $Cr^{3+}(aq) \rightarrow Cr(OH)_3(s) \rightarrow [Cr(OH)_6]^{3-}(aq)$  ▲ **Figure 11** *The reaction of $Cr^{3+}(aq)$ with NaOH*	$Cr^{3+}(aq) + 3OH^-(aq) \rightarrow Cr(OH)_3(s)$ green precipitate $Cr(OH)_3(s) + 3OH^-(aq) \rightarrow [Cr(OH)_6]^{3-}(aq)$ dark-green solution

Precipitation reactions with ammonia

You will recall from earlier in this topic the ligand substitution reactions of $[Cu(H_2O)_6]^{3+}(aq)$ and $[Cr(H_2O)_6]^{3+}(aq)$ with an excess of aqueous ammonia. In the first stage of these reactions precipitation takes place.

For copper the precipitation reaction is shown below:

$$Cu^{2+}(aq) + 2OH^-(aq) \rightarrow Cu(OH)_2(s)$$

Cu(OH)$_2$(s) is a blue precipitate which dissolves in excess ammonia to form a deep blue solution with the formula, [Cu(NH$_3$)$_4$(H$_2$O)$_2$]$^{2+}$(aq). You will find these colour changes in Figure 1.

For chromium the precipitation reaction is shown below:

$$Cr^{3+}(aq) + 3OH^-(aq) \rightarrow Cr(OH)_3(s)$$

Cr(OH)$_3$ is a green precipitate which dissolves in excess ammonia to form [Cr(NH$_3$)$_6$]$^{3+}$(aq) which is a purple solution. You will find these colour changes in Figure 6.

Fe^{2+}, Fe^{3+}, and Mn^{2+} react with an excess of aqueous ammonia in the same way as they react with aqueous sodium hydroxide, forming precipitates of Fe(OH)$_2$(s), Fe(OH)$_3$(s), and Mn(OH)$_2$(s). There is no further reaction with aqueous ammonia and so these precipitates do not dissolve.

Summary questions

1. **a** Write an ionic equation, with state symbols, for the reaction between Mn^{2+}(aq) and aqueous sodium hydroxide. *(1 mark)*
   b A solution of an iron salt reacts with aqueous sodium hydroxide to form a red-brown precipitate. Give the formula of the ion present in the solution. Write the electron configuration of this ion. *(1 mark)*
   c Why does an iron(II) hydroxide precipitate changes colour after standing in air. *(1 mark)*

2. When a small amount of ammonia solution is added to Cu^{2+}(aq), the reaction taking place is identical to that of Cu^{2+}(aq) with NaOH(aq).
   a Name, and give the formula of, the precipitate formed when aqueous sodium hydroxide (or a small amount of ammonia) is added Cu^{2+}(aq). *(1 mark)*
   b Write an ionic equation, with state symbols, for the formation of this product. *(2 marks)*
   c Cu^{2+}(aq) contains the [Cu(H$_2$O)$_6$]$^{2+}$ complex ion. When an excess of aqueous ammonia solution is added a ligand substitution reaction takes place. State the colour change you would expect in this reaction and give the formula of the coloured product formed. *(1 mark)*

3. When excess concentrated hydrochloric acid is added to an aqueous solution of copper(II) ions, the following equilibrium reaction occurs:
   $$[Cu(H_2O)_6]^{2+}(aq) + 4Cl^-(aq) \rightleftharpoons [CuCl_4]^{2-}(aq) + 6H_2O(l)$$
   a State the colours of [Cu(H$_2$O)$_6$]$^{2+}$(aq) and [CuCl$_4$]$^{2-}$(aq). *(2 marks)*
   b The [CuCl$_4$]$^{2-}$ ion formed is tetrahedral in shape. Draw a three-dimensional diagram to show the complex ion. State the bond angles and coordination number of this complex ion. *(3 marks)*
   c What type of reaction does the equation above represent? *(1 mark)*
   d State and explain the expected change in appearance if water is now added to the solution. *(2 marks)*

24.5 Redox and qualitative analysis

Specification reference: 5.3.1

Learning outcomes

Demonstrate knowledge, understanding, and application of:

→ the redox reactions and the accompanying colour changes for the Fe^{2+}/Fe^{3+} and $Cr^{3+}/Cr_2O_7^{2-}$ systems

→ the redox and disproportionation reactions of copper

→ qualitative analysis of ions on a test-tube scale.

Redox reactions involving transition metal ions

Reactions of iron(II), Fe^{2+}, and iron(III), Fe^{3+}

Oxidation of Fe^{2+} to Fe^{3+}

The redox reaction between iron(II) ions, $Fe^{2+}(aq)$, and manganate(VII) ions, MnO_4^-, in acid conditions is used as a basis for a redox titration.

In this reaction:

- Fe^{2+} is oxidised to Fe^{3+}
- MnO_4^- is reduced to Mn^{2+}.

The solution containing MnO_4^- ions is purple and is decolourised by $Fe^{2+}(aq)$ ions to form a colourless solution containing $Mn^{2+}(aq)$ ions. The equation for the reaction is shown below:

$$MnO_4^-(aq) + 8H^+(aq) + 5Fe^{2+}(aq) \rightarrow Mn^{2+}(aq) + 5Fe^{3+}(aq) + 4H_2O(l)$$
purple $\qquad\qquad\qquad\qquad\qquad\qquad$ colourless

Reduction of Fe^{3+} to Fe^{2+}

When a solution of $Fe^{3+}(aq)$ reacts with iodide ions, $I^-(aq)$, the orange-brown $Fe^{3+}(aq)$ ions are reduced to pale green $Fe^{2+}(aq)$ ions. Unfortunately this colour change is obscured by the the oxidation of iodide ions to form iodine, $I_2(aq)$, which has a brown colour. The equation for this reaction is shown below:

$$2Fe^{3+}(aq) + 2I^-(aq) \rightarrow 2Fe^{2+}(aq) + I_2(aq)$$
orange-brown $\qquad\qquad$ pale-green $\quad$ brown

▲ Figure 1 MnO_4^- (left) and Mn^{2+} (right)

In this reaction:

- Fe^{3+} is reduced to Fe^{2+}
- I^- is oxidised to I_2.

Synoptic link

For more information see Topic 23.2, Manganate(VII) redox titrations.

Using electrode potential to explain redox reactions

You can use the standard electrode potentials in Table 1 to explain why the two redox reactions of iron ions above take place.

▼ Table 1 Standard electrode potentials for iron redox reactions

Redox system	$E^{\ominus}$	
$I_2(aq) + 2e^- \rightleftharpoons 2I^-(s)$	+0.54 V	• More positive $E^{\ominus}$ value.
$Fe^{3+}(aq) + e^- \rightleftharpoons Fe^{2+}(aq)$	+0.77 V	• Equilibrium more likely to gain electrons, shift to the right and undergo reduction.
$MnO_4^-(aq) + 8H^+(aq) + 5e^- \rightleftharpoons Mn^{2+}(aq) + 4H_2O(l)$	+1.33 V	

Synoptic link

See also Topic 23.5 Predictions from electrode potential.

For the reaction of Fe^{2+} and MnO_4^- in acid conditions:

- $E^{\ominus}$ for MnO_4^-/Mn^{2+} is more positive than for Fe^{3+}/Fe^{2+}
- so Fe^{2+} is oxidised to Fe^{3+} and MnO_4^- is reduced to Mn^{2+}.

For the reaction of Fe^{3+} with I^-:

- $E^{\ominus}$ for Fe^{3+}/Fe^{2+} is more positive than for I_2/I^-
- so I^- is oxidised to I_2 and Fe^{3+} is reduced to Fe^{2+}.

Reactions of dichromate(VI), $Cr_2O_7^{2-}$, and chromium(III), Cr^{3+}

Reduction of $Cr_2O_7^{2-}$ to Cr^{3+}

Aqueous dichromate(VI) ions, $Cr_2O_7^{2-}$(aq), have an orange colour and aqueous chromium(III) ions, Cr^{3+}(aq), have a green colour. Acidified $Cr_2O_7^{2-}$(aq) ions can be reduced to Cr^{3+}(aq) ions by the addition of zinc as shown in Figure 2. The equation for the reaction is shown below:

$$Cr_2O_7^{2-}(aq) + 14H^+(aq) + 3Zn(s) \rightarrow 2Cr^{3+}(aq) + 7H_2O(l) + 3Zn^{2+}(aq)$$
  orange   green

With an excess of zinc, chromium(III) ions are reduced further to chromium(II), which is a pale blue colour (Figure 2).

$$Zn(s) + 2Cr^{3+}(aq) \rightarrow Zn^{2+}(aq) + 2Cr^{2+}(aq)$$
              green   pale blue

Table 2 shows the relevant standard electrode potentials for these two reactions.

> **Study tip**
>
> For practice, combine the half equations from Table 1 to give the overall equations shown for the reaction of Fe^{2+} with MnO_4^-/H^+ and the reaction of Fe^{3+} with I^-.

▲ **Figure 2** $Cr_2O_7^{2-}$ (left), Cr^{3+} (centre) and Cr^{2+} (right)

▼ **Table 2** *Standard electrode potentials for chromium redox reactions*

Redox system	$E^\ominus$/V	
$Zn^{2+}(aq) + 2e^- \rightleftharpoons Zn(s)$	−0.76	• More positive $E^\ominus$ value.
$Cr^{3+}(aq) + e^- \rightleftharpoons Cr^{2+}(aq)$	−0.41	• Equilibrium more likely to gain electrons, shift to the right and undergo reduction.
$Cr_2O_7^{2-}(aq) + 14H^+(aq) + 6e^- \rightleftharpoons 2Cr^{3+}(aq) + 7H_2O(l)$	+1.33	

In Table 2, zinc is the most powerful reducing agent and is capable of reducing both $Cr_2O_7^{2-}$ to Cr^{3+} and Cr^{3+} to Cr^{2+}. This explains why $Cr_2O_7^{2-}$ can be reduced down to Cr^{2+}.

Oxidation of Cr^{3+} to CrO_4^{2-}

Hot alkaline hydrogen peroxide, H_2O_2, is a powerful oxidising agent and can be used to oxidise chromium(III) in Cr^{3+} to chromium(VI) in CrO_4^{2-}. The overall equation is shown below.

$$3H_2O_2 + 2Cr^{3+} + 10OH^- \rightarrow 2CrO_4^{2-} + 8H_2O$$

Chromium is oxidised from +3 in Cr^{3+} to +6 in CrO_4^{2-}.
Oxygen is reduced from −1 in H_2O_2 to −2 in CrO_4^{2-}.

> **Synoptic link**
>
> You will recall the use of electrode potentials from Topic 23.4, Electrode potentials.

> **Synoptic link**
>
> The oxidation of Cr^{3+} to CrO_4^{2-} by alkaline hydroxide ion was discussed in detail in Topic 23.1, Redox reactions.

Reactions of copper(II), Cu^{2+}, and copper(I), Cu^+

Reduction of Cu^{2+} to Cu^+

When aqueous copper(II) ions, Cu^{2+}, react with excess iodide ions, I^-(aq), a redox reaction occurs.

- I^- is oxidised to brown iodine, I_2
- Cu^{2+} is reduced to Cu^+
- The Cu^+ forms a white precipitate of copper(I) iodide. The equation for the reaction is shown below:

$$2Cu^{2+}(aq) + 4I^-(aq) \rightarrow 2CuI(s) + I_2(s)$$
  pale blue   white precipitate brown

24.5 Redox and qualitative analysis

▲ Figure 3 *Copper(II) ions reacts with iodide ions to form copper(I) iodide and iodine*

If you look carefully in Figure 3, you can see the white precipitate of copper(I) iodide at the bottom of the brown solution that forms. The white colour is somewhat masked though by the brown iodine colour.

Disproportionation of Cu^+ ions

When solid copper(I) oxide, Cu_2O, reacts with hot dilute sulfuric acid, a brown precipitate of copper is formed together with a blue solution of copper(II) sulfate. In this reaction copper(I) ions, Cu^+, have been simultaneously oxidised and reduced. As the same element has been reduced and oxidised, this reaction is **disproportionation**.

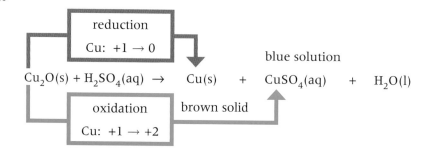

> **Synoptic link**
>
> See Topic 23.3, Iodine/thiosulfate titrations, for the use of the reaction between Cu^{2+} and I^- in analysis.

> **Synoptic link**
>
> See Topic 8.2, The halogens, for other examples of disproportionation.

Qualitative analysis

Identifying positive ions

Transition metal ions

Aqueous sodium hydroxide produces precipitates with aqueous transition metal ions. These reactions can be used for identifying transition metal ions in an unknown solution. The observations for these reactions are summarised in Topic 24.4, Table 1.

Ammonium ions, NH_4^+

When heated with hydroxide ions, NH_4^+ reacts to produce ammonia gas, NH_3.

$$NH_4^+(aq) + OH^-(aq) \rightarrow NH_3(g) + H_2O(l)$$

To test for the ammonium ion, aqueous sodium hydroxide, NaOH(aq), is heated gently with the solution being analysed. If ammonia is evolved, damp red pH indicator paper will turn blue, confirming the presence of NH_4^+ ions.

Identifying anions (negative ions)

Tests can be used to identify anions in solutions of unknown compounds. Where the tests are being carried out on the same solution it is important to carry them out in the same order as they are presented in Table 3.

> **Synoptic link**
>
> See Topic 8.3, Qualitative analysis, for further details of the cation test for ammonium ions, NH_4^+, and the anion tests for carbonate, CO_3^{2-}, sulfate, SO_4^{2-}, and halide ions, Cl^-, Br^-, I^-.

▼ **Table 3** *Tests for solutions containing one or more unknown anions*

Anion	Chemical test	Observations and equations
carbonate, CO_3^{2-}	add dilute nitric acid, HNO_3(aq)	effervescence as carbon dioxide is evolved: CO_3^{2-}(aq) + $2H^+$(aq) → CO_2(g) + H_2O(l)
sulfate, SO_4^{2-}	add Ba^{2+}(aq) ions	white precipitate of $BaSO_4$ is formed: Ba^{2+}(aq) + SO_4^{2-}(aq) → $BaSO_4$(s)
halide ions Cl^-, Br^-, or I^-	add Ag^+(aq) ions	• white precipitate of AgCl if Cl^- ions present, soluble in dilute NH_3(aq): Ag^+(aq) + Cl^-(aq) → AgCl(s) • cream precipitate of AgBr if Br^- ions present, soluble in concentrated NH_3(aq). • yellow precipitate of AgI if I^- ions present, insoluble in NH_3(aq).

Summary questions

1. **a** What is meant by the term disproportionation? *(1 mark)*
   b Use oxidation states to show that the following reaction is an example of disproportionation:
   $$Cl_2(g) + H_2O(l) \rightarrow HCl(aq) + HOCl(aq)$$
   (2 marks)

2. Describe a simple chemical test to distinguish between each of the following pairs of compounds:
   a NaCl and NaI *(1 mark)*
   b $CuSO_4$ and $FeSO_4$ *(1 mark)*
   c CuO and Cu_2O *(1 mark)*
   d $FeCl_2$ and $FeCl_3$ *(1 mark)*

3. Using the following redox systems, write equations for six reactions that can occur. *(6 marks)*

Redox system	$E^\ominus$ /V
Zn^{2+}(aq) + $2e^-$ ⇌ Zn(s)	−0.76
Cr^{3+}(aq) + e^- ⇌ Cr^{2+}(aq)	−0.41
$Cr_2O_7^{2-}$(aq) + $14H^+$(aq) + $6e^-$ ⇌ $2Cr^{3+}$(aq) + $7H_2O$(l)	+1.33
H_2O_2(aq) + $2H^+$(aq) + $2e^-$ ⇌ $2H_2O$(l)	+1.77

Chapter 24 Practice questions

Practice questions

1. Elements in the d-block of the periodic table form ions that combine with ligands to form complex ions. Most d-block elements are also classified as transition elements.

   a. Explain why two of the Period 4 d-block elements (Sc–Zn) are not also transition elements. In your answer you should link full electron configurations to your explanations. *(6 marks)*

   b. The cobalt(III) ion, Co^{3+}, forms a complex ion **A** with two chloride ligands and two 1,2-diaminoethane, $H_2NCH_2CH_2NH_2$, ligands. The structure of 1,2-diaminoethane is shown below.

   (i) Explain how ethanediamine is able to act as a bidentate ligand. *(2 marks)*

   (ii) Write the formula of complex ion **A**. *(1 mark)*

   (iii) What is the coordination number of cobalt in complex ion **A**? *(1 mark)*

   (iv) Complex ion **A** has *cis* and *trans* stereoisomers. One of these stereoisomers also has an optical isomer.
   Draw 3-D diagrams to show the three stereoisomers. *(3 marks)*

   c. The equilibrium reaction for the transport of oxygen by haemoglobin (Hb) in blood can be represented as **Equation 1**.

   $Hb(aq) + O_2(aq) \rightarrow HbO_2(aq)$ **Equation 1**

   Explain how ligand substitution reactions allow haemoglobin to transport oxygen in blood. *(2 marks)*

   OCR F325 June 14 Q5

2. Iron is heated with chlorine to form an orange–brown solid, **A**. Solid **A** is dissolved in water to form an orange–brown solution, **X**, containing the complex ion $[Fe(H_2O)_6]^{3+}$. Separate portions of solution **X** are reacted as shown in **Experiments 1–4** below.

   Experiment 1 Aqueous sodium hydroxide is added to solution **X**. An orange–brown precipitate **B** forms.

   Experiment 2 Excess zinc powder is added to solution **X** and the mixture is heated. The excess zinc is removed leaving a pale-green solution containing the complex ion **C** and aqueous Zn^{2+} ions.

   Experiment 3 An excess of aqueous potassium cyanide, KCN(aq), is added to solution **X**. The solution turns a yellow colour and contains the complex ion **E**. **E** has a molar mass of $211.8 \, g \, mol^{-1}$.

   Experiment 4 An aqueous solution containing ethanedioate ions, $(COO^-)_2$, is added to solution **X**. A coloured solution forms containing a mixture of optical isomers **F** and **G**. The structure of the ethanedioate ion is shown below.

   a. Write an equation for the formation of solid **A**. *(1 mark)*

   b. In **Experiment 1**, write an ionic equation for the formation of precipitate **B**. *(1 mark)*

   c. In **Experiment 2**:
   (i) write an equation for the formation of complex ion **C** *(2 marks)*
   (ii) state the type of reaction taking place. *(1 mark)*

   d. In **Experiment 3**:
   (i) write an equation for the formation of complex ion **E**
   (ii) state the type of reaction taking place. *(3 marks)*

   e. In **Experiment 4**, optical isomers **F** and **G** are formed.
   Show the 3-D shapes of **F** and **G**. In your diagrams, show the ligand atoms that are bonded to the metal ions and any overall charges. *(3 marks)*

f In a separate experiment, iron metal is heated with potassium nitrate, KNO_3, a strong oxidising agent. A reaction takes place and the resulting mixture is poured into water. A dark red solution forms containing ferrate(VI) ions. The ferrate(VI) ion has a 2– charge.

Suggest a possible formula for the ferrate(VI) ion. *(1 mark)*

OCR F325 June 2013 Q5

3 This question refers to chemistry of d-block elements in Period 4 (Sc–Zn).

a For each statement below, select the symbols of the correct element(s).

(i) The element that has atoms containing six electrons in the 3d sub-shell. *(1 mark)*

(ii) Two elements that have atoms with two unpaired d-electrons. *(2 marks)*

(iii) The element with ions that form a purple complex with excess ammonia *(1 mark)*

(iv) The element **X** that forms an oxide with the formula X_3O_4 with the molar mass of $228.7\,g\,mol^{-1}$. *(1 mark)*

(v) The element that has atoms with an average mass of $8.64 \times 10^{-23}\,g$. *(1 mark)*

b The flowchart below shows three reactions of the complex ion $[Cu(H_2O)_6]^{2+}$.

In the boxes below, write down the formulae of the species formed.

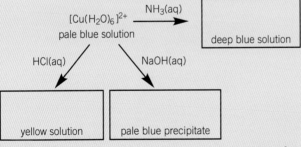

(3 marks)

c The answers to this question all refer to complex ions of nickel.

(i) State the shape of the complex ion $[Ni(H_2O)_6]^{2+}$. *(1 mark)*

(ii) What is the formula of the complex ion of Ni^{2+} containing six fluoride ligands? *(1 mark)*

(iii) Show the 3-D shapes of the stereoisomers of the complex ion $[Ni(en)_3]^{2+}$.

(en = $H_2NCH_2CH_2NH_2$) *(3 marks)*

OCR F325 Jan 2013 Q1

4 a A purple solid, **A**, containing 13.82% Fe, 4.49% H, 10.40% N and 71.28% O is dissolved in water to form a pale yellow solution, **B**. 40.12% of the mass of **A** is water. When aqueous sodium hydroxide is added to solution **B**, a brown precipitate, **C**, is formed. When NaSCN is added to **B**, one ligand in **B** is replaced resulting in the formation of a deep red solution containing a complex ion, **D**.

(i) Deduce the empirical formula of **A** and the number of waters of crystallisation in **A**, showing all working. *(4 marks)*

(ii) Identify **B**, **C**, and **D**. *(3 marks)*

b $NiSO_4 \cdot 6H_2O$ is dissolved in water forming a green solution containing the complex ion, $[Ni(H_2O)_6]^{2+}$. When the monodentate ligand NH_3 is added to $[Ni(H_2O)_6]^{2+}$, a blue complex **E** is formed where all six water ligands have been replaced. However when 1,2-diaminoethane, a bidentate ligand is added to $[Ni(H_2O)_6]^{2+}$, four water molecules are replaced resulting in the formation of a purple solution **F**.

(i) Give the formulae of the complex ions **E** and **F**. *(2 marks)*

(ii) Write an equation for the formation of **E** from $[Ni(H_2O)_6]^{2+}$. *(1 mark)*

(iii) Explain how 1,2-diaminoethane acts as a bidentate ligand. *(2 marks)*

(iv) **F** can form a number of stereoisomers.

Draw and label all of the stereoisomers of **F** and state the type of isomerism. *(4 marks)*

Module 5 Summary

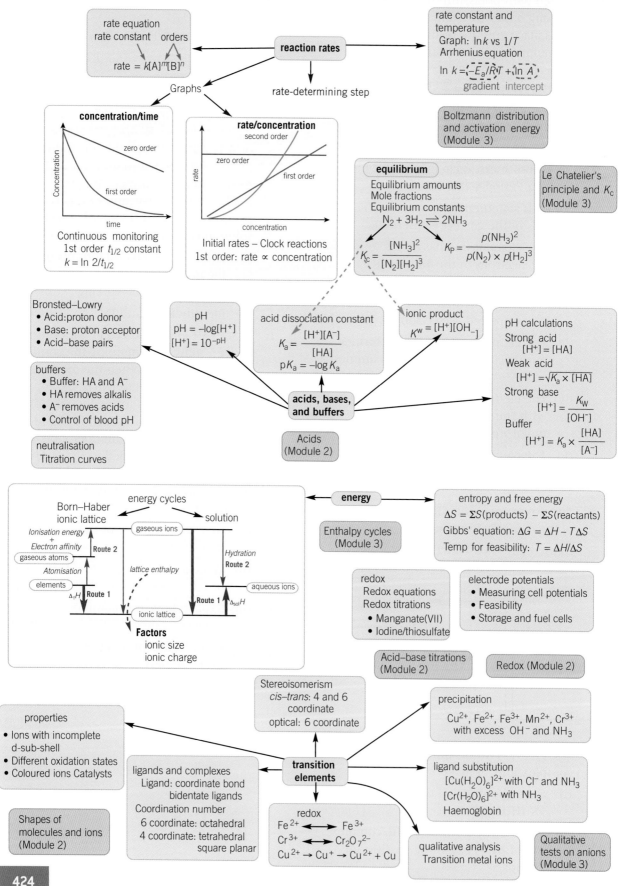

Oxalic acid

Oxalic acid (ethanedioic acid) is a dibasic acid found in plant matter such as spinach and rhubarb.

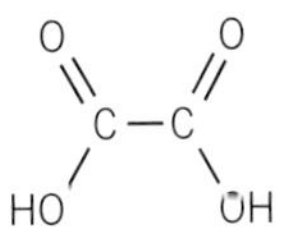

▲ Figure 1 *The structure of oxalic acid (ethandioic acid)*

▲ Figure 2 *Spinach leaves contain significant levels of oxalic acid*

The salts of oxalic acid contain oxalate (ethanedioate) ions, $C_2O_4^{2-}$. These ions are highly toxic and can cause health problems in lower concentrations when they form insoluble compounds with Fe^{2+} ions. Iron(II) oxalate is a cause of gout – an extremely painful complaint.

Although iron(II) oxalate is an insoluble ionic compound, oxalate ions will form a soluble complex ion with Fe^{3+} ions. This has the formula $[Fe(C_2O_4)_3]^{3-}$ and oxalate ion is acting as a bidentate ligand. Oxalate ions form particularly stable complexes with many metal ions – a process known as chelation.

The concentration of free oxalate ions in a solution can be found by a redox titration with manganate(VII) ions in acid conditions. The redox equilibria involved are:

$$2CO_2(aq) + 2e^- \rightarrow C_2O_4^{2-}(aq) \quad E^\ominus = -0.49\ V$$

$$MnO_4^-(aq) + 8H^+(aq) + 5e^- \rightarrow Mn^{2+}(aq) + 4H_2O(l)$$
$$E^\ominus = +1.51\ V$$

This titration can be used to analyse foods such as spinach to find the percentage of free oxalate ions in their leaves.

Although the reaction between manganate(VII) ions is quite slow, the product of the reaction, Mn^{2+} ions, can act as a catalyst for the process. This is known as autocatalysis, and means that the rate of the reaction changes in a rather unexpected way over the course of the reaction.

1. Oxalic acid is a dibasic acid. The K_a values for the two dissociations of oxalic acid are shown below.
   $K_a(H_2C_2O_4) = 5.9 \times 10^{-2}$ mol dm^{-3}
   $K_a(HC_2O_4^-) = 6.4 \times 10^{-5}$ mol dm^{-3}
   a Write equations for the two dissociation processes to which these K_a values relate.
   b i Calculate the pH of a 0.100 mol dm^{-3} solution of $NaHC_2O_4$, stating the assumptions that you make.
      ii Which of these assumptions would not be valid if you were trying to calculate the pH of 0.100 mol dm^{-3} $H_2C_2O_4$?

2. a Explain why the ethanedioate ion is able to act as a bidentate ligand.
   b Draw out a diagram to show of the structure of the complex ion $[Fe(C_2O_4)_3]^{3-}$. Comment on whether any stereoisomers will exist of this structure.

3. a Use electrode potential data to explain why acidified manganate(VII) ions are able to oxidise ethanedioate ions.
   b Write a balanced equation for this process.

Extension

1. The rate equation for a reaction can be deduced from knowledge of the steps in the mechanism. Research some mechanisms and the rate equation that can be deduced from these mechanisms. Comment on any situations in which the rate equation does not have the simple form that you have encountered in your A-level studies.

2. Prepare a summary of the factors that can lead to a reaction being considered feasible. You should consider factors affecting ΔH values, ΔS values, and how they are combined to produce an overall value of ΔG.

3. Research the anti-cancer drug oxaliplatin. Comment on the shape and bonding in the structure of the compound and use your knowledge of other anti-cancer drugs to explain a possible mechanism for its action.

Module 5 Practice questions

1. For a first order reaction, which row shows the correct changes as the reaction proceeds?

	Change in rate	Change in half life
A	remains the same	decreases
B	decreases	remains the same
C	remains the same	remains the same
D	decreases	decreases

(1 mark)

2. From a graph of ln k against $1/T$, what value is given by the gradient?

   A E_a
   B $-E_a$
   C $-E_a/R$
   D $E_a \times R$ (1 mark)

3. Nitrogen reacts with hydrogen to produce ammonia. $N_2(g) + 3H_2(g) \rightleftharpoons 2NH_3(g)$

   A mixture of 2.00 mol of N_2, 6.00 mol of H_2, and 2.40 mol of NH_3 is allowed to reach equilibrium in a sealed container of volume 1 dm³. At equilibrium, 2.32 mol of N_2 is present.

   What is the value of K_c under these conditions?

   A $\dfrac{1.76^2}{2.32 \times 6.96^3}$
   B $\dfrac{1.76^2}{2.32 \times 6.32^3}$
   C $\dfrac{1.76^2}{2.32 \times 6.96^3}$
   D $\dfrac{2.40^2}{2.32 \times 6.00^3}$ (1 mark)

4. $2SO_2(g) + O_2(g) \rightleftharpoons 2SO_3(g)$
   $\Delta H = -197\,\text{kJ mol}^{-1}$ (1 mark)

   The temperature of the equilibrium system is increased from 1000 K to 1100 K.
   Which row is correct?

A	SO_2 decreases	K_c increases
B	SO_2 increases	K_c decreases
C	SO_2 does not change	K_c does not change
D	SO_2 increases	K_c does not change

(1 mark)

5. What is the conjugate base of the acid HSO_4^-?

   A OH^-
   B H_2SO_4
   C SO_3
   D SO_4^{2-} (1 mark)

6. A solution of hydrochloric acid with a pH of 2.50 is made 10 times more dilute with water. What is the pH of the resulting solution?

   A 0.250
   B 2.50
   C 3.50
   D 12.50   (1 mark)

7. 100 cm³ of the 0.1 mol dm⁻³ solutions below are mixed. Which pair of solutions forms a buffer solution with an acidic pH?

   A hydrochloric acid and potassium chloride
   B methanoic acid and sodium methanoate
   C potassium hydroxide and potassium chloride
   D ammonia and ammonium chloride   (1 mark)

8. A buffer solution contains 0.10 mol of a weak acid, HA, (pK_a = 3.6) and 0.10 mol of the sodium salt of the acid, NaA, in a total volume of 1 dm³ of solution. What is the pH?

   A 7.0
   B 4.6
   C 3.6
   D 2.6   (1 mark)

9. Which reaction would be expected to have a positive entropy chnge?

   A $2Mg(s) + O_2(g) \rightarrow 2MgO(s)$
   B $NH_3(g) + HCl(g) \rightarrow NH_4Cl(s)$
   C $2H_2O_2(aq) \rightarrow 2H_2O(aq) + O_2(g)$
   D $2HCl(g) \rightarrow H_2(g) + Cl_2(g)$ (1 mark)

10. Nitrogen reacts with fluorine as below
    $N_2(g) + 3F_2(g) \rightarrow 2NF_3(g)$
    $\Delta H = -250\,\text{kJ mol}^{-1}$
    $\Delta S = -279\,\text{J K}^{-1}\,\text{mol}^{-1}$

    What is the value of ΔG in kJ mol⁻¹ for this reaction at 25 °C?

    A −333
    B −167
    C −243
    D 6725   (1 mark)

11. The redox reaction reaction takes place.
    $3Sn^{2+} + IO_3^- + 6H^+ \rightarrow 3Sn^{4+} + I^- + 3H_2O$
    Which element is being reduced?

    A Sn
    B I
    C O
    D H   (1 mark)

12. A standard cell is made based on the redox systems below.

    $Al^{3+}(aq) + 3e^- \rightleftharpoons Al(s)$ $E = -1.68\,\text{V}$
    $Sn^{2+}(aq) + 2e^- \rightleftharpoons Sn(s)$ $E = -0.14\,\text{V}$

When the cell is operating, which statement is true?

A Al is oxidised at one electrode and Sn is oxidised at the other.

B Al is oxidised at one electrode and Sn is reduced at the other.

C Al is reduced at one electrode and Sn is oxidised at the other.

D Al is reduced at one electrode and Sn is reduced at the other. *(1 mark)*

13 When aqueous barium nitrate is added to an aqueous solution of **X**, a white precipitate is formed. When aqueous sodium hydroxide was added to an aqueous solution of **X**, a green precipitate forms which is insoluble in an excess of aqueous sodium hydroxide.

Which compound could be **X**?

A $FeCl_2$ B $FeSO_4$
C $CrCl_3$ D $Cr_2(SO_4)_3$ *(1 mark)*

14 Which electron configuration is correct?

A Cu is $[Ar]3d^94s^2$ B Cu^+ is $[Ar]3d^94s^1$
C Cu^+ is $[Ar]3d^{10}$ D Cu^{2+} is $[Ar]3d^84s^1$ *(1 mark)*

15 In which ion does vanadium **not** have an oxidation state of +5.

A $[VO_3(OH)]^{2-}$ B $[VO(OH)_2]^-$
C $[V_4O_{12}]^{4-}$ D $[VO_4]^{3-}$ *(1 mark)*

16 Lattice enthalpy can be used as a measure of ionic bond strength. Lattice enthalpies can be determined indirectly using Born–Haber cycles.

The table below shows the enthalpy changes that are needed to determine the lattice enthalpy of lithium fluoride, LiF.

Enthalpy change	Energy / kJ mol^{-1}
1st electron affinity of fluorine	−328
1st ionisation energy of lithium	+520
atomisation of fluorine	+79
atomisation of lithium	+159
formation of lithium fluoride	−616

a Define the term *lattice enthalpy*. *(2 marks)*

b The diagram below shows an incomplete Born–Haber cycle that would allow the lattice enthalpy of lithium fluoride to be determined.

(i) On the four dotted lines, add the species present, including state symbols.

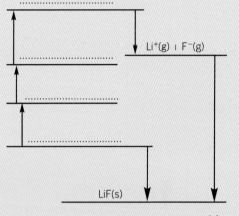

(4 marks)

(ii) Calculate the lattice enthalpy of lithium fluoride. *(2 marks)*

c The change that produces lattice enthalpy is spontaneous but has a negative entropy change.

Why is this change able to take place spontaneously? *(1 mark)*

d The lattice enthalpies of sodium fluoride, sodium chloride and magnesium fluoride are shown below.

Compound	Lattice enthalpy / kJ mol^{-1}
sodium fluoride	−918
sodium chloride	−780
magnesium fluoride	−2957

Explain the differences between these lattice enthalpies.

In your answer, your explanation should show how different factors affect lattice enthalpy. *(3 marks)*

F325 June 2012 Q1

Module 5 Practice questions

17 This question looks at energy changes in chemistry.

a Three processes are given below.

For each process, state and explain whether the change would be accompanied by an increase or decrease in entropy.

(i) $H_2O(l) \rightarrow H_2O(g)$ *(1 mark)*

(ii) $Zn(s) + 2HCl(aq) \rightarrow ZnCl_2(aq) + H_2(g)$ *(1 mark)*

(iii) $N_2(g) + 3H_2(g) \rightarrow 2NH_3(g)$ *(1 mark)*

b $N_2O_4(g)$ and $NO_2(g)$ form an equilibrium mixture at 25 °C.

$N_2O_4(g) \rightleftharpoons 2NO_2(g)$ $\Delta H = +58.0 \text{ kJ mol}^{-1}$
colourless   brown $\Delta S = +176 \text{ J K}^{-1} \text{mol}^{-1}$

(i) Calculate ΔG at 25 °C. *(1 mark)*

(ii) Calculate the temperature when the equilibrium will be midway between $N_2O_4(g)$ and $NO_2(g)$. *(2 marks)*

(iii) An equilibrium mixture of N_2O_4 and NO_2 is heated from 25 °C at constant pressure.

Explain what would happen to the colour of the equilibrium mixture as the mixture is heated. *(2 marks)*

18 Chemists and biochemists use pK_a values to compare the strengths of different acids. pK_a is a more convenient way of comparing acid strengths than K_a values.

pK_a values of several naturally occurring Brønsted–Lowry acids are shown in the table below.

Common name and source	Structural formula	pKa (at 25 °C)
benzoic acid	C_6H_5COOH	4.19
acetic acid	CH_3COOH	4.76
pyruvic acid	$CH_3COCOOH$	2.39
lactic acid	$CH_3CHOHCOOH$	3.86

a (i) What is meant by the term Brønsted–Lowry acid? *(1 mark)*

(ii) What is meant by the strength of an acid? Include an equation for one of the acids in the table in your answer. *(2 marks)*

(iii) What are the relative strengths of the four acids in the table? *(1 mark)*

(iv) Aqueous benzoic acid was mixed with aqueous lactic acid. An equilibrium mixture was formed containing conjugate acid–base pairs.

Complete the equilibrium below to show the components in the equilibrium mixture.

$C_6H_5COOH + CH_3CHOHCOOH \rightleftharpoons$ *(1 mark)*

b Aqueous pyruvic acid was reacted with an aqueous solution of calcium hydroxide.

(i) Write an equation for this reaction. *(1 mark)*

(ii) Write an ionic equation for this reaction. *(1 mark)*

c The pH of an acid solution can be calculated from its pK_a value.

Calculate the pH of a 0.0150 mol dm^{-3} solution of pyruvic acid at 25 °C.

Show all your working. Give your answer to **two** decimal places. *(4 marks)*

d Oxalic acid (ethanedioic acid), $C_2H_2O_4$, is present in the leaves of rhubarb plants.

Oxalic acid has two dissociations with $pK_a = 1.23$ and $pK_a = 4.19$.

(i) Draw the structure of oxalic acid. *(1 mark)*

(ii) Predict the equations that give rise to each dissociation. *(2 marks)*

OCR F325 Jun 11 Q4(a)–(d)

19 a Transition metals form complex ions with ligands, such as water, ammonia, and chloride ions. The complex ions may have different coordination numbers and shapes.

(i) Explain what is meant be the term *ligand*. *(1 mark)*

(ii) Explain what is meant by the term *coordination number*. *(1 mark)*

(iii) Some ligands are bidentate. Give an example of a bidentate ligand and a complex ion that it forms. *(2 marks)*

(iv) Platin is a complex ion that exist as stereoisomers. The molecular formula of platin is $PtCl_2N_2H_6$.

Draw and label the stereoisomers of platin. *(2 marks)*

b A series of reactions are carried out on compounds of transition metals.

Explain the following practical observations. For each different observation, state the type of reaction taking place, write the formula of any precipitate or complex ion formed and construct an equation.

(i) Solid copper(I) oxide, Cu_2O, is heated with dilute sulfuric acid. A pink–brown precipitate and a blue solution forms. *(4 marks)*

(ii) A solution containing $[Cr(H_2O)_6]^{3+}$ complex ions is added to two test-tubes. A few drops of aqueous sodium hydroxide are added to one test-tube. A grey-green precipitate forms.

An excess of aqueous sodium hydroxide is added to the other test-tube. A dark green solution is formed. *(6 marks)*

(iii) Concentrated hydrochloric acid is added to a solution containing aqueous $[Cu(H_2O)_6]^{2+}$ complex ions. A solution turns a yellow colour. *(3 marks)*

20 A student investigates the rate of the reaction of propanone, $(CH_3)_2C=O$, with iodine, I_2, in the presence of dilute acid, H^+.

$(CH_3)_2C=O + I_2 \rightarrow CH_3COCH_2I + HI$

A student carries out a series of reactions at the same temperature. The student's results are shown below.

$[(CH_3)_2C=O]/$ mol dm^{-3}	$[I_2]/$ mol dm^{-3}	$[H^+]/$ mol dm^{-3}	Initial rate / mol dm^{-3} s^{-1}
1.50×10^{-3}	0.0300	0.0200	2.1×10^{-9}
1.50×10^{-3}	0.0300	0.0400	4.2×10^{-9}
1.50×10^{-3}	0.0600	0.0400	4.2×10^{-9}
4.50×10^{-3}	0.0300	0.0200	6.3×10^{-9}

a Determine the orders, the rate equation, and the value of the rate constant, including units.

Show all your working. *(7 marks)*

b What would be the initial rate when the concentrations in **Experiment 1** are all tripled? *(1 mark)*

c The student proposes a two-step mechanism for the reaction with the rate-determining step as the first step.

(i) State what is meant by the term rate-determining step. *(1 mark)*

(ii) Suggest a possible two-step mechanism for this reaction. *(2 marks)*

(iii) Suggest, with an explanation, the role of $H^+(aq)$ ions in this reaction. *(2 marks)*

OCR 2816 Jan 2006 Q6

21 Methanol can be prepared from hydrogen and carbon monoxide. This is a reversible reaction.

$2H_2(g) + CO(g) \rightleftharpoons CH_3OH(g)$
$\Delta H = -129 \, kJ \, mol^{-1}$

At equilibrium the system contains 0.640 mol $H_2(g)$, 0.320 mol CO(g), and 0.240 mol CH_3OH. The equilibrium mixture has a pressure of $2.25 \times 10^4 \, kPa$.

a Calculate K_p. Show **all** your working. Include units in your answer. *(5 marks)*

b The chemist repeated the experiment three times using the same initial amounts of $H_2(g)$ and CO(g). The chemist makes one change to the conditions from the original experiment each time.

Complete the table to show the effect if any of each change. Use the words more, less or same.

Change	Equilibrium amount of CH_3OH	K_p	Time to reach equilibrium
Higher temperature			
Higher pressure			
Addition of a catalyst			

(3 marks)

MODULE 6
Organic chemistry and analysis

Chapters in this module
25 Aromatic compounds
26 Carbonyls and carboxylic acids
27 Amines, amino acids, and polymers
28 Organic synthesis
29 Chromatography and spectroscopy

Introduction

In this module you will build upon the knowledge and understanding gained during the first year of your studies of organic chemistry. You will learn about the physical and chemical properties of several new functional groups to extend your toolkit of organic reactions, so important for organic synthesis. You will also develop further analytical techniques for the identification of organic structures.

Aromatic compounds are chemicals containing a benzene ring. Many of our modern-day medicines are aromatic compounds such as aspirin, ibuprofen, and TCP. You will learn about the structure of benzene and study its important reactions in the production of new materials.

In **Carbonyls and carboxylic acids**, you will learn about reactions of ketones, aldehydes, and carboxylic acids. You will also learn about several new functional groups derived from carboxylic acids, esters, acyl chlorides, and amides.

Amines are organic bases, derived from ammonia. Amino acids are the building blocks of life and you will learn about their acid-base behaviour and ability to show optical isomerism. You will also learn about the important condensation polymers, polyamides, and polyesters and their uses in synthetic fibres and many modern plastics.

Organic synthesis allows you to acquire further practical techniques and to develop multi-stage reaction pathways to convert one organic molecule into another. You will also learn how carbon–carbon bond formation allows the carbon skeleton of complex organic structures to be synthesised.

In **Chromatography and spectroscopy**, you will learn about NMR spectroscopy and how it can be used, combined with the infrared spectroscopy and mass spectrometry, for the complete structure determination of organic compounds.

Knowledge and understanding checklist

From your previous studies you should be able to answer the following questions. Work through each point, using your notes and the support available on Kerboodle.

- ☐ Use IUPAC rules for naming organic compounds.
- ☐ Interpret and use general, structural, displayed, and skeletal formulae.
- ☐ Draw structural isomers and E/Z isomers from a molecular formula.
- ☐ Recognise alkanes, alkenes, alcohols, haloalkanes, aldehydes, ketones, and carboxylic acids.
- ☐ Understand the terms electrophile, nucleophile, addition, substitution, and elimination.
- ☐ Use curly arrows to write mechanisms for electrophilic addition and nucleophilic substitution.
- ☐ Describe the oxidation of primary, secondary, and tertiary alcohols.
- ☐ Know how to use Quickfit apparatus for distillation and reflux.
- ☐ Use infared spectra to identify O—H in alcohols, C=O is aldehydes and ketones, and C=O and O—H in carboxylic acids.
- ☐ Use mass spectra to identify the molecular ion peak and fragment peaks.

Maths skills checklist

In this module, you will need to use the following maths skills. You can find support for these skills on Kerboodle and through MyMaths.

- ☐ **Working with standard form, significant figures, and appropriate units.** You will need this to carry out all calculations in this module.
- ☐ **Changing the subject of an equation.** You will need this when carrying out structured and unstructured mole calculations.
- ☐ **Using ratios, fractions, and percentages.** You will need this when working with moles and equations using ratios, calculating percentage yields, and calculating atom economies.
- ☐ **Using angles and shapes in regular 2-D and 3-D structures.** You will need this when predicting the shapes of and bond angles in molecules.
- ☐ **Visualising and representing 2-D and 3-D forms including 2-D representations of 3-D objects.** You will need this when drawing different types of isomer.
- ☐ **Understand the symmetry of 2-D and 3-D shapes.** You will need this when describing the types of stereoisomerism shown by molecules.
- ☐ **Translating information between graphical, numerical, and algebraic forms.** You will need this when interpreting and analysing spectra.

MyMaths.co.uk
Bringing Maths Alive

25 AROMATIC COMPOUNDS
25.1 Introducing benzene

Specification reference: 6.1.1

Learning outcomes

Demonstrate knowledge, understanding, and application of:

→ the Kekulé and delocalised models for benzene

→ the nomenclature for systematically naming substituted aromatic compounds.

Benzene – the simplest aromatic hydrocarbon

Benzene, C_6H_6, was first discovered in 1825 by the English scientist Michael Faraday, who isolated and identified it in an oily residue from the gas that was used for street lighting. Benzene is:

- a colourless, sweet smelling, highly flammable liquid
- found naturally in crude oil, is a component of petrol, and also found in cigarette smoke
- classified as a carcinogen, that is, it can cause cancer.

A benzene molecule consists of a hexagonal ring of six carbon atoms, with each carbon atom joined to two other carbon atoms and to one hydrogen atom. Benzene is classed as an aromatic hydrocarbon or **arene**. The structure of benzene is shown in two ways (Figure 1).

You may think that benzene, with a name resembling an alkene, would react in the same way as the alkenes. However this is not the case.

Derivatives of benzene

Historically, aromatic was the term used to classify the derivatives of benzene, as many pleasant-smelling compounds contained a benzene ring. Many odourless compounds have been found to contain a benzene ring, yet the term aromatic is still used to classify these compounds.

The structures of benzaldehyde and thymol are shown in Figure 2.

▲ **Figure 1** *The structure of benzene. This book mainly uses the structure on the left – both are acceptable in exams*

Synoptic link

Revisit Chapter 13, Alkenes, to revise the structure of alkenes.

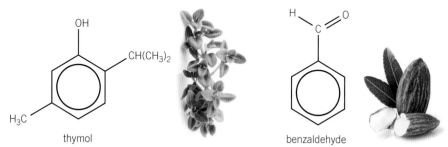

▲ **Figure 2** *The structure of benzaldehyde and thymol – both aromatic compounds containing a benzene ring. Each compound has a characteristic aroma. Benzaldehyde has the flavour of almonds (right) and thymol is found in the aromatic herb, thyme (left)*

Many aromatic compounds can be synthesised from benzene and you will meet some of these compounds and their methods of preparation in Topic 25.2.

The Kekulé and delocalised models of benzene

For many years, scientists attempted to establish a structure for benzene taking into account its molecular formula and the experimental evidence available at the time. Its molecular formula of C_6H_6 suggested a structure containing many double bonds or a structure containing double and triple bonds. Compounds containing multiple bonds were known to be very reactive, however benzene appeared unreactive.

The Kekulé model

In 1865, the German chemist Friedrich August Kekulé suggested that the structure of benzene was based on a six membered ring of carbon atoms joined by alternate single and double bonds (Figure 3). He claimed that he had thought of the ring shape of benzene whilst day-dreaming about a snake seizing its own tail.

Evidence to disprove Kekulé's model

Not all chemists accepted the Kekulé model of benzene as the structure is not able to explain all of its chemical and physical properties. This section outlines three pieces of evidence which appear to disprove Kekulé's model.

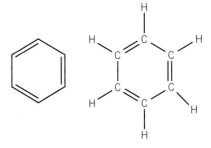

▲ Figure 3 *The skeletal and displayed structures of Kekulé's benzene*

1 The lack of reactivity of benzene

If benzene contained the C=C bonds shown in Figure 3, it should decolourise bromine in an electrophilic addition reaction. However:

- benzene does not undergo electrophilic addition reactions
- benzene does not decolourise bromine under normal conditions.

This has led scientists to suggest that benzene cannot have any C=C bonds in its structure.

2 The lengths of the carbon–carbon bonds in benzene

Using a technique called X-ray diffraction, it is possible to measure bond lengths in a molecule. When benzene was examined in 1929 by the crystallographer Kathleen Lonsdale, it was found that all the bonds in benzene were 0.139 nm in length. This bond length was between the length of a single bond, 0.153 nm, and a double bond, 0.134 nm.

3 Hydrogenation enthalpies

The Kekulé structure, containing alternate single and double bonds, could be given the name cyclohexa-1,3,5-triene to indicate the positioning of the double bonds.

If benzene did have the Kekulé structure, then it would be expected to have an enthalpy change of hydrogenation that is three times that of cyclohexene. When cyclohexene is hydrogenated (Figure 4) one double bond reacts with hydrogen. The enthalpy change of hydrogenation is -120 kJ mol^{-1}.

> **Synoptic link**
>
> Alkenes decolourise bromine water. This was covered in Topic 13.3, Reactions of alkenes, and Topic 13.4, Electrophilic addition in alkenes.

> **Study tip**
>
> Hydrogenation is the name given to the addition of hydrogen to an unsaturated compound.

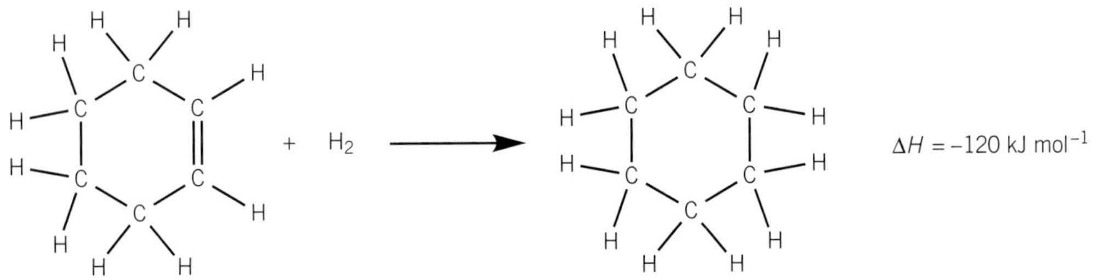

$\Delta H = -120 \text{ kJ mol}^{-1}$

▲ Figure 4 *The enthalpy change of hydrogenation of cyclohexene*

25.1 Introducing benzene

As the Kekulé structure is predicted to contain three double bonds the expected enthalpy change for reacting three double bonds with hydrogen would be 3 × −120 = −360 kJ mol^{-1} (Figure 5).

▲ **Figure 5** *The predicted enthalpy change of hydrogenation for cyclohexa-1,3,5-triene*

The actual enthalpy change of hydrogenation of benzene is only −208 kJ mol^{-1}. This means that 152 kJ mol^{-1} less energy is produced than expected. The actual structure of benzene is therefore more stable than the theoretical Kekulé model of benzene.

This information led scientists to propose the delocalised model of benzene.

The delocalised model of Benzene

The delocalised model of benzene was developed by scientists who decided that the experimental evidence was sufficient to disprove the Kekulé structure of benzene. The delocalised structure of benzene is shown in Figure 6. The main features of the delocalised model are listed below.

- Benzene is a planar, cyclic, hexagonal hydrocarbon containing six carbon atoms and six hydrogen atoms.
- Each carbon atom uses three of its available four electrons in bonding to two other carbon atoms and to one hydrogen atom.
- Each carbon atom has one electron in a p-orbital at right angles to the plane of the bonded carbon and hydrogen atoms.
- Adjacent p-orbital electrons overlap sideways, in both directions, above and below the plane of the carbon atoms to form a ring of electron density.
- This overlapping of the p-orbitals creates a system of **π-bonds (pi-bonds)** which spread over all six of the carbon atoms in the ring structure.
- The six electrons occupying this system of π-bonds (pi-bonds) are said to be delocalised.

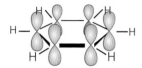
p-orbitals above and below plane of benzene ring

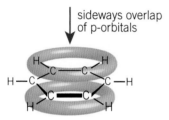
delocalised ring of electron density above and below the plane of benzene ring

▲ **Figure 6** *The delocalised structure of benzene forms when the p-orbitals overlap sideways forming a π-electron cloud above and below the plane of the carbon atoms*

> **Synoptic link**
>
> You learnt about the nature of the π-bond formed between two carbon atoms in alkenes in Topic 13.1, Properties of alkenes.

Naming aromatic compounds

In their names, some groups are shown as prefixes to benzene. These include short alkyl chains, halogens, and nitro groups.

Compounds with one substituent group

Aromatic compounds with one substituent group are monsubstituted. In aromatic compounds, the benzene ring is often considered to be the parent chain. Alkyl groups (CH_3, C_2H_5), halogens (F, Cl, Br, I), and nitro (NO_2) groups are all considered the prefixes to benzene (Figure 7).

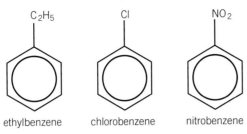

▲ **Figure 7** *The structures of ethylbenzene, chlorobenzene, and nitrobenzene*

When a benzene ring is attached to an alkyl chain with a functional group or to an alkyl chain with seven or more carbon atoms, benzene is considered to be a **substituent**. Instead of benzene, the prefix phenyl is used in the name (Figure 8).

▲ **Figure 8** *Examples of structures where the benzene ring is a substituent group*

As with all systems, there are some noticeable exceptions and unfortunately some names just have to be learnt and remembered. Benzoic acid (benzenecarboxylic acid), phenylamine, and benzaldehyde (benzenecarbaldehyde) are three common compounds you will encounter when studying aromatic chemistry (Figure 9).

Compounds with more than one substituent group

Some molecules may contain more than one substituent on the benzene ring, for example, disubstituted compounds have two substituents groups. The ring is now numbered, just like a carbon chain, starting with one of the substituent groups. The substituent groups are listed in alphabetical order using the smallest numbers possible.

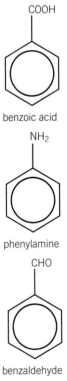

▲ **Figure 9** *The structures of benzoic acid, phenylamine, and benzaldehyde – all derivatives of benzene*

> **Worked example: Naming aromatic compounds**
>
> Name each of the following aromatic compounds.
>
> **Example 1**
> **Step 1:** The molecule is based on the methylbenzene with a bromine substituent.
>
> **Step 2:** When naming compounds, you always use the lowest combination of numbers possible. Therefore, bromine is on carbon-2
>
>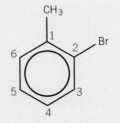
>
> ▲ **Figure 10** *2-bromomethylbenzene*
>
> **Step 3:** The compound is named 2-bromomethylbenzene.

Synoptic link

Revisit Topic 11.2, Nomenclature of organic compounds, if you need to revise the basics of naming aliphatic compounds.

25.1 Introducing benzene

Example 2
Step 1: Benzene is the parent molecule.

Step 2: The molecule contains two chlorine substituents: one on carbon-1 and one on carbon-4.

Step 3: The compound is called 1,4-dichlorobenzene.

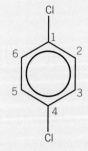

▲ **Figure 11** 1,4-dichlorobenzene

Example 3
Step 1: The molecule is based on methylbenzene with a chlorine substituent on carbon-3.

Step 2: The compound is called 3-chloromethylbenzene.

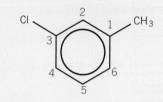

▲ **Figure 12** 3-chloromethylbenzene

Summary questions

1. State the empirical formula of benzene. *(1 mark)*
2. Explain how X-ray diffraction helped scientists to develop the delocalised model of benzene. *(3 marks)*
3. Explain why the delocalised model of benzene accounts for the observed stability of benzene better than the Kekulé model. *(3 marks)*
4. Draw the structures for the following aromatic compounds.
   a. 1,2,4-tribromobenzene *(1 mark)*
   b. 1,3-dinitrobenzene *(1 mark)*
   c. 2,4,6-trichloromethylbenzene *(1 mark)*
   d. phenylpropanone *(1 mark)*
   e. 2,3-dichlorobenzoic acid. *(1 mark)*
5. Name the following molecules.

25.2 Electrophilic substitution reactions of benzene

Specification reference: 6.1.1

Reactivity of benzene and its derivatives

Benzene and its derivatives undergo substitution reactions in which a hydrogen atom on the benzene ring is replaced by another atom or group of atoms. Benzene typically reacts with electrophiles and most of the reactions of benzene proceed by electrophilic substitution.

A typical equation representing electrophilic substitution is shown in Figure 1.

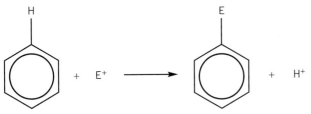

▲ **Figure 1** *In the equation above, the electrophile is E^+. In the reaction, E substitutes for a hydrogen atom on the benzene ring. This is a typical electrophilic substitution reaction*

Nitration of benzene

Benzene reacts slowly with nitric acid to form nitrobenzene (Figure 2). The reaction is catalysed by sulfuric acid and heated to 50 °C to obtain a good rate of reaction. A water bath is used to maintain the steady temperature.

In nitration, one of the hydrogen atoms on the benzene ring is replaced by a nitro, $-NO_2$, group.

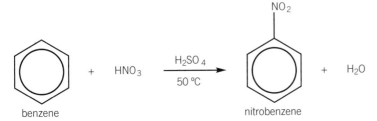

▲ **Figure 2** *The reaction of benzene with nitric acid*

If the temperature of the reaction rises above 50 °C, further substitution reactions may occur leading to the production of dinitrobenzene. This shows the importance of temperature control in the preparation of organic compounds.

Learning outcomes

Demonstrate knowledge, understanding, and application of:

→ electrophilic substitution in aromatic compounds
→ the mechanism for electrophilic substitution
→ the reaction of benzene and alkenes with bromine.

Synoptic link

You should recall the definitions of electrophile and substitution from Topic 13.4, Electrophilic addition in alkenes, and Topic 11.5, Introduction to reaction mechanisms.

Study tip

You should be able to define and give suitable examples of electrophiles and substitution reactions.

Synoptic link

You were first introduced to catalysts and their effect on reaction rates in Topic 10.2, Catalysts.

25.2 Electrophilic substitution reactions of benzene

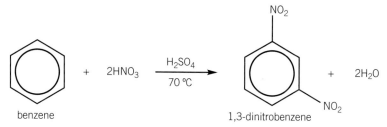

▲ **Figure 3** *Benzene can react with excess nitric acid above 50 °C to form 1,3-dinitrobenzene*

Nitrobenzene is an important starting material in the preparation of dyes, pharmaceuticals, and pesticides. It can be used as a starting material in the preparation of paracetamol (Figure 4).

The reaction mechanism for the nitration of benzene, electrophilic substitution, is shown in Figure 5.

The reaction involves an electrophile. However, nitric acid is not the electrophile involved in the mechanism. The electrophile in this reaction is the nitronium ion, NO_2^+, produced by the reaction of concentrated nitric acid with concentrated sulfuric acid (Step 1).

In the next step (Step 2) the electrophile, NO_2^+, accepts a pair of electrons from the benzene ring to form a dative covalent bond. The organic intermediate formed is unstable and breaks down to form the organic product, nitrobenzene, and the H^+ ion. A stable benzene ring is reformed.

Finally, the H^+ ion formed in Step 2 reacts with the HSO_4^- ion from Step 1 to regenerate the catalyst, H_2SO_4 (Step 3).

Step 1: $HNO_3 + H_2SO_4 \longrightarrow NO_2^+ + HSO_4^- + H_2O$

Step 2:

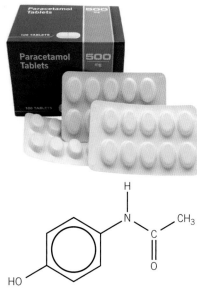

▲ **Figure 4** *Paracetamol is a common painkiller used to alleviate headache and symptoms of fever. Paracetamol can be synthesised from nitrobenzene*

Study tip
When drawing a mechanism you must ensure that your arrows start and end at the correct place.

In Figure 5, Step 2, you can see how precisely the curly arrows have been positioned.

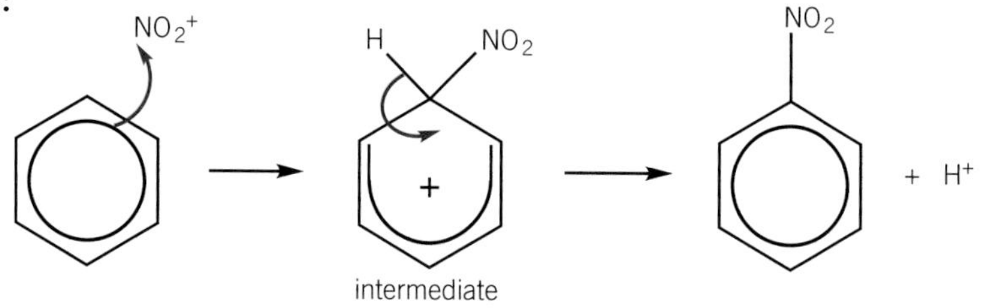

Step 3: $H^+ + HSO_4^- \longrightarrow H_2SO_4$

▲ **Figure 5** *The mechanism for the nitration of benzene*

Halogenation of benzene

The halogens do not react with benzene unless a catalyst called a **halogen carrier** is present. Common halogen carriers include $AlCl_3$, $FeCl_3$, $AlBr_3$, and $FeBr_3$, which can be generated *in situ* (in the reaction vessel) from the metal and the halogen.

Bromination of benzene

At room temperature and pressure and in the presence of a halogen carrier, benzene reacts with bromine in an electrophilic substitution reaction as shown in Figure 6.

Study tip
You may be asked to show the role of a catalyst.

From Figure 5, you would need to show Step 1 and Step 3.

In bromination, one of the hydrogen atoms on the benzene ring is replaced by a bromine atom.

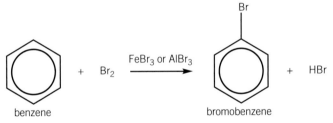

▲ **Figure 6** *The equation for the preparation of bromobenzene*

The reaction mechanism is electrophilic substitution as shown in Figure 7.

Benzene is too stable to react with a non-polar bromine molecule. The electrophile is the bromonium ion, Br^+, which is generated when the halogen carrier catalyst ($FeBr_3$ in this example) reacts with bromine in the first stage of the mechanism (Step 1).

In Step 2, the bromonium ion accepts a pair of electrons from the benzene ring to form a dative covalent bond. The organic intermediate is unstable and breaks down to form the organic product, bromobenzene, and an H^+ ion.

Finally, the H^+ formed in Step 2 reacts with the $FeBr_4^-$ ion from Step 1 to regenerate the $FeBr_3$ catalyst (Step 3).

Step 1: $Br_2 + FeBr_3 \longrightarrow FeBr_4^- + Br^+$

Step 2:

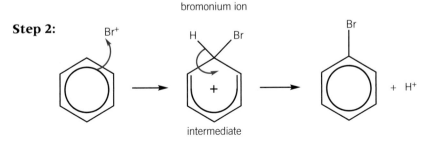

Step 3: $H^+ + FeBr_4^- \longrightarrow FeBr_3 + HBr$

▲ **Figure 7** *The mechanism for the bromination of benzene*

> **Study tip**
>
> You may be asked to show the role of the halogen carrier or how the electrophile is generated.
>
> You must include Step 1 and Step 3 in your response.

Chlorination of benzene

Chlorine will react with benzene in the same way as bromine and following the same mechanism. The halogen carrier used is $FeCl_3$, $AlCl_3$, or iron metal and chlorine, which react to make $FeCl_3$. The equation for the reaction is shown in Figure 8.

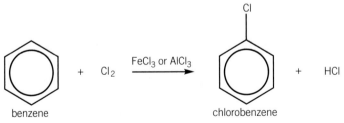

▲ **Figure 8** *Equation for the chlorination of benzene*

25.2 Electrophilic substitution reactions of benzene

Alkylation reactions

The alkylation of benzene is the substitution of a hydrogen atom in the benzene ring by an alkyl group. The reaction is carried out by reacting benzene with a haloalkane in the presence of $AlCl_3$, which acts as a halogen carrier catalyst, generating the electrophile.

Alkylation increases the number of carbon atoms in a compound by forming carbon–carbon bonds.

The reaction is sometimes called a Friedel-Crafts alkylation after the two chemists who first carried out the reaction. An example of a Friedel-Crafts alkylation is shown in Figure 9.

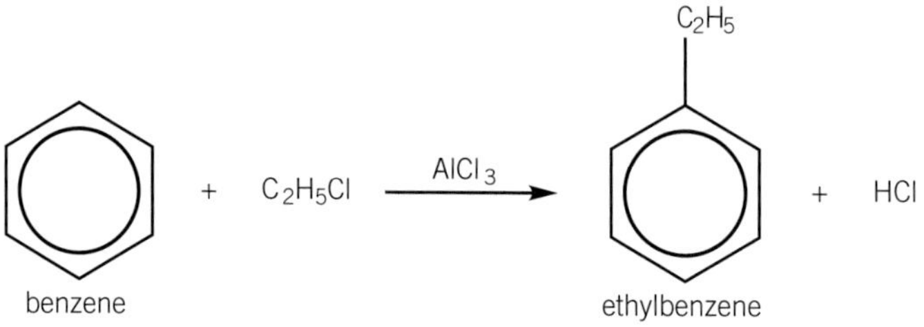

> **Synoptic link**
> Alkylation can be used to add a carbon chain to the benzene ring. This increases the number of carbon atoms in the structure. You will learn other methods of increasing the length of a carbon chain in Topic 28.1, Carbon–carbon bond formation.

▲ **Figure 9** *Equation for the alkylation of benzene*

Acylation reactions

When benzene reacts with an acyl chloride in the presence of an $AlCl_3$ catalyst, an aromatic ketone is formed. This is called an acylation reaction and is another example of electrophilic substitution. The reaction forms carbon–carbon bonds and is again useful in organic synthesis.

Ethanoyl chloride, CH_3COCl, is the first member of the acyl chloride homologous series. Phenylethanone in produced in the reaction between benzene and ethanoyl chloride (Figure 10) and is used in the perfume industry (Figure 11).

> **Study tip**
> You need to be able to interpret unfamiliar electrophilic substitution reactions for aromatic compounds or to work out an unfamiliar mechanism from given information.

> **Synoptic link**
> Acyl chlorides are discussed in more detail in Topic 25.4, Carboxylic acid derivatives.

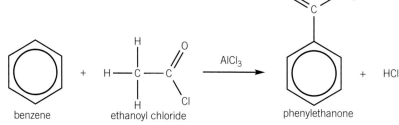

▲ **Figure 10** *The reaction between ethanoyl chloride and benzene, forming phenylethanone*

▲ **Figure 11** *Phenylethanone is an ingredient in fragrances that resemble almond, cherry, honeysuckle, jasmine, and strawberry*

Comparing the reactivity of alkenes with arenes

Alkenes decolourise bromine by an electrophilic addition reaction. One example of this reaction is the reaction between cyclohexene and bromine (Figure 12).

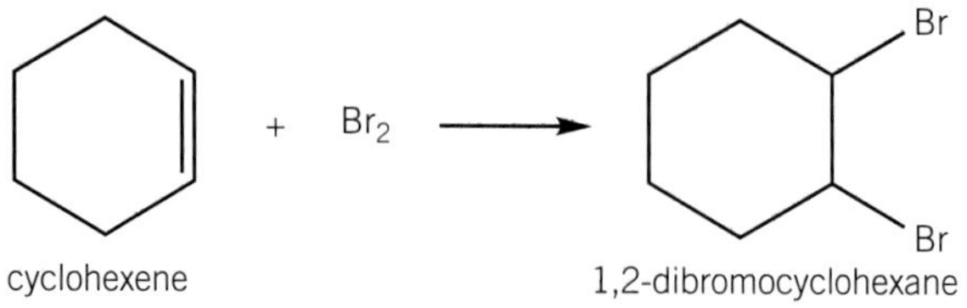

▲ **Figure 12** *The reaction between cyclohexene and bromine*

25 AROMATIC COMPOUNDS

In this reaction bromine adds across the double bond in cyclohexene.

1. The π-bond in the alkene contains *localised* electrons above and below the plane of the two carbon atoms in the double bond. This produces an area of high electron density.
2. The localised electrons in the π-bond induce a dipole in the non-polar bromine molecule making one bromine atom of the Br_2 molecule slightly positive and the other bromine atom slightly negative.
3. The slightly positive bromine atom enables the bromine molecule to act like an electrophile.

The mechanism for this reaction is electrophilic addition (Figure 13).

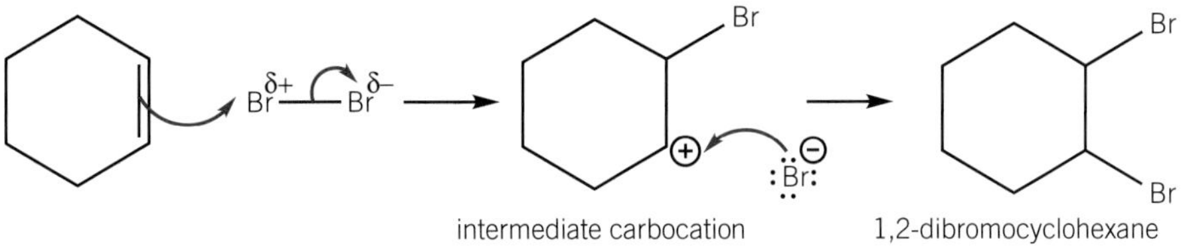

> **Synoptic link**
>
> The mechanism of electrophilic addition is covered in detail in Topic 13.4, Electrophilic addition of alkenes.

▲ **Figure 13** *The mechanism for the reaction between cyclohexene and bromine*

Unlike alkenes, benzene does not react with bromine unless a halogen carrier catalyst is present. This is because benzene has *delocalised* π-electrons spread above and below the plane of the carbon atoms in the ring structure. The electron density around any two carbon atoms in the benzene ring is less than that in a C=C double bond in an alkene.

When a non-polar molecule such as bromine approaches the benzene ring there is insufficient π-electron density around any two carbon atoms to polarise the bromine molecule. This prevents any reaction taking place.

The mechanism for the reaction of bromine with benzene in the presence of a halogen carrier is electrophilic substitution, as described in detail earlier in this topic.

> **Study tip**
>
> Alkenes react with bromine by electrophilic addition.
>
> Benzene reacts with bromine by electrophilic substitution.
>
> Make sure that you know the difference and learn the two mechanisms.

Summary questions

1. Benzene reacts by electrophilic substitution. Define the terms:
   a. electrophile *(1 mark)*
   b. substitution *(1 mark)*
2. Benzene reacts with chlorine by an electrophilic substitution mechanism. Outline this mechanism including how the electrophile is generated. *(4 marks)*
3. Benzene reacts with 2-chloropropane, in the presence of an $AlCl_3$ catalyst, by electrophilic substitution. Construct the equation for this reaction. *(2 marks)*
4. a. Deduce the structure of the organic product expected from the reaction of each of the following compounds with benzene in the presence of $AlCl_3$.
   i. $(CH_3)_2CHCOCl$
   ii. C_6H_5COCl *(2 marks)*
   b. Use curly arrows to show the mechanism for the reaction between $(CH_3)_2CHCOCl$ and benzene, clearly showing the role of the catalyst in the reaction *(4 marks)*

25.3 The chemistry of phenol
Specification reference: 6.1.1

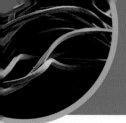

Learning outcomes
Demonstrate knowledge, understanding, and application of:
→ the acidity of phenol
→ electrophilic substitution reactions of phenol.

Phenols and phenol

Phenols are a type of organic chemical containing a hydroxyl, −OH, functional group directly bonded to an aromatic ring. The simplest member of the phenols, C_6H_5OH, has the same name as the group – phenol. In this topic you will study some of the reactions of phenol but any compound that contains an −OH group attached *directly* to the benzene ring will react in a similar way.

Some compounds, such as $C_6H_5CH_2OH$, contain an −OH group bonded to a carbon side chain, rather than the aromatic ring. These compounds are classified as alcohols rather than phenols.

Although alcohols and phenols have some common reactions, many reactions are different as the proximity of the delocalised ring influences the −OH group. It is important that you are able to recognise the difference between phenols and alcohols. Figure 2 shows three phenols and an aromatic alcohol.

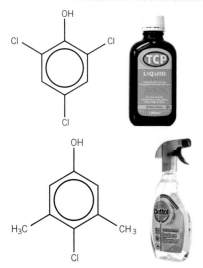

▲ **Figure 1** *Phenols are used in many everyday antiseptics. In the phenol groups of each structure, the −OH group is attached directly to the benzene ring*

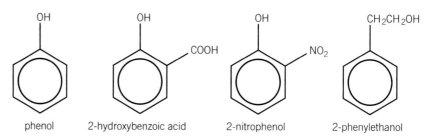

▲ **Figure 2** *The first three compounds all show the characteristic reactions of phenol, whereas the fourth compound, 2-phenylethanol, is an aromatic alcohol. The aromatic alcohol will react similarly to alcohols, rather than phenols*

 ## The manufacture of phenol

Phenol is an important chemical used in the production of disinfectants, detergents, plastics, paints, and even aspirin.

Original manufacture of phenol
Phenol used to be made from benzene, C_6H_6, using sulfuric acid and sodium hydroxide in a multi-stage process. The overall reaction is shown below.

$$C_6H_6 + H_2SO_4 + 2NaOH \rightarrow C_6H_5OH + Na_2SO_3 + 2H_2O$$

Current manufacture of phenol
Nowadays, the majority of phenol is manufactured from benzene and propene, C_3H_6, in a multistep reaction. The overall reaction is shown below.

$$C_6H_6 + C_3H_6 + O_2 \rightarrow C_6H_5OH + CH_3COCH_3$$

This reaction has an 86% yield of phenol from benzene. The other product is propanone, CH_3COCH_3 (also called acetone). Although propanone is a useful product, it is in less demand than phenol, and companies can find it difficult to make use of all the propanone produced.

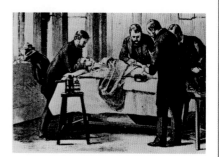

▲ **Figure 3** *Surgery in the nineteenth century used an antiseptic spray of phenol solution (seen here on the stool). Phenol harms the skin and has now been replaced by better antiseptics, but TCP (2,4,6-trichlorophenol), a related compound, is still used as an antiseptic in the home*

25 AROMATIC COMPOUNDS

Future manufacture of phenol

Research chemists are investigating another way of producing phenol using benzene and nitrogen(I) oxide, N_2O (also called nitrous oxide).

$$C_6H_6 + N_2O \rightarrow C_6H_5OH + N_2$$

This reaction has a 95% yield of phenol from benzene. Nitrous oxide is a gaseous waste product from the production of nylon and cannot be allowed to escape into the atmosphere, as it is a greenhouse gas.

1. Calculate the atom economy for the three methods for preparing phenol.
2. Calculate the mass of phenol that could be produced from 100 tonnes of benzene after taking into account the percentage yield for:
   a the current method
   b the possible future method.
3. Explain how the future method affects the atom economy for the production of nylon.

▲ **Figure 4** *Distillation units for separating phenol and propanone after reaction*

Phenol as a weak acid

Phenol is less soluble in water than alcohols due to the presence of the non-polar benzene ring. When dissolved in water, phenol partially dissociates forming the phenoxide ion and a proton (Figure 3). Because of this ability to partially dissociate to produce protons, phenol is classified as a weak acid. Similarly, other phenols act as weak acids.

Phenol is more acidic than alcohols but less acidic than carboxylic acids. This can be seen by comparing the acid dissociation constant K_a of an alcohol with a phenol and a carboxylic acid (Table 1).

- Ethanol does not react with sodium hydroxide (a strong base) or sodium carbonate (a weak base).
- Phenols and carboxylic acids react with solutions of strong bases such as aqueous sodium hydroxide.
- Only carboxylic acids are strong enough acids to react with the weak base, sodium carbonate.

A reaction with sodium carbonate can be used to distinguish between a phenol and a carboxylic acid – the carboxylic acid reacts with sodium carbonate to produce carbon dioxide, which is evolved as a gas.

Reaction of phenol with sodium hydroxide

Phenol reacts with sodium hydroxide to form the salt, sodium phenoxide, and water in a neutralisation reaction (Figure 4).

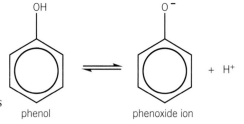

▲ **Figure 3** *The partial dissociation of phenol forms the phenoxide ion and a proton*

Synoptic link

For details of the acid dissociation constant K_a see Topic 20.3, The acid dissociation constant K_a.

▼ **Table 1** *The acid dissociation constants K_a of an alcohol, a carboxylic acid, and a phenol*

Compound	K_a (298 K) / mol dm^{-3}
ethanol	1.00×10^{-16}
phenol	1.26×10^{-11}
ethanoic acid	1.00×10^{-5}

Acidity increases as K_a value increases.

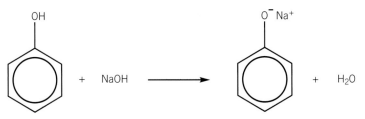

▲ **Figure 4** *The reaction of phenol with aqueous sodium hydroxide*

25.3 The chemistry of phenol

Electrophilic substitution reactions of phenol

Phenols are aromatic compounds and they undergo electrophilic substitution reactions. The reactions of phenol take place under milder conditions and more readily than the reactions of benzene.

Bromination of phenol

Phenol reacts with an aqueous solution of bromine (bromine water) to form a white precipitate of 2,4,6-tribromophenol (Figure 5 and Figure 6). The reaction decolourises the bromine water (orange to colourless). With phenol, a halogen carrier catalyst is *not* required and the reaction is carried out at room temperature.

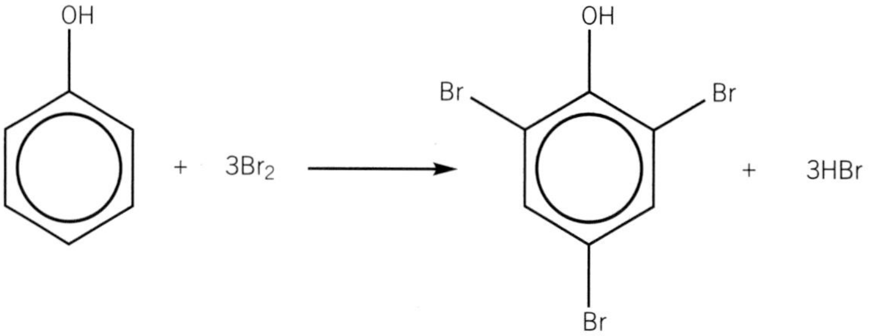

▲ **Figure 5** *When bromine water (orange) is reacted with phenol the bromine water is decolourised and a white precipitate is formed*

▲ **Figure 6** *Phenol reacts with aqueous bromine to form 2,4,6-tribromophenol*

Study tip

Addition of bromine water to phenol results in two observations — bromine decolourises and a white precipitate is formed.

You should always give both observations.

Nitration of phenol

Phenol reacts readily with dilute nitric acid at room temperature. A mixture of 2-nitrophenol and 4-nitrophenol is formed (Figure 7).

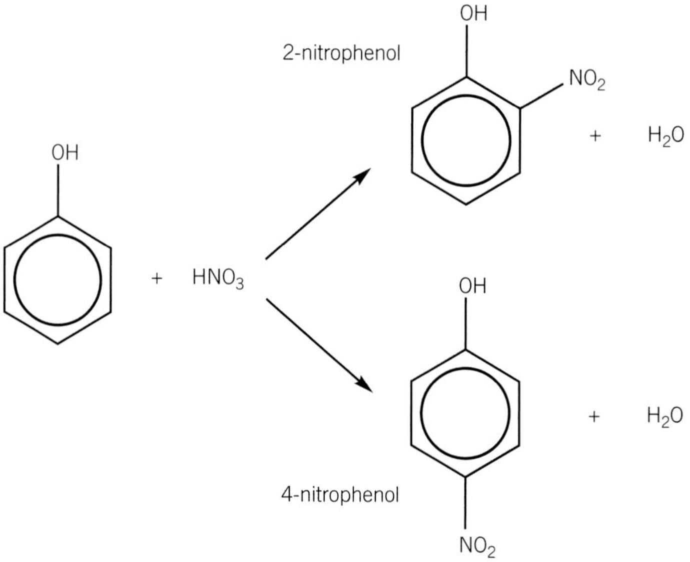

Study tip

The conditions for nitrating phenol are much milder than when nitrating benzene. Concentrated sulfuric acid is *not* required.

▲ **Figure 7** *The nitration of phenol produces a mixture of two isomers, 2-nitrophenol and 4-nitrophenol*

Synoptic link

See also the comparison in reactivity of bromine with benzene and alkenes in Topic 25.2, Electrophilic substitution reactions of benzene.

Comparing the reactivity of phenol and benzene

Bromine and nitric acid react more readily with phenol than they do with benzene. Phenol is nitrated with dilute nitric acid rather than needing concentrated nitric and sulfuric acids as with benzene.

The increased reactivity is caused by a lone pair of electrons from the oxygen p-orbital of the –OH group being donated into the π-system of phenol. The electron density of the benzene ring in phenol is increased. The increased electron density attracts electrophiles more strongly than with benzene.

The aromatic ring in phenol is therefore more susceptible to attack from electrophiles than in benzene. For bromine, the electron density in the phenol ring structure is sufficient to polarise bromine molecules and so no halogen carrier catalyst is required.

Summary questions

1 **a** Name each of the following molecules.

A 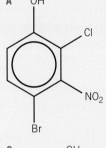   B CH₂CH₂CH₂OH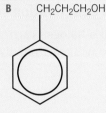

C 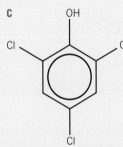   D

(4 marks)

b Identify the molecules that are phenols. (1 mark)

2 Write an equation to show the reaction of 2-hydroxyphenol with an excess of sodium hydroxide. (1 mark)

3 Outline how you would carry out a test to distinguish between solutions of pent-1-ene and phenol. (1 mark)

4 Explain why bromine reacts more readily with phenol than with benzene. (3 marks)

5 Phenol reacts with chloromethane to form 2-methylphenol. The reactivity of phenol means that a halogen carrier is not required. Using curly arrows, suggest a mechanism for this reaction.
(Hint: The initial attack is from the benzene ring to the electron-deficient carbon atom of chloromethane). (4 marks)

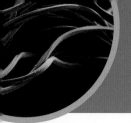

25.4 Directing groups
Specification reference: 6.1.1

Learning outcomes

Demonstrate knowledge, understanding, and application of:

→ the directing effects of electron donating and electron withdrawing groups in electrophilic substitution

→ the importance of directing groups in organic synthesis.

Further substitution

Phenol can undergo an electrophilic substitution reaction with nitric acid to form two isomers, 2-nitrophenol and 4-nitrophenol (Topic 25.3, The chemistry of phenol). Like phenol, many substituted aromatic compounds can undergo a second substitution – disubstitution. Some of these reactions take place more readily than benzene itself whereas other reactions take place less easily and require extreme conditions.

Activation and deactivation

Bromine requires a halogen carrier catalyst to react with benzene, whereas bromine will react rapidly with phenylamine (Figure 1).

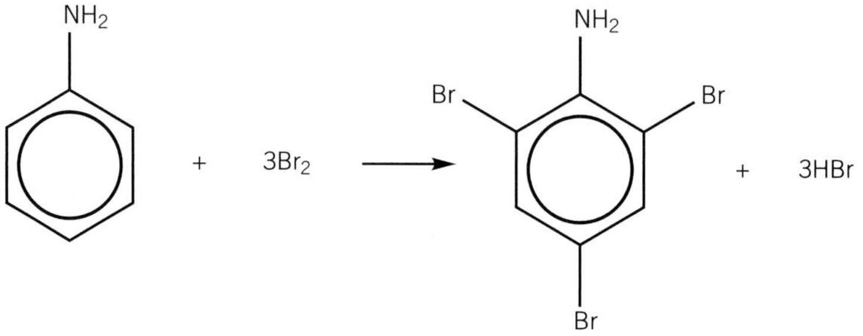

▲ **Figure 1** *Phenylamine reacts more readily than benzene with bromine to produce a multi-substituted product*

Nitrobenzene reacts slowly with bromine, requiring both a halogen carrier catalyst and a high temperature. The benzene ring in nitrobenzene is less susceptible to electrophilic substitution than benzene itself.

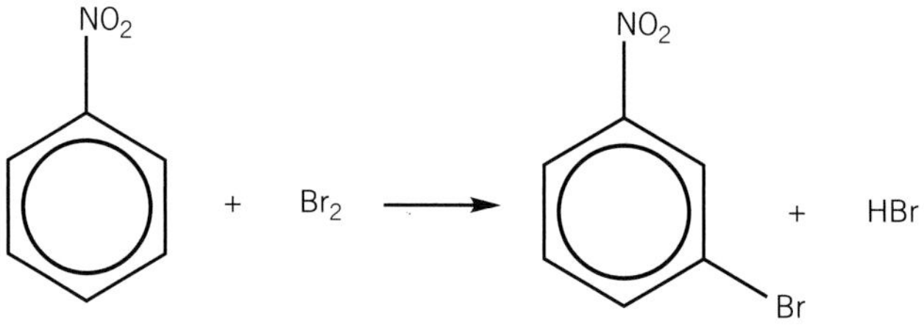

▲ **Figure 2** *The reaction of nitrobenzene with bromine produces only one product, 3-bromonitrobenzene*

In the first example shown in Figure 1, the $-NH_2$ group **activates** the ring as the aromatic ring reacts more readily with electrophiles. In the second example in Figure 2, the $-NO_2$ group **deactivates** the aromatic ring as the ring reacts less readily with electrophiles.

Not only does the rate of reaction and extent of substitution differ in the two examples but also the position of substitution on the benzene ring is also different.

AROMATIC COMPOUNDS   25

- The $-NH_2$ group directs the second substituent to positions 2 or 4.
- The $-NO_2$ group directs the second substituent to position 3.

> **Study tip**
>
> The $-NH_2$ group is said to be 2- and 4- directing.
>
> The $-NO_2$ group is said to be 3- directing.

Ortho, meta, and para

You may come across the terms *ortho*, *meta,* and *para* in more advanced textbooks and on the Internet. *Ortho* (*o-*), *meta* (*m-*), and *para* (*p-*) refer to positions on a benzene ring relative to the group at position 1.

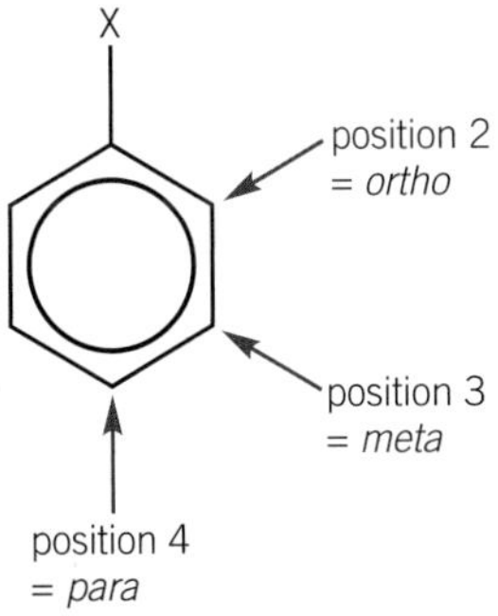

▲ **Figure 3** *The positions ortho, meta, and para can be used to identify positions on a benzene ring*

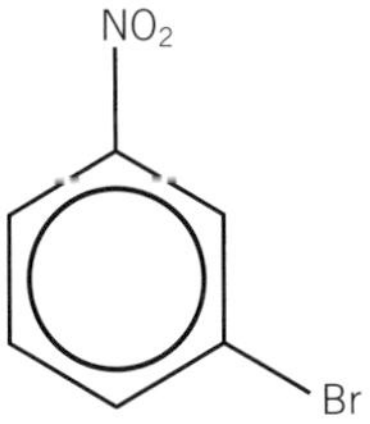

▲ **Figure 4** *3-bromonitrobenzene or meta-bromonitrobenzene or m-bromonitrobenzene*

For example, 3-bromonitrobenzene is also known as *meta*-bromonitrobenzene or more simply as *m*-bromonitrobenzene.

Name each of the following molecules using *ortho*, *meta*, and *para* names and the number-based IUPAC names.

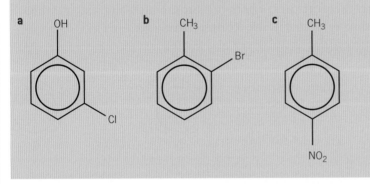

Directing effects

There are many different groups that can be attached to a benzene ring. Different groups can have a **directing effect** on any second substituent on the benzene ring.

All 2- and 4-directing groups (*ortho* and *para*-directors) are activating groups, with the exception of the halogens. All 3-directing groups (*meta* directors) are deactivating groups.

▼ **Table 1** *Table showing the effect of a first substituent on second substitutions*

2- and 4-directing (*ortho*-and-*para* directing)	3-directing (*meta* directing)
$-NH_2$ or $-NHR$	RCOR
$-OH$	$-COOR$
$-OR$	$-SO_3H$
$-R$ or $-C_6H_5$	$-CHO$
$-F, -Cl, -Br, -I$	$-COOH$
	$-CN$
	$-NO_2$
	$-NR_3^+$

25.4 Directing groups

Worked example: Predicting substitution products I

Benzoic acid, C₆H₅COOH, can be nitrated using a mixture of nitric acid, HNO₃, and sulfuric acid, H₂SO₄. Identify the mono-substituted organic product(s) of benzoic acid and write an equation for the reaction.

Step 1: As the –COOH group is a 3-directing group there is only one organic product and the NO₂ group will substitute to the 3-carbon position:

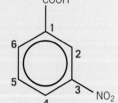

Step 2: Write the equation for the reaction.

Worked example: Predicting substitution products II

Methylbenzene reacts more readily than benzene with bromine, forming approximately 67% of one mono-substituted product of methylbenzene and 33% of a second mono-substituted product. Draw the structures of both organic products and suggest which product is formed in the greatest percentage.

Step 1: As the –CH₃ group is a 2- and 4-directing group, substitution of the bromine group can be at the 2-, 4-, or 6-positions.

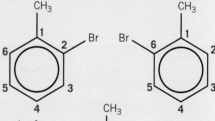

Step 2: Positions 2 and 6 are equivalent so you would expect approximately twice as much 2-bromomethylbenzene as 4-bromomethylbenzene.

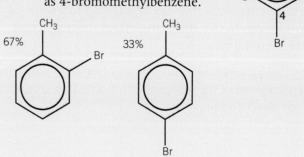

Study tip

Other factors can also affect the proportions of different substituted isomers. For example, a large substituted group may get in the way.

This is called a steric effect and forms part of university courses in chemistry.

AROMATIC COMPOUNDS
25

Using directing effects in organic synthesis

The directing effect of substituent groups can be used when planning an organic synthesis. When carrying out more than one electrophilic substitution reaction on an aromatic compound, you may have to consider the order in which the reactions are carried out to ensure the correct substitution pattern and that the required product is prepared.

> **Worked example: Planning an organic synthesis**
>
> Suggest a series of reactions that could be used to carry out the following synthesis. You should be able to carry out the reaction in two steps.
>
>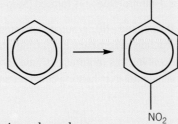
>
> **Step 1:** Identify the type of disubstitution.
>
> The required product is a 1,4-disubstituted product.
>
> Therefore the first substituent added must direct to the 4 position.
>
> **Step 2:** Identify the directing effect of the two substituents.
>
> - NO_2 directs the second substituent to carbon 3.
> - Cl directs the second substituent to carbons 2 and 4.
>
> Therefore Cl must be substituted first to give the correct direction for the second NO_2 substituent.
>
> **Step 3:** Write out the two steps with the reagents needed for the synthesis.
>
> benzene →(Cl₂/AlCl₃, Step 1)→ chlorobenzene →(HNO₃/H₂SO₄, Step 2)→ 4-chloro-nitrobenzene + 2-chloro-nitrobenzene
>
> **Step 4:** Is there is a mixture of organic products to separate?
>
> As Cl directs to positions 2 and 4, a mixture of the required 4- substituted product and the 2- substituted product would be obtained. These could be separated by distillation if they are liquids, or other methods including recrystallisation, if they are solids.

Synoptic link

For details of distillation, see Topic 16.1, Practical techniques in organic chemistry.

For recrystallisation, see Topic 28.2, Further practical techniques.

25.4 Directing groups

Making TNT

An application of directing groups can be seen in the reaction of methylbenzene (toluene) with an excess of concentrated nitric and sulfuric acid.

At 50 °C, toluene (methylbenzene) forms a monosubstituted product, 2-nitrotoluene (Step 1 in Figure 6).

▲ **Figure 5** *TNT is best known now as an explosive, but it was originally used as a yellow dye. TNT only found use as an explosive at the beginning of the twentieth century, when it was used in armour-piercing artillery shells*

The product of this reaction is less reactive than toluene because the $-NO_2$ group deactivates the benzene ring. However if the temperature of the reaction mixture is increased to 70 °C a disubstituted product, 2,4-dinitrotoluene is formed (Step 2 in Figure 6).

The second $-NO_2$ group deactivates the benzene ring further. Further reaction will produce 2,4,6-trinitrotoluene – TNT (Step 3 in Figure 6). However, this requires extreme conditions – a good job really as any accidental over-heating could result in an explosion.

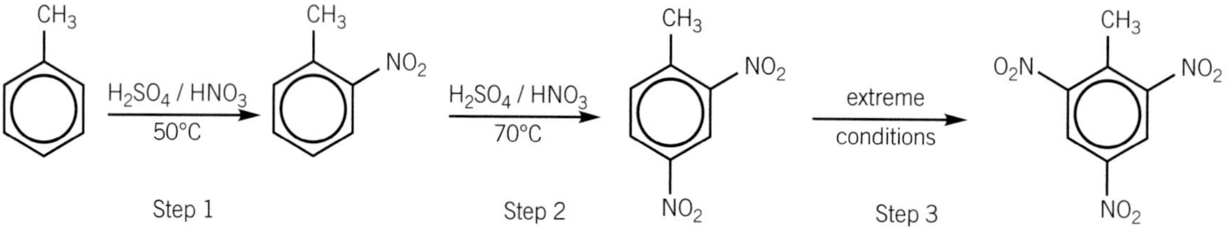

▲ **Figure 6** *The preparation of trinitrotoluene from benzene*

1. Write an equation for Step 2 in this synthesis.
2. Show the mechanism for the first step in this reaction. You do not need to show how the electrophile is prepared.

Summary questions

1. Identify each of the following compounds as containing 2-, 4-, or 3-directing groups.
   a $C_6H_5CH_2CH_3$ *(1 mark)*
   b C_6H_5Cl *(1 mark)*
   c C_6H_5COOH *(1 mark)*
   d $C_6H_5N^+(CH_3)_3$ *(1 mark)*

2. Identify and draw the structures of the product(s) of the following reactions.
   a $C_6H_5Br + HNO_3$ *(1 mark)*
   b $C_6H_5NO_2 + Br_2 + FeBr_3$ *(1 mark)*
   c $C_6H_5COOCH_2CH_3 + CH_3Cl + AlCl_3$ *(1 mark)*

3. Starting from benzene, suggest a two-step synthesis of:
   a 3-nitromethylbenzene *(2 marks)*
   b 2-nitromethylbenzene. *(2 marks)*

Practice questions

1 Salicylic acid is a naturally occurring carboxylic acid, widely used in organic synthesis.

   a The flowchart below shows some reactions of salicylic acid.

   (i) In the box below, draw the structure of the organic compound formed by **reaction 1**. *(1 mark)*

   (ii) Write a chemical equation to represent **reaction 2**. *(1 mark)*

   (iii) State the reagents and conditions in **reaction 3**. *(1 mark)*

   b Bromine reacts more readily with salicylic acid than with benzene.

   (i) Outline the mechanism for the bromination of salicylic acid shown in **reaction 2** in the flowchart. A halogen carrier is not required for this reaction. The electrophile is Br_2. *(4 marks)*

   (ii) Explain why bromine reacts more readily with salicylic acid than with benzene.
   In your answer, you should use appropriate technical terms, spelled correctly. *(3 marks)*

   OCR F324 June 2014 Q1 (a) (b)

2 **a** In the first stage of the synthesis of a dye, methyl 3-nitrobenzoate is formed. In this stage, methyl benzoate is nitrated by concentrated nitric acid, in the presence of concentrated sulfuric acid as a catalyst.

   (i) Outline the mechanism for this nitration of methyl benzoate. Show how H_2SO_4 behaves as a catalyst. *(5 marks)*

   (ii) State the name for this type of mechanism. *(1 mark)*

   OCR F324 Jan 2013 Q2(a)

3 Benzaldehyde, C_6H_5CHO, is the simplest aromatic aldehyde and has a characteristic smell of almonds.

   Benzaldehyde can be nitrated with a mixture of concentrated nitric acid and concentrated sulfuric acid to form 3-nitrobenzaldehyde.

   Explain, with the aid of curly arrows, the mechanism for the formation of 3-nitrobenzaldehyde. Your answer should clearly show the role of sulfuric acid as a catalyst. *(6 marks)*

   OCR F324 Jun 2012 Q4 (a)

4 Benzene and other arenes can be chlorinated to produce chloroarenes which are used in the manufacture of pesticides, drugs and dyes.

   a In 1865, Kekulé proposed a model for the structure and bonding of benzene, but there is considerable evidence to suggest that Kekulé's model may not be correct. Scientists have proposed alternative models for the structure and bonding of benzene. Explain the evidence that led scientists to doubt the model proposed by Kekulé. *(3 marks)*

b Chlorobenzene, C_6H_5Cl, is formed by the reaction of benzene and chlorine in the presence of a suitable catalyst, such as $AlCl_3$.

$$C_6H_6 + Cl_2 \rightarrow C_6H_5Cl + HCl$$

Outline the mechanism for the formation of chlorobenzene from benzene. Show how $AlCl_3$ behaves as a catalyst. *(6 marks)*

c Chlorobenzene reacts with trichloroethanal, Cl_3CCHO, to produce the pesticide DDT.

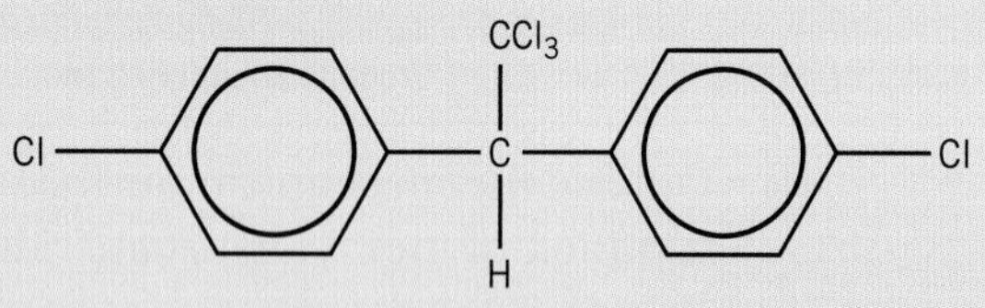

DDT

Construct an equation for the reaction of chlorobenzene with trichloroethanal to form DDT. *(2 marks)*

d Chlorobenzene can be nitrated to form a mixture of products.

Suggest why the reaction forms a mixture of products. *(1 mark)*

e Explain why phenol reacts more readily with chlorine than benzene reacts with chlorine.

In your answer, you should use appropriate technical terms, spelled correctly. *(3 marks)*

OCR F324 Jan 2010 Q1 and OCR F324 Jun 2011 Q1

5 a Benzene can be converted into benzenesulfonic acid, $C_6H_5SO_3H$, which is used in the manufacture of many detergents. The reaction between benzene and sulfuric acid is an electrophilic substitution reaction. Sulfur trioxide, SO_3, is the electrophile. Part of the mechanism for this reaction is shown below.

Complete the mechanism by drawing the intermediate and by adding curly arrows to show the movement of electron pairs in steps 1 and 2.

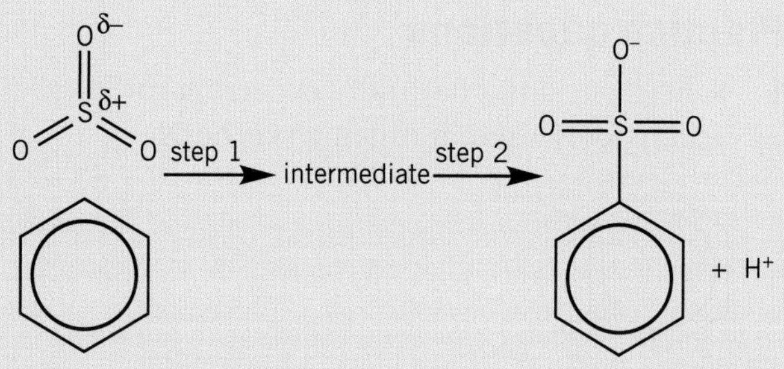

(4 marks)

b The painkiller paracetamol has the structure shown below

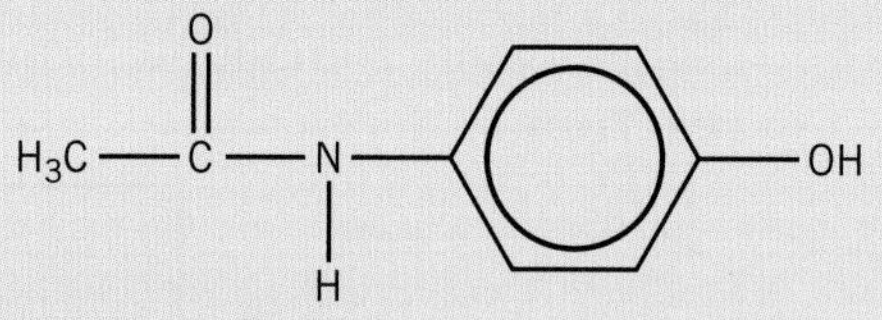

Separate samples of paracetamol are reacted with bromine, Br_2, and with cold sodium hydroxide, NaOH.

Draw the structures of possible organic products formed in each reaction.

(2 marks)

OCR F324 Jan 2012 Q2 (b) (c)

6 The mass spectrum of an aromatic hydrocarbon is found to have a molecular ion peak at *m/z* value of 106. Elemental analysis showed that the compound had the following composition by mass, C, 90.56% and H, 9.44%.

a Use the data to find the molecular formula of the aromatic compound.

(2 marks)

b A student realised that the calculated molecular formula could have a number of different structural isomers.

Draw all the possible structural isomers that contain a benzene ring. *(4 marks)*

c In the presence of concentrated nitric acid and concentrated sulfuric acid, one of the isomers, **A**, reacts to give one mono-nitrated compound, **B**.

Identify **A** and **B**. *(2 marks)*

AROMATIC COMPOUNDS 25

7 When monosubstituted aromatic compounds undergo further substitution, the position at which the new substituent attaches to the ring is directed by the original substituent. Use the table below to answer the questions which follow:

Original substituent	Position to which new substituent is directed
–COOH	3
–CH$_3$ (or any alkyl group)	2, 4
–NO$_2$	3
–Cl	2, 4
–COCH$_3$	3

   a (i) Draw the structure of the major monosubstituted product formed when phenylethanone, C$_6$H$_5$COCH$_3$, reacts with chlorine in the presence of AlCl$_3$. *(1 mark)*

      (ii) Show the mechanism for this reaction using curly arrows and illustrating the role of AlCl$_3$ in the reaction. *(5 marks)*

   b Identify the two monosubstituted isomers formed when methylbenzene reacts with chloroethane in the presence of AlCl$_3$ *(2 marks)*

8 Phenol is much more susceptible to electrophilic attack than benzene.

   a Explain why phenol reacts more readily with benzene with electrophiles. *(3 marks)*

   b Phenol reacts with dilute nitric acid to form a mixture of two monosubstituted isomers.

     Draw the structures of these two isomers and write an equation for the formation of one of these isomers. *(2 marks)*

9 Methylbenzene, C$_6$H$_5$CH$_3$, is an aromatic hydrocarbon, used widely as a solvent. It is readily nitrated and can form mono-, di-, or tri-nitromethylbenzenes.

   a 4-Nitromethylbenzene can be formed by the nitration of methylbenzene.

     Outline the mechanism for the formation of 4-nitromethylbenzene from methylbenzene using NO$_2^+$ as the electrophile. *(4 marks)*

   b There are six possible structural isomers of CH$_3$C$_6$H$_3$(NO$_2$)$_2$ that are dinitromethylbenzenes. Four of the isomers are shown below.

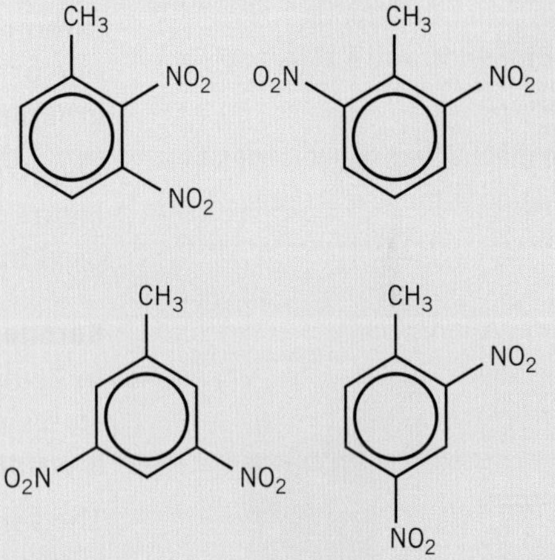

     Draw the structures of the other two isomers. *(2 marks)*

OCR F324 Jan 2011 Q1

26 CARBONYLS AND CARBOXYLIC ACIDS
26.1 Carbonyl compounds
Specification reference: 6.1.2

Learning outcomes

Demonstrate knowledge, understanding, and application of:

→ oxidation of aldehydes

→ reactions of carbonyl compounds with $NaBH_4$ and HCN

→ the mechanism of nucleophilic addition in these reactions.

Aldehydes and ketones

Aldehydes and ketones are important organic compounds that both contain the **carbonyl** functional group, C=O.

Aldehydes

In aldehydes the carbonyl functional group is found at the end of a carbon chain. The carbon atom of the carbonyl group is attached to one or two hydrogen atoms. In its structural formula, the aldehyde group is written as CHO.

The structure of the simplest aldehyde, methanal, HCHO, is shown in Figure 1. The common name for methanal is formaldehyde, which is used in solution to preserve biological specimens.

Ketones

In ketones, the carbonyl functional group is joined to two carbon atoms in the carbon chain. In its structural formula, the ketone group is written as CO.

The simplest ketone, propanone, CH_3COCH_3, contains three carbon atoms (Figure 2). The common name for propanone is acetone, which is used as an important industrial solvent and is also used in nail-varnish removers.

▲ **Figure 1** *Methanal, HCHO, is the simplest aldehyde. Biological specimens are preserved in methanal (formaldehyde)*

Naming carbonyl compounds

From the rules for naming organic compounds, you should be able to work out that the aldehyde and ketone functional groups are indicated by adding a suffix to the stem of the longest carbon chain:

- -al for an aldehyde
- -one for a ketone.

In an aldehyde, the carbon atom of the carbonyl group is always designated as carbon-1, whereas in a ketone the carbonyl carbon atom needs to be numbered.

▲ **Figure 2** *Propanone, CH_3COCH_3, with three carbon atoms, is the simplest ketone. It is an important industrial solvent and is also used in nail varnish remover*

🖩 Worked example: Naming an aldehyde

Use the naming rules for aldehydes to name the compound below.

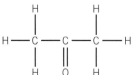

CARBONYLS AND CARBOXYLIC ACIDS 26

Step 1: There are eight carbons in the longest chain, so the stem is oct-.

Step 2: The –C=O functional group is on the end of the chain, an aldehyde, so the suffix is -al.

Step 3: The aldehyde carbon is carbon-1 but this is not included in the name.

Step 4: There is a suffix starting with a vowel so the alkane chain ending is shortened to -an-.

Step 5: The compound is octanal.

Synoptic link

For details of the rules for naming organic compounds, see Topic 11.2, Nomenclature of organic compounds.

Octanal in nature

Octanal, $CH_3(CH_2)_7CHO$, is a colourless fragrant liquid, one of a number of molecules responsible for the smell of oranges.

Octanal occurs naturally in citrus oils and is used commercially as a component of perfumes and in flavour production for the food industry.

▲ **Figure 6**

Researchers have found that the crested auklet, a native bird of Alaska, emits a strong tangerine-like perfume during the breeding seasons. Mating pairs spread this scent over each other as part of their breeding ritual.

▲ **Figure 7** Crested Aucklets

Use the Internet to find the other aldehydes that contribute to the smell of oranges. Draw the skeletal formulae of two of these aldehydes.

Worked example: Naming a ketone

Use the naming rules for ketones to name the compound below.

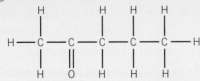

Step 1: There are five carbons in the longest chain, so the stem is pent-.

Step 2: There is a –C=O functional group: so the suffix is -one.

Step 3: The –C=O group is at carbon-2 so the suffix is -2-.

Step 4: There is a suffix starting with a vowel so the alkane chain ending is shortened to -an-.

Step 5: The compound is pentan-2-one.

Synoptic link

Look back at Topic 14.2, Reactions of alcohols, to revise how primary alcohols can be oxidised to form aldehydes and carboxylic acids, and how secondary alcohols can be oxidised to ketones.

Study tip

In step 2, numbering could be from either end of the molecule. But the name uses the lowest number, here 2 rather than 4

26
26.1 Carbonyl compounds

> **Study tip**
>
> Use [O] in brackets to represent an oxidation agent. It makes balancing the equations much easier too.

Oxidation of aldehydes

Aldehydes can be oxidised to carboxylic acids when refluxed with acidified dichromate(VI) ions, $Cr_2O_7^{2-}/H^+$, usually as a mixture of sodium or potassium dichromate(VI), $K_2Cr_2O_7$, and dilute sulfuric acid, H_2SO_4. The equation for the oxidation of butanal is shown in Figure 5.

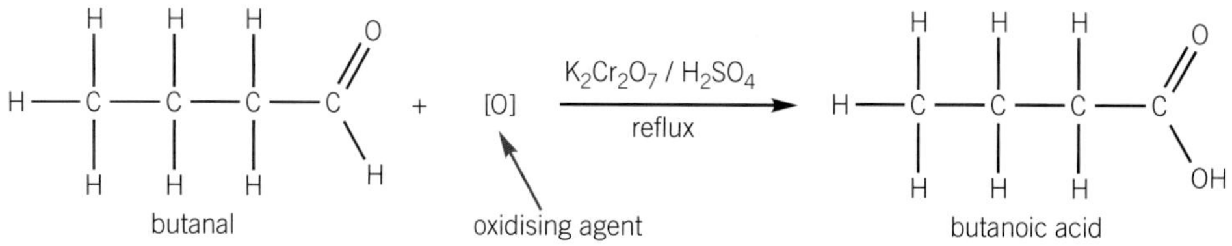

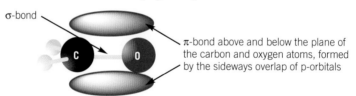

▲ **Figure 5** *The oxidation of butanal with excess acidified potassium dichromate forming butanoic acid*

▲ **Figure 6** *Aldehydes can be oxidised by acidified potassium dichromate(VI) – the dichromate solution changes colour from orange to green during the reaction*

Ketones do not undergo oxidation reactions. This lack of reactivity provides chemists with a way of distinguishing between aldehydes and ketones. In Topic 26.2, you will learn more about the chemical tests that can be used to show the presence of a carbonyl group, and to distinguish aldehydes from ketones.

Nucleophilic addition reactions of the carbonyl group

The C═O double bond

The reactivity of aldehydes and ketones is influenced by the nature of the carbon–oxygen double bond. The double bond is made up of both a σ-bond and a π-bond (Figure 7).

▲ **Figure 7** *The nature of the carbon–oxygen double bond*

You may think that carbonyl compounds would react in the same way as the alkenes, since carbonyl and alkene functional groups both contain a double bond. However, there is a significant difference in the nature of the double bond, which makes the reactions rather different.

- The C═C double bond in alkenes is *non-polar*.
- The C═O double bond in carbonyl compounds is *polar*.

Oxygen is more electronegative than carbon. The electron density in the double bond lies closer to the oxygen atom than to the carbon. This makes the carbon end of the bond slightly positive and the oxygen end slightly negative (Figure 8).

> **Synoptic link**
>
> Review the C═C double bond in alkenes from Topic 13.1, Properties of alkenes, which is also made up of a σ-bond and a π-bond.

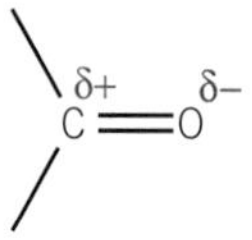

▲ **Figure 8** *The polar nature of the carbon–oxygen bond*

Due to the polarity of the C═O double bond, aldehydes and ketones react with some **nucleophiles**. A nucleophile is attracted to and attacks the slightly positive carbon atom resulting in addition across the C═O double bond.

The reaction type is **nucleophilic addition**.

CARBONYLS AND CARBOXYLIC ACIDS — 26

This is very different from the non-polar C=C double bond in alkenes which reacts with **electrophiles** by **electrophilic addition**.

Reaction of carbonyl compounds with NaBH$_4$

Sodium tetrahydridoborate(III), NaBH$_4$, is used as a reducing agent to reduce aldehydes and ketones to alcohols. The aldehyde or ketone is usually warmed with the NaBH$_4$ reducing agent in aqueous solution.

Reducing an aldehyde

Aldehydes are reduced to *primary* alcohols by NaBH$_4$. The reduction of butanal is shown in Figure 9.

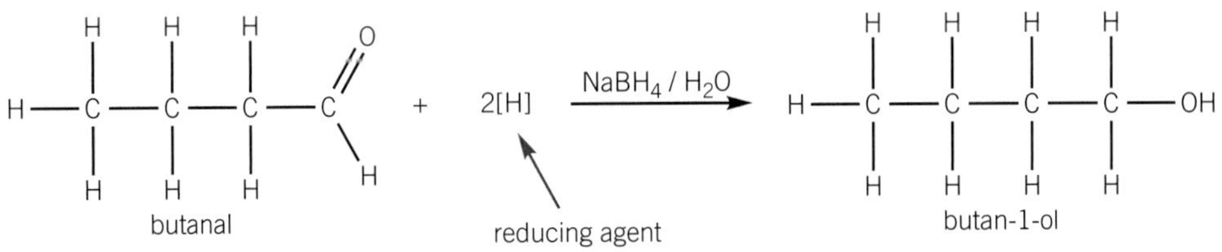

> **Synoptic link**
>
> You will recall from Topic 15.1, The chemistry of the haloalkanes, that a nucleophile is defined as an atom or group of atoms that is attracted to an electron-deficient carbon atom, where it donates a pair of electrons to form a new covalent bond.

▲ **Figure 9** *Butanal is reduced to butan-1-ol by NaBH$_4$*

Reducing a ketone

Ketones are reduced to *secondary* alcohols by NaBH$_4$. The reduction of propanone is shown in Figure 10.

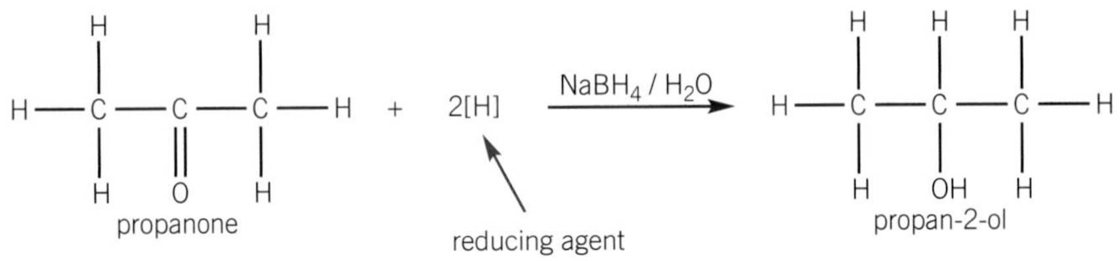

> **Study tip**
>
> Remember that reduction of organic compounds uses [H] to represent the reducing agent.

▲ **Figure 10** *Propanone is reduced to propan-2-ol, a secondary alcohol*

Reaction of carbonyl compounds with HCN

Hydrogen cyanide, HCN, adds across the C=O bond of aldehydes and ketones. Hydrogen cyanide is a colourless, extremely poisonous liquid that boils slightly above room temperature, so it cannot be used safely in an open laboratory. Sodium cyanide and sulfuric acid are used to provide the hydrogen cyanide in the reaction, but the reaction is still potentially very hazardous.

> **Study tip**
>
> The reaction with hydrogen cyanide is an *addition reaction* as two reactants combine to become one product.

The reaction is very useful as it provides a means of *increasing* the length of the carbon chain. The reaction of propanal with hydrogen cyanide is shown in Figure 11.

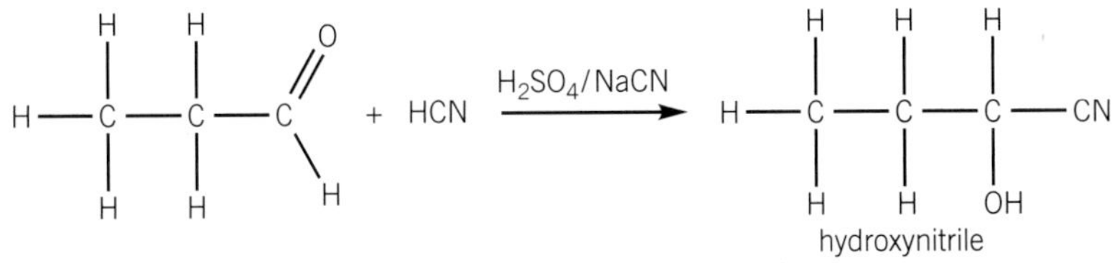

▲ **Figure 11** *Propanal reacts with hydrogen cyanide to form a hydroxynitrile*

26.1 Carbonyl compounds

The organic product formed from the reaction of an aldehyde with hydrogen cyanide contains two functional groups, a hydroxyl group, –OH, and a nitrile group, C≡N. These compounds are classed as hydroxynitriles (or cyanohydrins).

Mechanism for nucleophilic addition to carbonyl compounds

The carbon atom in the C=O double bond is electron deficient and attracts nucleophiles. Aldehydes and ketones both react by nucleophilic addition to form alcohols.

The mechanism for the reaction with $NaBH_4$

In this nucleophilic addition reaction, $NaBH_4$ can be considered as containing the hydride ion, :H$^-$, which acts as the nucleophile.

1. The lone pair of electrons from the hydride ion, :H$^-$, is attracted and donated to the δ+ carbon atom in the aldehyde or ketone C=O double bond.
2. A dative covalent bond is formed between the hydride ion and the carbon atom of the C=O double bond.
3. The π-bond in the C=O double bond breaks by heterolytic fission, forming a negatively charged intermediate.
4. The oxygen atom of the intermediate donates a lone pair of electrons to a hydrogen atom in a water molecule. The intermediate has then been protonated to form an alcohol.

The mechanism for the reaction of propanone with $NaBH_4$ is shown in Figure 12.

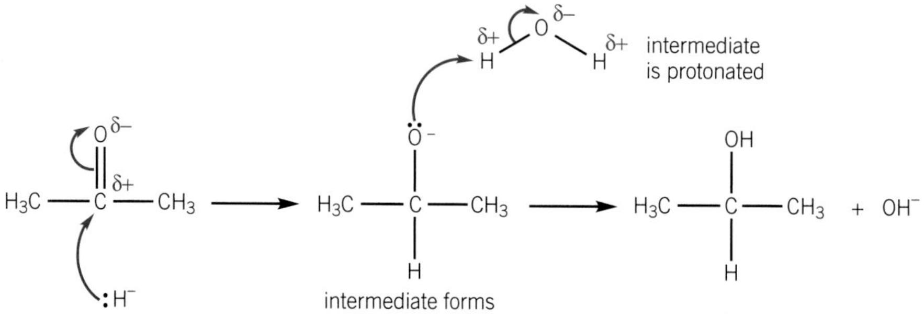

> **Study tip**
>
> Addition to carbonyl groups usually consists of two steps:
>
> 1. Nucleophilic attack on the carbonyl group to form a negatively charged intermediate.
> 2. Protonation of the intermediate to form an alcohol.

▲ **Figure 12** *Reduction of a carbonyl compound by nucleophilic addition*

The mechanism for the reaction with NaCN/H$^+$

In the first stage of the reaction of an aldehyde or ketone with NaCN/H$^+$, the cyanide ion, :CN$^-$, attacks the electron-deficient carbon atom in the aldehyde or ketone.

1. The lone pair of electrons from the cyanide ion, :CN$^-$, is attracted and donated to the δ+ carbon atom in the aldehyde or ketone C=O double bond. A dative covalent bond forms.
2. The π-bond in the C=O double bond breaks by heterolytic fission, forming a negatively charged intermediate.

> **Synoptic link**
>
> This reaction will be very useful in organic synthesis (Topic 28.1, Extending carbon chain length) as it provides a method of increasing the length of the carbon chain.

CARBONYLS AND CARBOXYLIC ACIDS   **26**

3. The intermediate is protonated by donating a lone pair of electrons to a hydrogen ion, to form the product.
4. The product is a hydroxynitrile.

The mechanism for the reaction of propanal with sodium cyanide in the presence of acid is shown in Figure 13. The second stage can also be drawn showing protonation by water, as in the mechanism with $NaBH_4$ in Figure 12.

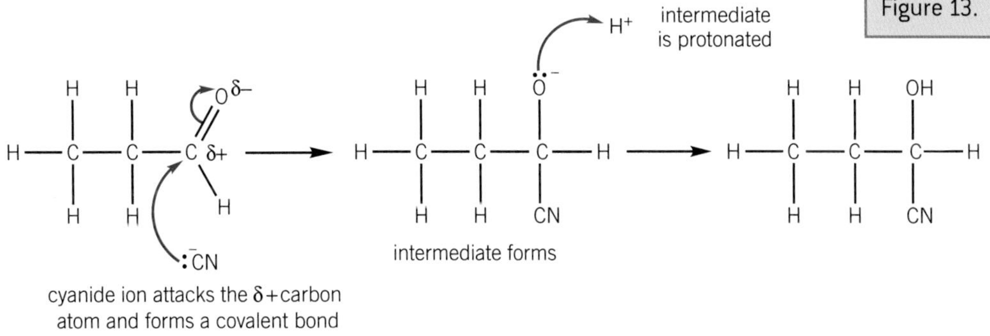

> **Study tip**
>
> Although the cyanide ion is usually written as CN^-, the negative charge is actually on the carbon atom. In the mechanism it is important to show the negative charge on the *carbon* atom, ^-CN, as shown in Figure 13.

▲ **Figure 13** *Reaction of propanal with cyanide ions to form a hydroxynitrile*

Summary questions

1. Draw the displayed formulae of the following molecules:
   a. hexan-3-one *(1 mark)*
   b. 2,3-dimethylnonanal *(1 mark)*
   c. 2-methylpentanal *(1 mark)*
   d. 3-ethyloctan-2-one. *(1 mark)*

2. Write the equations for the reduction of:
   a. propanal *(2 marks)*
   b. pentan-2-one. *(2 marks)*

3. Define the term nucleophile and explain why aldehydes and ketones undergo addition reactions. *(3 marks)*

4. Show the mechanism for the reaction between propanal and $NaBH_4$. *(4 marks)*

26.2 Identifying aldehydes and ketones
Specification reference: 6.1.2

Learning outcomes
Demonstrate knowledge, understanding, and application of:
→ using 2,4-dinitrophenylhydrazine to detect and identify carbonyls
→ using Tollens' reagent to identify aldehydes.

Study tip
You are not expected to know the structure of 2,4-DNP.

Detecting carbonyl compounds
A solution of 2,4-dinitrophenylhydrazine (2,4-DNP or 2,4-DNPH), also known as Brady's Reagent, is used to detect the presence of the carbonyl functional group in both aldehydes and ketones. The structure of 2,4-DNP is shown in Figure 1. In the presence of a carbonyl group, a yellow or orange precipitate called a 2,4-dinitrophenylhydrazone is produced.

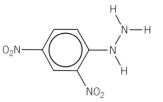

▲ Figure 1 Structure of 2,4-dinitrophenylhydrazine

Testing for the carbonyl group with 2,4-DNP
In practical work, 2,4-DNP is normally used dissolved in methanol and sulfuric acid as a pale orange solution called Brady's reagent. Solid 2,4-DNP can be very hazardous because friction or a sudden blow can cause it to explode.

Testing for the carbonyl group in aldehydes and ketones
The method for carrying out this test is outlined below.

1. Add 5 cm depth of a solution of 2,4-dinitrophenylhydrazine to a clean test-tube. This is in excess.
2. Using a dropping pipette, add three drops of the unknown compound. Leave to stand.
3. If no crystals form, add a few drops of sulfuric acid.
4. A yellow/orange precipitate (Figure 2) indicates the presence of an aldehyde or ketone.

The test tube containing the yellow/orange precipitate can then be put to one side and analysed later to identify the aldehyde or ketone that formed the precipitate.

▲ Figure 2 A yellow/orange precipitate formed in the reaction between an aldehyde or a ketone and 2,4-dinitrophenylhydrazine

Which of the following would produce an orange precipitate with 2,4-DNP?

A $CH_3CH_2CH_2CHO$
B $CH_3CH_2COCH_3$
C $CH_3CH_2CH_2CH_2OH$
D $(CH_3)_2CHCHO$
E $CH_3CHOHCH_3$

The reaction of aldehydes and ketones with 2,4-DNP
2,4-Dinitrophenylhydrazine is a derivative of hydrazine, H_2NNH_2. Hydrazine and substituted hydrazines react with the carbonyl group in a nucleophilic addition–elimination reaction. This reaction is often called a *condensation* reaction as water is eliminated. The equation for the reaction between propanone and 2,4-dinitrophenylhydrazine is shown in Figure 3.

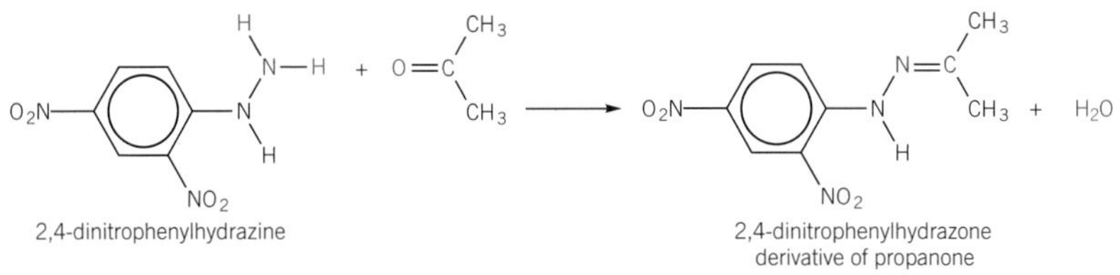

CARBONYLS AND CARBOXYLIC ACIDS 26

▲ Figure 3 *Reaction of propanone with 2,4-DNP*

When 2,4-DNP reacts with a carbonyl group, the NH_2 group adds across the C=O double bond, followed by removal of a water molecule.

> Write equations for the reactions of carbonyl compounds with the following hydrazine derivatives.
>
> a 2-methylpentan-3-one and hydroxylamine, NH_2OH
> b phenylethanone, $C_6H_5COCH_3$, and hydrazine, NH_2NH_2

Study tip

The equation for the reaction of an aldehyde or a ketone with 2,4-DNP is beyond the scope of the specification. However, you could be provided with the equation as additional information for an application-based question in an exam.

Distinguishing between aldehydes and ketones

Once a compound has been identified as containing a carbonyl group using 2,4-DNP, a fresh sample of the compound can be further classified as either an aldehyde or a ketone using Tollens' reagent – a solution of silver nitrate in aqueous ammonia. In the presence of an aldehyde group, a silver mirror is produced.

Testing for the aldehyde group with Tollens' reagent

Tollens' reagent has a short shelf-life and should be made up immediately before carrying out the test, as follows:

1 In a clean test tube, add 3 cm depth of aqueous silver nitrate, $AgNO_3(aq)$.
2 Add aqueous sodium hydroxide to the silver nitrate until a brown precipitate of silver oxide, Ag_2O, is formed.
3 Add dilute ammonia solution until the brown precipitate just dissolves to form a clear colourless solution. This is Tollens' reagent.

The method for carrying out the test for an aldehyde group is outlined below:

1 Pour 2 cm depth of the unknown solution into a clean test tube.
2 Add an equal volume of freshly prepared Tollens' reagent.
3 Leave the test tube to stand in a beaker of warm water at about 50 °C for about 10 to 15 minutes and then observe whether any silver mirror is formed (Figure 4).

◀ Figure 4 *Aldehydes react with Tollens' reagent to form a precipitate of silver (right) whereas ketones do not react (left)*

If the unknown solution contains an aldehyde then a silver mirror forms. If the unknown solution is a ketone then no reaction is observed.

1 Which of the following would produce a silver mirror with Tollens' reagent?
  A $CH_3CH_2CH_2CHO$ D $(CH_3)_2CHCHO$
  B $CH_3CH_2COCH_3$ E $CH_3CHOHCH_3$
  C $CH_3CH_2CH_2CH_2OH$

2 Write an equation for any reaction of an aldehyde with Tollens' reagent.

461

26.2 Identifying aldehydes and ketones

Study tip
Remember that oxidation of organic compounds uses [O] to represent the oxidising agent.

Synoptic link
You will cover recrystallisation and determination of melting points in Topic 28.2, Further practical techniques.

Tollens' reagent contains silver(I) ions, Ag^+ (aq), which act as an oxidising agent in the presence of ammonia. In the reaction, silver ions are reduced to silver as the aldehyde is oxidised to a carboxylic acid.

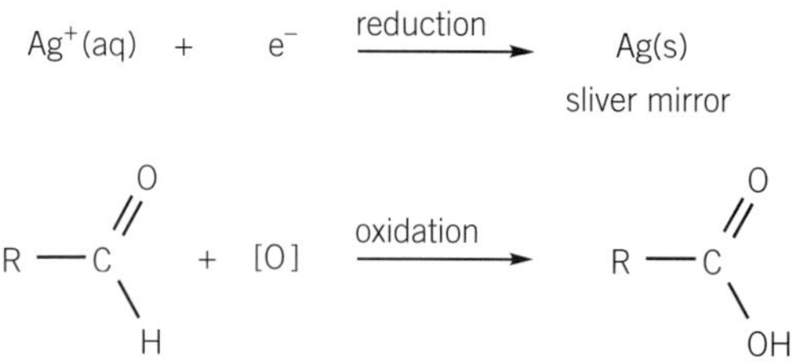

$$Ag^+(aq) + e^- \xrightarrow{reduction} Ag(s) \text{ silver mirror}$$

$$R-CHO + [O] \xrightarrow{oxidation} R-COOH$$

▲ **Figure 5** *Oxidation and reduction takes place in the reaction between an aldehyde and silver ions*

Identifying an aldehyde or ketone by melting point

The 2,4-dinitrophenylhydrazone precipitate formed in the 2,4-DNP test can be analysed to identify the carbonyl compound as follows:

1. The impure yellow/orange solid is filtered to separate the solid precipitate from the solution.
2. The solid is then recrystallised to produce a pure sample of crystals.
3. The melting point of the purified 2,4-dinitrophenylhydrazone is measured and recorded.

The melting point is then compared to a database or data table of melting points to identify the original carbonyl compound.

The table shows the melting points of the pure 2,4-DNP derivatives of several carbonyl compounds.

	2,4-DNP derivative melting point / °C
cyclopentanone	146
propanal	156
pentan-3-one	156
methanal	167

1. Draw the skeletal formulae for each carbonyl compound.
2. How would you then obtain a pure sample of the derivative for the melting point starting from the crude precipitate in the test tube?
3. Not all of these carbonyl compounds could be identified solely from the melting point. Which carbonyl compounds are the problem and what further test would be needed to distinguish between them?

Summary questions

1. Draw the structures of the two isomers with the molecular formula C_3H_6O that are carbonyl compounds and label as an aldehyde or a ketone. *(2 marks)*

2. **a** Describe a chemical test in which aldehydes and ketones react in the same way. State the expected observations. *(3 marks)*
   b Describe a chemical test in which aldehydes and ketones react differently. State the expected observations and write equations to show any reaction. *(4 marks)*

3. The three compounds below may be distinguished by simple chemical tests:
   $CH_3CH_2CH_2CHO$ $CH_3CH_2COCH_3$ $CH_3CH_2CH_2CH_2OH$
   Describe a sequence of tests that you could carry out to identify each compound. State all reagents used, any essential conditions, and expected observations. *(6 marks)*

26.3 Carboxylic acids
Specification reference: 6.1.3

The carboxyl group

A carboxylic acid is an organic acid which contains the **carboxyl** functional group. The carboxyl group (Figure 1) contains both a carbonyl group and a hydroxyl group.

You will find carboxylic acids in medicines, fruit juices, vinegar, and even in rhubarb leaves.

Although many carboxylic acids are useful in their own right, carboxylic acids and their derivatives are used in organic synthesis as starting materials or intermediates in the formation of more useful compounds. The painkiller, aspirin, can be synthesised from salicylic acid (Figure 2).

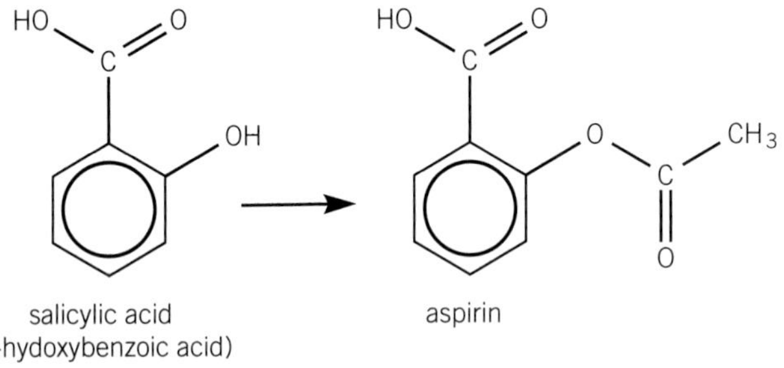

Learning outcomes

Demonstrate knowledge, understanding, and application of:
→ the solubility of carboxylic acids
→ reactions of carboxylic acids with metals and bases.

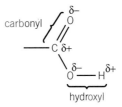

▲ **Figure 1** *The carboxyl group is made up of a carbonyl and a hydroxyl functional group, which give the carboxyl group its name*

▲ **Figure 2** *Aspirin can be made from salicylic acid*

Naming carboxylic acids

To name a carboxylic acid, the carboxylic acid (carboxyl) functional group is indicated by adding the suffix -oic acid to the stem of the longest carbon chain. The carboxyl carbon atom is always designated as carbon-1.

Study tip

Take special care here. The functional group is the *carboxyl* group, COOH. Don't be fooled into quoting both C=O and OH groups. The chemistry is changed entirely when the OH is attached to the carbon atom of the C=O.

> **Worked example: Naming a carboxylic acid**
>
> Name the carboxylic acid below: 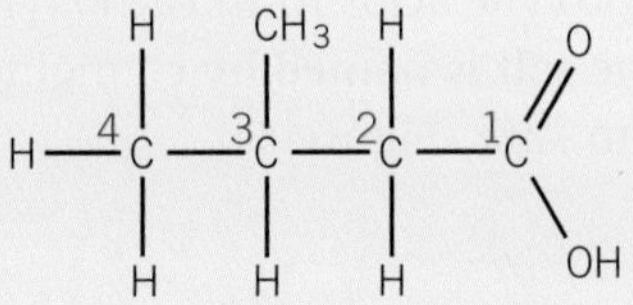
>
> **Step 1:** There are four carbons in the longest chain, so the stem is but-.
>
> **Step 2:** The –COOH functional group is on the end of the chain so the suffix is -oic acid.
>
> **Step 3:** There is a suffix that is followed with a vowel – the alkane chain ending is shortened to -an-.
>
> **Step 4:** There is a methyl group on carbon-3, so the prefix 3-methyl is added to the name.
>
> **Step 5:** The compound is 3-methylbutanoic acid.

Synoptic link

For details of the rules for naming organic compounds, see Topic 11.2, Nomenclature of organic compounds.

▲ **Figure 3** *3-methylbutanoic acid has a strong pungent cheesy or sweaty smell. It is a major component of the odour that can come from smelly feet*

463

26.3 Carboxylic acids

> **Synoptic link**
>
> For details of hydrogen bonding, see Topic 6.4, Hydrogen bonding.

Solubility of carboxylic acids

The C=O and O—H bonds in carboxylic acids are polar allowing carboxylic acids to form hydrogen bonds with water molecules (Figure 4).

- Carboxylic acids with up to four carbon atoms are soluble in water.
- As the number of carbon atoms increases, the solubility decreases as the non-polar carbon chain has a greater effect on the overall polarity of the molecule.

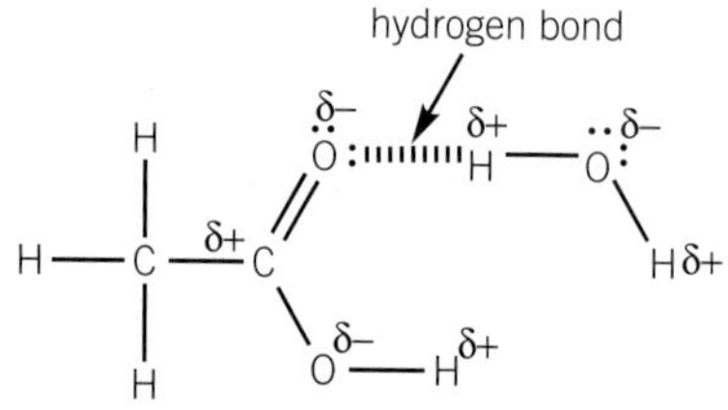

▲ **Figure 4** *Hydrogen bonds form between a carboxylic acid and water, making the carboxylic acid soluble in water*

Dicarboxylic acids have two polar carboxyl groups to form hydrogen bonds. They are solids at room temperature and dissolve readily in water.

Strength of carboxylic acids

Carboxylic acids are classified as weak acids. Methanoic acid, HCOOH, is the simplest carboxylic acid containing only one carbon atom. When dissolved in water, carboxylic acids partially dissociate:

$$HCOOH(aq) \rightleftharpoons H^+(aq) + HCOO^-(aq)$$

> **Synoptic link**
>
> For more details of the dissociation of weak acids and reactions of acids, see Topic 4.1, Acids, bases, and neutralisation, and Topic 20.1, Brønsted–Lowry acids and bases.

Acid reactions of carboxylic acids

Carboxylic acids take place in redox and neutralisation reactions:

- redox reactions with metals
- neutralisation reactions with bases (alkalis, metal oxides, and carbonates).

In these reactions, carboxylic acids form carboxylate salts. The **carboxylate ion** in the salt is named by changing the –ic acid ending of the carboxylic acid to –ate (Figure 5).

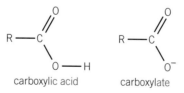

▲ **Figure 5** *Comparison of a carboxylic acid and a carboxylate ion*

▲ **Figure 6** *Magnesium metal reacts with carboxylic acids to produce hydrogen gas. The metal dissolves and there is an effervescence*

> **Synoptic link**
>
> For more details of redox reactions of acids with metals, see Topic 4.3, Redox,

Redox reactions of carboxylic acids with metals

Aqueous solutions of carboxylic acids react with metals in a redox reaction to form hydrogen gas and the carboxylate salt. In this reaction, you would observe the metal disappearing and effervescence as hydrogen gas is evolved. The reaction of propanoic acid with magnesium is shown below.

$$2CH_3CH_2COOH(aq) + Mg(s) \rightarrow (CH_3CH_2COO^-)_2Mg^{2+}(aq) + H_2(g)$$

26 CARBONYLS AND CARBOXYLIC ACIDS

Neutralisation reactions of carboxylic acids with bases

Carboxylic acids react with all bases – metal oxides, alkalis, and carbonates. These reactions are all examples of neutralisation.

Reaction with metal oxides

Carboxylic acids react with metal oxides to form a salt and water. The equation below shows the reaction of aqueous ethanoic acid with calcium oxide to form calcium ethanoate and water:

$$2CH_3COOH(aq) + CaO(s) \rightarrow (CH_3COO^-)_2Ca^{2+}(aq) + H_2O(l)$$

Reaction with alkalis

Carboxylic acids react with alkalis to form a salt and water. In this reaction, you may not see any reaction as two solutions react together to form an aqueous solution of the salt. Ethanoic acid reacts with sodium hydroxide to form sodium ethanoate and water.

Overall: $CH_3COOH(aq) + NaOH(aq) \rightarrow CH_3COO^-Na^+(aq) + H_2O(l)$

Ionic: $H^+(aq) + OH^-(aq) \rightarrow H_2O(l)$

Reaction with carbonates

When a carbonate is added to a carboxylic acid, carbon dioxide gas is evolved. If the carboxylic acid is in excess a solid carbonate would disappear. The equation below shows the reaction of ethanoic acid with aqueous sodium carbonate:

$$2CH_3COOH(aq) + Na_2CO_3(aq) \rightarrow 2CH_3COO^-Na^+(aq) + H_2O(l) + CO_2(g)$$

Test for the carboxyl group

The neutralisation reaction of carboxylic acids with carbonates, for example, sodium carbonate, provides chemists with a method of distinguishing carboxylic acids from any other organic compounds.

Carboxylic acids are the only common organic compounds sufficiently acidic to react with carbonates. This is especially useful for distinguishing carboxylic acids from phenols. Phenols are not acidic enough to react with carbonates.

> **Study tip**
>
> When writing the formula of the salt, you can include or omit charges. If you do show charges, you must show them all, otherwise the equation is unbalanced by charge. So both $(CH_3CH_2COO^-)_2 Mg^{2+}$ and $(CH_3CH_2COO)_2Mg$ are acceptable.

▲ Figure 7 Ethanoic acid reacting with solid sodium carbonate. The reaction produces a salt (soluble sodium ethanoate), carbon dioxide (the bubbles seen here), and water

> **Synoptic link**
>
> You will also encounter the sodium carbonate test for the carboxyl group in Topic 29.1, Tests for organic functional groups.

Summary questions

1. Draw the skeletal formula for the following carboxylic acids:
   a. 3-ethylhexanoic acid (1 mark)
   b. 2-chloro-3-methylpentanoic acid (1 mark)
   c. 2-bromobenzoic acid (1 mark)

2. Write balanced equations for the reactions between:
   a. propanoic acid and sodium carbonate (2 marks)
   b. butanoic acid and copper(II) oxide (2 marks)

3. Explain why ethanedioic acid is more soluble in water than ethanoic acid. (2 marks)

4. Write balanced equations, using structural formulae, for the reactions between:
   a. 2-bromobenzoic acid and magnesium (2 marks)
   b. propanedioic acid and sodium hydroxide (2 marks)

26.4 Carboxylic acid derivatives
Specification reference: 6.1.3

Learning outcomes
Demonstrate knowledge, understanding, and application of:
→ esterification
→ hydrolysis of esters
→ formation of acyl chlorides
→ use of acyl chlorides in synthesis.

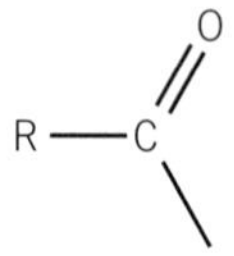

What is a derivative of a carboxylic acid?
A derivative of a carboxylic acid is a compound that can be hydrolysed to form the parent carboxylic acid. Carboxylic acid derivatives have a common sequence of atoms in their structure, known as an acyl group. The acyl group is shown in Figure 1.

Esters, acyl chlorides, acid anhydrides, and amides are all derivatives of carboxylic acids (Figure 2).

▲ **Figure 1** *The structure of the acyl group*

ester   acyl chloride   acid anhydride   amide

▲ **Figure 2** *Esters, acyl chlorides, acid anhydrides, and amides are all carboxylic acid derivatives. They all contain the acyl group (blue) attached to an electronegative atom or group of atoms (red)*

Esters
An ester is named after the parent carboxylic acid from which it is derived. To name an ester, remove the –oic acid suffix from the parent carboxylic acid and replace with –oate. The alkyl chain attached to the oxygen atom of the COO group is then added as the first word in the name (Figure 3).

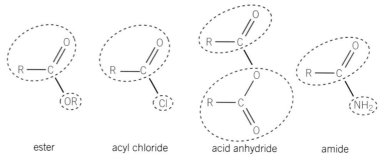

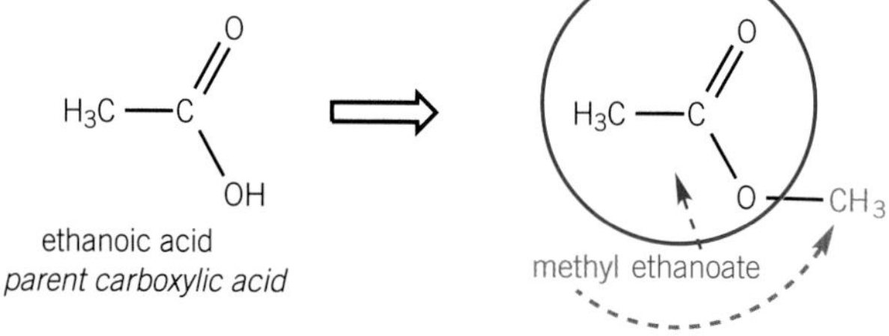

ethanoic acid
parent carboxylic acid

methyl ethanoate

▲ **Figure 3** *Naming esters*

Synoptic link
If you need some revision of how organic compounds are named, revisit Topic 11.2, Nomenclature of organic compounds.

Acyl chlorides
An acyl chloride is named after the parent carboxylic acid from which it is derived. To name an acyl chloride, remove the –oic acid suffix from the parent carboxylic acid and replace with –oyl chloride (Figure 4).

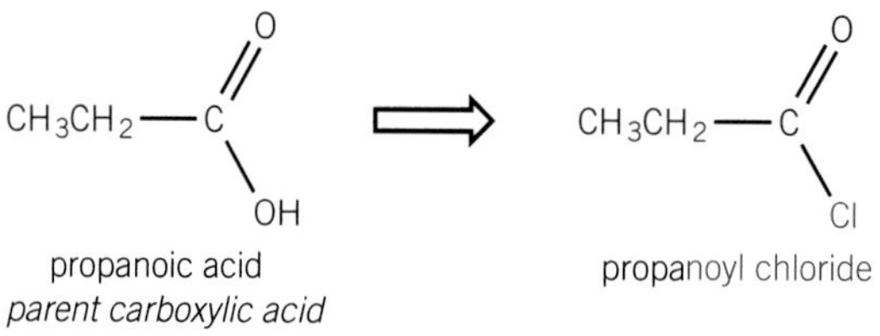

propanoic acid
parent carboxylic acid

propanoyl chloride   ethanoate

▲ **Figure 4** *Naming acyl chlorides*

CARBONYLS AND CARBOXYLIC ACIDS

Acid anhydrides

An acid anhydride is formed by removal of water from two carboxylic acid molecules, as shown in the formation of ethanoic anhydride (Figure 5).

$$H_3C-\underset{\underset{OH}{\|}}{\overset{O}{C}} \quad \underset{\underset{CH_3}{\|}}{\overset{O}{HO-C}} \Rightarrow H_3C-\underset{\underset{O}{\|}}{\overset{O}{C}}-\underset{\underset{CH_3}{\|}}{\overset{O}{C}} + H_2O$$

two molecules of ethanoic acid → ethanoic anhydride

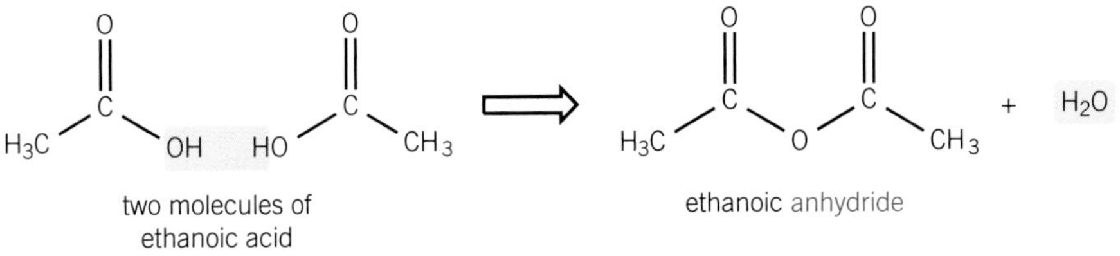

▲ **Figure 5** *Formation of an acid anhydride from a carboxylic acid*

Study tip
The acid anhydride looks quite complicated, however there is a clue to its structure in the name. The anhydride (which means without water) is two molecules of carboxylic acid without water.

Esterification

Esterification is the reaction of an alcohol with a carboxylic acid to form an ester. An alcohol is warmed with a carboxylic acid with a small amount of concentrated sulfuric acid, which acts as a catalyst. Esters are sweet-smelling liquids.

In the reaction shown in Figure 6, ethanol reacts with propanoic acid to produce an ester, ethyl propanoate, and water. The name of the ester is derived from the names of the two reactants.

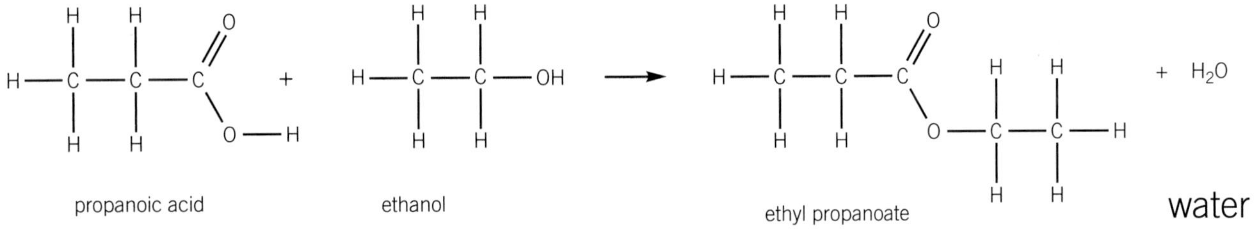

propanoic acid + ethanol → ethyl propanoate + H₂O

▲ **Figure 6** *The preparation of the ester ethyl propanoate*

The molecular formula for ethyl propanoate is $C_5H_{10}O_2$. It is a colourless liquid that smells like pineapples and is naturally produced in some fruits like kiwi fruit and strawberries.

Making ethyl propanoate

It is easy to make a sample of an ester in the laboratory.

1. Pour 2 cm³ of ethanol and 2 cm³ of propanoic acid into a boiling tube.
2. Carefully add a few drops of concentrated sulfuric acid to the mixture.
3. Place the boiling tube in a beaker of hot water at about 80 °C.
4. Leave the boiling tube in the water for about five minutes.
5. Pour the contents of the boiling tube into a beaker containing aqueous sodium carbonate. (Sodium carbonate is a weak base and neutralises the acids in the mixture. This removes the smell of any carboxylic acid that has not reacted, making it easier to detect the smell of the ester.)

You will notice that you now have oily drops of the ester floating on the water. You will also be able to smell the aroma of the ester you have made. (Don't make a habit of inhaling smells in the laboratory — some compounds are poisons or carcinogens.)

▲ **Figure 7** *Many esters have a fruity smell*

Write the equation for the reaction of propanoic acid with sodium carbonate.

Hydrolysis of esters

Esters can be hydrolysed by aqueous acid or alkali. **Hydrolysis** is the chemical breakdown of a compound in the presence of water or in aqueous solution.

Acid hydrolysis

Acid hydrolysis of an ester is the reverse of esterification.

To carry out acid hydrolysis:
- The ester is heated under reflux with dilute aqueous acid.
- The ester is broken down by water, with the acid acting as a catalyst.

The products of acid hydrolysis are a carboxylic acid and an alcohol, as shown in Figure 8.

▲ **Figure 9** *Methyl 2-hydroxybenzoate, also known as methyl salicylate (oil of wintergreen), is an organic ester naturally produced by many species of plants, particularly wintergreens*

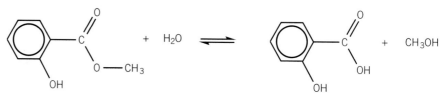

▲ **Figure 8** *The acid hydrolysis of the ester methyl 2-hydroxybenzoate*

Alkaline hydrolysis

Alkaline hydrolysis is also known as saponification and is irreversible. The ester is heated under reflux with aqueous hydroxide ions.

Figure 10 shows the alkaline hydrolysis of the ester methyl ethanoate to form a carboxylate ion and an alcohol.

▲ **Figure 10** *The alkaline hydrolysis of methyl ethanoate*

If aqueous sodium hydroxide is the alkali, the salt sodium ethanoate is formed during alkaline hydrolysis:

$$CH_3COOCH_3 + NaOH \rightarrow CH_3COO^-Na^+ + CH_3OH$$

Acyl chlorides and their reactions
Preparation of acyl chlorides

Acyl chlorides can be prepared directly from their parent carboxylic acid by reaction with thionyl chloride, $SOCl_2$. The other products of this reaction, SO_2 and HCl, are evolved as gases, leaving just the acyl chloride. The reaction should be carried out in a fume cupboard as the products are harmful.

The reaction for the formation of propanoyl chloride from propanoic acid is shown in Figure 11.

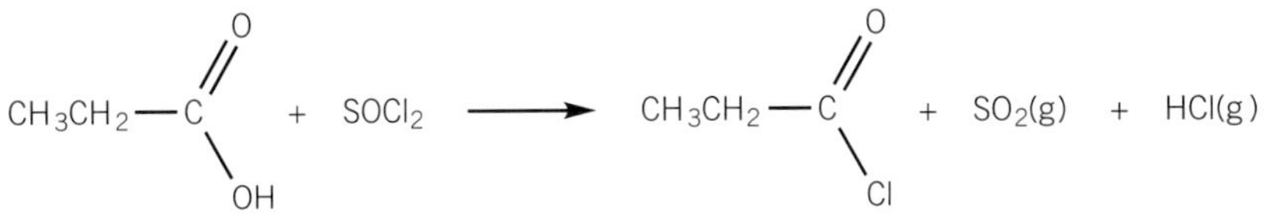

$CH_3CH_2-COOH + SOCl_2 \longrightarrow CH_3CH_2-COCl + SO_2(g) + HCl(g)$

propanoic acid → propanoyl chloride

▲ **Figure 11** *The preparation of propanoyl chloride from propanoic acid*

Reactions of acyl chlorides

Acyl chlorides are very reactive and are very useful in organic synthesis as they can be easily converted into carboxylic acid derivatives, such as esters and amides, with good yields. Acyl chlorides react with nucleophiles by losing the chloride ion whilst retaining the C=O double bond.

Reaction of acyl chlorides with alcohols to form esters

Acyl chlorides react with alcohols to form esters. The reaction of ethanoyl chloride with propan-1-ol is shown in Figure 12.

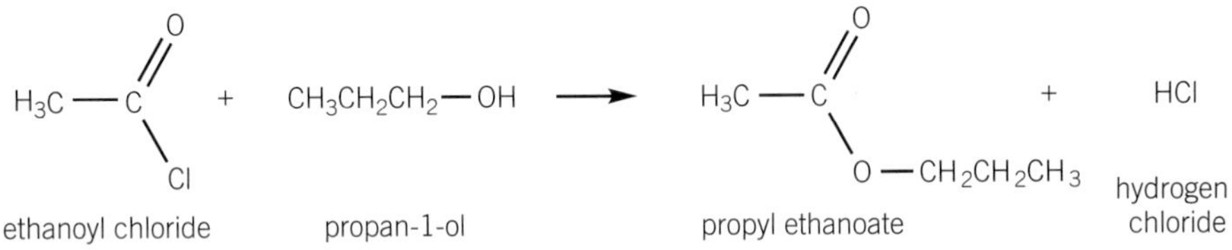

$H_3C-COCl + CH_3CH_2CH_2-OH \longrightarrow H_3C-COO-CH_2CH_2CH_3 + HCl$

ethanoyl chloride + propan-1-ol → propyl ethanoate + hydrogen chloride

▲ **Figure 12** *The reaction of an acyl chloride with an alcohol forming an ester*

Reaction of acyl chlorides with phenols to form esters

Carboxylic acids are not reactive enough to form esters with phenols. Acyl chlorides and acid anhydrides are both much more reactive than carboxylic acids and react with phenols to produce phenyl esters. Neither reaction needs an acid catalyst.

The equation for the formation of phenol ethanoate from ethanoyl chloride is shown in Figure 13.

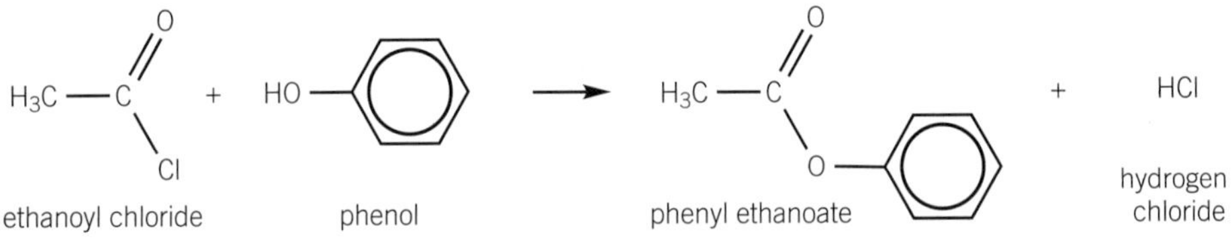

$H_3C-COCl + HO-C_6H_5 \longrightarrow H_3C-COO-C_6H_5 + HCl$

ethanoyl chloride + phenol → phenyl ethanoate + hydrogen chloride

▲ **Figure 13** *The preparation of phenyl ethanoate*

Reaction of acyl chlorides with water to form carboxylic acids

When water is added to an acyl chloride, a violent reaction takes place with the evolution of dense steamy hydrogen chloride fumes. A carboxylic acid is formed. The equation for the reaction of water with ethanoyl chloride is shown in Figure 14.

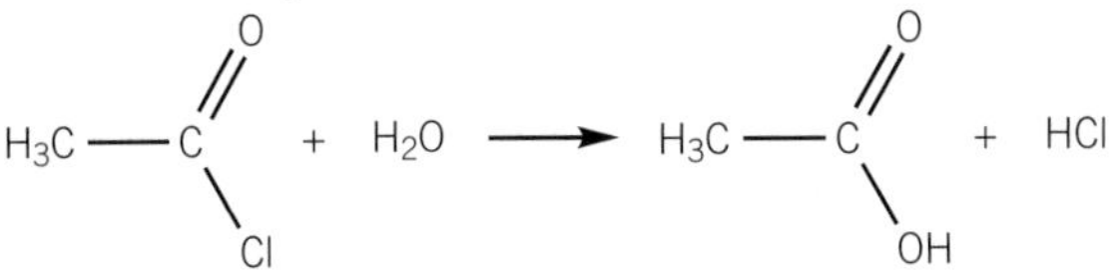

$H_3C-COCl + H_2O \longrightarrow H_3C-COOH + HCl$

▲ **Figure 14** *Ethanoyl chloride reacts vigorously with water to produce a carboxylic acid*

▲ **Figure 15** *When water is added carefully to ethanoyl chloride, steamy fumes of HCl are observed*

26.4 Carboxylic acid derivatives

> **Synoptic link**
>
> For details of primary, secondary, and tertiary amines, see Topic 27.1, Amines.

Reaction of acyl chlorides with ammonia and amines to form amides

Ammonia and amines can act as nucleophiles by donating the lone pair of electrons on the nitrogen atom to an electron-deficient species. Their reaction with acyl chlorides forms amides.

Ammonia reacts with acyl chlorides, forming a *primary* amide. In the primary amide the nitrogen atom is attached to *one* carbon atom. The reaction of ethanoyl chloride and ammonia is shown in Figure 16.

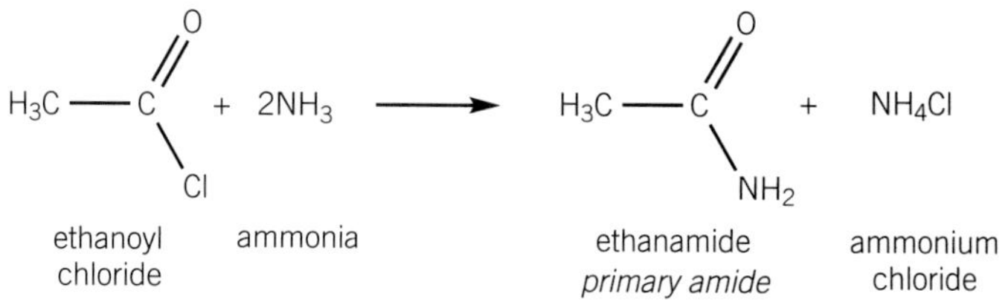

$$H_3C-COCl \; + \; 2NH_3 \longrightarrow H_3C-CONH_2 \; + \; NH_4Cl$$

ethanoyl chloride | ammonia | ethanamide primary amide | ammonium chloride

▲ **Figure 16** *Reaction between ethanoyl chloride and ammonia*

A primary amine reacts with an acyl chloride in the same way as ammonia to form a *secondary* amide. In the secondary amide the nitrogen atom is attached to *two* carbon atoms. The reaction of ethanoyl chloride and methylamine is shown in Figure 17.

$$H_3C-COCl \; + \; 2CH_3NH_2 \longrightarrow H_3C-CO-N(H)-CH_3 \; + \; CH_3NH_3^+Cl^-$$

ethanoyl chloride | methylamine primary amine | N-methylethanamide secondary amide | methylammonium chloride

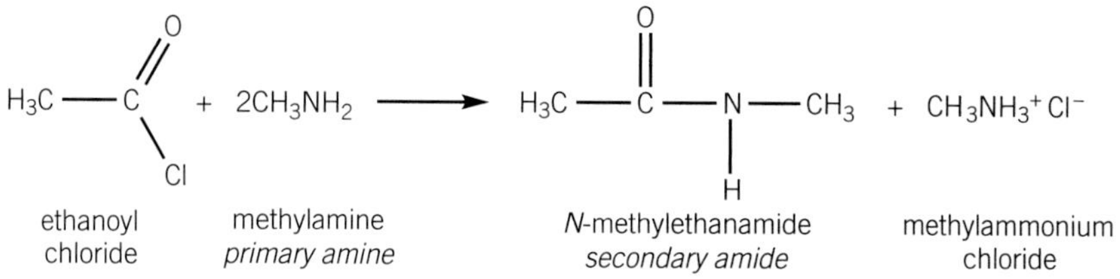

▲ **Figure 17** *The formation of a secondary amide from the reaction between an acyl chloride and an amine*

> **Study tip**
>
> The name N-methylethanamide shows us that the methyl group is attached to the nitrogen rather than a carbon atom of the main carbon chain.

Reactions of acid anhydrides

Acid anhydrides react in a similar way to acyl chlorides with alcohols, phenols, water, ammonia, and amines. They are less reactive than acyl chlorides and are useful for some laboratory preparations where acyl chlorides may be too reactive.

The equation for the formation of phenyl ethanoate from ethanoic anhydride and phenol is shown below.

$$2(CH_3CO)_2O \; + \; C_6H_5OH \rightarrow CH_3COOC_6H_5 \; + \; CH_3COOH$$

ethanoic anhydride | phenol | phenyl ethanoate | ethanoic acid

> ### The mechanism of the reaction between acyl chlorides and nucleophiles
>
> You have already seen how nucleophiles react with haloalkanes and carbonyl compounds. It should come as no surprise then that acyl chlorides also react with nucleophiles. Carbonyl compounds undergo addition whereas haloalkanes undergo substitution, so what about acyl chlorides which contain both C=O and Cl?
>
> The mechanism (Figure 18) shows water as the nucleophile. The mechanism has two steps – addition followed by elimination – containing features of both carbonyl and haloalkane chemistry.

CARBONYLS AND CARBOXYLIC ACIDS 26

Addition (Step 1)
The lone pair of electrons from the nucleophile is attracted to and donated to the δ+ carbon atom in the C=O group of the acyl chloride.

A dative covalent bond is formed between the nucleophile and the carbonyl carbon atom. The π-bond of the C=O group breaks, forming a negatively charged intermediate.

Elimination (Step 2)
A lone pair of electrons on oxygen reforms the C=O double bond, causing a chloride ion to be removed. A proton is also then lost to complete the elimination. For simplicity, the loss of Cl⁻ followed by H⁺ have been shown together in the second step of the mechanism.

The mechanism is called nucleophilic addition–elimination.

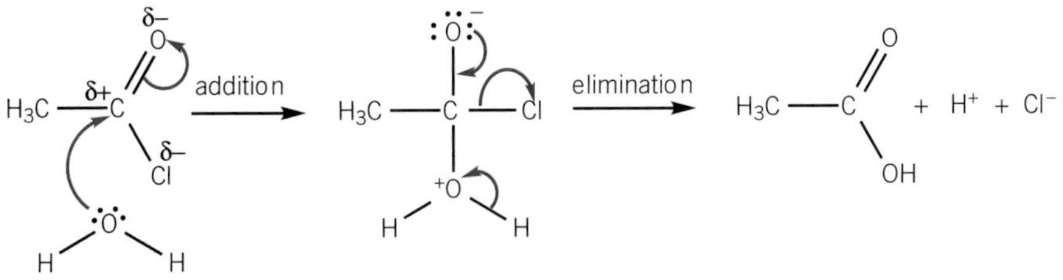

▲ Figure 18 *The mechanism of nucleophilic addition–elimination*

> 1 Describe how the shape around the central carbon atom changes during the different steps in the reaction mechanism.
> 2 Show the mechanism for the reaction of propanoyl chloride with ammonia.

Summary questions

1 State the name of each of the following compounds.
   a $CH_3CH_2CH_2COOH$ *(1 mark)*
   b $CH_3CH_2CH_2CONH_2$ *(1 mark)*
   c $(CH_3CH_2CO)_2O$ *(1 mark)*
   d $CH_3CH_2CH_2CH_2COCl$ *(1 mark)*

2 Write equations for the hydrolysis of propyl butanoate in both acid and alkaline conditions. *(4 marks)*

3 Write the equations for the reactions used to prepare a sample of:
   a propanoic acid from propanoyl chloride *(1 mark)*
   b butyl ethanoate from ethanoyl chloride *(1 mark)*
   c propanamide from propanoic acid.
      Hint: Two steps are required. *(2 marks)*

Practice questions

1. The following reactions were observed for a compound with molecular formula C_3H_6O.

   Test 1: The compound did not react with Tollens' reagent when heated

   Test 2: On addition 2,4-dinitrophenylhydrazine, a yellow–orange precipitate is formed

   Test 3: Reaction with $NaBH_4$ in the presence of water produced a colourless liquid.

   a Draw the displayed full structural formulae of two compounds of formula C_3H_6O. *(2 marks)*

   b What does the result of **Test 1** show? *(1 mark)*

   c The formation of a yellow–orange precipitate in **Test 2** is a positive test for a particular organic group.
   Identify this group. *(1 mark)*

   d Using the formula of the compound and the results of **Test 1** and **Test 2**, identify the compound by name. *(1 mark)*

   e Draw the mechanism for the reaction between the compound identified in **d** and $NaBH_4$. *(4 marks)*

2. An ester **D** is used as a solvent for paints and varnishes.

   a Ester **D** can be manufactured by heating an alcohol under reflux with ethanoic acid and a catalyst.

   (i) State a suitable catalyst for this reaction. *(1 mark)*

   (ii) Explain why the reaction is carried out under reflux. *(1 mark)*

   b Ester **D** has a structural formulae, $CH_3COOCH(CH_3)CH_2CH_3$.

   (i) Draw out the displayed formula of ester **D** and clearly display the functional group. *(1 mark)*

   (ii) State the name of the alcohol used to make ester **D**. *(1 mark)*

3. Esters are well known as compounds providing flavour in many fruits and the scent of some flowers. The ester $CH_3(CH_2)_2COOCH_3$ contributes to the aroma of apples.

   a Name the ester $CH_3(CH_2)_2COOCH_3$. *(1 mark)*

   b State the reagents and conditions for the hydrolysis of this ester in the laboratory to make a carboxylic acid. *(2 marks)*

   c Write a balanced equation for this hydrolysis of the ester. *(2 marks)*

4. Aspirin and paracetamol are common pain killers. The structures are shown below.

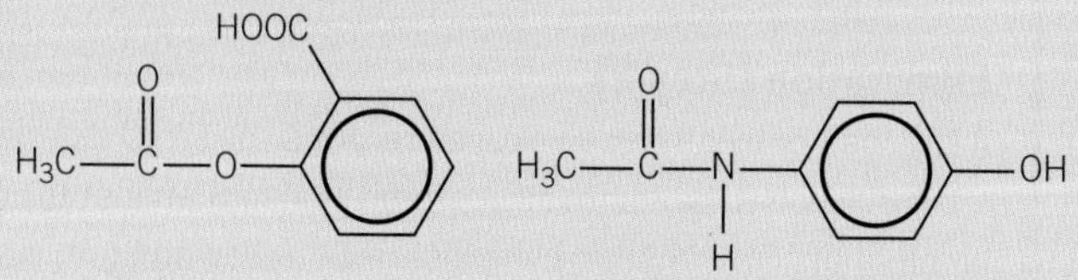

   aspirin   paracetamol

   a State the reagents, conditions, and write equations., showing organic compounds as structures, for the production of aspirin from ethanoyl chloride. *(3 marks)*

   b Aspirin is heated under reflux with aqueous sodium hydroxide.
   Write an equation for this reaction, showing organic compounds as structures. *(2 marks)*

5. Acyl chlorides, such as propanoyl chloride are useful in the synthesis of other organic compounds.

   Identify the structure of the organic products formed in the reactions between propanoyl chlorides and the following.

   a ethanol   b phenol
   c ammonia   d methylamine

   State the functional group present in each one of the organic products formed. *(4 marks)*

6. The following three carbonyl compounds are structural isomers of $C_5H_{10}O_2$.

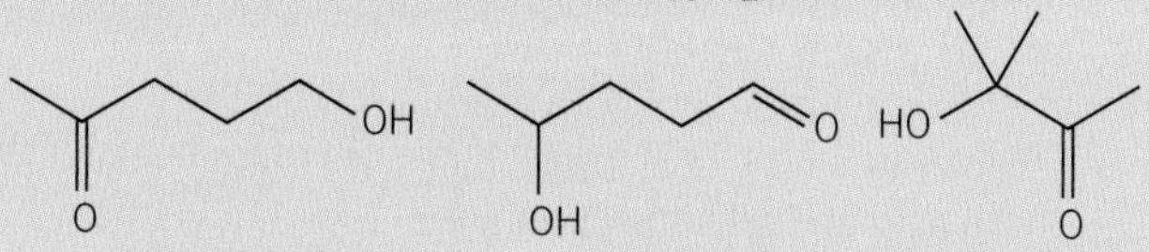

   compound **C**   compound **D**   compound **E**

   a Describe chemical tests that you could carry out in test-tubes to distinguish between compounds **C**, **D**, and **E**.
   Include appropriate reagents and any relevant observations. Also include equations showing structures for the organic compounds involved. *(4 marks)*

CARBONYLS AND CARBOXYLIC ACIDS — 26

b Aldehydes and ketones are both reduced by $NaBH_4$. When used in the presence of a $CeCl_3$ catalyst, $NaBH_4$ only reduces ketones. Compound **F** has the structural formula $CH_3COCH_2CH_2CHO$. It is reduced by $NaBH_4$ in the presence of a $CeCl_3$ catalyst to form one of the compounds **C**, **D**, or **E**.

Show the mechanism for this reduction of compound **F** and identify the product that is formed.

Use curly arrows and show relevant dipoles. You do not need to show the role of the $CeCl_3$ catalyst. *(4 marks)*

OCR F324 June 2014 Q2

7 Pyruvic acid, shown below, is an organic compound that has a smell similar to ethanoic acid. It is extremely soluble in water.

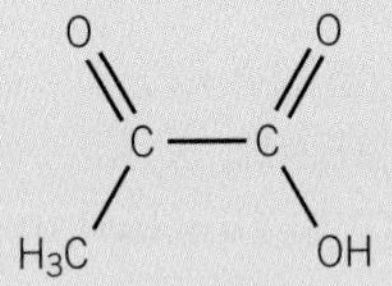

pyruvic acid

a Explain why pyruvic acid is soluble in water. Use a labelled diagram to support your answer. *(2 marks)*

b Pyruvic acid can be prepared in the laboratory by reacting propane-1,2-diol with excess acidified potassium dichromate(VI). The reaction mixture is heated under reflux.

Write an equation for this oxidation. Use **[O]** to represent the oxidising agent and show structural formulae for organic compounds. *(2 marks)*

c Pyruvic acid can also be reduced by $NaBH_4$ to form $CH_3CH(OH)COOH$.

Outline the mechanism for this reduction. Use curly arrows and show relevant dipoles. *(4 marks)*

d Compound **A**, shown below, is a structural isomer of pyruvic acid.

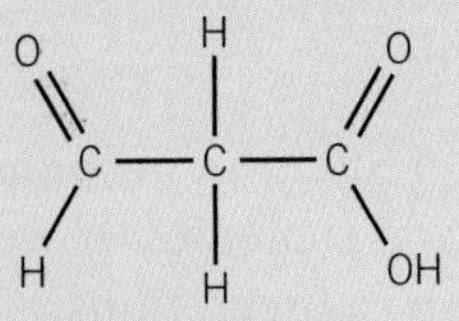

compound **A**

Describe a chemical test that could be carried out in a laboratory to distinguish between samples of pyruvic acid and compound **A**.

Your answer should include reagents, observations, the type of reaction and the organic product formed. *(4 marks)*

OCR F324 Jan 2013 Q1

8 Methylglyoxal, CH_3COCHO, is formed in the body during metabolism.

Describe **one** reduction reaction and **one** oxidation reaction of methylglyoxal that could be carried out in the laboratory.

Your answer should include reagents, equations and observations, if any. *(5 marks)*

OCR F324 Jun 2011 Q4

9 Safranal, shown below, is an aldehyde which contributes to the aroma of saffron.

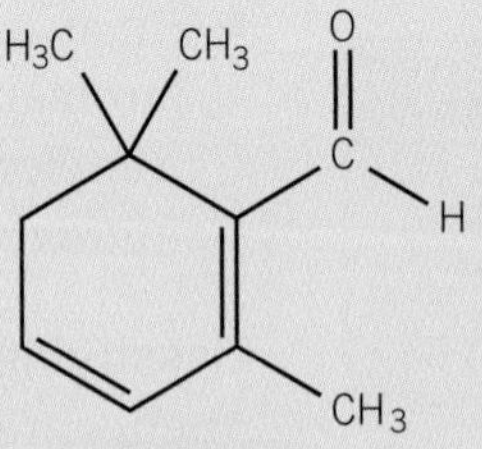

An undergraduate chemist investigated some reactions of safranal.

a She prepared a solution of Tollens' reagent and added a few drops of safranal. She then warmed the mixture for about five minutes in a water bath.

Describe what you would expect the chemist to see. State the type of reaction that the safranal undergoes.

Draw the structure of the organic product formed in this reaction. *(3 marks)*

b The chemist then reduced safranal using an aqueous solution of $NaBH_4$.

Outline the mechanism for this reaction. Use curly arrows and show any relevant dipoles. You can use the following structure to represent safranal.

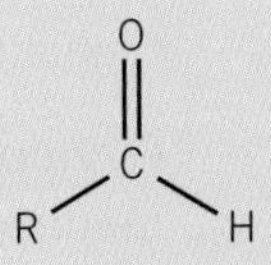

(4 marks)

OCR F324 Jan11 Q3

473

27 AMINES, AMINO ACIDS, AND POLYMERS

27.1 Amines

Specification reference: 6.2.1

Learning outcomes

Demonstrate knowledge, understanding, and application of:

→ amines as bases
→ preparation of amines
→ reactions of amines.

Study tip

When drawing the structure of an amine, you should show the lone pair.

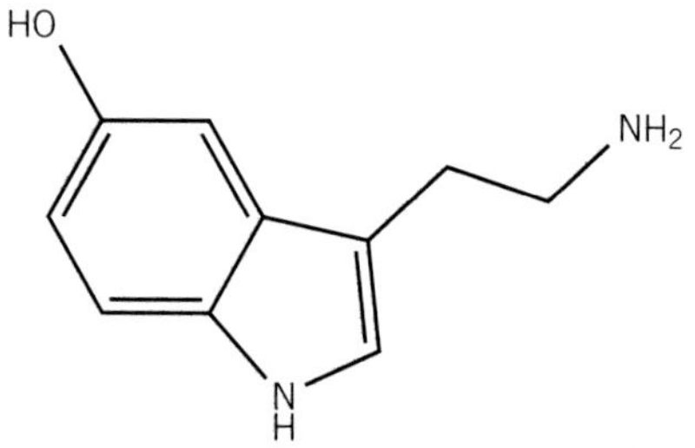

What are amines?

Amines are organic compounds, derived from ammonia, NH_3, in which one or more hydrogen atoms in ammonia have been replaced by a carbon chain or ring.

- In an **aliphatic amine**, the nitrogen atom is attached to at least one straight or branched carbon chain (alkyl group, R). Methylamine, CH_3NH_2, is the simplest aliphatic amine with one methyl group attached to the nitrogen atom.

- In an **aromatic amine**, the nitrogen atom is attached to an aromatic ring (aryl group, Ar). Phenylamine, $C_6H_5NH_2$ is the simplest aromatic amine, with a phenyl C_6H_5 group attached to the nitrogen atom.

Classifying amines

Amines are classified as primary, secondary, or tertiary by the number of alkyl or aryl groups attached to the nitrogen atom (Table 1).

▼ **Table 1** Primary, secondary, and tertiary aliphatic amines

Number of alkyl or aryl groups	1	2	3
Structure	R—N(H)—H	R—N(H)—H	R—N(R')—R''
	|	|	|
	H	R'	R'
Classification	primary	secondary	tertiary
Aliphatic	$CH_3CH_2CH_2NH_2$	$(C_2H_5)_2NH$	$(CH_3)_3N$
Aromatic	$C_6H_5NH_2$	$C_6H_5NH(CH_3)$	$C_6H_5N(CH_3)_2$

Amines occur commonly in nature with many being well known for their effects on the body.

Serotonin (Figure 1) acts as a neurotransmitter, responsible for the control of appetite, sleep, memory and learning, temperature regulation, muscle contraction, and depression.

Pseudoephedrine (Figure 2) is an active ingredient in decongestion medications such as in nose drops and in cold remedies. It works by shrinking nasal membranes and inhibiting nasal secretions.

▲ **Figure 1** The structure of serotonin

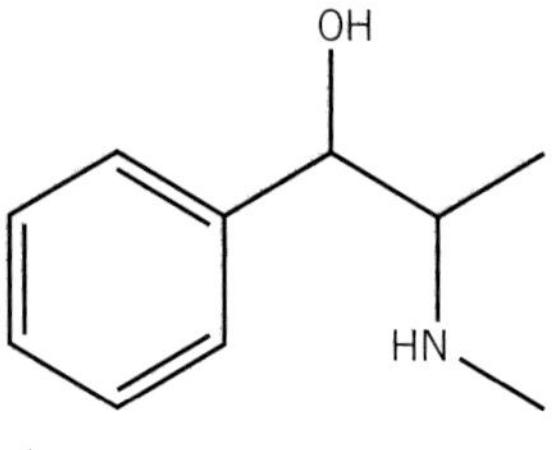

▲ **Figure 2** The structure of pseudoephedrine

474

AMINES, AMINO ACIDS, AND POLYMERS

Naming amines

There are many different ways of naming the amines. One way to name a **primary amine**, with the –NH_2 group on the end of the chain, is by adding the suffix -amine to the name of the alkyl chain (Figure 3).

Where a primary amine contains an amine group on any other carbon but carbon-1, the amine is named using the prefix amino- and a number is added to indicate the position of the amine group along that chain (Figure 4).

In **secondary** or **tertiary amines**, containing the same alkyl group, the prefixes di- or tri- are used to indicate the number of alkyl groups attached to the nitrogen atom, for example, $(CH_3)_2NH$ is dimethylamine.

When two or more different groups are attached to a nitrogen atom, the compound is named as a *N*-substituted derivative of the larger group, for example, $CH_3NHCH_2CH_2CH_3$ is *N*-methylpropylamine and $CH_3N(CH_2CH_3)CH_2CH_2CH_3$ is *N*-ethyl-*N*-methylpropylamine.

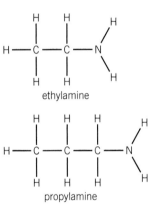

▲ **Figure 3** *The structures of the primary amines, ethylamine and propylamine*

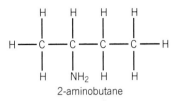

▲ **Figure 4** *The structure of 2-aminobutane*

The chemistry of decay

The smell of a decomposing organism, for example, rotting fish, is made up of all sorts of compounds, but amines and compounds of sulfur give the characteristic smell of decay. Most of the amines come from the breakdown of proteins, and two of the amines have such unpleasant odours that they have been named putrescine (1,4-diaminobutane), after the process of putrefaction, and cadaverine (1,5-diaminopentane), after the Latin-derived word for a corpse – cadaver.

Putrescine and cadaverine are also components of bad breath. Putrescine also has an industrial use – it is a raw material for producing some condensation polymers.

Draw the skeletal formulae of putrescine and cadaverine.

▲ **Figure 5** *Rotting fish release the amines putrescine and cadaverine whilst decomposing*

Reactions of amines

Amines as bases

Amines behave as bases in their chemical reactions as the lone pair of electrons on the nitrogen atom can accept a proton.

When an amine accepts a proton, a dative covalent bond is formed between the lone pair of electrons on the nitrogen atom and the proton (Figure 6).

> **Synoptic link**
>
> You first encountered dative covalent bonds in Topic 5.3, Covalent bonding.

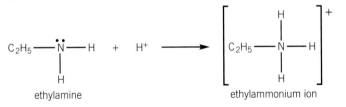

▲ **Figure 6** *Reaction of ethylamine as a base*

27.1 Amines

> **Study tip**
>
> The neutralisation of acids by amines is similar to the neutralisation of acids by ammonia:
>
> $NH_3 + HCl \rightarrow NH_4^+Cl^-$

> **Synoptic link**
>
> See Topic 15.1, The chemistry of haloalkanes, for details of nucleophilic substitution reactions of haloalkanes.

Salt formation

Amines are bases and they neutralise acids to make salts. For example, propylamine reacts with hydrochloric acid to form the salt propylammonium chloride.

$$CH_3CH_2CH_2NH_2 + HCl \rightarrow CH_3CH_2CH_2NH_3^+Cl^-$$

Ethylamine reacts with sulfuric acid to form ethylammonium sulfate.

$$2CH_3CH_2NH_2 + H_2SO_4 \rightarrow (CH_3CH_2NH_3^+)_2SO_4^{2-}$$

Preparation of aliphatic amines

Formation of primary amines

Ammonia has a lone pair of electrons on the nitrogen atom which allows ammonia to act as a nucleophile in a substitution reaction with a haloalkane. The product of this reaction is an ammonium salt. Aqueous alkali is then added to generate the amine from the salt.

The equations below show the formation of propylamine from 1-chloropropane and ammonia. Propylamine is a primary amine because there is *one* carbon atom attached to the nitrogen atom of the amine group.

salt formation:

$$CH_3CH_2CH_2Cl + NH_3 \rightarrow CH_3CH_2CH_2NH_3^+Cl^-$$
1-chloropropane   propylammonium chloride (salt)

amine formation:

$$CH_3CH_2CH_2NH_3^+Cl^- + NaOH \rightarrow CH_3CH_2CH_2NH_2 + NaCl + H_2O$$
propylammonium chloride   propylamine

For this reaction to occur, there are some essential conditions.

- Ethanol is used as the solvent. This prevents any substitution of the haloalkane by water to produce alcohols.
- Excess ammonia is used. This reduces further substitution of the amine group to form secondary and tertiary amines (see below).

Formation of secondary and tertiary amines

The reaction above is unsuitable for making a pure primary amine. The product still contains a lone pair of electrons on the nitrogen atom that can react further with a haloalkane to form a secondary amine. The product of this reaction is again an ammonium salt.

$$CH_3CH_2CH_2Cl + CH_3CH_2CH_2NH_2 \rightarrow (CH_3CH_2CH_2)_2NH_2^+Cl^-$$
                 propylamine   dipropylammonium chloride (salt)

The secondary amine is obtained from the salt by reacting the product with sodium hydroxide.

$$(CH_3CH_2CH_2)_2NH_2^+Cl^- + NaOH \rightarrow (CH_3CH_2CH_2)_2NH + NaCl + H_2O$$
dipropylammonium chloride   dipropylamine

Tertiary amines can also be formed by further reaction of the secondary amine. In this example, further substitution would form tripropylamine, $(CH_3CH_2CH_2)_3N$.

27 AMINES, AMINO ACIDS, AND POLYMERS

Preparation of aromatic amines

Phenylamine, $C_6H_5NH_2$, is made by the reduction of nitrobenzene, $C_6H_5NO_2$. Nitrobenzene is heated under reflux with tin and hydrochloric acid to form the ammonium salt, phenylammonium chloride, which is then reacted with excess sodium hydroxide to produce the aromatic amine, phenylamine. Tin and hydrochloric acid act as a reducing agent (Figure 7).

▲ **Figure 7** *The reduction of nitrobenzene to form phenylamine*

Perkin's Mauveine

Until the mid-19th century, most dyes were obtained from nature. Dyes such as indigo (blue-purple) and turmeric (orange) were extracted from plants whereas cochineal (pink) and shellfish purple (purple) were extracted from animals. Whilst these dyes could produce an amazing range of colours, they would often fade, run, or damage the materials they were meant to dye.

William Henry Perkin revolutionised the dye industry when he synthesised a purple substance, which he named mauveine. Mauveine was the first chemically produced dye which paved the way for the synthesis of many of the dyes used today. Mauvine is based upon the structure of phenylamine.

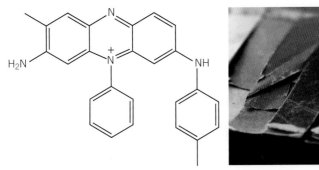

▲ **Figure 8** *The structure of Mauveine, together with some colour samples. Mauveine was the first synthetic dye, based on phenylamine*

What is the molecular formula of mauveine?

Summary questions

1. State whether the following amines are primary, secondary, or tertiary amines. Name each structure.

   a. $H_3C-\underset{\underset{H}{|}}{N}-CH_3$

   (1 mark)

   b. $H_3C-\underset{\underset{H}{|}}{N}-H$

   (1 mark)

   c. $H_3C-\underset{\underset{CH_3}{|}}{N}-C_2H_5$

   (1 mark)

2. Write equations for the reaction between an excess of ethylamine and:
   a. nitric acid *(1 mark)*
   b. sulfuric acid *(1 mark)*

3. The mechanism of the reaction between chloromethane and ethylamine is similar to that for a haloalkane and hydroxide ions. The lone pair on the nitrogen atom enables ethylamine to act as a nucleophile. Suggest a reaction mechanism for this reaction. *(3 marks)*

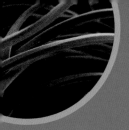

27.2 Amino acids, amides, and chirality

Specification reference: 6.2.2

Learning outcomes

Demonstrate knowledge, understanding, and application of:

→ the general formula for α-amino acids

→ reactions of the carboxylic acid group and amine groups in amino acids

→ structures of primary and secondary amides

→ optical isomerism and chirality.

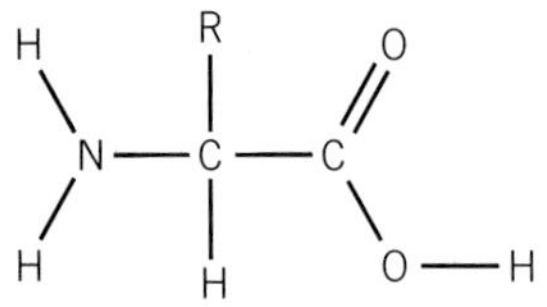

Amino acids

An amino acid is an organic compound containing both amine, NH_2, and carboxylic acid, COOH, functional groups. The body has 20 common amino acids that can be built into proteins. These amino acids are all α-amino acids in which the amine group is attached to the α-carbon atom – the second carbon atom, next to the carboxyl group. These 20 amino acids differ by the side chain, R, attached to the same α-carbon atom. The structure of an α-amino acid is shown in Figure 1.

The general formula of an α-amino acid can be written as $RCH(NH_2)COOH$. There are also less common amino acids in which the amine group is connected to the β-carbon atom (the third carbon) and the γ-carbon atom (the fourth carbon).

Since amino acids have both an acidic COOH and a basic NH_2 functional group, amino acids have similar reactions to both carboxylic acids and amines.

▲ **Figure 1** *The structure of an α-amino acid. All amino acids contain a basic amine group and an acidic carboxylic acid group. Each amino acid has a different R group*

Reactions of the amine group

The amine group is basic and reacts with acids to make salts. As such, amino acids will also react with acids to form salts. The amino acid alanine (2-aminopropanoic acid) reacts with hydrochloric acid to form an ammonium salt (Figure 2).

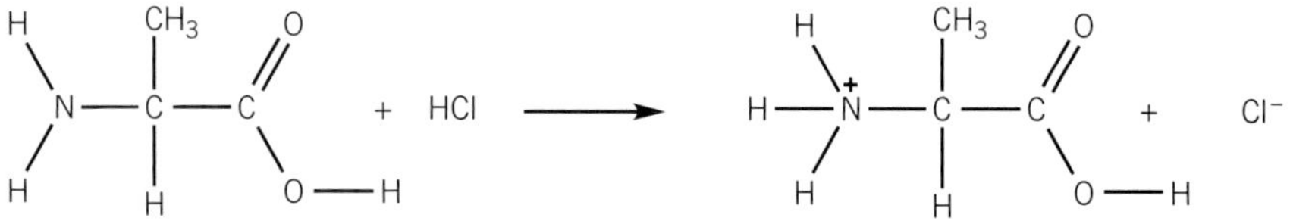

▲ **Figure 2** *The reaction of 2-aminopropanoic acid with hydrochloric acid*

Synoptic link

For details of reactions of salt formation and esterification of carboxylic acids, see Topic 26.3, Carboxylic acids, and Topic 26.4, Carboxylic acid derivatives.

Reactions of the carboxylic acid group in amino acids

The carboxylic acid group can react with alkalis to form salts and with alcohols to form esters.

Reaction with aqueous alkalis

An amino acid reacts with an aqueous alkali such as sodium or potassium hydroxide to form a salt and water. In Figure 3, glycine (aminoethanoic acid) reacts with sodium hydroxide to form a sodium salt.

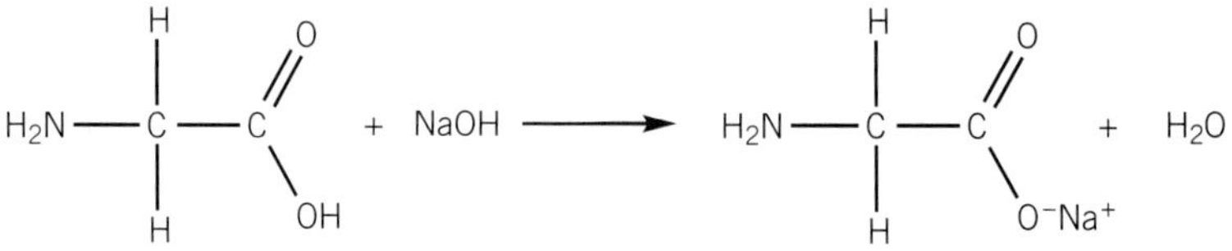

▲ **Figure 3** *An amino acid reacting with an aqueous base*

478

AMINES, AMINO ACIDS, AND POLYMERS

Esterification with alcohols

Amino acids, like carboxylic acids, are easily esterified by heating with an alcohol in the presence of concentrated sulfuric acid.

In Figure 4, the α-amino acid serine is reacted with excess ethanol and a small amount of a sulfuric acid. The carboxylic acid group is esterified, producing an ester. The acidic conditions protonate the basic amine group of the ester.

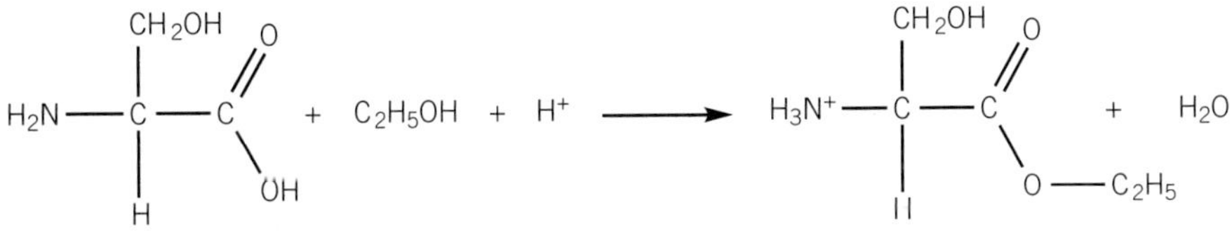

▲ **Figure 4** *An ester forms when the amino acid, serine, undergoes esterification*

Zwitterions and the isoelectric point

Within the structure of the amino acid, the basic amine group can accept a proton from the carboxylic acid group to form an ion containing both a positive and negative charge. This ion is known as a *zwitterion*. The zwitterion formed from the α-amino acid alanine is shown in Figure 5.

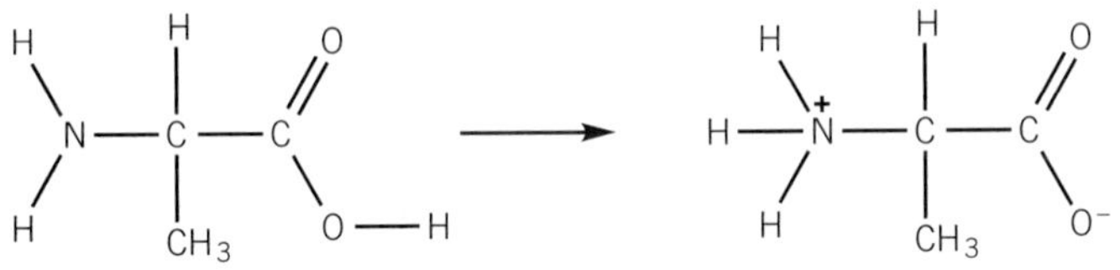

▲ **Figure 5** *The zwitterion formed from the α-amino acid alanine*

Zwitterions have no overall charge because the positive and negative charges cancel out. The isoelectric point is the pH at which the zwitterion is formed, and each amino acid has its own unique isoelectric point.

- If an amino acid is added to a solution with a pH greater than its isoelectric point, the amino acid behaves as an acid and loses a proton (Figure 6).
- If an amino acid is added to a solution with a pH lower than its isoelectric point, the amino acid behaves as a base and gains a proton (Figure 6)

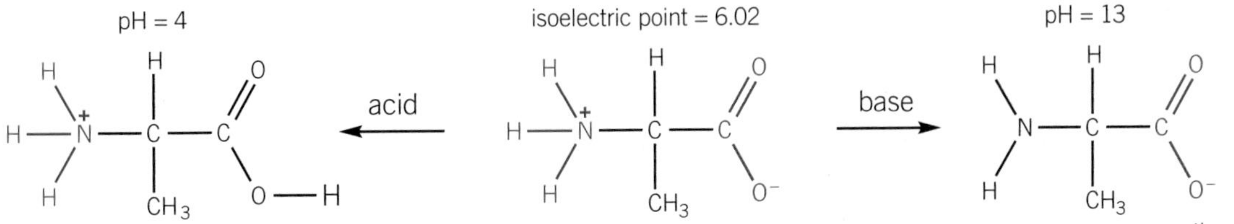

▲ **Figure 6** *Acid and base reactions of amino acids*

> The isoelectric points of valine, $(CH_3)_2CHCH(NH_2)COOH$, is 5.97 whereas that of aspartic acid, $HOOCCH_2CH(NH_2)COOH$, is 2.76. Draw the structures of valine and aspartic acid at pH 2.76 and at pH 12.20.

27.2 Amino acids, amides, and chirality

> **Synoptic link**
>
> For details of the preparation of amides from acyl chlorides, revisit Topic 26.4, Carboxylic acid derivatives.

Amides

Amides are the products of reactions of acyl chlorides with ammonia and amines. Amide groups are common in nature. For example, in proteins the amine and carboxylic acid groups of amino acids are bonded together to form amide groups.

As with amines, there are primary, secondary, and tertiary amides (Figure 7).

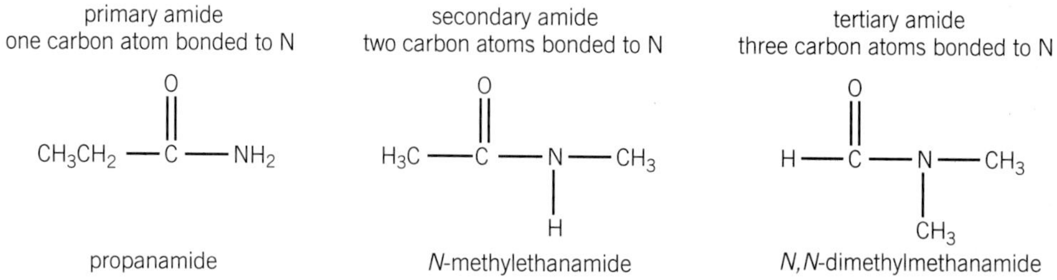

primary amide
one carbon atom bonded to N
$CH_3CH_2-C(=O)-NH_2$
propanamide

secondary amide
two carbon atoms bonded to N
$H_3C-C(=O)-N(H)-CH_3$
N-methylethanamide

tertiary amide
three carbon atoms bonded to N
$H-C(=O)-N(CH_3)-CH_3$
N,N-dimethylmethanamide

▲ **Figure 7** *The structure of a primary, secondary, and tertiary amide*

> **Synoptic link**
>
> See Topic 13.2, Stereoisomerism, for an introduction to stereoisomerism and E/Z isomerism.

> **Synoptic link**
>
> You have come across optical isomers in transition elements covered in Topic 24.3, Stereoisomerism in complex ions.

Stereoisomerism

You should recall that stereoisomers are compounds with the same structural formula but a different arrangement of atoms. Another type of stereoisomerism is **optical isomerism**.

Optical isomerism

Optical isomerism is found in molecules that contain a **chiral centre**. In organic chemistry, the chiral centre is a carbon atom that is attached to four different atoms or groups of atoms.

The presence of a chiral carbon atom in a molecule leads to the existence of two non-superimposable mirror image structures. These two molecules are known as optical isomers or **enantiomers**. For each chiral carbon atom, there is always one pair of optical isomers.

Chirality in α-amino acids

Chiral carbon atoms exist widely in naturally occurring organic molecules. For example, all sugars, proteins, and nucleic acids contain chiral carbon atoms.

With the exception of glycine, H_2NCH_2COOH, all α-amino acids, $RCH(NH_2)COOH$, contain a chiral carbon atom, with the α-carbon atom bonded to four different atoms or groups. In the structure, the chiral carbon atom is shown labelled with an asterisk *.

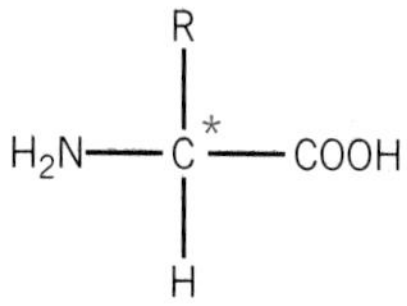

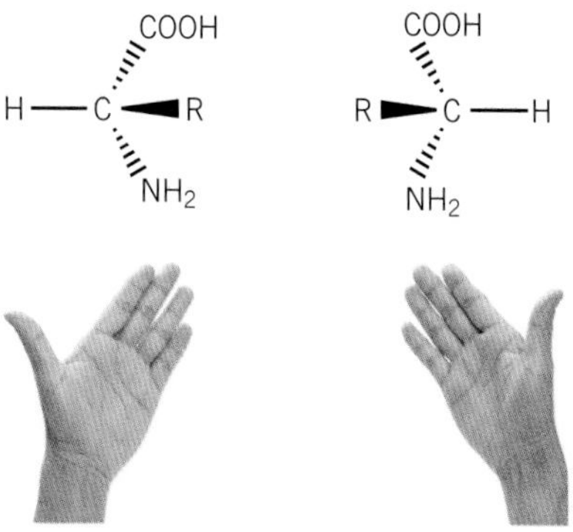

Chirality is not just reserved for carbon atoms and the term applies to any centre that holds attachments that can be arranged as two non-superimposable mirror image forms. Like a pair of hands, optical isomers can be considered as right- and left-handed forms and one optical isomer cannot be superimposed upon the other.

Figure 8 shows two hands above which are placed models of two optical isomers of an α-amino acid.

▲ **Figure 8** *Two optical isomers of an amino acid and a pair of hands – both examples of chirality*

AMINES, AMINO ACIDS, AND POLYMERS — 27

Drawing optical isomers

Optical isomers are drawn to show the 3D tetrahedral arrangement of the four different groups around the central chiral carbon atom. Once one isomer has been drawn, the other isomer is drawn as a mirror image, reflecting the first structure. The two optical isomers of butan-2-ol, $CH_3CH_2CH(OH)CH_3$, are shown in Figure 9.

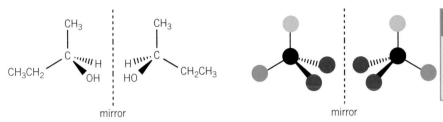

Synoptic link

See Topic 6.1 Shapes of molecules and ions for details of drawing 3-D shapes using wedges.

▲ Figure 9 Three-dimensional arrangement of the four different groups around a chiral carbon atom (left) and a simplified arrangement of two optical isomers (right) showing that they are non-superimposable mirror images of each other

Identifying a chiral carbon atom

You should be able to identify chiral centres in a molecule of any organic compound.

Optical isomerism – testing our senses

Our senses of taste and smell are regulated by chiral molecules, which interact with receptor sites in our mouths and noses. Optical isomers interact differently with our receptors, inducing different taste and smell sensations. The two optical isomers of the amino acid leucine, $(CH_3)_2CHCHCH_2CH(NH_2)COOH$, have different tastes, one being bitter whereas the other one is sweet. The bitter isomer is used as a food additive.

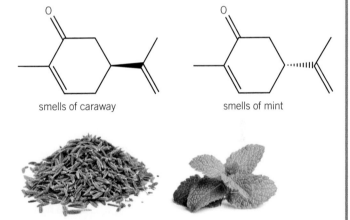

smells of caraway   smells of mint

▲ Figure 11 Caraway seeds (left) and mint leaves (right)

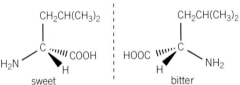

▲ Figure 10 The two optical isomers of leucine

Carvone a naturally occurring ketone which exists as two different optical isomers, depending on its source. One optical isomer smells of caraway seeds, the other of peppermint.

1. Limonene is responsible for the taste of oranges and lemons. Look up the structure of limonene and identify which optical isomers are responsible for each taste.
2. Optical isomers are also important in many drugs and medicine. Find out the effects of the optical isomers of the drugs:
   a. naproxen   b. propranolol   c. thalidomide.

Study tip

You may be asked to identify chiral centres in a molecule of any organic compound.

27.2 Amino acids, amides, and chirality

 Worked example: Identifying chiral centres in molecules

Identify any chiral carbon atoms, with an asterisk *, in the following molecules:

Step 1: Identify any chiral carbon atoms.

On the skeletal structures, remember that there will be hydrogen atoms also. You may find it easier to mark these on the structures.

Step 2: Identify the chiral carbon atoms as those that are attached to four different groups or atoms and label the chiral carbon atoms with an asterisk.

Here all groups have been shown at the chiral carbon atoms. Take care not to label the carbon atoms with two methyl groups. The hardest to see is in the right-hand structure where the side chain is attached to the ring.

Summary questions

1. Write equations for the reaction of 2-aminobutanoic acid with:
   a. sodium hydroxide *(1 mark)*
   b. ethanol in the presence of concentrated sulfuric acid *(1 mark)*
   c. nitric acid *(1 mark)*

2. Draw displayed formulae for the following molecules and label any chiral carbon atoms with an asterisk.
   a. $CH_3CH_2CHBrCH_2CH_2Cl$ *(1 mark)*
   b. $CH_3CH_2CH(OH)CH_3$ *(1 mark)*

3. For each of the molecules below, draw the three-dimensional arrangement of two optical isomers.
   a. 2-aminopentanoic acid *(1 mark)*
   b. 2-hydroxypropanoic acid *(1 mark)*

4. Write an equation for the reaction of aspartic acid, $HOOCCH_2CH(NH_2)COOH$, with an excess of methanol and a small quantity of concentrated sulfuric acid. *(2 marks)*

27.3 Condensation polymers
Specification reference: 6.2.3

Polyesters and polyamides

Alkenes can form polymers by addition polymerisation. In this topic, you will look at a different type of polymerisation called **condensation polymerisation**.

Polyesters and polyamides are two important condensation polymers made on an industrial scale. Carboxylic acids and their derivatives are common starting materials for their preparation.

Condensation polymerisation is the joining of monomers with loss of a small molecule, usually water (the condensation) or hydrogen chloride. In condensation polymerisation two different functional groups are needed.

Polyesters

When making a polyester, monomers are joined together by ester linkages in a long chain to form the polymer. Polyesters can be made from one monomer containing both a carboxylic acid and an alcohol group, or from two monomers – one containing two carboxylic acid groups and the other containing two alcohol groups.

Polyesters made from one monomer containing two different functional groups

For this type of a polyester formation, only *one* monomer is involved, containing both different functional groups, a hydroxyl group, –OH, and a carboxyl group, –COOH.

Poly(glycolic acid) (PGA) is the simplest polyester of this type, made by condensation polymerisation of its monomer, glycolic acid, $HOCH_2COOH$. Glycolic acid contains both a hydroxyl group and a carboxyl group.

Figure 1 shows the formation of a section of poly(glycolic) acid from its monomer. The carboxylic acid group in one molecule of glycolic acid reacts with the alcohol group of another molecule of glycolic acid to form the ester linkage and water.

▲ **Figure 1** *Condensation polymerisation of glycolic acid, $HOCH_2COOH$, showing three repeat units*

Learning outcomes

Demonstrate knowledge, understanding, and application of:

→ condensation polymerisation to form polyesters and polyamides

→ acid and base hydrolysis of the amide and ester group in polymers

→ identifying repeat units and monomers from both addition and condensation polymerisation.

Synoptic link

You covered addition polymerisation in Topic 13.5, Polymerisation in alkenes.

Study tip

You do not need to learn any specific examples of condensation polymerisation, but you do need to understand the principles of how the polymers are formed.

Study tip

When drawing a section of a polymer, it is important to leave open bonds at each end. The repeat unit is repeated time and time again along the whole polymer. It is usual to bracket the repeat unit as shown in Figure 1.

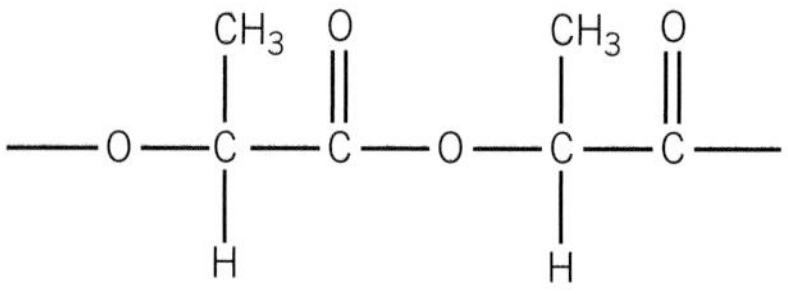

27.3 Condensation polymers

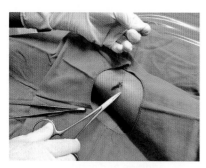

▲ Figure 2 *A section of the polymer, poly(lactic) acid (PLA)*

Lactic acid, HOCH(CH$_3$)COOH, also undergoes condensation polymerisation to form poly(lactic acid) (PLA) as shown in Figure 2. Both PLA and PGA are biodegradable polymers. Lactic acid is derived from maize, making its production much more sustainable than polymers derived from fossil fuels.

Polyesters made from two monomers each containing two functional groups

For this type of a polyester formation, *two* different monomers are involved, each with different functional groups.

- One monomer is a diol, with two hydroxyl groups.
- One monomer is a dicarboxylic acid, with two carboxyl groups.

Terylene is a condensation polymer made from the reaction between the two monomers, benzene-1,4-carboxylic acid, HOOCC$_6$H$_4$COOH, and ethane-1,2-diol, HO(CH$_2$)$_2$OH. The formation of Terylene from its monomers is shown in Figure 4. During the condensation reaction a hydroxyl group on the diol reacts with a carboxyl group on the dicarboxylic acid forming an ester linkage and water.

Terylene (polyethylene terephthalate) is often shortened to PET, and has many diverse uses ranging from clothing to plastic PET bottles. Polyesters are also used for electrical insulation.

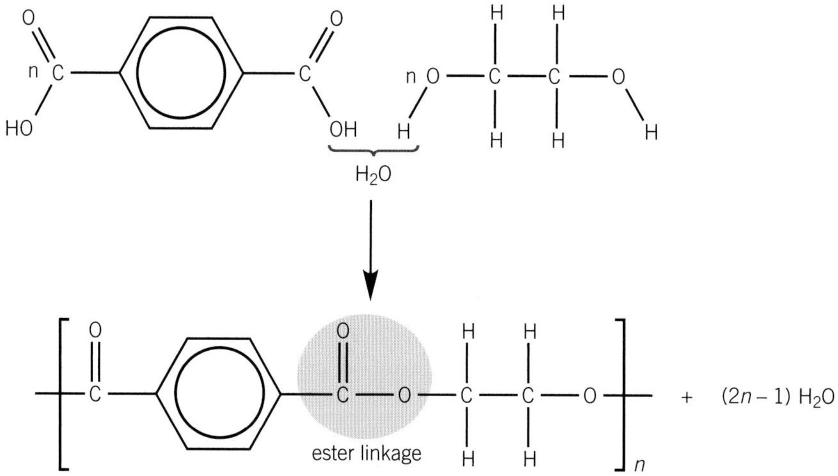

▲ Figure 3 *Sutures are one of modern medicine's most simple inventions, used to close wounds after surgery or injury, aiding the healing process without leaving scars. The suture is a thread of PLA or PGA fibre seen here held with a pair of scissor forceps. The suture biodegrades, softening in about a week and will dissolve completely in several months*

Study tip

A polyester can also be made using an diacyl chloride instead of a dicarboxylic acid. With an diacyl chloride, hydrogen chloride is lost, instead of water.

▲ Figure 4 *The formation of Terylene, a polyester*

Polyamides

Polyamides are condensation polymers formed when monomers are joined together by amide linkages in a long chain to form the polymer. As with polyesters, polyamides can be made from one monomer containing both a carboxylic acid (or acyl chloride) and an amine group – or from two monomers, one containing two carboxylic acid groups (or acyl chlorides) and the other containing two amine groups.

▲ Figure 5 *Items such as sports clothing, shirts, ties, and carpets are all made from Terylene*

Polyamides from one monomer with two functional groups

Amino acids contain both an amine group and a carboxylic acid group. Amino acids undergo condensation polymerisation to form polypeptides or proteins. A polypeptide or protein contains many different amino acids all linked together by amide bonds. When an amide bond is formed water is lost. Figure 6 shows a section of a polypeptide made from a number of amino acids.

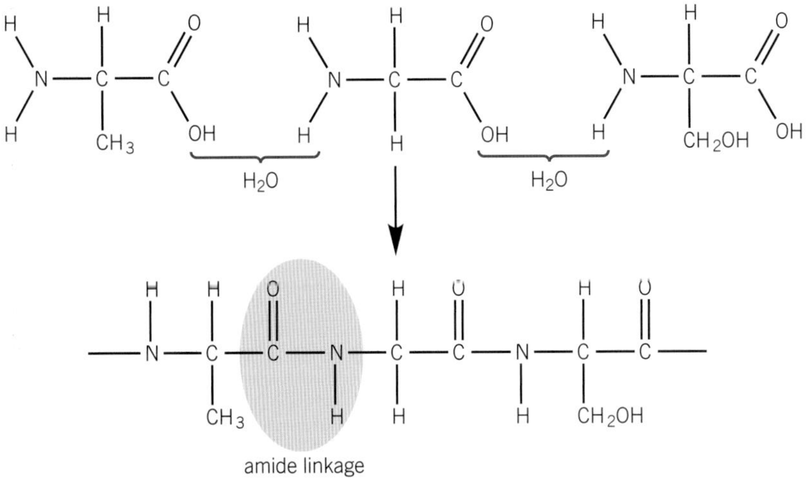

▲ **Figure 6** *The formation of amide bonds in protein formation*

▲ **Figure 7** *Nylon is a high strength fibre. It is used for making fishing nets, ropes, parachutes, and fabrics. It has also been used for classical guitar strings since the mid 1940s*

Polyamides from two monomers each with two functional groups

Polyamides can be made by the reaction of a dicarboxylic acid (or acyl chloride) with a diamine.

Figure 8 shows the formation of Nylon from the reaction of hexanedioic acid and 1,6-diaminohexane. Figure 9 shows the formation of nylon from the reaction of hexanedioyl chloride and 1,6-diaminohexane. Different types of Nylon can be made by varying the carbon-chain length. During the condensation reaction an amide bond forms between the amine on one monomer and the carboxyl (or acyl chloride) group on the other monomer.

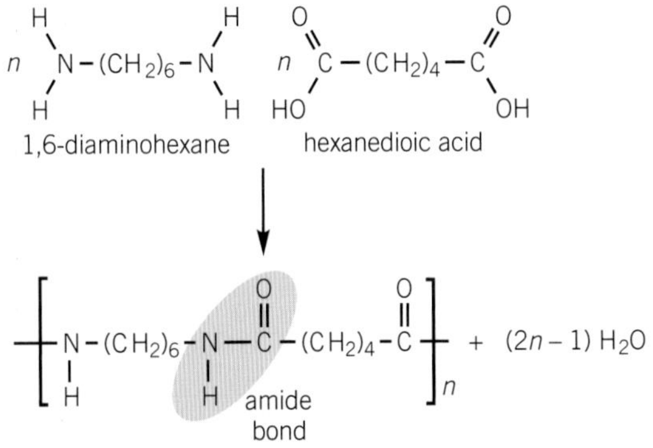

▲ **Figure 8** *Synthesis of Nylon from the reaction of a diamine with a dicaboxylic acid*

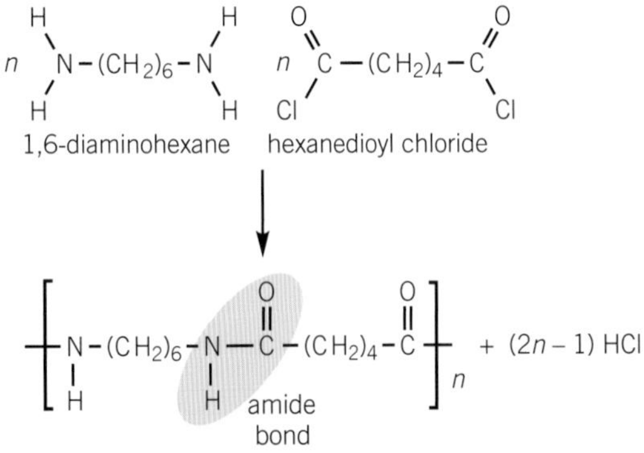

▲ **Figure 9** *Synthesis of Nylon from the reaction of a diamine with a diacyl chloride*

Hydrolysing condensation polymers

Polyesters and polyamides can be hydrolysed using hot aqueous alkali such as sodium hydroxide or by hot aqueous acid such as hydrochloric acid.

27.3 Condensation polymers

> **Synoptic link**
>
> See Topic 26.4 Carboxylic acid derivatives, for acid and alkaline hydrolysis of esters

Hydrolysing polyesters

Poly(Trimethylene Terephthalate) (PTT) is a polyester used in the manufacture of carpets and clothing fabrics. The acid and base hydrolysis of PTT is shown in Figure 9.

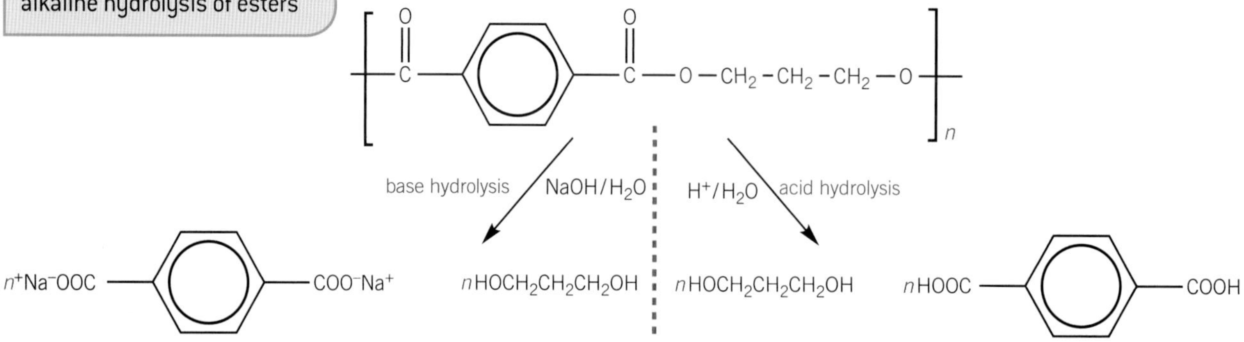

▲ **Figure 9** *The acid and base hydrolysis of PTT*

▲ **Figure 11** *Oven gloves containing the fire-resistant polyamide Nomex*

Hydrolysing polyamides

Nomex is a synthetic heat and fire-retardant polyamide, used in oven gloves and in the fire-protective suits worn by Formula 1 racing drivers and pit crew. The acid and base hydrolysis of Nomex is shown in Figure 10.

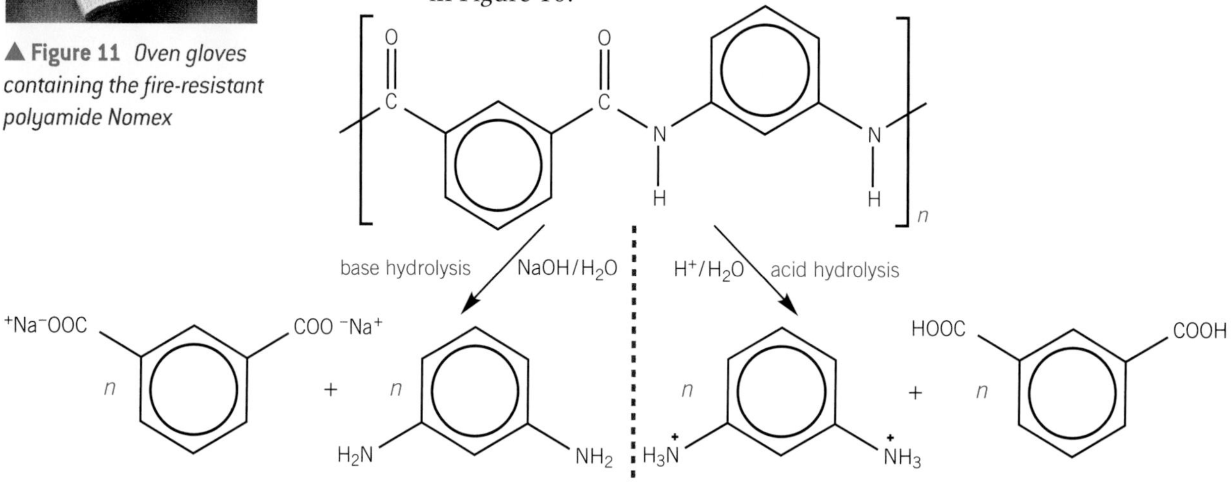

▲ **Figure 10** *The acid and base hydrolysis of Nomex*

Identifying monomers and repeat units in polymers

You should be able to identify monomers that could form polymer chains and to identify whether addition polymerisation or condensation polymerisation has taken place. Table 1 outlines what to look for when deciding the type of polymerisation.

▼ **Table 1** *Comparison of addition and condensation polymerisation*

Type	Characteristics	Example of monomer(s)
addition polymerisation	Monomer contains a C=C double bond. Backbone of polymer is a continuous chain of carbon atoms.	$ClCH=CHCH_3$
condensation polymerisation	Two monomers, each with two functional groups. One monomer with two different functional groups. Polymer contains ester or amide linkages.	$HOOCCH_2COOH$ and $HOCH_2CH_2OH$ H_2NCH_2COOH

AMINES, AMINO ACIDS, AND POLYMERS — 27

 Worked example: Identifying monomers and repeat units in polymers 1

Example 1
Identify the type of polymerisation, the repeat unit, and the monomer(s) for the polymer:

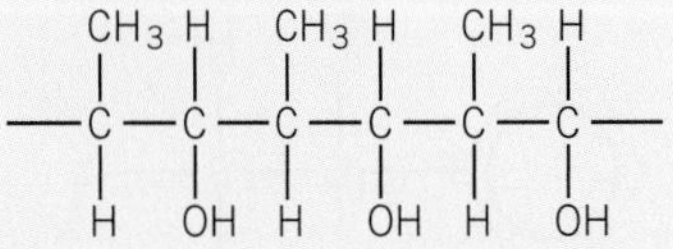

Step 1: The polymer backbone is a continuous chain of carbon atoms. Therefore it is an addition polymer.

Step 2: The repeat unit is

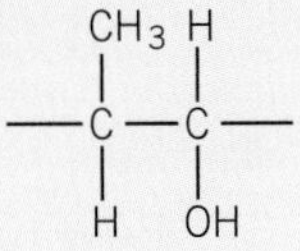

Step 3: The monomer is

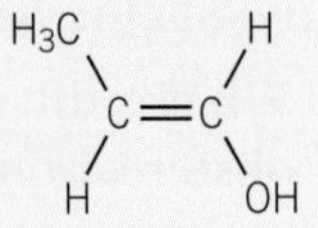

 Worked Example: Identifying monomers and repeat units in polymers 2

Example 2
Identify the type of polymerisation, the repeat unit and the monomer(s) for the polymer:

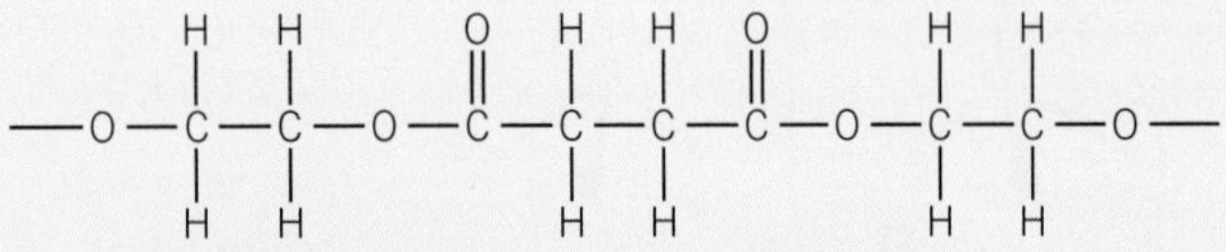

Step 1: The polymer contains ester linkages. Therefore is a condensation polymer.

Step 2: The repeat unit is

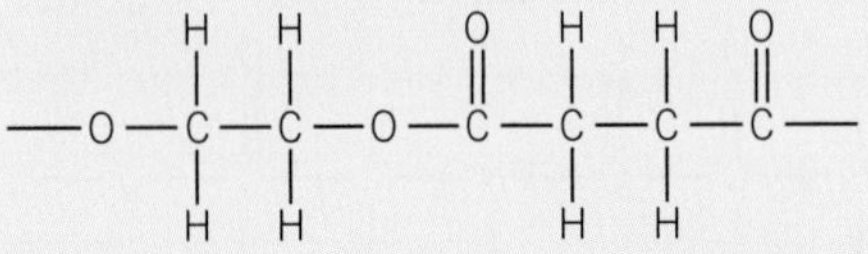

Step 3: The monomers are

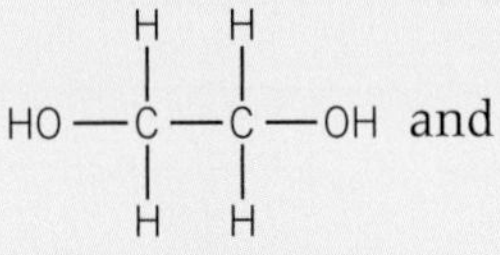

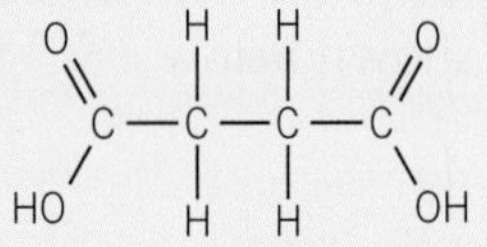

Summary questions

1. State the two functional groups that must be present in monomers to form:
   a. a polyamide *(2 marks)*
   b. a polyester. *(2 marks)*

2. For the monomer(s) below, state the type of polymerisation and draw one repeat unit of the polymer.
   a. *(2 marks)*
   b. *(2 marks)*
   c. *(2 marks)*

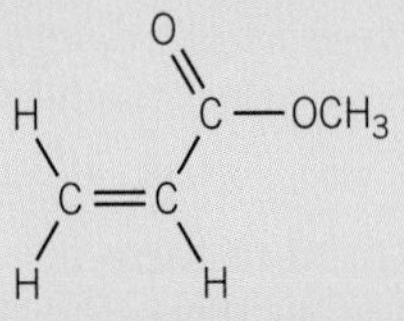

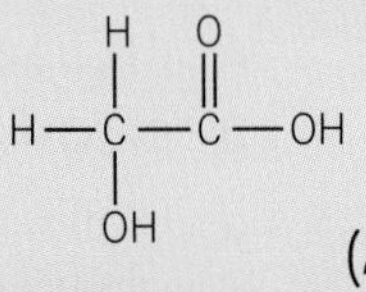

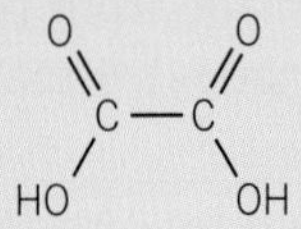

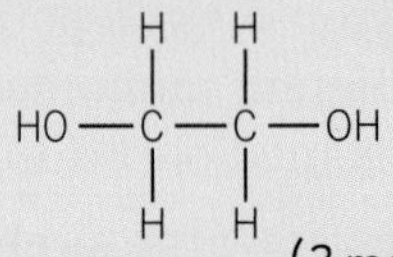

3. Draw structures for the products of acid and alkaline hydrolysis of the following polymers.
   a. *(4 marks)*
   b. *(4 marks)*

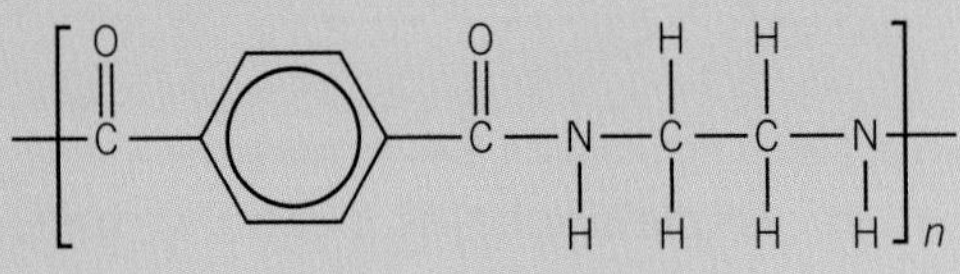

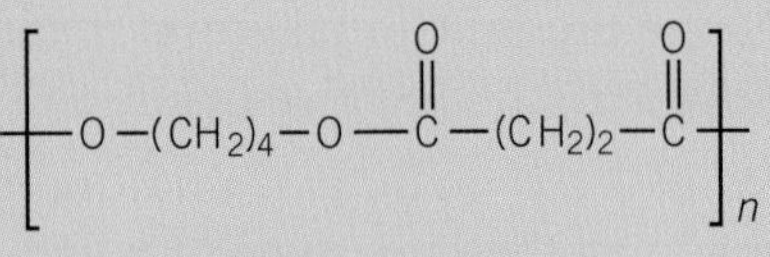

487

Chapter 27 Practice questions

Practice questions

1 Phenylalanine is a naturally occurring α-amino acid. Its structure is shown below.

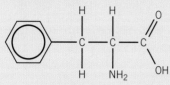

   a Phenylalanine has two stereoisomers.
   (i) What is the name of the type of stereoisomerism shown by phenylalanine? *(1 mark)*
   (ii) Complete the structures below to show the 3-D arrangement of the two stereoisomers of phenylalanine.

   mirror *(2 marks)*

2 Glutamic acid, an amino acid, can react with acids and bases as well as alcohols to form a variety of products. The structure of glutamic acid is shown below.

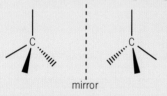

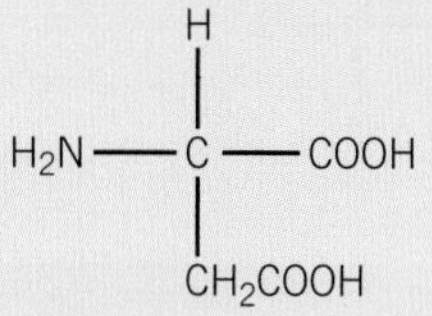

Draw the structures for the reaction of glutamic acid and the following substances.
   a NaOH(aq) *(1 mark)*
   b HCl(aq) *(1 mark)*
   c C_2H_5OH and concentrated sulfuric acid *(1 mark)*

3 Nylon-6,6 is a polymer made from two different monomers: hexane-1,6-diamine and hexanedioic acid.
   a Draw the **skeletal** formula for each monomer. *(2 marks)*
   b Draw the structural formula for the repeat unit of nylon-6,6. *(2 marks)*
   c Name the functional group made during polymerisation to give nylon-6,6. *(1 mark)*
   d Wool is a protein. It is a natural polymer made by the same type of polymerisation as nylon-6,6. A section of the polymer chain in a protein is shown below.

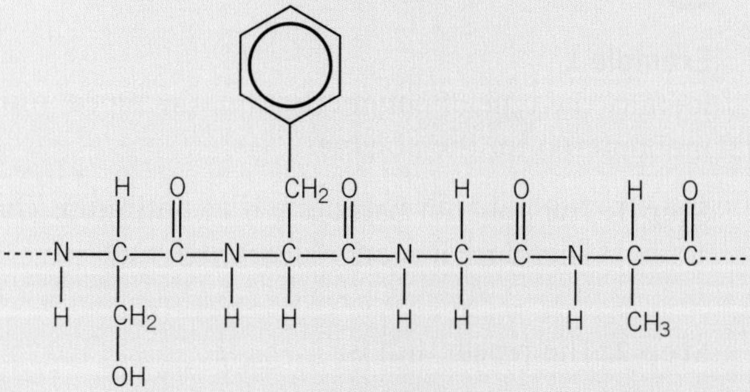

   (i) How many monomer units does this section contain? *(1 mark)*
   (ii) Draw a displayed formula for **one** of the monomer molecules that was used to form this section. *(1 mark)*
   (iii) The protein section shown above is reacted with hot aqueous acid.
   Draw the structures of the products of this reaction. *(4 marks)*

4 Aspartame is a low calorie sweetener introduced in 1983. The structure of aspartame is shown below.

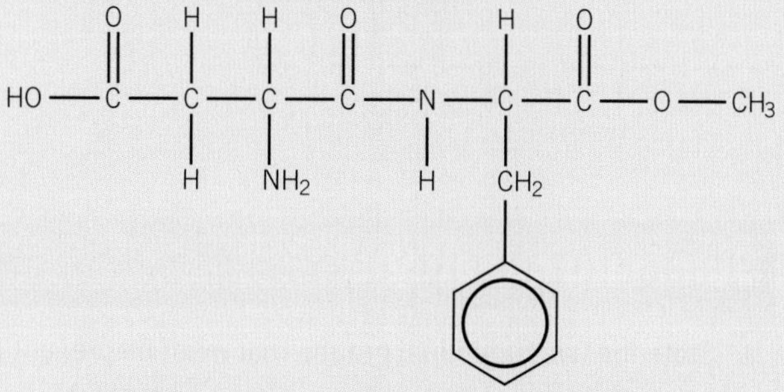

Aspartame can be hydrolysed by aqueous HCl and aqueous NaOH.
Predict the products of the hydrolysis of aspartame in
   a HCl(aq) *(3 marks)*
   b NaOH(aq) *(3 marks)*

5 a Salicylic acid can be used to form a condensation polymer similar to Terylene®.

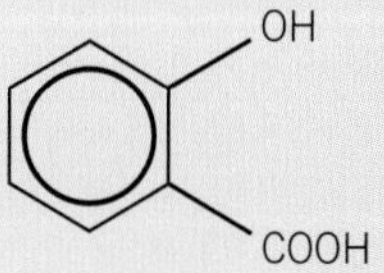

salicylic acid

   (i) Explain what is meant by the term condensation polymer. *(1 mark)*

(ii) The repeat unit of Terylene® is shown below.

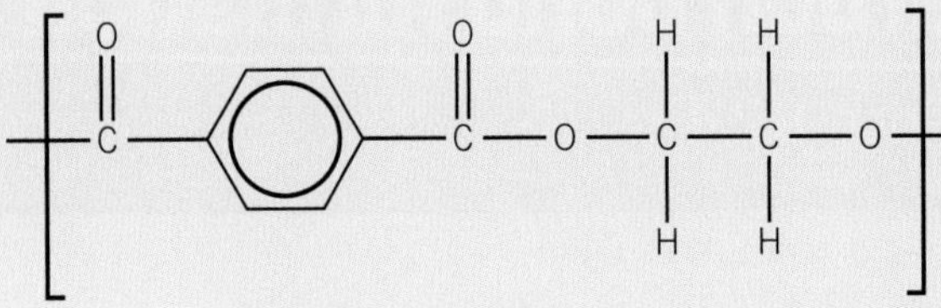

[structure of Terylene repeat unit]

Draw the skeletal formulae of **two** monomers that can be used to form Terylene®. *(2 marks)*

(iii) Salicylic acid reacts with 3-hydroxypropanoic acid to form a mixture of condensation polymers. To form one polymer, the two monomers react in equal quantities.

Draw the repeat unit of this polymer, displaying the link between the monomer units. *(2 marks)*

OCR F324 June 2014 Q1(d)

6 Penicillamine is an α-amino acid that is used as a drug to treat rheumatoid arthritis. The structure of penicillamine is shown below.

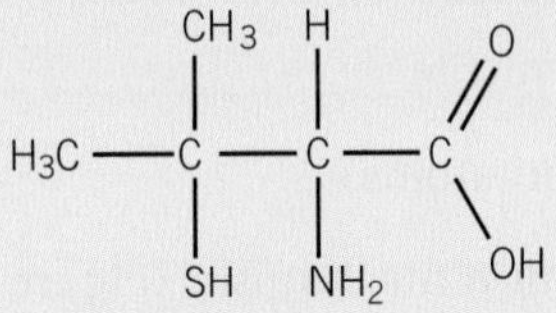

[structure of penicillamine]

a Explain why penicillamine is described as an α-amino acid. *(1 mark)*

b Penicillamine reacts with dilute aqueous acid. Draw the structure of the product formed in this reaction. *(1 mark)*

c Penicillamine exists as optical isomers. Draw the two optical isomers of penicillamine. *(2 marks)*

d Penicillamine is able to undergo condensation polymerisation.

Draw **two** repeat units of the polymer formed from penicillamine. *(2 marks)*

OCR F324 Jan 2013 Q3

7 This question looks at different types of condensation polymers – polyesters and polyamides.

a Polyester **A** is a degradable polymer prepared by bacterial fermentation of sugars.

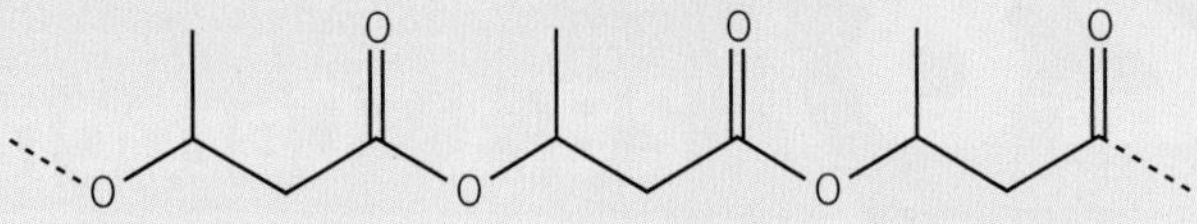

[structure of polyester A]

When polyester **A** is hydrolysed with aqueous acid, compound **B** is formed.

Draw the skeletal formula of compound **B**. *(1 mark)*

b Nylon-4,6 is a polyamide that can be prepared by reacting 1,4-diaminobutane, $H_2N(CH_2)_4NH_2$, with hexanedioic acid, $HOOC(CH_2)_4COOH$.

(i) $H_2N(CH_2)_4NH_2$ can be synthesised from 1,4-dichlorobutane, $Cl(CH_2)_4Cl$.

State the reagents and conditions required for this synthesis. *(1 mark)*

(ii) $H_2N(CH_2)_4NH_2$ can act as a base and forms salts with dilute acids.

Explain how an amine can act as a base.

Write the formula of the salt formed when $H_2N(CH_2)_4NH_2$ reacts with an **excess** of dilute hydrochloric acid. *(2 marks)*

(iii) Draw the repeat unit of nylon-4,6. Clearly display the bonding that links the two monomers. *(3 marks)*

OCR F324 Jun 2012 Q2

8 A student was investigating the reactions and uses of organic amines.

a The student found that amines such as ethylamine, $C_2H_5NH_2$, and phenylamine, $C_6H_5NH_2$, both behave as bases.

Explain why amines can behave as bases. *(1 mark)*

b The student reacted an excess of $C_2H_5NH_2$ with two different acids.

Write the formulae of the salts that would be formed when an **excess** of $C_2H_5NH_2$ reacts with

(i) sulfuric acid *(1 mark)*

(ii) ethanoic acid. *(1 mark)*

c Phenylamine, $C_6H_5NH_2$, can be prepared from nitrobenzene.

State the reagents and conditions. Write the equation and state the type of reaction. *(3 marks)*

28 ORGANIC SYNTHESIS
28.1 Carbon–carbon bond formation

Specification reference: 6.2.4

Learning outcomes

Demonstrate knowledge, understanding, and application of:

→ the use of C—C bond formation in synthesis to increase the length of a carbon chain

→ the formation and reaction of nitriles

→ alkylation and acylation of benzene.

Carbon–carbon bond formation in synthesis

Reactions that form carbon–carbon bonds have a central importance to organic chemistry as they provide a means of synthesising new compounds containing more carbon atoms. In this topic you will be looking at reactions that can be used to lengthen a carbon chain, to add a side chain to a benzene ring and to introduce new functional groups which can then be reacted further.

Formation of nitriles

The nitrile group is the organic functional group –CN.

Nitriles from haloalkanes

Nitriles can be formed by reacting haloalkanes with sodium cyanide, NaCN, or potassium cyanide, KCN, in ethanol. In this reaction the length of the carbon chain is increased.

$$CH_3CH_2CH_2Cl + KCN \rightarrow CH_3CH_2CH_2CN + KCl$$

1-chloropropane   butanenitrile

(3 carbon atoms)   (4 carbon atoms)

The reaction follows a nucleophilic substitution mechanism (Figure 1).

Synoptic link

You should recall the mechanism of nucleophilic substitution from Topic 15.1, The chemistry of the haloalkanes.

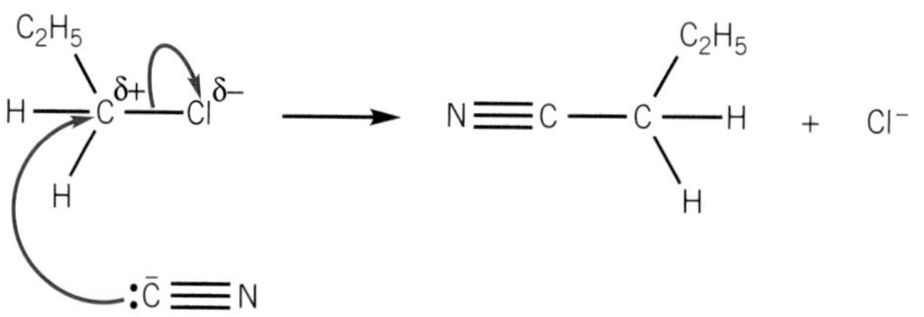

▲ **Figure 1** *Formation of a nitrile by a nucleophilic substitution reaction of a haloalkane*

Nitriles from aldehydes and ketones

Aldehydes and ketones will react with hydrogen cyanide in a nucleophilic addition reaction. The reaction forms a carbon–carbon bond, thus increasing the number of carbon atoms in the molecule.

$$CH_3COCH_3 + HCN \rightarrow CH_3C(OH)(CN)CH_3$$

The product is known as a hydroxynitrile (or cyanohydrin).

Hydrogen cyanide is far too poisonous to use, and an increased reaction rate can be obtained in the presence of cyanide ions. A mixture of sodium cyanide and sulfuric acid is used, which improves safety and increases reaction rate. The reaction mechanism is shown in Figure 2.

Synoptic link

The reactions of aldehydes and ketones with hydrogen cyanide were discussed in Topic 26.1, Carbonyl compounds.

ORGANIC SYNTHESIS 28

▲ **Figure 2** Nucleophilic addition mechanism for the reaction of a carbonyl compound with HCN

Hydroxynitriles in nature

A species of millipede, *Apheloria corrugata*, stores 2-hydroxy-2-phenylethanenitrile in one of its body segments. When the millipede is attacked, the 2-hydroxy-2-phenylethanenitrile is released from a storage chamber into a reaction chamber where it mixes with an enzyme. The reaction between the enzyme and the hydroxynitrile releases very toxic hydrogen cyanide, together with benzaldehyde (Figure 3). The millipede secretes this toxic mixture from glands located in its body segments. An adult millipede can produce 0.6 mg of hydrogen cyanide, which is enough to kill ants and deter most of their other enemies.

> What mass of 2-hydroxy-2-phenylethanenitrile must the millipede break down to produce 0.6 mg of HCN?

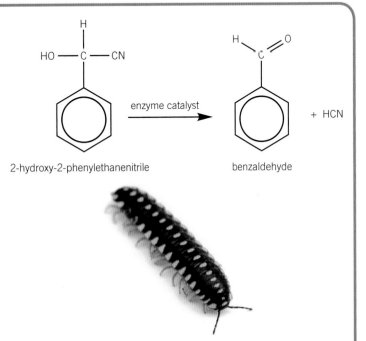

▲ **Figure 3** *Apheloria corrugata* uses an enzyme to break down a hydroxynitrile to form deadly hydrogen cyanide

Nitriles are useful intermediates in the synthesis of other organic compounds such as amines and carboxylic acids. The reduction and hydrolysis reactions of nitriles are described below.

Reduction of nitriles

Nitriles can be reduced to amines by reacting with hydrogen in the presence of a nickel catalyst.

$$CH_3CH_2C\equiv N + 2H_2 \xrightarrow{\text{Ni catalyst}} CH_3CH_2CH_2NH_2$$
propanenitrile → propylamine

Hydrolysis of nitriles

Nitriles undergo hydrolysis to form carboxylic acids by heating with dilute aqueous acid, for example, HCl(aq).

$$CH_3CH_2CH_2C\equiv N + 2H_2O + HCl \xrightarrow{\text{heat}} CH_3CH_2CH_2COOH + NH_4Cl$$
butanenitrile → butanoic acid

28.1 Carbon–carbon bond formation

> **Synoptic link**
>
> In Topic 25.2, Electrophilic substitution reactions of benzene, you have already met alkylation and acylation as methods for increasing the number of carbon atoms in an aromatic compound.

Forming carbon–carbon bonds to benzene rings

Alkylation

Alkylation is a reaction that transfers an alkyl group from a haloalkane to a benzene ring. The reaction takes place in the presence of a Friedel–Crafts catalyst such as aluminium chloride. The ethylation of benzene is shown in Figure 4.

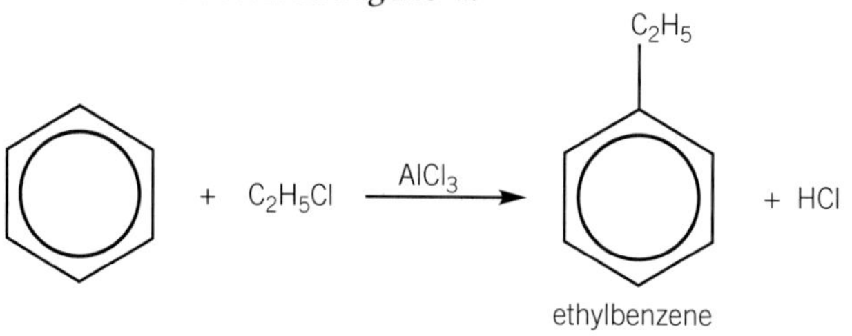

▲ **Figure 4** *The ethylation of benzene*

Acylation

When benzene reacts with an acyl chloride in the presence of an aluminium chloride catalyst, a ketone is formed. This reaction is known as acylation. This reaction is useful in synthesis as the organic product undergoes the typical reactions of a ketone. The acylation of benzene is shown in Figure 5.

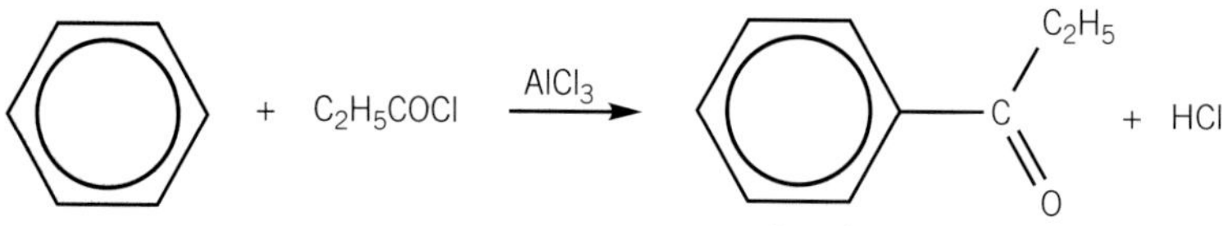

▲ **Figure 5** *acylation of benzene to form phenylpropanone*

➕ Organometallic compounds in carbon–carbon bond formation

Compounds that contain a carbon–metal bond are known as organometallic compounds. Organometallic compounds are very important for forming carbon–carbon bonds in organic compounds to increase the length of a carbon chain. Organometallic compounds containing magnesium or lithium are often used for this purpose.

- An organometallic compound provides a source of a nucleophile called a carbanion, which contain a carbon atom with a negative charge.
- The carbanion reacts with a positively charged or $\delta+$ carbon atom to form a new carbon–carbon bond.

Grignard reagents

Organometallic compounds called Grignard reagents are made by reacting magnesium with an alkyl or aryl haloalkane dissolved in an ether solvent.

$$CH_3CH_2Br + Mg \longrightarrow CH_3CH_2MgBr$$

bromoethane → ethylmagnesium bromide (Grignard reagent)

This Grignard reagent reacts as the carbanion, $CH_3CH_2^-$.

28 ORGANIC SYNTHESIS

Aldehydes and ketones can be reacted with Grignard reagents, followed by dilute aqueous acid, to form alcohols with a longer carbon chain. For example, butanal, $CH_3CH_2CH_2CHO$, can form a secondary alcohol:

$CH_3CH_2CH_2-C(=O)H$ →(1. CH_3CH_2MgBr; 2. H^+/H_2O)→ $CH_3CH_2CH_2-C(OH)(H)-CH_2CH_3$

aldehyde → secondary alcohol (carbon–carbon bond)

The mechanism for this reaction can be shown using a carbanion. This is similar to nucleophilic addition of carbonyl compounds by hydride ions, H^-, or cyanide ions, CN^-.

$CH_3CH_2CH_2-C^{\delta+}(H)=O^{\delta-}$ with $^-CH_2CH_3$ carbanion attacking → intermediate forms $CH_3CH_2CH_2-C(H)(CH_2CH_3)-O^-$ → H^+ intermediate is protonated → $CH_3CH_2CH_2-C(H)(CH_2CH_3)-OH$

$CH_3CH_2^-$ carbanion attacks the $\delta+$ carbon atom and forms a C–C bond

1. A chemist needs to prepare this organic compound from benzaldehyde, C_6H_5CHO. Draw the structure of the Grignard reagents that could be used in this reaction.

   (structure: $C_6H_5-C(OH)(H)-C(H)(CH_3)-CH_2CH_3$)

2. Outline the mechanism for the reaction above
3. Grignard reagents also react with carbon dioxide. Predict the organic product formed when carbon dioxide reacts with the Grignard reagent, CH_3-MgBr, followed by dilute aqueous acid.

Synoptic link
See Topic 26.1 for details of nucleophilic substitution by hydride and cyanide ions.

Summary questions

1. Write equations for the reaction of:
   a. chloroethane with potassium cyanide *(1 mark)*
   b. ethanal with hydrogen cyanide. *(1 mark)*

2. Outline the stages in the synthesis of $CH_3CH(OH)COOH$ from CH_3CHO. State the reagents, essential conditions and provide an equation for each step. *(6 marks)*

3. Outline the stages in the synthesis to the right.

   The CH_2Br group is 2,4-directing and you can assume that the main substitution is at the 4-position. Write equations for each step. *(3 marks)*

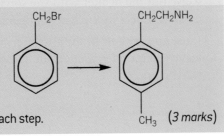

28.2 Further practical techniques
Specification reference: 6.2.5

Learning outcomes
Demonstrate knowledge, understanding, and application of:
→ organic preparation including use of Quickfit apparatus for distillation and heating under reflux
→ purification of an organic solid including the use of filtration under reduced pressure, recrystallisation, and measurement of a melting point.

Synoptic link
Distillation and heating under reflux were covered in Topic 16.1, Practical techniques in organic chemistry.

Preparing, purifying, and determining the melting point an organic solid are important practical skills. You have already seen how to use Quickfit apparatus for distilling and heating under reflux.

Preparation of an organic solid
There are a number of experiments that you may carry out to prepare an organic solid. In the preparation of aspirin, you reflux salicylic acid, ethanoic anhydride, and glacial ethanoic acid using Quickfit apparatus. When preparing benzoic acid by alkaline hydrolysis, you use Quickfit apparatus to reflux methyl benzoate with aqueous sodium hydroxide, followed by acidification. In both preparations a solid forms which has to be separated from the reaction mixture before it can be purified.

Filtration under reduced pressure
Filtration under reduced pressure is a technique for separating a solid product from a solvent or liquid reaction mixture. You will need the following apparatus to carry out this technique:
- Buchner flask
- Buchner funnel
- pressure tubing
- filter paper
- access to filter or vacuum pump.

1. Connect one end of the pressure tubing to the vacuum outlet or to the filter pump whilst attaching the other end of the rubber tubing to the Buchner flask.
2. Fit the Buchner funnel to the Buchner flask ensuring that there is a good tight fit. This is usually obtained using a Buchner ring or a rubber bung (Figure 1).
3. Switch on the vacuum pump, or the tap, to which your filter pump is attached.
4. Check for good suction by placing your hand across the top of the funnel.
5. Place a piece of filter paper inside the Buchner funnel and wet this with the same solvent used in preparing your solid. You should see the paper being sucked down against the holes in the funnel.
6. To filter your sample, slowly pour the reaction mixture from a beaker into the centre of the filter paper.
7. Rinse out the beaker with the solvent so that all of the solid crystals collect in the Buchner funnel.
8. Rinse the crystals in the Buchner funnel with more solvent and leave them under suction for a few minutes so that the crystals start to dry.

▲ Figure 1 Buchner flask and funnel connected to filter pump on standard water tap

ORGANIC SYNTHESIS 28

▲ **Figure 2** *The stages of filtering under reduced pressure*

Recrystallisation

The solid product obtained after filtration will contain impurities which can be removed by carrying out recrystallisation. Purification by recrystallisation depends upon the desired product and the impurities having different solubilities in the chosen solvent.

1. Pour a quantity of the chosen solvent into a conical flask. If the solvent is flammable, warm the solvent over a water bath. If the solvent is water, place the conical flask on a tripod and gauze over a Bunsen and warm the solvent.
2. Tip the impure sample into a second conical flask or beaker.
3. Slowly add the solvent to the impure sample until it dissolves in the solvent. You should add the minimum volume of solvent needed to dissolve the solid.

solvent added to sample　　　　sample dissolved　　　　crystals form on cooling

▲ **Figure 3** *The stages of recrystallisation. Minimum volume of hot solvent to crude sample (left), solid dissolved (centre), and fine crystals of the solid formed on cooling (right)*

4. Once the solid has dissolved, allow the solution to cool. Crystals of the desired product should form in the conical flask or beaker. When no more crystals form, filter the crystals under reduced pressure to obtain the dry crystalline solid.

28.2 Further practical techniques

▲ Figure 4 *Closing the end of a capillary tube by heating in a Bunsen burner. The tube is rotated quickly in the hot Bunsen flame to seal one end of the tube. Only the edge of the flame is used so that only the tip of the capillary tube is heated. Otherwise the tube might bend as this tube has started to do*

▲ Figure 5 *The sample is forced up the capillary tube by dipping the open end of the tube into the sample*

▲ Figure 6 *Sample tube prepared and ready for insertion into melting point apparatus*

Melting point determination

Chemists determine the melting point of solids to identify whether a solid compound is pure. A pure organic substance usually has a very sharp melting range of one or two degrees. The melting range is the difference between the temperature at which the sample starts to melt and the temperature at which melting is complete. If the compound contains impurities, the solid melts over a wide range of temperatures. An impure sample also has a lower melting point than a pure sample.

1. Before taking the melting point of a solid you should ensure that the sample is completely dry and free flowing.
2. Take a glass capillary tube or melting point tube. Hold one end of the capillary tube in the hot flame of a Bunsen burner. Rotate the tube in the flame until the end of the tube is sealed (Figure 4).
3. The capillary tube is allowed to cool, and is then filled with crystals to about 3 mm depth. This is usually carried out by pushing the open end of the capillary into the solid sample to force some of the solid into the tube (Figure 5).
4. Once you have prepared your sample you will need to take its melting point. In schools and colleges, one of two methods will be available depending on the apparatus available.

Using electrically heated melting point apparatus

1. Place the capillary tube containing the sample into a sample hole and a 0–300 °C thermometer in the thermometer hole of the melting point apparatus (Figure 7).
2. Using the rapid heating setting, start to heat up the sample whilst observing the sample through the magnifying window.

▲ Figure 7 *Electrically operated melting point apparatus*

3. Once the solid is seen to melt, record the melting point. Allow the melting point apparatus to cool.
4. Prepare a second sample in a new capillary tube and place in the melting point apparatus and again heat up the sample.
5. As the melting point is approached, set to low and raise the temperature slowly whilst observing the sample. An accurate determination of the melting point can then be obtained.

ORGANIC SYNTHESIS   28

Using an oil bath or Thiele tube method

1. Set up the Thiele tube or oil bath (Figure 8).
2. Attach the capillary tube containing the sample to a thermometer using a rubber band.
3. Insert the thermometer through a hole in the cork if using a Thiele tube or clamp the thermometer if using an oil bath. The end of the thermometer and the end of the capillary tube should dip into the oil (Figure 9).
4. Using a micro-burner, slowly heat the side arm of the Thiele tube or the oil bath whilst observing the solid. When the solid starts to melt, remove the heat and record the temperature at which all of the solid has melted. It is important to heat the oil slowly when approaching the melting point, and it is advisable to repeat the melting point determination a second time to ensure that you obtain an accurate value.

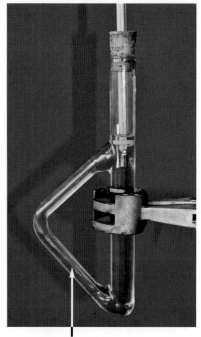

heat with microburner

▲ **Figure 8** *The standard set-up of a Thiele tube melting point apparatus*

Summary questions

1. Why is it important to heat the sample slowly when carrying out a melting point determination? *(1 mark)*

2. When carrying out recrystallisation it is important to allow the mixture to cool back to room temperature before re-filtering the sample. What effect would not cooling the mixture have on the final yield of product? Explain your reasoning. *(2 marks)*

3. A student recrystallised a sample of an organic solid and took the melting point of the recrystallised product.

   The student's results are shown below.

   Sample 1 melting point before recrystallisation 126–130 °C

   Sample 2 melting point after recrystallisation 134–136 °C

   Why are the melting points of Sample 1 and 2 different? *(2 marks)*

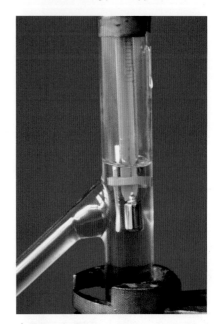

▲ **Figure 9** *The thermometer and melting point tube can be seen dipping into the oil at the top of the Thiele tube.*

28.3 Further synthetic routes

Specification reference: 6.2.5

Learning outcomes

Demonstrate knowledge, understanding, and application of:

→ identifying functional groups in organic molecules

→ predicting the reactions of organic molecules

→ multi-stage synthetic routes for preparing organic compounds.

Identifying functional groups in organic molecules

In this topic, you will revisit all of the functional groups you have studied, identify multiple functional groups in organic molecules, and solve multi-stage synthesis problems.

Worked example: Aspartame

Aspartame (Figure 1) is used as an artificial sweetener. Identify the functional groups present in aspartame.

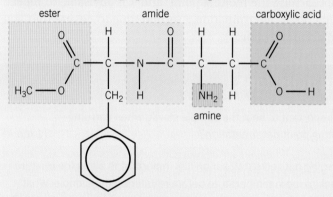

▲ Figure 1 *The structure of aspartame*

Aspartame contains an ester functional group, an amide, an amine, and a carboxylic acid functional group (Figure 1).

Worked example: Capsaicin

Capsaicin (Figure 3) is commonly used in food products to give them added spice. Identify the functional groups present in the structure.

▲ Figure 3 *The structure of capsaicin*

The molecule contains a phenol group, an amide, and an alkene (Figure 3). The molecule also contains an ether which is beyond the scope of the specification.

▲ Figure 2 *Aspartame used as an artificial sweetener*

▲ Figure 4 *Chillies owe their warming properties to capsaicin*

ORGANIC SYNTHESIS 28

The reactions of the main functional groups

The flow chart in Figure 5 links together all of the functional groups that you have studied so far in your course.

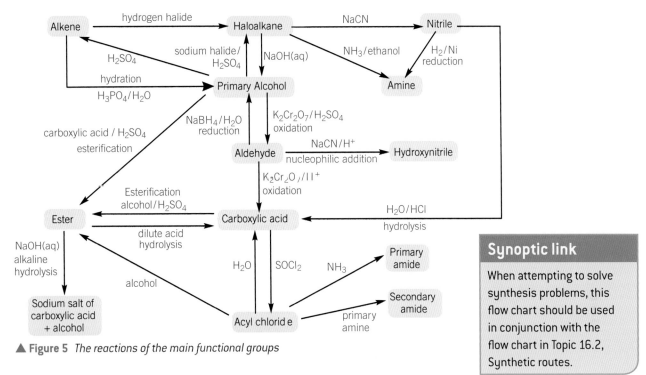

▲ **Figure 5** *The reactions of the main functional groups*

Synoptic link

When attempting to solve synthesis problems, this flow chart should be used in conjunction with the flow chart in Topic 16.2, Synthetic routes.

The reactions of benzene and its derivatives

The flow chart in Figure 6 summarises the common reactions of benzene and its compounds.

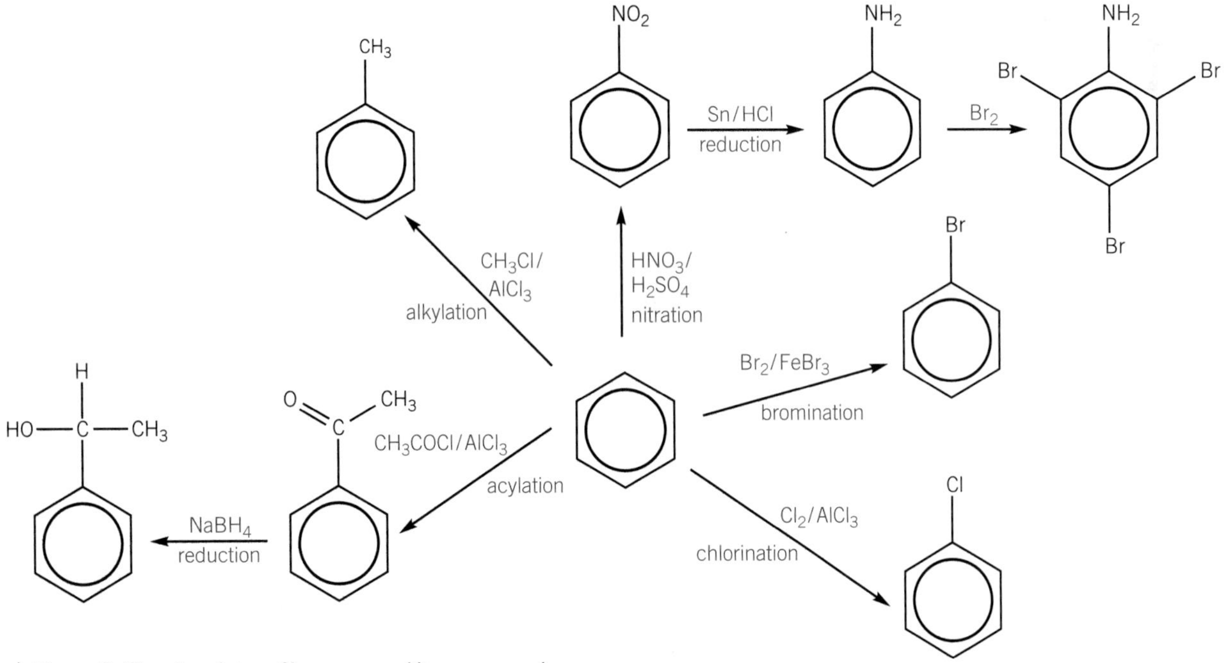

▲ **Figure 6** *The chemistry of benzene and its compounds*

28.3 Further synthetic routes

The reactions of phenol and its derivatives

The reactions of phenol are summarised in Figure 7.

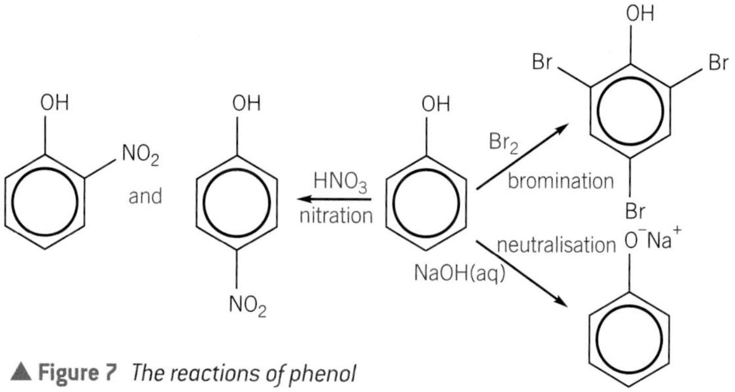

Figure 7 *The reactions of phenol*

Organic synthesis

The term target molecule is used to indicate the molecule that you are trying to obtain. In order to devise an organic synthesis you will need to identify the functional groups in the target molecule and in the starting material. You then select a sequence of steps to convert the starting material into the target molecule. Sometimes it is easier to work backwards from the target molecule.

> **Study tip**
> You need to be able to suggest reagents, conditions, and equations for converting one compound into another in multi-stage syntheses.

> **Study tip**
> Both molecules contain the same number of carbon atoms. The synthesis will involve changing functional groups only.

> **Study tip**
> The steps are easier to work out by working back from the target molecule

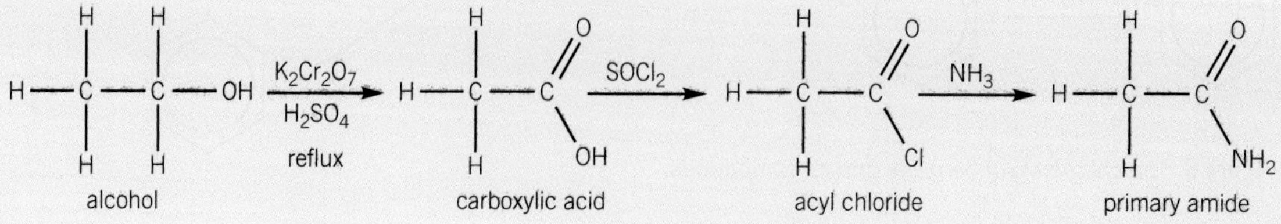

Worked example: Synthesis of ethanamide

Devise a flow chart to show how a sample of ethanamide could be prepared starting from chloroethane. Show the reagents you would use to carry out each step in your flow chart.

Stage 1: Identify the functional groups in the starting and target molecules.

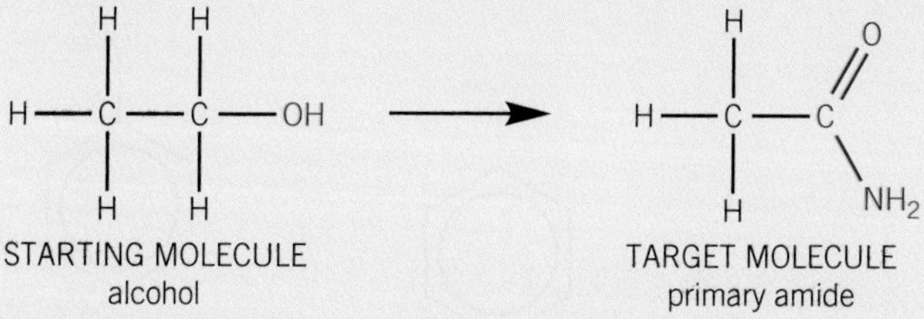

STARTING MOLECULE — alcohol → TARGET MOLECULE — primary amide

Stage 2: Identify a sequence of chemical reactions that could convert the functional groups in the starting molecule into the functional groups of the target molecule
- Primary amides can be prepared from acyl chlorides.
- Acyl chlorides can be made from carboxylic acids.
- An alcohol can be oxidised to a carboxylic acid.

Stage 3: The complete flow chart is shown below showing the conversion of each functional group.

alcohol $\xrightarrow[\text{reflux}]{K_2Cr_2O_7 / H_2SO_4}$ carboxylic acid $\xrightarrow{SOCl_2}$ acyl chloride $\xrightarrow{NH_3}$ primary amide

ORGANIC SYNTHESIS 28

Worked example: Synthesis of butanoic acid

Devise a synthesis of butanoic acid starting from propanal. State the reagents and conditions and give suitable equations for each step.

Stage 1: Identify the functional groups in the starting and target molecules

STARTING MOLECULE
aldehyde (CH₃CH₂CHO)

TARGET MOLECULE
carboxylic acid with extra carbon atom (CH₃CH₂CH₂COOH)

> **Study tip**
>
> The target molecule contains one extra carbon atom – this is a good clue that you will have to use a synthetic route containing the reaction of a haloalkane with NaCN in ethanol to lengthen the carbon chain.

Stage 2: Identify a sequence of chemical reactions that could convert the functional groups of the starting molecule into the functional groups of the target molecule.

- Nitriles can be used to form carboxylic acids.
- A carbon chain can be lengthened using NaCN/ethanol to form a nitrile.
- Haloalkanes can be prepared from alcohols.
- An aldehyde can be reduced to a primary alcohol.

> **Study tip**
>
> The steps are easier to work out by working back from the target molecule

Stage 3: Write out the synthesis.

Step 1: Reagents $NaBH_4$ in H_2O

$CH_3CH_2CHO + 2[H] \rightarrow CH_3CH_2CH_2OH$

Step 2: Reagents HBr (made from NaBr and H_2SO_4), heat

$CH_3CH_2CH_2OH + NaBr + H_2SO_4 \rightarrow CH_3CH_2CH_2Br + NaHSO_4 + H_2O$

Step 3: Sodium cyanide in ethanol

$CH_3CH_2CH_2Br + NaCN \rightarrow CH_3CH_2CH_2C\equiv N + NaBr$

Step 4: Heat with dilute aqueous HCl

$CH_3CH_2CH_2C\equiv N + 2H_2O + HCl \rightarrow CH_3CH_2CH_2COOH + NH_4Cl$

501

28 28.3 Further synthetic routes

▲ Figure 7 *Organic chemist carrying out a synthesis*

Study tip

In this synthesis, the ketone group must be attached to the benzene ring using a 3-directing group (the nitro group).

Had the nitro group been reduced to an amine first, the ketone would have been substituted at the 2- and 4-positions ($-NH_2$ is a 2- and 4-directing group).

Worked example: Synthesis of 1-(3-aminophenyl)ethanol

1-(3-aminophenyl)ethanol can be prepared from benzene as shown below. Complete a reaction scheme to show this synthesis.

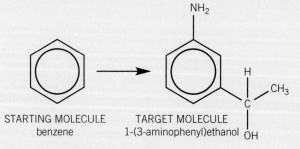

STARTING MOLECULE — benzene

TARGET MOLECULE — 1-(3-aminophenyl)ethanol

Step 1: Identify the functional groups in the starting and target molecules.

Starting material is benzene.

Target molecule contains an amine group and a side chain added to the benzene ring.

Step 2: Look back at Figure 6 to identify the reactions that take place:

- Nitration of benzene with nitric acid, then reduction of the NO_2 to form the NH_2 group.
- A nitro-group directs other groups to carbon number 3.
- Acylation of benzene to add a ketone group. The ketone group can then be reduced to an alcohol.

Step 3: Draw out the reaction scheme.

ORGANIC SYNTHESIS 28

Summary questions

1. Identify the functional groups in the following molecules.

   a [structure: aspirin-like molecule with COOH and ester groups on benzene ring] *(2 marks)*

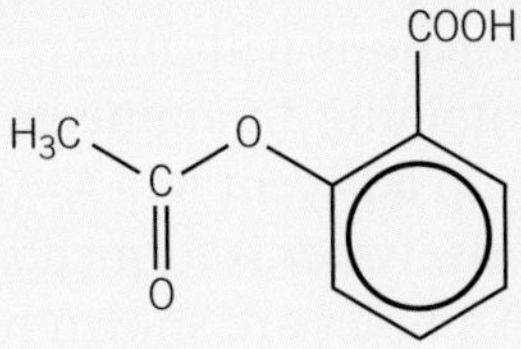

   b [structure: paracetamol-like molecule with amide and OH on benzene ring] *(2 marks)*

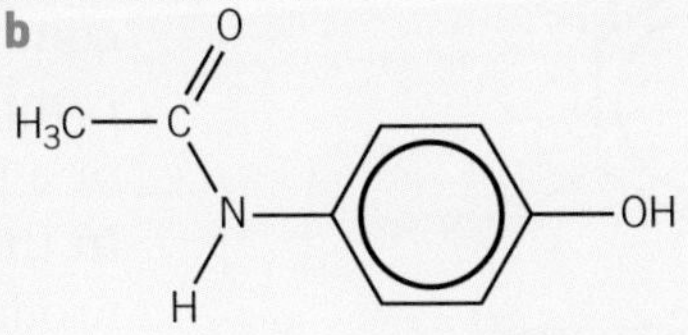

2. Outline the steps required to convert chloroethane into each of the following compounds. For each step state the reagents, conditions, and write an equation.

   a ethanoic acid *(2 marks)*
   b 2-hydroxypropanenitrile *(2 marks)*
   c propylamine *(2 marks)*

3. Using flow charts giving the essential conditions for each step, show how the following organic syntheses can be carried out.

   a $CH_3CH_2Br \rightarrow CH_3CH_2NHCOCH_3$ *(4 marks)*

   b [structure: phenyl-COCH$_2$CH$_3$ → 3-amino substituted benzene with CH=CHCH$_3$ group] *(3 marks)*

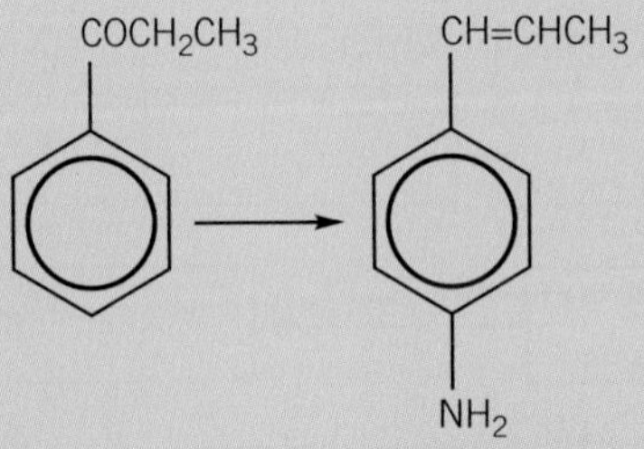

4. Two compounds that contribute to the flavour of foods are shown below

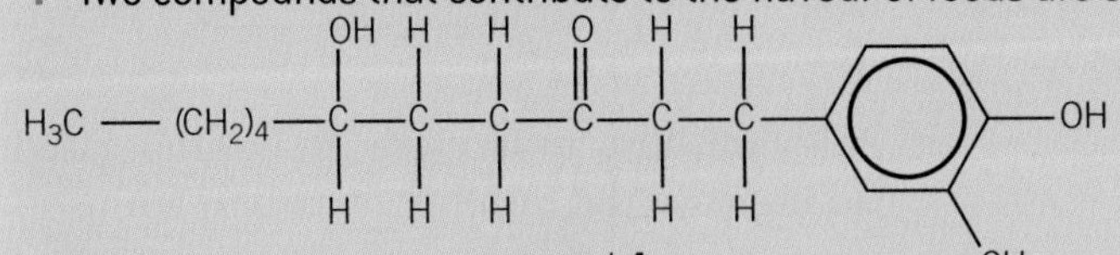

   [structure of compound A: $H_3C-(CH_2)_4-CH(OH)-CH_2-CH_2-CO-CH_2-CH_2$-(phenol with CH$_3$ substituent)]
   compound **A**

   [structure of compound B: $H_3C-(CH_2)_4-CH=CH-CH_2-CO-CH_2-CH_2$-(phenol with CH$_3$ substituent)]
   compound **B**

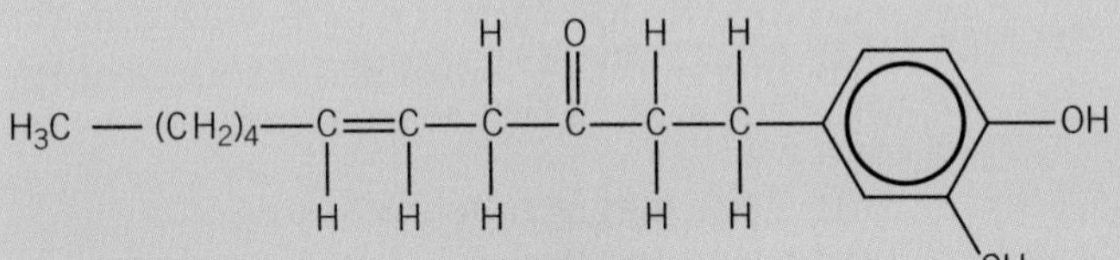

   a Name one functional group common to **A** and **B**. *(1 mark)*
   b Name a functional group present in **A** but not in **B** *(1 mark)*
   c Chose a chemical reagent for each of the following situations and state what you would expect to see.
      i A reagent that reacts with both **A** and **B**.
      ii A reagent that reacts with **B** but not with **A**.
      iii A reagent that reacts with **A** but not with **B**. *(3 marks)*
   d **A** could be converted into **B** in the laboratory. State reagents and conditions for this reaction. *(2 marks)*

Chapter 28 Practice questions

Practice questions

1. Mesalazine is a drug that can be synthesised from salicylic acid in two steps.

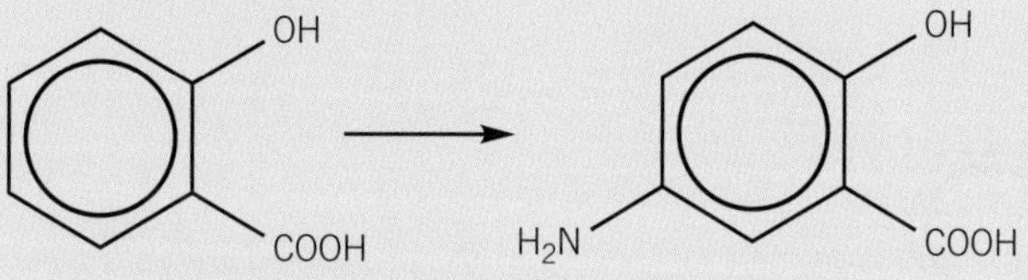

   Suggest a **two-step** synthesis to prepare mesalazine from salicylic acid. For **each** step state the reagents used and write a chemical equation. *(4 marks)*

   OCR F324 June 2014 Q1(c)

2. Oxandrolone is a type of synthetic drug called an anabolic steroid, prescribed to promote muscle growth. The structure of oxandrolone is shown below.

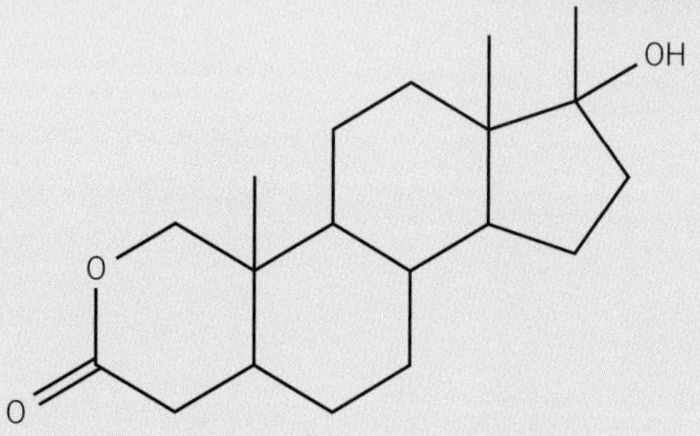

   a. What are the functional groups in oxandrolone? *(2 marks)*
   b. Compound **C** below is an intermediate formed during the synthesis of oxandrolone.

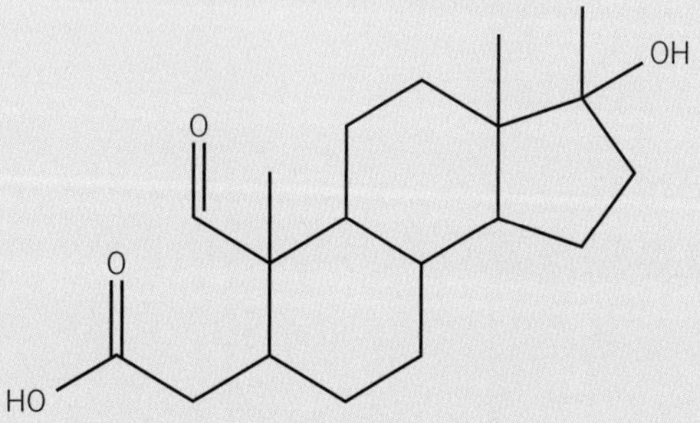

   Suggest a two-step synthesis of oxandrolone from compound **C**. For each step of the synthesis:
   - state the reagents and any conditions
   - state the functional groups that would react and those that would form.

   (4 marks)

   F324 June 2012 Q3(c)

3. Benzaldehyde, C_6H_5CHO, is the simplest aromatic aldehyde and has a characteristic smell of almonds.

   a. The aldehyde group takes part in condensation reactions with many compounds containing an amine group or a methyl group adjacent to a C=O. In these reactions, water is formed as a product. Two examples are shown below.

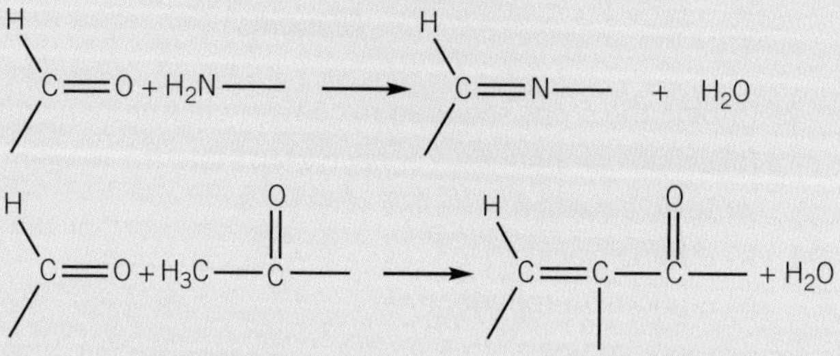

   Predict the organic products formed in the following condensation reactions of benzaldehyde.

   In each reaction, an excess of benzaldehyde is used. Draw the structure of each organic product.

   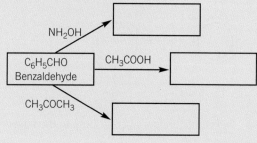

   (3 marks)

   b. Alkyllithium compounds, RLi, can be used to increase the number of carbon atoms in an organic compound. Different alkyl groups, R, add carbon chains with different chain lengths. RLi provides a source of R^- ions, which act as a nucleophile.

   (i) The diagram below shows an incomplete mechanism for the reaction of RLi with benzaldehyde, followed by reaction with aqueous acid.
   - Complete, using curly arrows and relevant dipoles, the mechanism for **stage 1**.
   - Give the structure of the intermediate and the organic product.

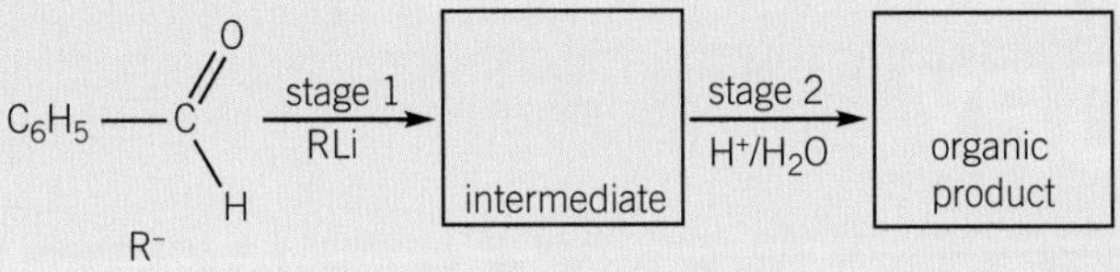

ORGANIC SYNTHESIS 28

C$_6$H$_5$—C(=O)H R⁻ —[stage 1, RLi]→ intermediate —[stage 2, H⁺/H$_2$O]→ organic product

(4 marks)

(ii) A chemist needs to prepare the organic compound below from benzaldehyde.

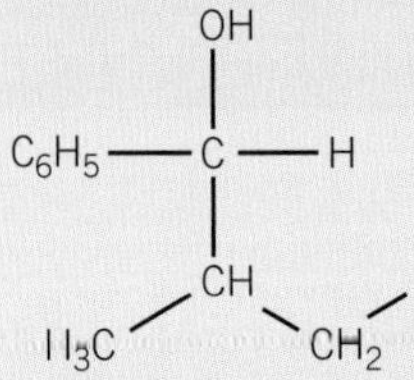

C$_6$H$_5$—C(OH)(H)—CH(CH$_3$)—CH$_2$CH$_3$ (with H$_3$C branch)

Draw the structure of the alkyllithium compound needed for this synthesis. *(1 mark)*

OCR F324 June 2012 Q4(c)

4 A food scientist synthesised the ester shown below, for possible use as a flavouring.

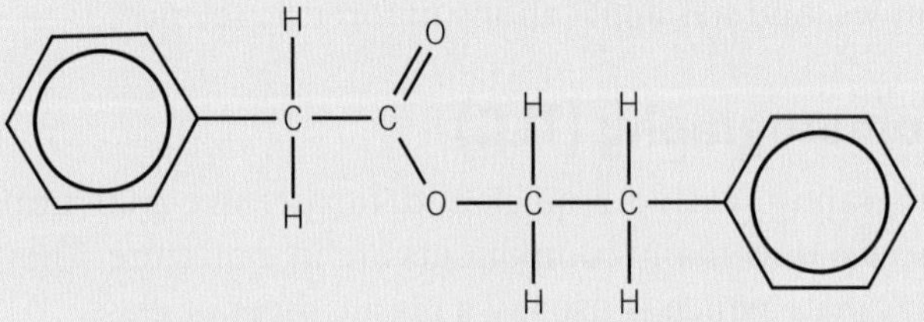

C$_6$H$_5$—CH$_2$—C(=O)—O—CH$_2$—CH$_2$—C$_6$H$_5$

The **only** organic compound available to the food scientist was phenylethanal, C$_6$H$_5$CH$_2$CHO.

Explain how the food scientist was able to synthesise this ester using only phenylethanal and standard laboratory reagents. *(7 marks)*

OCR F324 June 10 Q4 (c)

5 A student carried out an experiment to prepare a sample of phenylethanamide by reacting phenylamine with ethanoic anhydride as shown in the equation below.

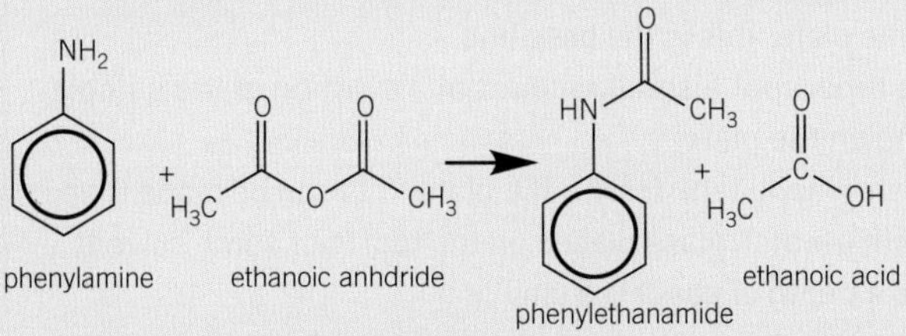

phenylamine + ethanoic anhydride → phenylethanamide + ethanoic acid

In the reaction the two reagents are shaken together in a conical flask and impure solid is formed. The solid is filtered by vacuum filtration and recrystallised using hot water. After recrystallisation, the pure product is filtered again and washed with ice-cold water.

A student carried out this experiment using 0.120 g phenylamine and collected 0.142 g of the phenylethanamide product.

a Calculate the percentage yield obtained. *(3 marks)*

b What would you expect if the student added too much hot water during the preparation? *(2 marks)*

c How would the melting point of the crude product compare to that of the purified sample of phenylethanamide. *(2 marks)*

d During filtration, why is it important to only wash your solid with ice-cold solvent? *(1 mark)*

6 Devise a series of steps to synthesise the ester, ethyl 4-nitrobenzaldehyde from 4-nitrobenzene. For each step suggest reagents, conditions and provide an equation.

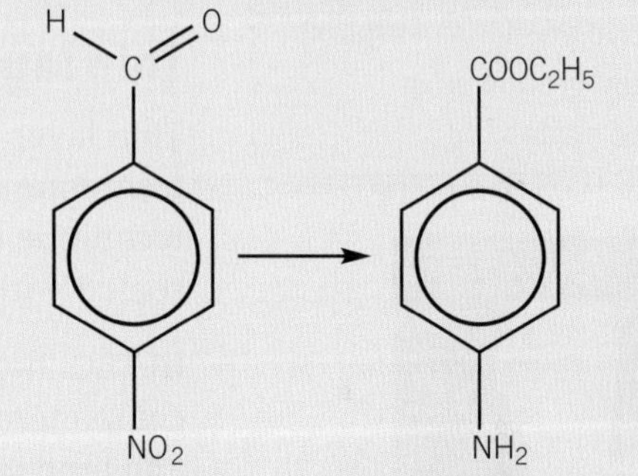

4-nitrobenzaldehyde → ethyl 4-aminobenzoate *(6 marks)*

7 Outline how you could prepare a sample of compound **D** starting with methylbenzene. Provide reagents and conditions for each step and show the structures of products for each step.

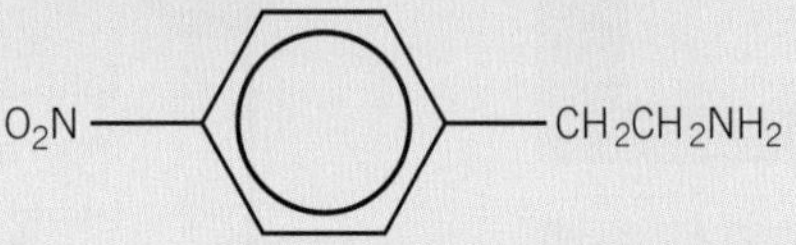

O$_2$N—C$_6$H$_4$—CH$_2$CH$_2$NH$_2$

Compound **D** *(8 marks)*

505

29 CHROMATOGRAPHY AND SPECTROSCOPY

29.1 Chromatography and functional group analysis

Specification reference: 6.3.1

Learning outcomes

Demonstrate knowledge, understanding, and application of:

→ one-way TLC chromatograms in terms of R_f values

→ gas chromatograms in terms of retention times and the proportions of the components of a mixture

→ qualitative analysis of organic functional groups on a test-tube scale.

Chromatography

Chromatography is used to separate individual components from a mixture of substances. All forms of chromatography have a **stationary phase** and a **mobile phase**.

- The stationary phase does not move and is normally a solid or a liquid supported on a solid.
- The mobile phase does move, and is normally a liquid or a gas.

Chromatography can be used in the analysis of drugs, plastics, flavourings, air samples, and has applications in forensic science.

Thin layer chromatography (TLC)

Thin layer chromatography (TLC) is a quick and inexpensive analytical technique that indicates how many components are in a mixture. The technique uses a TLC plate which is usually a plastic sheet or glass, coated with a thin layer of a solid **adsorbent** substance – usually silica.

In TLC, the adsorbent is the stationary phase. The different components in the mixture have different affinities for the absorbent and bind with differing strengths to its surface. **Adsorption** is the process by which the solid silica holds the different substances in the mixture to its *surface*. Separation is achieved by the relative adsorptions of substances with the stationary phase

▲ **Figure 1** *TLC plate marked with pencil base line (top), spotting on base line (middle) and run the plate in TLC tank or beaker (bottom)*

 Carrying out TLC

1. Take a TLC plate. Using a pencil, draw a line across the plate about 1 cm from one end of the plate. This is the base line.
2. Using a capillary tube, spot a small amount of a solution of the sample onto the base line on the plate.
3. Prepare a chromatography tank for the TLC plate. This can be made from a small beaker with a watch glass placed on the top. Pour some solvent into the beaker to a depth of about 0.5 cm.
4. Place the prepared TLC plate in the beaker, making sure that the solvent does not cover the spot. Cover the beaker with the watch glass and leave it undisturbed on the bench. The solvent will rise up the TLC plate (Figure 1).

CHROMATOGRAPHY AND SPECTROSCOPY 29

5. Allow the solvent to rise up the plate until it is about 1 cm below the top of the plate. Remove the plate from the beaker and immediately mark the solvent front with a pencil. Allow the plate to dry.

6. If there are any visible spots, circle them with a pencil. Alternatively hold a UV lamp over the plate and circle any spots that you can see. Sometimes the plate is sprayed with a chemical or a locating agent, such as iodine, to show the position of the spots that may be invisible to the naked eye.

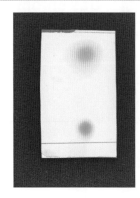

◀ **Figure 2** *Developed TLC plate of a mixture of amino acids. The two spots correspond to the amino acids leucine and lysine. Here the locating agent, ninhydrin, has been used, which is specific to amino acids and related compounds*

In the steps above, why is it important that the solvent depth is less than 1 cm?

Interpretation of a TLC plate

Thin layer chromatograms are analysed by calculating the value for the **retention factor** R_f for each component. Each component can be identified by comparing its R_f value with known values recorded using the same solvent system, and absorbent.

To calculate the R_f value you use the formula:

$$R_f = \frac{\text{distance moved by the component}}{\text{distance moved by the solvent front}}$$

Worked example: TLC

A mixture of amino acids is analysed by TLC. The TLC plate produced is shown in Figure 3. Identify the amino acids in the sample. R_f values are provided in Table 1.

Step 1: On the developed TLC plate, measure the distance moved by each component (*x*) and the distance moved by the solvent front from the sample line (*y*).

Step 2: Work out the R_f value for each component as $\frac{x}{y}$.

Step 3: Match the R_f value with the known values shown in Table 1.

▼ **Table 1** R_f values for six amino acids

Amino acid	R_f value
alanine	0.33
aspartic acid	0.24
valine	0.44
leucine	0.61
cysteine	0.37
isoleucine	0.53

Blue spot: $R_f = \frac{x}{y} = \frac{2.82}{4.63} = 0.61$
- The blue spot matches the R_f of leucine.

Green spot: $R_f = \frac{x}{y} = \frac{2.45}{4.63} = 0.53$
- The green spot is isoleucine.

Red spot: $R_f = \frac{x}{y} = \frac{2.04}{4.63} = 0.44$
- The red spot is valine.

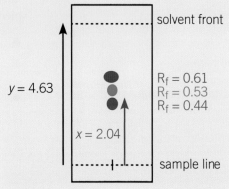

▲ **Figure 3** *TLC plate containing a mixture of substances*

It is common to run a TLC of a sample alongside pure samples of compounds that may be present. It is then easy to identify the amino acids in the unknown sample visually, without needing to calculate any R_f values.

29.1 Chromatography and functional group analysis

▲ Figure 5 *The inside of a gas chromatograph showing the copper column wound into a coil*

Gas chromatography

Gas chromatography (Figure 4) is useful for separating and identifying volatile organic compounds present in a mixture.

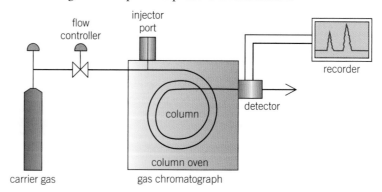

▲ Figure 4 *Schematic diagram of gas chromatography*

The stationary phase is a high boiling liquid adsorbed onto an inert solid support. The mobile phase is an inert carrier gas such as helium or neon.

A small amount of the volatile mixture is injected into the apparatus, called a gas chromatograph. The mobile carrier gas carries the components in the sample through the capillary column which contains the liquid stationary phase absorbed onto the solid support.

The components slow down as they interact with the liquid stationary phase inside the column. The more soluble the component is in the liquid stationary phase, the slower it moves through the capillary column.

The components of the mixture are separated depending on their solubility in the liquid stationary phase. The compounds in the mixture reach the detector at different times depending on their interactions with the stationary phase in the column. The compound retained in the column for the shortest time has the lowest retention time and is detected first. The **retention time** is the time taken for each component to travel through the column.

Interpretation of a gas chromatogram

Each component is detected as a peak on the gas chromatogram. Two pieces of information can be obtained from a gas chromatogram.

- Retention times can be used to identify the components present in the sample by comparing these to retention times for known components.
- Peak integrations (the areas under each peak) can be used to determine the concentrations of components in the sample.

CHROMATOGRAPHY AND SPECTROSCOPY 29

Retention times

Figure 6 shows a gas chromatogram run to investigate the presence of drugs in a urine sample. The analysis results show that five drugs were present in the sample, identified from their retention times. You can see from the relative sizes of the peaks that there is a different proportion of each drug.

Concentration of components

The concentration of a component in a sample is determined by comparing its peak integration (peak area) with values obtained from standard solutions of the component. The procedure is outlined below.

1. Prepare standard solutions of known concentrations of the compound being investigated.
2. Obtain gas chromatograms for each standard solution.
3. Plot a calibration curve (Figure 7) of peak area against concentration. This is called external calibration and offers a method for converting a peak area into a concentration.
4. Obtain a gas chromatogram of the compound being investigated under the same conditions.
5. Use the calibration curve to measure the concentration of the compound.

Figure 7 shows gas chromatogram readings for four different concentrations of caffeine and the resulting calibration curve. It is then just a matter of running the sample through the gas chromatograph and comparing the compounds peak area with the calibration curve.

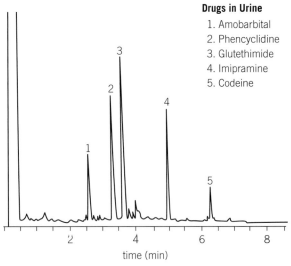

Drugs in Urine
1. Amobarbital
2. Phencyclidine
3. Glutethimide
4. Imipramine
5. Codeine

▲ **Figure 6** *Gas chromatogram of drugs in urine sample*

> **Synoptic link**
>
> See Topic 18.2, Rate graphs and orders, for use of a calibration curve to determine concentrations using a colorimeter.

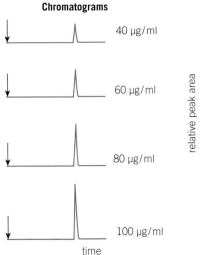
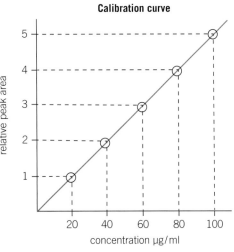

▲ **Figure 7** *External calibration technique to obtain a calibration curve for caffeine. The calibration curve can be used to determine concentrations from peak integrations made by the gas chromatograph*

29.1 Chromatography and functional group analysis

Qualitative analysis of organic functional groups

Reactions can be carried out on a test-tube scale to identify the functional groups present in organic compounds. You have already met many tests for different functional groups throughout the course. These tests are summarised in Table 2 below.

▼ **Table 2** *Tests for functional groups*

Functional group	Chemical test	Observation
alkene	add bromine water drop-wise	bromine water decolourised from orange to colourless 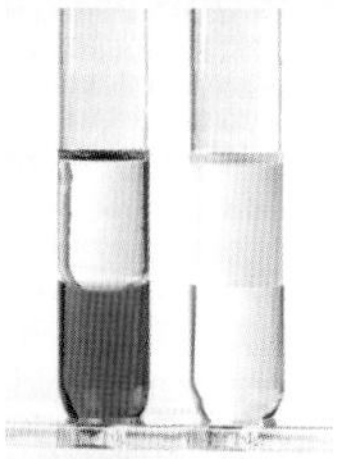
haloalkane	add silver nitrate and ethanol and warm to 50 °C in a water bath	chloroalkane – white precipitate bromoalkane – cream precipitate iodoalkane – yellow precipitate 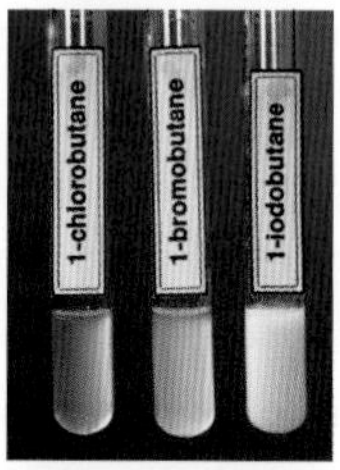
carbonyl	add 2,4-dinitrophenylhydrazine	orange precipitate 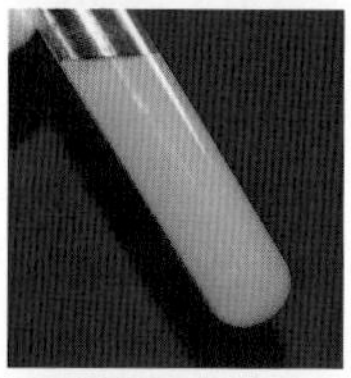
aldehyde	add Tollens' reagent and warm	silver mirror 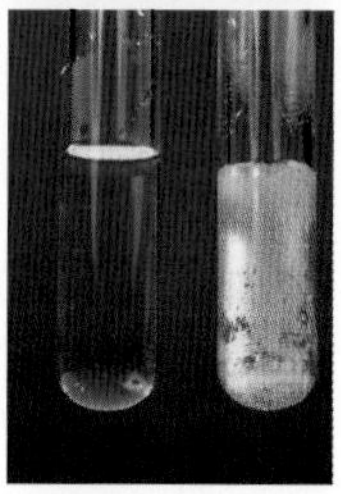
primary and secondary alcohol, and aldehyde	add acidified potassium dichromate(VI) and warm in a water bath	colour change from orange to green 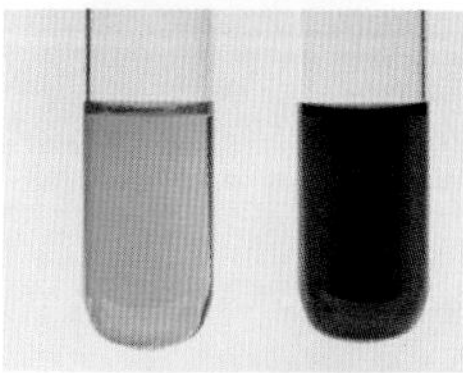
carboxylic acid	add aqueous sodium carbonate	effervescence 

29 CHROMATOGRAPHY AND SPECTROSCOPY

Identifying a phenol

Phenols are acidic compounds which can be tested using pH indicator paper. Phenols are not as acidic as carboxylic acids and do not react with sodium carbonate.

Phenol undergoes an electrophilic substitution reaction with bromine at room temperature. When bromine is added to an aqueous solution of phenol, the bromine is decolourised and a white precipitate is formed (Figure 8).

Name the white precipitate that is formed.

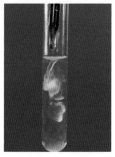

▲ Figure 8 *A white precipitate is formed when bromine is added to phenol*

Summary questions

1 The gas chromatogram in Figure 9 was obtained from a sample of wine.
   Use the retention times in Table 3 to identify peaks 4, 7, and 8 in the sample.

Compound	Retention time / min
ethanal	2.0
ethanol	4.0
propan-1-ol	8.2
ethanoic acid	19.5
2-methylbutan-1-ol	12.0

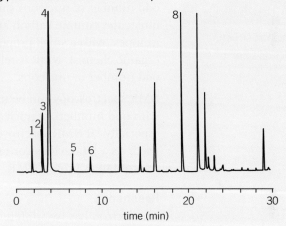

▲ Figure 9 *Gas chromatogram of wine sample*   (3 marks)

2 A sample of paint was analysed by TLC. The TLC plate is shown below.
   Calculate the R_f values for the components in the paint.   (3 marks)

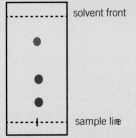

▲ Figure 10 *A TLC plate for a sample of paint*

3 Describe chemical tests that will enable you to distinguish between:
   a an aldehyde and a ketone   (1 mark)
   b a phenol and an alkene   (1 mark)
   c a ketone and a secondary alcohol.   (1 mark)

29.2 Nuclear Magnetic Resonance (NMR) spectroscopy

Specification reference: 6.3.2

Learning outcomes

Demonstrate knowledge, understanding, and application of:

→ NMR spectroscopy

→ use of tetramethylsilane (TMS)

→ the need for deuterated solvents, for example, $CDCl_3$.

Synoptic link

See Topic 5.1, Electron structure, for details of electron spin.

NMR spectroscopy

Since its invention about 70 years ago, **nuclear magnetic resonance** (NMR) spectroscopy has revolutionised the analysis of organic compounds. The technique uses a combination of a very strong magnetic field and radio frequency radiation. With the right combination of magnetic field strength and frequency, the nuclei of some atoms absorb this radiation. The energy for the absorption can be measured and recorded as an NMR spectrum.

What is nuclear magnetic resonance?

Nuclear spin

You will remember that electrons have a property called spin. The nucleus also has a **nuclear spin**, that is significant if there is an odd number of nucleons (protons and neutrons). Almost all organic molecules contain carbon and hydrogen, mostly as the ^1H and ^{12}C isotopes, with a small proportion (1.1%) of the ^{13}C isotope. So for the organic chemist, NMR is relevant for ^1H and ^{13}C, the isotopes with an odd number of nucleons.

NMR spectroscopy can be used to detect isotopes of other elements with odd numbers of nucleons, such as ^{19}F and ^{31}P. However ^{13}C and especially ^1H NMR spectroscopy are the commonest forms of analysis used. As a ^1H nucleus consists of just a proton, ^1H NMR is usually referred to as proton NMR.

Resonance

An electron has two different spin states. The nucleus also has two different spin states and these have different energies. With the right combination of a strong magnetic field and radio frequency radiation, the nucleus can absorb energy and rapidly flips between the two spin states. This is called **resonance** and the whole process gives the name 'nuclear magnetic resonance'.

The NMR spectrometer

Radio frequency radiation has much less energy than the infrared radiation used in IR spectroscopy. The frequency required for resonance is proportional to the magnetic field strength and it is only in strong and uniform magnetic fields that this small quantity of energy can be detected. Typically a very strong super-conducting electromagnet is used, cooled to 4 K by liquid helium.

Figure 1 shows a typical NMR spectrometer. You can see the large cylinder which houses the electromagnet, cooled by liquid helium. For organic chemistry, most routine spectrometers operate at

▲ Figure 1 *A researcher adding a sample to a nuclear magnetic resonance (NMR) spectrometer*

radio frequencies of 100, 200, or 400 MHz. Universities have NMR spectrometers but they are far too expensive to be viable for schools and colleges. They are also found in hospitals as MRI (magnetic resonance imaging) body scanners, a technique that uses the same technology.

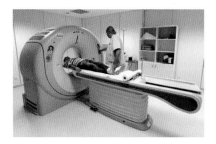

▲ Figure 2 *An MRI body scanner, essentially a large horizontal NMR spectrometer*

Chemical shift and TMS

In an organic molecule, every carbon and hydrogen atom is bonded to other atoms. All atoms have electrons surrounding the nucleus, which shifts the energy and radio frequency needed for nuclear magnetic resonance to takes place. The frequency shift is measured on a scale called **chemical shift** δ, in units of parts per million (ppm).

Tetramethylsilane (TMS), $(CH_3)_4Si$, is used as the standard reference chemical against which all chemical shifts are measured. TMS is given a chemical shift value of 0 ppm.

The amount of chemical shift is determined by chemical environment, especially the presence of nearby electronegative atoms (Figure 3).

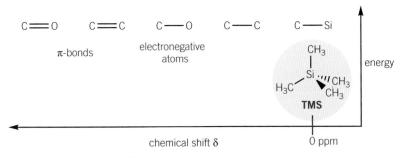

▲ Figure 3 *Chemical shift*

> **Study tip**
>
> In an NMR spectrum, the chemical shift in ppm increases in value from right to left.
>
> So the TMS peak, at 0 ppm, is alway found at the right-hand end of the chemical shift scale.

So depending on the chemical environment, nuclear magnetic resonance requires a different energy and frequency, producing absorption peaks at chemical shifts. This means that the carbon and hydrogen arrangement in a molecule can be mapped out without needing to carry out conventional chemical tests and without destroying the organic compound under test.

Running the spectrum

In an NMR spectrometer, the sample is dissolved in a solvent and placed in a narrow NMR sample tube, together with a small amount of TMS (Figure 4).

The tube is placed inside the NMR spectrometer, where it is spun to even out any imperfections in the magnetic field within the sample. The spectrometer is zeroed against the TMS standard and the sample is given a pulse of radiation containing a range of radio frequencies, whilst maintaining a constant magnetic field. Any absorptions of energy resulting from resonance are detected and displayed on a computer screen (Figure 5). After analysis, the sample can be recovered by evaporation of the solvent.

▲ Figure 4 *NMR sample tubes containing samples for analysis by NMR*

29.2 Nuclear Magnetic Resonance (NMR) spectroscopy

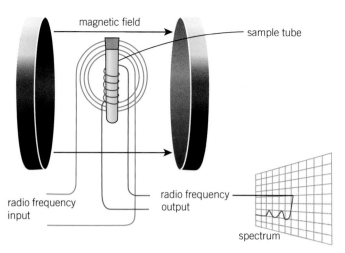

▲ Figure 5 *The set-up within an NMR spectrometer*

Synoptic link
See Topic 2.1, Atomic structure and isotopes, for information about deuterium.

Deuterated solvents
Molecules of most common solvents contain carbon and hydrogen atoms, which will produce a signal in both ^{13}C and ^1H NMR spectra. A deuterated solvent is usually used in which the ^1H atoms have been replaced by ^2H atoms (deuterium, D). Deuterium produces no NMR signal in the frequency ranges used in ^1H and ^{13}C NMR spectroscopy. Deuterated trichloromethane, $CDCl_3$, is commonly used as a solvent in NMR spectroscopy, but this will still produce a peak in a carbon-13 NMR spectrum. The computer usually filters out this peak before displaying the spectrum.

Summary questions

1 What is the standard reference chemical for chemical shift measurements? *(1 mark)*

2 Why are deuterated solvents used? Give an example of a deuterated solvent. *(2 marks)*

3 What is meant by chemical shift? *(1 mark)*

29.3 Carbon-13 NMR spectroscopy

Specification reference: 6.3.2

Carbon-13 NMR spectroscopy

A carbon-13 NMR spectrum provides two important pieces of information about a molecule:

- the *number* of different carbon environments – from the number of peaks
- the *types* of carbon environment present – from the chemical shift.

Figure 1 shows chemical shift values for carbon atoms in different environments. The chemical shift δ is referenced against TMS at δ = 0 ppm. The chemical shift range of about 220 ppm is sufficiently wide to separate carbon atoms that have only slightly different environments.

Learning outcomes

Demonstrate knowledge, understanding, and application of:

→ carbon-13 NMR spectroscopy
→ the number of carbon environments
→ the different types of carbon environment
→ predictions about possible structures for an unknown molecule.

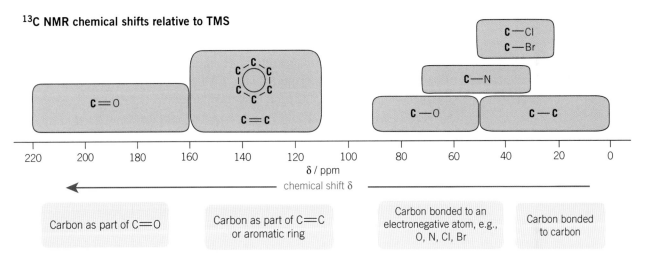

▲ **Figure 1** *Carbon-13 NMR chemical shifts*

There are four main types of carbon atom that absorb over different chemical shift ranges, as shown by the labels below the chemical shift axis. The chemical shifts may also be outside of these ranges, depending on the solvent, concentration, and substituents.

The chemical environment of a carbon atom is determined by the position of the atom within the molecule.

- Carbon atoms that are bonded to different atoms or groups of atoms have different environments and will absorb at different chemical shifts.
- If two carbon atoms are positioned symmetrically within a molecule, then they are *equivalent* and have the *same* chemical environment. They will then absorb radiation at the same chemical shift and contribute to the same peak.

Study tip

You do not need to memorise the chemical shifts of different groups as you will be provided with them on the *Data Sheet*.

29.3 Carbon-13 NMR spectroscopy

Carbon-13 NMR spectra for propanal and propanone

Propanal and propanone are structural isomers of C_3H_6O. Propanal and propanone produce different ^{13}C NMR spectra. The NMR spectra and the reasons for the difference are described below.

1 Propanal, CH_3CH_2CHO

There are three carbon atoms and these are clearly in three different environments as shown below.

- Carbon-**1** is part of the CHO functional group.
- Carbon-**2** is part of a CH_2 group between a CH_3 group and an aldehyde group.
- Carbon-**3** is part of a CH_3 group bonded to CH_2CHO.

2 Propanone, CH_3COCH_3

There are three carbon atoms but only two different environments as shown below.

- Carbon-**1** is part of the C=O group.
- The other two carbon atoms, both labelled **2**, have the same environment. They are positioned symmetrically and can both be described in an identical way – each carbon atom is part of a CH_3 group bonded to $COCH_3$.

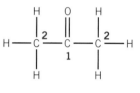

The carbon-13 NMR spectra for propanal and propanone are shown below. The peaks have been identified from the chemical shift values in Figure 1.

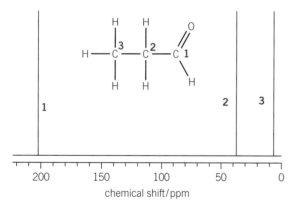

Propanal, CH_3CH_2CHO
- Carbon-**1** at δ ~ 203 ppm for **C**HO
- Carbon-**2** at δ ~ 37 ppm for **C**—C=O
- Carbon-**3** at δ ~ 8 ppm for C—**C**H$_3$

▲ Figure 2 *Carbon-13 NMR spectrum of propanal*

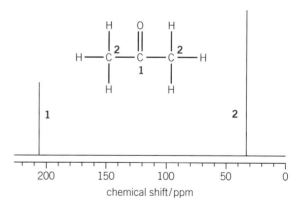

Propanone, CH_3COCH_3
- Carbon-**1** at δ ~ 205 ppm for **C**=O
- Carbon-**2** at δ ~ 32 ppm for two carbon atoms of the type: **C**H$_3$C=O

▲ Figure 3 *Carbon-13 NMR spectrum of propanone*

In the carbon-13 NMR spectrum for propanal, carbon-2 (CH_2) was assigned to the peak at δ ~ 37 ppm. Although, this carbon atom falls into the same broad C—C environment as carbon-3 (CH_3), it is nearer to the oxygen atom connected to carbon–1 so it is likely to be shifted more than carbon-3.

29 CHROMATOGRAPHY AND SPECTROSCOPY

Interpreting carbon-13 NMR spectra

The worked examples below show how carbon atoms can be matched to different carbon-13 NMR spectra, based on the number of peaks. Chemical shifts then allow the peaks to be assigned.

Worked example: Isomers of alcohols

The two spectra below are for two isomers of $C_4H_{10}O$ that are alcohols. Identify the isomers that give each spectrum and assign each peak.

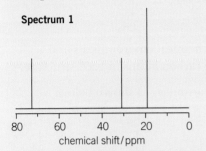
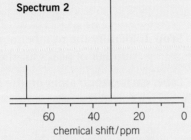

Step 1: Draw out the isomers of $C_4H_{10}O$ that are alcohols and identify the number of chemical environments for the carbon atoms and the expected number of peaks.

A
4 environments
4 peaks

B
4 environments
4 peaks

C
2 environments
2 peaks

D
3 environments
3 peaks

> **Study tip**
> The number of peaks is the same as the number of environments.

Step 2: Match the structures to the spectra from the number of peaks.

- Spectrum 1 has three peaks and three environments which matches the three environments in structure **D**.

- Spectrum 2 has two peaks and two environments which matches the two environments in structure **C**.

Step 3: Assign the peaks from the chemical shifts in Figure 1.

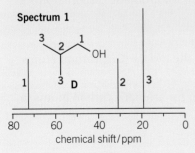
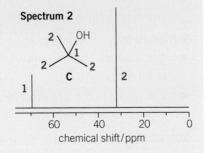

29.3 Carbon-13 NMR spectroscopy

Worked example: Aromatic isomers

For each isomer below, predict the number of peaks in its carbon-13 NMR spectrum and the approximate chemical shifts for each peak.

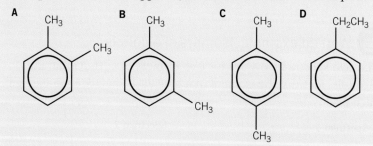

Step 1: Identify the number of chemical environments and number of peaks for the carbon atoms in each structure.

Label each carbon atom on the formulae, as shown below.

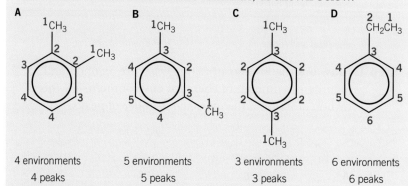

A	B	C	D
4 environments	5 environments	3 environments	6 environments
4 peaks	5 peaks	3 peaks	6 peaks

> **Study tip**
> The number of peaks is the same as the number of environments.

Step 2: Predict the approximate chemical shifts.

Structure A   four peaks
- *One* peak at δ = 0–50 ppm for the *two* C—**CH**$_3$ carbon atoms in environment **1**.
- *Three* peaks at δ = 110–160 ppm for aromatic carbon atoms in environments **2**, **3**, and **4**.

Structure B   five peaks
- *One* peak at δ = 0–50 ppm for the *two* C—**CH**$_3$ carbon atoms in environment **1**.
- *Four* peaks at δ = 110–160 ppm for aromatic carbon atoms in environments **2**, **3**, **4**, and **5**.

Structure C   three peaks
- *One* peak at δ = 0–50 ppm for the *two* C—**CH**$_3$ carbon atoms in environment **1**.
- *Two* peaks at δ = 110–160 ppm for aromatic carbon atoms in environments **2** and **3**.

Structure D   six peaks
- *Two* peaks at δ = 0–50 ppm for the *one* C—**CH**$_3$ carbon atom in environment **1** and the *one* C—**CH**$_2$—C carbon atom in environment **2**.
- *Four* peaks at δ = 110–160 ppm for aromatic carbon atoms in environments **3**, **4**, **5**, and **6**.

CHROMATOGRAPHY AND SPECTROSCOPY 29

The worked example for aromatic isomers shows just how useful carbon-13 NMR spectroscopy can be. The structures are so similar that it would be very difficult to distinguish between them by other means. But from the carbon-13 NMR spectra, you just need to count the number of peaks. It may not always be as easy and you may find that different isomers give the same number of peaks, but you can often identify some of the isomers using this technique.

Further analysis of a carbon-13 NMR spectrum

The carbon-13 NMR spectrum of propan-1-ol is shown in Figure 4 with the three peaks for the three carbon environments labelled.

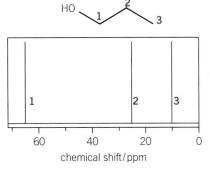

▲ Figure 4 *Carbon-13 NMR spectrum of propan-1-ol*

Looking at the chemical shifts in Figure 1, carbon-**1** is clearly in a C—O environment giving a peak in the range 50–90 ppm, but carbon atoms **2** and **3** both match the same C—C environment in the range 0–50 ppm. So how do you know that the peaks are this way round?

Carbon-**2** is closer than carbon-**3** to the oxygen atom connected to carbon-**1**. The proximity of an electronegative atom or multiple bond is an important factor in shifting a peak to a higher δ value. You saw this idea in the spectrum of propanal earlier in this topic.

This prediction agrees with the actual NMR spectrum, but you do need to be very careful as other factors such as *concentration* and *interactions with the solvent* may also affect the chemical shift.

Sketch carbon-13 NMR spectra to predict the number of peaks and the order of peaks for the following compounds.

a butan-1-ol   b pentan-3-one

Summary questions

Use the ^{13}C chemical shifts in Figure 1 to help you answer these questions.

1 For each compound below, predict the number of ^{13}C peaks and the chemical shift of each peak:
   a CH_3COOH (2 marks)   b CH_3CHO (2 marks)
   c CH_3NHCH_3 (2 marks)

2 For each compound below, predict the number of ^{13}C peaks and the chemical shift of each peak:
   a $CH_3COOCH_2CH_3$ (2 marks)   b $CH_3CH=CHCH_3$ (2 marks)
   c $HOCH_2CH_2CHO$ (2 marks)

3 Draw the structures of the following and label each carbon environment. Predict the number of peaks in the carbon-13 NMR spectra of each isomer.
   a the three aromatic isomers of nitromethylbenzene   (3 marks)
   b 2,3-dimethylphenol   (3 marks)
   c 3,5-dimethylphenol   (3 marks)
   d 1,3,5-trimethylbenzene   (3 marks)

29.4 Proton NMR spectroscopy

Specification reference: 6.3.2

Learning outcomes

Demonstrate knowledge, understanding, and application of:

→ proton NMR spectroscopy
→ the different types of proton environment
→ the relative numbers of each type of proton
→ spin–spin splitting
→ identification of O—H and N—H protons.

Proton NMR spectroscopy

A proton NMR spectrum provides four important pieces of information about a molecule. Firstly, the spectrum provides similar information to a carbon-13 NMR spectrum but for protons:

- the *number* of different proton environments – from the number of peaks
- the *types* of proton environments present – from the chemical shift.

There are the two extra pieces of information:

- the *relative numbers* of each type of proton – from integration traces or ratio numbers of the relative peak areas
- the number of non-equivalent protons *adjacent* to a given proton – from the spin–spin splitting pattern.

Figure 1 shows chemical shift values for protons in different environments. The chemical shift δ is referenced against TMS at δ = 0 ppm. The chemical shift range of about 12 ppm is much narrower than the carbon-13 range of about 220 ppm.

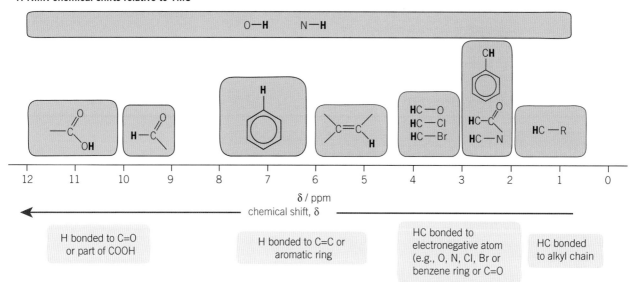

▲ **Figure 1** *Proton NMR chemical shifts*

Study tip

You do not need to memorise the chemical shifts of different groups. You will be provided with these on the *Data Sheet*.

As with carbon-13 NMR spectra, there are broad categories for types of proton as the chemical shift value increases. These are shown in Figure 1 below the chemical shift axis. Factors such as solvent, concentration, and substituents may move a peak outside of these ranges.

29 CHROMATOGRAPHY AND SPECTROSCOPY

Equivalent and non-equivalent protons

For carbon-13 NMR, you saw that carbon atoms in the same chemical environment absorb at the same chemical shift value. These carbon atoms are equivalent to one another. For proton NMR, the same principle applies:

- If two or more protons are *equivalent*, they will absorb at the same chemical shift, increasing the size of the peak.
- Protons of different types have different chemical environments and are *non-equivalent* – they absorb at different chemical shifts.

A good way of visualising equivalent and non-equivalent protons is to look for any plane of symmetry. The examples below show different molecules containing more than one CH_2 group.

Butanoic acid

In butanoic acid, $CH_3CH_2CH_2COOH$, the two CH_2 groups are clearly in *different* environments (Figure 2). Each CH_2 group is connected to different groups on one side and the protons in the two CH_2 groups are non-equivalent. There is no plane of symmetry and there are four proton environments giving four peaks in the proton NMR spectrum.

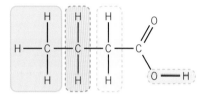

$CH_3CH_2CH_2COOH$
4 environments
4 peaks

▲ **Figure 2** *Non-equivalent protons in butanoic acid*

Butanedioic acid

In butanedioic acid, $HOOCCH_2CH_2COOH$, the two CH_2 groups are in the *same* environment (Figure 3). Each CH_2 group is connected to a CH_2 group on one side and a COOH group on the other side. There is a plane of symmetry between the two CH_2 group and the protons in the two CH_2 groups are equivalent. There are two proton environments giving two peaks in total:

- one peak for the two equivalent COOH protons
- one peak for the four equivalent protons in the two CH_2 groups.

The proton NMR spectrum therefore has two peaks, one with twice the area of the other.

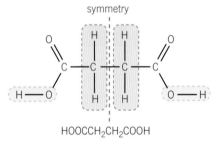

$HOOCCH_2CH_2COOH$

▲ **Figure 3** *Equivalent protons in butanedioic acid*

Relative numbers of each type of proton

In a carbon-13 NMR spectrum, the peak area is not directly related to the number of carbon atoms responsible for the peak. This all changes for proton NMR – the ratio of the relative areas under each peak gives the ratio of the number of protons responsible for each peak.

The NMR spectrometer measures the area under each peak as an **integration trace** (mathematically, integration means the area under a curve). The integration trace is shown either as an extra line on the spectrum or as a printed number of the relative peak areas. This provides invaluable information for identifying an unknown compound.

The proton NMR spectra of methyl chloroethanoate, $ClCH_2COOCH_3$, shows an integration trace (the blue line) in the peak area ratio of 2 : 3 for the two protons in CH_2 and three protons in CH_3 (Figure 4).

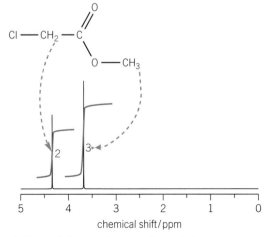

▲ **Figure 4** *Integration trace for methyl chloroethanoate, $ClCH_2COOCH_3$*

521

Spin–spin coupling

So far, you can see that a proton NMR spectrum reveals a large amount of information about the structure of molecules. A proton NMR peak can also be split into sub-peaks or splitting patterns. These are caused by the proton's spin interacting with the spin states of nearby protons that are in different environments. This provides information about the number of protons bonded to adjacent carbon atoms.

The $n + 1$ rule

The splitting of a main peak into sub-peaks is called **spin–spin coupling** or **spin–spin splitting**, and the number of sub-peaks is one greater than the number of adjacent protons causing the splitting. This sounds complicated but the pattern is easy to see using the $n + 1$ **rule**.

- For a proton with n protons attached to an adjacent carbon atom, the number of sub-peaks in a splitting pattern = $n + 1$

When analysing spin–spin splitting, you are really seeing the number of hydrogen atoms on the immediately *adjacent* carbon atom. The possible splittings for different sequences of protons in an organic molecule are shown in Table 1.

> **Study tip**
>
> Spin–spin splitting occurs only if adjacent protons are in a different environment from the protons being split. Refer back to the section on equivalent and non-equivalent protons, earlier in this topic.

▲ Table 1 *Spin–spin splitting patterns*

n	$n + 1$	Splitting pattern	Relative peak areas within splitting	Pattern	Structural feature
0	1	singlet	1		no H on adjacent atoms
1	2	doublet	1 : 1		adjacent CH
2	3	triplet	1 : 2 : 1		adjacent CH_2
3	4	quartet	1 : 3 : 3 : 1		adjacent CH_3

▲ Figure 5 *Heptet splitting of the CH group in $CH(CH_3)_2$ caused by six adjacent protons in the two CH_3 groups*

The patterns shown above are very common in NMR spectra. This degree of splitting continues with more adjacent protons. Another common pattern is for $CH(CH_3)_2$ where a CH proton has six protons on the adjacent carbon atoms. This gives the heptet (7) splitting pattern shown in Figure 5.

More complex splitting

In $CH(CH_3)_2$ above, the two CH_3 groups are in the same environment. Sometimes, adjacent protons may have different environments. In molecules such as $CH_3CH_2CH_2COOH$, the central $-CH_2-$ would be split differently by the CH_3 and CH_2 protons. The resulting splitting would then show as a multiplet. More advanced work allows multiplets to be analysed.

> **Study tip**
>
> You will not need to interpret more complex splitting beyond observing a multiplet.

Aromatic protons

From Figure 1, aromatic protons are expected to absorb in the range δ = 6.2–8.0 ppm. Splitting does occur but this can be difficult to interpret. You are only expected to interpret aromatic protons as groups of protons often forming one or more mutiplets.

Splitting patterns

The relative peak areas within spin–spin coupling follow a pattern called Pascal's triangle. If you take A level Mathematics, you may have come across this pattern. From Table 1, as the number of sub-peaks increases, the new extra peak has a relative area equal to the sum of the peak areas immediately above:

```
                        1
Pascal's             1    1
triangle           1   2   1
                 1   3   3   1
```

Predict the relative peak areas in the splitting patterns for a quintet (5), hextet (6), and heptet (7).

Spin–spin coupling occurs in pairs

In an NMR spectrum, if you see one splitting pattern there must always be another – splitting patterns occur in pairs because each proton splits the signal of the other. This can make it very easy to spot a structural feature when analysing a molecule. There are several very common splitting pairs that you may see in a spectrum and these are shown in Figure 6 below, together with the structural feature causing the splitting. You don't need to learn these as they can all be worked out using the $n + 1$ rule. But you will quickly come to recognise the triplet/quartet combination for a CH_3CH_2 combination because it is so common.

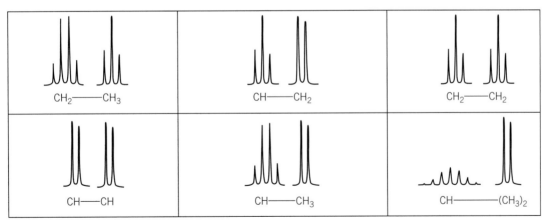

▲ Figure 6 *Common pairs of splitting patterns and their structural features*

Hydroxyl and amino protons

Organic compounds may contain protons that are not bonded to carbon atoms, for example, organic compounds often contain –OH and –NH protons.

29.4 Proton NMR spectroscopy

NMR spectrum of CH₃OH

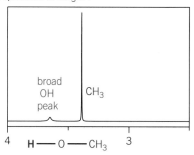

NMR spectrum of CH₃OH with D₂O added

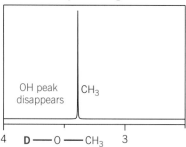

▲ **Figure 7** *NMR spectra of methanol, CH_3OH, run without and with D_2O*

Synoptic link

In Topic 29.2, Nuclear magnetic resonance (NMR) spectroscopy, you came across deuterated solvents.

The functional groups involved include:
- alcohols, RO**H**, phenols, ArO**H**, and carboxylic acids, RCOO**H**
- amides, RN**H₂**, amides, RCON**H₂**, and amino acids, RCH(N**H₂**)COOH.

In solution, NH and OH protons may be involved in hydrogen bonding and the NMR peaks are often broad and of variable chemical shift. In Figure 1, you can see that OH and NH peaks can occur at almost any chemical shift. Carboxyl COOH protons are more predictable, absorbing at 10–12 ppm.

The broadening of the peaks also means that OH and NH protons are not usually involved in spin–spin coupling. All this can make assigning OH and NH protons difficult.

Proton exchange

Chemists have devised a technique called proton exchange for identifying –OH and –NH protons:

1. A proton NMR spectrum is run as normal.
2. A small volume of deuterium oxide, D_2O, is added, the mixture is shaken and a second spectrum is run.

Deuterium exchanges and replaces the OH and NH protons in the sample with deuterium atoms. For example, the following equilibrium is set up with methanol:

$$CH_3OH + D_2O \rightleftharpoons CH_3OD + HOD$$

So the second spectrum is essentially being run on CH_3OD. As deuterium does not absorb in this chemical shift range, the OH peak disappears. You can see this for methanol in Figure 7.

Summary questions

Use the proton chemical shifts given in Figure 1 to help you answer these questions.

1. For each compound below, predict:
   i the number of peaks
   ii the relative peaks areas
   iii any difference in the presence of D_2O.
   a CH_3COOH *(3 marks)*
   b CH_3CHO *(3 marks)*
   c CH_3COCH_3 *(3 marks)*
   d $CH_3COOCH_2CH_3$ *(3 marks)*
   e $H_2NCH(CH_3)COOH$ *(3 marks)*

2. For each compound below, predict:
   i the number of peaks
   ii the type of proton and chemical shift
   iii the relative peaks areas
   a $HOCH_2CH_2OH$ *(3 marks)*
   b $HOCH_2CH_2CH_2OH$ *(3 marks)*
   c $(CH_3)_2CHOH$ *(3 marks)*

3. a Name the structural sequence indicated by the presence of the following splitting patterns in a proton NMR spectrum:
   i triplet and quartet *(1 mark)*
   ii quartet and doublet *(1 mark)*
   iii heptet and doublet *(1 mark)*
   iv doublet and triplet *(1 mark)*
   v two triplets *(1 mark)*
   vi two doublets *(1 mark)*
   b Predict the splitting patterns in the proton NMR spectra of the three compounds in question 2. *(6 marks)*

29.5 Interpreting NMR spectra
Specification reference: 6.3.2

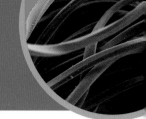

Interpreting proton NMR spectra

There are no set rules about how to identify a compound from a proton NMR spectrum. Although the worked examples follow a step-by-step procedure, you can solve spectral problems in almost any order. With one spectrum, you might start with the splitting patterns (especially with a triplet/quartet combination), for another you might start by looking at chemical shifts. It depends on which piece of evidence is the first that you see and there are countless ways of solving these problems.

> ### Learning outcomes
> Demonstrate knowledge, understanding, and application of:
> → analysis of proton NMR spectra of an organic molecule
> → predictions about possible structures for the molecule
> → prediction of a carbon-13 or proton NMR spectrum for a given molecule.

 Worked example: Isomer of $C_4H_8O_2$

The proton NMR spectrum below is for an isomer of $C_4H_8O_2$. The numbers are the relative peak areas. Analyse the spectrum to identify the compound.

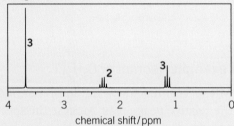

Step 1: Analyse the types of proton present and how many of each type.

There are *three* peaks so there are *three* types of proton.
- From the relative *peak areas*, the ratio of types of protons is 3 : 2 : 3.
- So the peaks are 3.7 ppm CH_3, 2.3 ppm CH_2, 1.1 ppm CH_3.

Step 2: Analyse the splitting patterns to find information about adjacent protons.
- The singlet at 3.7 ppm indicates no adjacent H atoms by the $n + 1$ rule (0 + 1 = 1).
- The triplet at 1.1 ppm indicates an adjacent CH_2 by the $n + 1$ rule (2 + 1 = 3).
- The quartet at 2.3 ppm indicates an adjacent CH_3 by the $n + 1$ rule (3 + 1 = 4).
- The combination of a triplet $\delta = 1.1$ ppm and a quartet at $\delta = 2.3$ ppm indicates CH_3CH_2.

Step 3: Using the data in Topic 29.4 Figure 1, analyse the chemical shifts for the types of proton.
- The CH_3 peak at $\delta = 3.7$ ppm has the type **H—C—O**, supporting the sequence CH_3—O.
- The CH_2 peak at $\delta = 2.3$ ppm has the type **H—C=O**, supporting the sequence CH_2—C=O.
- The CH_3 peak at $\delta = 1.1$ ppm has the type **H—C—R**, supporting the sequence CH_3—C.

Step 4: Combine the information to suggest a structure.

Combining the evidence from Steps 1–3, the correct structure is $CH_3CH_2COOCH_3$ (methyl propanoate), which has the given formula $C_4H_8O_2$. The spectrum below shows how the splitting patterns, relative peak areas and chemical shifts provide the evidence for the identification.

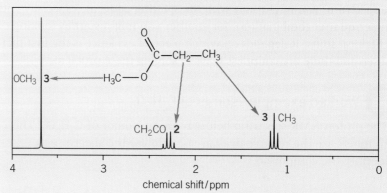

Worked example: Isomer of C_3H_7Cl

The proton NMR spectrum below is for an isomer of C_3H_7Cl. The numbers are the relative peak areas. Analyse the spectrum to identify the compound.

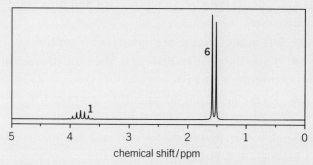

Step 1: Analyse the types of proton present and how many of each type.

There are *two* peaks so there are *two* types of proton.

- From the relative peak areas, the ratio of types of protons is 1 : 6.
- So the peaks are 3.8 ppm CH and 1.6 ppm 2 × CH_3.

Step 2: Analyse the splitting patterns to find information about adjacent protons.

- The heptet at 3.8 ppm indicates 6 H atoms on adjacent C atoms by the $n + 1$ rule ($6 + 1 = 7$) which must be two CH_3 groups as $(CH_3)_2$.
- The doublet at 1.6 ppm indicates an adjacent CH by the $n + 1$ rule ($1 + 1 = 2$).
- The combination of a doublet at $\delta = 1.6$ ppm and a heptet at $\delta = 3.8$ ppm indicates $CH(CH_3)_2$.

Step 3: Using the data in Topic 29.4, Figure 1 to analyse the chemical shifts for the types of proton.

- The CH peak at δ = 3.8 ppm has the type **H—C—Cl**, supporting the sequence CH—Cl.
- The (CH$_3$)$_2$ peak at δ = 1.6 ppm has the type **H—C—R**, supporting the sequence C(CH$_3$)$_2$.

Step 4: Combine the information to suggest a structure.

Combining the evidence from Steps 1–3, the correct structure is (CH$_3$)$_2$CHCl (2-chloropropane), which has the given formula C$_3$H$_7$Cl. The spectrum below shows and how the splitting pattern, relative peak areas and chemical shifts provide the evidence for the identification.

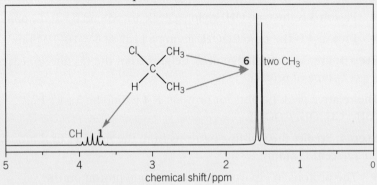

Predicting NMR spectra

Using the information from Topics 29.3 and 29.4, you can now predict both carbon and proton NMR spectra.

 Worked example: Predicting a carbon-13 NMR spectrum

Predict the carbon-13 NMR spectra of 2-hydroxypropanoic acid, CH$_3$CHOHCOOH.

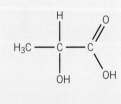

Step 1: Draw out the structure

Step 2: Identify the number of chemical environments.

There are three types of carbon atoms giving three peaks.

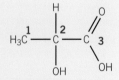

Step 3: Using a data sheet in Topic 29.3, Figure 1, predict the chemical shifts.

- Carbon-1 is the type **C—C**, giving a peak at δ = 0–50 ppm.
- Carbon-2 is the type **C—O**, giving a peak at δ = 50–90 ppm.
- Carbon-3 is the type **C=O**, giving a peak at δ = 160–220 ppm.

Checking the prediction

The actual carbon-13 NMR spectrum of CH$_3$CHOHCOOH is shown to the right, which is in good agreement with the prediction.

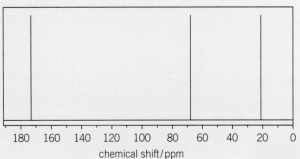

29.5 Interpreting NMR spectra

Worked example: Predicting a proton NMR spectrum

Predict the proton NMR spectra of 2-hydroxypropanoic acid, CH₃CHOHCOOH.

Step 1: Draw the structure and identify the number of chemical environments.

This is the same molecule as in the carbon-13 NMR prediction. There are four types of protons which should give four peaks.

Step 2: Use the data in Topic 29.4, Figure 1, predict the chemical shifts.

- Protons **1** are the type **H**—C—R, giving a peak at δ = 0.5–2.0 ppm.
- Proton **2** is the type **H**—C—O, giving a peak at δ = 3.0–4.2 ppm.
- Proton **3** is the type O**H**, giving a peak anywhere: δ = 0–12 ppm.
- Proton **4** is the type COO**H**, giving a peak at δ = 10–12 ppm.

Step 3: Predict the relative peak heights from the number of each type of proton.

The relative peak heights for protons **1–4** would be 3 : 1 : 1 : 1 (CH₃, CH, OH, COOH)

Step 4: Predict the splitting patterns from the number of H atoms on adjacent C atoms.

- The protons in environment **1** have one proton on the adjacent C atom (i.e. CH) giving a doublet by the n + 1 rule (3 + 1 = 2).
- The protons in environment **2** have three protons on the adjacent C atom (i.e. CH₃), giving a quartet by the n + 1 rule (3 + 1 = 2).
- The OH and COOH protons would give broad peaks and would not show splitting.

Checking the prediction

The actual proton NMR spectrum of CH₃CHOHCOOH is shown below, which is in good agreement with the prediction.

Notice that the quartet at δ = 4.2 for the CH group is at the upper end of our predicted chemical shift range of 3.0–4.2 ppm. If you look at the structure, this CH group is adjacent to both an O atom and a C=O group, two groups that increase the chemical shift.

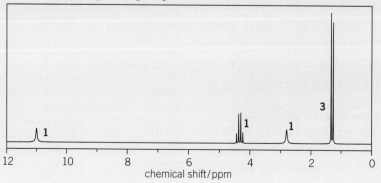

Summary questions

1. Predict the proton NMR spectrum of $CH_3COOCH_2CH_3$. *(3 marks)*

2. **a** Identify and explain the peaks in the NMR spectrum for $HOC(CH_3)_2COCH(CH_3)_2$ below.

   The splitting pattern for the peak at $\delta = 1.8$ ppm has also been shown magnified.

   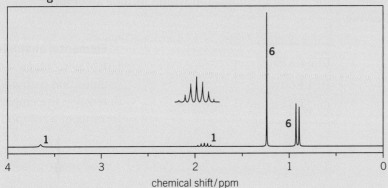

   (4 marks)

   b The proton NMR spectrum below is for an isomer of C_4H_8O. The numbers are the relative peak areas. Analyse the spectrum to identify the compound.

   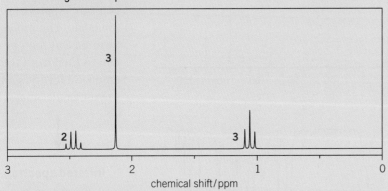

   (4 marks)

3. Predict the differences between the proton NMR spectra of the following.
   a $CH_3COCH_2CH_2CH_3$ and $CH_3CH_2COCH_2CH_3$ *(4 marks)*
   b $CH_3CH_2COOCH_3$ and $CH_3COOCH_2CH_3$ *(4 marks)*

29.6 Combined techniques

Specification reference: 6.3.2

Learning outcomes

Demonstrate knowledge, understanding, and application of:

→ deduction of the structures of organic compounds from different analytical data including:
  i   elemental analysis
  ii mass spectra
  iii IR spectra
  iv NMR spectra.

Structure determination

In practice, organic chemists use and interpret information from a variety of sources when determining the structure of an organic molecule.

A typical sequence for identification would include all of the following.

C: 64.82%
H: 13.60%
O: 21.58%

Elemental analysis
Use of percentage composition by mass to determine the empirical formula of a compound.

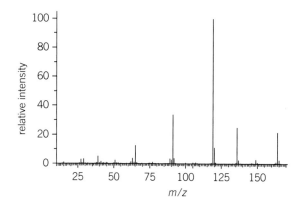

Mass spectra
Use of the molecular ion peak to determine the molecular mass and fragment ions to determine parts of the molecule. The molecular formula can then be determined from the empirical formula and the molecular mass.

Synoptic link

You need to be familiar with these techniques. You have covered NMR spectroscopy in this chapter, but if you need to revise any of the other techniques, use the following topics:

- Topic 3.1, Amount of substance and the mole, for elemental analysis to determine the empirical formula
- Topic 17.1, Mass spectrometry, for Mass spectra
- Topic 17.2, Infrared spectroscopy, for IR spectra.

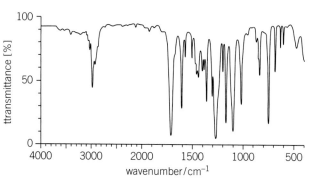

Infrared spectra
Use of absorption peaks to identify bonds present and functional groups

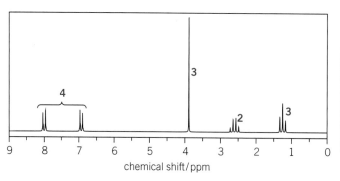

NMR spectra
To determine the number and types of carbon and hydrogen atoms from the chemical shifts of peaks and the order of atoms within molecules from splitting patterns.

530

29 CHROMATOGRAPHY AND SPECTROSCOPY

Analysis of an unknown compound

Chemical analysis of an unknown compound gave the percentage composition by mass: C, 73.17%; H, 7.32%; O, 19.51%.

The mass spectrum, infrared spectrum, carbon-13, and proton NMR spectra are shown in Figures 2–5.

Analyse this information to suggest the structural formula of the compound.

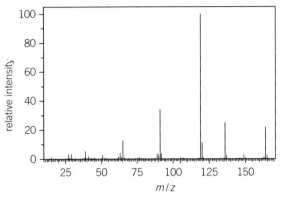

▲ **Figure 1** Mass spectrum of unknown compound

▲ **Figure 2** Infrared spectrum of unknown compound

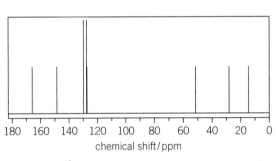

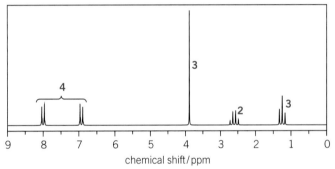

▲ **Figure 3** ^{13}C NMR spectrum of unknown compound

▲ **Figure 4** Proton NMR spectrum of unknown compound

Step 1: Determine the empirical formula from the elemental analysis data.

Convert percentage by mass to amounts in moles: C, 73.17%; H, 7.32%; O, 19.51%.

$$n(C) = \frac{73.17}{12.0} = 6.0975 \text{ mol} \quad n(H) = \frac{7.32}{1.0} = 7.32 \text{ mol} \quad n(O) = \frac{19.51}{16.0} = 1.2194 \text{ mol}$$

Find the smallest whole number ratio and the empirical formula.

$$n(C) : n(H) : n(O) = \frac{6.0975}{1.2194} : \frac{7.32}{1.2194} : \frac{1.2194}{1.2194} = 5 : 6 : 1$$

Empirical formula = C_5H_6O

Step 2: Determine the molecular formula using the mass spectrum and the empirical formula.

- Use the m/z value for the molecular ion peak to determine the molecular formula.
- From the mass spectrum, molecular ion peak is at m/z = 164; so molecular mass = 164

Synoptic link

See Topic 3.2, Determination of formulae, for details of empirical formula and molecular calculations.

See Topic 17.2 Chapter 17, Mass spectrometry, for molecular ion peak.

29.6 Combined techniques

> - Find the molecular mass by comparing the molecular mass with empirical formula mass
>
> Relative mass of empirical formula $= 12.0 \times 5 + 1.0 \times 6 + 16.0 \times 1$
> $= 82.0$
>
> Number of C_5H_6O units in one molecule $= \dfrac{164}{82.0} = 2$
>
> Molecular formula $= C_5H_6O \times 2 = C_{10}H_{12}O_2$
>
> **Step 3:** Identify the functional groups using the infrared spectrum and the Data sheet.
>
> Peak at 1710 cm^{-1} indicates the presence of a C=O group in an aldehyde, ketone or ester (cannot be a carboxylic acid as no broad peak present at 2500–3300 cm^{-1} for O—H).
>
> **Step 4:** Analyse the ^{13}C NMR spectrum.
>
> There are 8 peaks and 8 types of carbon atom (8 carbon environments).
>
> **Step 5:** Analyse the proton NMR spectrum.
>
> - Analyse the types of proton present and how many of each type.
>
> There are four peaks so there are four types of proton.
>
> From the relative peak areas, the ratio of protons from the left is 4 : 3 : 2 : 3.
>
> So the peaks are:
>
> 6.9–8.0 ppm: **4 aromatic (Ar)** protons **4** : 3 : 2 : 3
>
> 3.8 ppm: CH$_3$ 4 : **3** : 2 : 3
>
> 2.6 ppm: CH$_2$ 4 : 3 : **2** : 3
>
> 1.2 ppm: CH$_3$ 4 : 3 : 2 : **3**
>
> - Analyse the splitting patterns to find information about adjacent protons.
>
> The triplet at 1.2 ppm indicates an adjacent CH$_2$ by the $n + 1$ rule (2 + 1 = 3).
>
> The quartet at 2.6 ppm indicates an adjacent CH$_3$ by the $n + 1$ rule (3 + 1 = 4).
>
> The singlet at 3.8 ppm indicates no adjacent H atoms by the $n + 1$ rule (0 + 1 = 1).
>
> The combination of a triplet $\delta = 1.2$ ppm and a quartet at $\delta = 2.6$ ppm indicates CH$_3$CH$_2$.
>
> - Use the data in Topic 29.6, Figure 1, to analyse the chemical shifts for the types of proton.
>
> The CH$_3$ peak at $\delta = 3.8$ ppm has the type H—C—O, supporting the sequence O—CH$_3$.
>
> The CH$_2$ peak at $\delta = 2.6$ ppm has the type H—C=O or H—C—Ar, supporting the sequence CH$_2$—C=O or CH$_2$—Ar.
>
> The CH$_3$ peak at $\delta = 1.2$ ppm has the type H—C—R, supporting the sequence CH$_3$—C.

Study tip

In this example, the identification of the functional group is uncertain from the IR spectrum.

Study tip

The peaks at 6.9–8.0 are grouped together as one type of carbon atom. The reason for the two groupings within the 4 protons will be revealed later.

Study tip

Look again at the aromatic protons in the 1H NMR spectrum.

They appear to be split into two doublets.

This splitting pattern is characteristic of 1,4-disubstitution.

1,2 and 1,3 disubstitutions produce more complex multiplets.

From your knowledge of splitting patterns, you should be able to see why.

Step 6: Combine the information to suggest a structure.

From the evidence, the CH_3CH_2 group could be adjacent to either a carbonyl or aromatic ring.

The ^{13}C NMR spectrum suggests that the aromatic ring is 1,4-disubstituted.

From all the evidence, the compounds could have one of two structures below:

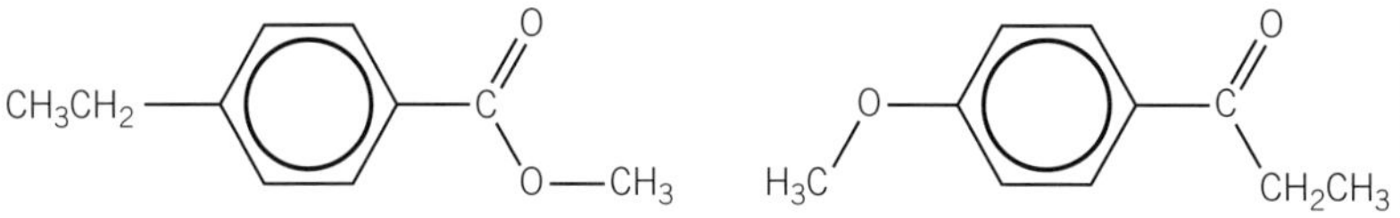

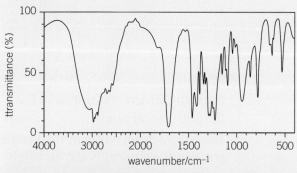

▲ **Figure 5**

1. Use the ^{13}C NMR spectrum to show that the aromatic ring must be 1,4-disubstituted rather than structures 1,2- or 1,3-disubstituted.
2. Describe a simple chemical test that would confirm which of the two structures above is likely to be correct.

Synoptic link

See Topic 29.3, ^{13}C NMR spectroscopy. for details of ^{13}C NMR spectra of disubstitutued aromatic compounds.

Summary questions

Chemical analysis of an unknown compound gave the percentage composition by mass: C, 62.07%; H, 10.34%; O, 27.59%. The molecular ion peak is at m/z = 116. The infrared and ^{13}C NMR spectra are shown in Figures 7–8.

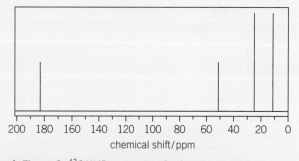

▲ **Figure 7** *The infrared spectrum of unknown compound*

▲ **Figure 8** *^{13}C NMR spectrum of unknown compound*

1. Determine the empirical and molecular formulae of the compound. *(2 marks)*

2. Analyse the infrared and ^{13}C NMR spectra to suggest two possible structures for the compound. *(8 marks)*

3. Explain how the compound could be identified from the proton NMR spectrum of the compound *(4 marks)*

Chapter 29 Practice questions

Practice questions

1. α-Amino acids are found in human sweat. A student had read that chromatography could be used to separate and identify the amino acids present in human sweat.

   a The student used thin-layer chromatography (TLC) to separate the α-amino acids in a sample of human sweat and discovered that three different α-amino acids were present.

   (i) State the property of α-amino acids that allow them to be separated by TLC. *(1 mark)*

   (ii) The chromatogram was treated to show the positions of the separated α-amino acids.

   Explain how the student could analyse the chromatogram to identify the three α-amino acids that were present. *(2 marks)*

   (iii) Several α-amino acids have structures that are very similar.

   Suggest why this could cause problems when using TLC to analyse mixtures of α-amino acids. *(1 mark)*

   F324 Jan 2010 Q3(a)

2. Proton NMR spectroscopy can be used to distinguish between the two esters below.

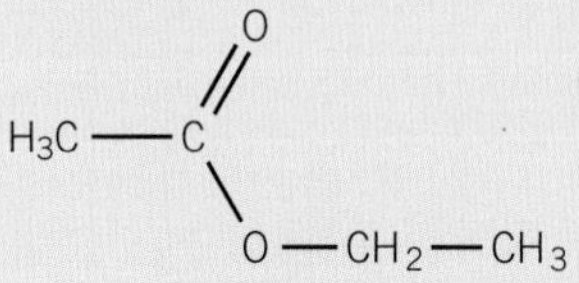

   $H_3C-C(=O)-O-CH_2-CH_3$ $H_3C-CH_2-C(=O)-O-CH_3$

   A 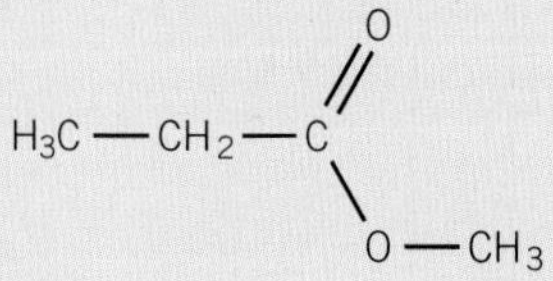   B

   a Complete the table for the peaks in the ^{1}H NMR spectrum of **A**.

Proton environment	Chemical shift/ppm	Splitting	Relative peak area
CH_3–CO			
O–CH_2			
C–CH_3			

   (3 marks)

   b How would the ^{1}NMR spectrum of **B** differ from the ^{1}NMR spectrum of **A**? *(1 mark)*

3. A student was given three compounds, an aldehyde, a ketone, and a carboxylic acid.

   a The student carried out the same two chemical tests on each compound. This allowed her to distinguish between all three compounds.

   • Describe two suitable tests that the student could have used.

   • Show how the observations would allow her to distinguish between the compounds. *(4 marks)*

   b Explain how infrared spectroscopy could be used to confirm which compound is a carboxylic acid. *(1 mark)*

   c The aldehyde has the molecular formula $C_5H_{10}O$. The ^{1}H NMR spectrum of the aldehyde contains a doublet at δ = 0.9 ppm with a relative peak area of six compared with the aldehyde proton as well as a number of other peaks.

   Analyse this information to deduce the structure of the aldehyde. Explain your reasoning. *(4 marks)*

   d The ketone also has the molecular formula $C_5H_{10}O$. There are three structural isomers of this formula that are ketones.

   (i) Two of these isomers are shown below.

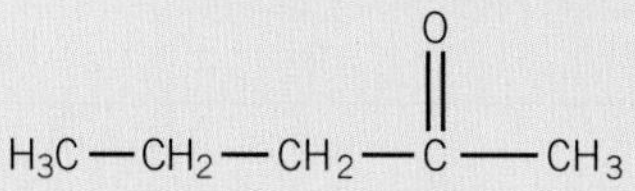

   $H_3C-CH_2-CH_2-C(=O)-CH_3$ $CH_3-CH(CH_3)-C(=O)-CH_3$

   ketone 1 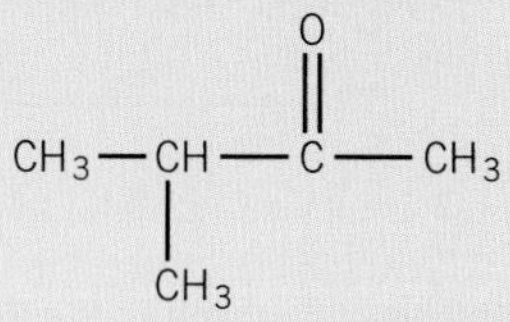   ketone 2

   Draw the structural formula of the third structural isomer. *(1 mark)*

   (ii) The ^{13}C NMR spectrum of the ketone given to the student is shown below.

   Use the spectrum to identify the ketone. Explain your reasoning.

   Identify the carbon responsible for the peak at δ = 210 ppm.

   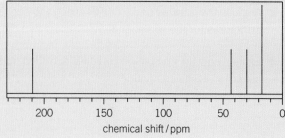

   (3 marks)

   F324 June 2010 Q3

CHROMATOGRAPHY AND SPECTROSCOPY

4 An industrial chemist discovered five bottles of different chemicals (three esters and two carboxylic acids) that were all labelled $C_5H_{10}O_2$.

The different chemicals had the structural formulae below.

CH₃CH₂COOCH₂CH₃   (CH₃)₃CCOOH
CH₃COOCH(CH₃)₂   (CH₃)₂CHCH₂COOH
(CH₃)₂CHCOOCH₃

a The chemist used infrared and ^{13}C NMR spectroscopy to identify the two carboxylic acids and to distinguish between them.

How do the spectra allow the carboxylic acids to be identified and distinguished from one another? *(3 marks)*

b The chemist analysed one of the esters by ^{1}H NMR spectroscopy. The spectrum is shown below.

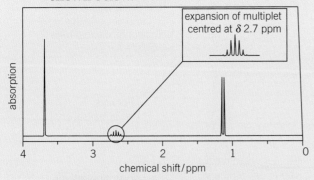

Analyse the splitting patterns and the chemical shift values to identify the ester.

Give your reasoning.

In your answer, you should use appropriate technical terms, spelt correctly. *(6 marks)*

F324 Jan 2010 Q4

5 Compounds **A** and **B** are structural isomers.

Elemental analysis gave the following percentage composition by mass of A and B:
C, 40.68%   H, 5.08%   O, 54.24%.

The infrared and mass spectra of compounds of **A** and **B** are very similar. The spectra for **A** are shown below.

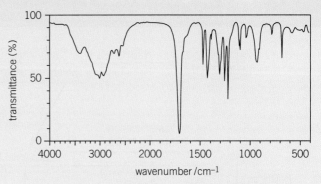

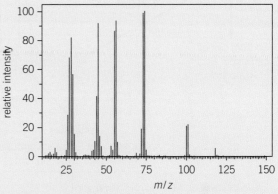

The ^{13}C NMR spectra of **A** and **B** are shown below.

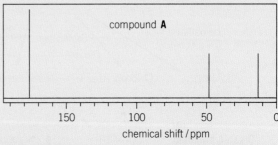

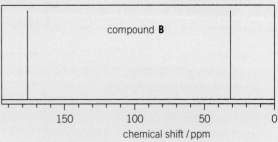

a Using the analytical data, determine the structures of compounds **A** and **B**.
Show all your reasoning. *(6 marks)*

b Predict the ^{1}H NMR spectra of **A** and **B**.
Show all your reasoning. *(6 marks)*

c What difference, if any, would there be if the ^{1}H NMR spectra were run with a few drops of D₂O added? Explain your answer *(1 mark)*

Module 6 Summary

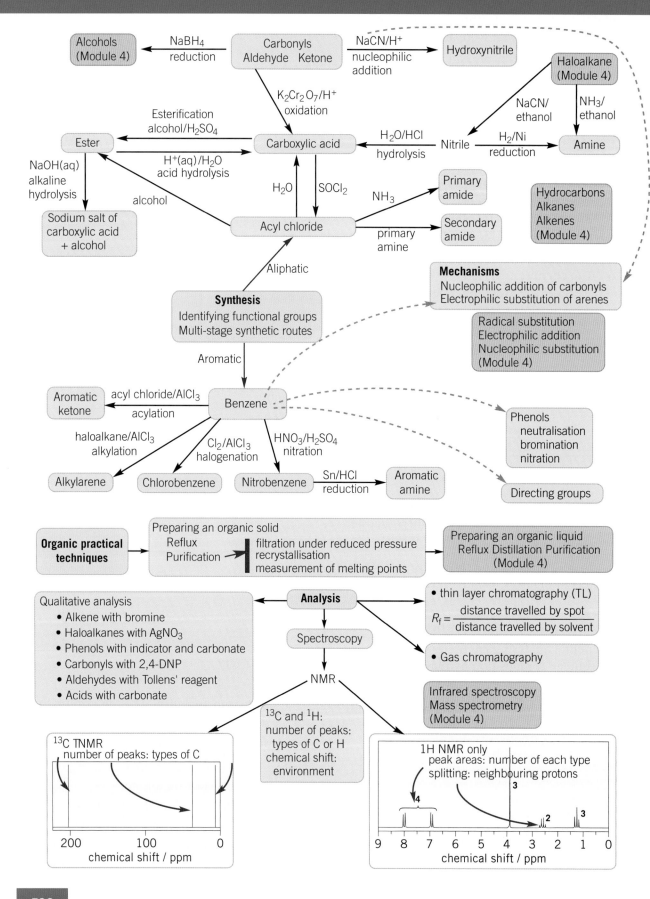

Antibiotics

Antibiotics are substances that destroy bacteria and are one of the cornerstones of modern medicine.

Synthetic organic chemists play an important role in developing new semi-synthetic antibiotics. These make use of structural components from existing antibiotic molecules, combined with a molecule synthesised by chemists.

One example is ampicillin, which was developed in the 1960s but is still important in treating many infections today.

▲ **Figure 1** *The structure of the semi-synthetic antibiotic, ampicillin*

Ampicillin, like all semi-synthetic penicillins, is made using the naturally occurring compound 6-APA (6-aminopenicillanic acid), which can be extracted from penicillin mould.

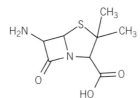

▲ **Figure 2** *Penicillin mould, from which 6-APA can be extracted*

▲ **Figure 3** *6-APA, one of the starting materials for the synthesis of ampicillin*

The other starting material is usually 2-amino-2-phenylethanoic acid, which is classified as an α-amino acid, although it is not one of the 20 naturally occurring amino acids found in all living organisms.

1. Name as many functional groups as posisble in the ampicillin molecule. (Some will be unfamilliar to you.)
2. a Draw out the structure of 2-amino-2-phenylethanoic acid.
   b Explain why it is classified as an α-amino acid.
   c 2-Amino-2-phenylethanoic acid exists as stereoisomers. Describe the type of stereoisomerism and explain why this molecule exists as stereoisomers.
3. 6-APA extracted from penicillium mould can be analysed using various spectroscopic methods to check its identify.
   a How many different peaks would be seen in the carbon-13 spectrum of 6-APA?
   b Predict the *m/z* value of the M⁺ ion in the mass spectrum of 6-APA
   c Predict the peaks in the infrared spectrum of 6-APA resulting from the main functional groups present in the molecule. Give the approximate wavenumber range for each peak.
4. Describe a three-step synthesis of ampicillin, starting from 2-amino-2-phenylethanol, and making use of 6-APA and any necessary inorganic reagents.

Extension

1. Amide groups in a penicillin molecule can be hydrolysed by acid or alkali in a similar way to the hydrolysis of amide groups in a polyamide. Draw the structures of the organic products of the hydrolysis of ampicillin using **a** hydrochloric acid **b** sodium hydroxide solution.
2. The pattern of heights of the peaks in a multiplet (e.g., a triplet) can be predicted using Pascals triangle (Topic 29.4, Proton NMR spectroscopy). By considering the possible spins of the coupled protons as ↑ or ↓, explain why the heights of the peaks in a signal from a proton coupled to a CH₂ group follow this pattern, and so are in the ratio 1:2:1.

Module 6 Practice questions

1. Which is the correct systematic name of this compound?

   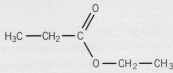

   A ethyl propanoate
   B ethyl ethanoate
   C propyl ethanoate
   D propyl pentanoate (1 mark)

2. What is the number of structural isomers that are carboxylic acids with the formula $C_5H_{10}O_2$?
   A 2   B 3   C 4   D 5   (1 mark)

3. The thin layer chromatography plate shown below has a polar stationary phase. It was developed using hexane as the solvent.

   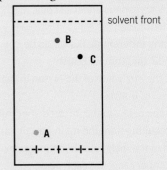

   Which sample has the most polar molecules?
   A sample A
   B sample B
   C sample C
   D There is not enough information to determine which sample has the most polar molecules. (1 mark)

4. Which statements about $CH_3CH=CHCH_2CHO$, is **not** true?
   A It turns acidified potassium dichromate(VI) a green colour.
   B It has an infrared peak at about 1700 cm^{-1}.
   C It has stereoisomers.
   D It reacts with a carboxylic acid form an ester. (1 mark)

5. Which compound has two peaks in its ^{13}C NMR spectrum?
   A 1-chloropropan-2-ol
   B 3-bromopropan-1-ol
   C 2-iodopropan-2-ol
   D 1-bromopropan-1-ol (1 mark)

6. Which compound has a proton NMR spectrum that contains a doublet?
   A 2-methylbutan-2-ol
   B hexan-3-one
   C 2-bromobutane
   D propyl methanoate (1 mark)

7. P is $HOCH_2CH(OH)CHO$.
   Q is $HOCH_2COCH_2OH$
   Which statement is **not** correct?
   A P and Q both react with 2,4-dinitrophenylhydrazine.
   B P and Q both react with sulfuric acid and potassium dichromate(VI).
   C P and Q both react with Tollens' reagent.
   D P and Q both react with $NaBH_4$. (1 mark)

8. What type of mechanism takes place when methylbenzene reacts with bromine in the presence of $FeBr_3$?
   A Nucleophilic addition
   B Electrophilic addition
   C Nucleophilic substitution
   D Electrophilic substitution (1 mark)

9. The reaction pathway for the synthesis of paracetamol, a mild painkiller, is provided below.

   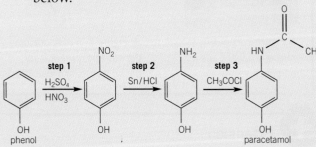

   Which step or steps in this synthesis involve(s) a reduction reaction?
   A step 1 only
   B step 2 only
   C steps 1 and 3 only
   D steps 1, 2 and 3 (1 mark)

10 Cinnamic acid is an organic substance that partly contributes to the flavour of oil of cinnamon. A structure of cinnamic acid is given below.

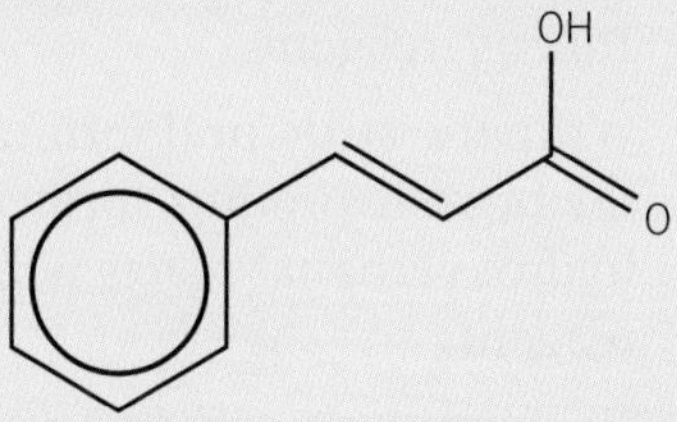

Which row shows whether cinnamic acid would react with the reagents?

	$CH_3CH_2Cl/AlCl_3$ catalyst	Br_2(aq)	CH_3OH/H_2SO_4 catalyst
A	yes	yes	yes
B	yes	no	yes
C	no	yes	yes
D	no	yes	no

(1 mark)

11 The structure of cholesterol is shown below.

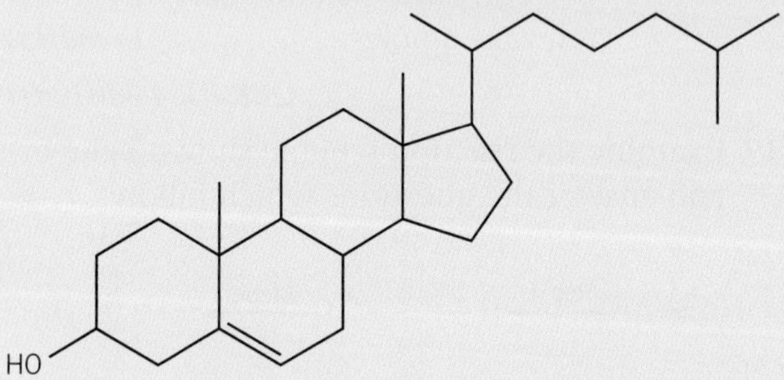

How many chiral centres are there in one molecule of cholesterol?

A 4 B 5 C 8 D 9 *(1 mark)*

12 Compound **R** has the molecular formula $C_5H_{10}O$

Compound **T** is synthesised from compound **R** as below.

$R \xrightarrow{H_2SO_4/K_2Cr_2O_7 \text{ reflux}} S \xrightarrow{\text{ethanol}/H_2SO_4 \text{ reflux}} T$

What formula could be **T**?

A $CH_3COO(CH_2)_4CH_3$
B $CH_3CH_2COO(CH_2)_3CH_3$
C $CH_3(CH_2)_3COCH_2CH_3$
D $CH_3(CH_2)_3COOCH_2CH_3$

(1 mark)

13 An excess of calcium carbonate is added to $100\,cm^3$ of $1\,mol\,dm^{-3}$ citric acid.

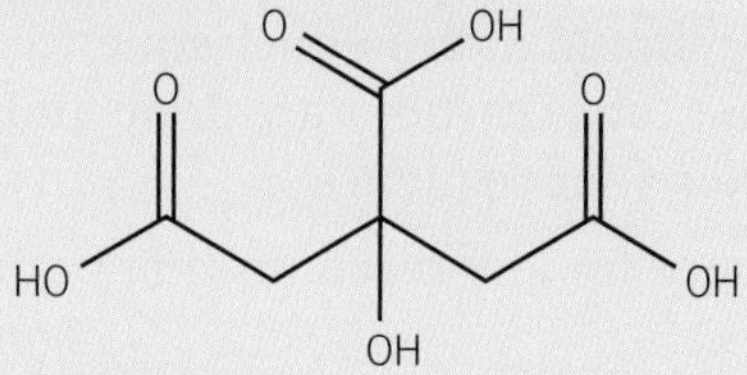

Assuming that no CO_2 dissolves, what volume of CO_2(g), measured at room temperature and pressure, is formed?

A $2.4\,dm^3$ B $4.8\,dm^3$
C $7.2\,dm^3$ D $9.6\,dm^3$ *(1 mark)*

14 How many of the compounds below react with aqueous sodium hydroxide to form the sodium salt of a carboxylic acid?

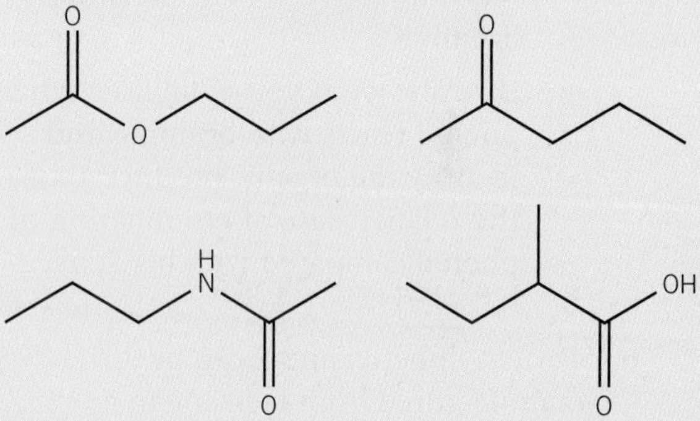

A 1 B 2 C 3 D 4 *(1 mark)*

15 A section of a condensation polymer made from two monomers is shown below.

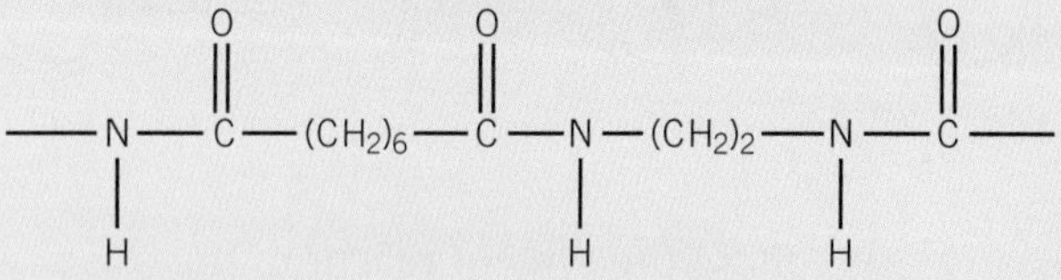

What is the repeat unit?

A $-[NHCO(CH_2)_6CONH(CH_2)_2NHCO]-$
B $-[NHCO(CH_2)_6CONH(CH_2)_2NH]-$
C $-[CO(CH_2)_6CONH(CH_2)_2NHCO]-$
D $-[CO(CH_2)_6CONH(CH_2)_2NH]-$ *(1 mark)*

16 The four compounds below (**A–D**) are in bottles from which the labels have fallen off.

A $CH_3CH_2CH_2COOH$ B C_6H_5OH
C $CH_3CH_2CH_2CHO$ D $CH_3CH_2CHCH_2$

Devise a scheme, using test-tube reactions, to identify the contents of the bottles.

Module 6 Practice questions

You may choose any appropriate reagents to use in the identification.

Your scheme should indicate the sequence in which the test tube reactions are done and the reasons for your sequence.

Write equations for any reactions that occur.
(9 marks)

17 Benzene and phenol are used as starting materials for making aromatic compounds.

a Benzene and phenol both react with bromine. The reaction between benzene and bromine requires a catalyst. The reaction between phenol and bromine does not require a catalyst.

(i) Identify a suitable catalyst for the reaction between benzene and bromine. (1 mark)

(ii) Describe what you would see when phenol reacts with bromine and identify the organic product. Explain the relative ease of bromination of phenol compared with benzene.
(6 marks)

b 4-Methylphenylamine can be manufactured from benzene in a multistep synthesis, as shown below.

(i) State the type of reaction, the reagents and conditions and write a balanced equation for **Step 1**, **Step 2** and **Step 3**. (9 marks)

(ii) Describe, with the aid of curly arrows, the mechanism for **Step 1**. Include an equation for the formation of the electrophile. (4 marks)

OCR 2814 Jun 2010 Q1

18 Hydroxyethanal, $HOCH_2CHO$, is the simplest molecule that contains both an aldehyde group and an alcohol group. A biochemist investigated some redox reactions of hydroxyethanal.

a The biochemist reacted hydroxyethanal with Tollens' reagent.

(i) State what the biochemist would see when hydroxyethanal reacts with Tollens' reagent. (1 mark)

(ii) Write the structural formula of the organic product formed when hydroxyethanal reacts with Tollens' reagent. (1 mark)

b The biochemist also reacted hydroxyethanal with acidified dichromate by heating under reflux.

Write an equation for this oxidation. Use [O] to represent the oxidising agent.
(2 marks)

c The biochemist then reduced hydroxyethanal using aqueous $NaBH_4$.

(i) Write the structural formula of the organic product. (1 mark)

(ii) Outline the mechanism for this reduction.
Use curly arrows and show any relevant dipoles. (4 marks)

OCR F324 Jan10 Q2

19 Examine the reaction scheme shown below and answer the questions which follow

a (i) For **stage 1**, state the reagents, the type of reaction and write a balanced equation. (3 marks)

(ii) Describe, with the aid of curly arrows, the reaction mechanism in **stage 1**. Show any relevant dipoles. (4 marks)

b For **stage 2**, state the reagents, the type of reaction and write a balanced equation.
(3 marks)

c Identify and show the structures for compounds **A**, **B**, and **C**. (3 marks)

Module 6 Organic chemistry and analysis

20 This question looks at the synthesis of organic compounds.

   a Identify by name and skeletal formula, the alcohol and carboxylic acid needed to be reacted in the presence of concentrated sulfuric acid to make each ester.

   (i) methyl propanoate (2 marks)

   (ii) ethyl hexanoate. (2 marks)

   b With the aid of reagents and conditions, and equations, describe how the following compounds could be synthesised from either methyl propanoate or 2-chloropropan-1-ol.

   (i) sodium propanoate (1 step) (1 mark)

   (ii) 2-chloropropanoyl chloride (2 steps) (4 marks)

   (iii) 2-aminopropanal (2 steps) (4 marks)

21 Compound **A**, has the molecular formula $C_4H_{10}O$.

   Compound **A** reacts with concentrated sulfuric acid to form a mixture of three isomers, **B**, **C**, and **D** with molecular formula C_4H_8.

   Compound **B** does not have *E/Z* isomers.

   Compounds **C** and **D** are *E/Z* isomers of one another.

   Compound **A** reacts with NaCl and sulfuric acid to produce compound **E**, C_4H_9Cl.

   Compound **E** is reacted with potassium cyanide in ethanol to produce the compound **F**.

   Compound **F** is warmed with dilute hydrochloric acid to form compound **G**.

   Compound **A** reacts with compound **G** in the presence of sulfuric acid to produce a sweet smelling compound **H**, $C_9H_{16}O_2$.

   Draw the structures for compounds **A–H**.
   (8 marks)

22 A chemist prepares and analyses some esters.

   a The chemist prepares an ester of propan-2-ol, $CH_3CH(OH)CH_3$, by reacting $CH_3CH(OH)CH_3$ with excess ethanoic anhydride, $(CH_3CO)_2O$.

   Using structural formulae, write an equation for the reaction of propan-2-ol and ethanoic anhydride. (2 marks)

   b A sample contains a mixture of two esters contaminated with an alkane and an alcohol.

   The chemist attempts to separate the four organic compounds in the mixture using GC. The column contains a liquid alkane which acts as the stationary phase.

   (i) By which property are the organic compounds in a mixture separated?
   (2 marks)

   (ii) Suggest how well these four compounds would be separated using the alkane stationary phase.

   In your answer, include some indication of the relative length of the retention times. Explain your answer. (2 marks)

   c One of the esters in a perfume is separated by GC and then analysed by proton NMR spectroscopy.

   The results are shown below.

   Elemental analysis by mass

   C, 66.63%; H, 11.18%; O, 22.19%

   Mass spectrum

   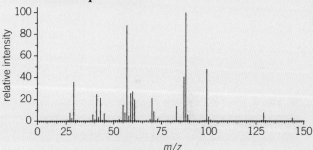

   Proton NMR spectrum

   The numbers by each peak are the relative peak areas.

   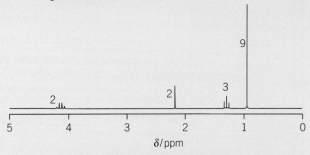

   Use the results to identify the ester. Show **all** your reasoning. (10 marks)

   OCR 2814 Jan 2012 Q4

UNIFYING CONCEPTS

Introduction to unifying concepts

A unifying concept in chemistry links together more than one area of chemistry within a single theme. For example, a chemistry theme might include material from the three traditional branches of chemistry – inorganic, physical, and organic chemistry – and could include practical processes and techniques. A synoptic question might assess the theme within all of these areas, some of these areas or even introduce new unfamiliar aspects of chemistry or novel real life scenarios that you have not encountered before.

Unified chemistry

In the unified chemistry exam paper, question styles will include short answer questions (structured questions, problem solving, calculations, practical) and extended response questions. Chemistry is an evolving science and new applications of theory are discovered each year. As such, it is impossible to predict the kind of questions that may be set on this paper. The section that follows contains a number of context questions which test key concepts but also your ability to think deeper about how the concepts you have studied can be applied in unfamiliar situations

Practicing the questions will help you to develop your problem solving skills, however it is very unlikely that you will see questions exactly the same as these on any of the synoptic papers set.

Table 1 shows how just one part of a module, Acids, can link with other sections. The practical skills from Module 1 also link to all other modules.

▼ **Table 1** *Table to show how a topic within a module links to other modules*

Learning outcome	Links with other modules		What you might be asked
2.1.4 Acids	Neutralisation and redox	2.1.4 Acids 2.1.5 Redox 1.1.3 Analysis	Construct equations for acid reactions and carry out titration calculations.
	Enthalpy changes	3.2.1 Enthalpy changes 1.1.1 Planning 1.1.3 Analysis	Plan an experiment that would allow you to calculate an enthalpy change of neutralisation.
	Acid–base equilibria, pH and buffers	5.1.3 Acids, bases and buffers	Calculate pH values for acids, bases and buffers.
	Carboxylic acids and their derivatives, and phenols	6.1.3 Carboxylic acids and esters 6.1.1 Aromatic chemistry	Develop synthetics routes for organic compounds with different functional groups using carboxylic acids as the starting material.

Analysing and answering a synoptic question

Let's look at a question and the knowledge and understanding you would need to answer it. Links with different topics are shown, with each number in square brackets representing a different topic.

Electrophiles[3] were covered in Topic 25.2, Electrophilic substation reactions of benzene, but sulfuric acid as an electrophile was not covered.

Oxidising agents[2] were covered in Topic 23.1, Redox reaction.

Acids[1] were covered in Chapter 4, Acids and redox, and in Chapter 20, Acids, bases, and pH.

1 In its reactions, sulfuric acid, H_2SO_4, can behave as an acid, an oxidising agent, an electrophile, and a dehydrating agent. The displayed formula of sulfuric acid is shown below.

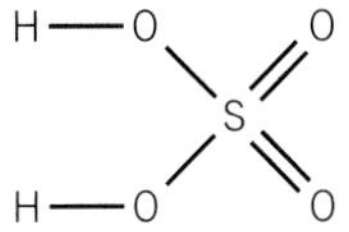

Oxidation number[1] was covered in Topic 4.3, Redox. Oxygen has an oxidation number of −2. Hydrogen has an oxidation number of +1. The molecule has no overall charge.
$(4 \times -2) + (2 \times +1) = -6$

a What is the oxidation number of sulfur in sulfuric acid?
(1 mark)

Answer to part (a)

+6 ✓

b The boiling point of sulfuric acid, at 270 °C, is much higher than expected.
Explain why. Show a diagram in your answer, including relevant dipoles.
(3 marks)

Intermolecular forces[4] are largely responsible for physical propertied such as boiling points, and were covered in Topic 6.3, Intermolecular forces.

Answer to part (b)

Hydrogen bonds exist between molecules of sulfuric acid ✓

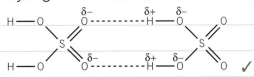

Hydrogen bonding is a strong intermolecular force and energy needs to be supplied to break the hydrogen bonds, increasing the boiling point. ✓

c Dilute sulfuric acid takes part in acid–base reactions.
Write a balanced equation for the reaction of dilute sulfuric acid with an excess of solid sodium carbonate. (1 mark)

Balanced equations[5] were covered in Topic 2.3, Formulae and equations

Answer to part (c)

$H_2SO_4 + Na_2CO_3 \rightarrow Na_2SO_4 + CO_2 + H_2O$ ✓

Analysing and answering a synoptic question

d Sulfuric acid is bibasic. When sulfuric acid is reacted with sodium hydroxide, in a 1:1 molar ratio, the acid salt $NaHSO_4$ is formed.

$$H_2SO_4(aq) + NaOH(aq) \rightarrow NaHSO_4(aq) + H_2O(l)$$

The resulting solution contains the hydrogensulfate(VI) ion, $HSO_4^-(aq)$ which behave as a weak acid.

$$HSO_4^-(aq) \rightleftharpoons H^+(aq) + SO_4^{2-}(aq) \qquad K_a = 1.2 \times 10^{-2} \text{ mol dm}^{-3}$$

A student reacts 25.0 cm³ of 0.160 mol dm⁻³ H_2SO_4 with 25.0 cm³ 0.160 mol dm⁻³ NaOH(aq) to form a solution containing sodium hydrogensulfate(VI) and water only.

(i) Calculate the pH of the resulting solution. *(3 marks)*

Answer to part (d)(i)

Sulfuric acid has been reacted with sodium hydroxide in a 1:1 molar ratio.

So $n(HSO_4^-) = n(H_2SO_4) = \dfrac{0.160 \times 25.0}{1000} = 4.00 \times 10^{-3}$ mol

The total volume is now 50.0 cm³.

Concentration, c, of $HSO_4^- = \dfrac{4.00 \times 10^{-3} \times 1000}{50.0} = 0.0800$ mol dm⁻³ ✓

For $HSO_4^- \quad K_a = \dfrac{[H^+(aq)][SO_4^{2-}(aq)]}{[HSO_4^-(aq)]} \sim \dfrac{[H^+(aq)]^2}{[HSO_4^-(aq)]}$

> First work out the **concentration**[6] of hydrogensulfate(VI) ions in the resulting solution, covered in Topics 3.3, Moles and volumes

$[H^+(aq)]^2 = K_a \times [HSO_4^-(aq)]$

$[H^+(aq)] = \sqrt{K_a \times [HSO_4^-(aq)]} = \sqrt{(1.2 \times 10^{-2} \times 0.0800)}$

$= 0.0310$ mol dm⁻³ ✓

$pH = -\log[H^+(aq)] = -\log(0.0310) = 1.51$ ✓

> Then calculate **pH for a weak acid**[7] using the approximations learnt in Topic 20.4, The pH of weak acids.

> pH[7] was covered in Chapter 20, Acids, bases, and pH.

(ii) The student measures the pH of the solution formed and found that it had a higher pH value than calculated in (i).
Suggest an explanation for the difference. *(2 marks)*

Answer to part (d)(ii)

The K_a value of HSO_4^- showed that HSO_4^- is a 'strong' weak acid and the approximation that the equilibrium concentration of $HSO_4^-(aq)$ is the same as the undissociated HSO_4^- is not valid. ✓

This results in $[H^+(aq)]$ being less than calculated using the approximation and the pH higher ✓

---

e Concentrated sulfuric acid acts as an oxidising agent in many redox reactions. For example, concentrated sulfuric acid oxidises hydrogen iodide to iodine. The sulfuric acid is reduced to hydrogen sulphide, H_2S. Water is also formed.
Construct a balanced equation for this redox reaction of sulfuric acid with iodide ions.
Include all changes in oxidation number in your answer.

(3 marks)

Unifying concepts

> Remember, **Oxidation numbers**[1] were covered in Topic 4.3, Redox.

Answer to part (e)

$$H_2SO_4 + HI \rightarrow H_2S + I_2 + H_2O$$

+6   −2   oxidation number change: −8

      −1   0   oxidation number change: +1

Balance ONLY the species that contain the elements that have changed oxidation number.

There is one sulfur atom on both sides with an oxidation number change of −8 (+6 to −2)

To match the decrease of −8 for sulfur, you need 8 atoms of I on both sides for a total increase of +8.

So HI needs to be multiplied by 8 and I_2 needs to be multiplied by 4.

$$H_2SO_4 + 8HI \rightarrow H_2S + 4I_2 + H_2O$$

Balance any remaining atoms

As the oxidation numbers are now balanced, the hydrogen and oxygen atoms are balanced by including a 4 in front of the H_2O.

$$H_2SO_4 + 8HI \rightarrow H_2S + 4I_2 + 4H_2O$$

> **Balanced equations**[5] were covered in Topic 2.3, Formulae and equations

f Concentrated sulfuric acid reacts with benzene in an electrophilic substitution reaction

$$C_6H_6 + H_2SO_4 \rightarrow C_6H_5SO_3H + H_2O$$

In the reaction, two molecules of sulfuric acid first react to form the SO_3H^+ electrophile.
The SO_3H^+ electrophile then reacts with benzene to form the sulfonic acid.
Outline a mechanism for this reaction, including an equation for formation of the SO_3H^+ electrophile. *(5 marks)*

> Although this reaction is not in the specification, the equations and mechanism are virtually identical as for the nitration of benzene. H_2SO_4 and SO_3H^+ replaces HNO_3 and NO_2^+. The **nitration of benzene**[4] was covered in covered in Topic 25.2, Electrophilic substation reactions of benzene.

Answer to part (f)

$$2H_2SO_4 \rightarrow SO_3H^+ + HSO_4^- + H_2O \checkmark$$

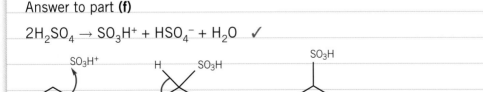

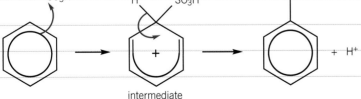

Curly arrow ✓   Intermediate ✓   Curly arrow ✓   Formation of H^+ ✓

Analysing and answering a synoptic question

g Concentrated sulfuric acid dehydrates many organic compounds, forming water as one of the products. For example, sulfuric acid dehydrates alcohols by eliminating water to form an alkene and water. Three examples of sulfuric acid dehydrating organic compounds are shown below.
- Sulfuric acid dehydrates methanoic acid to form a gas, **A**, with the same molar mass as ethene.
- Sulfuric acid dehydrates cyclopentanol to form a liquid, **B** (M_r = 68.0)
- Sulfuric acid dehydrates ethane-1,2-diol to form a compound **C** (M_r = 88.0)

Suggest the identity of **A**, **B** and **C**. Write equations for each reaction, using structural formulae for organic compounds.

(6 marks)

Answer to part (g)

M_r of ethane, C_2H_4 = 24 + 4 = 28

A: CO (M_r = 12 + 16 = 28) ✓

HCOOH → CO + H_2O ✓

B = (cyclopentene)   ✓   (M_r C_5H_8 = 12 × 5 + 1 × 8 = 68)

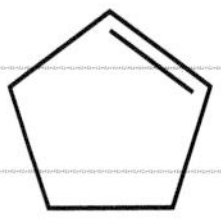

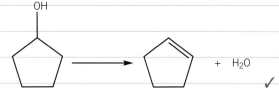

+ H_2O ✓

M_r of ethane-1,2-diol, $HOCH_2CH_2OH$ = 62

62 × 2 − 88 = 36 which suggests that two $HOCH_2CH_2OH$ molecules react together with elimination of 2H_2O molecules.

2$HOCH_2CH_2OH$ → $C_4H_8O_2$ + 2H_2O ✓

A possible structure for structure **C** is

 ✓

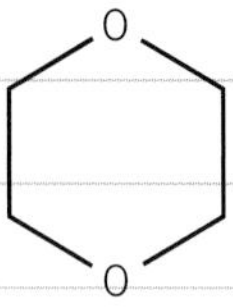

Practice questions

1 Amines and amides are organic bases derived from ammonia. The structure of propylamine is shown below.

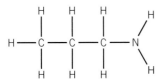

a 1 mol of propylamine reacts with 1 mol phosphoric(V) acid, H_3PO_4, to form a salt.

(i) Write the equation for this reaction. *(1 mark)*

(ii) State and explain how the bond angle around the nitrogen atom changes during this reaction. *(4 marks)*

b When propylamine dissolves in water, it forms a weakly alkaline solution as shown in the following equilibrium.

$$CH_3CH_2CH_2NH_2 + H_2O \rightleftharpoons CH_3CH_2CH_2NH_3^+ + OH^-$$

This equilibrium can be expressed as a base dissociation constant K_b given by the expression below.

$$K_b = \frac{[CH_3CH_2CH_2NH_3^+][OH^-]}{[CH_3CH_2CH_2NH_2]}$$

Phenylamine has a pK_b value of 3.32. Calculate the pH of a solution of 0.0200 mol dm^{-3} propylamine. *(4 marks)*

c Propylamine reacts with platinum(II) iodide, PtI_2, to form a mixture of compounds. Two of these compounds are stereoisomers with four-coordinate geometry.

Draw and label the structures of these two stereoisomers. *(3 marks)*

d The amide, N-phenylbenzamide, can be synthesised from benzoic acid and benzene by a multi-stage synthesis.

Complete the flow chart for this synthesis.
- In the boxes, draw the structures of organic compounds formed in the synthesis.
- Give the reagents required for each step. *(6 marks)*

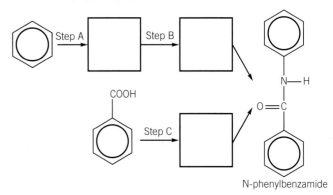

N-phenylbenzamide

(6 marks)

2 a Compound **A** is a straight-chain compound containing two different functional groups. Compound **A** forms an orange crystalline derivative with 2,4-dinitrophenylhydrazine.

Compound **A** is completely oxidised by acidified potassium dichromate(VI) to form compound **B**.

Compound **B** reacts with hydrogen bromide forming compound **C**.

0.203 g of compound **B** reacts completely with 35.0 cm^3 of 0.100 mol dm^{-3} NaOH. In this reaction 1 mol of compound **B** reacts with 2 mol of NaOH.

Compound **B** can form both a condensation polymer with ethane-1,2-diol as well as forming an addition polymer.

(i) Identify compounds **A**, **B**, and **C**.
Using all of the information in the question, explain your reasoning. *(9 marks)*

(ii) Draw one repeat unit of the condensation polymer formed between compound **B** and ethane-1,2-diol. *(1 mark)*

(iii) Draw two repeat units of the addition polymer formed from compound **B**. *(1 mark)*

(iv) Compounds **B** and **C** both show stereoisomerism.
Draw and label the stereoisomers of **B** and **C**. *(4 marks)*

b The acid hydrolysis of compound **D** produces compounds **E**, **F**, and **G** in a molar ratio of 1:1:1.

Compound **E** contains a chiral centre and has the following percentage composition by mass: C, 40.5%; H, 7.9%; N, 15.7%; O, 35.9%.

Compound **F** is an alcohol with an enthalpy change of combustion of -2006 kJ mol^{-1}. Complete combustion of 0.750 g of compound **F** raised the temperature of 500 g of water by 12.0°C. Compound **F** is oxidised and the organic product is distilled off as it forms. This product does not react with Tollens' reagent.

Compound **G**, reacts with aqueous sodium carbonate, forming carbon dioxide gas. In its mass spectrum, compound **G** has a molecular ion peak at $m/z = 88$. The ^{13}C NMR and ^{1}H NMR spectra of compound **G** are shown below.

^{13}C NMR spectrum

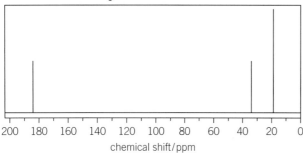

^{1}H NMR spectrum

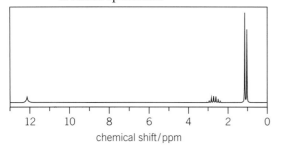

- Identify compounds **E**, **F**, and **G**.
- Draw a full displayed formula for compound **D**. *(12 marks)*

3 a When hydrogen gas is passed over heated sodium, a grey solid, **A**, is formed, which conducts electricity when molten.

Solid **A** reacts with water to form a colourless gas, **B** and an alkaline solution **C**.

Solid **A**, reacts with BCl$_3$ to produce a reducing agent **D** and one other product.

Compound **D** can be used to reduce carbonyl compounds.

  (i) Identify compounds **A**, **B**, **C**, and **D**. *(4 marks)*

  (ii) Write an equation for the reaction of **A** with BCl$_3$. *(1 mark)*

b Trichloroethanal, Cl$_3$CCHO, reacts with cyanide ions, :CN$^-$, in the presence of dilute aqueous acid.

  (i) What type of reaction is taking place? *(1 mark)*

  (ii) Outline the mechanism for this reaction. *(4 marks)*

c Hydrogen reacts with halogens to produce hydrogen halides. The table below shows the boiling points of the hydrogen halides and the bond enthalpies of the hydrogen–halide bonds.

Hydrogen halide	HF	HCl	HBr	HI
Boiling point / °C	20	−85	−67	35
Bond enthalpy / kJ mol^{-1}	568	432	366	298

  (i) Explain the trend in boiling points from HCl to HI and suggest why the boiling point of HF does not fit in with this trend. *(4 marks)*

  (ii) The hydrogen halides all dissolve in water to form strongly acidic solutions.

  Identify the strongest acid and explain your answer. *(2 marks)*

4 Read the section below and use your knowledge and understanding of chemistry to answer the questions that follow.

Compound **X** is a dicarboxylic acid, $M_r = 90$, present in many foods such as cocoa, green vegetables and rhubarb. Compound **X** may be harmful to humans because it reacts with calcium ions in the bloodstream to form solid particles. These are then deposited in the kidneys as kidney stones.

Analysis of a kidney stone showed that it was made up almost entirely of a single compound **Y** with the percentage composition by mass:
Ca, 31.3%; C, 18.7%; O, 50.0%.

Some people develop kidney stones more readily than others. Compound **X** is very soluble in water and people susceptible to kidney stones are advised to drink large quantities of water to flush this compound from their bodies.

  a Calculate the empirical formula of compound **Y**, the main compound in kidney stones. *(2 marks)*

  b A kidney stone with a mass of 2.0 g was removed from a patient.
  Calculate the number of calcium ions that have been removed from the bloodstream to form this kidney stone. *(3 marks)*

  c Deduce the molecular formula of compound **X** and suggest its structural formula. *(2 marks)*

  d Explain why the carboxylic acid compound **X** is very soluble in water. *(2 marks)*

  e A student carries out a titration to determine the concentration of a solution of compound **X**. The student adds 25.0 cm³ of the solution of **X** to a conical flask and adds dilute sulfuric acid. This mixture is titrated against 0.0200 mol dm⁻³ potassium manganate(VII), $KMnO_4$. 14.25 cm³ this $KMnO_4$ solution was added before the solution remained permanent pink.
  Calculate the concentration of the solution of compound **X**. *(3 marks)*

The following equations may be useful.
$$MnO_4^-(aq) + 8H^+(aq) + 5e^- \rightarrow Mn^{2+}(aq) + 4H_2O(l)$$
$$C_2O_4^{2-}(aq) \rightarrow 2CO_2(g) + 2e^-$$

5 A student reacts 4.23 g of phenol with nitric acid and sulfuric acid at 100 °C to form a yellow crystalline solid, compound **B** with the following percentage composition by mass:
C, 31.44%; H, 1.31%; N, 18.34%; O, 48.91%.

The percentage yield of compound **B** was 65.8%.

The mass spectrum and ¹³C NMR spectrum of compound **B** are shown below.

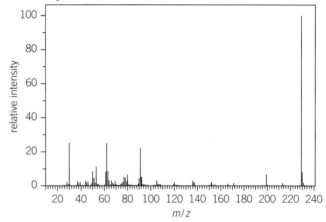

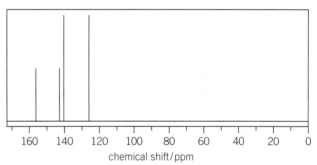

  a Calculate the mass of compound **B** prepared by the student.
  Give your answer to **three** significant figures. *(2 marks)*

  b Determine the molecular formula of compound **B**.
  Show your working. *(2 marks)*

  c Suggest a possible structure for compound **B**.
  Explain your reasoning *(3 marks)*

Practice questions

6 This question looks at different synthetic routes in organic chemistry.

Each synthesis requires more than one step.

In the formulae for synthesis 2, C_6H_5 = phenyl group.

Synthesis 1 Synthesis of $CH_3CH_2NH_2$ starting from $CH_2=CH_2$

Synthesis 2 Synthesis of $C_6H_5COOCH(CH_3)_2$ starting from $C_6H_5CH_2OH$

Synthesis 3 Synthesis of CH_3CH_2COOH starting from CH_3CH_2Br

For each synthesis, describe the steps required, show the structural formulae for any organic intermediate compounds and include details of reagents and conditions. *(9 marks)*

7 The statue of liberty consists of an iron frame covered in 28 123 kg of copper plates.

Although the statue would be expected to be copper-coloured, it is actually green due to weathering. The copper metal has reacted with oxygen, carbon dioxide and water in the atmosphere. This green colour is an approximate 1:1 mixture of copper(II) carbonate and copper(II) hydroxide.

When copper(II) carbonate is heated, it decomposes to form a black solid, copper(II) oxide and a colourless gas which has a density of 1.83×10^{-3} g cm^{-3} at room temperature and pressure. The black solid is reacted with dilute hydrochloric acid to form a green solution. Copper(II) hydroxide reacts with dilute nitric acid to produce a blue solution.

a Calculate the number of copper atoms in the plates covering the statue of liberty. *(2 marks)*

b Write equations for the reactions of copper and copper compounds described in the passage above. *(4 marks)*

c (i) The blue solution formed from copper(II) hydroxide and nitric acid contains a complex ion. What is the formula of the complex ion? *(1 mark)*

(ii) Suggest why the solution formed from copper(II) hydroxide and hydrochloric acid is green. *(1 mark)*

d Use the data below to calculate the lattice enthalpy of copper(II) oxide. *(1 mark)*

Enthalpy change	kJ mol^{-1}
Atomisation of copper	+339
First ionisation energy of copper	+745
Second ionisation energy of copper	+1960
Atomisation of oxygen	+248
First electron affinity of oxygen	−141
Second electron affinity of oxygen	+791
Formation of copper(II) oxide	−155

e The standard electrode potentials and enthalpy changes of reactions for copper(II) and zinc ions are given in the table below.

	$E^\ominus$/V	$\Delta H^\ominus$/kJ mol^{-1}
$Zn^{2+}(aq) + 2e^- \rightleftharpoons Zn(s)$	−0.76	+152.4
$Cu^{2+}(aq) + 2e^- \rightleftharpoons Cu(s)$	+0.34	−64.4

Copper metal is frequently used with zinc metal in electrochemical cells.

(i) Write the equation for the cell reaction taking place in a zinc/copper cell and calculate the standard cell potential. *(2 marks)*

(ii) How will the following changes influence the cell potential?
- Increasing the concentration of the Zn^{2+} ions.
- Increasing the temperature of the Cu^{2+}/Cu half cell. *(4 marks)*

f The standard entropies of $Cu(s)$, $Cu^{2+}(g)$, $Zn(s)$ and $Zn^{2+}(aq)$ are given in the table below.

Substance	Cu	Cu^{2+}	Zn	Zn^{2+}
S / J K^{-1} mol^{-1}	33.2	−99.6	41.6	−112.1

(i) Calculate the standard entropy change for the reaction given in **(e)(i)**. *(1 mark)*

(ii) Show that the reaction is feasible at 25 °C. *(3 marks)*

8 Compound **A** was analysed in the laboratory and was shown to have the percentage composition by mass: Na, 21.60%; Cl, 33.33%; O, 45.07%.

On gentle heating, compound **A** formed sodium chlorate(VII), $NaClO_4$, and compound **B** in a 3:1 molar ratio.

On strong heating, 0.2205 g $NaClO_4$ was broken down into compound **B** and oxygen gas.

An aqueous solution of compound **B** formed a white precipitate, C, with aqueous silver nitrate.

  a Identify substances **A–C**. Show your working *(5 marks)*

  b Write balanced equations for all reactions that took place. *(3 marks)*

  c Calculate the mass of **B** formed from 0.2205 g of $NaClO_4$, and the volume of oxygen formed at RTP *(4 marks)*

9 A student carried out an investigation to identify a volatile liquid using a gas syringe.

A sample of a volatile liquid was drawn up into a hypodermic needle and small syringe. The mass of the syringe and liquid was recorded.

The gas syringe is placed inside a heating jacket and the temperature of the apparatus is allowed to reach a constant 100 °C. The volatile liquid is injected into the gas syringe.

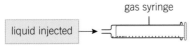

The empty hypodermic syringe is reweighed.

The liquid vaporises and the final volume of gas inside the syringe is recorded. The pressure is also recorded.

The following results were obtained.

Mass of hypodermic syringe and liquid	45.369 g
Mass of hypodermic syringe	45.220 g
Volume of air in gas syringe at start	4 cm³
Volume in gas syringe at end	81 cm³
Temperature of heating jacket	100 °C
Pressure	100 kPa

  a Calculate the molar mass of the volatile liquid. *(4 marks)*

  b Describe how you could use spectroscopy to confirm the molar mass of the volatile liquid. *(2 marks)*

  c A sample of the volatile liquid was heated with acidified potassium dichromate(VI). The colour changed from orange to green and a neutral organic product was formed.

When heated with concentrated sulfuric acid, the volatile liquid formed a compound, which decolourised bromine water.

Suggest a structural formula for the volatile liquid and write equations for the two reactions above. *(3 marks)*

Glossary

acid A species that releases H⁺ ions in aqueous solution.

acid dissociation constant K_a The equilibrium constant that shows the extent of dissociation of a weak acid.
For a weak acid HA(aq), $K_a = \dfrac{[H^+(aq)][A^-(aq)]}{[HA(aq)]}$

acid–base pair A pair of two species that transform into each other by gain or loss of a proton.

activation energy The minimum energy required to start a reaction by the breaking of bonds.

actual yield The amount of product obtained from a reaction.

addition polymerisation Formation of a very long molecular chain, by repeated addition reactions of many unsaturated alkene molecules (monomers).

addition reaction A reaction in which two reactants join together to form one product.

adsorption The process that occurs when a gas or liquid or solute is held to the surface of a solid.

alicyclic Containing carbon atoms joined together in a ring that is not aromatic.

aliphatic Containing carbon atoms joined together in straight or branched chains.

alkali A type of base that dissolves in water forming hydroxide ions, OH⁻(aq) ions.

alkanes The hydrocarbon homologous series with single carbon-to-carbon bonds and the general formula: C_nH_{2n+2}.

alkenes The hydrocarbon homologous series with at least one double carbon-to-carbon bond.

alkyl group A side chain formed by removing a hydrogen atom removed from an alkane parent chain, for example, CH_3, C_2H_5; any alkyl group is often shown as R.

alkynes The hydrocarbon homologous series with at least one triple carbon-to-carbon bond.

amount of substance The quantity whose unit of the mole, used as a means of counting any species such as atoms, ions and molecules.

anhydrous Containing no water molecules.

anion A negatively charged ion with more electrons than protons.

aromatic Containing one or more benzene rings.

atom economy (Sum of molar masses of desired products)/(sum of molar masses of all products) Å~100%.

atomic (proton) number The number of protons in the nucleus of an atom.

atomic orbital A region around the nucleus that can hold up to two electrons, with opposite spins.

average bond enthalpy The average enthalpy change that takes place when breaking by homolytic fission 1 mol of a given type of bond in the molecules of a gaseous species.

Avogadro constant N_A The number of atoms per mole of the carbon-12 isotope. (6.02×10^{23} mol⁻¹)

base A compound that neutralises an acid to form a salt

binary compound A compound containing two elements only

bond angle The angle between two bonds at an atom

bond dissociation enthalpy The enthalpy change that takes place when breaking by homolytic fission 1 mol of a given bond in the molecules of a gaseous species.

bonded pair A pair of electrons shared between two atoms to make a covalent bond

Brønsted-Lowry acid A species that is a proton, H⁺, donor.

Brønsted-Lowry base A species that is a proton, H⁺, acceptor.

buffer solution A system that minimises pH changes on addition of small amounts of an acid or a base.

carbocation An ion that contains a positively charged carbon atom.

catalyst A substance that increases the rate of a chemical reaction without being used up in the process; a catalyst provide an alternative route for the reaction with lower activation energy.

cation A positively charged ion with fewer electrons than protons.

chain reaction A reaction in which the propagation steps release new radicals that continue the reaction.

chemical shift δ A scale, in ppm, that compares the frequency of an NMR absorption with the frequency of the reference TMS at $\delta = 0$ ppm.

chiral carbon A carbon atom attached to four different atoms or groups of atoms.

chromatogram A visible record showing the result of separation of the components of a mixture by chromatography.

***cis–trans* isomerism** A special type of *E/Z* isomerism in which there are two non-hydrogen groups and two hydrogen atoms around the C=C double bond: the *cis* isomer (*Z* isomer) has H atoms on each carbon on the same side; the *trans* isomer (*E* isomer) has H atoms on each carbon on different sides.

closed system A system isolated from its surroundings.

collision theory Two reacting particles must collide for a reaction to occur, and must be in the correct orientation and have sufficient energy to overcome the activation energy of the reaction.

complex ion A transition metal ion bonded to ligands by coordinate bonds (dative covalent bonds).

concentration The amount of solute, in moles, dissolved in $1 dm^3$ ($1000 cm^3$) of solution.

condensation reaction A reaction in which two small molecules react together to form a larger molecule with elimination of a small molecule such as water.

conjugate acid A species that releases a proton to form a conjugate base.

conjugate base A species that accepts a proton to form a conjugate acid.

coordinate bond A shared pair of electrons in which the bonded pair has been provided by one of the bonding atoms only; also called a dative covalent bond.

coordination number The total number of coordinate bonds formed between a central metal ion and ligands.

covalent bond The strong electrostatic attraction between a shared pair of electrons and the nuclei of the bonded atoms.

dative covalent A shared pair of electrons in which the bonded pair has been provided by one of the bonding atoms only; also called a coordinate bond.

dehydration An elimination reaction in which water is removed from a saturated molecule to make an unsaturated molecule.

delocalised electrons Electrons that are shared between more than two atoms.

desorption Release of an adsorbed substance from a surface.

dipole A separation in electrical charge so that one atom of a polar covalent bond, or one end of a polar molecule, has a small positive charge $\delta+$ and the other has a small negative charge $\delta-$.

displacement reaction A reaction in which a more reactive element displaces a less reactive element from an aqueous solution of its ions.

displayed formula A formula showing the relative positioning of all the atoms in a molecule and the bonds between them.

disproportionation A redox reaction in which the same element is both oxidised and reduced.

dynamic equilibrium The equilibrium that exists in a closed system when the rate of the forward reaction is equal to the rate of the reverse reaction and concentrations do not change.

***E/Z* isomerism** A type of stereoisomerism in which different groups attached to each carbon of a C=C double bond may be arranged differently in space because of the restricted rotation of the C=C bond.

electron configuration A shorthand representation that shows how electrons occupy sub-shells in an atom.

electronegativity A measure of the attraction of a bonded atom for the pair of electrons in a covalent bond.

electrophile An atom (or group of atoms) which is attracted to an electron-rich centre or atom, where it accepts a pair of electrons to form a new covalent bond.

electrophilic addition An addition reaction in which the first step is attack by an electrophile on a region of high electron density.

electrophilic substitution A type of substitution reaction in which an electrophile is attracted to an electron-rich centre or atom, where it accepts a pair of electrons to form a new covalent bond.

elimination reaction The removal of a molecule from a saturated molecule to make an unsaturated molecule.

empirical formula The formula that shows the simplest whole-number ratio of atoms of each element present in a compound.

Glossary

enantiomers Stereoisomers that are non-superimposable mirror images of each other; also called optical isomers.

end point The point in a titration where the indicator changes colour; the end point indicates when the reaction is just complete.

endothermic reaction A reaction in which the enthalpy of the products is greater than the enthalpy of the reactants, resulting in heat being taken in from the surroundings (ΔH is positive).

enthalpy H The heat content that is stored in a chemical system.

enthalpy change ΔH The difference between the enthalpy of the products and the enthalpy of the reactants.

enthalpy cycle A diagram showing alternative routes between reactants and products which allows the indirect determination of an enthalpy change from other known enthalpy changes using Hess's law.

enthalpy profile diagram A diagram for a reaction to compares the enthalpy of the reactants with the enthalpy of the products.

entropy The used for the dispersal of energy and disorder within the chemicals making up the chemical system.

equilibrium constant K_c A measure of the position of equilibrium; the magnitude of an equilibrium constant indicates whether there are more reactants or more products in an equilibrium system.

equivalence point The point in a titration at which the volume of one solution has reacted exactly with the volume of the second solution.

esterification A reaction in which a carboxylic acid reacts with an alcohol to form an ester and water.

exothermic reaction A reaction in which the enthalpy of the products is smaller than the enthalpy of the reactants, resulting in heat loss to the surroundings (ΔH is negative).

fingerprint region An area of an infrared spectrum below 1500 cm^{-1} that gives a characteristic pattern for different compounds.

first electron affinity The enthalpy change that takes place when one electron is added to each atom in one mole of gaseous atoms to form one mole of gaseous 1− ions.

first ionisation energy The energy required to remove one electron from each atom in one mole of gaseous atoms of an element to form one mole of gaseous 1+ ions.

fractional distillation The separation of components in a liquid mixture by their different boiling points into fractions with different compositions.

fragment ions Ions formed from the breakdown of the molecular ion in a mass spectrometer.

fragmentation The process in mass spectrometry that causes a positive ion to spilt into smaller pieces, one of which is a positive fragment ion.

free energy change ΔG The balance between enthalpy, entropy and temperature for a process given by $\Delta G = \Delta H - T\Delta S$. A process is feasible when $\Delta G < 0$.

functional group The part of the organic molecule responsible for its chemical reactions.

general formula The simplest algebraic formula of a member of a homologous series. For example, the general formula of the alkanes is C_nH_{2n+2}.

giant covalent lattice A three-dimensional structure of atoms, bonded together by strong covalent bonds.

giant ionic lattice A three-dimensional structure of oppositely charged ions, bonded together by strong ionic bonds.

giant metallic lattice A three-dimensional structure of positive ions and delocalised electrons, bonded together by strong metallic bonds.

group A vertical column in the periodic table. Elements in a group have similar chemical properties and their atoms have the same number of outer shell electrons.

half-life The time taken for the concentration of a reactant to decrease by half.

hess's Law If a reaction can take place by more than one route and the initial and final conditions are the same, the total enthalpy change is the same for each route.

heterogeneous catalysis A reaction in which the catalyst has a different physical state from the reactants; frequently reactants are gases whilst the catalyst is a solid.

heterogeneous equilibrium An equilibrium in which the species making up the reactants and products have different physical states.

heterolytic fission The breaking of a covalent bond forming a cation (positive ion) and an anion (negative ion).

homogeneous catalysis A reaction in which the catalyst and reactants are in the same physical state, which is most frequently the aqueous or gaseous state.

homogeneous equilibrium An equilibrium in which all the species making up the reactants and products have the same physical state.

homologous series A series of organic compounds with the same functional group but with each successive member differing by CH_2.

homolytic fission The breaking of a covalent bond with one of the bonded electrons going to each atom, forming two radicals.

hydrated A crystalline compound containing water molecules.

hydrocarbon A compound of hydrogen and carbon only.

hydrogen bond A strong dipole-dipole attraction between an electron-deficient hydrogen atom of –NH, –OH or HF on one molecule and a lone pair of electrons on a highly electronegative atom containing N, O or F on a different molecule.

hydrolysis A reaction with water that breaks a chemical compound into two compounds, the H and OH in a water molecule becomes incorporated into the two compounds.

induced dipole–dipole interaction attractive forces between induced dipoles in different molecules – also called London Forces.

infrared spectroscopy An instrumentation method of analysis that identifies bonds from absorption of the infrared radiation of different wavelengths.

initial rate of reaction The change in concentration of a reactant or product per unit time at the start of the reaction: $t = 0$.

initiation The first stage in a radical reaction in which radicals starts when a covalent bond is broken by homolytic fission of a covalent bond.

intermediate A species formed during a reaction that reacts further and is not present in the final products.

intermolecular force An attractive force between molecules. Intermolecular forces can be London forces, permanent dipole-dipole interactions or hydrogen bonding.

ion A positively or negatively charged atom or a (covalently bonded) group of atoms (a polyatomic ion), where the number of electrons is different from the number of protons.

ionic bonding ionic bonding The electrostatic attraction between positive and negative ions.

ionic product of water K_w The product of the ions formed in the partial dissociation of water, given by $K_w = [H^+(aq)] [OH^-(aq)]$.

isotopes Atoms of the same element with different numbers of neutrons and different masses.

lattice enthalpy The enthalpy change that accompanies the formation of one mole of an ionic compound from its gaseous ions under standard conditions.

Le Chatelier's principle When a system in dynamic equilibrium is subjected to a external change, the system readjusts itself to minimise the effect of the change and to restore equilibrium.

ligand A molecule or ion that can donate a pair of electrons to the transition metal ion.

ligand substitution A reaction in which one or more ligands in a complex ion are replaced by different ligands.

limiting reagent The reactant that is not in excess, which will be used up first and stop the reaction.

lone pair An outer shell pair of electrons that is not involved in chemical bonding.

mass number The sum of the number of protons and neutrons in the nucleus – also known as nucleon number.

metallic bond The electrostatic attraction between positive metal ions and delocalised electrons.

mobile phase The phase that moves in chromatography.

molar gas volume V_m The volume per mole of gas molecules at a stated temperature and pressure

molar mass M The mass per mole of a substance, in units of g mol^{-1}.

Glossary

mole The amount of any substance containing as many elementary particles as there are carbon atoms in exactly 12 g of the carbon-12 isotope, that is, 6.02×10^{23} particles.

molecular formula molecular formula A formula that shows the number and type of atoms of each element present in a molecule.

molecular ion The positive ion formed in mass spectrometry when a molecule loses an electron.

molecule The smallest part of a covalent compound that can exist while retaining its chemical identity, consisting of two or more atoms covalently bonded together.

monomer A small molecule that combines with many other monomers to form a polymer.

neutralisation A chemical reaction in which an acid and a base react together to produce a salt.

nomenclature A system of naming compounds.

non-polar With no charge separation across a bond or in a molecule.

nucleophile An atom (or group of atoms) which is attracted to an electron-deficient centre or atom, where it donates a pair of electrons to form a new covalent bond.

nucleophilic substitution A reaction in which a nucleophile is attracted to an electron-deficient carbon atom, and replaces an atom or group of atoms on the carbon atom.

optical isomers Stereoisomers that are non-superimposable mirror images of each other; also called 'enantiomers'.

order The power to which the concentration of a reactant is raised in the rate equation.

overall order The sum of the individual orders of reactants in the rate equation: $m + n$.

oxidation Loss of electrons or an increase in oxidation number.

oxidation number A measure of the number of electrons that an atom uses to bond with atoms of another element. Oxidation numbers are derived form a set of rules.

oxidation state The oxidation number.

oxidising agent A reagent that oxidises (takes electrons from) another species.

π-bond A bond formed by the sideways overlap of two p-orbitals, with the electron density above and the plane of the bonding atoms.

partial dissociation The splitting of some of a species in solution into aqueous ions.

Pauling electronegativity value A value assigned as a measure of the relative attraction of a bonded atom for the pair of electrons in a covalent bond.

percentage yield % yield
$$= \frac{\text{actual amount, in mol, of product}}{\text{theoretical amount, in mol, of product}} \times 100$$

period A horizontal row of elements in the periodic table. Elements show trend in properties across a period.

periodicity A repeating trend in properties of the elements across each period of the periodic table.

permanent dipole A small charge difference that does not change across a bond, with δ+ and δ– partial changes on the bonded atoms: the result of the bonded atoms having different electrongativities.

Permanant dipole–dipole interaction An attractive force between permanent dipoles in neighbouring polar molecules.

pH The expression, pH = –log[H⁺(aq)].

polar (*molecule*) With δ+ and δ– charges at different ends of the molecule.

polar covalent bond A bond with a permanent dipole, having δ+ and δ– partial changes on the bonded atoms.

polar molecule A molecule with an overall dipole, having taken into account any dipoles across bonds and the shape of the molecule.

polyatomic ion An ion containing more than one atom.

polymer An large molecule formed from many thousands of repeat units of smaller molecules known as monomers.

position of equilibrium The relative quantities of reactants and products, indicating the extent of a reversible reaction at equilibrium.

precipitation reaction The formation of a solid from a solution during a chemical reaction. Precipitates are often formed when two aqueous solutions are mixed together.

primary On a carbon atom at the end of a chain.

Glossary

primary alcohol An alcohol in which the OH group is attached to a carbon atom that is attached to two or three hydrogen atoms.

principal quantum number n A number representing the relative overall energy of each orbital, which increases with distance from the nucleus. The sets of orbitals with the same n-value are referred to as electron shells or energy levels.

propagation The steps that continue a free radical reaction, in which a radical reacts with a reactant molecule to form a new molecule and another radical, causing a chain reaction.

proton number The number of protons in the nucleus of an atom; also known as atomic number.

radical A species with an unpaired electron

rate constant k The constant that links the rate of reaction with the concentrations of the reactants raised to the powers of their orders in the rate equation.

rate equation For a reaction: $A + B \rightarrow C$ with orders m for A and n for B, the rate equation is given by: rate = $k[A]^m[B]^n$

rate of reaction The change in concentration of a reactant or a product in a given time.

rate-determining step The slowest step in the reaction mechanism of a multi-step reaction.

reaction mechanism The sequence of bond breaking and bond-forming steps that shows the path taken by electrons during a reaction.

redox reaction A reaction involving reduction and oxidation.

reducing agent A reagent that reduces (adds electron to) another species.

reduction Gain of electrons or a decrease in oxidation number.

reflux The continual boiling and condensing of a reaction mixture back to the original container to ensure that the reaction takes place without the contents of the flask boiling dry.

relative atomic mass A_r The weighted mean mass of an atom of an element compared with one-twelfth of the mass of an atom of carbon-12.

relative formula mass The weighted mean mass of the formula unit of a compound compared with one-twelfth of the mass of an atom of carbon-12.

relative isotopic mass The mass of an atom of an isotope compared with one-twelfth of the mass of an atom of carbon-12.

relative molecular mass M_r The weighted mean mass of a molecule of a compound compared with one-twelfth of the mass of an atom of carbon-12.

repeat unit A specific arrangement of atoms that occurs in the structure over and over again. Repeat units are included in brackets outside of which is the symbol n.

retention time In gas chromatography, The time for a component to pass from the column inlet to the detector.

reversible reaction A reaction that takes place in both forward and reverse directions.

R_f value $R_f = \dfrac{\text{distance moved by component}}{\text{distance moved by solvent}}$

σ-bond A bond formed by the overlap of one orbital from each bonding atom, consisting of two electrons and with the electron density centred around a line directly between the nuclei of the two atoms.

salt The product of a reaction in which the H^- ions from the acid are replaced by metal or ammonium ions.

saturated Containing single bonds only.

saturated hydrocarbon A hydrocarbon with single bonds only.

second electron affinity The enthalpy change that takes place when one electron is added to each ion in one mole of gaseous 1– ions to form one mole of gaseous 2– ions.

second ionisation energy The energy required to remove one electron from each ion in one mole of gaseous 1+ ions of an element to form one mole of gaseous 2+ ions.

secondary On a carbon atom to which two carbon chains are attached.

secondary alcohol An alcohol in which the –OH group is attached to a carbon atom that is attached to two carbon chains and one hydrogen atom.

shell A group of atomic orbitals with the same principal quantum number, n. Also known as a main energy level.

shielding effect The repulsion between electrons in different inner shells. Shielding reduces the net attractive force between the positive nucleus on the outer shell electrons.

Glossary

simple molecular lattice A three-dimensional structure of molecules, bonded together by weak intermolecular forces.

skeletal formula A simplified organic formula, with hydrogen atoms removed from alkyl chains, leaving just a carbon skeleton and associated functional groups.

specific heat capacity c The energy required to raise the temperature of 1 g of a substance by 1 °C.

spectator ions Ions that are present but take no part in a chemical reaction.

spin-spin coupling In an NMR spectrum, the interaction between spin states of non-equivalent nuclei that results in the splitting of a signal.

standard conditions A pressure of 100 kPa, a stated temperature, usually 298K (25 °C) and a concentration of 1 mol dm^{-3} (for reactions with aqueous solutions).

standard electrode potential $E^\ominus$ The e.m.f. of a half-cell compared with a standard hydrogen half-cell, measured at 298 K with solution concentrations of 1 mol dm^{-3} and a gas pressure of 100 kPa.

standard enthalpy change of atomisation The enthalpy change that takes place when one mole of gaseous atoms forms from the element in its standard state.

standard enthalpy change of combustion $\Delta_c H^\ominus$ The enthalpy change that takes place when one mole of a substance reacts completely with oxygen under standard conditions, all reactants and products being in their standard states.

standard enthalpy change of formation $\Delta_f H^\ominus$ The enthalpy change that takes place when one mole of a compound is formed from its constituent elements in their standard states under standard conditions.

standard enthalpy change of hydration The enthalpy change that takes place when one mole of isolated gaseous ions is dissolved in water forming one mole of aqueous ions under standard conditions.

standard enthalpy change of neutralisation $\Delta_{neut} H^\ominus$ The enthalpy change that accompanies the reaction of an acid by a base to form one mole of H$_2$O(l), under standard conditions, with all reactants and products in their standard states.

standard enthalpy change of reaction $\Delta_r H^\ominus$ The enthalpy change that accompanies a reaction in the molar quantities expressed in a chemical equation under standard conditions, all reactants and products being in their standard states.

standard enthalpy change of solution The enthalpy change that takes place when one mole of a compound is completely dissolved in water under standard conditions.

standard solution A solution of known concentration.

standard state The physical state of a substance under standard conditions of 100 kPa and a stated temperature (usually 298 K).

stationary phase The phase that does not moves in chromatography.

stereoisomers Compounds with the same structural formula but with a different arrangement of the atoms in space.

stoichiometry The ratio of the amount, in moles, of each substance in a chemical equation (essentially the ratio of the balancing numbers).

strong acid An acid that dissociates completely in solution.

structural formula A formula showing the minimal detail for the arrangement of atoms in a molecule.

structural isomers Molecules with the same molecular formula but with different structural formulae.

sub-shell A group of orbitals of the same type within a shell.

substitution reaction A reaction in which an atom or group of atoms is replaced with a different atom or group of atoms.

surroundings Everything that is not the chemical system.

system The chemicals involved in the reaction.

termination The step at the end of a radical substitution when two radicals combine to form a molecule.

tertiary alcohol An alcohol in which the –OH group is attached to a carbon atom that is attached to three carbon atoms and no hydrogen atoms.

theoretical yield The yield resulting from complete conversion of reactants into products.

Glossary

thermal decomposition The breaking up of a chemical substance with heat into at least two chemical substances.

titre The volume added from the burette when the volume of one solution has exactly reacted with eth other solution.

transition element A d-block element which forms an ion with an incomplete d-sub-shell.

unsaturated containing a multiple carbon-carbon bond.

volatility The ease at which a liquid turns into a gas. Volatility increases as boiling point decreases.

water of crystallisation Water molecules that are bonded into a crystalline structure of a compound.

weak acid An acid that dissociates only partially in solution.

Answers

2.1

> **Heavy water**
>
> i 18   ii 20   iii 22

Melting point and boiling point would be higher; density would be greater.

1. **a** 6 p$^+$, 6 n, 6 e$^-$ **b** 6 p$^+$, 7 n, 6 e$^-$
   c 6 p$^+$, 8 n, 6 e$^-$
2. **a** Nuclei contain different numbers of neutrons.
   b The nuclei contain the same number of protons.
3. **a** 7 p$^+$, 8 n, 7 e$^-$ **b** 47 p$^+$, 62 n, 47 e$^-$
   c 82 p$^+$, 125 n, 82 e$^-$
4. **a** 19 p$^+$, 22 n, 18 e$^-$ **b** 16 p$^+$, 18 n, 18 e$^-$
   c 24 p$^+$, 29 n, 21 e$^-$
5. **a** ^{7}Li has one more neutron
   b ^{18}O^{2-} has two more electrons
   c ^{40}Ca^{2+} has one more proton

2.2

> **Relative atomic masses – time for change?**
>
> **a** ^{10}B: 20%, ^{11}B: 80%
> **b** ^{10}B: 17%, ^{11}B: 83%

1. **a** Mass of an isotope relative to $\frac{1}{12}$th of the mass of an atom of carbon-12
   b The weighted mean mass of an atom of an element relative to $\frac{1}{12}$th of the mass of an atom of carbon-12.
2. **a** 39.14   **b** 121.86   **c** 20.18
3. **a** 34.45   **b** Compared with 35.48 obtained from using 35 and 37, the difference is significant to two decimal places but both calculations would give the same value, 35.5, to one decimal place.

2.3

1. **a** K$_2$O   **b** MgI$_2$ **c** Ca$_3$P$_2$
   d Fe(OH)$_3$ **e** (NH$_4$)$_2$CO$_3$ **f** Fe(NO$_3$)$_2$
2. **a** Aluminium nitride   **b** ammonium phosphate   **c** iron(III) sulfate
3. **a** 4NH$_3$(g) + 5O$_2$(aq) → 4NO(aq) + 6H$_2$O(l)
   b C$_6$H$_{14}$(g) + 9$\frac{1}{2}$O$_2$(g) → 6CO$_2$(g) + 7H$_2$O(l)
   c Al$_2$O$_3$(s) + 2H$_3$PO$_4$(aq) → 2AlPO$_4$(aq) + 3H$_2$O(l)
   d 3Zn(s) + 8HNO$_3$(aq) → 3Zn(NO$_3$)$_2$(aq) + 2NO(g) + 4H$_2$O(l)
4. **a** 6Mg(s) + P$_4$(s) → 2Mg$_3$P$_2$(s)
   b 2Fe(s) + 3Cu(NO$_3$)$_2$(aq) → 2Fe(NO$_3$)$_3$(aq) + 3Cu(s)
   c 2Pb(NO$_3$)$_2$(s) → 2PbO(s) + 4NO$_2$(g) + O$_2$(g)

3.1

1. **a** 0.300 mol   **b** 5.00 mol   **c** 7.50 × 10^{-3} mol
   d 8.33 × 10^{-5} mol   **e** 4.66 × 10^{-4} mol
2. **a** 7000 g   **b** 9.45 g   **c** 4.9 g
   d 1.325 g   **e** 0.747 g
3. **a** 28 g mol^{-1} **b** 74 g mol^{-1} **c** 85 g mol^{-1}

3.2

> **Formula of a hydrated salt**
>
> CoCl$_2$•6H$_2$O

1. **a** i 64.1   ii 284.0   iii 100.5
   b i 184.1   ii 80.0   iii 342.3
2. x = 6; Ni(NO$_3$)•6H$_2$O
3. C$_2$H$_4$ (28), C$_3$H$_6$ (42), C$_4$H$_8$ (56), C$_5$H$_{10}$ (70), C$_6$H$_{12}$ (84), C$_7$H$_{16}$ (98), C$_8$H$_{16}$ (112)
4. The first salt would not lose all H$_2$O so x would appear to be smaller than it should be.

   The second salt loses more mass than just water so x would appear to be greater than it should be.

3.3

> **Finding a relative molecular mass**
>
> 160

> **Real gases**
>
> High pressure: gas molecules are close together and the volume of the molecules becomes significant compared with the volume of the container.
>
> Low temperatures. Gas molecules slow down and have less energy than higher temperatures. Intermolecular forces may then become significant.

1. **a** 0.250 mol   **b** 0.00200 mol
2. **a** 20.0 g dm^{-3} **b** 157.5 g dm^{-3}

Answers

3 **a** 60.0 mol **b** 0.0300 mol **c** 1.42×10^{-3} mol
4 **i** 23.5 dm^3 **ii** 27.8 dm^3
5 **a** 192 cm^3 **b** 48.0 cm^3 **c** 15.0 cm^3
6 60

3.4

Identifying an unknown metal

magnesium

1 a $2Ca(s) + O_2(g) \rightarrow 2CaO(s)$
  b 3.2 g Ca and 1.28 g O_2 completely react together to form 4.488 g CaO **c** 840 cm^3
2 a 100% **b** 51.1%
3 % yield: 61.0%; atom economy 22.7%
4 a $2SO_2(g) + O_2(g) \rightarrow 2SO_3(g)$
  b 180 cm^3 $SO_2(g)$ and 90 cm^3 $O_2(g)$
  c i O_2 in excess as 2 mol SO_2 react with only 1 mol O_2 **ii** 150 cm^3 SO_3 as SO_2 is the limiting reagent
5 a $CH_4(g) + H_2O(g) \rightarrow 3H_2(g) + CO(g)$
  $\dfrac{3 \times 2.0}{3 \times 2 + 28.0} \times 100 = 17.6\%$
  b $\dfrac{13.5}{18.75} \times 100 = 72.0\%$
6 a $Cr(s) + 2HCl(aq) \rightarrow CrCl_2(aq) + H_2(g)$
  b 540 cm^3; **c** 0.30 mol dm^{-3}
7 $n(H_2) = 3.0 \times 10^{-3}$ mol; $n(X) = 2.0 \times 10^{-3}$ mol; $M(X) = 27$ g mol^{-1}; **X** = Al

4.1

Dissociation in sulfuric acid

$H_3PO_4(aq) \rightarrow H^+(aq) + H_2PO_4^-(aq)$

$H_2PO_4^-(aq) \rightleftharpoons H^+(aq) + HPO_4^{2-}(aq)$

$HPO_4^{2-}(aq) \rightleftharpoons H^+(aq) + PO_4^{3-}(aq)$

1 a completely dissociates **b** partially dissociates
2 a $HNO_3(aq) \rightarrow H^+(aq) + NO_3^-(aq)$
  b $CH_3CH_2COOH(aq) \rightleftharpoons H^+(aq) + CH_3CH_2COO^-(aq)$
3 a $MgO(s) + 2HCl(aq) \rightarrow MgCl_2(aq) + H_2O(l)$ magnesium chloride
  b $2NaOH(aq) + H_2SO_4(aq) \rightarrow Na_2SO_4(aq) + 2H_2O(l)$ sodium sulfate
  c $ZnCO_3(s) + 2HNO_3(aq) \rightarrow Zn(NO_3)_2(aq) + H_2O(l) + CO_2(g)$ zinc nitrate
  d $NaOH(aq) + CH_3COOH(aq) \rightarrow CH_3COONa(aq) + H_2O(l)$ sodium ethanoate
4 a $H_2CO_3(aq) + NaOH(aq) \rightarrow NaHCO_3(aq) + H_2O(l)$
  $H_2CO_3(aq) + 2NaOH(aq) \rightarrow Na_2CO_3(aq) + 2H_2O(l)$
  b One hydrogen atom has been replaced by a metal ion. The other hydrogen atom can still behave as an acid.
  c The reaction produces bubbles of carbon dioxide which make the cake rise.

4.2

Preparing standard solutions

1 Titre would be less as solution is more dilute.
2 Solution used in first titration would be more dilute than solution used later. First titre would need less solution.

The acid–base titration procedure

1 Titre would be less as less solution has been added than the correct pipette volume.
2 No effect as volume is measured by difference between two readings.

Identification of a carbonate

97.2

1 a i 2.75×10^{-3} **ii** $NaOH(aq) + HNO_3(aq) \rightarrow NaNO_3(aq) + H_2O(l)$ **iii** 0.118 mol dm^{-3}
  b $2KOH(aq) + H_2SO_4(aq) \rightarrow K_2SO_4(aq) + 2H_2O(l)$; 0.0587 mol dm^{-3}
2 $n(HCl) = 5.30 \times 10^{-4}$ mol
  $n(X(OH)_2) = 2.65 \times 10^{-4}$ mol in 25.0 cm^3
  $n(X(OH)_2) = 2.65 \times 10^{-3}$ mol in 250 cm^3
  $M(X(OH)_2) = 74.0$ g mol^{-1};
  Formula = $Ca(OH)_2$
3 $n(NaOH) = 0.00250$ mol
  $n(H_2C_2O_4) = 0.00125$ mol in 23.8 cm^3
  $n(H_2C_2O_4) = 0.0131$ mol in 250 cm^3
  $M(H_2C_2O_4 \cdot xH_2O) = 126$;
  $n = 2$;
  Formula = $H_2C_2O_4 \cdot 2H_2O$

Answers

4.3

1. **a** Ag$^+$: +1; **b** F$_2$: 0
   c NaClO$_3$: Na +1, Cl +5, O −2.
   a H$_2$S: −2; **b** SO$_4^{2-}$: +6; **c** Na$_2$S$_2$O$_3$: +2.
   a Mg: 0 → +2; H: +1 → 0.
   b Mg oxidised; H reduced.

5.1

Electrons in shells

2. $n = 5$: 50, $n = 6$: 72, $n = 7$: 98

1. **a** 2 **b** 6 **c** 32 **d** 2
2. **a** C: 1s^{2}2s^{2}2p^2 **b** S: 1s^{2}2s^{2}2p^{6}3s^{2}3p^4
   c K: 1s^{2}2s^{2}2p^{6}3s^{2}3p^{6}4s^1
   d Co: 1s^{2}2s^{2}2p^{6}3s^{2}3p^{6}3d^{7}4s^2
   e As: 1s^{2}2s^{2}2p^{6}3s^{2}3p^{6}3d^{10}4s^{2}4p^3
3. **a** Mg^{2+}: 1s^{2}2s^{2}2p^6
   b P^{3-}: 1s^{2}2s^{2}2p^{6}3s^{2}3p^6
   c Br$^-$: 1s^{2}2s^{2}2p^{6}3s^{2}3p^{6}3d^{10}4s^{2}4p^6
   d Fe^{2+}: 1s^{2}2s^{2}2p^{6}3s^{2}3p^{6}3d^6
4. **a** Si: [Ne]3s^{2}3p^2 **b** Cl$^-$: [Ne]3s^{2}3p^6
   c Mn^{2+}: [Ar]3d^5 **d** Ga: [Ar]3d^{10}4s^{2}4p^1

5.2

Ions bones and teeth answers

1. Ca$_5$(PO$_4$)$_3$OH
   five Ca^{2+} gives 10+, three PO$_4^{3-}$ +
   one OH$^-$ gives −10 ions balance.
   Ca$_5$(PO$_4$)$_3$F
   five Ca^{2+} gives 10+, three PO$_4^{3-}$ +
   one F$^-$ gives −10 ions balance.
2. The ionic attraction in the lattice is between highly charged Ca^{2+} and PO$_4^{3-}$ ions. This leads to strong attraction which is difficult for polar water molecules to break down.
3. A fluoride ion is a fluoride atom that has gained one electron to give the stable noble gas electron structure. Fluorine has two fluorine atoms bonded by a covalent bond in an F$_2$ molecule. F$_2$ is one of the most reactive substances known.

1. **a,b,c**

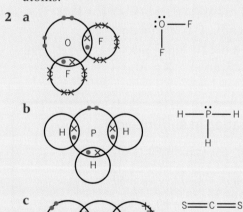

2. Strong electrostatic attraction between oppositely charged ions. High temperature needed to provide sufficient energy to overcome the attractions.

3. Polar water molecules are attracted towards ions on the surface of the ionic lattice. Water molecules bond to the ions, weakening ionic bonding. Ionic bonds are broken. Ions become surrounded by water molecules and break free from the lattice.

5.3

1. Strong electrostatic attraction between a shared pair of electrons and the nuclei of the bonded atoms.

2. **a, b, c** (dot-and-cross diagrams and displayed formulae for OF$_2$, PH$_3$, CS$_2$)

562

Answers

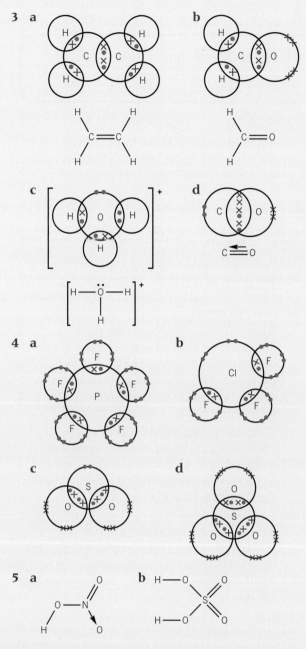

6.1

1. **a** BeI$_2$ linear, 180° **b** BCl$_3$ trigonal planar, 120° **c** SiH$_4$; tetrahedral, 109.5° **d** H$_2$C=O trigonal planar, 120° **e** CS$_2$ linear, 180°

2. **a** C: tetrahedral, 109.5° O: non-linear, 104.5°
   b trigonal planar, 120° **c** non-linear, 120°
   d pyramidal, 107°

3. **a** BF$_3$: 120°, NH$_3$: 107°
   b All bond angles 109.5°. BF$_3$ has three bonded pairs around boron. NH$_3$ has three bonded pairs and one lone pair around nitrogen. In F$_3$BNH$_3$, boron and nitrogen now have four bonded electrons around boron and around nitrogen.

6.2

1. **a** Measure of the attraction of a bonded atom for the pair of electrons in a covalent bond.
   b Shared pair of electrons where the electron pair is not shared equally between the two bonded atoms.
   c Charge separation across a bond with one atom having a δ+ charge and one atom a δ− charge.

2. Ionic → KF, Na$_2$O, Al$_2$O$_3$, LiBr, SiO$_2$, NO$_2$, Br$_2$ and PH$_3$ → Covalent

3. **a** H(δ+), S(δ−); Be(δ+), Br(δ−); N(δ−), H(δ+); B(δ+), F(δ−); Si(δ+), Cl(δ−); H(δ+), C(δ−) and C(δ+), Cl(δ−); H(δ+), C(δ−) and C(δ+), O(δ−); P(δ+), F(δ−); S(δ+), F(δ−)
   b Polar: All non-symmetrical molecules, H$_2$S, NH$_3$, CHCl$_3$, H$_2$C=O; Non polar: all symmetrical molecules where individual dipoles cancel, BeBr$_2$, BF$_3$, SiCl$_4$, PF$_5$, SF$_6$

6.3

1. Fluctuation in the electron density around a molecule creates an instantaneous dipole in a molecule. The instantaneous dipole induces a dipole in a neighbouring molecule.

2. **a** Weak intermolecular forces are broken by the energy present at low temperatures.
   b There is little interaction between the molecules in the lattice and the polar solvent molecules.
   c There are no mobile charged particles.

3. **a** Giant Ionic lattice. High melting and boiling point. Strong electrostatic attraction between oppositely charged ions. High temperature is needed to provide sufficient energy to overcome the attractions.

   In solid state, ions are fixed in the lattice and cannot move so there is no conductivity. In liquid and aqueous states, the ions are free to move and conduct electricity.

   Charged ions can interact with polar solvent molecules and dissolve.

   b Simple molecular lattice. Low melting and boiling point. Weak intermolecular forces are broken by the energy present at low temperatures.

   No conductivity as there are no mobile charged particles.

   When added to a non-polar solvent, intermolecular forces form between the simple molecular structure and the solvent. CCl$_4$ dissolves.

Answers

6.4

> ### Boiling points of hydrides and intermolecular forces
>
> 1. **a** water: −75 °C, hydrogen fluoride: −90 °C, ammonia: −100 °C
>   **b** All three have hydrogen bonding which is a stronger intermolecular force than other dipole interactions. Greater energy needed to overcome the intermolecular forces, so boiling points are much higher.

2. Increase in electrons increases the strength of the London forces.

3. From Period 3 to Period 6, difference in electronegativity between hydrogen and element decreases, decreasing the permanent dipole–dipole interactions. Number of electrons increases, increasing London forces. Boiling point increases so London forces are stronger and more significant than the permanent dipole–dipole interactions.

> ### Hydrogen bonding in DNA
>
> 1. Two purine bases would be too close together. Two pyramidine bases would be too far apart.
> 2. $4^{3\,000\,000\,000}$

1. Ice is less dense than liquid water because hydrogen bonds hold molecules apart in open lattice structure. Higher melting and boiling point than expected because appreciable energy is needed to break the hydrogen bonds.

2. H_2O and CH_3OH

3. **a**

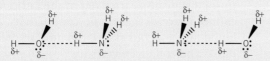

   b

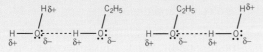

7.1

> ### Focus of ekasilicon
>
> **a** 71 **b** ECl_3

1. **a** Sr **b** As   2. ns^2np^3

3. Group 18(0). Group 18 contains the very unreactive noble gases, and no elements from Group 18 had been discovered at that time.

7.2

1. $S(g) \rightarrow S^+(g) + e^-$ 2. $Al(g) \rightarrow Al^+(g) + e^-$
   $S^+(g) \rightarrow S^{2+}(g) + e^-$

3. As each electron is removed, the outer shell is drawn closer to the nucleus. Nuclear attraction is greater and more energy is needed to remove the next electron.

4. **a** Nuclear charge increases from Na to Ar, and outer electrons are in the same shell with similar shielding. The atomic radius decreases. This results in an increase in nuclear attraction on the outer electrons and an increase in first ionisation energy.

   b Sharp drop reflects addition of a new shell with a resulting increase in distance and shielding. This decreases the nuclear attraction on the outer electrons, decreasing the first ionisation energy.

   c Ionisation energy decreases from He to Ne to Ar due to increase in the number of shells, so increasing atomic radius and shielding. This causes a decrease in nuclear attraction on outer electrons, decreasing the first ionisation energy.

5. Silicon there is a large increase between the fourth and fifth ionisation energies, so the fifth electron is removed from an inner shell. Therefore, there are four electrons in the outer shell, so Group 14(4).

6. **a** The 3p sub-shell in aluminium has a higher energy level than the 3s sub-shell in magnesium. The 3p electron easier to remove.

   b Phosphorus has three electrons in the 3p sub-shell, one electron in each 3p orbital. Sulfur has four electrons in the 3p sub-shell, two electrons paired in one orbital and one electron in each of the other two 3p orbitals. The paired electrons in sulfur repel one another making it eaer to remove one of these electrons than an unpaired electron.

Answers

7.3

> **Graphene and graphite**
>
> a The structure in carbon fibre is the hexagonal arrangement as in graphene and graphite.
>
> b Carbon fibre are bundled together, either as sheet, bonded together with resin, as as a fibres that are woven together.

1 Strong electrostatic attraction between cations (positive ions) and delocalised electrons. The delocalised electrons can move across a potential difference.

2 A simple molecular lattice has London forces between molecules. A giant covalent lattice has covalent bonds between atoms.

3 a The Group 4 element, germanium has a giant metallic lattice rather than a giant covalent lattice in carbon and silicon.

  b The Group 5 element, arsenic has a giant covalent lattice whereas in Period 2 and 3, the Group 5 element has a simple molecular lattice.

8.1

1 Group 2 metals add electrons to other species.

2 a Mg changes from 0 to +2, H changes from +1 to 0 b Mg oxidised, H reduced

3 Down Group 2, the total energy from 1st and 2nd ionisation energies decreases as the nuclear attraction on the outer electrons decreases because of increased atomic radius and increased shielding. It therefore becomes easier to remove the electrons and the reactivity increases.

4 Group 2 oxides react with water forming the metal hydroxide. Hydroxide ions in solution cause alkalinity. Down Group 2, the solubility of the metal hydroxide increases, increasing the pH and alkalinity.

8.2

> **Halide ions as reducing agents**
>
> a $2H^+ + H_2SO_4 + 2I^- \rightarrow SO_2 + I_2 + 2H_2O$
> $4H^+ + SO_2 + 4I^- \rightarrow S + 2I_2 + 2H_2O$
> $2H^+ + S + 2I^- \rightarrow H_2S + I_2$
>
> b S: +6 to +4, I: −1 to 0
>   S: +4 to 0, I: −1 to 0
>   S: 0 to −2, I: −1 to 0
>
> c $8H^+ + H_2SO_4 + 8I^- \rightarrow SO_2 + 4I_2 + 4H_2O$

1 Boiling point increases due to increase in electrons down the group. This increases the strength of the London forces and more energy needed to break the intermolecular forces.

2 a $Cl_2(aq) + 2KBr(aq) \rightarrow 2KCl(aq) + Br_2(aq)$
    ionic equation $Cl_2(aq) + 2Br^-(aq) \rightarrow 2Cl^-(aq) + Br_2(aq)$

  b $Cl_2(aq) + MgI_2(aq) \rightarrow MgCl_2(aq) + I_2(aq)$
    ionic equation $Cl_2(aq) + 2I^-(aq) \rightarrow 2Cl^-(aq) + I_2(aq)$

3 $3Cl_2(aq) + 6NaOH(aq) \rightarrow NaClO_3(aq) + 5NaCl(aq) + 3H_2O(l)$

There are six Cl atoms in Cl_2. Five Cl atoms has been reduced from 0 to −1 in NaCl.

One Cl has been oxidised to from 0 to +5 in $NaClO_3$. So Cl has been both oxidised and reduced.

8.3

> **A barium meal – making use of precipitation reactions**
>
> Barium sulfate is very insoluble in water, so passes through the body without interacting with it.

1 Add $AgNO_3(aq)$ to an aqueous solution of each.
NaCl forms a white precipitate which dissolves in dilute $NH_3(aq)$.

NaBr forms a cream precipitate which dissolves in concentrated $NH_3(aq)$.

NaI forms a yellow precipitate which is insoluble in concentrated $NH_3(aq)$.

2 Dilute nitric acid reacts with carbonate ions and no precipitate of barium carbonate will then form.

3 If sulfuric acid is used, sulfate ions are added and will show up in the sulfate test with $Ba^{2+}(aq)$. If hydrochloric acid is used, chloride ions are added and will show up in the halide test with $Ag^+(aq)$.

9.1

> **What about calories?**
>
> 1 a The value of the specific heat capacity of water. b 957.22 kJ
>
> 2 Assuming mass of cup of tea = 200 g
> $957220 = m \times 4.18 \times 85$
> $m = 2694$
> $\frac{2694}{200} = 13.5$ cups of tea

Answers

1

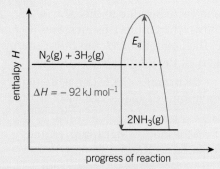

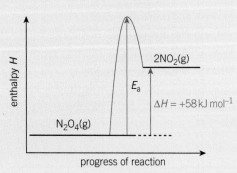

2 a $C(s) + 2H_2(g) \rightarrow CH_4(g)$
   b $\frac{1}{2}N_2(g) + O_2(g) \rightarrow NO_2(g)$;

3 a $H_2S(g) + 1\frac{1}{2}O_2(g) \rightarrow H_2O(l) + SO_2(g)$
   b $Al(s) + \frac{3}{4}O_2(g) \rightarrow \frac{1}{2}Al_2O_3(s)$

9.2

Spirit burners
a Final answer is 514.
b Final answer is 520.

Cooling curves
$\Delta_r H = -493\,kJ\,mol^{-1}$

1 a 8.99 kJ   b 8.36 kJ
2 $-2280\,kJ\,mol^{-1}$
3 a $C_2H_5OH(l) + 3O_2(g) \rightarrow 2CO_2(g) + 3H_2O(l)$;
   b $-906\,kJ\,mol^{-1}$

9.3

Bond enthalpies and combustion
1 Extra bonds broken: C—C, 2 C—H, 1.5 O=O
   Extra bonds made: 2 C=O, 2 O—H
   $\sum$(bonds broken – bond made) = 1920 – 2538
   $= -618\,kJ\,mol^{-1}$

2 Enthalpy change = $-658 + (19 \times -618)$
   $= 12\,400\,kJ\,mol^{-1}$

3 Actual bond enthalpies for bonds broken and made may be different from average bond enthalpies.
   Alcohols and water are liquids under standard conditions. Bond enthalpies apply to gases.

1 Average bond enthalpy is the energy required to break one mole of a type of bond in gaseous molecules. The bond is in different environments.

2 a $-130\,kJ\,mol^{-1}$
   b The bonds broken are stronger than the bonds made.

3 a $345\,kJ\,mol^{-1}$ b $469\,kJ\,mol^{-1}$

9.4

Unfamiliar enthalpy cycle
1 $A = -168\,kJ\,mol^{-1}$, $B = -285\,kJ\,mol^{-1}$,
   $C = -393\,kJ\,mol^{-1}$, $D = -54\,kJ\,mol^{-1}$
2 $-792\,kJ\,mol^{-1}$

1 $\Delta_c H^\ominus$ (C(s)) is for combustion of 1 mol C(s) with 1 mol $O_2(g)$ to form 1 mol $CO_2(g)$.
   $\Delta_f H^\ominus$ ($CO_2(g)$) is for formation of 1 mol $CO_2(g)$ from 1 mol C(s) and 1 mol $O_2(g)$. Both processes represent the same reaction and equation and the enthalpy changes will have the same value.

2 a $-175\,kJ\,mol^{-1}$;   b $154\,kJ\,mol^{-1}$;
   c $-295\,kJ\,mol^{-1}$

3 a $-203\,kJ\,mol^{-1}$;   b $-279\,kJ\,mol^{-1}$;

10.1

Monitoring the production of a gas using gas collection
a $2.5\,cm^3\,s^{-1}$ b $1.5\,cm^3\,s^{-1}$

Answers

1. Temperature, surface area, and a catalyst. When concentration of a reactant is increased, the rate of reaction increases. Increase in concentration means more reactant molecules in the same volume. The reactant molecules are closer together and collide more frequently.
2. Gas collection or mass loss
3. a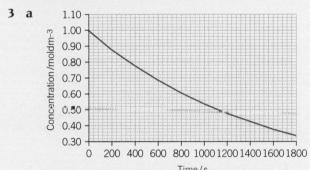

   b Initial rate = $\frac{1}{1600}$ = 6.25×10^{-4} mol dm^{-3}s^{-1}

   Rate at 1000s = $\frac{0.58}{1700}$ = 3.41×10^{-4} mol dm^{-3}s^{-1}

10.2

> **Heterogeneous catalysts and atmospheric pollution**
>
> $2CO(g) + 2NO(g) \rightarrow 2CO_2(g) + N_2(g)$

> **Autocatalysis**
>
> 1 Mn^{2+}
> 2 $2Mn^{3+} + C_2O_4^{2-} \rightarrow 2CO_2 + 2Mn^{2+}$

1. A homogeneous catalyst has the same physical state as the reactants. A heterogeneous catalyst has a different physical state from the reactants.
2. Lowers the activation energy by providing an alternative route for the reaction. There is no effect on ΔH.
3. a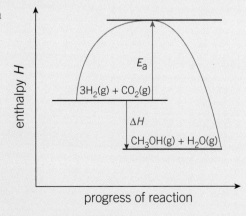

   b 274 kJ mol^{-1}

10.3

1. Minimum energy required to start a reaction by the breaking of bonds.
2. Rate of reaction decreases as fewer molecules have energy greater than the activation energy. Collisions are less frequent as the molecules are moving slower but this is a much smaller effect than the decrease in molecules exceeding the activation energy.
3. a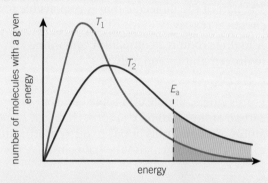

   b More particles have energy greater than or equal to the activation energy. More collisions are then able to result in a reaction.

   c A catalyst provides an alternative reaction pathway with lower activation energy. Therefore a greater number of molecules have energy greater than or equal to the activation energy. More collisions are then able to result in a reaction.

10.4

> **Making ammonia with the Haber process**
>
> Low temperature shifts equilibrium in the exothermic direction – to the right and towards ammonia.
>
> High pressure shifts equilibrium to the right because there are fewer molecules (two ammonia on the right-hand side compared to one nitrogen and three hydrogen, four in total, on the left-hand side).

1. a moves left b moves right c No change
2. a goes right b goes left
3. a low temperature and high pressure
   b high temperature and low pressure

Answers

10.5

1 a $K_c = \dfrac{[CH_3COOC_2H_5][H_2O]}{[CH_3COOH][C_2H_5OH]}$

  b 1.56

2 $K_c = \dfrac{[HI]^2}{[H_2][I_2]}$

  0.163

3 $K_c = \dfrac{[SO_3]^2}{[SO_2]^2[O_2]}$

  11.7

11.1

1 The part of the molecule that is responsible for chemical reactivity

2 A hydrocarbon is a compound containing hydrogen and carbon only

3 $C_5H_{11}OH$

4 Forms single, double and triple bonds to other carbon atoms. Carbon can form chains of carbon atoms. Carbon forms bonds to other atoms such as oxygen, nitrogen and the halogens.

11.2

1 a chloroalkane b ketone
  c alcohol d bromalkane

2 a chloroethane b propanone
  c 3-methylbutan-2-ol d 2,3-dibromobutane

3 a–f (displayed formulae)

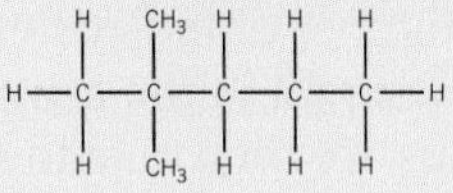

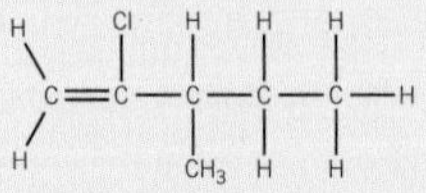

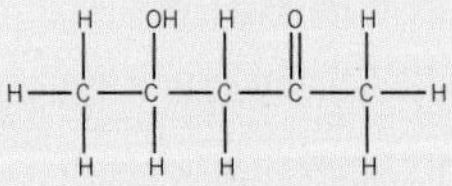

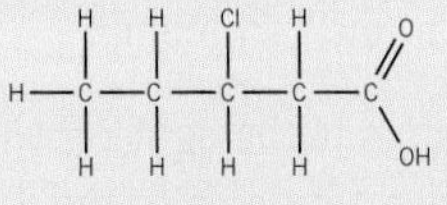

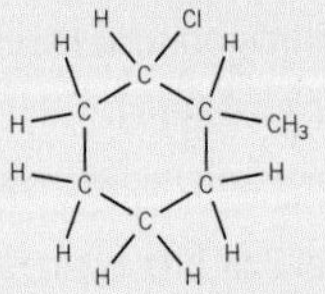

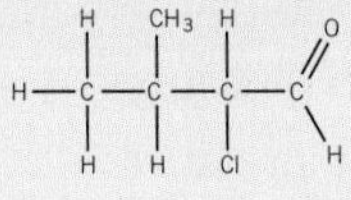

11.3

1 a C_3H_6O; $C_6H_{12}O_2$ b CH; C_6H_6
  c CH_3O; $C_2H_6O_2$ d C_2H_4Cl; $C_4H_8Cl_2$

2 Displayed formulae:

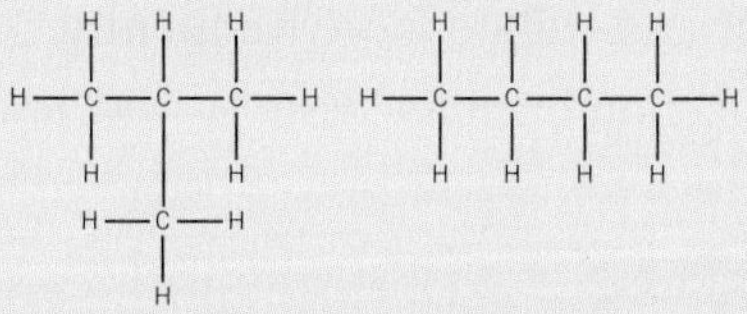

Structural formulae: $(CH_3)_3CH$; $CH_3CH_2CH_2CH_3$

3 a   b 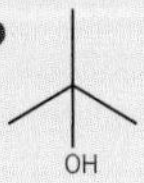   c

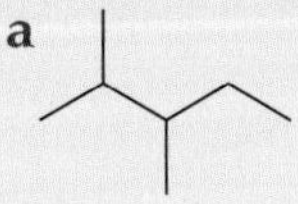

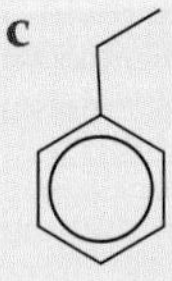

11.4

Detecting isomerism by smell

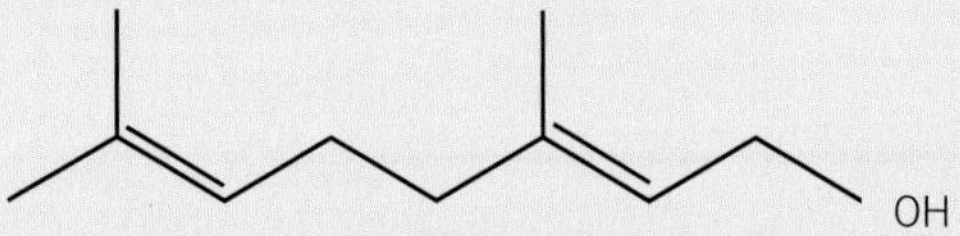

geraniol

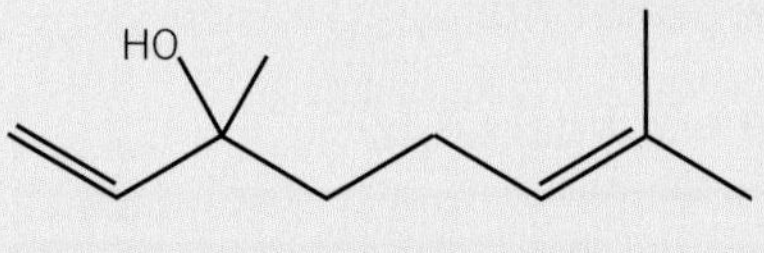

linalool

1 a   b

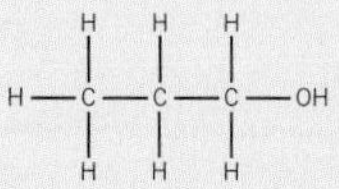

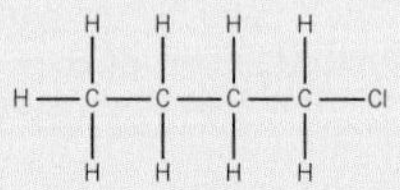

2 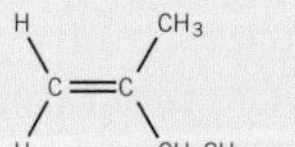 2-methylbut-1-ene 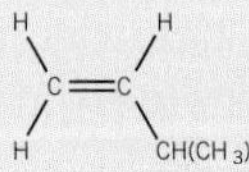   3-methylbut-1-ene 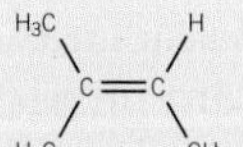   2-methylbut-2-ene

3 butanoic acid 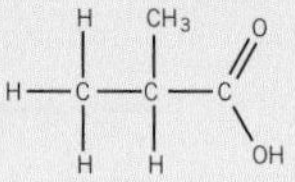   methylpropanoic acid

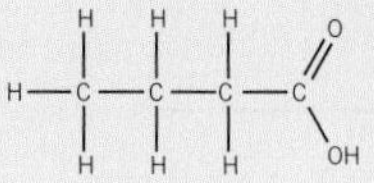

4

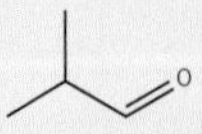

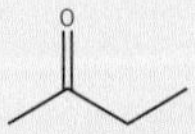

Answers

11.5

Curly arrows and homolytic fission

H₃C—H → H₃C• + •H (methane showing homolytic fission of a C—H bond)

1. The breaking of a covalent bond where one of the bonding atoms takes both electrons from the bond to form ions.
2. A species with an unpaired electron. The breaking of a covalent bond where each of the bonded atoms takes one electron from the bond to form radicals.
   Cl—Cl → Cl• + Cl•
3. **a** Substitution **b** Addition **c** Elimination **d** Addition

12.1

1. Fractional distillation. Different fractions in crude oil have different boiling points. Increase the temperature of the system so each fraction vaporises separately from the other fractions and can be collected.
2. There is more surface contact between straight-chain isomer so it has more/stronger London forces between the chains. More energy required to break these London forces.
3. As you go down the chain length increases, there are more points of contact between the molecules and stronger London forces. More energy required to break these London forces.
4. Single covalent bonds between the C—C and the C—H bonds are σ-bonds. σ-bond is the overlap of orbitals, one from each bonding atom. Each carbon atom is bonded to four other atoms by covalent bonds. The shape around each carbon is atom is tetrahedral. The four bonding pairs of electrons repel equally to give a bond angle of 109.5° around the carbon atom.

12.2

Methane

Methane produced from decomposition of organic waste and plant material is burnt as a fuel rather than released into the atmosphere. This then reduces the amount of methane released into the atmosphere.

The silent killer

$CH_4(g) + 1\frac{1}{2}O_2(g) \rightarrow CO(g) + 2H_2O(l)$ or
$CH_4(g) + O_2(g) \rightarrow C(s) + 2H_2O(l)$

1. complete $C_{12}H_{26}(g) + 18\frac{1}{2}O_2(g) \rightarrow 12CO_2(g) + 13H_2O(l)$
   incomplete $C_{12}H_{26}(g) + 12\frac{1}{2}O_2(g) \rightarrow 12CO(g) + 13H_2O(l)$
2. Alkanes are fairly unreactive due to the presence strong carbon to hydrogen and carbon to carbon single bonds which have no or little polarity.
3. $C_6H_{14}(g) + 6\frac{1}{2}O_2(g) \rightarrow 6CO(g) + 7H_2O(l)$
4. $C_3H_8(g) + Br_2(g) \rightarrow C_3H_7Br(g) + HBr(g)$
5. Initiation: $Cl_2 \rightarrow 2Cl•$
   Propagation: $C_2H_6 + Cl• \rightarrow C_2H_5• + HCl$
   $C_2H_5• + Cl_2 \rightarrow C_2H_5Cl + Cl•$
   Termination: $Cl• + Cl• \rightarrow Cl_2$
   $C_2H_5• + C_2H_5• \rightarrow C_4H_{10}$
   $C_2H_5• + Cl• \rightarrow C_2H_5Cl$
6. $C_3H_6Cl_2 + Cl• \rightarrow C_3H_5Cl_2• + HCl$
   $C_3H_5Cl_2• + Cl_2 \rightarrow C_3H_5Cl_3 + Cl•$

13.1

1. Compound with one or more double or triple bonds
2. C_9H_{18}. Non-4-ene
3. A σ-bond is the result of the head on overlap of orbitals whereas a π-bond is formed from the sideways overlap of p-orbitals.
4. In an alkene, there are three bonding regions about the carbon atom. These repel each other equally resulting in a trigonal planar shape. In an alkane, there are four bonding pairs of electrons on the carbon atom. These repel each other equally resulting in a tetrahedral shape. The bond angle for a trigonal planar shape is 120°, whereas for a tetrahedral shape the bond angle is 109.5°
5. Structural diagrams of but-1-ene, but-2-ene, and 2-methylpropene.

569

Answers

13.2

Solving the mystery of maleic and fumaric acid

1. Fumaric acid
   empirical: CHO, molecular: $C_4H_4O_4$
   Maleic acid
   empirical: CHO, molecular: $C_4H_4O_4$
2. In maleic acid the two COOH groups are on the same side and are able to react together to form a five-membered ring by loss of water.
   In fumaric acid, it is impossible to form a five-membered ring at the COOH groups are on opposite sides of the molecule and they cannot then react react with one another.

1. Molecules with the same structural formula but a different arrangement of the atoms in space.
2. A carbon to carbon double bond and each carbon atom in the double bond must be attached to two different groups.
3. **a** and **b** have no *E/Z* isomers but **c**, **d**, and **e** can form *E/Z* isomers.
4. **c** (*E*)-2,3-dichlorobut-2-ene, (*Z*)-2,3-dichlorobut-2-ene
   d (*E*)-2,3-dichlorobut-2-ene-1,4-diol, (*Z*)-2,3-dichlorobut-2-ene-1,4-diol
   e (*E*)-hex-3-ene, (*Z*)-hex-3-ene

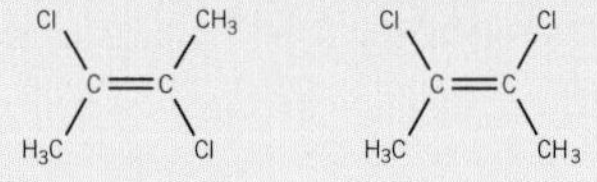

13.3

Focus on margarine

1. stearic acid 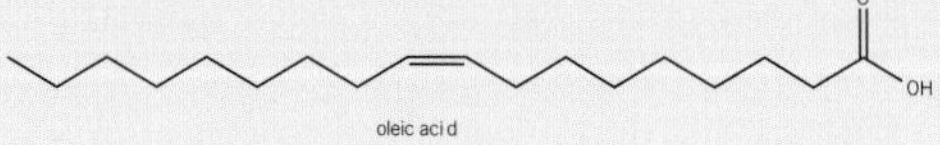
   oleic acid 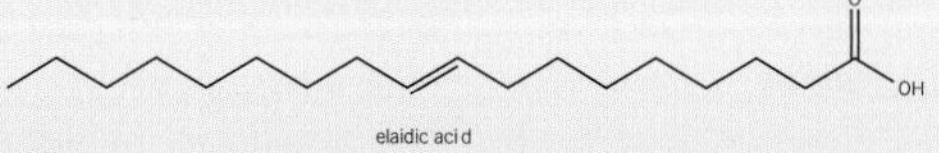
   elaidic acid
2. Stearic acid and elaidic acid have a more regular shape and molecules can pack together more effectively than oleic acid. Therefore there will be more surface contact between molecules and greater London forces.

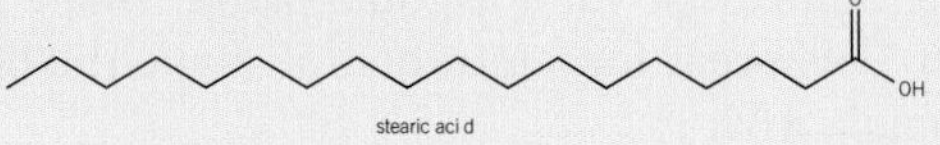

1. (pent-2-ene + H_2 → pentane)
2. 2,3-dibromo-3-methylpentane structure
3. 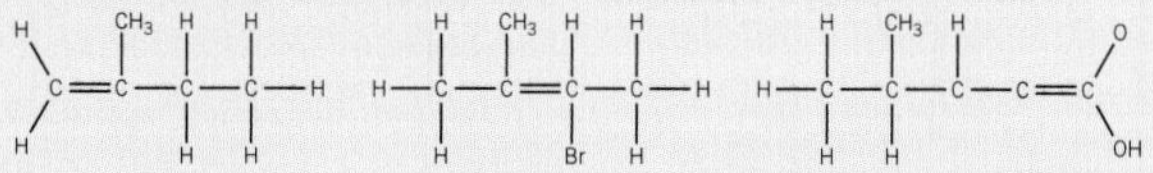 2-methylbut-1-ene, 2-methylbut-2-ene, 3-methylbut-1-ene
4. **a** 3-bromopentane, 2-bromopentane
   b (pentene + HBr → 2-bromopentane)
5. **a** cyclopentadiene + 2Br$_2$ → tetrabromo product
   b 2-methylbuta-1,3-diene + 2H$_2$ → 2-methylbutane
6.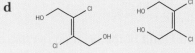

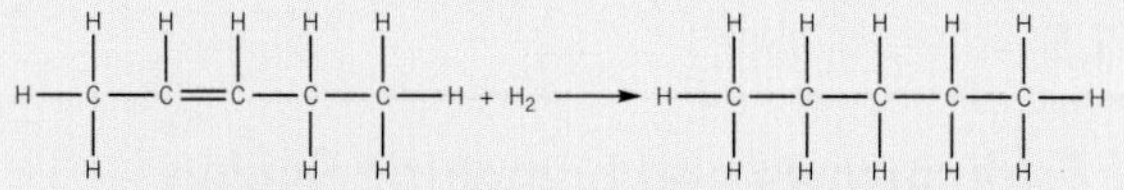

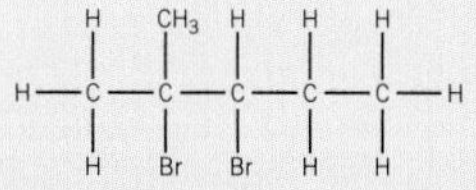

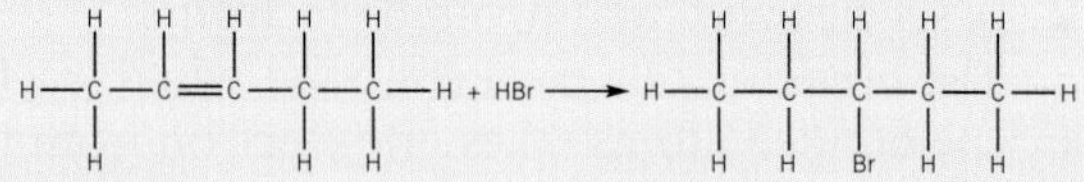

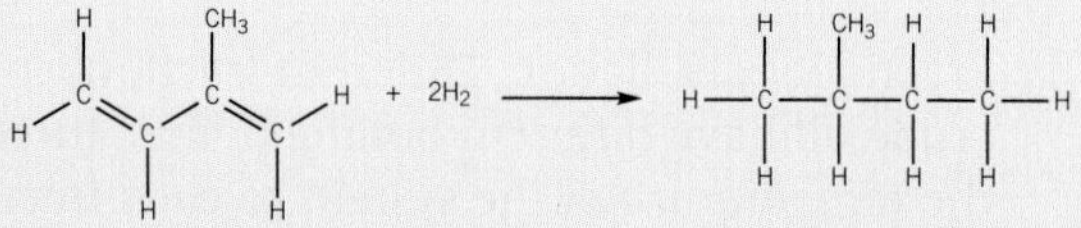

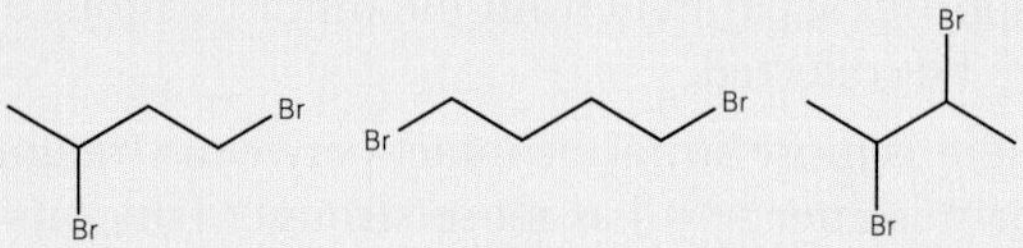

13.4

1. An electron pair acceptor.

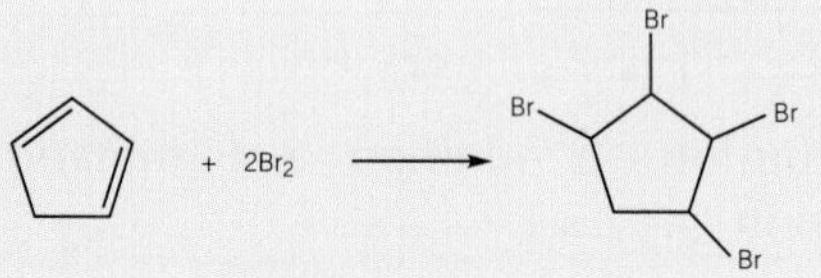

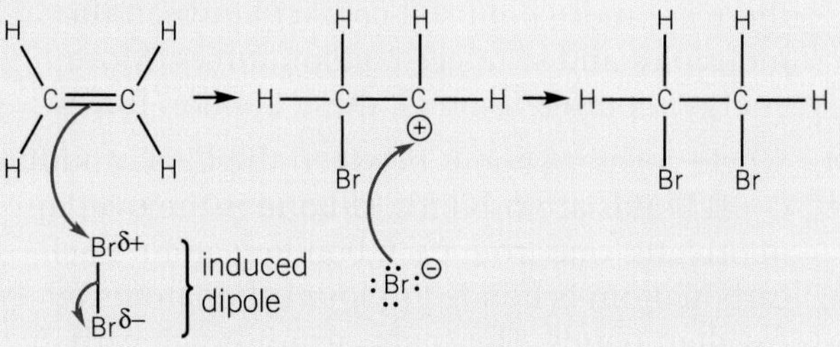

   (mechanism: ethene + Br$_2$ with induced dipole Br$^{\delta+}$–Br$^{\delta-}$, :Br:$^-$)

2. Bromine approaches double bond. High electron density of the double bond repels the electrons in the bromine molecule, inducing a dipole. Slightly positive end of the Br$_2$ molecule is able to accept an electron pair and react.

3. **a** 2-bromobutane
   b 2-bromoheptane

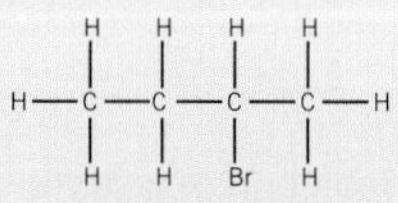

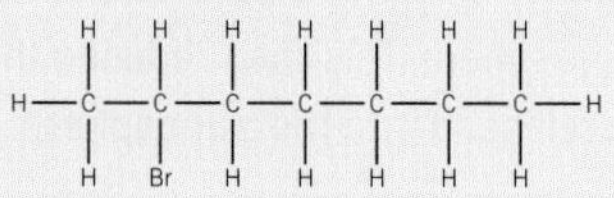

570

Answers

c 2-bromo-2-methylbutane

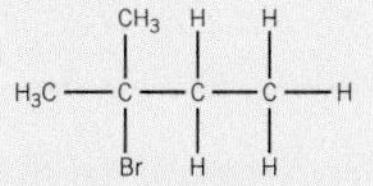

d 1-bromo-1-methylcyclohexane

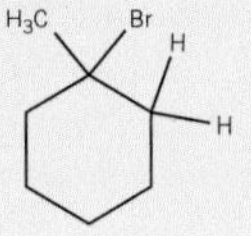

4

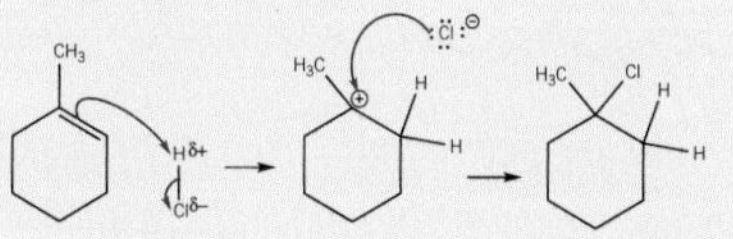

13.5

> ### The history of polyethene
> 1 Linear chains in HDPE can pack closer together than branched chains, so more molecules in a given amount. There is more surface contact between the HDPE molecules giving greater London forces and increased strength.
> 2 7142–17857

1 a Small molecule that combines with many other monomers to form the polymer.
  b Arrangement of atoms that occurs in the polymer structure over and over again.
2 HCl, CO, or Cl_2
3 a 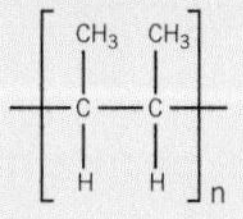
4 a 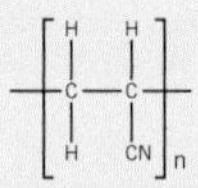   b 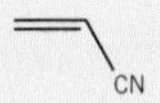
5

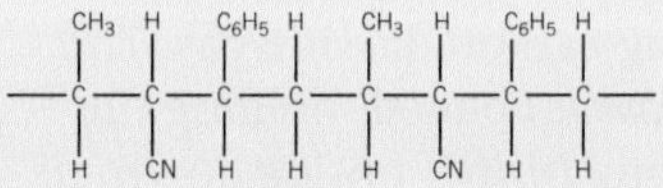

14.1

> ### Ethylene glycol to the rescue
> Ethylene glycol forms hydrogen bonds with water molecules.

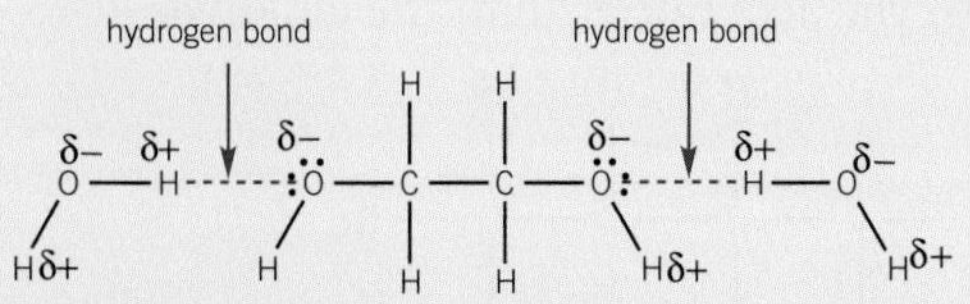

1 a i pentan-2-ol ii propan-2-ol
     iii 2-methylbutan-1-ol
  b i secondary ii primary iii primary

2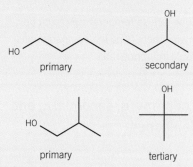
primary   secondary
primary   tertiary

3 As the chain length increases the boiling point increases due to stronger/more London forces. More energy is required to break these additional/stronger London forces.

4 Hydrogen bond

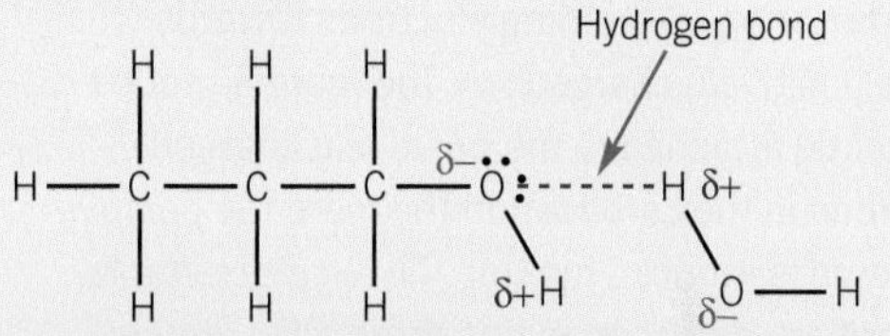

5 Propane-1,2,3-triol has three –OH groups in its structure. Each OH groups is able to form hydrogen bonds with water molecules. The more hydrogen bonds formed between a molecule and water the greater the solubility.

14.2

1 a 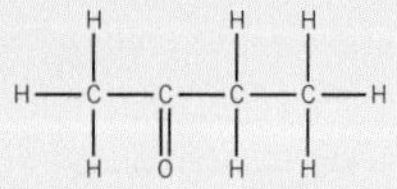
  b 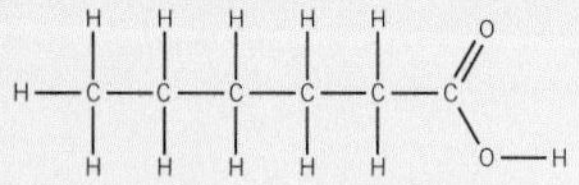
  c 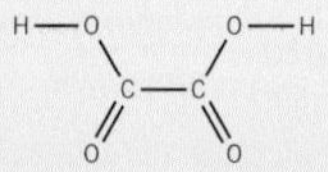
  d

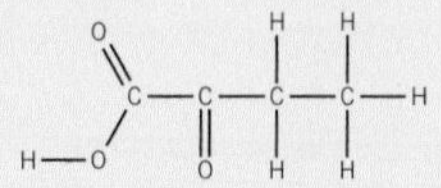

2 $CH_3CH_2CH_2CH_2OH + NaI + H_2SO_4 \rightarrow$
   $CH_3CH_2CH_2CH_2I + NaHSO_4 + H_2O$
   name: 1-iodobutane
   reaction: substitution

3 a   b

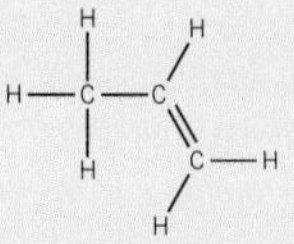

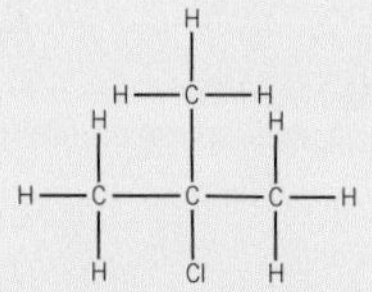

571

Answers

15.1

> **Hydrolysis of haloalkanes**
>
> You are comparing rate so temperatures must be constant. Otherwise there is no fair comparison as temperature would influence rate.

> **Hydrolysis of primary, secondary, and tertiary haloalkanes**
>
> 1. [mechanism diagram showing hydrolysis of tertiary haloalkane via carbocation]
>
> 2. Alkyl groups reduce the charge on the carbocation by pushing electrons towards the positively charged ion. The more R-groups directly attached to the positively charged carbon in the carbocation the more the positive charge is reduced making the ion more stable. A tertiary carbocation has three alkyl groups directly attached to the positive carbon atom but a primary carbocation has only one alkyl group.

1. Halogen atoms are more electronegative than carbon atoms. The electrons in the C–Hal bond are attracted towards the halogen making the bond polar.

2. [mechanism diagram showing hydrolysis of chloroethane with OH⁻]

3. Fastest Rate B, C, A Slowest Rate

4. **a** [structure of CH₃CH₂CH₂-O-CH₂CH₃]

   b [mechanism diagram showing reaction of bromoethane with ethoxide giving diethyl ether + Br⁻]

15.2

> **The end of the road for organohalogen compounds?**
>
> [structures of a CFC and a brominated cyclic compound, plus PVC polymer repeating unit]

1. Refrigerants, air conditioning units, and aerosol sprays.

2. Stratosphere

3. It is thought that depletion of the ozone layer will allow more harmful UV-b to penetrate the troposphere leading to increased incidence of skin cancer and genetic damage.

4. $HCClF_2 \rightarrow HCF_2\bullet + Cl\bullet$
   Propagation step 1 $Cl\bullet + O_3 \rightarrow ClO\bullet + O_2$
   Propagation step 2 $ClO\bullet + O \rightarrow Cl\bullet + O_2$

5. Step 1 $OH\bullet + O_3 \rightarrow O_2H\bullet + O_2$
   Step 2 $O_2H\bullet + O_3 \rightarrow OH\bullet + 2O_2$
   Overall $2O_3 \rightarrow 3O_2$

16.1

1. Calcium chloride, calcium sulfate, or magnesium sulfate

2. Add aqueous sodium carbonate and shaking the mixture in the separating funnel.

   Any acid present will react with sodium carbonate releasing carbon dioxide gas. The tap needs to be slowly opened, holding the stoppered separating funnel upside down, to release any gas pressure that may build up.

   The aqueous sodium carbonate layer and organic layer are run off and the organic layer washed with water before running both layers off into two separate flasks.

3. Refluxing a primary alcohol would form a carboxylic acid rather than the aldehyde.

4. **a** [structure of 1-bromobutane: H-C-C-C-C-Br with H substituents]

b $CH_3CH_2CH_2CH_2OH + H_2SO_4 + NaBr \rightarrow$
  $CH_3CH_2CH_2CH_2Br + NaHSO_4 + H_2O$

  c The lower layer will contain the bromobutane which is more dense than water.

16.2

> **Does nature really provide a cure for everything?**
>
> 1 carboxylic acid, COOH
> 2 Aspirin: $C_9H_8O_4$; Ibuprofen: $C_{13}H_{18}O_2$
> **Step 1**: NaOH(aq), reflux
> $RCH_2Br \rightarrow RCH_2OH$
>
> **Step 2**: H_2SO_4 and $K_2Cr_2O_7$, reflux
> $RCH_2OH \rightarrow RCOOH$

> **Ozonolysis**
>
> Step 1: React with ozone
>
> Step 2: Reflux with acidified potassium dichromate(VI)

1 a alkene, alcohol, and aldehyde.
   b alcohol and carboxylic acid

2 Sodium chloride and H_2SO_4 Conditions: Reflux

3 **Synthesis 1**
   Single step: H_2O(g) and H_3PO_4 catalyst at high temperature
   but-2-ene, $CH_3CH=CHCH_3 \rightarrow$
   butan-2-ol, $CH_3CHOHCH_2CH_3$

   Synthesis 2
   Step 1: NaCl and H_2SO_4, reflux
   but-2-ene, $CH_3CH=CHCH_3 \rightarrow$
   2-chlorobutane, $CH_3CHClCH_2CH_3$
   Step 2: NaOH(aq), reflux
   2-chlorobutane, $CH_3CHClCH_2CH_3 \rightarrow$
   butan-2-ol, $CH_3CHOHCH_2CH_3$

4 Reaction 1: H_2O(g) + H_3PO_4 at high temperature
   Equation $CH_2=CHCH_2OH + H_2O \rightarrow$
   $CH_2CH(OH)CH_2OH$

   Reaction 2: HBr
   Equation $CH_2=CHCH_2OH + HBr \rightarrow$
   $CH_3CH(Br)CH_2Br$

   Reaction 3: H_2/Ni catalyst
   Equation $CH_2=CHCH_2OH + H_2 \rightarrow CH_3CH_2CH_2OH$

17.1

> **Using the M and M + 1 peaks**
>
> 9

> **Drug testing in sport**
>
> 277

1 a $C_9H_{20} \rightarrow C_9H_{20}^+ + e^-$
   b $CH_3CH_2CH_2CH_2OH \rightarrow$
   $CH_3CH_2CH_2CH_2OH^+ + e^-$

2 The M + 1 peak arises because there is a small percentage of carbon-13 in the sample and this adds one to the molecular mass of the molecular ion.

3 Compound A M^+ peak is at 72 so the molecular mass is 72.

   Compound B: M^+ peak at 100 so the molecular mass is 100

4 m/z = 29: $CH_3CH_2^+$
   m/z = 31: CH_2OH^+

5 a $CH_3CH_2CH_2^+$ and CH_3CO^+
   b $CH_3CH_2CH_2CH_2COCH_3^+ \rightarrow$
   $CH_3CH_2CH_2^+ + CH_2COCH_3\bullet$
   $CH_3CH_2CH_2CH_2COCH_3^+ \rightarrow$
   $CH_3CH_2CH_2CH_2\bullet + CH_3CO^+$

17.2

1 Stretching and bending.

2 a O–H at 3200–3600 cm^{-1}
   b O–H at 2500–3300 cm^{-1} and C=O at 1630–1820 cm^{-1}
   c O–H at 3200–3600 cm^{-1} and C=O at 1630–1820 cm^{-1}
   d C=O at 1630–1820 cm^{-1} O–H at 2500–3330 cm^{-1} and O–H at 3200–3600 cm^{-1}

3 a X is O–H in carboxylic acids and Y is C=O
   b X is C–H and Y is C=O and Z is C–O
   c X is O–H in alcohols and Y is C–H
   d X is C–H and Y is C=O

Answers

18.1

1. **a** **A:** 0, **B:** 2, **C:** 1
   b overall order: 3
2. **a** **i** no change **ii** rate doubles
   iii rate quadruples
   b Rate increases by a factor of 8
3. **a** **A:** 1, **B:** 0, **C:** 1
   b overall order: 2
   c rate = $k[\mathbf{A}][\mathbf{C}]$
   d Units of k = $dm^3\ mol^{-1}\ s^{-1}$
4. rate = $k[\mathbf{A}]^2[\mathbf{B}]$
   $k = 27.0\ dm^6\ mol^{-2}\ s^{-1}$

18.2

> **Analysing by colorimetry**
>
> Monitor the concentration of $H^+(aq)$ ions with time with a pH meter.

> **Exponential decay in medicine**
>
> 1 Salbutamol ~ 1/64 remains;
>   Salmeterol ~ 1/4 remains
> 2 Salbutamol when quick action is required during an asthma attack; Salmeterol slower release taken for longer term relief, e.g. just before sleep.

1. Rate constant = gradient of the straight line graph.
2. Reaction 1, $0.0151\ s^{-1}$, Reaction 2, $4.20 \times 10^{-3}\ s^{-1}$
3. **a**

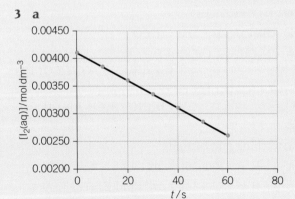

   b Zero order: a straight-line graph with a negative gradient.
4. **a** $t_{1/2}$ is constant at 50 s; rate = $k[N_2O]$
   b When t = 70 s, rate = gradient of tangent = $0.00524\ mol\ dm^{-3}\ s^{-1}$
   c **i** At 70 s, $[N_2O]$ = 0.400 mol dm^{-3} and
   k = rate/$[N_2O]$ = 0.0054/0.400 = $0.0135\ s^{-1}$
   ii $k = \ln 2/t_{1/2} = \ln 2/52 = 0.0133\ s^{-1}$

18.3

> **Log–Log graphs**
>
> 1 Gradient = order = 2
> 2 Intercept = Log(rate) = 0.9
>   rate constant = $10^{0.9}$ = 7.9

> **Iodine clock procedure**
>
> 1 Thiosulfate ions react with iodine forming I^- ions which are not coloured. As soon as all the thiosulfate has reacted, iodine ions will start to build up in solution producing the colour.
> 2 The amount of iodine is removed each time by the thiosulfate is constant. This means that rate is proportional to $1/t$.

1. Zero order 0 is a horizontal line, 1st order is a straight line starting at the origin, 2nd order is an upward curve, starting at the origin.
2. For a first order reaction, gradient gives the rate constant k.
3. **a** $1/t$ values in s^{-1}: 0.030, 0.023, 0.017, 0.013, 0.005
   b

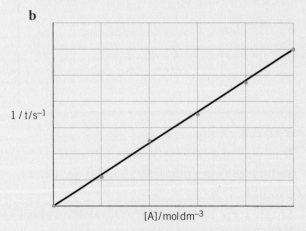

   c 1st order with respect to **A**.

18.4

1. A one step reaction would require 4 molecules colliding at the same time which is very unlikely
2. **a** rate = $k[H_2O_2(aq)][Br^-(aq)]$
   b $2H_2O_2(aq) \rightarrow 2H_2O(l) + O_2(g)$
   c BrO^-
   d Catalyst as it is used in first step but regenerated in second step. Overall, it is not used up.
3. $N_2O(g) \rightarrow N_2(g) + O(g)$ slow
   $N_2O(g) + O(g) \rightarrow N_2(g) + O_2(g)$ fast

18.5

1. A graph of ln k against $1/T$ is plotted. The gradient of the straight line graph is equal to $-E_a/R$. The intercept with the y axis is ln A.

2. **a** k increases **b** k decreases **a** k increases

3.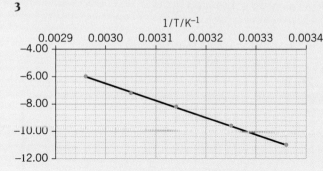

$E_a = 103$ kJ mol^{-1}, $A = 3.98 \times 10^{30}$ s^{-1}

19.1

> **Determining K_c from experimental results**
>
> 1. Set up another identical experiment at start and sample after 2 weeks. If the result is the same, then it is safe to assume that equilibrium had been established after 1 week.
> 2. $K_c = 4.6$

1. **a** $\dfrac{[HBr(g)]^2}{[H_2(g)][Br_2(g)]}$ No units

   b $\dfrac{[CH_3OH(g)]}{[CO(g)][H_2(g)]^2}$ dm^6 mol^{-2}

   c $K_c = [CO_2(g)]$ mol dm^{-3}

2. **a** $n(H_2) = 0.210$ mol, $n(HI) = 0.480$ mol
   b $K_c = 18.3$ (no units)

3. **a** $n(N_2) = 2.40$ mol, $n(H_2) = 4.00$ mol
   b $K_c = 0.0807$ dm^6 mol^{-2}

19.2

> **Dissolving oxygen in the blood**
>
> 1. The amount of oxygen that can dissolve in the blood is proportional to the partial pressure of O_2. So as the partial pressure of oxygen decreases with altitude, the amount of oxygen that dissolves in the blood also decreases.

2. **a** 0.34 atm
   b $0.34 \times 0.21 = \sim 0.07$ atm
   c With pure oxygen on Everest, $p(O_2) = 34$ kPa which is about 50% more than at sea level. But at sea level, $p(O_2)$ of pure O_2 would be 1 atm which would dissolve far too much oxygen in the blood (about 500% more than from air at sea level).

1. **a** N_2: $x = 0.75$, $p = 240$ kPa
      O_2: $x = 0.25$, $p = 80$ kPa
   b O_2: $x = 0.2$, $p = 100$ kPa
      N_2: $x = 0.5$, $p = 250$ kPa
      H_2: $x = 0.3$, $p = 150$ kPa

2. **a** $K_p = \dfrac{p_{H_2} \times p_{I_2}}{p_{HI}} = 8.64$

   b $K_p = \dfrac{p_{NO_2} \times p_{O_2}}{p_{NO_2}^2} = 53$ kPa

   c $K_p = \dfrac{p_{SO_3}^2}{p_{SO_2}^2 \times p_{O_2}} = 0.16$ kPa^{-1}

3. **a** 1.5 mol SO_2, 0.75 mol O_2, 0.5 mol SO_3.
      Total = 2.75 mol
   b $x(SO_2) = 0.545$; $x(O_2) = 0.273$
      $x(SO_3) = 0.182$
      $p(SO_2) = 491$ kPa
      $p(O_2) = 245$ kPa
      $p(SO_3) = 164$ kPa
   c 4.55×10^{-4} kPa^{-1}

19.3

1. Reaction A: endothermic
   K_c increases with increasing temperature.
   Reaction B: exothermic
   K_c decreases with increasing temperature.

2. **a** $K_c = \dfrac{[C(aq)]}{[A(aq)][B(aq)]^2}$

   b **i** K_c increases (reaction is endothermic)
      ii No effect (K_c only changes with temperature).

   c Blue colour first deepens with increased [C]. Then, blue colour decreases in intensity as increased [C] shifts equilibrium to left. Final blue colour is more intense than before addition of C.

3. **a** $K_p = \dfrac{p(H_2) \times p(CO_2)}{p(H_2O) \times p(CO)}$

   b Equilibrium amount of CO_2 = 0.6 mol.
      Total number of moles = 1.0 + 1.0 + 0.6 + 0.6
      = 3.2

Answers

If total pressure = P,

$$K_p = \frac{(0.6/3.2 \times P) \times (0.6/3.2 \times P)}{(1.0/3.2 \times P) \times (1.0/3.2 \times P)} = 0.36$$

c i No effect. There is the same number of gaseous moles on both sides. K_p does not change as K_p is only affected by temperature and not pressure.

ii Percentage of hydrogen decreases as forward reaction is exothermic and equilibrium position moves left. K_p decreases as forward reaction is exothermic.

4 Increasing pressure increases all concentrations. As there are more concentration terms on the bottom of the expression, the ratio of products to reactants becomes less than K_c and the reaction is no longer at equilibrium. The concentrations of reactants decreases and products increases until the ratio once again is equal to K_c. The overall effect shifts the equilibrium to the right and increases the yield of NH_3.

20.1

1 a i A proton donor; ii A proton acceptor
   b i NO_3^- ii Br^- iii $CH_3CH_2COO^-$
2 a $CH_3COOH(aq) + H_2O(aq) \rightleftharpoons$
       $H_3O^+(aq) + CH_3COO^-(aq)$
       Acid 1   Base 2   Acid 2   Base 1
   b $NH_3(aq) + H_2O(l) \rightleftharpoons NH_4^+(aq) + OH^-(aq)$
       Base 2   Acid 1   Acid 2   Base 1
   c $HCO_3^-(aq) + HCl(aq) \rightleftharpoons H_2CO_3(aq) + Cl^-(aq)$
       Base 2   Acid 1   Acid 2   Base 1
3 a $2Al(s) + 3H_2SO_4(aq) \rightarrow Al_2(SO_4)_3(aq) + 3H_2(g)$
       $2Al(s) + 6H^+(aq) \rightarrow 2Al^{3+}(aq) + 3H_2(g)$
   b $2HNO_3(aq) + FeO(s) \rightarrow Fe(NO_3)_2(aq) + H_2O(l)$
       $2H^+(aq) + FeO(s) \rightarrow Fe^{2+}(aq) + H_2O(l)$
   c $2HCl(aq) + Ca(OH)_2(aq) \rightarrow$
       $CaCl_2(aq) + 2H_2O(l)$
       $H^+(aq) + OH^-(aq) \rightarrow H_2O(l)$
   d $2H_3PO_4(aq) + 3Na_2CO_3(aq) \rightarrow$
       $2Na_3PO_4(aq) + 3H_2O(l) + 3CO_2(g)$
       $2H^+(aq) + CO_3^{2-}(aq) \rightarrow H_2O(l) + CO_2(g)$

20.2

1 a 1,000,000 times
   b pH 2
   c i pH 2 ii pH 13 iii pH 4
   d i 10^{-3} mol dm^{-3} ii 10^{-6} mol dm^{-3}
      iii 10^{-12} mol dm^{-3}

2 a pH 2.60 b pH 5.09 c pH 7.43
   d pH 9.35 e pH 11.91 f pH −0.36
3 a 1.48×10^{-3} mol dm^{-3}
   b 1.23×10^{-8} mol dm^{-3}
   c 5.37×10^{-13} mol dm^{-3}
   d 2.04×10^{-10} mol dm^{-3}
   e 1.95×10^{-5} mol dm^{-3}
   f 12.02 mol dm^{-3}

4 a i pH = 1.78 ii pH = 2.90
   b volume = 41.67 cm^3

20.3

Weak acids in wine

1 [structural formula equilibria shown]

2 $C_3H_4OH(COOH)_3(aq) \rightleftharpoons$
    $H^+(aq) + C_3H_4OH(COOH)_2(COO^-)(aq)$
   $C_3H_4OH(COOH)_2(COO^-)(aq) \rightleftharpoons$
    $H^+(aq) + C_3H_4OH(COOH)(COO^-)_2(aq)$
   $C_3H_4OH(COOH)(COO^-)_2(aq) \rightleftharpoons$
    $H^+(aq) + C_3H_4OH(COO^-)_3(aq)$

1 a $K_a = \dfrac{[H^+(aq)][NO_2^-(aq)]}{[HNO_2(aq)]}$ mol dm^{-3}

   b $K_a = \dfrac{[H^+(aq)][C_6H_5COO^-(aq)]}{[C_6H_5COOH(aq)]}$ mol dm^{-3}

   c $K_a = \dfrac{[H^+(aq)][HS^-(aq)]}{[H_2S(aq)]}$ mol dm^{-3}

2 a i pK_a 7.02 ii pK_a 3.15 iii pK_a 7.20;
   b i 1.2×10^{-2} mol dm^{-3}
      ii 2.0×10^{-8} mol dm^{-3}
      iii 1.7×10^{-5} mol dm^{-3}

3 a $K_a = \dfrac{[H^+(aq)][H_2PO_4^-(aq)]}{[H_3PO_4(aq)]}$

      $K_a = \dfrac{[H^+(aq)][HPO_4^{2-}(aq)]}{[H_2PO_4^-(aq)]}$

      $K_a = \dfrac{[H^+(aq)][PO_4^{3-}(aq)]}{[HPO_4^{2-}(aq)]}$

   b From 1st to 2nd to 3rd dissociations of H_3PO_4, K_a decreases and pK_a increases

Answers

20.4

> **Checking the approximations**
> 1. Using quadratic, pH = 2.21.
>   Using the approximation, pH = 2.20.
>   Therefore approximation is justified.
> 2. Approximation is valid when $K_a < 10^{-2}$ mol dm^{-3}

1. a i $\dfrac{[H^+(aq)][CN^-(aq)]}{[HCN(aq)]}$

   ii $\dfrac{[H^+(aq)][C_6H_5COO^-(aq)]}{[C_6H_5COOH(aq)]}$

   b i 1 The dissociation of water is negligible, i.e. $[H^+]_{eqm} \sim [A^-]_{eqm}$
      2 The concentration of the undissociated acid is much greater than the H^+ concentration at equilibrium, i.e., $[HA]_{start} \sim [HA]_{eqm}$.

      ii $[H^+(aq)] = \sqrt{K_a \cdot [HA(aq)]}$

2. a i pH = 2.44   ii pH = 2.83
   b i $K_a = 1.83 \times 10^{-4}$ mol dm^{-3}, $pK_a = 3.74$
      ii $K_a = 1.19 \times 10^{-5}$ mol dm^{-3}, $pK_a = 4.92$

3. a The second dissociation of H_2SO_4 is incomplete.
   b Calculate $[H^+]$ from first and second dissociations of sulfuric acid (strong and weak acids).
      Add the $[H^+]$ concentrations together and calculate the pH using $-\log[H^+]$
   c There is a large concentration of H^+ from the first dissociation of sulfuric acid already so $[H^+(aq)] \gg [A^-(aq)]$. To solve the pH,
   $1.2 \times 10^{-2} = \dfrac{(0.01 + x)x}{(0.01 - x)}$

20.5

> **pOH**
> 1. a 13.10   b 11.51
> 2. a 5.75×10^{-4} mol dm^{-3} b 2.40×10^{-4} mol dm^{-3}

1. a $K_w = [H^+(aq)][OH^-(aq)]$; b 10^{-10} mol dm^{-3}
   c K_w has two concentrations multiplied together.
2. a pH = 12.08; b pH = 11.13; c pH = 13.24
3. a 0.0447 mol dm^{-3} b 3.72×10^{-3} mol dm^{-3}
   c 0.0338 mol dm^{-3}
4. a i pH = 6.52   ii pH = 7.27
   b Water is neutral as $[H^+(aq)] = [OH^-(aq)]$.
   c Endothermic as K_w increases with temperature.

21.1

1. a 4.82
   b $CH_3(CH_2)_2COOH(aq) \rightleftharpoons$
      $H^+(aq) + CH_3(CH_2)_2COO^-(aq)$
   c The H^+ reacts with OH^- added.
      $CH_3(CH_2)_2COOH$(dissociates, moving the position of equilibrium to the right to restore most of the removed $H^+(aq)$ ions
2. a pH = 4.20   b pH = 3.50   c pH = 4.80
3. a pH = 4.56   c pH = 5.24

21.2 Answers

> **The Henderson–Hasselbalch equation**
> a $pH = 3.75 + \log\dfrac{0.750}{0.250} = 4.23$
> b $pH = 3.83 + \log\dfrac{0.400}{0.800} = 3.53$
> c $pH = 3.08 + \log\dfrac{0.625}{0.475} = 3.20$

1. a i $H_2CO_3(aq) \rightleftharpoons H^+(aq) + HCO_3^-(aq)$
      ii $H^+(aq)$ ions react with the $HCO_3^-(aq)$, the equilibrium position shifts to the left, removing most of the $H^+(aq)$ ions.
2. a i pH 7.40, $[H^+] = 3.98 \times 10^{-8}$ mol dm^{-3}
         pH 6.92, $[H^+] = 1.20 \times 10^{-7}$ mol dm^{-3}
      ii 3 times more concentrated.
3. a ratio = 4.46/1
   b i $K_a\ 10^{-7.40} \times 20 = 7.96 \times 10^{-7}$ mol dm^{-3}
      ii $HCO_3^-/H_2CO_3 = 7.96 \times 10^{-7}/10^{-7.20}$
          $= 12.6 : 1$

21.3

> **Using a pH meter**
> At pH = 5, $[H^+(aq)] = 1 \times 10^{-5}$ mol dm^{-3};
> At pH = 4.65, $[H^+(aq)] = 2.24 \times 10^{-5}$ mol dm^{-3}.
>
> pH % error = 7.5%; $[H^+(aq)]$ % error = 104%.

1. The equivalence point of the titration is the volume of one solution that exactly reacts with the volume of the other solution.
   The end point of a titration shows an indicator colour that is in between the two extreme colours when the indicator contains equal concentrations of HA and A^-.
2. a $HA(aq) \rightleftharpoons A^-(aq) + H^+(aq)$
   b i Yellow   ii Red   iii Orange
3. The starting pH and first section will be at higher pH and the sharp vertical section would be at 12.5 cm^3.

Answers

22.1

1 a i first ionisation energy of calcium
  ii second electron affinity of sulfur
  iii enthalpy change of formation of calcium chloride

  b i $Mg^{2+}(g) + O^{2-}(g) \rightarrow MgO(s)$
  ii $F(g) + e^- \rightarrow F^-(g)$
  iii $\tfrac{1}{2}O_2(g) \rightarrow O(g)$

2 a A: enthalpy change of formation of sodium bromide
  B: first electron affinity of bromine
  C: enthalpy change of atomisation of sodium
  D: first ionisation energy of sodium
  E: enthalpy change of atomisation of bromine

b

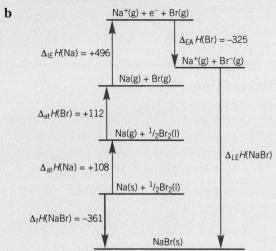

c -752 kJ mol^{-1}

3 a

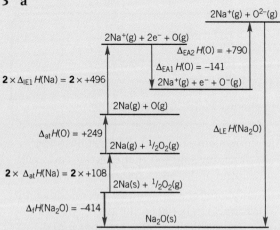

b -2520 kJ mol^{-1}

22.2

Experimental determination of the enthalpy change of solution
a $+28.7$ kJ mol^{-1} **b** -41.4 kJ mol^{-1}

Cold packs – where does the heat go?
The reaction in the cold pack is endothermic as more energy is required to break bonds than is given out when new bonds form. The energy is taken from the immediate surroundings which are the part of the person in contact with the cold pack.

1 a The standard enthalpy change of solution is the enthalpy change that takes place when one mole of a solute dissolves in a solvent.

  b The enthalpy change of hydration is the enthalpy change that accompanies the dissolving of one mole of gaseous ions in water to form aqueous ions.

2 a Lattice enthalpy of KF; Enthalpy change of hydration of K^+. Enthalpy change of hydration of F^-

  b $KF(s) + aq \rightarrow K^+(aq) + F^-(aq)$

  c

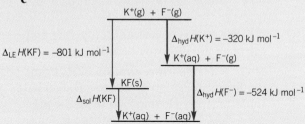

  d -43 kJ mol^{-1}

3 $q = 55.18 \times 4.18 \times 29.0/1000 = 6.6889196$ kJ
  $n(LiF) = 5.18/25.9 = 0.200$ mol
  $\Delta_{sol}H(LiF) = 6.5735934/0.200 = -33.4$ kJ mol^{-1}

4 a

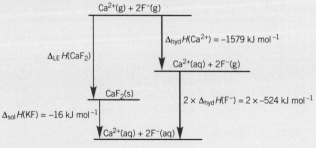

  b $\Delta_{LE}H(CaF_2) = -2611$ kJ mol^{-1}

578

22.3

> **Ionic liquids**
>
> **a** The ions are very different sizes and they are unable to pack together tightly into an ionic lattice
>
> **b** $C_6H_{11}N_2Cl$
>
> **c** There are many possible uses including synthesis of drugs, cellulose processing, algae processing, dispersants, gas handling, nuclear fuel reprocessing, waste recycling and batteries.

1 The smaller the ionic size, the greater the charge density and attraction for water molecules, and the more exothermic the hydration enthalpy.

The larger the ionic charge, the greater the charge density and attraction for water molecules, and the more exothermic the hydration enthalpy.

2 a Na^+ ions are smaller and has a greater charge density than K^+ ions. In NaCl, there is a greater attraction between ions and the lattice enthalpy of NaCl is more exothermic.

b Mg^{2+} ions are smaller and have a greater charge than Na^+ ions, so Mg^{2+} ions have a greater charge density than Na^+ ions. In $MgCl_2$, there is a greater attraction between ions and the lattice enthalpy of $MgCl_2$ is more exothermic.

3 a Na^+ ions are smaller than K^+ ions and have greater attraction.

Cl^- ions are smaller than Br^- ions and have greater attraction.

So in each pair, there is the ion with greater attraction and the ion with less attraction. It is not possible to predict their relative effects on the overall attraction.

b F^- ions are smaller than O^{2-} ions, so have the greater attraction in terms of ionic size.

F^- ions have a smaller charge than O^{2-} ions, so will less attraction.

So in each pair, the two comparisons oppose one another and it is difficult to know which has the greater effect.

22.4

1 a decrease as liquid is less disordered than gas

b increase as a gas is produced

c decrease as there are fewer moles of gas produced

2 a -509 kJ mol^{-1} **b** -391 kJ mol^{-1}

3 $+230$ J K^{-1} mol^{-1}

22.5

> **Diamonds are forever … or are they?**
>
> The rate is very slow as there is a large activation energy.

1 a $\Delta G = \Delta H - T\Delta S$

b $\Delta G < 0$

c The rate may be very slow

2 $\Delta S = +215$ J mol^{-1} K^{-1}

$\Delta G = +91 - 298 \times 0.215 = +26.93$ kJ mol^{-1}

As $\Delta G > 0$, reaction is not feasible at 25 °C.

Minimum temperature for feasibility = 91/0.215 = 423 K = 150 °C

3 a $\Delta S = +356$ J mol^{-1} K^{-1}, $\Delta H = +154$ kJ mol^{-1}

b $\Delta G = +47.9$ kJ mol^{-1} so reaction is not feasible at 25 °C

c Minimum temperature = 432.6 K which is 160 °C

23.1

1 a Oxidised: Mg from 0 to +2, Reduced: Zn from +2 to 0

Oxidising agent: Zn^{2+}, Reducing agent: Mg

b Oxidised: Cu from 0 to +2, Reduced: S from +6 to +4

Oxidising agent: H_2SO_4, Reducing agent: Cu

c Oxidised: Fe from +2 to +3, Reduced: N from +3 to +2

Oxidising agent: NO_2^-, Reducing agent: Fe^{2+}

2 a $6Fe^{2+} + Cr_2O_7^{2-} + 14H^+ \rightarrow 6Fe^{3+} + 2Cr^{3+} + 7H_2O$

b $3Zn + 2VO_2^+ + 8H^+ \rightarrow 3Zn^{2+} + 2V^{2+} + 4H_2O$

c $5H_2O_2 + 2MnO_4^- + 6H^+ \rightarrow 5O_2 + 2Mn^{2+} + 8H_2O$

3 a $2MnO_4^- + 6H^+ + 5H_2S \rightarrow 2Mn^{2+} + 5S + 8H_2O$

b $BrO_3^- + 5Br^- + 6H^+ \rightarrow 3Br_2 + 3H_2O$

4 a $MnO_2 + 4H^+ + 2e^- \rightarrow Mn^{2+} + 2H_2O$

b $NO_3^- + 4H^+ + 3e^- \rightarrow NO + 2H_2O$

5 a $8HI + H_2SO_4 \rightarrow 4I_2 + H_2S + 4H_2O$

b $PbO_2 + 4HCl \rightarrow PbCl_2 + Cl_2 + 2H_2O$

Answers

23.2

> **Manganate(VII) titrations**
>
> percentage uncertainty = $(2 \times 0.05/22.30) \times 100$
> = 0.45%

> **Analysing the percentage purity of an iron(II) compound**
>
> $n(MnO_4^-) = 1.44 \times 10^{-4}$ mol
>
> $n(Fe^{2+}) = 7.20 \times 10^{-4}$ mol
>
> percentage purity = 13.2%

1. $MnO_4^-(aq) + 8H^+(aq) + 5Fe^{2+}(aq) \rightarrow Mn^{2+}(aq) + 5Fe^{3+}(aq) + 4H_2O(l)$
2. Hydrochloric acid contains chloride ions that would be oxidised instead.
3. 28.50 cm^3
4. a $Cr_2O_7^{2-} + 14H^+ + 6Fe^{2+} \rightarrow 2Cr^{3+} + 6Fe^{3+} + 7H_2O$
   b 1 mole
5. 1.19412 g of Fe so % by mass = 18.5%
6. Molar mass = 287.8 g mol^{-1}, $x = 6$
   Formula = $Fe(NO_3)_2 \cdot 6H_2O$

23.3

> **Analysis of oxidising agents**
>
> The blue–black colour is a complex of starch and iodine. If added too early, the starch-iodine complex may precipitate out of solution, preventing some of the iodine reacting with the thiosulfate. Close to the end-point, the iodine concentration is low enough for the complex not to precipitate out.

> **Analysis of household bleach**
>
> 1 1 mol NaOCl contains 1 mol NaClO
>   100 cm^3 of the bleach contains 0.0605 mol NaClO
>   Mass of NaClO is 100 cm^3 bleach
> $= 0.0605 \times M(NaClO) = 0.0605 \times 74.5 = 4.5$ g.
> 2 $n(S_2O_3^{2-}) = 1.74 \times 10^{-3}$ mol
> $n(ClO^-) = 8.70 \times 10^{-4}$ mol
>   Concentration of ClO$^-$ in bleach = 0.870 mol dm^{-3}

1. $2S_2O_3^{2-}(aq) + I_2(aq) \rightarrow 2I^-(aq) + S_4O_6^{2-}(aq)$
2. All of the oxidising agent reacts with the excess KI producing iodine, I_2, for the titration.
3. 20.00 cm^3
4. 1 mol IO_3^- produces 3 mol I_2 which reacts with 2 mol $S_2O_3^{2-}$.
   Therefore 1 mol IO_3^- is equivalent to 6 mol $S_2O_3^{2-}$
5. In the titration, $n(S_2O_3^{2-}) = n(Cu^{2+})$
   $= 5.65175 \times 10^{-4}$ mol
   In 250.0 cm^3 solution, $n(Cu^{2+})$
   $= 5.65175 \times 10^{-3}$ mol
   Molar mass = $1.365/(5.65175 \times 10^{-3})$
   = 241.5 g mol^{-1}
   Mass of H_2O = 241.5 – 187.5 = 54 g (equivalent to $3H_2O$)
   Formula = $Cu(NO_3)_2 \cdot 3H_2O$

23.4

> **Measuring standard cell potentials**
>
> 1 a From the Mg electrode
>   b oxidation: $Mg(s) \rightarrow Mg^{2+}(aq) + 2e^-$
>   reduction: $Fe^{3+}(aq) + e^- \rightarrow Fe^{2+}(aq)$
>   overall: $Mg(s) + 2Fe^{3+}(aq) \rightarrow Mg^{2+}(aq) + 2Fe^{2+}(aq)$
> 2 1.0 mol dm^{-3} FeSO$_4$ and 0.5 mol dm^{-3} Fe$_2$(SO$_4$)$_3$

1. a electrons; b ions
2. a The standard electrode potential is the e.m.f. of a half-cell connected to a standard hydrogen half-cell under standard conditions.
   b 298 K, solution concentrations of 1 mol dm^{-3} and a pressure of 100 kPa.
3.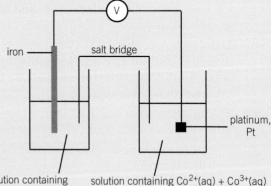
4. a i 1.02 V   i 1.22 V   i 1.56 V
   b oxidation: $Cu \rightarrow Cu^{2+} + 2e^-$
      reduction: $Cl_2 + 2e^- \rightarrow 2Cl^-$
      $Cu + Cl_2 \rightarrow Cu^{2+} + 2Cl^-$
      oxidation: $Al \rightarrow Al^{3+} + 3e^-$
      reduction: $Fe^{2+} + 2e^- \rightarrow Fe$
      $2Al + 3Fe^{2+} \rightarrow 3Fe + 2Al^{3+}$
      oxidation: $Zn \rightarrow Zn^{2+} + 2e^-$

reduction: $Ag^+ + e^- \rightarrow Ag$

$Zn + 2Ag^+ \rightarrow 2Ag + Zn^{2+}$

5 $H^+(aq), H_2(g)/Pt(s)$ 0 V

$Pb^{2+}(aq)/Pb(s)$ –0.13 V

$Ni^{2+}(aq)/Ni(s)$ –0.25 V

$Cd^{2+}(aq)/Cd(s)$ –0.40 V

$Co^{3+}(aq), Co^{2+}(aq)/Pt(s)$ +1.81 V

23.5

> **Standard electrode potentials and ΔG**
>
> a $Fe(s) + Cd^{2+}(aq) \rightarrow Fe^{2+}(aq) + Cd(s)$
> $E^\ominus_{cell}$ –0.04 V
> b $\Delta G^\ominus = -nFE^\ominus_{cell} = -2 \times 96500 \times 0.04 = -7720$ J
> $= -7.72$ kJ
>
> From $\Delta G^\ominus$, the reaction is feasibly but its small negative value suggest that the feasibility may be susceptible to small changes in conditions.

1 The reaction may have a slow rate with high activation energy.

   The concentration may not be 1 mol dm⁻³, changing the E value.

2 a i $Br_2(l)$ ii $U(s)$
   b $U(s)$ and $V^{2+}(aq)$ will react

   $3Fe^{3+}(aq) + U(s) \rightarrow 3Fe^{2+}(aq) + U^{3+}(aq)$

   $Fe^{3+}(aq) + V^{2+}(s) \rightarrow Fe^{2+}(aq) + V^{3+}(aq)$

3 $6Fe^{2+}(aq) + Cr_2O_7^{2-}(aq) + 14H^+(aq) \rightarrow$
       $6Fe^{3+}(aq) + 2Cr^{3+}(aq) + 7H_2O(l)$

   $2Fe^{2+}(aq) + Cl_2(g) \rightleftharpoons 2Fe^{3+}(aq) + 2Cl^-(aq)$

   $3Cl_2(g) + 2Cr^{3+}(aq) + 7H_2O(l) \rightarrow$
       $6Cl^-(aq) + Cr_2O_7^{2-}(aq) + 14H^+(aq)$

23.6

> **Lithum-ion and lithium-ion polymer cells**
>
> a +1.16 V
> b From +4 in CoO_2 to +3 in $LiCoO_2$

1 A primary cell is just used once and cannot be recharged. A secondary cell can be recharged.

2 A fuel cell uses the energy from the reaction of a fuel with oxygen to create a voltage. A fuel cell can be run continuously provided that the fuel and oxygen are continually being supplied to the cell.

3 a $2NiO(OH) + Cd + 2H_2O \rightarrow$
          $2Ni(OH)_2 + Cd(OH)_2$
   b Cd oxidised from 0 to +2; Ni reduced from +3 to +2

   c $E^\ominus$ = +0.49 V
   d $2Ni(OH)_2 + Cd(OH)_2 \rightarrow$
          $2NiO(OH) + Cd + 2H_2O$ (reverse of (a))

4 a $CH_3OH + 1½O_2 \rightarrow CO_2 + 2H_2O$
   b $CH_3OH + H_2O \rightarrow 6H^+ + 6e^- + CO_2$

24.1

1 Transition elements form compounds in which the transition element has different oxidation states.

   They form coloured compounds.

   Transition elements and their compounds can act as catalysts.

2 Scandium is a d-block metal as it has its highest energy electrons occupying the 3d sub-shell, $1s^22s^22p^63s^23p^64s^23d^1$. Scandium only forms the Sc^{3+} ion which has the electron configuration of $1s^22s^22p^63s^23p^6$ with an empty 3d sub-shell. As a transition element must form an ion with a partially filled d sub-shell, scandium is not a transition element.

3 a +3   b +6   c +4   d +2

4 Colour in transition metals is associated with the presence of a partially filled d-orbitals. Zinc forms one ion, Zn^{2+}, with a full 3d sub-shell. As there is no partially filled d-orbital, there is no colour.

24.2

> **Colours in transition metal chemistry**
>
> a red
> b yellow

1 a A complex ion consists of a central metal ion bonded to ligands by coordinate bonds.

   b The coordination number is the number of coordinate bonds attached to the central metal ion.

   c A ligand is defined as a molecule or ion that donates a pair of electrons to a central metal ion to form a coordinate or dative covalent bond.

2 $[Cr(H_2O)_4Cl_2]^+$

   The ion is octahedral

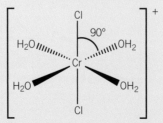

Answers

3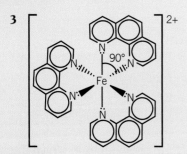

24.3

The role of cis–trans isomerism in medicine

cyclohexane-1,2-diamine

oxalate ion

1 a
cis [Co complex with OH₂, H₂O, NH₃, H₃N, NH₃, NH₃]⁺

b
trans [Ru complex with Cl, Cl, and two ethylenediamine ligands]

2 *Cis*-platin works by forming a platinum complex inside of a cell which binds to DNA and prevents the DNA of the cell from replicating. Activation of the cell's own repair mechanism eventually leads to apoptosis, or systematic cell death.

3
trans [Co(oxalate)₂(OH₂)₂]⁻

cis [Co(oxalate)₂(OH₂)₂]⁻ (two forms shown)

cis has optical isomers

24.4

1 a $Mn^{2+}(aq) + 2OH^-(aq) \rightarrow Mn(OH)_2(s)$

b Fe^{3+}, $1s^2 2s^2 2p^6 3s^2 3p^6 3d^5$

c The iron(II) hydroxide turns orange-brown as it is oxidised by the air to form iron(III) hydroxide.

2 Copper(II) hydroxide, $Cu(OH)_2$

b $Cu^{2+}(aq) + 2OH^-(aq) \rightarrow Cu(OH)_2(s)$

c Pale blue solution forms a deep dark-blue solution of $[Cu(NH_3)_4(H_2O)_2]^{2+}(aq)$.

3 a $[Cu(H_2O)_6]^{2+}(aq)$ is pale blue and $[CuCl_4]^{2-}(aq)$ is yellow.

b $[CuCl_4]^{2-}$ tetrahedral structure

Bond angle = 109.5°

The coordination number is 4.

c Ligand substitution

d Colour change from yellow, through green, to blue. Addition of water moves the position of equilibrium to the left to minimise the increase in water.

24.5

1 A reaction in which the same element is simultaneously oxidised and reduced.

Chlorine has been reduced from 0 in Cl_2 to -1 in HCl and has been oxidised from 0 in Cl_2 to $+1$ in HOCl. This is disproportionation as the same element, chlorine, has been both oxidised and reduced.

2 a Add $AgNO_3(aq)$, NaCl gives a white precipitate whereas NaI gives a yellow precipitate

b Add NaOH(aq), $CuSO_4$ would give a pale-blue precipitate whereas $FeSO_4$ gives a pale-green precipitate.

c Heat with dilute H_2SO_4. CuO forms a blue solution whereas Cu_2O forms a blue solution and a brown copper solid

d Add NaOH(aq), $FeCl_2$ forms a pale-green precipitate whereas $FeCl_3$ forms a brown precipitate.

3 $Zn(s) + H_2O_2(aq) + 2H^+(aq) \rightarrow$
$$Zn^{2+}(aq) + 2H_2O(l)$$

$2Cr^{2+}(aq) + H_2O_2(aq) + 2H^+(aq) \rightarrow$
$$2Cr^{3+}(aq) + 2H_2O(l)$$

Answers

$2Cr^{3+}(aq) + H_2O(l) + 3H_2O_2(aq) \rightarrow Cr_2O_7^{2-}(aq) + 8H^+(aq)$

$3Zn(s) + Cr_2O_7^{2-}(aq) + 14H^+(aq) \rightarrow 3Zn^{2+}(aq) + 2Cr^{3+}(aq) + 7H_2O(l)$

$6Cr^{2+}(aq) + Cr_2O_7^{2-}(aq) + 14H^+(aq) \rightarrow 8Cr^{3+}(aq) + 7H_2O(l)$

$Zn(aq) + 2Cr^{3+}(aq) \rightarrow Zn^{2+}(aq) + 2Cr^{2+}(aq)$

25.1

1. CH

2. Using x-ray diffraction, it is possible to measure the lengths of the bonds in a molecule.

   All carbon-to-carbon bonds have the same length: 0.139 nm, between the bond length of a single C–C bond: 0.153 nm and the bond length of a C=C double bond: 0.134 nm. This illustrates that benzene does not have single and double bonds but bonds intermediate in length.

3. The delocalised model has the pi-bond electron density spread out rather than having concentrated areas of electron density from separate double bonds as in the Kekule structure. Compounds containing delocalised electrons are more stable than those that do not have delocalised electrons.

4. a) 1,2,4-tribromobenzene structure
   b) 1,3-dinitrobenzene structure
   c) 2,4,6-trichlorotoluene structure
   d) phenyl ethanal chain structure
   e) 2,3-dichlorobenzoic acid structure

5. a) 1,2-dimethylbenzene
   b) 2-bromophenylamine
   c) 4-chlorobenzoic acid

25.2

1. i) An electron pair acceptor.
   ii) An atom or group replaces another atom or group

2. Step 1: $Cl_2 + FeCl_3 \rightarrow FeCl_4^- + Cl^+$

   Step 2: [mechanism diagram showing Cl+ attack on benzene, intermediate cation, then loss of H+ giving chlorobenzene]

   Step 3: $FeCl_4^- + H^+ \rightarrow FeCl_3 + HCl$

3. [benzene + $(CH_3)_2CHCl$ with $AlCl_3$ → isopropylbenzene + HCl]

4. a) i) phenyl isopropyl ketone structure
      ii) diphenyl ketone (benzophenone) structure

   b) Step 1: $H_3C-CH(CH_3)-COCl + AlCl_3 \rightarrow H_3C-CH(CH_3)-C^+=O + AlCl_4^-$

      Step 2: [mechanism diagram showing +COCH(CH_3)_2 attack on benzene, intermediate, loss of H+]

      Step 3: $AlCl_4^- + H^+ \rightarrow AlCl_3 + HCl$

25.3

The manufacture of phenol

1. Method 1: 36.7%, Method 2: 61.8%, Method 3: 77.0%

2. a) 104 tonnes b) 114 tonnes.

3. By finding a use for the N_2O waste product, the percentage yield increases.

Answers

1 a **A:** 4-bromo-2-chloro-3-nitrophenol
      B: 3-phenylpropan-1-ol
      C: 2,4,6-trichlorophenol
      D: 4-ethylphenol
   b **A**, **C**, and **D** are phenols

2 [benzene-1,2-diol + 2NaOH → disodium benzene-1,2-diolate + 2H$_2$O]

3 Add bromine water, both would decolourise the bromine water but phenol would produce a white precipitate.

4 Lone pair of electrons on the oxygen atom of the –OH group in phenol is delocalised into the ring. This increases the electron density in the ring in phenol. Phenol is more able to polarise bromine than benzene. In benzene all the electrons are delocalised throughout the structure and benzene is not able to polarise bromine.

5 [Mechanism: phenol + CH$_3$–Br (δ+/δ−) → arenium intermediate → 2-methylphenol + H$^+$]

25.4 Answers

Ortho, meta, and para

a *meta*-chlorophenol   3-chlorophenol
b *ortho*-bromomethylbenzene
   2-bromomethylbenzene
c *para*-nitromethylbenzene   4-nitromethylbenzene

Making TNT

1 [methylbenzene + HNO$_3$, H$_2$SO$_4$, 50°C → 2-nitromethylbenzene + H$_2$O]

2 [Mechanism: methylbenzene + NO$_2^+$ → arenium intermediate → 2-nitrophenol + H$^+$]

1 a 2,4-directing   b 2,4-directing
   c 3-directing   d 3-directing

2 a [1-bromo-2-nitrobenzene and 1-bromo-4-nitrobenzene]
   b [1-nitro-3-bromobenzene]
   c [ethyl 3-methylbenzoate]

3 a [benzene + HNO$_3$, H$_2$SO$_4$, 50°C → nitrobenzene + H$_2$O]

   [ethylbenzene + CH$_3$Cl, AlCl$_3$ → 3-methyl... (nitro compound shown) + HCl]

   b [benzene + CH$_3$Cl, AlCl$_3$ → methylbenzene + HCl]

   [ethylbenzene + HNO$_3$, H$_2$SO$_4$, 50°C → nitroethylbenzene + H$_2$O]

   [benzene + C$_2$H$_5$Cl, AlCl$_3$ → ethylbenzene + HCl]

   [ethylbenzene + HNO$_3$, H$_2$SO$_4$, 50°C → 2-nitroethylbenzene + H$_2$O]

584

Answers

26.1

Octanal in nature

Several aldehydes are present including hexanal and decanal, shown below.

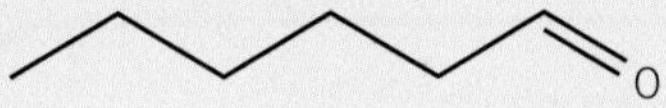

hexanal

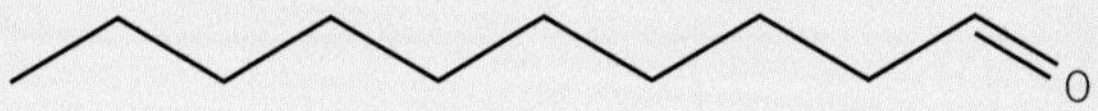

decanal

1 a [structure: hexan-3-one]

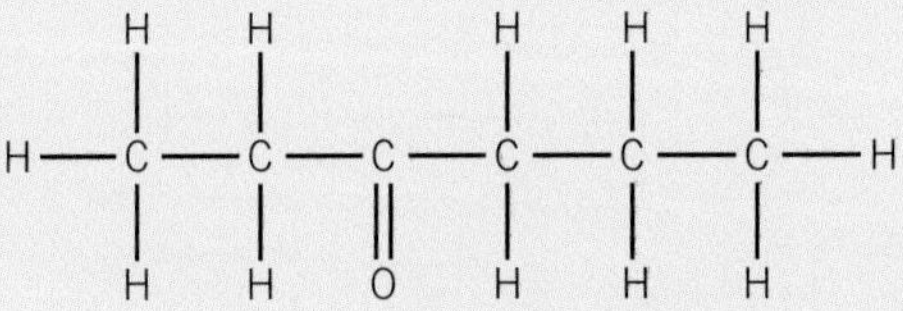

b [structure: 2,3-dimethyloctanal]

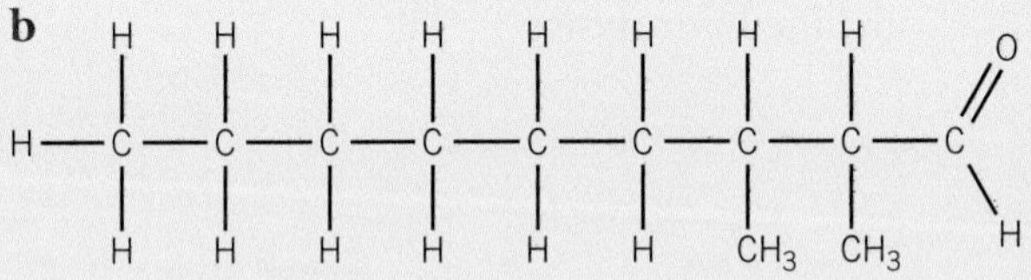

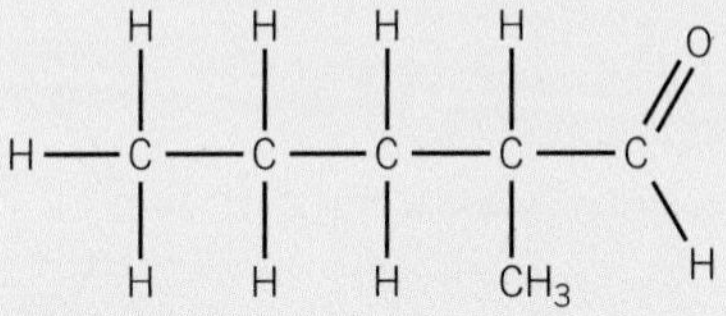

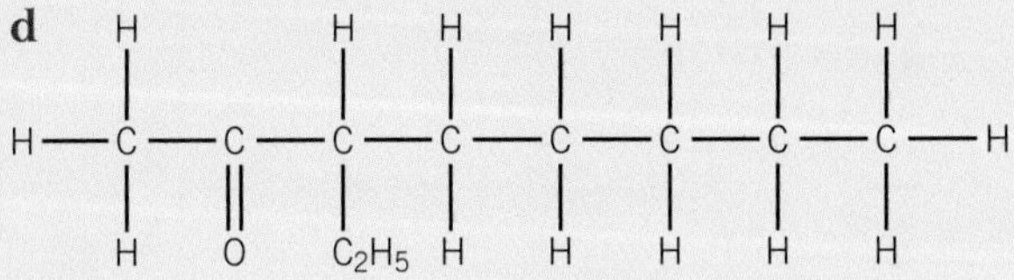

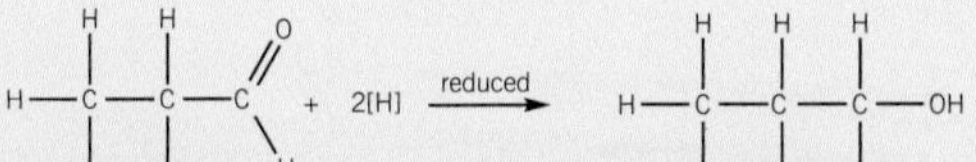

c [structure: 2-methylpentanal]

d [structure: 3-ethyloctan-2-one? — shown with C₂H₅ branch]

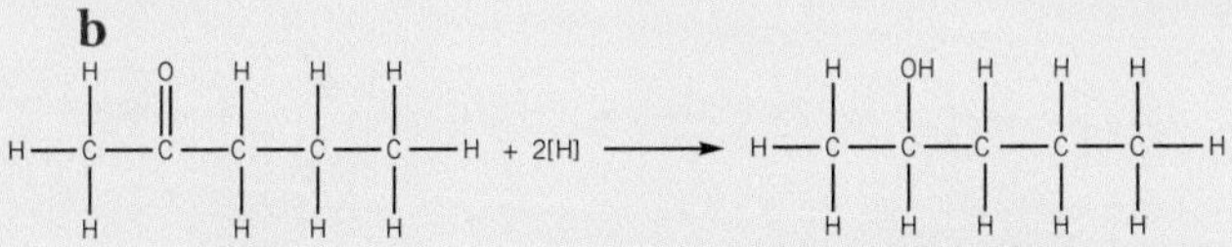

2 a propanal + 2[H] → propan-1-ol (primary alcohol)

b pentan-2-one (ketone) + 2[H] → pentan-2-ol (secondary alcohol)

3 A nucleophile is an electron pair donor. Aldehydes undergo addition reactions because they are unsaturated and contain a carbon to oxygen double bond.

4 [mechanism diagram showing nucleophilic addition of H⁻ to carbonyl, followed by protonation by water, giving alcohol + OH⁻]

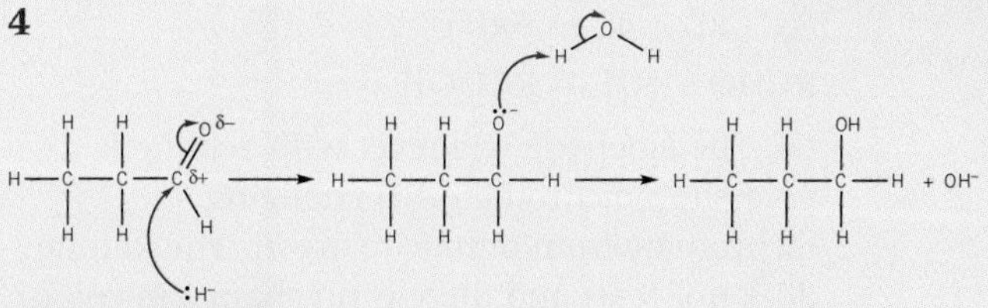

26.2 Answers

Testing for the carbonyl groups in aldehydes and ketones

$CH_3CH_2CH_2CHO$, $CH_3CH_2COCH_3$, $CH_3CH(CHO)CH_3$

The reaction of aldehydes and ketones with 2,4-DNP

a $H_3C-CH(CH_3)-CO-CH_2-CH_2-CH_3 + NH_2OH \longrightarrow H_3C-CH(CH_3)-C(NOH)-CH_2-CH_2-CH_3 + H_2O$

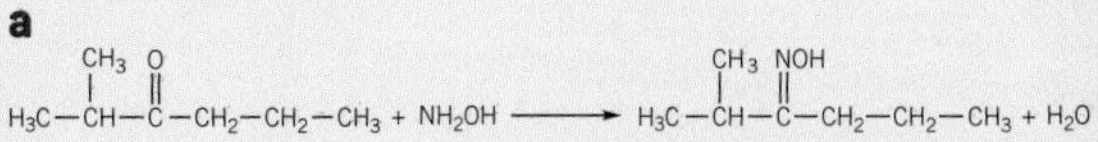

b PhCOCH₃ + NH₂NH₂ ⟶ PhC(NNH₂)CH₃ + H₂O

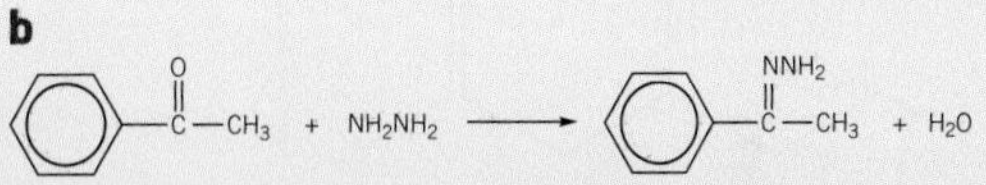

Application: Testing for the aldehyde group with Tollens' reagent

$CH_3CH_2CH_2CHO$, $CH_3CH(CHO)CH_3$

Identifying an aldehyde or ketone by melting point

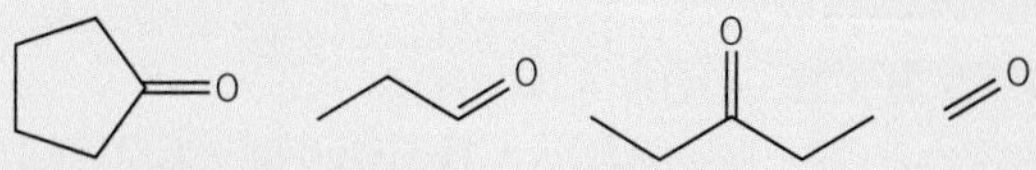

cyclopentanone   propanone   pentan-3-one   methanal

1 [structures: ketone (propanone) and aldehyde (propanal)]

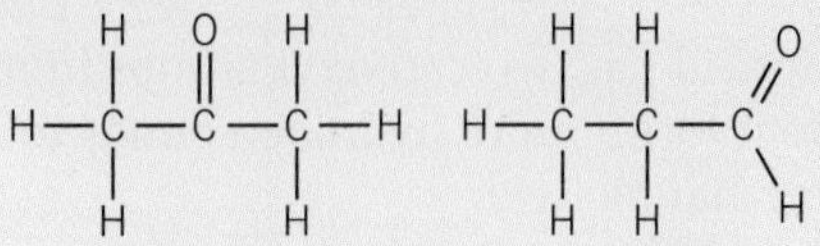

2 a Aldehydes and ketones both react with 2,4-dinitrophenylhydrazine to give a yellow-orange precipitate

b Only an aldehyde reacts with Tollens' reagent to give a silver mirror. The ketone will not react.

585

Answers

$Ag^+(aq) + e^- \rightarrow Ag(s)$

$RCHO + [O] \rightarrow RCOOH$

OR The aldehyde will react with acidified potassium dichromate(VI) changing dichromate from orange to green. The ketone does not react and there is no change in colour.

3 Add 2,4-dinitrophenylhydrazine to each of the three solutions. The compound that does not give a yellow precipitate is the alcohol, $CH_3CH_2CH_2CH_2OH$.

Take fresh sample of the compounds that give yellow precipitates. Add Tollens' reagent to each compound at a temperature of 50 °C in a water bath. The compound that gives a silver mirror is the aldehyde $CH_3CH_2CH_2CHO$. The ketone $CH_3CH_2COCH_3$ does not react.

26.3

1 a [structure: 3-ethylhexanoic acid type carboxylic acid]

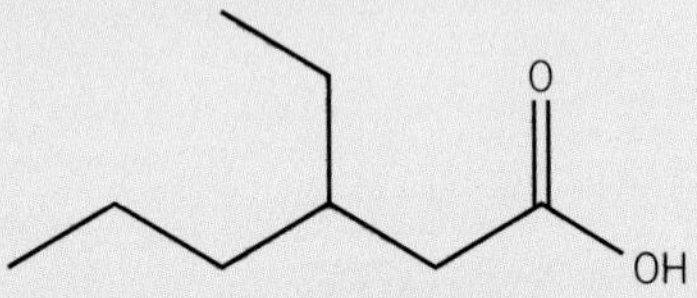

b [structure: 2-chloro-3-methylpentanoic acid type]

c [structure: 2-bromobenzoic acid]

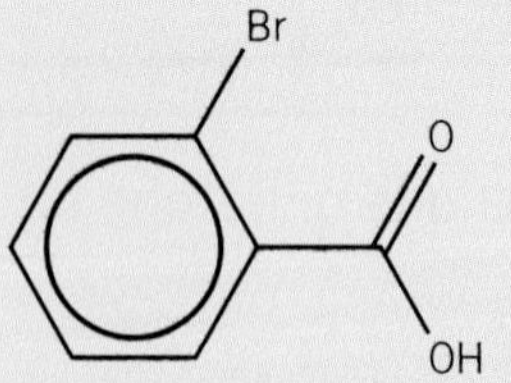

2 a $2CH_3CH_2COOH + Na_2CO_3 \rightarrow$
$2CH_3CH_2COONa + CO_2 + H_2O$

b $2CH_3CH_2CH_2COOH + CuO \rightarrow$
$(CH_3CH_2CH_2COO)_2Cu + H_2O$

3 Ethanedioic acid has two carboxyl groups in its structure so it forms more hydrogen bonds to water than ethanoic acid.

4 a [structure: 2-bromobenzoic acid + Mg → magnesium salt + H₂]

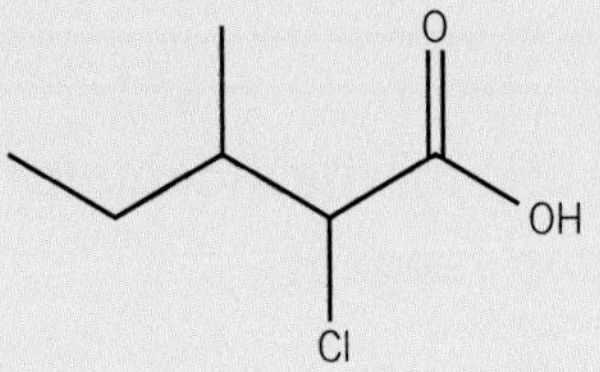

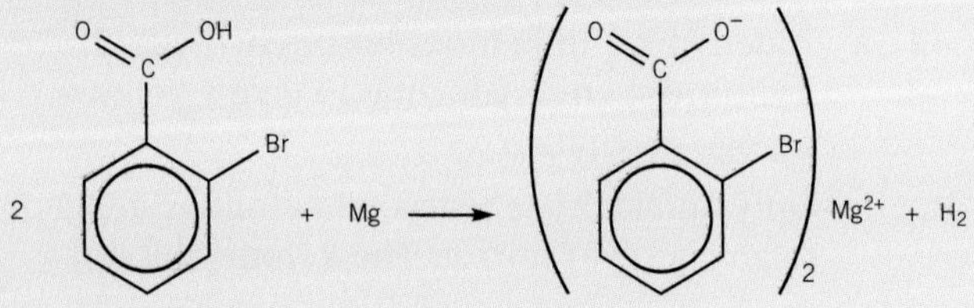

b [structure: propanedioic acid + 2NaOH → disodium salt + 2H₂O]

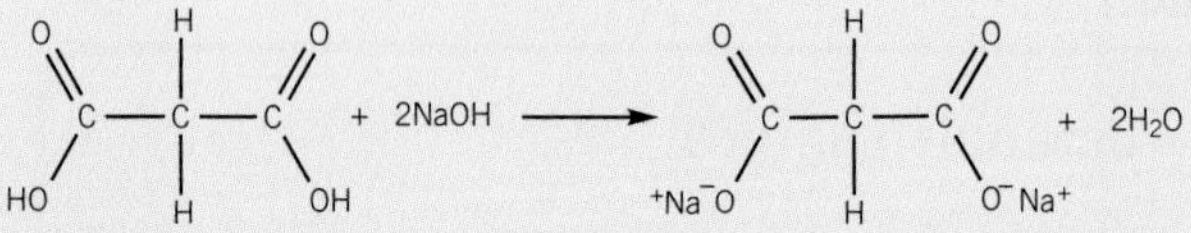

26.4

> **Making ethyl propanoate**
>
> $2CH_3CH_2COOH + NaCO_3 \rightarrow 2CH_3CH_2COONa + CO_2 + H_2O$

1 a butanoic acid;   b butylamide;
   c propanoic anhydride;   d pentanoyl chloride

2 acid hydrolysis

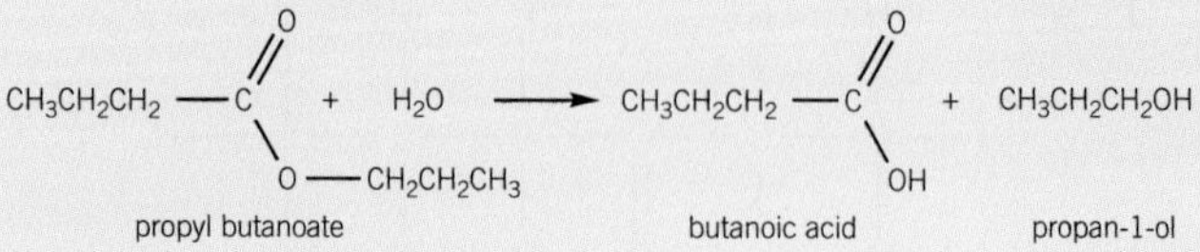

$CH_3CH_2CH_2-C(=O)-O-CH_2CH_2CH_3 + H_2O \rightarrow CH_3CH_2CH_2-C(=O)-OH + CH_3CH_2CH_2OH$

propyl butanoate → butanoic acid + propan-1-ol

alkaline hydrolysis

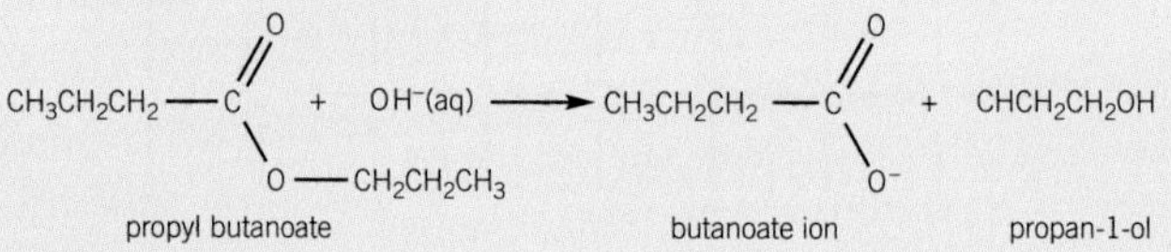

$CH_3CH_2CH_2-C(=O)-O-CH_2CH_2CH_3 + OH^-(aq) \rightarrow CH_3CH_2CH_2-C(=O)-O^- + CHCH_2CH_2OH$

propyl butanoate → butanoate ion + propan-1-ol

3 a $CH_3CH_2-C(=O)Cl + H_2O \rightarrow CH_3CH_2-COOH + HCl$

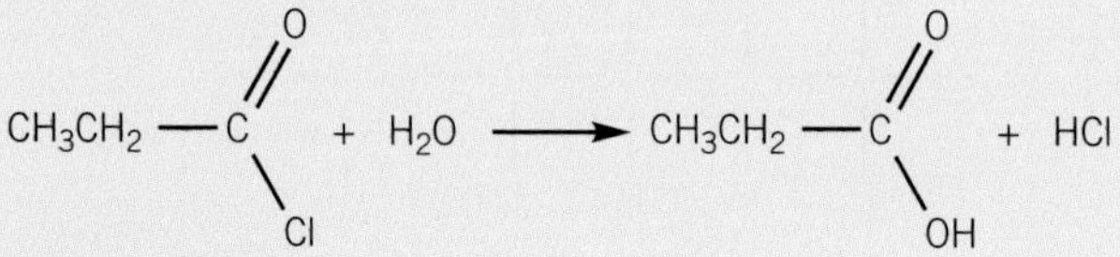

b $H_3C-C(=O)Cl + CH_3CH_2CH_2CH_2OH \rightarrow H_3C-C(=O)OCH_2CH_2CH_2CH_3 + HCl$

c $CH_3CH_2-COOH + SOCl_2 \rightarrow CH_3CH_2-C(=O)Cl + SO_2 + HCl$

$CH_3CH_2-C(=O)Cl + 2NH_3 \rightarrow CH_3CH_2-C(=O)NH_2 + NH_4Cl$

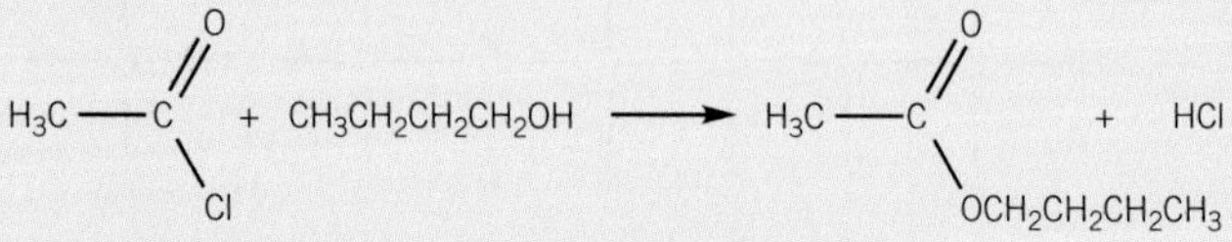

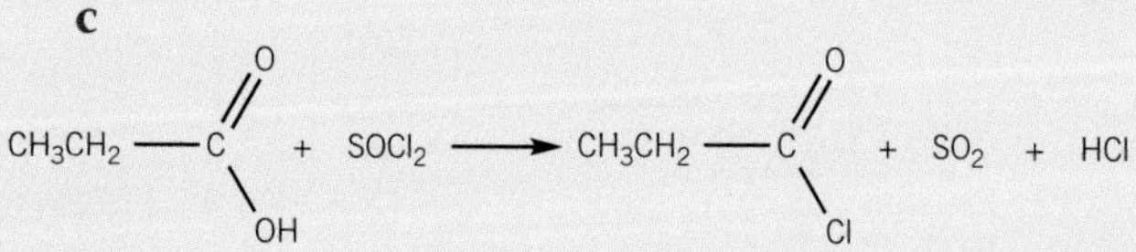

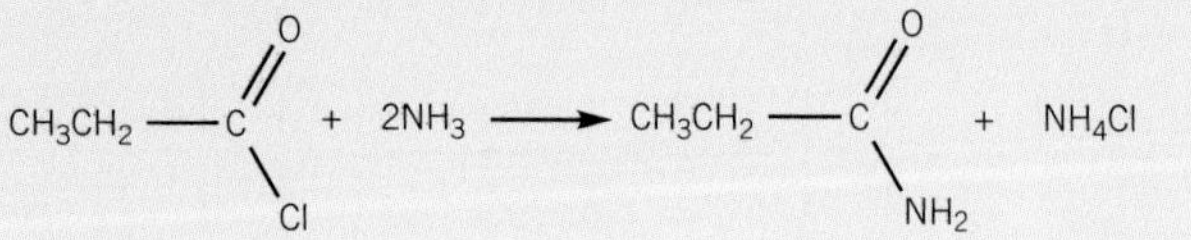

Answers

The mechanism of the reaction between acyl chlorides and nucleophiles

1. In the acyl chloride, the shape is trigonal planar as there are 3 areas of electron density surrounding the central carbon atom.

   In the intermediate, the shape becomes tetrahedral as there are now four electron pairs surrounding the central carbon atom.

   In the carboxylic acid product, the shape once again becomes trigonal planar as there are 3 areas of electron density surrounding the central carbon atom.

2. [mechanism diagram]

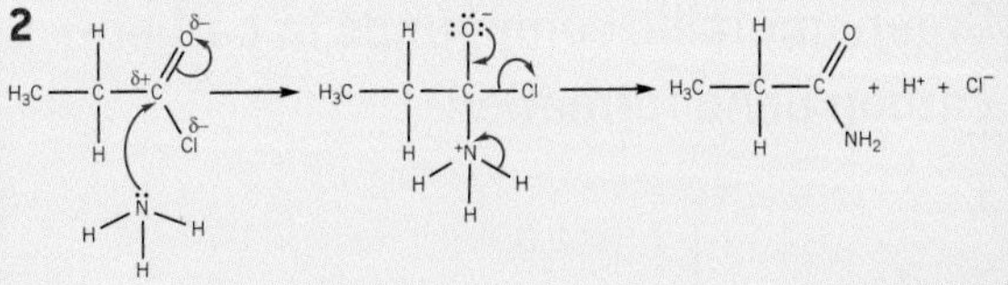

27.1

Chemistry of Decay

1. putrescine: H_2N—(chain)—NH_2 ; cadaverine: H_2N—(chain)—NH_2

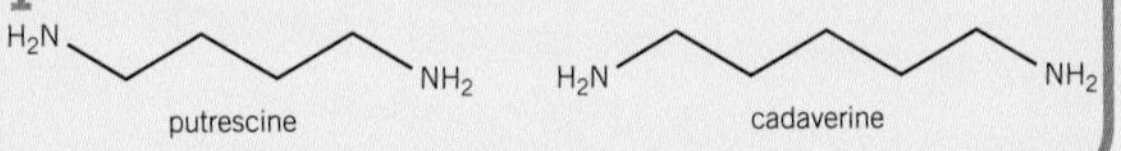

Perkin's Mauveine

$C_{26}H_{23}N_4^+$

1. a secondary: dimethylamine
   b primary: methylamine
   c tertiary: N-methyl-N-ethylamine

2. a $CH_3CH_2NH_2 + HNO_3 \rightarrow CH_3CH_2NH_3^+ NO_3^-$
   b $2CH_3CH_2NH_2 + H_2SO_4 \rightarrow (CH_3CH_2NH_3^+)_2 SO_4^{2-}$

3. [mechanism diagram]

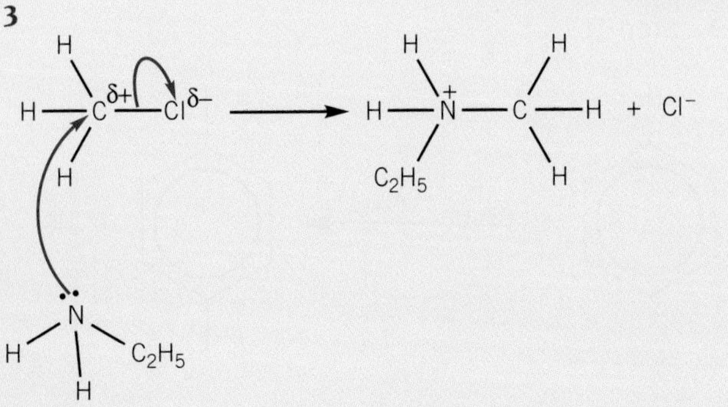

27.2

Zwitterions and the isoelectric point

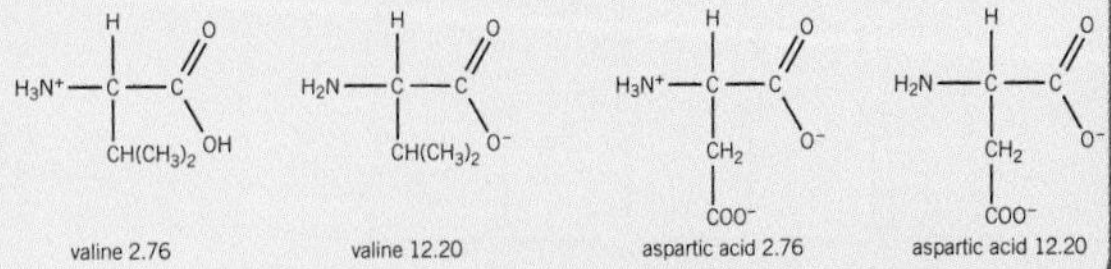

valine 2.76, valine 12.20, aspartic acid 2.76, aspartic acid 12.20

Optical isomerism – testing our senses

1.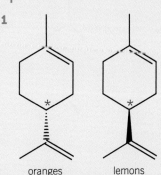

   oranges / lemons

2. Naproxen: One isomer treats arthritis pain; the other isomer causes liver poisoning.
   Propranolol: One isomer treat heart disease. Both have a local anesthetic effect.
   Thalidomide: One isomer is a sedative effective against morning sickness; the other isomer causes birth defects.

1. a [amino acid + NaOH → sodium salt + H_2O]
   b [amino acid + C_2H_5OH → ester + $2H_2O$]
   c [amino acid + HNO_3 → protonated amine + NO_3^-]

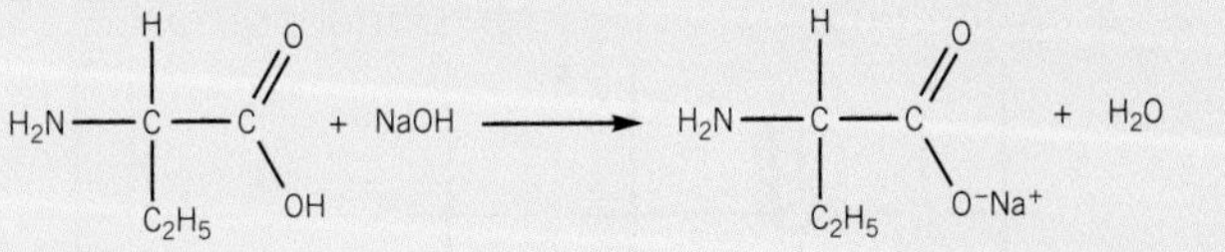

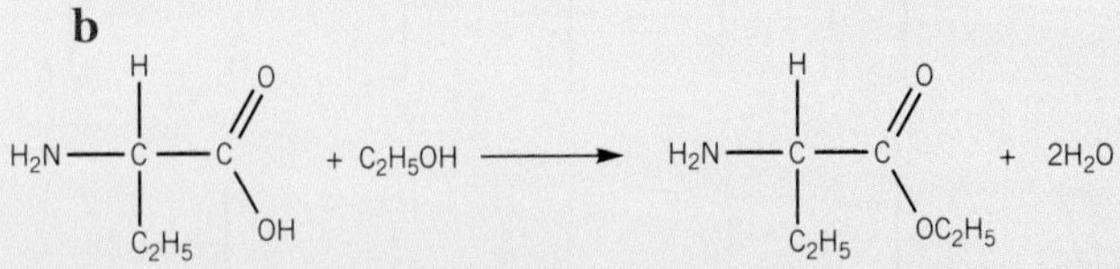

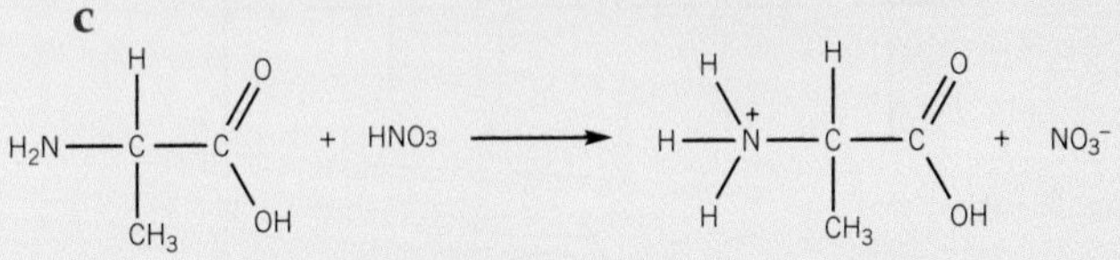

Answers

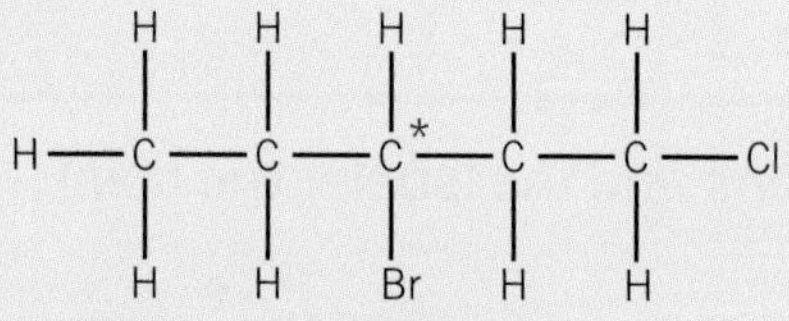

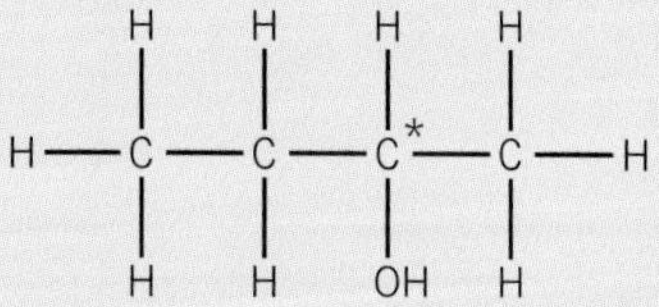

2 a [structure: H-C(H)(H)-C(H)(H)-C*(Br)(H)-C(H)(H)-C(H)(H)-Cl]

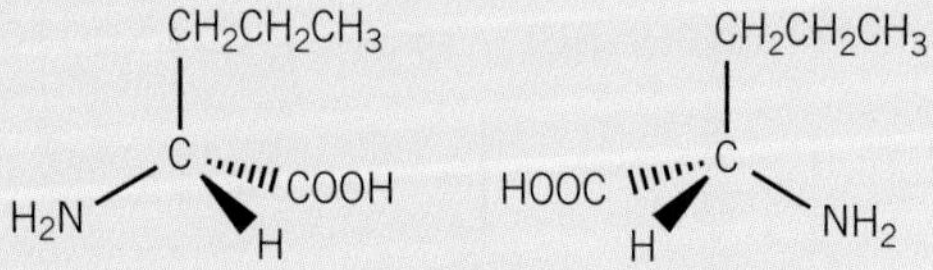

b [structure: H-C(H)(H)-C(H)(H)-C*(OH)(H)-C(H)(H)-H]

3 a [two enantiomers with CH₂CH₂CH₃, NH₂/H₂N, COOH, H]

b [two enantiomers with CH₃, OH/HO, COOH, H]

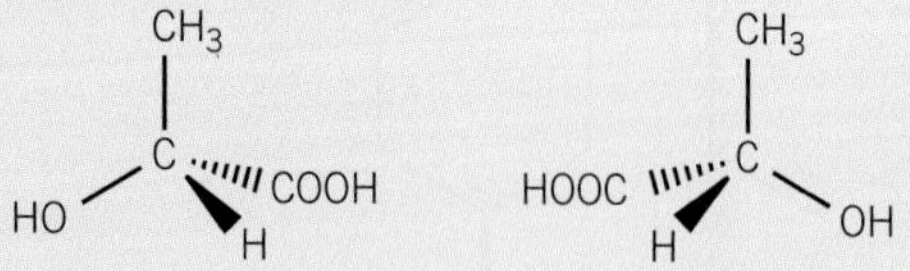

4
$$H_3N^+-C(H)(CH_2COOH)-COOH + 2CH_3OH \xrightarrow{H_2SO_4} H_3N^+-C(H)(CH_2COOCH_3)-COOCH_3 + 2H_2O$$

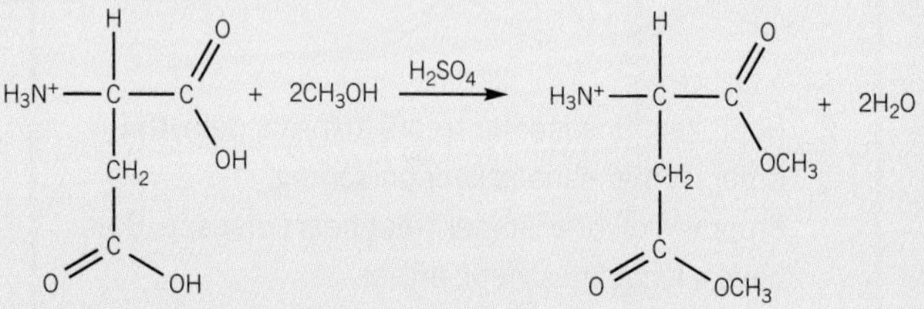

27.3

1 a amine and carboxylic acid (or acyl chloride);
  b alcohol and carboxylic acid

2 a Addition   **b** Condensation

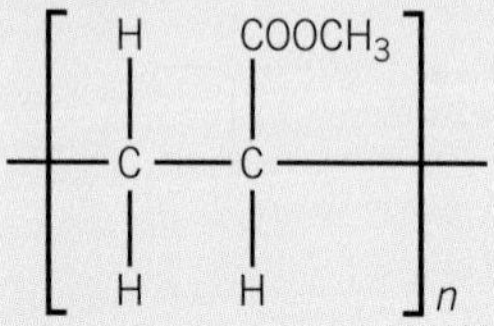

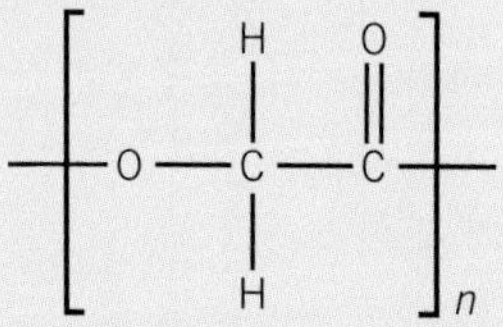

[polymer structures: -[CH(COOCH₃)-CH₂]ₙ- and -[O-CH₂-CO]ₙ-]

  c Condensation

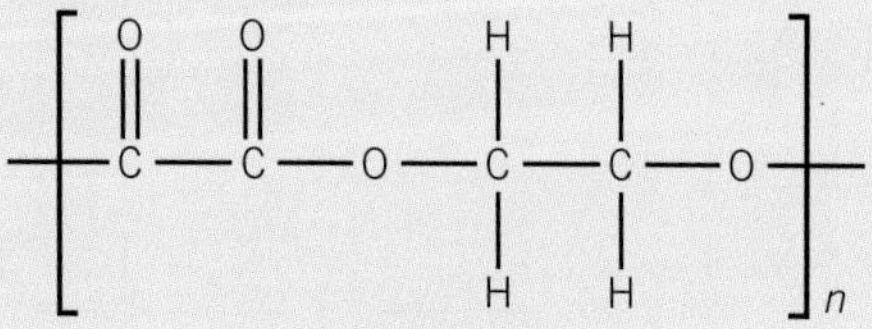

[polymer: -[CO-CO-O-CH₂-CH₂-O]ₙ-]

3 a

Acid hydrolysis HOOC—⌬⌬—COOH   H₃N⁺—(CH₂)₂—N⁺H₃

Alkaline hydrolysis ⁻OOC—⌬⌬—COO⁻   H₂N—(CH₂)₂—NH₂

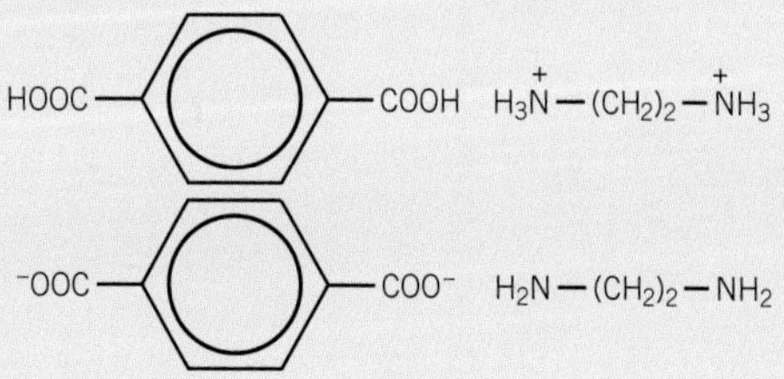

b

Acid hydrolysis   HO—(CH₂)₄—OH   HOOC—(CH₂)₂—COOH

Alkaline hydrolysis   HO—(CH₂)₄—OH   ⁻OOC—(CH₂)₂—COO⁻

28.1

> **Hydroxynitriles in nature**
>
> $n(HCN) = 0.006/27.0 = 2.22 \times 10^{-4}$ mol
>
> mass of 2-hydroxy-2-phenylethanenitrile
> $= 133 \times 2.22 \times 10^{-4} = 0.03$ g

> **Organometallic compounds in carbon-carbon bond formation**
>
> **1** [structure: CH₃CH₂-C(H)(CH₃)-Li]
>
> **2** [mechanism showing addition of butyl group to propanal carbonyl, via intermediate, to give 2-hydroxyhexane (H₃C-C(OH)(H)-CH₂CH₂CH₂CH₃)]
>
> **3** CH₃COOH

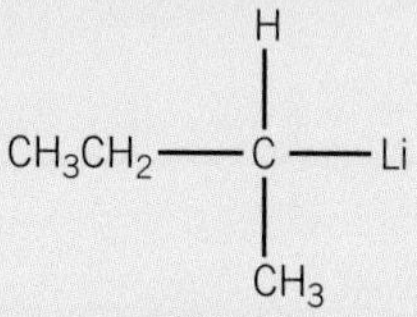

1 a $CH_3CH_2Cl + KCN \rightarrow CH_3CH_2CN + KCl$
  b $CH_3CHO + HCN \rightarrow CH_3CH(OH)CN$

2 $CH_3CHO + HCN \rightarrow CH_3CH(OH)CN$
  Reagent: NaCN/H₂SO₄
  Conditions: Warm
  $CH_3CH(OH)CN + 2H_2O + HCl \rightarrow$
  $\qquad CH_3CH(OH)COOH + NH_4Cl$
  Reagent: Aqueous acid
  Condition: Heat

3 Step 1:

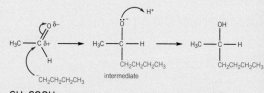

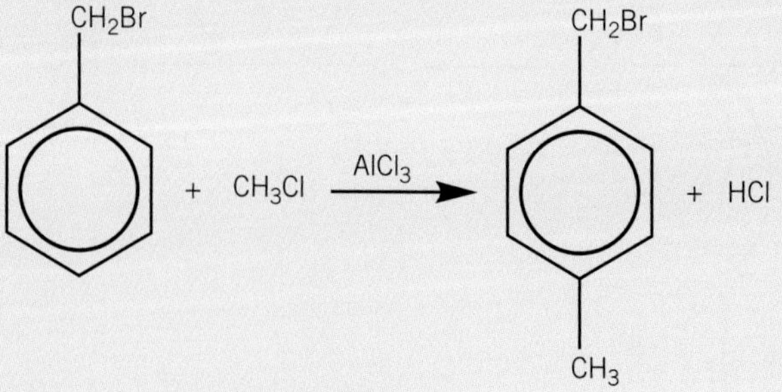

Answers

Step 2:

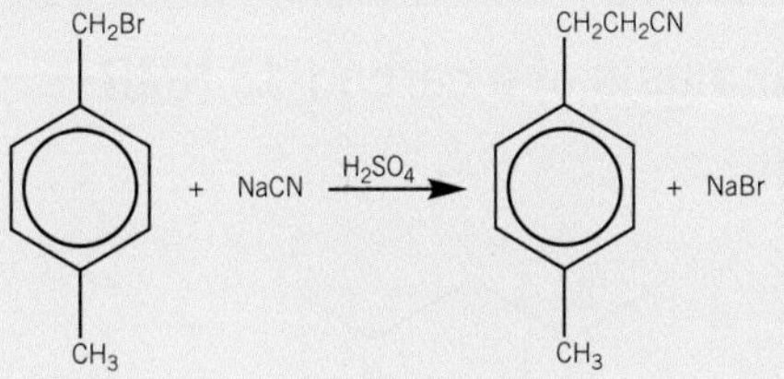

[Diagram: CH₂Br-C₆H₄-CH₃ + NaCN → (H₂SO₄) → CH₂CH₂CN-C₆H₄-CH₃ + NaBr]

Step 3:

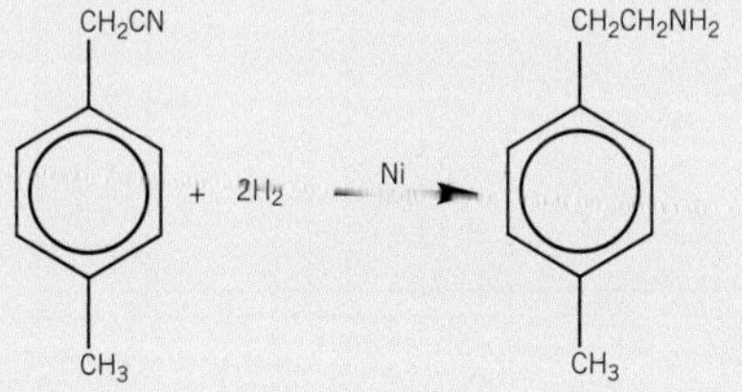

[Diagram: CH₂CN-C₆H₄-CH₃ + 2H₂ → (Ni) → CH₂CH₂NH₂-C₆H₄-CH₃]

4 a

[Diagram: benzene + (CH₃)₂CHCl → (AlCl₃) → isopropylbenzene + HCl]

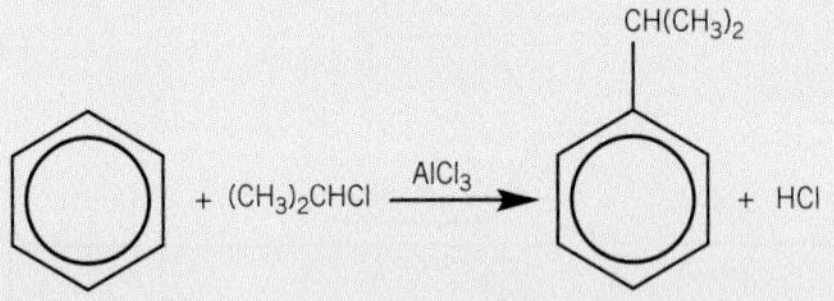

b Step 1: $(CH_3)_2CHCl + AlCl_3 \rightarrow (CH_3)_2CH^+ + AlCl_4^-$

Step 2:

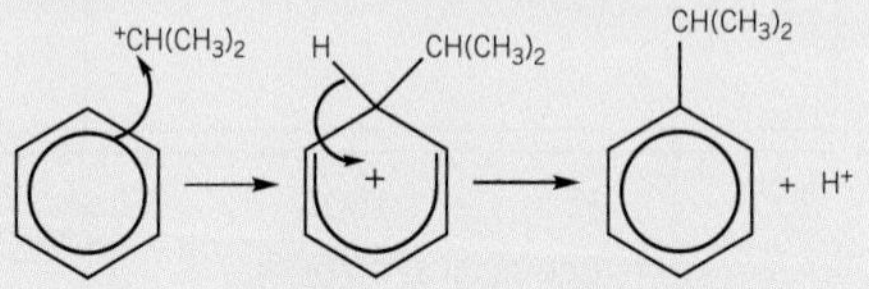

[Mechanism diagram showing electrophilic aromatic substitution with +CH(CH₃)₂ attacking benzene to form arenium intermediate then CH(CH₃)₂-benzene + H⁺]

Step 3: $AlCl_4^- + H^+ \rightarrow AlCl_3 + HCl$

28.2

1 The sample melts over a range of temperatures. Heating too quickly will result in the thermometer recording a higher temperature than the actual melting range of the sample.

2 If the solvent is still warm some of the solid will still be dissolved in the solvent and so not all of the solid will have crystallised. The yield would be lower because some of the solid is still dissolved in the solvent.

3 There are fewer impurities in sample 2, so the melting point is higher and sharper.

4 **a** carbonyl/ketone/phenol
 b alcohol
 c i 2,4-dinitrophenyl hydrazine – orange precipitate
 ii hydrogen bromide – addition of bromine to carbon-carbon double bond
 iii potassium dichromate and sulfuric acid – ketone formed
 d concentrated sulfuric acid and reflux

28.3

1 **a** Carboxylic acid and ester
 b Primary amide and phenol

2 **a** Step 1 – Aqueous sodium hydroxide
   $CH_3CH_2Cl + NaOH \rightarrow CH_3CH_2OH + NaCl$
   Step 2 – Acidified potassium dichromate(VI) / Reflux
   $CH_3CH_2OH + 2[O] \rightarrow CH_3COOH + H_2O$

 b Step 1 – Aqueous sodium hydroxide
   $CH_3CH_2Cl + NaOH \rightarrow CH_3CH_2OH + NaCl$
   Step 2 – Acidified potassium dichromate / Distil
   $CH_3CH_2OH + [O] \rightarrow CH_3CHO + H_2O$
   Step 3 – NaCN / H₂SO₄
   $CH_3CHO + HCN \rightarrow CH_3CH(OH)CN$

 c Step 1 NaCN in ethanol
   $CH_3CH_2Cl + NaCN \rightarrow CH_3CH_2CN + NaCl$
   Step 2 H₂/Ni
   $CH_3CH_2CN + 2H_2 \rightarrow CH_3CH_2CH_2NH_2$

3 **a** $CH_3CH_2Br \xrightarrow{NH_3/ethanol} CH_3CH_2NH_2 \xrightarrow{CH_3COCl} CH_3CH_2NHCOCH_3$

b

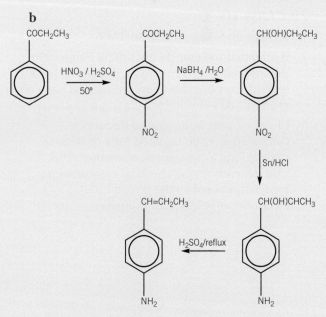

Answers

29.1

> **Carrying out TLC**
>
> So that the solvent does not wash the compounds in the spot off the TLC plate.

> **Identifying a phenol**
>
> 2,4,6-tritribromphenol

1. 4: ethanol; 7: 2-methylbutan-1-ol; 8: ethanoic acid
2. purple 0.74, red 0.38 and blue 0.17
3. a Warm with Tollens' reagent. An aldehyde gives a silver mirror but the ketone does not react
   b Add bromine water, both will decolourise the bromine water but the phenol will also give a white precipitate also. Alternatively, both could be tested with pH indicator. The phenol would show an weakly acidic pH but the alkene would be neutral.
   c Add 2,4-DNP, ketone will produce an orange precipitate but the secondary alcohol does not react. Alternatively, warm both with acidified potassium dichromate(VI). With a secondary alcohol, the colour changes from orange to green. With a ketone, the colour stays orange.

29.2 Answers

1. Tetramethylsilane, $(CH_3)_4Si$ (TMS)
2. Deuterated solvents do not produce signals in the frequency ranges used in ^1H and ^{13}C NMR spectroscopy because deuterium atoms have an even number of nucleons. Example is $CDCl_3$.
3. Chemical shift is the shift in frequency, compared to TMS, required for a nucleus to undergo nuclear magnetic resonance. The shift depends on the chemical environment, especially caused by the proximity of electronegative atoms or π-bonds.

29.3 Answers

Further analysis of a carbon-13 NMR spectrum

a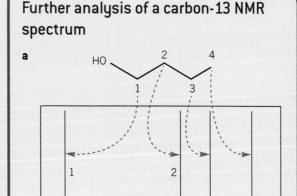

b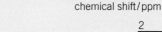
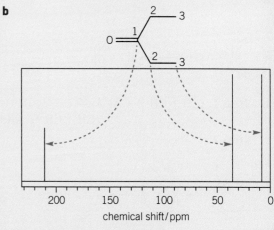

1. a CH_3COOH: 2 peaks, **C**H_3–C at 0–50 ppm; **C**OOH at 160–220 ppm
   b CH_3CHO: 2 peaks, **C**H_3–C at 0–50 ppm; **C**HO at 160–220 ppm
   c CH_3NHCH_3: 1 peak **C**–N at 30–70 ppm
2. a $CH_3COOCH_2CH_3$: 4 peaks, **C**H_3–C at 0–50 ppm; **C**OO at 160–220 ppm; O–**C**H_2 at 50–90 ppm; –O**C**H_2–**C**H_3 at 0–50 ppm
   b $CH_3CH=CHCH_3$: 2 peaks, one peak for 2 C atoms of type C=C at 110–160 ppm; one peak for 2 C atoms for type **C**H_3–C at 0–50 ppm
   c $HOCH_2CH_2CHO$: 3 peaks, **C**H_2–O at 50–90 ppm; **C**H_2–C at 0–50 ppm; **C**HO at 160–220 ppm

Answers

3 a

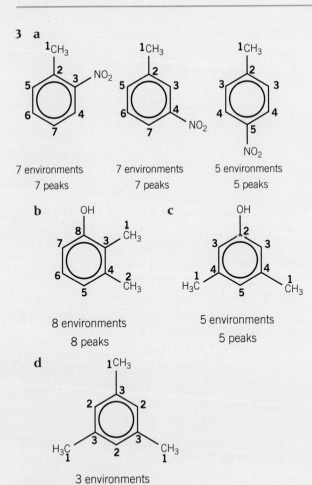

7 environments
7 peaks

7 environments
7 peaks

5 environments
5 peaks

b 8 environments
8 peaks

c 5 environments
5 peaks

d 3 environments
3 peaks

29.4 Answers

Splitting patterns

quintet, 1 : 4 : 6 : 4 : 1; hextet 1 : 5 : 10 : 10 : 5 : 1; heptet 1 : 6 : 15 : 20 : 15 : 6 : 1.

1 a CH_3COOH **i** 2 peaks, **ii** Ratio 3 : 1 **iii** 1 peak disappears with D_2O

b CH_3CHO **i** 2 peaks, **ii** Ratio 3 : 1 **iii** No difference with D_2O

c CH_3COCH_3 **i** 1 peak, **ii** Ratio 1 **iii** No difference with D_2O

d $CH_3COOCH_2CH_3$ **i** 3 peaks, **ii** Ratio 3 : 2 : 3. **iii** No difference with D_2O

e $H_2NCH(CH_3)COOH$ **i** 4 peaks **ii** Ratio 2 : 1 : 3 : 1 **iii** 2 peaks disappear with D_2O

2 a $HOCH_2CH_2OH$ **i** 2 peaks **ii** $HOCH_2CH_2OH$ at 3.0–4.2 ppm, **HO**CH_2CH_2OH at 0–12 ppm. **iii** Ratio 2 : 1

b $HOCH_2CH_2CH_2OH$: **i** 3 peaks
ii **HO**$CH_2CH_2CH_2$**OH** at 0–12 ppm
$HOCH_2CH_2CH_2OH$ at 3.0–4.2 ppm
$HOCH_2$**CH$_2$**CH_2OH at 0.5–2.0 ppm
iii Ratio 1 : 2 : 1 (for 2 : 4 : 2)

c $(CH_3)_2CHOH$ **i** 3 peaks **ii** $(CH_3)_2CHOH$ at 0.5–2.0 ppm, $(CH_3)_2$**CH**OH at 3.0–4.2 ppm $(CH_3)_2CHOH$ at 0–12 ppm **iii** Ratio 6 : 1 : 1

3 a **i** CH_3CH_2 **ii** CH_3CH **iii** $(CH_3)_2CH$ **iv** CH_2CH **v** CH_2CH_2 **vi** $CHCH$

b **i** $HOCH_2CH_2OH$ no splitting because CH_2 groups are equivalent in same environment

ii $HOCH_2$**CH$_2$**CH_2OH split into a triplet $(2 + 1 = 3)$, $HOCH_2CH_2CH_2OH$ split into a quintet $(4 + 1 = 5)$

iii $(CH_3)_2CHOH$ split into a doublet $(1 + 1 = 2)$, $(CH_3)_2$**CH**OH split into a heptet $(6 + 1 = 7)$

29.5 Answers

1

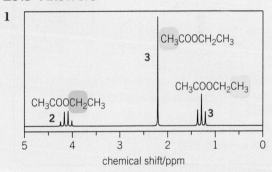

2 a

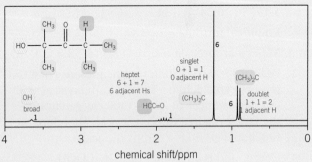

b

Answers

3 a CH₃COCH₂CH₂CH₃ has 4 peaks for
CH₃COCH₂CH₂CH₃, CH₃COC**H₂**CH₂CH₃,
CH₃COCH₂C**H₂**CH₃, CH₃COCH₂CH₂C**H₃**.
Ratio 3 : 2 : 2 : 3

CH₃CH₂COCH₂CH₃ has 2 peaks for
CH₃CH₂COCH₂C**H₂**, CH₃C**H₂**COC**H₂**CH₃.
Ratio 3 : 2 (6 : 4)

b CH₃CH₂COOCH₃ has a **CH₃** peak at 0.5–2.0, a **CH₂**C=O peak at 2.0–3.0 ppm and a O**CH₃** peak at 3.0–4.2 ppm (ratio in order of increasing chemical shift 3 : 2 : 3)

CH₃COOCH₂CH₃ has a **CH₃** peak at 0.5–2.0 ppm, a **CH₃**C=O peak at 2.0–3.0 ppm, an O**CH₂** peak at 3.0–4.2 ppm (ratio in order of increasing chemical shift 3 : 3 : 2)

29.6 Answers

Analysis of an unknown compound

1 1,4- gives 8 types of carbon atom and 8 peaks. Both 1,2- and 1,3- give 10 peaks. (See below)

2 Add 2,4-DNP. An orange precipitate would suggest the presence of a C=O group in an aldehyde or ketone and would confirm the right-hand structure. No precipitate would suggest the left-hand structure.

1 C : H : O = 62.07/12.0 : 10.34/1.0 : 25.59/16.0 = 5.1725 : 10.34 : 1.724 = 3 : 6 : 1

Empirical formula = C_3H_6O

Molecular formula = $C_3H_6O \times 116/58 = C_6H_{12}O_2$

2 Peak at 1700 cm⁻¹ indicates the presence of a C=O group in an aldehyde, ketone, carboxylic acid or ester. Broad peak at 2500–3300 cm⁻¹ indicates the O–H peak in a carboxylic acid. Therefore the functional group is a carboxylic acid.

Compound is an isomer of $C_5H_{11}COOH$.

As the ¹³C NMR spectrum has 4 peaks, the correct structure should be one of the two structures with 4 types of carbon atom.

3 Predicted proton NMR spectrum

For the first structure, there are four types of proton giving four peaks.

Protons in environment **1** are the type **H**–C–R and adjacent to CH₂.
triplet at δ = 0.5–2.0 ppm

Protons in environment **2** are the type **H**–C–R and adjacent to CH₃ and CH.
multiplet at δ = 0.5–2.0 ppm

Protons in environment **3** are the type **H**C–C=O and adjacent to 2 × CH₂ = 4 H.
quintet at δ = 2.0–3.0 ppm

Proton **4** is the type COO**H** and would give a peak at δ = 10–12 ppm

Relative peak heights will be: 6 : 4 : 1 : 1 for 2 × CH₃ : 2 × CH₂ : CH : COOH

For the second structure, there are three types of proton giving three peaks.

Protons in environment **1** (3CH₃ groups) are the type **H**–C–R and adjacent to a C with no H.
singlet at δ = 0.5–2.0 ppm

Protons in environment **2** are the type **H**–C=O and adjacent to a C with no H.
singlet at δ = 2.0–3.0 ppm

Proton **4** is the type COO**H** and would give a peak at δ = 10–12 ppm

Relative peak heights will be: 9 : 2 : 1 for 3 × CH₃ : CH₂ : COOH

Index

acid anhydrides 466–467, 470–471
acid–base equilibria 310–312
acid–base reactions 40–47
acid–base titrations 340–343
acid dissociation constant K_a 319–321
acid hydrogen fuel cells 396
acid hydrolysis of esters 468
acids 40–47, 310–325
    alkali reactions 314
    Brønsted–Lowry theory 310–312
    carbonate reactions 117, 119, 120, 314
    condensation polymer hydrolysis 485–486
    conjugate base pairs 310–311
    dissociation 40, 42, 319–325
    Group 2 reactions 109–111
    hydrogen ions 313–325
    metal oxide reactions 314
    metal redox reactions 313
    mono-/di-/tribasic 312–313
    neutralisation 41–47, 110–111, 128, 133, 312–314
    pH scale 315–318
    redox reactions 51, 313
    solutions 327–328
    strong and weak 40, 42
    see also carboxylic acids
activation of bromine 446–447
activation energy 126, 145, 149, 151–153
acylation of benzene 440, 492
acyl chlorides 178, 466, 468–470
    amidations 470

carboxylic acid formation 469
    esterifications 469
    formation 468–469
addition–elimination reactions 470–471
addition polymerisation 486
addition reactions 185, 207–217
    bromine/cyclohexene 440–441
    electrophilic 440–414
    nucleophilic 456–459, 490–491
adsorption 150, 506
alcohols 173, 177, 222–231
    amino acid esterifications 479
    bond angles 201
    carbon-13 NMR 517
    chemical tests 510
    classification 224
    dehydration 229
    eliminations 186
    esterifications 228, 467, 469
    from alkenes 209–210
    from haloalkane hydrolysis 233
    infrared spectra 256–261
    naming 222
    oxidation 226–227
    properties 223–224
    reactions 226–229
    structural isomerism 183
    substitutions 229
aldehydes 226, 246, 454–462
    identification 460–462, 510
    melting points 462
    molecular formulae 178
    naming 454–455
    nitrile formation 490–491
    oxidations 456
    reductions 457–458

alicyclic hydrocarbons 174, 176, 181
aliphatic hydrocarbons 174–175
alkali hydrogen fuel cells 396
alkalis 40–41
    acid neutralisation 314
    carboxylic acid reactions 465, 478
    condensation polymer hydrolysing 485–486
    ester hydrolysis 468
    solutions 327–329
alkanes 173, 174–176, 190–199
    boiling points 191–192
    bonds 190, 193
    branched 175–176, 182, 191–192
    chain length 175–176, 191
    combustion 193–195
    halogenation reactions 195–197
    molecular shapes 190–192
    naming 174–176
    nitrile formation 490
    properties 190–192
    reactivity 190, 193–197
    structural isomerism 182
alkenes 174, 200–221, 497–498
    additions 207–214
    branched 200, 206
    chemical tests 510
    differences to arene reactivity 440–441
    double bonds 200–201, 203
    electrophilic addition 211–214
    halogen reactions 208–209, 212–213
    hydration 209–210
    hydrogenation 207–208
    hydrogen halide reactions 209, 211–212

isomeric products 213–214
    naming 176, 204–206
    polymerisation 215–219
    preparation from alcohols 229
    properties 200–206
    reactivity 207–219, 500
    stereoisomerism 203–206
alkylations of benzene 440, 492
alkyl groups 174, 175
alkynes 174, 177
amide bonds 485
amides 480, 497–498
    formation from acyl chlorides 470
    polymers 484–487
    proteins 485
    reactivity 498
amines 173, 178, 246, 474–489, 497–498
    acyl chloride reactivity 470
    amino acids 478–482
    as bases 475–476
    classification 474
    esterifications 479
    isomerism 480–482
    naming 475
    polymers 484–487
    reactivity 498
    salt formation 476
amino acids 478–482
    chirality 480
    esterification 479
    polymerisation 485
    as zwitterions 479
α-amino acids 478–482
amino groups 523–524
ammonia 71, 118, 159, 328–329
    acyl chloride reactivity 470
    ligand substitution 413–415
    transition metal precipitations 416–417
ammonium ions 420

Index

amount of substances 20–37
concentrations 26–27, 33
formula determination 22–25
gases 27–28, 33
moles 20–21
reactions 32–37
volumes 26–28
analytical methods
carbon-13 NMR 515–519, 527
colorimetry 277–278
combined 530–533
functional groups 509–510
gas chromatography 508–511
infrared spectroscopy 256–261, 530
iodine/thiosulfate redox titrations 381–385
manganate(VII) redox titrations 376–379
mass spectrometry 530
nuclear magnetic resonance spectroscopy 512–533
pH meters 340
proton NMR 520–529
qualitative 420–421
redox titrations 376–385
thin layer chromatography 506–508
titrations 340–343
anhydrous salts 24–25, 245
anions 11, 117–120, 243, 420–421
aqueous solutions
acids/alkalis 327–329
amount of substances 26–27
electrical conductivity 61–62
enthalpy changes 352–361
equilibria 294–297, 304–305
halogens 113
hydronium ions 311–312, 326

aromatic compounds (arenes) 432–453
alcohols 442
amines 474, 477
carbon-13 NMR 518–519
differences to alkene reactivity 440–441
naming 434–436
proton NMR spectroscopy 523
aromatic hydrocarbons 174, 181
Arrhenius equation 289–291
atmosphere 150, 237–238, 256–257, 258
atom economy of reactions 35–36
atomic mass 12–14, 20–21
atomic numbers 9, 93
atomic radius 96
atomic structure 8–11
atomisation, standard enthalpy changes 347
average bond enthalpies 67, 134–136
Avogadro constant 20
Avogadro's hypothesis 27

balanced equations 17, 32, 35
bases 40–47, 110–111, 310–312, 314–315, 326–329
amines 475–476
Brønsted–Lowry theory 310–312
conjugate acid pairs 310–311
neutralisation 314
pH scale 315–318
batteries 394–395
benzaldehyde 432
benzene 432–441, 444–445
acylations 440, 492
alkylations 440, 492
comparison to phenol 444–445
delocalisation 434
derivatives 432, 434–441

electrophilic substitutions 437–441
halogenations 438–439
hydrogenation 433–434
nitrations 437–438
reactivity 433–434, 437–441, 498
benzene rings 174, 181
bidentate ligands 406, 411
binary compounds 15, 16
biodegradable polymers 219
blocks of the periodic table 58, 94–95
blood 299, 337–338, 415
boiling points
alcohols 223
alkanes 191–192
entropy 362
giant covalent lattices 103
halogens 112
hydrides 82
hydrogen chloride 79
intermolecular forces 77, 79, 82
ionic compounds 61
metals 102
simple molecular substances 79
water 82
Boltzmann distributions 152–153, 289
bond angles 70–73, 201
bond breaking 134, 184, 185, 238
heterolytic fission 184, 211, 212
homolytic fission 184, 185, 195–196, 238
bond enthalpies 67, 134–136
bond frequencies 257–261
bonding 59–69
alkanes 190, 193
alkenes 200–201, 203
benzene 434
carbonyl groups 456
bond polarity 74–75
bonds
carbon-13 chemical shifts 515
proton spectra 520–524

strengths 61, 67, 77, 234–235, 237–238
π-bonds 200–201, 203, 434, 456
σ-bonds 190, 193, 456
Born–Haber cycles 346–351
Brady's reagent 460–461
branched molecules
alkanes 175–176, 182, 191–192
alkenes 200, 206
naming 175–176
bromide test 118
brominations 195–197, 208–209, 211–213, 438–439, 444
Brønsted–Lowry acids and bases 310–312
buffer solutions 332–339
blood 337–338
Henderson–Hasselbalch equation 338–339
pH 333–339
preparation 332–333
burettes 44

Cahn-Ingold–Prelog nomenclature 205–206
calculation
cell potentials 390
enthalpy changes 346–357
entropy changes 363–364
lattice enthalpy 348–351
carbocations 205–208
carbon-13 NMR spectroscopy 515–519, 527
carbon 103, 104, 172, 480–482
carbonates 42, 117, 119, 120, 314, 465
carbon–carbon bonds 433, 490–492
carbon–carbon double bonds 200–201, 203, 409
carbon dioxide 35–36, 71–72, 76, 193–194, 415
carbon–halogen bonds 234–235, 237–238

Index

carbon–hydrogen bond spectra 258
carbonyl compounds 454–473
  carboxylic acids 463–471
  chemical tests 510
  hydrogen cyanide additions 457–459
  identification 460–462, 510
  melting points 462
  naming 454–455, 463–464, 466–467
  nitrile formation 490–491
  nucleophilic additions 456–457
  reductions 457–458
carboxyl groups 463, 465
carboxylic acids 246, 463–471, 497–498
  acid reactions 464–465
  alkali reactivity 478
  amino acids 478–479
  chemical tests 510
  derivatives 466–471
  esterifications 228, 467
  infrared spectra 256–261
  molecular formulae 178
  naming 463–464
  neutralisation 464–465
  preparation 227
  reactivity 498
  redox 464
  solubility 464
catalysts 149–151, 153, 159, 306, 403–404
catalytic converters 150
cations 15, 120, 211–214, 420
cell potentials 388–390
chain elongation 457–458
chain length 175–176, 191
charge effects of ions 359
charges 8–9, 11, 15, 96
chemical shifts 513
  carbon-13 NMR 515
  prediction 527–528
  proton NMR spectroscopy 520–524

chemical tests
  alcohols 510
  aldehydes 510
  aldehydes/ketone 461–462
  alkenes 510
  carbonyl compounds 460–462, 510
  carboxyl groups 465
  carboxylic acids 510
  haloalkanes 510
  phenols 510
chiral centres 477–482
chirality 480–482
chlorination of benzene 439
chlorine 65, 113–115
chlorofluorocarbons (CFCs) 237–238
chromatography 506–511
cis-platin 412
cis–trans isomerism 203–205, 409–412
clock reactions 284–286
collision theory 145
colorimetry 277–278
coloured compounds of transition metals 403, 408
combined analytical techniques 530–533
combustion 127–128, 130–131, 136, 193–195, 226
complementary colours 408
complete combustion 193–194
complete oxidation 227
complex ions 405–417
  cis–trans isomerism 409–411
  formation 405
  ligand substitution 413–415
  optical isomerism 411–412
  shapes 407, 409–412
  stereoisomerism 409–412
compounds
  enthalpy change of formation 127

  formulae and equations 15–18
  Group 2 elements 110–111
  oxidation numbers 48
concentration
  calculating from titres 45–47
  dynamic equilibrium 155–156, 160–161
  equilibrium constants 294–297, 304–305
  gas chromatography 509
  moles 26–28
  reaction rates 144–145
concentration–time graphs 277–281
  half-lives 279–280
  order of reaction 278
condensation polymers 483–487
condensers 243–244
configuration of electrons 57–58
conjugate acid–base pairs 310–311, 332–333, 337–338
conservation of energy 124
Contact process 403
continuous monitoring 277–278
coordinate bonds 66–67, 405–407
coordination numbers 405–407
covalent bonds 63–67
  alkane σ-bonds 194, 197
  bond strengths 77
  electronegativity and polarity 74–76
  homolytic/heterolytic fission 184–185
  infrared spectroscopy 256–261
  organic compounds 172
  vibrations 256
covalent molecules
  dot-and-cross diagrams 64, 66–67
  formulae and equations 15–18

  properties of simple molecular substances 79–80
  structures 103–105
crude oil 190, 191
cyanations, carbonyl compounds 457–459
cyclic organic compounds 174, 176, 181, 200
cyclohexene
  bromine additions 440–441
  enthalpy change of hydrogenation 433–434
  hydrogenation 433–434

dative covalent bonds 66–67, 405–407
d-block elements 400–423
  electronic configuration 400–401
  general properties 400
  see also transition elements
deactivation of bromine 446–447
decay 279–280, 475
decolourising of bromine 440–441
decomposition of hydrogen peroxide 369, 404
dehydration
  alcohols 229
  organic liquids 245
delocalised electrons 70, 101, 434
derivatives of benzene 432, 434–441
derivatives of carboxylic acids 466–471
desorption from catalysts 150
determination
  cell potentials 388–389
  electrode potentials 387–388
  K_c 295–297
  K_p 299–301
  melting points 495–496
deuterated solvents 514
deuterium 14, 15, 514, 524
diatomic molecules 21, 22
dibasic acids 312–313

Index

dilutions, pH 317–318
2,4-dinitrophenylhydrazine (2,4-DNP) 460–461
dipoles 75–77, 212
  see also London forces; polar bonds
directing effects, phenol 447–450
directing groups, phenol 446–450
displayed formulae 64, 180–181
disproportionations
  copper(I) ions 420
  halogens 114–115
dissociation of acids 40, 42
dissociation constants K_a 319–325
  pK_a 319–321
  strong acids 319–320
  weak acids 319–325
dissolving
  enthalpy changes 352–357, 360
  ionic compounds 354–355, 360
  standard enthalpy change of solution 352–355
distillation 191, 226, 243–245
disubstitutions, phenol 446–450
DNA hydrogen bonding 83
2,4-DNP (2,4-dinitrophenyl-hydrazine) 460–461
d-orbitals 55–59, 95
dot-and-cross diagrams 59–60, 64, 66–67
double bonds 66, 200–201, 203, 258–260
drugs 111, 246–247, 255, 276
drying organic liquids 245
dynamic equilibrium 154–161
  catalyst effects 159
  concentration effects 155–156
  equilibrium constant 160–161
  le Chatelier's principle 154–159

pressure effects 158–159
temperature effects 156–158

electrical conductivity
  aqueous solutions 61–62
  carbon structures 103, 104
  giant covalent lattices 103, 104
  ionic compounds 61–62
  metals 101, 102
  salt bridges 387
  simple molecular substances 80
electrochemical cells 386
electrode potentials 386–396
  cell potentials 388–390
  feasibility 392–393
  free energy 393
  fuel cells 396
  half-cells 386
  redox 388, 391–393, 418
  standard 387–388
  storage cells 394–395
electron density
  benzene 433–434, 444–445
  carbonyl groups 456
  phenol 444–445
electronegativity 74
electronic configuration
  atoms 57–58
  d-block elements 400–403
  transition elements 401–403
electrons
  atomic structures 54–58
  bonding 59–69
  carbonyl groups 456
  delocalisation 434
  ionic structures and charges 11, 15
  lone pairs 64, 67, 70–71, 81
  oxidation/reduction 372
  pairs 70–73, 184–185
  properties 8–11, 12
  redox reactions 50–51

shells and orbitals 54–58
electron shielding 96
electrophilic additions 211–214, 440–441
electrophilic substitutions 437–441, 444–445
elemental analysis 530
elements
  atomic numbers 9
  in equations 17
  oxidation numbers 46
elimination reactions 186, 229
empirical formulae 22, 179
endothermic reactions
  dissolving 354–355
  enthalpy 124, 125, 134–135, 355
  entropy and free energy 368–369
  equilibria 156–159, 303–304
end points, titrations 341–342
energy
  activation energy 126
  calories 128
  conservation 124
  see also enthalpy
enthalpy H 124–143
enthalpy change ΔH 346–361, 365–371
  of atomisation 347
  average bond enthalpies 135–136
  Born–Haber cycles 346–351
  catalysts 149
  of combustion 127–128, 130–131, 136
  definition 124
  dynamic equilibrium 156–159
  exothermic/endothermic changes 124–126, 132–133, 355
  of formation 127, 347
  of hydration 354–357, 360
  indirect determination using Hess' law 137–140

measurement 129–133
  of neutralisation 128, 133
  of reaction 126–127, 132
  in solution 352–361
  of solution 352–355
  standard enthalpy changes 126–128
enthalpy cycles 137–140
enthalpy profile diagrams 124–126, 135, 137, 149
entropy changes ΔS 362–371
  calculation 363–364
  prediction 362–363
environmental issues
  atmospheric pollution 258
  catalytic converters 150
  global warming 193, 256–257
  organohalogens 237–239
  polymers/plastics 218–219
equations 15, 17–18, 33–34, 36, 114
equilibria 294–309
  catalyst effects 306
  heterogeneous 295, 301
  homogeneous 295, 300
  K_c 294–297, 304–305
  K_p 298–301, 305–306
  position 302–306
  see also dynamic equilibrium
equilibrium constant K_c 160–161, 294–297, 304–305
equilibrium constant K_p 298–301, 305–306
equilibrium constants K 294–309
  concentration 294–297, 304–305
  partial pressure 298–301, 305–306
  position of equilibrium 302–306
  temperature 302–304
equilibrium law 160

Index

equivalence points, titrations 342–343
equivalent protons 521
esterifications 228, 246, 248–249, 467, 469, 479
esters 183, 466–468, 497–498
   formation 228, 246, 248–249, 467, 469, 479
   hydrolysis 468
   molecular formulae 178
   polymers 483–484
   reactivity 498
ethers 497–498
ethylation of benzene 492
excess reagents 34–35
exothermic reactions
   dissolving 354–355
   enthalpy 124, 125, 131, 134–135
   equilibria 156–159, 302–303
exponential decay 279–280
E/Z isomerism 197–200

fats
   hydrogenations 404
feasibility 365–369, 392–393
feedstock recycling 218–219
filtration under reduced pressure 493–494
fingerprint region of spectra 261
first electron affinity 348
first ionisation energy 96–98, 109–110, 348
first order reactions 273, 278–280, 282
flame retardants 237, 239
f-orbitals 55–58, 95
formulae 15–17
   determination 22–25
   displayed 64, 180–181
   empirical 22, 179
   ionic compounds 16–17
   molecular 22, 179, 182–183
   organic compounds 179–183

four-coordinate complexes 407
fractional distillation 191
fragmentation peaks 253–254
fragment ions 253–254
free energy ΔG 365–371, 393
fuel cells 396
fuels 190, 193, 218, 396
functional groups
   carbon-13 spectra 515–519
   chemical tests 510
   identification 246, 258–260, 497–502
   infrared spectroscopy 256–261
   medicines 247
   organic compounds 173, 177–178
   proton NMR spectra 520–524
   qualitative analysis 509–510
   reactivities 498
   structural isomerism 182–183
   synthetic routes 248–250

gas chromatography 508–511
gases
   dynamic equilibrium 158–159
   entropy 363
   mole fractions 298–301
   partial pressures 298–301
   reaction rates 146–147
   tests for anions 117, 120
   volumes and moles 27–28
general formulae of organic compounds 173
giant lattice structures 60, 102–105
Gibbs free energy ΔG 365–369
   electrode potentials 393
global warming 193, 256–257
graphene/graphite 104

greenhouse gases 183, 244–245
Grignard reagents 492
Group 2 elements 108–111
groups of periodic table 93–95, 98, 109–116

Haber process 159, 403
haemoglobin 415
half-cells 386
half-equations 372, 374–375
half lives 279–280
halides 113–114, 116, 118–120
haloalkanes 232–236
   carbon–halogen bond strength 234–235
   chemical tests 510
   functional group 177, 246
   hydrolysis 233–236, 287–288
   naming 232
   nitrile formation 490
   nucleophilic substitution 233
   preparation from alcohols 229
   preparation from alkanes 195–197
   preparation from alkenes 208–209, 211–214
   reactivity 232–233
   substitution reactions 185, 233
halogenations 208–209, 211–213, 438–439
halogens 95, 112–116
heat
   enthalpy change 124–125, 129–133
   see also temperature
heating under reflux 227, 242–243
Henderson–Hasselbalch equation 338–339
Hess' law 137–140
heterogeneous catalysts 150, 404
heterogeneous equilibria 295, 301

heterolytic fission 184, 211, 212
homogeneous catalysts 149–150, 404
homogeneous equilibria 295, 300
homologous series of hydrocarbons 173, 174, 222
homolytic fission 184, 185, 195–196, 238
hydrated salts 24–25
hydration
   alkenes 209–210
   enthalpy changes 354–357, 360
hydrides 82
hydrocarbons
   basic concepts 170–189
   formulae 179–181
   nomenclature 174–178
   saturation 207, 208, 211
   structural isomerism 182–183
hydrogenations
   alkenes 207–208
   benzene 433–434
   cyclohexene 433–434
   vegetable fats 404
hydrogen bonding 77, 81–83
hydrogencarbonate–carbonic acid buffer system 337–338
hydrogen chloride 75, 78–79
hydrogen cyanide 457–459
hydrogen fuel cells 396
hydrogen halides 209, 211–212
hydrogen ions
   acids 313–325
   converting concentration to pH 316
hydrogen isotopes 10, 11
hydrogen peroxide 369, 404
hydrolysis
   condensation polymers 486
   esters 468

Index

haloalkanes 233–236, 287–288
nitriles 491
hydronium ions 311–312
hydroxides
  acid neutralisation 41
  ammonium ion test 120
  chlorine reaction 115
  Group 2 elements 110
hydroxyl functional group 177, 222, 523–528
hydroxynitriles 457–458, 491

ice, density 14, 81–82
identification
  alcohols 510
  alkenes 510
  carbonyl compounds 460–462, 510
  carboxylic acids 510
  chiral centres 481–482
  haloalkanes 510
  monomers/repeat units 486–487
  organic compounds 510
  phenols 510
  unknown compound structures 531–532
incomplete combustion 194
indicators 44, 341–343
indirect determination of enthalpy changes 137–140
induced dipole–dipole interactions (London forces) 77–78, 104, 191–192
induced dipoles 212
industrial process sustainability 36–37
infrared (IR) radiation effects 256–261
infrared (IR) spectroscopy 256–261, 530
initial rates 275–276, 284–286
instantaneous dipoles 77
integration traces 521
intermediate molecules in synthesis 249
intermolecular forces 77–80, 104, 191–192

iodine clocks 284–286
iodine/thiosulfate redox titrations 381–385
ionic compounds
  dissolving 354–355, 360
  dot-and-cross diagrams 59–60
  formation 40–41, 51, 109
  formulae and equations 16–17
  hydrated/anhydrous salts 24–25
  hydration 354–355, 360
  lattice enthalpy 346–351, 358–359
  melting points 358–359
  properties 60–62
  solubility in polar solvents 76
  in solution 352–355
  structures 60
ionic equations 114
ionic product of water K_w 326–329
ion/ion half cells 386
ionisation energies 96–100, 109–110
ions
  charge effects 359
  complex 405–417
  conjugate acid–base pairs 310–311
  d-block elements 401
  electron configuration 58
  halides 113–116
  heterolytic fission of covalent bonds 184
  mass spectrometry 252–253
  mixture analysis 120
  oxidation numbers 48
  polyatomic 16
  precipitation reactions 415–417
  qualitative analysis 117–120, 420–421
  shapes 73
  size effects 358
  stability 213–214
  structures and charges 11, 15
isoelectric point 479

isomerism
  carbon-13 NMR 517
  complex ions 409–412
  optical 203, 411–412, 480–482
  stereo 203–206, 409–412, 480–482
  structural 182–183, 197, 253
isotopes 9–14, 252
IUPAC names of compounds 174–178
K_a see dissociation constants
K_c see equilibrium constant
Kelvin 129
ketones 227, 246, 250, 258, 259, 454–462
  identification 460–462
  melting points 462
  molecular formulae 178
  naming 455
  nitrile formation 490–491
  reductions 457–458
kinetic energy 289
kinetics 368

lattice enthalpy 346–351, 358–359
lattices 60, 79, 102–105
law of conservation of energy 124
le Chatelier's principle 154–159
ligands 405–417
  bidentate 406, 411
  isomerism 410–411
  monodentate 406, 410–411
  substitution 413–415
light 195, 219, 237–238, 408
lithium-ion cells 395
lithium-ion polymer cells 395
logarithmic Arrhenius equation 290–291
log–log graphs 283
London forces 77–78, 104, 191–192

$M^+/M+1$ peaks 252, 254

manganate(VII) redox titrations 376–379
Markownikoff's rule 213–214
mass
  enthalpy calculations 129, 353–354
  molar mass 25
  relative mass of atoms 12–14
  subatomic particles 8–9, 12
  volume relationship 26, 28
mass number 10, 13
mass spectrometry 252–255, 530
melting points
  aldehydes/ketones 462
  carbonyl compounds 462
  determination 495–496
  entropy 362
  giant covalent lattices 103
  intermolecular forces 77, 79, 82
  ionic compounds 61, 358–359
  metals 102
  periodic trends 105
  prediction 359
  simple molecular substances 79
  unsaturated versus saturated hydrocarbons 208
  water 82
metallic bonding 101
metal/metal ion half-cells 386
metal oxide–acid reactions 314, 465
metals
  acid reactions 51
  acid redox reactions 313
  bonding and structures 101–102
  carboxylic acid reactivity 464
  Group 2 108–111
  ion formation 15
  ionic bonding 59–62
  properties 101, 102

Index

transition 15, 95, 102
versus non-metals 101
mobile phases 506
molar concentration 26–27
molar mass 21
molar volume 28
molecular covalent bonding 63–64
molecular formulae 22, 179, 182–183
molecular ions 252–254
molecular masses 20–21, 252
molecular shapes 70–71, 190–192
mole fractions 300–301
moles
    formula determination 22–25
    gases 27–28
    mass relationship 20–21
    reacting quantities 32–36
    solutions 26–27
    volume relationships 26–28
monobasic acids 312–313
monodentate ligands 406, 410–411
monomers 215–217, 483–487
multiplets 522–523
multi-step reactions 287–288

$n+1$ rule 522–523
neutralisation 340–343
    acids 41–47, 110–111, 128, 133, 312–314
    amines 476
    bases 314
    carboxylic acids 464–465
    indicators 341–343
neutrons 8–12
nitrations 437–438, 444
nitriles 177, 246, 486–487
NMR 512–533
nomenclature
    alcohols 222
    amines 475
    aromatic compounds 434–436

carbonyl compounds 454–455, 463–464, 466–467
carboxylic acids 463–464
E/Z isomerism 204–206
haloalkanes 232
organic compounds 174–178
non-equivalent protons 521
non-metals
    covalent bonding 63–67
    giant structures 103–105
    halogens 112–116
    ionic bonding 59–62
    properties 103–105
    simple molecular structures 103, 105
non-polar bonds 74–76
non-polar molecules 75, 76, 79–80, 193
non-polar solvents 80
nuclear charge 8–9, 96
nuclear magnetic resonance (NMR) spectroscopy 512–533
    aromatic compounds 518–519, 523
    carbon-13 515–519, 527
    equivalent/non-equivalent protons 521
    integration traces 521
    interpretation 515–519, 525–527
    multiplets 522–523
    predicting spectra 527–528
    proton exchange 524
    proton spectra 520–529
    spin–spin coupling 522–523
nuclear spin 512
nucleon number 10
nucleophiles 232–233
nucleophilic addition–eliminations 470–471
nucleophilic additions 456–459, 490–491
nucleophilic substitutions 233, 490
nucleus of atoms 8–9

octahedral complexes 410–412
octahedral molecules 72, 73
OIL RIG 372
optical isomerism 203, 411–412, 480–482
orbitals of electrons 54–58
order of reactions 272–274, 278–283
organic acids 203
organic chemistry 170–261, 430–535
    acid anhydrides 466–467, 470–471
    acyl chlorides 466, 468–470
    aldehydes 454–462
    alkenes 497–498
    amides 480, 497–498
    amines 474–489, 497–498
    amino acids 478–482
    analytical techniques 506–535
    aromatic compounds 432–453
    basic concepts 172–188
    benzene 432–441, 444–445, 492, 498
    carbon–carbon bond formation 490–492
    carbonyl compounds 454–473
    carboxylic acids 463–471, 497–498
    chromatography 506–511
    combined analysis 530–533
    condensation polymers 483–487
    directing effects 447–450
    esters 466–468, 497–498
    ethers 497–498
    functional group identification 497–502
    ketones 454–462, 490–491
    melting point determination 495–496

naming 434–436, 447, 454–455, 463–464, 466–467, 475
nitriles 490–491
nomenclature 174–178
organometallic compounds 492
phenol/phenols 442–451, 499
reaction mechanisms 184–187
separation methods 493–494
stereoisomerism 480–482
structure determination 530–532
organic solids, preparation 493–496
organic synthesis 242–233, 490–505
    benzene additions 492
    carbon–carbon bond formation 490–492
    directing groups 447–450
    functional groups 248, 497–502
    Grignard reagents 492
    melting point determination 495–496
    nitriles 490–491
    organic solids 493–496
    phenol 442–443
    preparation of liquids 242–244
    problem solving 248–249
    purification 244–245
    recrystallisation 494
    reduced pressure filtrations 493–494
    target molecules 247, 499–501
organohalogens 237–239
organometallic compounds 492
overall order of reactions 273–274
oxidation numbers 48–51, 372–374, 402–403
oxidations 50–51, 113, 226–227, 388, 456

Index

oxidation states of transition metals 402–403
oxides 41, 110
oxidising agents 113, 372, 381–385
oxygen 99, 100, 108, 299, 415
oxygen–hydrogen bonds 258–260
ozone 150, 237–238

partial pressure 296–301, 305–306
Pascal's triangle 523
Pauling electronegativity values 74
peak integration 508–509
percentage yield 34, 36
periodicity 92–107, 358–360
periodic table 9, 14, 15, 92–107
   blocks 60, 94–95
   groups 93–95
   history 92–93
   metal/non-metal transition 101, 105
   names and numbers 95
   periods 93–94
   reactivity trends within groups 108–116
permanent dipoles 75–76
permanent dipole–dipole interactions 77–79
petrol 190, 193
pH 315–329
   blood 337–338
   buffer solutions 333–339
   converting to H^+ 316
   dilutions 317–318
   dissociation constants 319–325
   Henderson–Hasselbalch equation 338–339
   indicators 44, 341–343
   meters 340
   scale 315–318
   strong acids 316–317
   strong bases 326–328
   titrations 340–341
   weak acids 322–325
   weak bases 328–329

phenol/phenols 442–451
   brominations 444
   chemical tests 510
   comparison to benzene 444–445
   directing groups 446–450
   disubstitutions 446–450
   electrophilic substitutions 444–445
   esterifications 469
   manufacturing 442–443
   nitrations 444
   reactivity 443, 499
   sodium hydroxide reactivity 443
   as a weak acid 443
photodissociation 238
physical chemistry 270–429
   acids 310–325
   Arrhenius equation 289–291
   bases 310–312, 314–315, 326–329
   buffer solutions 332–339
   concentration 294–297, 304–305
   d-block elements 400–423
   enthalpy 346–361
   entropy 362–371
   equilibria 294–309
   free energy 365–371
   hydronium ions 311–312, 326
   mole fractions 298–301
   neutralisation 340–343
   order of reactions 272–276, 278–283
   partial pressure 298–271, 305–306
   pH 315–329
   pOH 329
   rate-determining steps 287–288
   rates of reactions 272–293
   transition elements 400–423
π-bonds 200–201, 203, 434, 456

pK_a 319–321
planar structures 104
pOH 329
polar bonds 75, 76, 456, 464
polar molecules 75–76, 78–80, 211, 223–224, 232
polar solvents 61, 76, 80
pollution 150, 218, 258
polyamides 484–487
polyatomic ions 16, 17, 50, 59, 70
polyesters 483–484, 486
poly(ethene) 215–216
polyethylene terephthalate (PET) 484
polymers 215–219
   addition-type 486
   condensation-type 483–487
polypeptide formation 485
polyunsaturated alkenes 200
p-orbitals 55–58, 95, 100
position of equilibrium 154–161, 302–306
potassium dichromate 226–227, 456
precipitate tests 116–120
precipitation reactions 415–417
prediction
   carbon-13 NMR spectra 527
   entropy changes 362–363
   melting points 359
   proton NMR spectra 528
   redox reactions 374–375, 391–393
   solubility 360
prefix, organic compound names 174, 177–178
pressure
   dynamic equilibrium 158–159
   equilibrium constants 298–301, 305–306
   reaction rate effects 146
   room temperature and pressure 28
   standard conditions 126

primary alcohols 224, 226–227, 243, 510
primary amides 470, 480
primary amines 474–476
primary carbocations 207–208
primary cells 394
primary haloalkanes 232, 234–236
principal quantum numbers 54
proteins 485
proton exchange 524
proton NMR spectroscopy 520–529
   chemical shifts 520–524
   equivalent/non-equivalent protons 521
   integration traces 521
   interpretation 525–527
   spin–spin coupling 522–523
protons 8–12
   converting concentration to pH 316
   equivalent/non-equivalent 521
   splitting 522–523
pure covalent bonds 75
purification 244, 493–494
purity analysis 377–378

qualitative analysis 117–120
   ammonium test 120
   anions 117–120
   carbonate test 117, 119, 120
   cations 120
   functional groups 509–510
   halide test 116, 118–120
   ion mixtures 120
   sequence of testing 119–120
   sulfate test 117, 119, 120
   transition metal ions 420–421
quantitative analysis 43–47, 117
quantities, equations and reactions 32–36

Index

quickfit apparatus 242, 243

radicals 184, 196, 238, 253
rate–concentration graphs 282–286
rate constant k 273–274
    Arrhenius equation 289–291
    first order reactions 279–280
    half-lives 279–280
    initial rates 284–286
    logarithmic Arrhenius equation 290–291
    log–log graphs 283
    rate–concentration graphs 282–286
    temperature effects 289–291
    units 274
rate-determining steps 287–288
rate equations 273–276, 280, 282–286
rates of reactions 272–293
    Arrhenius equation 289–291
    clock reactions 284–286
    colorimetry 277–278
    concentration–time graphs 277–281
    half-lives 279–280
    initial 275–276, 284–286
    log–log graphs 283
    order of reaction 272–276, 278–283
    rate–concentration graphs 282–286
    rate constants 273–274, 279–280, 282–286
    rate-determining steps 287–288
    rate equations 273–276, 280, 282–286
    reaction mechanisms 287–288
    temperature effects 289–291
reaction mechanisms 287–288
reaction rates 144–153
    Boltzmann distribution 152–153

bond strengths 234–235
catalysts 149–151, 153
definition 144
factors affecting 145–146
following progress 146–148
gases 146–147
haloalkane hydrolysis 234–236
reaction mechanisms 236
reactions
    atom economy 35–36
    Born–Haber cycles 346–351
    enthalpy change 126–127, 132
    entropy 362–364
    feasibility 365–369
    mechanisms 184–187, 228, 236
    quantities 32–36
    yield 34
reactivity
    acyl chlorides 469
    alkanes 190, 193
    alkenes 207–219, 498
    amides 498
    amines 470, 498
    aromatic compounds 467–471
    benzene 433–434, 437–441, 498
    carboxylic acids 464, 478, 498
    esters 498
    haloalkanes 232–233
    phenol/phenols 443, 499
    trends 108–123
rechargeable batteries 394–395
recrystallisation 494
recycling polymers 218
redistillation 225
redox 40–51, 372–399
    acids and metals 313, 464
    carboxylic acids with metals 464
    disproportionations 420
    electrode potentials 386–396, 418
    equations 372–375

fuel cells 396
Group 2 elements 108–109
half-equations 372, 374–375
halogens 112–116
iodine/thiosulfate titrations 381–385
manganate(VII) titrations 376–379
OIL RIG 50–51, 372
oxidation numbers 372–374
prediction 391–393
primary cells 394
rechargeable batteries 394–395
titrations 376–385
transition metals 418–420
reduced pressure filtrations 493–494
reducing agents 108, 116, 372, 376–379
reductions 50–51, 388, 457–458, 477, 491
reflux heating 227, 242–243
relative atomic mass 13–14
relative formula mass 23
relative isotopic mass 12–14
relative masses 8, 12–14, 23
relative molecular mass 23
repeat units 483–487
repulsion of electron pairs 70–71
resonance 512
retention factors 507–508
retention times 508–509
reversible reactions 154–161
room temperature and pressure (RTP) 29, 126, 131, 327–328
rotation around hydrocarbon bonds 190, 201, 203

salt bridges 387
saturated fats 208
saturated hydrocarbons 173, 207–209, 211

secondary alcohols 224, 225, 227, 250, 510
secondary amides 470, 480
secondary amines 474–476
secondary carbocations 213–214
secondary cells 394
secondary haloalkanes 232, 236
second electron affinity 350–351
second ionisation energy 96–97, 109–110
second order reactions 273, 278, 282
separation of organic synthesis products 244–245, 493–494
shapes of molecules and complexes 70–73, 190–192, 407, 409–412
shells of electrons 55–57, 96–100
σ-bonds 190, 193, 456
silver mirror test 461–462
simple molecular structures 79–80, 103, 105
single covalent bonds 64–66
six-coordinate complexes 407
size effects of ions 358
skeletal formulae 180–181
sodium hydroxide 115, 120, 328, 415–416, 443
solubility
    alcohols 223–224
    carboxylic acids 464
    giant covalent lattices 103
    ionic compounds 61
    metals 102–103
    non-polar solvents 82
    polar solvents 76
    prediction 360
    simple molecular substances 79–80
solutions
    acids/alkalis 327–329
    amount of substances 26–27

601

Index

electrical conductivity 61–62
enthalpy changes 352–361
equilibria 294–297, 304–305
halogens 113
hydronium ions 311–312, 326
solvents 61, 76, 80, 237, 514
solvent separation 413–494
s-orbitals 55–58, 95, 100
specific heat capacity 129
spectator ions 313
spectroscopy 252–261
spin of nuclei 512
spin–spin coupling 522–523
splitting of resonances 522–523
square planar complexes 407, 409–410
standard conditions 126, 131, 327–328
standard electrode potentials 387–388
standard enthalpy changes 126–128
  of atomisation 347
  of formation 347
  of solution 352–355
standard entropy 363–364
standard solutions 27, 43
starting molecules for organic synthesis 248–250
states of matter 362–363
stationary phases 506
steam, alkene reactions 209–210
stems of organic compound names 174, 175
stereoisomerism 203–206, 409–412, 480–482
stoichiometry 32
storage cells 394–395
strong acids 40, 42, 316–317, 318–320, 342
strong bases 326–328, 342
structural formulae 180, 182–183

structural isomers 182–183, 197, 253
structure determination 530–532
subatomic particles 8–12
subshells of electrons 55, 99–100
substituents, aromatic compounds 434–436
substitution reactions 185, 195–197, 229, 233, 490
successive ionisation energies 96–97
suffixes of organic compound names 174, 175, 177–178
surroundings, energy transfer 124–125, 129–133
sustainability 35–36, 151
symmetrical molecules 75–76

target molecules 247–250, 499–501
temperature
  Boltzmann distribution 152–153
  calculating energy changes 129
  dynamic equilibrium 156–158
  entropy 362–363
  equilibrium constants 302–304
  feasibility 365–369
  rates of reactions 289–291
reaction rates 145, 153
room temperature 28
standard conditions 126
tertiary alcohols 224, 225, 227
tertiary amines 474–476
tests for ions 116–120
tetrahedral complexes 407
tetrahedral ions 73
tetrahedral molecules 70–72
tetramethylsilane (TMS) 513
theoretical yield 34
Thiele tubes 496

thin layer chromatography (TLC) 506–508
three-dimensional shapes of molecules 70
titrations 43–47, 340–343, 376–385
TLC (thin layer chromatography) 506–508
TMS (tetramethylsilane) 513
TNT (trinitrotoluene) 450
Tollens' reagent 461–462
transition elements 400–423
  definition 401–402
transition metals 15, 95, 102
  catalysis 403–404
  coloured compounds 403, 408
  complex ions 405–417
  oxidation states 402–403
  precipitation reactions 415–417
  qualitative analysis 420–421
  redox 418–420
transport of oxygen in blood 299
tribasic acids 312–313
trigonal planar structures 73
triple covalent bonds 66

ultraviolet (UV) radiation 195, 237–238
units
  K_c 294
  rate constants 274
unknown compounds, analysis 531–532
unknown enthalpy changes 355–357
unsaturated fats 208

van der Waals' forces 78
variable oxidation states of transition metals 402–403
voltaic cells 386
volumes
  gases 27–28, 33

mass relationships 26–28
reactions 33
solutions 26–27, 33
volumetric flasks 43

waste polymers 218
water
  acid–base equilibria 326–328
  acyl chloride reactivity 469
  alkene steam reaction 209–210
  boiling/melting points 82
  chlorine reaction 121
  of crystallisation 24–25
  density of ice 10, 81–82
  Group 2 reactions 109
  hydrogen bonding 81–82
  hydronium ions 311–312, 326
  ionic product 326–329
  molecular shape and bond angle 71
  polarity 76
  properties 81–82
  removal from organic synthesis products 244–245
wavenumbers 257–258
weak acid buffers 332–333
weak acids 40, 42
  buffer solutions 332–333
  carboxylic acids 464–465
  dissociation constants 319–325
  partial neutralisation 332–333, 336
  pH 322–325
  phenol 443
  titrations 342
  wine 320–321
weak bases 328–329, 342
weak intermolecular forces 79

yield of reactions 34, 36

zero order reactions 272–273, 278, 282
zwitterions 479

Periodic table

(1)	(2)												(3)	(4)	(5)	(6)	(7)	(0)
1																		**18**
1 **H** hydrogen 1.0																		2 **He** helium 4.0
	2											**13**	**14**	**15**	**16**	**17**	
3 **Li** lithium 6.9	4 **Be** beryllium 9.0												5 **B** boron 10.8	6 **C** carbon 12.0	7 **N** nitrogen 14.0	8 **O** oxygen 16.0	9 **F** fluorine 19.0	10 **Ne** neon 20.2
11 **Na** sodium 23.0	12 **Mg** magnesium 24.3	**3**	**4**	**5**	**6**	**7**	**8**	**9**	**10**	**11**	**12**	13 **Al** aluminium 27.0	14 **Si** silicon 28.1	15 **P** phosphorus 31.0	16 **S** sulfur 32.1	17 **Cl** chlorine 35.5	18 **Ar** argon 39.9
19 **K** potassium 39.1	20 **Ca** calcium 40.1	21 **Sc** scandium 45.0	22 **Ti** titanium 47.9	23 **V** vanadium 50.9	24 **Cr** chromium 52.0	25 **Mn** manganese 54.9	26 **Fe** iron 55.8	27 **Co** cobalt 58.9	28 **Ni** nickel 58.7	29 **Cu** copper 63.5	30 **Zn** zinc 65.4	31 **Ga** gallium 69.7	32 **Ge** germanium 72.6	33 **As** arsenic 74.9	34 **Se** selenium 79.0	35 **Br** bromine 79.9	36 **Kr** krypton 83.8
37 **Rb** rubidium 85.5	38 **Sr** strontium 87.6	39 **Y** yttrium 88.9	40 **Zr** zirconium 91.2	41 **Nb** niobium 92.9	42 **Mo** molybdenum 95.9	43 **Tc** technetium	44 **Ru** ruthenium 101.1	45 **Rh** rhodium 102.9	46 **Pd** palladium 106.4	47 **Ag** silver 107.9	48 **Cd** cadmium 112.4	49 **In** indium 114.8	50 **Sn** tin 118.7	51 **Sb** antimony 121.8	52 **Te** tellurium 127.6	53 **I** iodine 126.9	54 **Xe** xenon 131.3
55 **Cs** caesium 132.9	56 **Ba** barium 137.3	57–71 lanthanoids	72 **Hf** hafnium 178.5	73 **Ta** tantalum 180.9	74 **W** tungsten 183.8	75 **Re** rhenium 186.2	76 **Os** osmium 190.2	77 **Ir** iridium 192.2	78 **Pt** platinum 195.1	79 **Au** gold 197.0	80 **Hg** mercury 200.6	81 **Tl** thallium 204.4	82 **Pb** lead 207.2	83 **Bi** bismuth 209.0	84 **Po** polonium	85 **At** astatine	86 **Rn** radon
87 **Fr** francium	88 **Ra** radium	89–103 actinoids	104 **Rf** rutherfordium	105 **Db** dubnium	106 **Sg** seaborgium	107 **Bh** bohrium	108 **Hs** hassium	109 **Mt** meitnerium	110 **Ds** darmstadtium	111 **Rg** roentgenium	112 **Cn** copernicium		114 **Fl** flerovium		116 **Lv** livermorium		

Key
atomic number
Symbol
name
relative atomic mass

57 **La** lanthanum 138.9	58 **Ce** cerium 140.1	59 **Pr** praseodymium 140.9	60 **Nd** neodymium 144.2	61 **Pm** promethium 144.9	62 **Sm** samarium 150.4	63 **Eu** europium 152.0	64 **Gd** gadolinium 157.2	65 **Tb** terbium 158.9	66 **Dy** dysprosium 162.5	67 **Ho** holmium 164.9	68 **Er** erbium 167.3	69 **Tm** thulium 168.9	70 **Yb** ytterbium 173.0	71 **Lu** lutetium 175.0
89 **Ac** actinium	90 **Th** thorium 232.0	91 **Pa** protactinium	92 **U** uranium 238.1	93 **Np** neptunium	94 **Pu** plutonium	95 **Am** americium	96 **Cm** curium	97 **Bk** berkelium	98 **Cf** californium	99 **Es** einsteinium	100 **Fm** fermium	101 **Md** mendelevium	102 **No** nobelium	103 **Lr** lawrencium